Advances in Distributed System Reliability

ADVANCES in DISTRIBUTED SYSTEM RELIABILITY

SURESH RAI and
DHARMA P. AGRAWAL

IEEE Computer Society Press Tutorial

Advances in Distributed System Reliability

Suresh Rai
Louisiana State University

Dharma P. Agrawal
North Carolina State University

This tutorial text is dedicated to our families:
Jayanti, Manjul, and Shiwali Rai
Purnima, Sonali, and Braj Agrawal

IEEE Computer Society Press
Los Alamitos, California

Washington • Brussels • Tokyo

Published by

IEEE Computer Society Press
10662 Los Vaqueros Circle
P.O. Box 3014
Los Alamitos, CA 90720-1264

Cover designed by Jack I. Ballestero

Printed in United States of America

IEEE Computer Society Order Number 1907
Library of Congress Number 89-45997
IEEE Catalog Number EH0300-4
ISBN 0-8186-8907-2 (Case)
ISBN 0-8186-5907-6 (Microfiche)

Additional copies may be ordered from:

IEEE Computer Society
10662 Los Vaqueros Circle
P.O. Box 3014
Los Alamitos CA 90720-1264

IEEE Service Center
445 Hoes Lane
P.O. Box 1331
Piscataway, NJ 08855-1331

IEEE Computer Society
13, Avenue de l'Aquilon
B-1200 Brussels
BELGIUM

IEEE Computer Society
Ooshima Building
2-19-1 Minami-Aoyama,
Minato-Ku
Tokyo 107 JAPAN

THE INSTITUTE OF ELECTRICAL AND ELECTRONICS ENGINEERS, INC.

Preface

The objective of this tutorial is to provide a literacy forum for exchanging information among planners and design engineers of parallel and distributed computing networks, computer systems analysts, specialists in computer reliability and maintainability, and other computer engineering professionals. The tutorial emphasizes the importance of such a close interaction and the impact of reliability on parallel/distributed computing networks. It also covers traditional strategies for performability evaluation and case studies.

This tutorial is intended for system designers, application engineers, scientists, researchers working in universities or with government agencies/organizations, and students, who would like to know more about the reliability of parallel and distributed computing networks. Some background in computer communications, switching theory, and probability will be assumed.

Viewpoint

Recent advances in VLSI circuitry have tremendous impact on implementing a fairly complex process on a single chip. This development has led to increased use of stand-alone work stations connected in the form of a powerful distributed system. Potential benefits offered by such distributed systems include better cost performance resulting from exploiting parallelism in most of the algorithms, enhanced fault tolerance, a high degree of modularity, increased system throughput, and efficient sharing of resources. In this tutorial text, we focus on the reliability issues in such systems because the computation of system reliability metrics has become an integral part of the system designer's task.

We have considered all of the three broad categories—namely, closely coupled, loosely coupled, and barely coupled types—of distributed systems. The key issue distinguishing these systems is the grain of computation, which can be roughly expressed as the computation time divided by the communication time. If this ratio is below 10, we have a closely coupled system. If the ratio is between 10 and 100, we have a loosely coupled system. If the ratio is above 100, the system is barely coupled.

From the reliability aspect, a distributed system is modeled as a collection of resourses (computers, programs, datafiles, devices, etc.) connected by an arbitrary communication network and controlled by a distributed operating system. This text investigates the following issues:

- multiprocessor and multiterminal reliability
- multimode and dependent-failure analysis
- performability analysis
- task-based reliability and software reliability

Six chapters, some of which are further divided into subchapters, address these problems. Each chapter or subchapter consists of two to six papers that illustrate the conceptual and research issues of the topic. The chapter organization reflects continuity; special precautions have been taken to ensure a smooth transition from one subject to another. Each chapter begins with introductory remarks addressed to both novice and advanced readers. At the end of the tutorial text, a bibliography listing relevant journals, books, and research papers provides interested readers with further references in the area.

Chapter Descriptions

Chapter 1 introduces the topic and provides a critical assessment of current network models useful for evaluation reliability. It is suggested that readers refer to a companion tutorial text, *Distributed Computing Network Reliability,* published by the IEEE Computer Society, for a preliminary discussion and introductory comments on reliability issues. Reliability of multiprocessor systems is the theme in Chapter 2. Here we have considered multiprocessors with processor, memory, switch (PMS) notation and single- and redundant-path multistage interconnection networks (MINs) as examples.

Chapter 3 concentrates on multiterminal reliability analysis of distributed processing systems. Note that in all such systems—like distributed database systems, teamwork, distributed data processing systems, and computer communication networks—*system (or application) survivability* is best characterized by some multiterminal reliability measure. In fact, the single fact of reliable terminals could not be used to compute the survivability of a system because various events depend on one another.

Chapter 4 discusses multimode and dependent-failure issues in reliability analysis. These assumptions provide more flexible and realistic models of communication network reliability. Degraded communication capabilities between different locations are much more common than complete communication outages in most commercial networks. Further, there are enough causes for the failure of common modes that dependent failures could be said to be the dominant causes of downtime or degraded performance. A similar behavior studied in computing systems is considered in Chapter 5 in which we develop a unified model of reliability and performance.

The tutorial text concludes in Chapter 6 with subchapters discussing such topics as task-based reliability, software reliability, and case studies.

Acknowledgments

The authors thank various groups and individuals for making this undertaking possible. The IEEE Computer Society Press Editorial Board and other IEEE Computer Society Press staff have always provided necessary help and guidance. We appreciate their efforts. We sincerely thank the Department of Electrical and Computer Engineering at North Carolina State University, for constant support. We thank Peggie Dixon for her painstaking and meticulous secretarial work. We also thank T.Y. Chung, a graduate student, who helped us collect some of the reprints of papers for this tutorial text, which we hope will be helpful to the reliability-conscious community. Finally, we thank our parents who encouraged us to be inquisitive in life, making this tutorial text possible.

Suresh Rai
Dharma P. Agrawal

Advances in Distributed System Reliability

Table of Contents

Chapter 1: Introduction

Distributed systems span a wide spectrum in the design space, and forming a very specific definition for a distributed system is difficult. We will consider the following three broad categories:

- closely coupled systems
- loosely coupled systems
- barely coupled systems

The key characteristic distinguishing these systems is the grain of computation, which can be expressed roughly as the computation time divided by the communication time.

In practice, the amount of time required for communication is determined by the interconnection media, the communication hardware, and the operating system. In a system consisting of a large number of CPU boards on a single backplane with shared memory, one processor may, in microseconds, write a word in another processor's memory. On the other hand, processors that communicate over a local area network (LAN) by message-passing typically require milliseconds to send a message and get a response. Finally, when a wide-area network (WAN) is being used, a communication time of hundreds of milliseconds or more is not abnormal. Thus, if the grain of computation ratio is below 10, we have a closely coupled system; if it is between 10 and 100, we have a loosely coupled system; if it is above 100, the system is barely coupled. A general survey on distributed systems is given by Tanenbaum and van Renesse in their paper in the journal *Computing Surveys*, (17, Dec. 1985, 419-470).

The topology or interconnection network defines the layout of communication links and processing elements (PEs) and determines the data paths that may be used between any pair of PEs in a distributed system. Several schemes have been suggested to classify the topology. They are distinguished as static vs. dynamic, link-oriented vs. bus-oriented; and so on. Some typical examples of interconnection topology include linear array, ring, star, tree, mesh, D-dimensional W-wide hypercube, cube-connected cycle (CCC), and single- and redundant-path MINs. The choice of a topology depends on a variety of factors such as diameter, average distance, expandability, availability, capacity, flexibility, fault tolerance, problem/algorithm mapping characteristics, and centralized/distributed control strategy.

Network metrics such as diameter and average distance are commonly used as evaluation measures of an inter-connection scheme. Larger diameter means that messages have to go through a relatively larger number of interfaces and links; therefore, the message delay will increase. The average distance characterizes the average message delay when a PE communicates with an arbitrary destination PE and, in a way, determines the throughput of the network. This also depends on the total number of I/O ports or channels per processor. A network that has a low average distance may require an unreasonable number of communication ports for each computer.

Expandability is a measure of the ability to match the interconnection structure to increased processing requirements. Each new application of the processing architecture should not involve a complete redesign effort in either hardware or communication software. Capacity describes the network's ability to increase communication bandwidth as more processors are added to the system. The flexibility of a network is indicated by its ability to adapt to change in the data or processor communication flow. For example, when a primary candidate is occupied, the network should be able to reroute the data to an available processor without incurring excessive overhead.

Fault-resilient behavior of a network is generally specified by edge- and node-connectivity. This behavior provides information regarding edge- and node-disjoint paths, hence on a possible number of link and/or node failures that the system could tolerate.

Reliability is an equally important characteristic that should be examined for different topologies. Reliability is mathematically defined as the probability that the system will perform satisfactorily in a given time, assuming that operation commences successfully. For distributed system networks, reliability translates into finding alternate communication paths between a pair of nodes and seeing whether at least one of them is operational. The greater the number of alternate paths, the more reliable is the network. Interested readers are referred to this companion tutorial text, *Distributed Computing Network Reliability*, noted previously for a preliminary discussion on reliability issues, specifically terminal and network reliability. This text provides information regarding advances in the reliability of distributed systems.

In organizing this tutorial, we have carefully considered the continuity of the subject matter and have concentrated on just a few topics in each chapter or subchapter. Toward this end, this chapter provides a graph model of multistage interconnection networks so that the existence of a path

EH0300-4/90/0000/0001$01.00

between a source-destination pair in a finite number of passes could be found out. In later chapters, this model has been used in computing the reliability and the availability of the network. Appropriate modeling of a complete system is a very complex issue and computing its reliability belongs to NP-hard class of problems.

Another paper in this chapter provides a critique of major weaknesses of current techniques for analyzing the reliability of computer communication networks. Most weaknesses result from making some unrealistic assumptions in the model(s) used for predicting reliability. This paper not only gives a conceptual view of the difficulties faced in reliability modeling, but also suggests important areas where future research on the reliability of telecommunication networks is needed. Chapters 3 through 5 address some issues raised in this paper. Also, the multiprocessor reliability is considered in Chapter 2, and specific application examples and case studies are given in Chapter 6.

Reprinted from *IEEE Transactions on Computers*, Volume C-34, Number 3, March 1985, pages 255–266.

Dynamic Accessibility Testing and Path Length Optimization of Multistage Interconnection Networks

DHARMA P. AGRAWAL, SENIOR MEMBER, IEEE, AND JA-SONG LEU, STUDENT MEMBER, IEEE

Abstract **— The emergence of multiple processor systems has seen the increased use of multistage interconnection networks (MIN's), built with several stages of 2-input 2-output switching elements (SE's). The connectivity and fault tolerance of these networks are important problems as MIN's are expected to be the heart of these systems. This paper employs a versatile graph model of an SE that could represent all possible stuck type terminal faults at the control lines and input/output data lines. This technique leads to a graph model of a given MIN, amenable to testing of its dynamic full access (DFA) capability. The basic strategy of employing adjacency and reachability matrices enables testing under various combinations of multiple faults. Simulation of various networks is carried out to evaluate the average path lengths which illustrates the effect of connection pattern on the network performance. A design methodology for implementing a class of 2^n-input 2^n-output networks with m stages ($m < n$) of 2×2 MIN's is also outlined so that DFA capability and the maximum availability could be ensured. Optimality of such a network under the presence of faults is verified by the simulation results that show a negligible increase in the average path length.**

Index Terms **— Adjacency matrix, average distance, connectivity, dynamic full access capability, graph model, multistage interconnection networks, reachability matrix, stuck-at faults.**

I. INTRODUCTION

RECENT advances in VLSI technology have encouraged the use of multiprocessor and multicomputer systems with a large number of processing elements (PE's) and memory modules (MM's). In such systems, various techniques are utilized to support restructurable data paths between the PE's and MM's. Thus, the intercommunication is becoming an increasingly complex but inevitable issue. Several design techniques for the interprocessor communication have been reviewed [1] and some of them have been constructed. The current trend is to employ multistage interconnection networks (MIN's) which requires segmentation of the network into several stages, with each stage partially satisfying the input–output connection requirements.

Various design issues of MIN's have been covered in the literature. The main emphasis has been in finding their equivalence and nonequivalence and comparing their permutation capabilities [2]–[8], designing for conflict-free permutations [9], [10], algorithmic adaptability characteristics [11], and understanding relative advantages and disadvantages of MIN's [12]. But not much attention has been paid to the aspect of fault tolerance which is crucial to the successful operation of a multiple processor system. It has become extremely important, as MIN's are considered the heart of parallel systems. This paper is concerned with the fault-tolerant capability of MIN's when employed to provide interprocessor communication in a multiprocessor environment. The network should be implemented in such a way that a noncatastrophic fault may not force a complete shut-off and the system should continue working with reduced capacity. This graceful degradation characteristic is not only dependent on the processors failure, but also on the MIN's. Similarily, the path length required to provide a logical link between two processors indicates the time delay involved in transmitting the information from a source to any desired destination. Thus, average path lengths in both MIN's and single-stage interconnection networks (SSIN's) [13] seem to be a good representative of their communication delays.

The type of multiple processor systems we are concerned with is shown in **Fig. 1** [1], [14]. In this system model, the PE's with their private memory modules provide desired parallelism while PE-to-PE transfer is achieved through the MIN Thus, all PE's are connected to both sides of the network so each PE can transmit data via the network input side while the output is useful in receiving data from another PE.

A brief introduction of the MIN's and an overview of existing testing techniques is covered in Section II. Our fault model including multiple stuck-at-faults at input–output lines and control lines is described in Section III. A generalized test procedure and the use of adjacency and reachability matrices are illustrated in Section IV. Section V outlines a procedure for computing the static average path length under various faulty conditions. Section VI provides an insight to the optimum design methodology of MIN's so that a larger number of faults can be tolerated. Finally, concluding remarks are included in Section VII.

II. MULTISTAGE INTERCONNECTION NETWORKS AND EXISTING TESTING TECHNIQUES

Several of the proposed MIN's have been designed using 2×2 SE as a basic building block. A simple representation of such an SE is shown in **Fig. 2(a)** and its two possible states are illustrated in Fig. **2(b) and (c).** The parallel and cross connection of an SE is determined by the logical level applied at its control line. An N-input N-output MIN (for simplicity, N is assumed to be some power of 2; say $N = 2^n$) is constructed in several stages with each stage consisting of $N/2$ SE's. The links between successive stages of the network are assigned in such a way that each input could be connected to as many outputs as possible. An n-stage network can provide a path between any of the N-inputs and to each one of the N-outputs and such a MIN has been reproduced from [2] in Fig. 2(d). The control lines are not shown for clarity of the diagram.

The problem of communication and data transfer rate tends to be increasingly critical when either the network load is heavy, or when some of the lines happen to be faulty. A link between any input and an arbitrary output cannot be established if a conflict occurs or a required line happens to be faulty. In such situations, if data are to be transferred between any two PE's of Fig. 1, the data may have to be passed to one or more intermediate PE's before reaching the destination PE. These considerations encouraged us to look into MIN's with m stages ($m \leq n$).

The fault diagnosis for a class of switching networks has been described by Operfman and Tsao-Wu [15]. They utilize a sequence of tests to ensure the correct operation of the two possible states of each SE. An unalterable state of an SE means that the control lines are permanently stuck at 0 or 1. Shen and Hayes [16] have used a graph model for demonstrating the effect of faults at the control lines. In their fault model, they consider the output side to be fed back to the corresponding input lines, as rerouting of data through the PE is possible. They have modeled an SE by a node and each interconnection link is represented by a directed edge connecting the nodes. A faulty SE is indicated by a node partitioned into two sections. Their main concern is to test whether, under a given fault condition, each input of the network could be logically connected to any one of the network outputs in a finite number of passes. They define this property as the dynamic full access (DFA) capability. The control line faults are said to be critical if they destroy the DFA characteristics. The same model has been used [17] in analyzing the fault tolerance of various redundant MIN's. The major shortcoming is to assume the presence of a fault(s) only at the control input(s). In Section III, we propose a graph model which overcomes this limitation.

Narraway and So [18] have considered a diagnosis model for a general switching network constructed with k-input k-output switches ($k \geq 2$). Their basic strategy is to use known good connecting data paths in identifying good switches and use this progressively in defining a faulty connecting path. They do not consider faults at the control lines. In another recent paper, Wu and Feng [19] have described a simple testing technique for MIN's. In their paper, they have illustrated 16 different possible states of an SE and they consider only two states (parallel or cross connected) as fault-free situations and all others are interpreted as faults in the SE's. In their novel scheme, they require only four sequences for testing any MIN. In fact, only two complementary sequences are needed and are repeated for two different control settings of the network, one for parallel-connected and another for cross-connected modes. Each sequence is selected in such a way that a one-out-of-two code is used in space. This means that only one input (and hence one output) of each SE is made one while the other remains zero.

It is well known that the probability of logical faults inside an IC chip is quite small and most of the faults occur at the pins [20]. Thus, it is more important to consider stuck-at faults at inputs, outputs, and control lines of the SE's. Wu and Feng's algorithm can still test faults at the input and output line(s) of SE(s). Further utilization of the same sequences in testing control line fault has been covered in [21]. Another model for the control line faults in Omega [22] and other networks has recently been considered in [23]. Their opti-

Manuscript received January 5, 1984; revised July 27, 1984. An earlier version of this work was presented at the Fourth International Conference on Distributed Computing Systems, San Francisco, CA, May 14–18, 1984.

The authors are with the Department of Electrical and Computer Engineering, North Carolina State University, Raleigh, NC 27695.

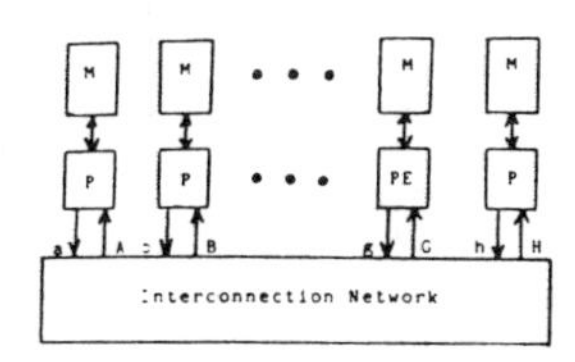

Fig. 1. Multiple processor system organization.

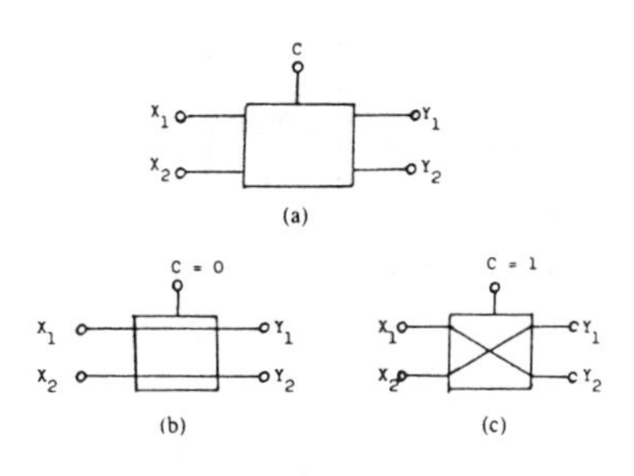

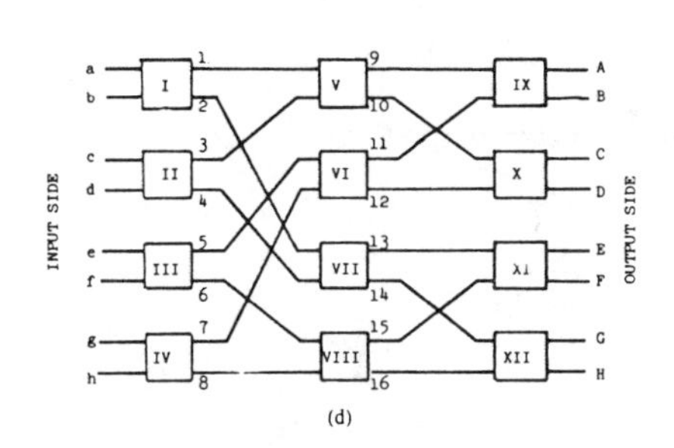

Fig. 2. (a) Switching element. (b) Parallel connection with $C = 0$. (c) Cross connection with $C = 1$. (d) Baseline network for $N = 8$ ($n = 3$).

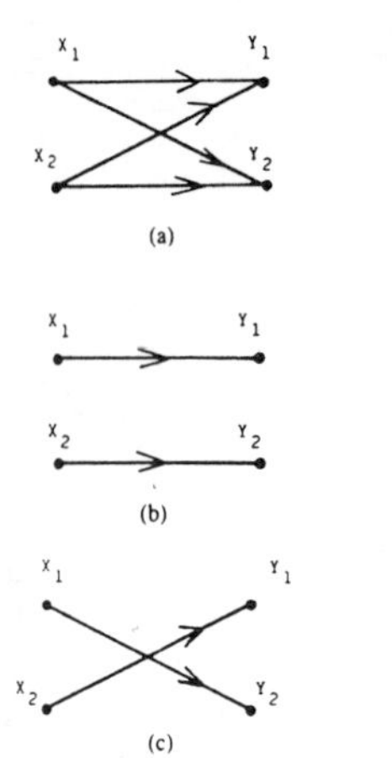

Fig. 3. (a) Graph model of the SE of Fig. 1(a). (b) Graph model of the SE when C S-a-0. (c) Graph model of the SE when C s-a-1.

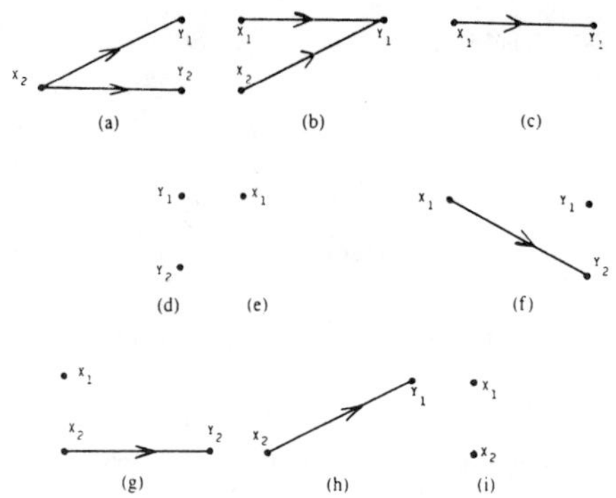

Fig. 4. (a) Graph model of the SE when X_1 s-a-α. (b) Graph model of the SE when Y_2 s-a-α. (c) Graph model of the SE when X_2 s-a-α and Y_2 s-a-β. (d) Graph model of the SE when X_1 s-a-α and X_2 s-a-β. (e) Graph model of the SE when X_2 s-a-α, Y_1 s-a-β, and Y_2 s-a-γ. (f) Graph model of the SE when control s-a-1 and X_2 s-a-α. (g) Graph model when control s-a-0 and Y_1 s-a-α. (h) Graph model when control s-a-1, X_1 s-a-α, and Y_2 s-a-β. (i) Graph model when control s-a-α, Y_1 s-a-β, and Y_2 s-a-γ.

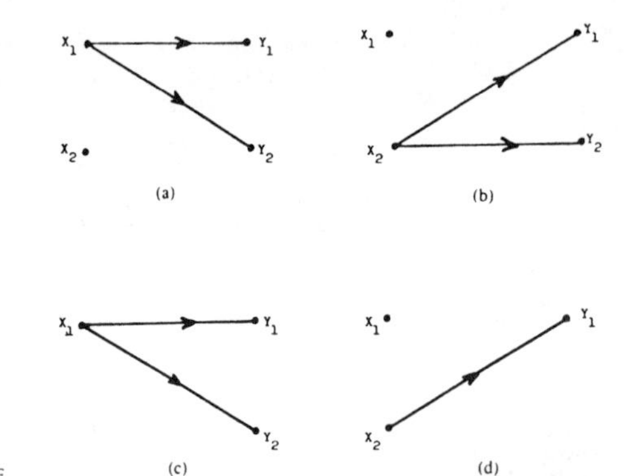

Fig. 5. (a) Graph model of an SE for Omega network when s-a upper broadcast. (b) Graph model when s-a lower broadcast. (c) Graph model when s-a upper broadcast and X_2 s-a-γ. (d) Graph model when s-a lower broadcast and Y_2 s-a-γ.

mum test sequence allows location of single-stage multiple faults and is helpful in removing the faulty SE's, if a single IC chip constitutes a stage. In another recent work [24], design procedures for $(2n - 1)$ stage Benes type network [15] that could tolerate single and multiple faults at the control lines have been outlined. A comprehensive review of testing techniques has appeared in a recent publication [25]. A comparison of various redundant networks has been given in [26]. In this paper, we evaluate the DFA capabilities of n-, single, and m-stage $(m < n)$ MIN's under stuck-at faults at the control lines as well as at the inputs and outputs of the SE's.

III. Fault Model

The model for a 2-input 2-output SE of Fig. 2(a) is shown in Fig. 3(a) [5], [6]. This graph model is based on the connectivity property between the input lines X_1 and X_2 and output lines Y_1 and Y_2 of the SE. Each one of them is assigned a node. As X_1 (X_2) of these [Fig. 2(a)] can be connected to either Y_1 or Y_2 (Y_2 or Y_1), these are shown by the directed edges in the graph model of Fig. 3(a). It is worth mentioning that the graph model of Fig. 3(a) [6] resembles the regular SW Banyan network [27] in appearance. But, instead of using the SE of Fig. 2(a), the Banyan is designed with individually controlled paths and its graph shows the actual link pattern. Similarly, the graph model of cross-bar switches [28] utilizes one edge for each cross-point and a single switch fault removes an edge in the graph. In this way, our model of a 2-input 2-output SE is altogether different from the Banyan network.

Considering the SE of Fig. 2(a) again, it could be in either of the two operating modes, if no fault is present. But when the control line is stuck-at-zero (s-a-0), the graph of **Fig. 3(a)** is reduced to as shown in Fig. 3(b). A similar modification is shown in **Fig.** 3(c) whenever the control line is s-a-1. The stuck-at faults at the input and output lines of an SE have to be treated in a different way. If a line is faulty (s-a-α, $\alpha \in 0, 1$), then the line cannot be used to transmit any data. This is reflected in the graph model by removing the corresponding node and hence eliminating all the edges connecting the node. This is shown in **Fig. 4(a)** for s-a fault at one of the input lines while **Fig.** 4(b) illustrates the model when a fault is present at one of the output lines. Multiple faults are represented by α, β, and $\gamma (\alpha, \beta, \gamma \in 0, 1)$. Models for the two possible double faults are shown in Fig. 4(c) and (d). Faults at the two inputs and outputs lead to the model shown in Fig. 4(e), while faults at all the input and output lines of an SE eliminate all the nodes from the model. If the control line is stuck-at-0 or -1, and one of the input or output lines is also s-a-α, then the graph models are reduced as shown in Fig. 4(f) and (g). Two possible combinations of faults with the control line s-a-α and two of the input/output lines s-a-α are given in Fig. 4(h) and (i). In this way, Figs. 3(b) and (c) and 4(a)–(i) represent a reduced graph model under all single and multiple faults in an SE.

The model of Fig. 3(a) could also be used [5], [6] to represent the upper and lower broadcast used in Omega network [22]. In the case of upper broadcast, the upper input X_1 is sent to both the outputs Y_1 and Y_2 while the lower broadcast provides $Y_1 = Y_2 = X_2$. The graph models of Figs. 3(b) and (c) and 4 remain valid for various faults in the SE for Omega network. The additional faulty situations and the corresponding reduced graphs for SE which lower and upper broadcast, are shown in **Fig.** 5.

In this way, once the SE's have been appropriately modeled, the analysis is similar for SE's with and without broadcast capability. For conciseness of the text, we will be considering the MIN's implemented with SE's having only two valid states, i.e., without having any broadcast facility.

IV. Graph Model of a MIN and Its DFA Capability

Before we go any further, let us define two matrices, the adjacency matrix and the reachability matrix, obtained from the graph model we illustrated earlier. The adjacency matrix A of a graph is the $N \times N$ matrix $[a_{ij}]$ with $a_{ij} = 1$ if there is a connecting link from node i to node j in the graph; otherwise, $a_{ij} = 0$. **Fig. 6(a)** shows the adjacency matrix of the bipartite directed graph [29] of a 2*2 SE of Fig. 2(a). When the control line is s-a-0 or s-a-1, the matrices are as shown in Fig. 6(b) and (c), respectively.

The reachability matrix R of a graph is defined as an $N*N$ matrix $[r_{ij}]$ with $r_{ij} = 1$ if node j is reachable from node i, and $r_{ij} = 0$ otherwise. Here, the information we need from the reachability matrix is whether an input port of MIN can reach an output port or not, and thereafter, it is not necessary to retain any connectivity information from the input nodes to the intermediate nodes [e.g., node numbers 1–16 in **Fig. 7(a)**]. The graph model of a 2*2 SE given in Fig. 3(a) can be used to obtain the graph model for the baseline net-

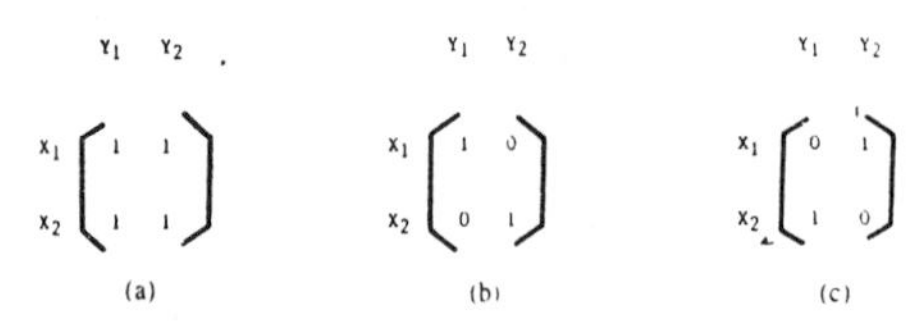

Fig. 6. (a) Adjacency matrix of the SE of Fig. 2(a). (b) Adjacency matrix of the SE of Fig. 2(b). (c) Adjacency matrix of the SE of Fig. 2(c).

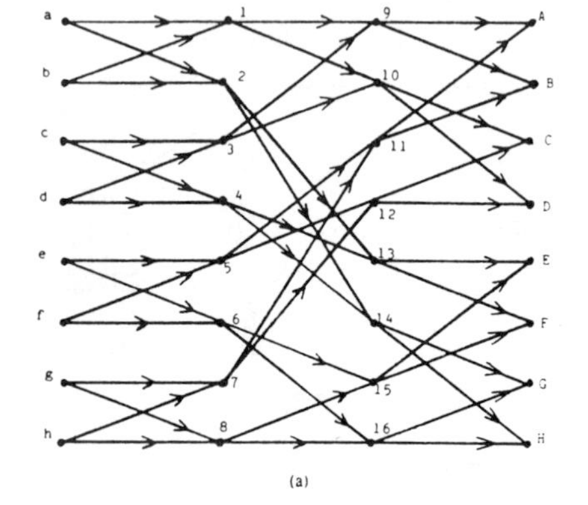

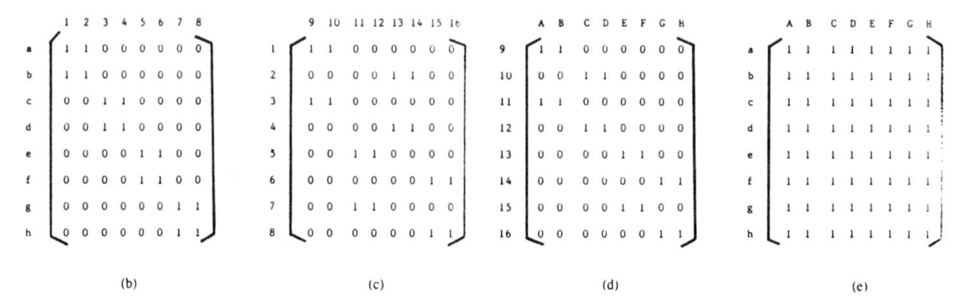

Fig. 7. (a) Graph model for the baseline network of Fig. 2(d). (b) A_1, first-stage adjacency matrix for MIN of Fig. 7(a). (c) A_2, second-stage adjacency matrix for MIN of Fig. 7(a). (d) A_3, third-stage adjacency matrix for MIN of Fig. 7(a). (e) R, reachability matrix for MIN of Fig. 7(a).

work of Fig. 2(d) and is shown in Fig. 7(a). The adjacency matrices of each of the three stages are shown in Fig. 7(b), (c), and (d), respectively.

Corollary 1: In an m-stage MIN, let A_i be the adjacency matrix of the bipartite directed graph of the ith stage representing its connectivity, then R, the reachability matrix from input nodes to output nodes of the MIN, could be given by

$$R = \prod_{i=1}^{m} A_i .$$

Proof: The proof is obvious.

Now we can compute the reachability matrix R by multiplying the adjacency matrices A_1, A_2, and A_3 and the resulting R matrix is shown in Fig. 7(e). In a similar way, the model could be obtained for a MIN with any number of stages and with any number of arbitrarily located faults and the R-matrix could be used to test for the DFA capability.

For example, consider the multiple single faults of **Fig. 8(a)**. As shown in the corresponding graph model of **Fig. 8(b)**, node 11 and 8 edges have been eliminated because of the faults. The adjacency matrices for stage 1, 2, 3 are given in Fig. 8(c), (d), and (e), respectively. The reachability matrix, $R = A_1 \cdot A_2 \cdot A_3$ given in Fig. 8(f) shows that not all input nodes can reach all the output nodes. If multiple passes are allowed (i.e., output can be fed back to the input side), then additional nodes can be accessed in successive passes.

Corollary 2: Let R be the reachability matrix of a MIN, then the reachability in K passes (and hence, its DFA) could be given by

$$R_k = R^k.$$

Proof: The proof is self-explanatory.

DFA is defined as a property that provides each input of the network to be connected to any one of its outputs in a finite number of passes (and hence, any PE to any other PE). The R-matrix of Fig. 8(f) shows that the network of Fig. 8(a) does not allow all input nodes to be connected to each one of the output nodes. But by multiplying A three times, it could be observed [Fig. 8(h)] that three passes through the network of Fig. 8(a) are good enough to provide communication paths from any input to any one of the output nodes. Hence, the DFA property is satisfied. It may be noted that its DFA could not be retained for other combinations of faults. For example, an additional s-a-0 at the control input of SE II in the network of Fig. 8(a) leads to a graph model **of Fig. 9(a)**. Th**is** graph is clearly divided into two unconnected subgraphs, one consisting of the input nodes b and c and the output nodes A, B, and C, and the other containing the rest of the input and output nodes. This could also be seen from the reachability matrix of Fig. 9(b) which could be partitioned as two smaller nonzero submatrices as shown in the figure. This will be true for any such case if R could be partitioned as [29].

$$R = \left[\begin{array}{c|c} R' & 0 \\ \hline 0 & R'' \end{array}\right]$$

where R' and R'' are the two nonzero submatrices. This observation could easily be verified by obtaining the reachability matrices in two, three, and four passes of the network and is shown in Fig. 9(c), (d), and (e), respectively.

As $R_6 = R_5 = R_4$, it is clear that no matter how many times we multiply, we will never be able to get any better result. This means the network no longer possesses the DFA capability.

Theorem 1: In an N-input N-output MIN, the DFA characteristic is ascertained for single or multiple faults at the control line and/or input and output lines of one or more SE's if there exists a reachability matrix R_k $(1 \leq k \leq N)$ in k passes, such that $r_{ij}^k = 1$ for all i,j; $1 \leq i,j \leq N$.

Proof: If a MIN has the DFA property in a minimum of k passes then $R_j = R_{j-1}$ $1 \leq j \leq k$. This means that any input node of the network could reach at least one output node (different than itself as feedback is assumed) after the first pass, and thereafter it will be able to reach at least one additional output node in successive passes. Thus, each input node should be able to reach all N output nodes in at most N passes if the network has the DFA capability. This means that an R_k $(1 \leq k \leq N)$ must exist such that $r_{ij}^k = 1$ for all i,j, $1 \leq i,j \leq N$. When this is satisfied, we will be able to connect any input node to any one of the output nodes in at most k passes. Q.E.D.

Corollary 3: If there are any stuck type faults at the input side or the output side of a MIN (hereinafter called primary inputs and outputs, respectively), then the DFA property cannot be provided.

Proof: If there is a fault at any one of its output lines, then it is obvious that nothing can be transmitted on that line, and hence the corresponding PE cannot receive any data. In terms of the graph model, the output line cannot be accessed by any one of the inputs. Similarly, if there is a fault at any one of the input lines, then it cannot communicate to the output lines, and hence the corresponding PE cannot send any data. Q.E.D.

Theorem 2: The faults in the MIN may destroy its DFA capability if and only if there exist at least two reachability matrices R^l and R^{l-1} for $1 \leq l < N$, such that $r_{ij}^l = r_{ij}^{l-1}$ for all i,j; $1 \leq i,j \leq N$ and at least one $r^l = 0$ for any i,j; $1 \leq i,j \leq N$.

Proof: If some faults in the MIN cause it to lose the DFA property, then there remains at least one pair of input and output nodes i and j that cannot be connected in a finite number of steps and is indicated by an entry of $r^k = 0$. Moreover, if $R^{l-1} = R^l$ then all successive powers of R (i.e., R^{l+2} and so on) will remain equal to R^l. This means that the connection from i to j can never be provided for any number of passes. Q.E.D.

Corollary 4: In an N-input N-output MIN, R^{p+1}, the reachability matrix within $p + 1$ passes is equal to R^p for all $p \geq N$.

Proof: As per Theorem 1, if there exists a path from an input to an output node, then the maximum number of passes required is equal to N. Hence, after N passes, we ought to

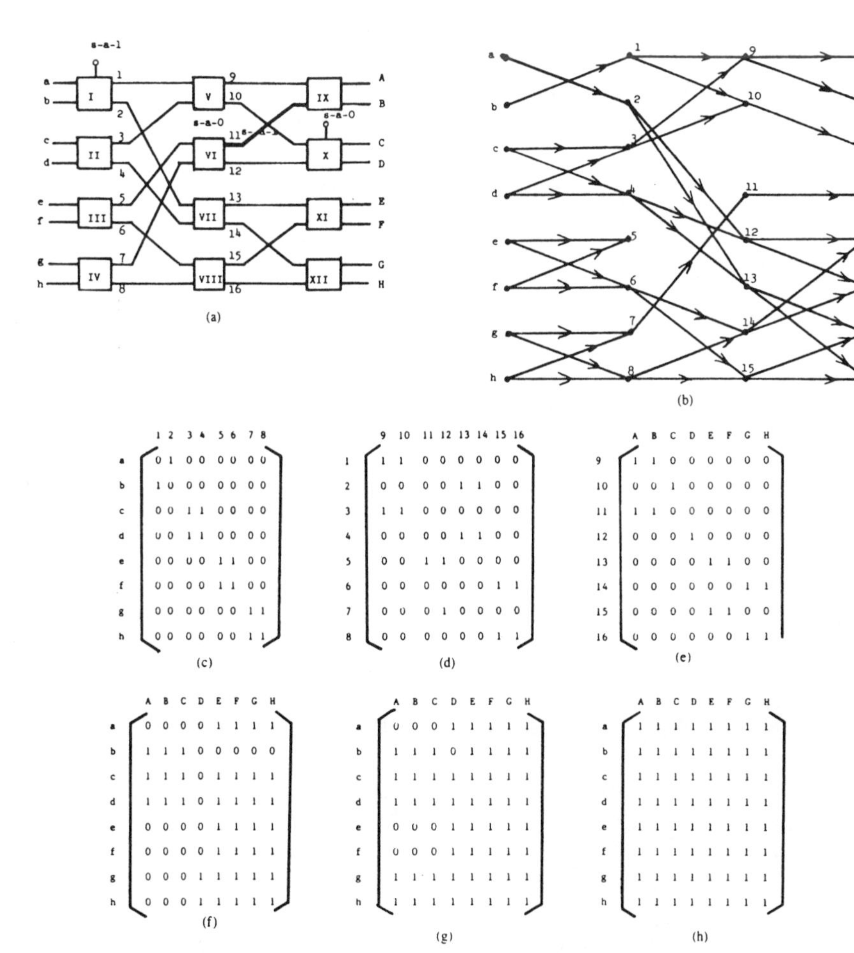

Fig. 8. (a) Baseline network with faults. (b) Graph model of the baseline network of Fig. 8(a). (c) Adjacency matrix in first stage. (d) Adjacency matrix in second stage. (e) Adjacency matrix in third stage. (f) R, the reachability matrix. (g) R^2 for Fig. 8(f). (h) R^3 for Fig. 8(f).

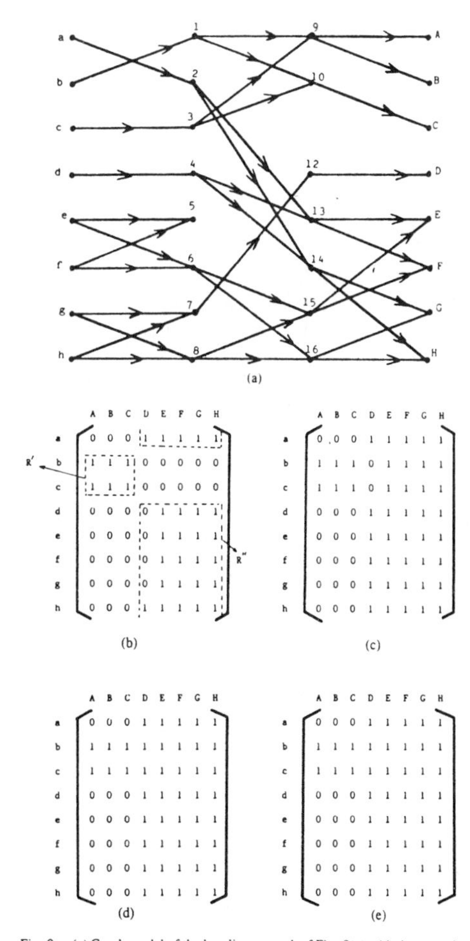

Fig. 9. (a) Graph model of the baseline network of Fig. 8(a) with the control input of SE-II s-a-0. (b) R, the reachability matrix. (c) R^2 for Fig. 9(b). (d) R^3 for Fig. 9(b). (e) $R^4 = R^3$ for Fig. 9(b).

have the corresponding a_{ij}^N element of R_N as 1 and any further pass cannot modify the reachability elements, and hence a_{ij} elements. Hence, R_{N+1} ought to remain the same as R_N. Q.E.D.

V. Distance Matrix and Average Path Length of the MIN's

The distance matrix D [29] of a MIN is defined as an $N \times N$ matrix $[d_{ij}]$ with the entries

$$d_{ij} = \begin{cases} 0 \text{ when } i = j \\ \text{the least } l \text{ (if any) such that } r_{ij}^l = 1 \text{ in } R_l;\ 1 \leq l \leq N \\ \infty \text{ otherwise.} \end{cases}$$

The entries indicate the number of the passes needed for a request to reach the destination. **Fig. 10(a), (b), and (c)** shows the distance matrices of the networks of Figs. 7(a), 8(a), and 9(a), respectively.

Corollary 5: If no conflict in path length is assumed for the random requests, then the average of the minimum path length (called the static average and represented by SAV), can be computed from the distance matrix D as follows:

$$\text{SAV} = \frac{\delta}{N^2} \sum_{i=1}^{N} \sum_{j=1}^{N} d_{ij}$$

where δ represents the delay time needed in each pass.

After we have the entries for the reachability matrices of a MIN, it is easy to set the distance matrix and calculate the average path lengths. Once we have all these results, we may examine the question of whether the network is the best from the SAV viewpoint or not. In other words, could we have a better SAV by changing the link connection patterns or doing something else? This aspect could be easily examined at least for some of networks like single-stage interconnection network (SSIN) [4], [13]. The minimization of path lengths in an SSIN has been covered in [13] and a few rules have been suggested for defining the connection pattern in an SSIN. But the effect of faults on the SAV has never been considered.

A simulation program is implemented to observe the performance of a 16 input 16 output SSIN (one stage of the Omega network [22] configuration) and the modified version as shown **in Fig. 11(a)** and (b). **Fig. 12** indicates the effect of changing the connection pattern when no faults are present. The impact of a single fault on the SAV is shown **in Fig.** 13 and the modified version is seen to provide better performance. The next question to be addressed is whether such conclusions are valid for a general MIN with several stages. Some conclusive simulation results are obtained from the computer program. Under no faults, **Fig. 14 sho**ws the effect of changing connection pattern from a 2-stage 16-input

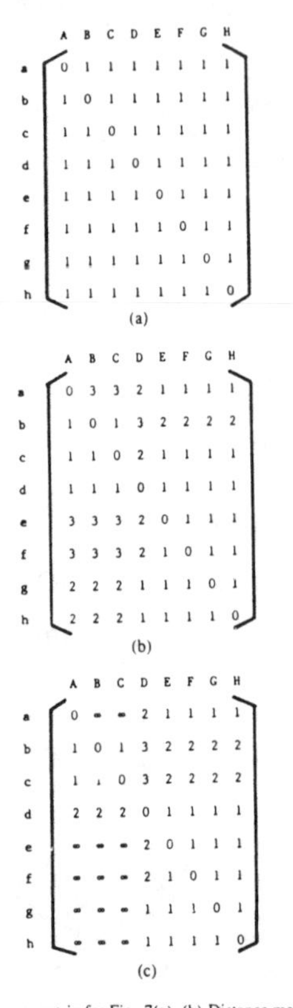

Fig. 10. (a) Distance matrix for Fig. 7(a). (b) Distance matrix for Fig. 8(a). (c) Distance matrix for Fig. 9(a).

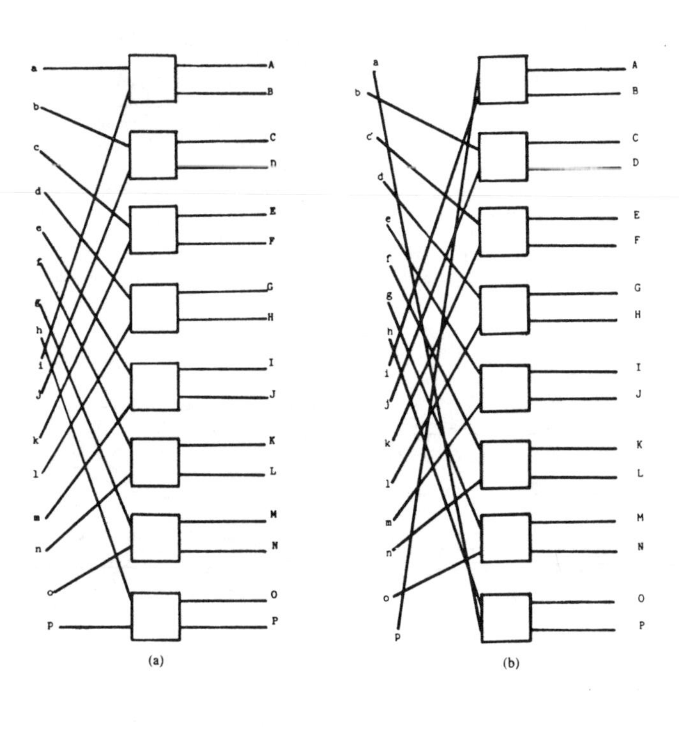

Fig. 11. (a) A 16-input 16-output Omega type SSIN. (b) Modified version of SSIN.

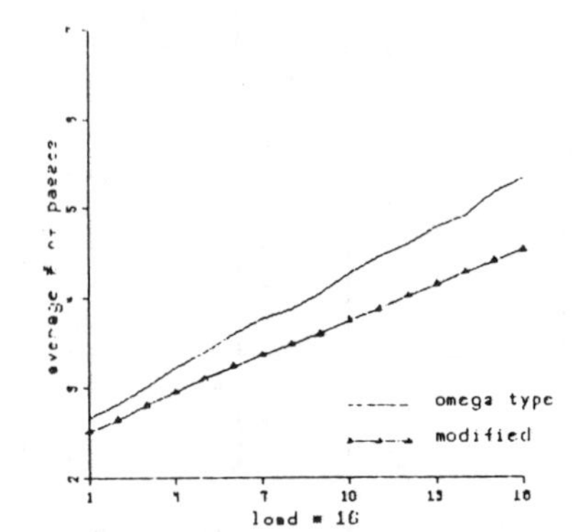

Fig. 12. The effect of changing connection patterns when no fault is presented in 16-input 16-output SSIN.

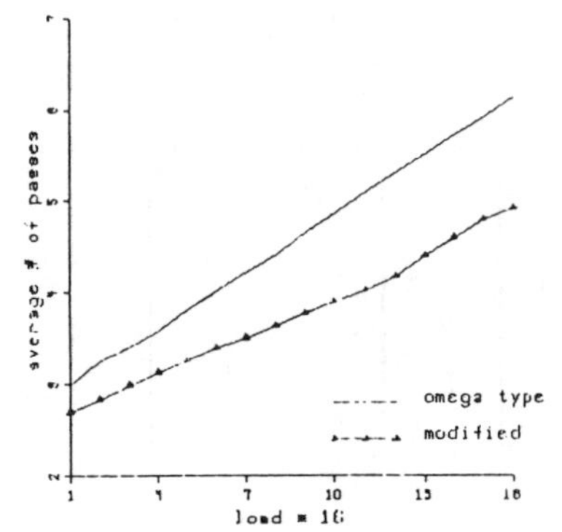

Fig. 13. The effect of changing connection pattern when a single fault occurs in a control line of 16-input 16-output SSIN.

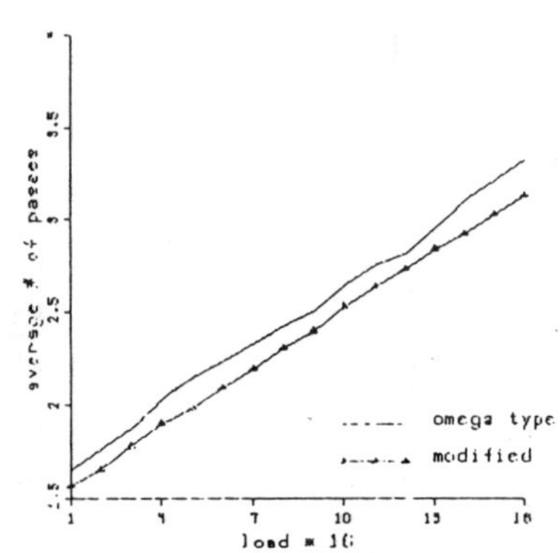

Fig. 14. The effect of changing connection pattern when no fault occurs in the 2-stage 16-input 16-output network.

16-output Omega-type network to a modified version. **Figs. 15, 16, and 17** show the performances of these 2-stage networks when a single fault occurs in control line at the first stage, second stage, or at a link connecting the two stages, respectively.

VI. Optimum Design of a MIN for DFA Capability

The m-stage MIN design with $m = n$ has been widely covered in the literature [2]–[4], [7]–[12]. These networks are designed in such a way that there is a one-to-one correspondence between input and output nodes, i.e., from each input line, there is a unique path to each one of the output lines, and full connectivity requirements are satisfied. Systematic ways of designing these networks have also been described [6]. For $m > n$, alternate paths between each input–output pair provides redundancy and recent work [26] provides a detailed account of their fault-tolerant capabilities. Our main concern is to describe a design methodology for a class of MIN's, with $m < n$, so that the availability and graceful degradation of the parallel computing system could be optimized. In other words, the network could be designed such that the DFA property could be retained for as many faults as possible. Although it may be possible to devise other schemes too, the proposed methodology does provide a certain degree of optimality from the DFA view point.

The design procedure is based on set theory. The two steps are as follows.

A) Partition the N $(= 2^n)$ inputs and outputs into $N/(2^m)$ sets with each set consisting of 2^m inputs and 2^m outputs.

B) Design the m-stages of the network such that 2^m-inputs of one set could be connected to 2^m-outputs of another set.

Thus, each set will consist of m stages, with each stage formed with 2^{m-1} SE's. In this way, each of the subnetworks becomes fully connected network of size 2^m inputs and 2^m outputs. One such example for $m = 2$ and $n = 4$ is shown in **Fig. 18,** wherein 16 inputs and 16 outputs are divided into four subsets with each group consisting of four elements. The input subsets are (a, b, c, d), (e, f, g, h), (i, j, k, l), and (m, n, o, p), while the outputs are divided into (A, B, C, D), (E, F, G, H), (I, J, K, L), and (M, N, O, P) subsets. This satisfies part A) of the design procedure. Part B) is assured by assigning output nodes in such a way that there is no common alphabet between the inputs and outputs of each subset. In other words, the outputs from the input node subset (a, b, c, d) are not connected to the output node subset (A, B, C, D). It must be remembered that for the multiple processor system of Fig. 1, both nodes a and A are logically the same, as the PE works as a link between the input–output pair a and A. Hence, no advantage is gained by connecting nodes a and A through the MIN. Such a design procedure for an arbitrary value of $m < n$ is shown **in Fig. 19.** Before we consider the optimality of our design procedure, three lemmas are in order.

Lemma 1: In the fault-free partition of $m \cdot 2^{m-1}$ SE's connecting 2^m inputs and corresponding outputs, any input can access any one of its outputs in just one pass.

Proof: The design methodology described earlier makes each of the partitioned networks a MIN with 2^m inputs–2^m outputs. Moreover, the R-matrix of each partition could be seen to contain all one elements and hence DFA is satisfied in only one pass. Hence, any of its inputs can access all of its output lines in one pass. Q.E.D.

Lemma 2: In each partitioned group consisting of $m \cdot 2^{m-1}$ SE's and connecting $2m$ inputs and the corresponding output lines, if some or all SE's have single faults (except at the primary input and output lines), then each primary input line can be connected to at least one of the primary output lines.

Proof: From Fig. 3(b) and (c), it is obvious that a fault at the control line of an SE allows each input to be connected to one output. A single fault at one of the inputs of the SE allows the other input to be connected to both the outputs [Fig. 4(a)] and a single fault at the output side of the SE permits both inputs to be connected to the nonfaulty output [Fig. 4(b)]. It may be noted from Figs. 3 and 4 that a fault at an input (output) line is reflected as a fault at the corresponding output (input) line. Hence, simultaneous faults at an input and an output line of the same SE are considered a multiple fault. As the primary input and output lines of the MIN are assumed to be fault free, one input can be connected to at least one output line. This can also be proved using the adjacency matrices for each stage and by using the resultant one-pass reachability matrix. Q.E.D.

Lemma 3: A special case of Lemma 2 arises when the control lines of all the switches are stuck-at-zero or -one; then the graph model for a partitioned network will contain 2^m

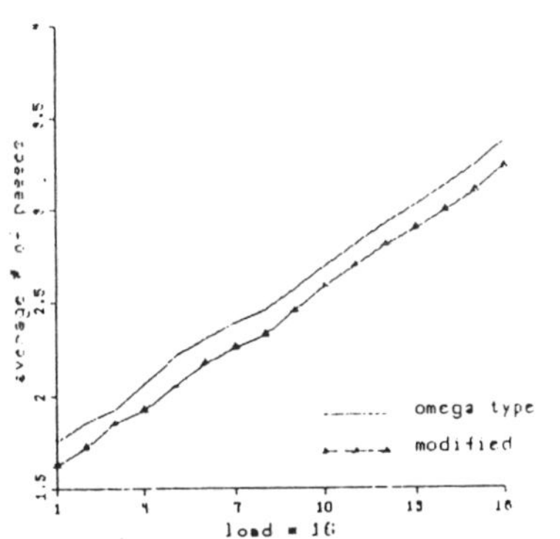

Fig. 15. The effect of changing connection pattern when a single fault occurs in the control line at the first stage of a 2-stage 16-input 16-output network.

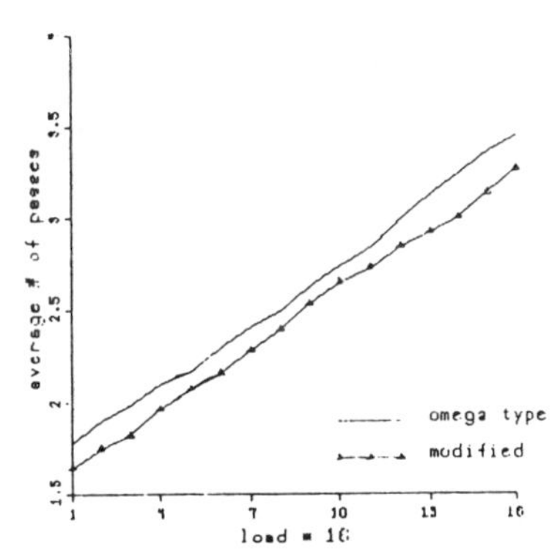

Fig. 16. The effect of changing connection pattern when a single fault occurs in the control line at the second stage of 2-stage 16-input 16-output network.

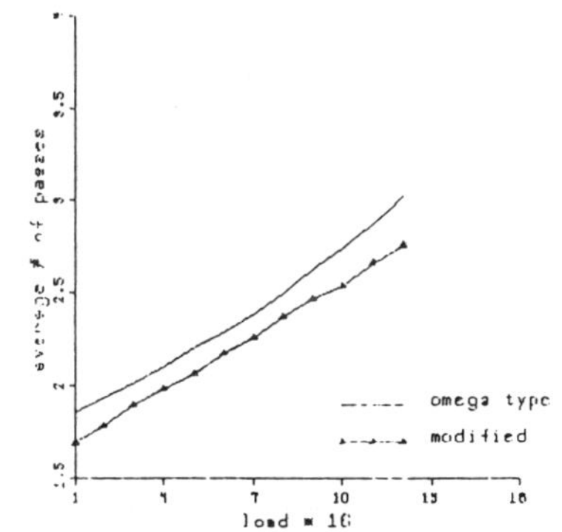

Fig. 17. The effect of changing connection pattern when a single fault occurs in the link between 2 stages of 2-stage 16-input network.

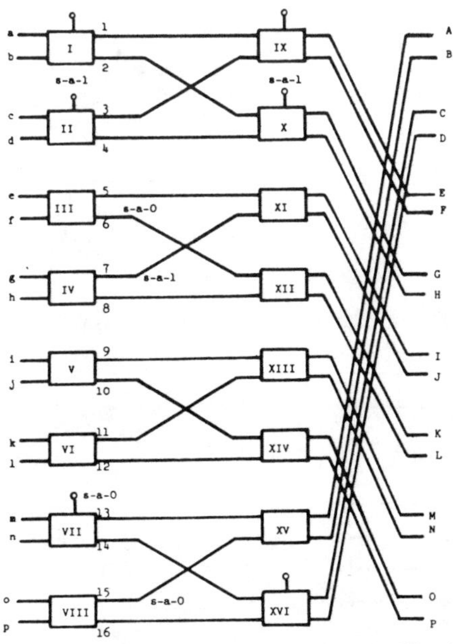

Fig. 18. Two-stage MIN with single stuck type faults at various SE's (only one group of 4-SE's, V, VI, VIII, and XIV, not faulty).

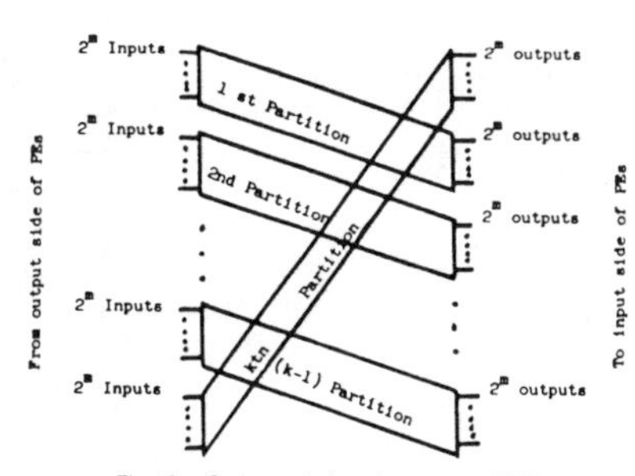

Fig. 19. Optimum design of an m-stage MIN.

unconnected subgraphs, with each one directed from one primary input to only one of the primary outputs.

Proof: From Fig. 3(b) and (c), whenever the control line of an SE is faulty, there is a one-to-one connection. This means only one input is connected to one and only one of the outputs, and hence the adjacency matrix for each stage will have only one nonzero element for each row and each column. Hence, the overall R-matrix will also have only one "1" entry in each row and each column. This would lead to an overall one-to-one connection, with 2^m unconnected subgraphs for each group. Q.E.D.

The optimality of the design in terms of DFA capability is demonstrated by the following theorem.

Theorem 3: If a 2^n input–2^n output MIN is implemented by m-stages ($m < n$) according to the design procedures A) and B), then DFA capability is ensured for multiple single faults provided that $m \cdot 2^{m-1}$ SE's constituting one partition of 2^m inputs and 2^m outputs are assumed fault free. The upper bound for the number of passes required to provide the DFA is (2^{K-1}) where $K = 2^{n-m}$.

Proof: The graph models for possible single faults have been given in Figs. 3(b) and (c) and 4(a) and (b). The connectivity consideration (and the adjacency matrix), is important for the DFA capability. A typical m-stage MIN is shown in Fig. 19. Then the worst case fault could be said to be present when the SE's of all the $(K - 1)$ partitions are faulty. For simplicity (and without losing the generality) let us assume that the kth partition is healthy. The reachability matrix R for each partition could be obtained and the R-matrix for the first $(K - 1)$ partitions would satisfy Lemma 2 while Lemma 1 is applicable to the last partition. As interpreted earlier, in the first $(K - 1)$ partitions, one pass would allow any input to be connected to at least one of the outputs; while the Kth partition would allow any input to be connected to any one of its outputs in one pass.

Let us assume that we started access from one of the inputs of the first partition. Under the assumed faults, the first pass will allow access to at least one of its output nodes and the feedback path through the corresponding PE takes us to a node of the second partition. The second pass takes us to one of the outputs of third partition and so on. In the worst case, it will take $(K - 1)$ passes before we reach the Kth partition. As this group contains all healthy SE's and the corresponding R matrix contains all "1" elements, the next pass can take us to all the output nodes of the Kth partition. Now, the feedback path through PE's takes us back to the first partition and access to all 2^m inputs of the first partition is possible. The next pass provides access to all 2^m outputs of the second partition. If this process is continued, a total of $(K - 1 + K) = (2K - 1)$ passes is required to access any one of the output nodes. Q.E.D.

Lemma 4: The maximum number of tolerable control line faults, under the conditions given in Theorem 3, is $m \cdot (2^{n-1} - 2^{m-1})$.

Proof: For a given m-stage 2^n input–2^n output MIN, the number of SE's in one stage $= 2^{n-1}$, so the total number of SE's $= m \cdot 2^{n-1}$. The number of partitions $= 2^{n-m}$ and the number of SE's in one partition $= m \cdot 2^{m-1}$. According to the statement in Theorem 3, only one partition is assumed to be fault free while control lines in all other partitions may be faulty. Therefore, the maximum number of tolerable control line faults $= m \cdot (2^{n-1} - 2^{m-1})$. Q.E.D.

Lemma 5: The maximum number of tolerable link line faults under the conditions given in Theorem 3 is $(m - 1) \cdot (2^{n-1} - 2^{m-1})$.

Proof: For a given m-stage 2^n-input–2^n output MIN, there are $(m - 1)$ intermediate connections between the stages and 2^n link lines for each stage. Hence, there is a total of $(m - 1)2^n$ intermediate links (except the primary inputs and outputs). Each partition will consist of $(m - 1)2^m$ intermediate links. As only partition is assumed to be fault free and a single link failure per SE is allowed in all other partitions, the total number of tolerable link faults $= 1/2(m - 1)(2^n - 2^m) = (m - 1)(2^{n-1} - 2^{m-1})$. Q.E.D.

Since the number of faults tolerable in our design is fairly close to the number of SE's in the network, we could claim that our methodology provides a close to optimal solution. As an example, the graph model of the MIN shown in Fig. 18 which contains several single faults is provided in Fig. 20. The reachability matrix in the first pass is given in Fig. 21. Under the random request loading and close-to-finish arbitration [30], the computer simulation mentioned earlier provides the performances of the MIN of Fig. 18 and is shown in Fig. 22. The average path lengths are computed for no fault case and when a single fault is present at either the control line of either stage or at a link connecting the two stages. The resulting curves indicate that a single fault has a very marginal increase on the average time delay and could be considered to be a very valuable simulation result to support our claim that our design is good and close to optimal from a fault-tolerance viewpoint. It has not been possible to compare our design methodology to others, as, to the best of our knowledge, there does not exist any such technique in the literature.

Corollary 6: The restriction imposed by Theorem 3 is not a necessary condition for the DFA property.

Proof: Theorem 3 is sufficient for ensuring DFA, but not necessary. The faults may be such that the R-matrix elements may contain arbitrary 1's and the multiple pass (hence, multiplication of the R matrix to itself) may lead to an R' matrix with all "1" entries. Q.E.D.

One such exception is illustrated in Fig. 23, which can be said to possess a high degree of fault tolerance. Thus, our design procedure is very useful in implementing a MIN with the DFA capability in the presence of faults. Theorem 3 identifies the set of single faults at the SE's so that it is possible to ascertain the DFA characteristic even without obtaining a graph model and without performing a lot of connectivity and reachability computation.

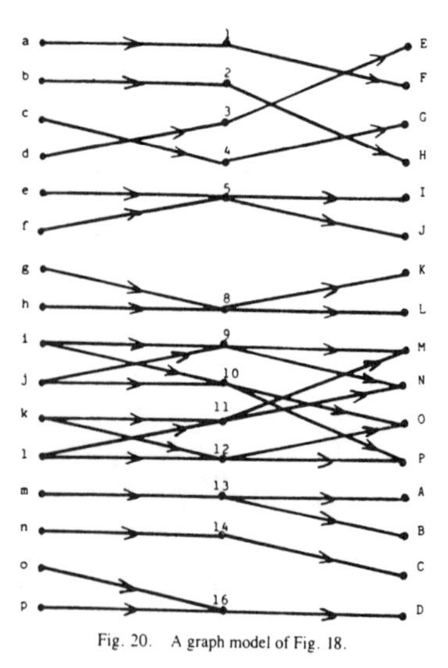

Fig. 20. A graph model of Fig. 18.

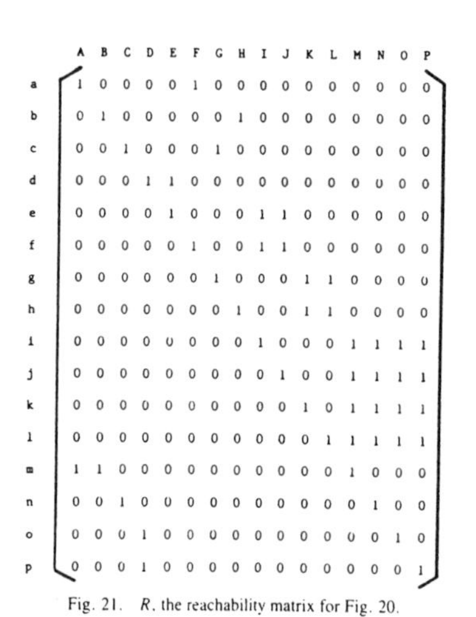

	A	B	C	D	E	F	G	H	I	J	K	L	M	N	O	P
a	1	0	0	0	0	1	0	0	0	0	0	0	0	0	0	0
b	0	1	0	0	0	0	0	1	0	0	0	0	0	0	0	0
c	0	0	1	0	0	0	1	0	0	0	0	0	0	0	0	0
d	0	0	0	1	1	0	0	0	0	0	0	0	0	0	0	0
e	0	0	0	0	1	0	0	0	1	1	0	0	0	0	0	0
f	0	0	0	0	0	1	0	0	1	1	0	0	0	0	0	0
g	0	0	0	0	0	0	1	0	0	0	1	1	0	0	0	0
h	0	0	0	0	0	0	0	1	0	0	1	1	0	0	0	0
i	0	0	0	0	0	0	0	0	1	0	0	0	1	1	1	1
j	0	0	0	0	0	0	0	0	0	1	0	0	1	1	1	1
k	0	0	0	0	0	0	0	0	0	0	1	0	1	1	1	1
l	0	0	0	0	0	0	0	0	0	0	0	1	1	1	1	1
m	1	1	0	0	0	0	0	0	0	0	0	0	1	0	0	0
n	0	0	1	0	0	0	0	0	0	0	0	0	0	1	0	0
o	0	0	0	1	0	0	0	0	0	0	0	0	0	0	1	0
p	0	0	0	1	0	0	0	0	0	0	0	0	0	0	0	1

Fig. 21. R, the reachability matrix for Fig. 20.

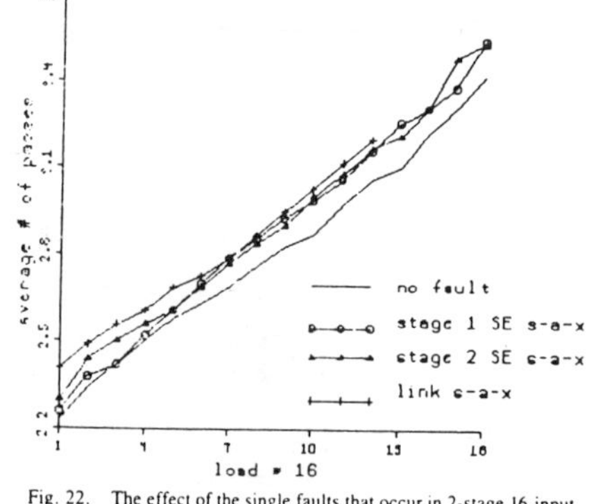

Fig. 22. The effect of the single faults that occur in 2-stage 16-input 16-output network.

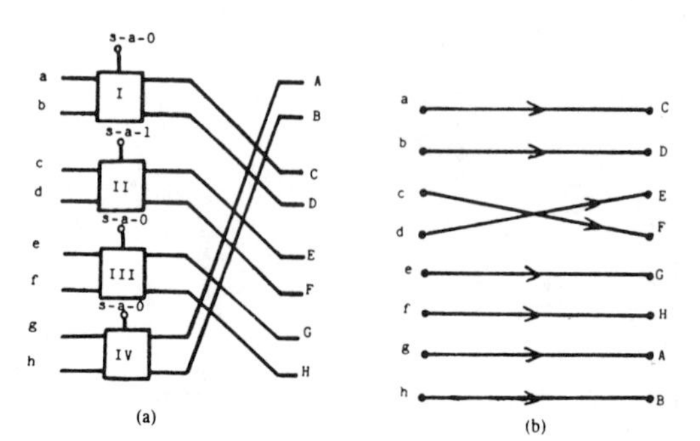

Fig. 23. (a) One-stage MIN with 4 faults. (b) Graph model for Fig. 23(a).

VII. Concluding Remarks

The fault model of an SE is used to model MIN's, and an adjacency matrix and reachability matrices are employed to provide a systematic procedure of testing the network's DFA capability. The versatility of the fault model enables us to test the network under multiple stuck type faults both at the control lines and the input–output lines of the SE's. In addition, the R-matrix in successive passes is also useful in computing the average path length. The network design procedure enables the MIN to posses maximum fault tolerance, and hence in turn optimizes the availability of the system. The selection of connection pattern is seen to be a key issue in minimizing the path lengths in both the SSIN's and MIN's which could also be used as an index for the performance. If we consider the fault tolerance as well as the minimization of the path lengths simultaneously, then the optimization problem of the MIN becomes extremely complex and we hope to present additional results in the near future.

References

[1] D. P. Agrawal and T. Y. Feng, "A study of communication processor systems," Rome Air Devel. Center, Tech. Rep. RADC-TR-79-310, Dec. 1979.

[2] C. L. Wu and T. Y. Feng, "On a class of multistage interconnection networks," *IEEE Trans. Comput.*, vol. C-29, pp. 694–702, Aug. 1980.

[3] H. J. Siegel, "The theory underlying the partitioning of permutation networks," *IEEE Trans. Comput.*, vol. C-29, pp. 791–801, Sept. 1980.

[4] D. K. Pradhan and K. L. Kodandapani, "A uniform representation of single- and multi-stage interconnection networks used in SIMD machines," *IEEE Trans. Comput.*, vol. C-29, pp. 777–791, Sept. 1980.

[5] D. P. Agrawal, "On graph theoretic approach to n- and $(2n - 1)$-stage interconnection networks," in *Proc. 19th Annu. Allerton Conf. Commun. Contr. Comput.*, Sept. 30–Oct. 2, 1981, pp. 559–568.

[6] ——, "Graph theoretic analysis and design of multistage interconnection networks," *IEEE Trans. Comput.*, vol. C-32, pp. 637–648, July 1983.

[7] L. N. Bhuyan and D. P. Agrawal, "Design and performance of a general class of interconnection networks," in *Proc. 1982 Int. Conf. Parallel Processing*, Aug. 24–27, 1982, pp. 2–9; also in *IEEE Trans. Comput.*, vol. C-32, pp. 1081–1090, Dec. 1983.

[8] D. P. Agrawal and S. C. Kim, "On non-equivalent multistage interconnection networks," in *Proc. 10th Int. Conf. Parallel Processing*, Aug. 25–28, 1981, pp. 234–237.

[9] M. A. Abidi and D. P. Agrawal, "On conflict-free permutations in multistage interconnection network," *J. Digital Syst.*, vol. V, no. 2, pp. 115–134, Summer 1980.

[10] ——, "Two single pass permutations in multistage interconnection networks," in *Proc. 1980 Conf. Inform. Sci. Syst.*, Mar. 26–28, 1980, pp. 516–522.

[11] D. A. Pauda, D. J. Kuck, and D. H. Lawrie, "High-speed multiprocessors and compilation techniques," *IEEE Trans. Comput.*, vol. C-29, pp. 763–776, Sept. 1980.

[12] C. L. Wu and T. Y. Feng, "The reverse-exchange interconnection network," *IEEE Trans. Comput.*, vol. C-29, pp. 801–811, Sept. 1980.

[13] J. E. Wirsching and T. Kishi, "Minimization of path lengths in single stage connection networks," in *Proc. 3rd Int. Conf. Distrib. Comput. Syst.*, Oct. 18–22, 1982, pp. 563–571.

[14] C. L. Wu, T. Feng, and M. Lin, "Star: A local network for real-time management of imagery data," *IEEE Trans. Comput.*, vol. C-31, pp. 923–933, Oct. 1982.

[15] D. C. Operferman and N. T. Tsao-Wu, "On a class of rearrangeable switching networks, Part II: Enumeration studies and fault diagnosis," *Bell Syst. Tech. J.*, pp. 1601–1618, May/June 1971.

[16] J. P. Shen and J. P. Hayes, "Fault tolerance of a class of connecting networks," in *Proc. 7th Symp. Comput. Arch.*, La Baule, France, May 6–8, 1980, pp. 61–71.

[17] J. P. Shen, "Fault tolerance analysis of several interconnection networks," in *Proc. 1982 Int. Conf. Parallel Processing*, Aug. 24–27, 1982, pp. 102–112.

[18] J. J. Narraway and K. M. So, "Fault diagnosis in inter-processor switching networks," in *Proc. Int. Conf. Circ. Comput.*, Oct. 1–3, 1980, pp. 750–753.

[19] C. L. Wu and T. Y. Feng, "Fault-diagnosis for a class of multistage interconnection networks," in *Proc. 1979 Int. Conf. Parallel Processing*, Aug. 21–24, 1979, pp. 269–278.

[20] D. P. Siewioriek *et al.*, "A case study of C*mmp, Cm*, and C*vmp: Part I—Experiences with fault tolerance in multiprocessor systems," *Proc. IEEE*, vol. 66, pp. 1178–1200, Oct. 1978.

[21] D. P. Agrawal, "Automated testing of computer networks," in *Proc. 1980 Int. Conf. Circ. Comput.*, Oct. 1–3, 1980, pp. 717–720.

[22] D. K. Lawrie, "Access and alignment of data in an array processor," *IEEE Trans. Comput.*, vol. C-24, pp. 1145–1155, Dec. 1975.

[23] K. M. Falavarianai and D. K. Pradhan, "Fault-diagnosis of parallel processor interconnection networks," in *Proc. 1981 Fault Tolerant Comput Symp.*, June 1981.

[24] S. Sowrirajan and S. M. Reddy, "A design for fault-tolerant full connection networks," *1980 Conf. Inform. Sci. Syst.*, pp. 536–540.

[25] D. P. Agrawal, "Testing and fault-tolerance of multistage interconnection networks," *IEEE Computer*, vol. 15, pp. 41–53, Apr. 1982.

[26] D. P. Agrawal and D. Kaur, "Fault tolerant capabilities of redundant multistage interconnection networks," in *Proc. Real-time Syst. Symp.* Arlington, VA, Dec. 6–8, 1983, pp. 119–127.

[27] L. R. Goke and G. J. Lipovski, "Banyan networks for partitioning of the multiprocessor systems," in *Proc. 1st Annu. Symp. Comput. Arch.*, Dec 1973, pp. 21–28.

[28] G. M. Masson, G. C. Gingher, and S. Nakamura, "A sampler of circuit switching networks," *IEEE Computer*, vol. 12, pp. 32–48, June 1979.

[29] F. Harary, *Graph Theory*. Reading, MA: Addison-Wesley, 1972.

[30] P. Y. Chen, P. C. Yew, and D. Lawrie, "Performance of packet switching in buffered single-stage shuffle-exchange networks," in *Proc. 3rd Int. Conf. Distrib. Comput. Syst.*, Oct. 18–22, 1982, pp. 622–627.

Dharma P. Agrawal (M'74–SM'79) was born in Balod, India, on April 12, 1945. He received the B.E. degree in electrical engineering from the Ravishankar University, Raipur, India, in 1966, the M.E. (hons.) degree in electronics and communication engineering from the University of Roorkee, Roorkee, India, in 1968, and the D. Sc. Tech. degree from the Federal Institute of Technology, Lausanne, Switzerland, in 1975.

He has been a Member of the Faculty in the M.N. Regional Engineering College, Alahabad, India, the University of Roorkee, the Federal Institute of Technology, Southern Methodist University, Dallas, TX, and Wayne State University, Detroit, MI. Currently, he is with the North Carolina State University, Raleigh, as a Professor in the Department of Electrical and Computer Engineering. His research interests include parallel/distributed processing, computer architecture, fault tolerance, and information retrieval.

Dr. Agrawal has served as a Referee for various reputed journals and international conferences. He was a Member of Program Committees for the COMPCON Fall of 1979, the Sixth IEEE Symposium on Computer Arithmetic, and Seventh Symposium on Computer Arithmetic. During the years 1980–1983 he served as a Member and the Secretary of the IEEE Computer Society Publications Board, and has been awarded the Society's "Certificate of Appreciation" for his services. Currently, he is the Chairman of the Rules of Practice Committee of the Publications Board. He was the Treasurer of the IEEE-CS Technical Committee on Computer Architecture and as the Program Chairman for the Thirteenth International Symposium on Computer Architecture held in June 1984. He has been a Co-Guest Editor of the *IEEE Transactions on Computers* Special Issue on Computer Arithmetic and is an Editor of the new *Journal on Parallel and Distributed Computing* published by Academic Press. He is also a Distinguished Visitor of the IEEE Computer Society. He is listed in Who's Who in the Midwest, the 1981 Outstanding Young Men of America, and in the Directory of World Researchers' 1980's subjects published by the International Technical Information Institute, Tokyo, Japan. He is a member of the ACM, SIAM, and Sigma Xi.

Ja-Song Leu (S'84) was born in Yun-Lin, Taiwan, on October 15, 1957. He received the B.E. degree in EECS in 1980 from Chung-Yuan College, Taiwan, the M.S. degree in computer studies from North Carolina State University, Raleigh, in 1983, and is now a Research Assistant working towards the Ph.D. degree in the Department of Electrical and Computer Engineering, North Carolina State University. His current interests include parallel/distributed processing and computer communication.

Reprinted from *IEEE Journal on Selected Areas in Communications*, Volume SAC-4, Number 7, October 1986, pages 1168–1173.

Current Telecommunication Network Reliability Models: A Critical Assessment

JOHN D. SPRAGINS, SENIOR MEMBER, IEEE, JAMES C. SINCLAIR, STUDENT MEMBER, IEEE, YONG J. KANG, AND HOSSEIN JAFARI

***Abstract*—This paper is a critique of major weaknesses of current techniques for analyzing the reliability of computer communication networks. Several of the most important factors influencing the reliability seen by users of such systems have been almost completely ignored or modeled in a rudimentary fashion. Ignoring such factors may be partially justified by the inherent complexity of the problems being studied, which often precludes exact computations being feasible for networks of any size, even with simplifying assumptions. Nevertheless, the results of models which ignore such factors are questionable even if they are computable. The focus of models of such networks should be on the effects of system malfunctions, degraded performance, excessive delays, etc., on system users. New approximate modeling techniques appear to be the most realistic approach to studying such problems.**

Introduction

An unfortunate percentage of the papers on telecommunication network reliability and/or availability are described by a paper written by Rosanoff a few years ago [1], [2], "Rigorous argument from inapplicable assumptions produces the world's most durable nonsense." Although this criticism may seem harsh, papers based on models ignoring some of the most important factors influencing reliability or availability do not produce answers which can be trusted. If the analysis is, nonetheless, rigorous, at least a few such papers tend to be durable in the sense that they become standard references for other workers. This does not mean that they yield the right answers, however.

Modern telecommunication networks are based on merging the technologies of the computer and communications fields. As such, they use a variety of technologies, and are subject to a wide variety of types of failures. The user of such a network normally does not know, or care, whether network unavailability results from hardware malfunctions, software failures, protocol deadlocks, excessive network congestion, error recovery delays, or other phenomena.

Manuscript received December 9, 1985; revised May 12, 1986. This paper was presented in part at IEEE GLOBECOM '84, Atlanta, GA, November 1984, and at the '85 Pacific Computer Communications Symposium, Seoul, South Korea, October 1985.

J. D. Spragins, J. C. Sinclair, and Y. J. Kang are with the Department of Electrical and Computer Engineering, Clemson University, Clemson, SC 29631.

H. Jafari was with the Department of Electrical and Computer Engineering, Clemson University, Clemson, SC 29631. He is now with AT&T Bell Laboratories, Holmdel, NJ 07733.

IEEE Log Number 8609806.

Some standard definitions of reliability or availability, such as those based on the probability that all components of a system are operational at a given time, can be dismissed as irrelevant when studying large telecommunication networks. Many telecommunication networks are so large that the probability they are operational according to this criterion may be very nearly zero; at least one item of equipment may be down essentially all of the time. The typical user, however, does not see this unless he or she happens to be the unlucky person whose equipment fails; the system may still operate perfectly from this user's point of view. A more meaningful criterion is one based on the reliability seen by typical system users. The reliability apparent to system operators is another valid, but distinct, criterion. (Since system operators commonly consider systems down only after failures have been reported to them, and may not hear of short self-clearing outages, their estimates of reliability are often higher than the values seen by users.)

Difficulties in reliability modeling are not unique to telecommunication networks, as is indicated by a paper by Harris and O'Connor [3]. "Reliability engineers and their associates in logistics and statistics have worked on the reliability prediction problem for the past 40 years or so. Unfortunately, the results have been disappointing. Systems all too often achieve reliabilities markedly different (usually lower) than those predicted. The credibility of reliability engineers has inevitably suffered as they have argued over the merits and failings of different methods apparently without managing to make any significant improvement in the precision of their predictions."

Even techniques for estimating reliability published in standard handbooks exhibit extreme variability in their results. This is pointed out by a recent paper comparing the results of calculations made according to formulas in five standard handbooks (published in the U.S., Great Britain, France and Japan) plus reliability predictions for the same equipment from two suppliers [4]. The predicted failure rates for a large memory board ranged from 0.6 percent per year to 3713 percent per year, a factor of over 6000 in variability!

Since essentially all telecommunication systems are repairable systems, they fall into a class of systems for which reliability theory techniques are surprisingly incomplete. The general neglect of this area in the reliability theory literature is described in a paper by Ascher and

Feingold [5]. "It is empirically obvious that most "real world" systems are designed to be repairable. . . . However, with the partial exception of probabilistic cost models and reliability growth models, the entire area of repairable systems has been seriously neglected in the reliability literature. For example, when one of us remarked to a colleague that, to a first approximation, almost nobody in the reliability field understands anything about repairable systems, he replied that that assessment would hold as a second approximation as well!"

Predicting the reliability of telecommunication networks is especially difficult, with virtually all problems of interest in this field classified as NP-hard problems [6]–[11], even with the simplifying assumptions commonly made. This implies that computational requirements for general algorithms appear to inevitably explode, for at least some cases, as network size grows. Many current day networks are well beyond the size where precise models appear to be feasible. Some justification for using unrealistic assumptions may be obtained from this fact; if reasonably simple models cannot be solved, using more elaborate models may be questionable. On the other hand, models which ignore the dominant causes of system down time are questionable even if they yield computable formulas.

Important Research Areas

Some important areas where future research on the reliability of telecommunication networks is needed, include the following:

1) development of more efficient computational algorithms;

2) exploiting network routing algorithms and similar protocols to help develop more realistic, and possibly simpler, reliability models;

3) modeling the effects of statistical dependencies among failures of different network components;

4) development of models reflecting the impact of several factors which are difficult to quantify but have major impact on network reliability;

5) developing better models for software reliability/availability and measuring their effect on overall network reliability or availability;

6) adequately modeling nodal reliabilities or availabilities including effects of both hardware and software failures;

7) finding techniques for including the impact of protocol related factors such as deadlocks, routing, flow control, congestion and error recovery delays on network performance and perceived reliabilities and availabilities;

8) development of unified reliability and performance models.

More Efficient Computational Algorithms

The fact that a problem is NP-hard does not preclude major improvements in computational efficiency resulting from new algorithms. An example is a radically improved algorithm recently developed by Shier and Whited [12] which has reduced computational time for analyzing the reliability of a network in the form of a dodecahedron (20 nodes, 30 edges)—apparently the most complex network yet analyzed precisely (with simplifying assumptions)—from a previous best time of 43 min (on an IBM 4341) [J. S. Provan, personal communication] to a 4 min (on an IBM 3081) [D. R. Shier, personal communication]. The new algorithm also provides more comprehensive results; the 43-min calculation referenced involved only computing reliability between a specified source node and a specified destination node, while the 4-min calculation involved computing reliability between the same source node and each of the other 19 nodes in the network. Further, the new algorithm produces an iterative series of approximations to each of the reliabilities computed, with each approximation a lower bound to reliability and monotonic convergence to the exact value. In each case studied so far, terminating the iteration after times far smaller than required for precise solution would produce reliability estimates virtually indistinguishable from the precise values.

Exploiting Network Protocols to Simplify Models

Recent work by the authors and their students indicates that computation of the reliability of some real networks may be far simpler than the NP-hard theory would indicate. SNA is a good example; since a maximum of eight possible routes between a given source and destination pair are defined at system generation time [13], there is no exponential increase in complexity with network size. The routing algorithms used with other types of networks seem to allow reasonable computation algorithms, also. Possible problems with this approach could, however, be indicated by recent research by Provan and Ball [14], who have shown that polynomial time algorithms for computing network reliability from previously determined network paths do not exist for general networks although polynomial time algorithms based on network cuts do exist. This indicates it may be difficult to exploit network routing algorithms, which normally search for paths, in order to simplify network reliability calculations for general network configurations. Still more recent research by Shier *et al.* [15] has indicated that polynomial time algorithms based on paths do exist for planar networks, though. Since essentially all commercial networks are planar, this class should be broad enough to cover such networks.

The only network studied in this manner in any detail so far is TELENET, for which a polynomial time algorithm for computation of a lower bound on network reliability has been obtained by neglecting some fine details of the routing algorithm [16]. Good approximate models which realistically reflect the factors most important in determining network reliability or availability are another approach to general networks and should be much more realistic than exact analyses based on unrealistic assumptions. The authors and their colleagues are among several groups currently looking into this.

Statistical Dependencies Between Failures

The first named author has been studying statistical dependencies between failures of different components for over ten years [17]–[29]. His original motivation was the discovery, by his colleagues when he worked for a major computer manufacturer, that dependent failures of different components of telecommunication networks were often the dominant factor causing network downtime.

Dependent failures in telecommunication networks commonly arise from use of common equipment, sometimes in portions of networks that appear at first glance to be disjoint. When common equipment fails, this may cause failures of all lines using this equipment. An example occurs in networks using telephone facilities. Although most models of such networks are based on an assumption that backup communication lines and the lines being backed up fail independently, in almost all such cases the lines share the same multipair cables at least between each end point and the closest telephone switching office, with other common equipment in switching offices and elsewhere. It is not normally economically feasible for telephone companies to provide completely independent facilities. Other networks are even more strongly impacted by dependent failures, with tactical communication networks among the most affected. Such networks often use large numbers of radio links in limited geographical areas, with various links affected by common weather disturbances, interference effects caused by their proximity to one another, common multipath effects (when a wave of aircraft come over the horizon or similar phenomena occur), as well as by possible enemy actions such as jamming.

A variety of approaches to this problem are discussed in the references listed. Most of the work has concentrated on simply finding techniques for adequately representing the effects of dependent failures, regardless of computability. The most successful techniques are discussed in [24]–[26]. Although these methods allow representation of completely general types of failure dependencies, they require exponentially growing numbers of parameters and extremely long computational times for networks of any size. More recent work [26]–[28] attacks the computability problems; approximations to the general models have proved to be computable in times close to those required by some of the faster independent failure models developed to date. The accuracy of these approximations is currently being studied.

Recent work by Lam and Li [30] has focused on an alternative approach to this problem. Their "event network model" incorporates the effects of events which explicitly model the mechanisms causing dependent failures into the computation of reliability. The relationship between their model and those in [17]–[29] is currently being investigated.

Factors that are Difficult to Quantify

Several difficult to quantify factors can have major impact on reliability. A simple example is provided by a leased circuit in the United Kingdom discussed in [31]. This leased circuit "showed an error rate varying over the range 10^{-4}–10^{-7}. The daily average varied with the day of the week from 2×10^{-5} on Wednesdays to 3×10^{-7} on Thursdays for no obvious reason." Temporal variations in line error rate or similar reliability measures of this magnitude can have major impact on telecommunication network reliability, but it is extremely rare to find reliability models incorporating such factors. No satisfactory method for quantifying such variations, aside from measurements taken on individual network components, is available.

A variety of factors in this class are mentioned by Giloth and Frantzen [32]. These factors include "environment, software errors, craft errors, program retrofits, planned downtime, system size, system growth, etc." Similarly, the authors have seen claims of as much as a factor of three or four variation in average down time depending on the identity of the system repairperson. Recent studies of the reliability of certain types of telecommunication systems, in particular stored-program controlled switching systems, indicate that close to half of the system outages currently being experienced in such systems are due to procedural errors, such as taking incorrect actions during repair, growth and update procedures [33]–[34]. All factors of this type are extremely difficult to quantify, but models ignoring them are suspect. Some models which attempt to include such human performance factors are given in [34]–[35].

An interesting approach to including additional factors important in analyzing reliability is given in a recent paper by Martella, Pernici and Schreiber [36]. This paper involves analysis of the reliability of a distributed transaction system including such parameters as the availability of data items needed by the read-set and the write-set for a transaction, state-dependent and workload-dependent transition rates between different states, and concurrency control for the database process. Their work constitutes a promising beginning for studying such factors, but numerous issues are still open.

Software Reliability

In a large percentage of today's telecommunication networks, software failures cause more down time than do hardware failures. Although it is dangerous to cite numbers as typical, one example of the frequency of software failures is a command and control software system which experienced 25 critical failures during 1000 h of CPU operation, corresponding to a mean time between critical failures of 40 h [37]. Software reliability, unfortunately, is far less successfully modeled with the techniques currently known than is hardware reliability.

Software reliability theory is in a very early development stage. Since a large percentage of the persons developing software reliability techniques have previously worked on hardware reliability, it is not surprising that most software reliability models are derivatives of hardware models. There are important differences between hardware and software which imply that distinct techniques are needed, however.

References [38]–[41] are recommended for additional information on software reliability models. A few papers discussing combined hardware/software models are [42]–[47], but the results obtained are extremely limited. The inadequacy of such models is indicated by [41], which includes the results of a test of the applicability of seven standard software reliability models to a major C^3I system. Model success percentages, as measured by chi-square tests, ranged from 0 percent to 53 percent!

Protocol Performance Factors

Excessive delays due to the interaction of network congestion with network protocols can, in many instances, have as serious an impact on perceived system reliability as actual outages. Although some models for evaluating customer acceptance of telephone systems have been developed [49], no models which really examine the impact of delays on the data communications user have been located by the authors. The impact of delays on the user is, in most situations, a very nonlinear function of the actual delays. In fact, there are some reasonably well-defined psychological thresholds on delay, such that delays less than the threshold value are acceptable but delays greater than the threshold become very disturbing [50]. A typical threshold is one approximately equal to two seconds for intensive interaction with a teleprocessing system; most persons are not able to maintain the same intensity of interaction with delays greater than two seconds that they can with shorter delays. The delays which are tolerable also vary with the type of task; after task steps resulting in "psychological closure," such as dialing the last digit in a telephone number, more delay is acceptable than after task steps without psychological closure. Including such factors in telecommunications network reliability models, however, is far beyond the current state of the art.

The most spectacular congestion-related factors experienced in telecommunication systems are system deadlocks, where the system comes to an abrupt halt because of some protocol induced state [51]–[53]. Although it should be impossible to reach deadlock states with well-designed communications protocols, they do in fact occur. New protocol design techniques which avoid deadlocks are being developed, however.

Excessive delays due to network congestion are more common than deadlocks. Such delays are normally modeled by queueing theory techniques [54]–[55], but such techniques become extremely complex for large networks. Typically, only mean delays are studied, if even they can be obtained. The probabilities of delays exceeding thresholds, such as should really be computed if the impact on the user is to be studied, are seldom available.

Other papers which treat problems similar to that proposed here from a reliability viewpoint are [56]–[60]. The models in these papers are very limited, however.

Unified Reliability and Performance Models

The relationship between component failures and performance degradations in the network is the least adequately studied factor we will discuss. Most network reliability models simply provide measures of network connectivity. Experience with current commercial telecommunications networks suggests that connectivity failures are relatively rare and affect, at most, a small percentage of network users at any one time. The failure of one or more network components may, however, lead to increases in network congestion and delay and decreases in the available throughput. These phenomena can be experienced simultaneously by all users of the network and affect the way in which the reliability of the network is perceived.

In recent years, efforts have been made to develop a unified framework for the study of reliability and performance. This is one of the major emphases of current work by the authors [61]. Much of the original work was developed within the context of degradable computing systems, but the concepts transfer nicely to the study of telecommunication networks. Beaudry [62] defined a set of performance-related reliability measures based on the concept of computation capacity, which can be defined as the amount of computation which can be performed by the computing system per unit time. Meyer [63] used this concept to develop a model which relates the state of the computing system to the level of performance achievable by a system in that state. The model assumes that there is a finite number of states in which the system can be found, and that transitions of the system between the various states can be described by a Markov chain. In this case, measures such as performability, which may be defined as the probability that the system is able to achieve a given level of performance, and effectiveness, which is the expected value of the level of performance achieved by the system, can be obtained.

In applying these concepts to the study of telecommunication networks, three issues seem to be crucial. The first is selection of an appropriate measure of performance. While either delay or throughput can be studied separately, in practice the goals of minimizing delay and maximizing throughput conflict with each other. A measure of performance which allows the study of tradeoffs between delay and throughput is network power, originally defined by Giessler *et al.* [64].

The second and third issues, which are closely related, are definition of the network states and evaluation of the performance achieved by each state. Since the total number of possible network states explodes with size of the network, it is infeasible to perform an exact analysis for each possible state. Three possible solutions to the problem exist. The first is to accept a rough approximation of the performance in each state. The second is to limit the analysis to some subset of the possible state space, possibly the most likely states as suggested by Li and Sylvester [65]. The third approach, developed by Deuermeyer [66], involves the use of simulation to select a random sample of system states for analysis.

Summary

This paper has summarized some major problems still to be solved before telecommunication network reliability and availability models are adequate. Although progress has been made on a few of these problems, models which really portray the impact various factors of the type discussed have on system users, and are at the same time computable, are not now available. Good approximate modeling techniques appear, to the authors, to be the only realistic approach to the problems listed.

References

[1] R. A. Rosanoff, "A survey of modern nonsense as applied to matrix computations," *Tech. Papers for Meeting, AIAA/ASME 10th Structures, Structural Dynamics and Materials Conf.*, New Orleans, LA, Apr. 1969.

[2] J. Spragins, "Analytical queueing models: Guest editor's introduction," *Comput.*, vol. 13, no. 4, Apr. 1980.

[3] N. Harris and P. D. T. O'Connor, "Reliability prediction: Improving the crystal ball," *Proc. 1984 Annual Reliability and Maintainability Symp.*, San Francisco, CA, 1984.

[4] J. L. Spencer, "The highs and lows of reliability predictions," *Proc. 1986 Annual Reliability and Maintainability Symp.*, Las Vegas, NV, 1986.

[5] H. E. Ascher and H. Feingold, "Repairable systems reliability: Future Research topics," in *Reliability in Electrical and Electronic Components and Systems*, E. Langer and J. Møltoft, Eds. Amsterdam, The Netherlands: North-Holland, 1982.

[6] M. O. Ball, "Computing network reliability," *Oper. Res.*, vol. 17, no. 4, July-August 1979.

[7] —, "Complexity of network reliability computations," *Networks*, vol. 10, pp. 153-165, 1980.

[8] —, "An overview of the computational complexity of network reliability analysis," *Working Paper Series MS/S 84-010*, College of Business and Management, University of Maryland, Feb. 1984.

[9] E. Hansler, "Reliability analysis of large complex systems," in *Proc. Fifth European Conf. Electrotechnics-EUROCON '82*, 1982.

[10] E. Lawler, *Combinatorial Optimization*. New York: Holt, 1976.

[11] L. G. Valiant, "The complexity of enumeration and reliability problems," *SIAM J. Computing*, vol. 11, pp. 298-313, 1982.

[12] D. R. Shier and D. E. Whited, "Algebraic methods applied to network reliability problems," Dep. Math. Sci., Clemson University, Clemson, SC, Tech. Rep. 486, Aug. 1985.

[13] IBM Corporation, *System Network Architecture: Technical Overview*, Manual GC30-3073-0, File No. S370/4300/8100-30, SLSS No. 5743-SNA, 1982.

[14] J. S. Provan and M. O. Ball, "Computing network reliability in time polynomial in the number of cuts," *Oper. Res.*, vol. 32, pp. 516-526, May-June 1984.

[15] D. R. Shier *et al.*, "Reliability of planar graphs," presented at the ORSA-TIMS Nat. Symp., Atlanta, GA, Oct. 1985.

[16] J. C. Sinclair, "Reduced complexity algorithm for the computation of a lower bound in the reliability of a virtual circuit network," Comput. Commun. Sys. Lab., Dep. Elec. Comput. Eng., Clemson University, Clemson, SC, Oct. 31, 1985.

[17] J. Spragins, "Reliability models for computer communication systems," in *Proc. Seventh Asilomar Conf. Circuits, Systems and Computers*, Pacific Grove, CA, Nov. 1973.

[18] —, "Birth-and-death equation approach to reliability theory problems," Tech. Rep. TR 27.0141, IBM Corporation, Systems Communications Division, Research Triangle Park, NC, Aug. 1975.

[19] —, "Dependent failures in data communication systems," *IEEE Trans. Commun.*, vol. COM-25, no. 12, pp. 1494-1499, Dec. 1977.

[20] —, "Reliability problems in data communications systems," in *Proc. Fifth ACM/IEEE Data Commun. Symp.*, Snowbird, UT, Sept. 1977.

[21] J. Spragins and J. Assiri, "Communication network reliability calculations with dependent failures," in *Proc. IEEE 1980 Nat. Telecommun. Conf.*, Houston, TX, Nov.-Dec. 1980.

[22] J. A. Assiri, "Development of dependent failure reliability models for distributed communication networks," Ph.D. dissertation, Oregon State Univ., Corvallis, OR, June 1980.

[23] J. D. Spragins, J. D. Markov, M. W. Doss, S. A. Mitchell, and D. C. Squire, "Communication network availability predictions based on measurement data," *IEEE Trans. Commun.*, vol. COM-29, no. 10, Oct. 1981.

[24] S. N. Pan and J. D. Spragins, "Dependent failure availability models for tractical communications networks," Final Rep. Naval Ocean Syst. Center, Contract N66001-82-C-0118, Aug. 1982.

[25] S. N. Pan, "Dependent failure availability models for tactical communications networks," Ph.D. dissertation, Clemson University, Clemson, SC, Aug. 1982.

[26] S. N. Pan and J. D. Spragins, "Dependent failure reliability models for tactical communications networks," in *Proc. IEEE 1983 Int. Conf. Commun.*, Boston, MA, June 1983.

[27] J. D. Spragins, "A fast algorithm for computing availability in networks with dependent failures," in *Proc. IEEE INFOCOM '84*, San Francisco, CA, Apr. 1984.

[28] G. A. Knab, "Comparative analysis of reliability models for computer communication networks with dependent failures," M.S. thesis, Clemson Univ., Clemson, SC, Aug. 1984.

[29] D. R. Shier and J. D. Spragins, "Exact and approximate dependent failure reliability models for telecommunication networks," in *Proc. IEEE INFOCOM '85*, Washington, DC, Mar. 1985.

[30] Y. F. Lam and V. O. K. Li, "Reliability modeling and analysis of communication networks with dependent failures," in *Proc. IEEE INFOCOM '85*, Washington, DC, Mar. 1985.

[31] D. W. Davies and D. L. A. Barber, *Communication Networks for Computers*. New York: Wiley, 1973.

[32] P. K. Giloth and K. D. Frantzen, "Design for quality and certification of a digital switch," in *Proc. IEEE Global Telecommun. Conf.*, Atlanta, GA, Nov. 1984.

[33] S. R. Ali, "Study of total system outage data for SPC switching systems," presented at IEEE Workshop on the Reliability of Comput. Controlled Commun. Syst., La Sapiniere, P.Q., Canada, Oct. 1985.

[34] W. J. Martin, III, "Including procedural errors in system reliability estimates," presented at IEEE Workshop on the Reliability of Comput. Controlled Commun. Syst., La Sapiniere, P.Q., Canada, Oct. 1985.

[35] B. S. Dillon and S. N. Rayapati, "Human performance reliability modeling," *Proc. 1986 Annual Reliability and Maintainability Symp.*, Las Vegas, NV, Jan. 1986.

[36] G. Martella, B. Pernici and F. A. Schreiber, "An availability model for distributed transaction systems," *IEEE Trans. Software Eng.*, vol. SE-11, no. 5, pp. 483-491, May 1985.

[37] M. Lipow, "On software reliability: A preface by the guest editor," *IEEE Trans. Reliability*, vol. R-28, no. 3, Aug. 1979.

[38] B. Littlewood, "How to measure software reliability and how not to," *IEEE Trans. Reliability*, vol. R-28, no. 2, June 1979.

[39] M. Shooman, "Operational testing and software reliability estimation during program development," *Rec. 1973 IEEE Symp. Comput. Software Reliability*, New York, NY, Apr.-May 1973.

[40] —, *Software Engineering*. New York: McGraw-Hill, 1982.

[41] J. E. Angus, "The application of software reliability models to a major C^3I system," in *Proc. 1984 Annual Reliability and Maintainability Symp.*, San Francisco, CA, Jan. 1984.

[42] J. L. Romeu and K. A. Dey, "Classifying combined hardware/software *R* models," in *Proc. 1984 Annual Reliability and Maintainability Symp.*, San Francisco, CA, Jan. 1984.

[43] A. L. Goel and J. Soenjoto, "Models for hardware-software system operational-performance evaluation," *IEEE Trans. Reliability*, vol. R-30, no. 3, Aug. 1981.

[44] —, "Hardware-software availability: A cost based trade-off study," in *Proc. 1983 Annual Reliability and Maintainability Symp.*, Orlando, FL, Jan. 1983.

[45] R. Piskar, "Hardware/software availability for a phone system," in *Proc. 1983 Annual Reliability and Maintainability Symp.*, Orlando, FL, Jan. 1983.

[46] W. E. Thompson, "System hardware and software reliability analysis," in *Proc. 1983 Annual Reliability and Maintainability Symp.*, Orlando, FL, Jan. 1983.

[47] J. E. Angus and L. E. James, "Combined hardware/software reliability models," in *Proc. 1982 Annual Reliability and Maintainability Symp.*, Los Angeles, CA, Jan. 1982.

[48] B. S. Liebesman and M. Tortorella, "Reliability of a class of telephone switching systems," in *Proc. 1982 Annual Reliability and Maintainability Symp.*, Los Angeles, CA, Jan. 1982.

[49] B. W. Kort, "Models and methods for evaluating customer acceptance of telephone connections," in *Proc. IEEE GLOBECOM '83*, San Diego, CA, Nov.-Dec. 1983.

[50] R. B. Miller, "Response time in man-computer conversational transactions," TR 00.1660-1, IBM Syst. Development Division, Poughkeepsie Lab., Poughkeepsie, NY, Jan. 1968.

[51] E. G. Coffman, M. J. Elphick, and A. Soshani, "System deadlocks," *Commuting Surveys*, vol. 3, no. 2, June 1971.

[52] H. Opderbeck and L. Kleinrock, "The influence of control procedures on the performance of packet-switched networks," in *Rec. 1974 Nat. Telecommun. Conf.*, San Diego, CA, Dec. 1974.

[53] A. Danthine and E. Eschenauer, "Influence on packet node behavior of the internode protocol," *IEEE Trans. Commun.*, vol. COM-24, no. 6, June 1976.

[54] L. Kleinrock, *Queueing Systems, Volume II: Computer Applications.* New York: Wiley, 1976.

[55] H. Kobayashi, *Modeling and Analysis: An Introduction to System Performance Evaluation Methodology.* Reading, MA: Addison-Wesley, 1978.

[56] Y. J. Park and S. Tanaka, "Reliability evaluation of a network with delay," *IEEE Trans. Reliability*, vol. R-28, no. 4, Oct. 1979.

[57] G. Barberis and U. Mazzei, "Traffic-based criteria for reliability and availability analysis of computer networks," in *Rec. 1977 Int. Conf. Commun.*, vol. 2, Chicago, IL, June 1977.

[58] F. Shatwan and D. G. Smith, "The reliability of a computer communication network under a given level of traffic," in *Proc. 2nd Int. Network Planning Symp.: Networks '83*, 1983.

[59] D. F. Lazaroiu and E. Staicut, "Congestion-reliability-availability relationship in packet-switching computer networks," *IEEE Trans. Reliability*, vol. R-32, no. 4, pp. 354–357, Oct. 1983.

[60] —, "A Markov model for availability of a packet-switching computer network," *IEEE Trans. Reliability*, vol. R-32, no. 4, pp. 358 and 365, Oct. 1983.

[61] J. C. Sinclair, "Performability analysis of virtual circuit based networks including routing and flow control effects," Comput. Commun. Syst. Lab., Dep. Elec. Comput. Eng., Clemson University, Clemson, SC, Oct. 31, 1985.

[62] M. D. Beaudry, "Performance-related reliability measures for computing systems," *IEEE Trans. Comput.*, vol. C-27, pp. 540–547, 1978.

[63] J. F. Meyer, "On evaluating the performability of degradable computing systems," *IEEE Trans. Comput.*, vol. C-29, no. 8, pp. 720–731, Aug. 1980.

[64] A. Geissler, J. Hänle, A. König, and E. Pade, "Free buffer allocation–An investigation by Simulation," *Comput. Networks*, vol. 1, no. 3, pp. 191–204, July 1978.

[65] Y. O. K. Li and J. A. Sylvester, "Performance analysis of networks with unreliable components," *IEEE Trans. Commun.*, vol. COM-32, pp. 1105–1110.

[66] B. L. Deuermeyer, "A new approach for network reliability analysis," *IEEE Trans. Reliability*, vol. R-31, no. 4, pp. 350–354, Oct. 1982.

[67] J. Spragins, "Limitations of current telecommunication network reliablity models," in *Proc. IEEE GLOBECOM '84*, Atlanta, GA, Nov. 1984.

[68] J. Spragins, Y. J. Kang, and H. Jafari, "State of the art in telecommunication network reliability modeling," *Proc. Pacific Comput. Commun. Symp.*, Seoul, Republic of Korea, Oct. 1985.

John D. Spragins (SM'84), for a photograph and biography, see this issue, p. 1098.

James C. Sinclair (S'84) received the B.S. and M.S. degrees in electrical engineering from Virginia Tech, Blacksburg, VA, in 1981 and 1982, respectively.

In January 1983, he joined the Computer Communications Systems Laboratory at Clemson University, Clemson, SC, as a Research Assistant, and is currently working towards the Ph.D. degree. His doctoral research is focused on network reliability and performance modeling.

Mr. Sinclair is a member of Tau Beta Pi and Eta Kappa Nu.

Yong J. Kang received the B.S. degree in electronics from Sogang University, Seoul, Korea, in 1976, and the M.S. and Ph.D. degrees in electrical and computer engineering from Oregon State University, Corvallis, OR, in 1979 and 1981, respectively.

In 1977 he served as a Laboratory Instructor, Electronics Department, Sogang University. From 1980 to 1984, he was a Technical Staff member and Technical Director, Advanced Control Technology, Inc., Albany, OR. In 1984 he became an Assistant Professor, Department of Electrical and Computer Engineering, Clemson University, Clemson, SC. Since the beginning of 1986 he has been on leave to serve as an Executive Director, Gold Star Cable Co., Seoul, Korea. His research interests are in the areas of computer communication networks, distributed processing systems, factory automation systems, and network operating systems.

Hossein Jafari received the B.S. degree in electronics from Tehran Polytechnique, Tehran, Iran, in 1966, the M.E.E. degree in computer engineering from North Carolina State University, Raleigh, NC, in 1972, and the Ph.D. degree in computer engineering from Oregon State University, Corvallis, OR, in 1977.

He is currently a member of Technical Staff of Operations Systems Planning Center at AT&T Bell Laboratories. Prior to joining AT&T Bell Laboratories, he was a principle member of Technical Staff of Advanced Research Group at ITT Telecom, Raleigh, NC, and on the faculty with the Department of Electrical and Computer Engineering, Clemson University, SC, respectively, during the 1984–1986 period. From 1982 to 1984 he worked in Iran at International Digital Systems (IDS) as a directing manager; and from 1978 to 1982 at Iranian International Engineering Company as a Computer Communications Senior Advisor and Project Manager.

Chapter 2: Multiprocessor System Reliability

As research and development of multiprocessing systems continues, more and more attention is being focused on the reliability of the connections between the functional components of the system. This chapter considers reliability efforts in multiprocessors described by the PMS notation, as well as in single- and redundant-path multistage interconnection networks (MINs).

When calculating a reliability measure for a system of components, three items of information are necessary:

1. the reliabilities of the individual components in the system
2. the physical or logical interconnection structure of these components, which enables a particular characteristic to exist in the system and thereby has a major impact on its reliability
3. the requirements or constraints on the system, in terms of sets of minimum numbers of working components, which define the conditions under which the system is considered to be operational

Most researchers use the physical interconnection structure and the system's operational requirements to generate an intermediate representation of a fault tree or reliability graph. Then the traditional reliability concerns are "What is the probability that any two nodes or vertices of the graph are able to communicate at any time?" or "If the links are subject to stochastic failure, what is the probability that enough links are operational to preserve at least a spanning tree of the network?" and so on. The companion tutorial text, *Distributed Computing Network Reliability,* describes various reliability measures and discusses methods of deriving reliability functions for arbitrary interconnection structures.

Traditional programs for calculating reliability have addressed the reliability issue largely at the level of fault trees and probabilistic graphs with homogeneous vertices and arcs, where the vertices, the arcs, or both are prone to fail. Moreover, most examples (in reliability literature) deal with the reliability of loosely coupled systems as geographically distributed computer communication networks. PMS representation, developed at the Carnegie Mellon University (CMU) describes computers and multiprocessor structures at a higher level of design and has been used at the Digital Equipment Corporation as an aid to system planning. In the case of PMS systems, the attributes of vertices in the interconnection graph could adequately represent the heterogeneous components of the system. Thus, the PMS design model for a multiprocessor system for reliability computation must be labeled to reflect this heterogeneity. Furthermore, the criteria for system functionality typically require that a certain minimum number of an assortment of components be functional and be capable of intercommunication. Kini and Siewiorek (1982) address this aspect of reliability computation in multiprocessor systems.

Recently there have been many reliability analyses and comparisons of single- and redundant-path MINs. Network reliability is the probability of maintaining full access capability throughout the network. Terminal reliability is the probability that at least one path exists from a particular processor to a particular memory (or memories). Reliability analysis of the MINs has focused in one of three areas:

1. the average number of switch failures that can be tolerated
2. the mean time to failure
3. the reliability of the terminals, a measure often used in the analysis of packet switching networks

Generally, a graph-theoretic concept and combining method are used to evaluate the reliability of multiprocessor structures such as crossbar switches, time-shared buses, multiport memories, and MINs. But most of the current algorithms for analyzing the reliability of MINs are either specific to particular areas of investigation or apply to the type of network that is being analyzed.

EH0300-4/90/0000/0017$01.00 © 1990 IEEE

Reprinted from *IEEE Transactions on Computers*, Volume C-31, Number 8, August 1982, pages 752–771.

Automatic Generation of Symbolic Reliability Functions for Processor-Memory-Switch Structures

VITTAL KINI, MEMBER, IEEE, AND DANIEL P. SIEWIOREK, FELLOW, IEEE

***Abstract*—Calculation of the reliability of computer system architectures with built-in redundancy, such as multiprocessors, is gaining in importance. The task of computing the reliability function for arbitrary Processor-Memory-Switch (PMS) interconnection structures, however, is tedious and prone to human error. Existing reliability computation programs make one of two assumptions:**

- **That the case analysis of success states of the system has been carried out. Such analysis must be done manually. In this instance, input to the program is usually in the form of an intermediate representation (e.g., fault tree, reliability graph).**
- **That the interconnection structure is a member of, or can be partitioned into, some limited class of structures for which a parametric family of equations exists (e.g., *N*-modular redundant systems, hybrid redundant systems).**

This paper represents a first step in the development of a methodology for automating the computation of symbolic reliability functions for arbitrary interconnection structures at the PMS level. The work reported here automates the task of case analysis and problem partitioning in the hard-failure reliability computation for PMS structures. As a consequence, the user is freed to focus almost wholly on specifying the reliability computation problem. The advantages of such an approach are: 1) utility to a larger class of users, not necessarily expert in reliability analysis, and 2) a lower probability of human error in the computation.

A program named ADVISER (Advanced Interactive Symbolic Evaluator of Reliability) was constructed as a research vehicle. ADVISER accepts as inputs

1) the interconnection graph of the PMS structure, and

2) a succinct statement of the operational requirements on the structure in the form of a regular expression.

Each component in the system, which may have internal redundancy, is represented by a symbol. The operational requirements in the case of a multiprocessor architecture may be, for example, "two processors and four memory boxes and one I/O channel." ADVISER considers communication structures in the PMS system (e.g., buses, crosspoint switches) in addition to the explicitly stated requirements to determine how the interconnection structure affects the system reliability. The program's output is a symbolic reliability equation for the system, subject to the given requirements. This paper describes the ADVISER program and methodology.

***Index Terms*—Automatic generation of symbolic reliability functions, processor-memory-switch (PMS) structures, reliability computation program, symbolic hard-failure reliability functions.**

Manuscript received October 6, 1981; revised January 12, 1982. This work was supported by the National Science Foundation under Grant GJ 32758X and the Office of Naval Research under Contracts NR-048-645 and N00014-77-C-0103.

V. Kini is with the Information Sciences Institute, University of Southern California, Marina del Rey, CA 90291.

D. P. Siewiorek is with the Department of Computer Science and Electrical Engineering, Carnegie-Mellon University, Pittsburgh, PA 15213.

I. INTRODUCTION

WITH the advent of practical multiprocessor architectures in recent years the growing complexity of systems has increased the importance of system reliability as a design parameter. Computation of system reliability metrics has therefore become part of the catalog of system design tasks. Various efforts have been reported in the literature and are in progress to provide designers with reliability design tools that will make the task of computing system reliability easier and more efficient. Computing system reliability for complex multiprocessor architectures can be very tedious and prone to error, sometimes even for experienced reliability analysts. Software tools that currently exist to help estimate or calculate system reliability usually assume an understanding of reliability analysis techniques and are usually more in the nature of computational aids once preliminary system decomposition and analysis has been manually achieved.

The work reported in this paper was prompted by the question, "Is it possible to build reliability design aids that will assume the burden of a significant portion of the system analysis effort, leaving mainly the system reliability specification task to the designer?" The result of this effort was the ADVISER program, which accepts the interconnection structure of the architecture at the Processor-Memory-Switch (PMS) level [1] and a simple set of operational requirements on the architecture. It then produces the symbolic form of the system hard-failure reliability function under the given requirements. For the present version of the program perfect coverage was assumed (see Section IV). The program attempts to efficiently analyze, using the divide-and-conquer paradigm, the various possible *classes* of cases of system success using information gleaned from the interconnection structure of the system. The current scope, capabilities, and user-interface of ADVISER are modest, but the methodology underlying its design shows promise for building more sophisticated future versions.

Section II of this paper summarizes previous work. Sections III and IV describe the motivations for the present work and its underlying assumptions and concepts. Section V presents a broad overview of the process by which the input PMS structure and the operational requirements are analyzed in ADVISER to produce the symbolic hard-failure reliability functions for the structure. Finally, Section VI shows an example of the use of ADVISER.

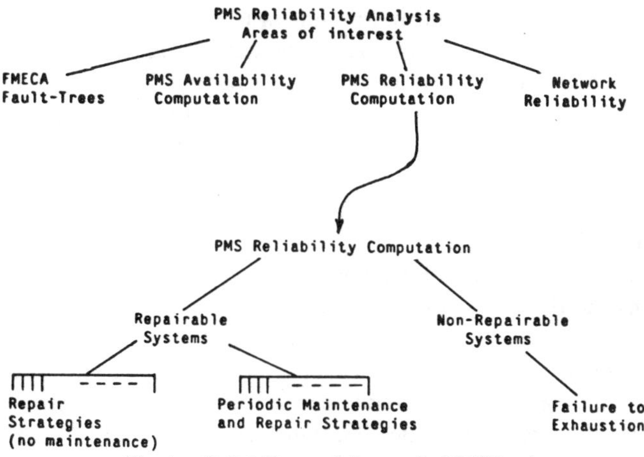

Fig. 1. Reliability modeling at the PMS level.

II. PREVIOUS WORK

Computer systems may be studied at various levels of detail. In their book on computer structures, Bell and Newell [1] proposed four broad levels at which attention is usually focused. There are, respectively, 1) the gate level, 2) the register-transfer level, 3) the software level, and 4) the Processor-Memory-Switch (PMS) level. The work reported here deals with the PMS level of detail at which computer systems are viewed as assemblages or networks of components that process and/or store information. This is also often termed the system architecture level. Fig. 1 shows the areas of reliability assessment that are of broad interest at the PMS level. The categories in the figure are not necessarily disjoint and serve only to grossly characterize the distribution of work reported in the literature. Failure Modes, Effects, and Criticality Analysis (FMECA) is described in Greene [2]. An introduction to fault-tree analysis may be found in Barlow and Lambert [3]. Although not dealing with computing systems, one of the best known uses of fault-trees in system reliability analysis is the Rasmussen Report [4]. Lapp and Powers [5] describe recent work that developed a methodology for automatic synthesis of fault-trees for chemical engineering systems.

The term "network reliability problem" is variably applied in the literature to two distinct kinds of network problems. The first variety deals with graph models of computer communication networks as in Wilkov [6] and Hansler [7]. The second variety addresses two-terminal directed networks, which are essentially reliability graphs. Such a graph is *not* necessarily a model of the physical interconnection structure of the system, but is rather a *representation* of it which characterizes the system's reliability. *In this paper we use the term "network reliability problem" exclusively in the sense of Wilkov* [6]. The network reliability problem is concerned with calculating fairly simple measures of reliability for a system. Typically, the system is a computer communication network and the vertices of the interconnection graph denote the computers while the arcs denote the communication links. Either arcs or vertices, or both, are assumed to fail stochastically. Typically although not always, all vertices are considered homogeneous with identical probabilities of failure. Arcs are usually treated the same. Two common reliability measures computed for such a system are [6]:

- the probability that some specific pair of vertices will have at least one communication path between them at all times,
- the probability that the operative arcs always contain a spanning tree of the network.

Despite their seeming simplicity these types of network reliability calculation problems have been shown to be NP-hard in the case of general networks [8], [9].

In the work reported here the problem formulation is most similar to that of network reliability problems as in Wilkov. The difference is that components in the network are not homogeneous (processors, memories, switches, etc., are connected by the network), and the operational requirements on the network are much more complicated than those noted above.

A. Extant Reliability Calculation Programs

In this section we shall review a few representative extant computer programs for system reliability computation. Their characteristics and intent will be briefly discussed with a view to setting up a framework within which to classify ADVISER.

1) Reliability Estimation: Nelson, Batts, and Beadles [10] describe a program that computes the bounds on system reliability given its reliability graph (which must be manually derived from the PMS or structural graph). An upper bound on system reliability is given as the sum of the probabilities of functioning of the path sets of the reliability graph. A lower bound for the system reliability is obtained by taking the first two terms of the finite series

$$\Pr\{E_1 \cup E_2 \cup \cdots \cup E_n\} = \sum_i \Pr\{E_i\} - \sum_{i<j} \Pr\{E_i \cap E_j\} + \sum_{i<j<k} \Pr\{E_i \cap E_j \cap E_k\} \cdots + (-1)^{n-1}\Pr\{E_1 \cap E_2 \cap \cdots \cap E_n\}$$

where the event E_i is the functioning of all components in the ith tie set. Increasingly tighter upper and lower bounds are obtained by taking more terms of the above expression. Similar bounds can be obtained on the *un*reliability by considering E_i to be the event that all components fail in the ith minimal cut set of the reliability graph. The existence of the system reliability graph is assumed. Components are assumed to have constant reliabilities. Matrix methods are used to generate the minimal cut sets of the graph. Bounds based on the tie sets are recommended in the low reliability region and those based on cut sets are recommended in the high reliability region.

2) Reliability Block Diagram Representation: The exact combinatorial system reliability derived from the reliability block diagram representation is the subject of Fleming [11], Chelson [12], and Kim *et al.* [13]. All assume that the system reliability graph (in the form of a reliability block diagram) has been previously derived by the analyst. Fleming [11] describes the RELCOMP program, which computes the system reliability and MTBF. It accepts a purely series reliability block diagram. RELCOMP assumes that the system is composed of independent subsystems which fall into one of eight categories provided for (e.g., standby redundant configuration, actively redundant configuration). The corresponding eight commonly used reliability equations are built into the program. Both exponential and Weibull failure distributions are represented in the equation repository.

Chelson [12] describes a program that accepts a form of block diagram able to represent systems with standby redundancy. More than one block may represent a given system component (these are called equivalent blocks). Exponential failure distributions are assumed throughout and different failure rates may be assigned to spares and active modules. The switches representing the recovery capability of the system may be modeled as imperfect. The program constructs a probability tree for the system and computes the reliability from it.

Kim *et al.* [13] describe a method for computing reliability from nonseries-parallel reliability block diagrams. Their procedure consists of three steps: 1) reduction of all series and parallel connections until the block diagram cannot be reduced further (Krishnamurthy [14] describes another reduction method), 2) enumeration of all paths from source vertex to sink vertex in the block diagram, and 3) computation of system reliability from the path reliabilities using an operation which counts the probability of a given component only once in each product term. Matrix methods are used to compute the path sets for the block diagram.

More recently, work has been reported on the use of reliability graphs to produce symbolic system reliability functions [15], [16]. Satyanarayana and Hagstrom [17] describe an algorithm which efficiently produces a symbolic expression for two measures for a probabilistic directed network, namely, 1) *source-to-multiple-terminal reliability:* the probability that a given node can reach every other node in the network, and 2) *overall reliability:* the probability that any node in the network can reach every other node in the network. ADVISER considers a different and more complicated problem: that of finding the probability that certain minimum numbers and types of nodes are functional in the network and can communicate. However, the low-level operations on the intermediate representation created internally to ADVISER are similar to computation of the reliability of two-terminal networks. Hence, the results of Satyanarayana and Hagstrom could be applicable, with modifications, within the ADVISER framework and would probably serve to make the program more efficient.

3) Hybrid-Redundant System Analysis: Another class of programs for system reliability analysis focus on weak points of the purely combinatorial analysis technique, i.e., the inability to deal with systems containing varieties of dynamic redundancy [18]. In such systems the switching in of spares to replace failed modules is viewed as an imperfect process contrary to the assumptions of static reliability models. In such "staged" systems the so-called *coverage factor*, or the probability of system recovery after a fault, is of central importance since the system reliability has been shown to be very sensitive to the factor [19].

The early effort in this instance was the REL program, which was succeeded by REL70 [20] written in APL. Bouricius *et al.* derived basic equations for systems with standby sparing, largely under the assumption of constant failure rates for all system components. The coverage factor C was included in these equations and it was shown that assuming perfect coverage ($C = 1$), even when coverage was in fact "near" perfect ($C = 0.99$), could produce gross errors. The results of this work were incorporated as an equation repository in REL70 to analyze memory and processor subsystems of a typical computer.

Mathur [21] describes a computer program named CARE, which improved on REL70. Systems being analyzed were viewed as cascades of independent hybrid-redundant subsystems. Again, a repository of equations was built into the program for the analysis of each of such subsystems. Equations developed in Bouricius *et al.* [20] were also included. The system reliability was taken to be the product of the independent subsystem reliabilities. The latest version of the program, CARE III, developed by the Raytheon Corporation, is considerably more complex. A Markov process approach has been incorporated into the program along with decomposition methods that agglutinate states to reduce the large state space of a complex model. Time-dependent parameters for transitions between the states of the Markov model (i.e., a nonhomogeneous Markov model) are handled in cases of nonrepairable systems. Since ultrareliable systems are the subject of CARE III much attention has been paid to reducing numerical error.

More recently, Ng and Avizienis [22], [23] developed a unified reliability model for fault-tolerant systems. This model is based on a Markov process view of the graceful degradation of dynamically redundant systems. Various earlier reliability equations derived for different types of static and dynamic redundant systems are available as special cases of the unified model [23]. In addition, the model is extended to repairable systems under a restricted model of the repair process. Degradation under transient faults is also modeled by the same Markov techniques. The ARIES program embodies the results of the unified model. However, the model is still restricted in its applicability to those types of systems which are decomposable into cascades of independent hybrid-redundant subsystems.

Landrault and Laprie [24] describe the SURF program, which views repairable systems as being governed by nonexponential failure processes. In cases where the nonexponential distribution being considered is related to the exponential (e.g., Gamma, Erlang), the Coxian device of stages [25] is used to judiciously introduce series of fictitious states (with exponentially distributed transition times among them) so as to convert the non-Markov process to a Markovian one. For some problems, semi-Markov processes, which suppose the existence of a finite number of instants possessing the property of independence on past history (i.e., an imbedded Markov chain), are also used.

4) PMSL: A quite different view of PMS systems is contained in Knudsen [26] who reports on PMSL, a language and a system for describing arbitrary PMS structures. The notation developed by Knudsen is similar to its progenitor, the PMS notation of Bell and Newell [1]. Programmed in SNOBOL, PMSL was a powerful description facility which allowed users to construct interconnection models of arbitrary PMS structures. During the construction, the program did various attribute checks on the structure for legality of interconnections. The PMSL system, although more in the nature of a PMS-database manipulation system, also allowed the user to compute the combinatorial reliability of the PMS structure input to it. However, the program suffered from very rudimentary reliability calculation facilities. Reliability calculation was applicable only to uniprocessor structures, and enumeration of system success states was used as the (inefficient) computation method. PMSL is included in this survey of reliability calculation programs because the level of detail in its model of PMS structures is similar to that in ADVISER, although the instruments provided to manipulate PMS descriptions were more powerful in PMSL. In PMSL, PMS structures are viewed as being hierarchical, and components in them are described by a list of attribute-value pairs.

5) *Automatic Fault-Tree Synthesis:* We end this brief survey of PMS reliability computation programs with an example from the field of chemical engineering. Although not entirely relevant to computer systems, this example is important since it is a step toward the eminently desirable goal of easier and less error-prone reliability computation for complex systems. Lapp and Powers [5], [27] describe the FTS program, which constructs the fault-tree representation of a complex chemical engineering process from a much simpler logical model of the process. The program contains hazard models of commonly used pieces of equipment within the process (e.g., valves, pumps, sensors, reactors). The user constructs a logical flow diagram of the process, labeled with various process parameters, and the program uses its database of hazard models and information to synthesize the fault tree for the process. Cut set analysis may then be used to compute the probability of the top event or system failure [28].

III. Motivation

In the construction of ADVISER the goal was to produce a reliability calculation program capable of computing the symbolic reliability function for an arbitrary PMS interconnection structure, given a simple statement of the operational requirements placed on the structure. Therefore, the following ends were pursued.

- The program should require only a modicum of information from the user as input, i.e., the specification of the problem should be simplified. Ideally, the existence of a program such as ADVISER should cause the focus of attention to shift to the design of formal languages that clearly and unambiguously *specify* reliability problems.
- The program should attempt to assume most of the analysis of an interconnection structure before computing its reliability function. This will make it attractive to the user who is less experienced in reliability analysis, and will reduce the chances of human error entering into the computation.
- The program should output the system reliability function in symbolic form in order that arbitrary failure distributions for the individual component reliabilities may be experimented with by the user. This also allows the reliability function to be manipulated by symbolic manipulation systems such as MACSYMA [29].

The major emphasis in ADVISER is the avoidance of manual construction of a reliability graph or equivalent representation for the system; ADVISER is therefore preferable to programs such as those described in Chelson [12] and Kim *et al.* [13]. *Also, since knowledge of the physical interconnection structure provides information about the structural dependence (as distinct from statistical dependence) of components in determining the system reliability, the operational requirements on the structure can be expressed very simply in terms of a few key classes of system components. Further information can then be deduced from the interconnection topology. This leads to succinct statements of minimal system requirements in the ADVISER paradigm.*

In our investigations we were interested primarily in systems which could not be partitioned into independent hybrid-redundant subsystems as assumed in Mathur [21] and Ng [23]. Examples of such systems are the Pluribus [30], Cm* [31], and Tandem-16 [32] multiprocessors in which fault recovery and reconfiguration is done largely by software or firmware. This is not to exclude the possibility that, for example, one of the processors within a multiprocessor such as Tandem-16 could be constructed for reliable operation by internally using hybrid redundancy. The difference is one of the level of detail at which the system is being studied.

Network reliability problems, as defined in Section II and addressed in Hansler [7], for example, were only of marginal interest in our investigations. The reason is that PMS structures are generally more closely coupled than computer communication networks and the operational requirements on them are usually more complex than in the kinds of problems studied in Hansler [7], it is shown in Kini [33] that ADVISER can be used for a subclass of the latter kind of problem.

The ADVISER program is aimed more toward solving the common problem of deriving the combinatorial reliability of complex interconnection structures under various operational requirements, particularly in the context of comparative reliability studies of PMS interconnection structures. A possible use of ADVISER would be in an iterative design study of a candidate PMS interconnection structure wherein the structure topology is perturbed, components are added or deleted, etc., until the appropriate reliability is achieved.

IV. Underlying Assumptions and Concepts

The general case of deriving reliability functions for arbitrary interconnection structures of components is a task that is difficult to program. Much depends on the semantics of the behavior of the components in the structure, how their tasks interrelate, whether their probabilities of functioning are statistically independent, and so on. Some idealizations become necessary in order to make the problem tractable for an initial attempt.

A comparison of two PMS interconnection-structure designs is valid, as long as the metric used is consistent across the space of designs being considered. For this reason we decided to study the hard-failure reliability of a system unencumbered by such issues as: 1) the effects of policy decisions regarding manner of use, 2) software reliability, 3) transient failures, and 4) statistical dependence of component failures in any form. The comparisons would therefore take into account the best possible reliability performance of each PMS structure being considered. The following assumptions were made.

1) To begin, failure processes in individual components of the structure were assumed to be stochastically independent. Since PMS structures are the focus of the study, there seems to be justification in making this assumption. For example, the typical components being considered, such as processors and memories, are generally physically separated. This does not, however, imply that failures of different system components affect the system uniformly; clearly, the topology of the interconnection in the structure has a bearing on the question.

2) *Only the hard-failure reliability functions of components under perfect coverage [19] were treated.* However, the methodology used does not prevent the inclusion of coverage in the model, and we estimate the effort required to do so would be moderate. The effects of transient failure mechanisms were not considered.

3) Components in the PMS structure were assumed to be binary state objects, i.e., either "failed" or "working." The emphasis was on success probabilities of components, so that all reliability functions are expressed in these terms.

4) The graph of the interconnections of the PMS structure was modeled as a nondirected graph. Vertices in the graph correspond to the components in the structure and the functionality of the components is lumped into these vertices. Each nondirected arc of the graph is considered perfectly reliable and simply represents the capability of information to flow between its two end vertices. Thus, the failure of a component is assumed to be equivalent to removing the corresponding vertex and all of its incident arcs from the graph.

5) *It is assumed that in order for an assemblage of information-processing components to comprise a useful functioning system, some distinguished set of critically important system components must be able to communicate among themselves. In other words, information should be capable of flowing between any two components from the distinguished set whether via other distinguished components or any other components in the structure, or both.* This will henceforth be referred to as the *Communication Axiom.* It is a cornerstone of the ADVISER methodology and is elaborated upon below.

We first introduce some concepts basic to our discussion of the Communication Axiom and then formally state the axiom. Throughout the rest of this paper the terms "system success" and "component success" will be used interchangeably with the terms "system is functional" and "component is functional," respectively. In any system that we may consider, some subset of the total set of components in the system will be distinguished in that their correct functioning is of vital importance to system success, e.g., the CPU in a uniprocessor system. More accurately there will be a set of generic classes or *types* of components of vital importance (e.g., processors, memories). Also, a certain minimum number of components drawn from each distinguished generic class or type, must be functional for system success. These distinguished component types will be termed *critical component types* (*CCT's*). Each such type is taken to constitute a class of *identical* components and the members of these classes will be termed *critical components.* The need to consider members of a component type class to be identical stems from the desire to detect possible symmetries in the interconnection structure which could help ease the analysis problem (see Sections V-A.1 and V-B.1). All components that are not critical in the PMS structure will be termed *auxiliary components.* There may be system success states in which auxiliary components are not functional; however, there are *no* system success states in which critical components are *not* functional.

A minimum number of critical components from each CCT are required for system success. Together they constitute a *minimal critical resource set* (henceforth MCRS). The set is minimal in the sense that, although the system *may* function if all components in an MCRS are functional (depending on the status of the auxiliary components in the structure), the structure is guaranteed to fail if any component of this MCRS fails. In other words, the success of an MCRS is a necessary, although not sufficient, condition for system success. This concept is not to be confused with a *minimal system success state* in which the failure of any one functioning component, whether critical or auxiliary, causes system failure (cf. the *minimal cut vector* in the terminology of Barlow and Proschan [34]). The latter would be a stronger condition on minimality.

If there is redundancy in the supply of critical components configured in the structure, then there will be more than one minimal critical resource set. Each such set will, in general, be included in one or more system success states, again depending on the disposition of the auxiliary components.

It appears fundamental that, in order for an information processing system to do useful work, there must be pathways, or channels, of information flow between the components of an MCRS of that system. This basic rule is tacitly assumed during hand calculation of reliability of PMS structures. We contend that the reliability of PMS structures may be computed by a program using the following simple paradigm. The user inputs

1) component type classifications,

2) graph of the PMS interconnections, and

3) a Boolean statement of which component types are critical component types and how these are related in determining system reliability.

The last of these three items is what we shall term the *minimal requirements* on the system. An example of such a minimal requirement phrased in English might be "at least one processor *and* at least one memory *and* (at least one disk *or* at least two tape units) must be functional" where the *or* is an Inclusive-OR. A regular expression is used to convey such a requirement. The program would employ the minimal requirement and the interconnection graph of the system in the context of the Communication Axiom. The component types referred to in the minimal requirements would be labeled critical component types by default and the rest labeled auxiliary types. The minimal requirements would be used to generate all the MCRS's of the system. For a given MCRS, the Communication Axiom and the interconnection graph then identify sets of paths between pairs of vertices in the graph that represent the components of the MCRS. A path is deemed functional iff all the components along that path are functional. The Communication Axiom implies that components in the

MCRS must be part of a connected graph of functional paths for a reliable system.

We now define the Communication Axiom as follows.

Let $T = \{t_i\}, i \in \{1, 2, \cdots, n\}$ be the set of component types specified in the *minimal requirements* input. The set T is then, by default, the set of critical component types.

Let $Q_i = \{q_{ji}\}, j \in \{1, 2, \cdots, m_i\}$ be the set of all identical components of type t_i present in the structure.

Let $T' \subseteq T$ and let $M_k \subseteq Q_k, t_k \in T'$.

If $M' = \cup_k M_k$ is a set of critical components such that the Boolean statement of minimal requirements is satisfied minimally, then M' is a Minimal Critical Resource Set (MCRS).

A simple path p_{ab} between any two vertices v_a and v_b in G is said to be a *functional path* iff all components represented by vertices along that path are functional.

Let $V_{M'}$ be the set of vertices in G that represent the components in M', and MCRS of G.

Definition 1: We define a *Communicability Graph* or *K-Graph*, $G_K(V_K, E_K)$, for M' as follows.

- There is a bijective mapping from the vertex set V_K to the vertex set $V_{M'}$.
- A *Communicability Edge* or *K-edge* $(v'_K, v''_K) \in E_K$ will exist iff at least one functional path exists between the vertices in G that are the images, under the bijective mapping, of v'_K and v''_K in G_K, respectively.

Axiom 1—Communication Axiom: For any MCRS M' of the system represented by G, if the components in M' are all functional, then the system will be functional iff the K-graph of M' is connected.

In order to intuitively understand how the Communication Axiom is used in the program, the reader may think of each MCRS as a skeleton of critical components which must be "fleshed out" with a set of paths in the graph between the vertices of the skeleton, so as to form a connected graph. These paths will allow communication among the components in the MCRS. Each such possible fleshing out of the skeleton will correspond to one minimal success state of the system. *Furthermore, each set of paths chosen to flesh out the skeleton will identify the other (auxiliary) components along those paths that must be functional for the vertices of the MCRS to communicate.* This method of identifying the additionally necessary components has an important and useful side effect from the viewpoint of the user of the program. Consider a component in the structure whose component type is *not* referred to in the requirements expression. It may be the case that this component is required to be functional in every system success state, i.e., it is truly a "critical" component although it has been labeled auxiliary by default. An example of such a component might be a memory controller that lies on the path to a memory required for system success by a requirement of the type shown above. Although the memory controller is not referred to in the requirements, thereby not explicitly making it "critical," the strategy of fleshing out the MCRS with paths from the graph will always find the memory controller to be necessary at all times, since all paths to the memory pass through it. Typically, therefore, few component types will need to be explicitly labeled critical by inclusion in the requirements expression since other critical components will be deduced from the interconnection structure by applying the Communication Axiom.

On the basis of the foregoing discussion it is evident that each MCRS will be part of possibly several system success states, depending on how many combinations of paths can be found to flesh out the skeleton it provides. For each MCRS a reliability expression will be generated to account for all the probabilities of all the functional states of which the MCRS could be a part. These reliability expressions will henceforth be referred to as *Intermediate Results* or *Partial Results.* Intermediate results relating to all possible MCRS's will finally be combined in conjunction or disjunction, as appropriate, to generate the system reliability expression.

The use of the Communication Axiom, in the manner referred to in the paradigm outlined above, initially appears sufficient to derive the reliability function for many cases of arbitrary PMS interconnection structures. However, constraints beyond those implied by the Communication Axiom are sometimes posed during calculation of PMS system reliability by the special types of behavior exhibited by various system components. For example, a crosspoint switch, unlike a bus, generally allows communication only between components connected to distinct sides of the switch and not among components connected to the same side. It would be impractical to include in the program all the semantics of various types of special behavior ever to be encountered, although this might be reasonable for a limited set of special component types. We postulated that the inclusion of three further types of simple modeling information as additional input *side-constraints* would enable the program to handle a majority of cases. This keeps the model and the required operations simple while providing a useful tool in the program. The three types of additional information, noted below, are discussed in Section V-A.3:

- the internal port-connection matrix of a component,
- the possibility of intracomponent-type information transfer,
- the required clustering of functioning critical components in parts of the PMS structure.

V. Overview of Adviser

We now describe the process by which a PMS description and the associated reliability requirements upon it are operated on by ADVISER to produce a symbolic reliability function.

Three items of information are needed to calculate a reliability measure for a system of components:

1) the reliabilities of the individual components in the system,

2) the physical or logical connections of those components which give the system its particular existence and define its reliability, and

3) the operational requirements placed on the system that affect its perceived reliability (for, clearly, a multiprocessor will be more reliable in the case of a task that requires any m of its processors to be functional as opposed to the case where a task requires only any $n \neq m$).

The above three kinds of information are entered into the program in the input phase (Sections V-A.1–V-A.3). After the detection of symmetric substructures in the interconnection graph (Section V-B.1) the graph is segmented (Section V-B.2). The program then enumerates the various ways of satisfying the minimal requirements by choosing critical components from the segments (Section V-C.1). A symbolic reliability expression is generated at each step if the choice satisfies the Communication Axiom and side-constraints (Section V-C.2). These intermediate expressions are combined to form the system reliability function. The latter is printed out symbolically (Section V-D). Fig. 2 illustrates the structure of the ADVISER program and its various phases, which will be discussed in the following.

A. Program Inputs

1) Declaration of Component Types: The first input to the program is a list of *types* of components that comprise the PMS structure yet to be described and for which the reliability function is to be computed. Each type-declaration will contain information about the reliability function for a component of that type, whether it be a function type known to the program or some user-defined function elsewhere. The type-declaration will contain a "print-name" to represent the component when the reliability function is printed out. When the interconnection structure is defined the components comprising it will each be assigned a type to be selected only from this list of type declarations. Components of like type are then assumed, in the current implementation of ADVISER, to be identical in a reliability sense. In other words, they are drawn from the same population. Another option is to consider components of like type to be identical only for the purposes of symmetry detection but to let them have possibly different reliability functions. This would require trivial changes to ADVISER. Table I shows some representative type declarations. The unique type numbers in the first column are assigned by the program. Components labeled as belonging to type 3 are of type Cent. Proc (central processors). The symbol representing their reliability will print out as PC in the system reliability function that is eventually produced. All Cent.Proc's are declared to be identical and to have exponential reliability functions with a failure rate of 200.1/mh (failures per million hours). All type 4 components, likewise, are of the class IO.Proc (input/output processors) and have Weibull reliability functions, each with a scale factor of 385.3/mh and a shape factor of 0.86.

Assigning a type to each component in a PMS structure may be viewed as labeling the vertex representing that component in the interconnection graph. This information is used by the program as a constraint in detecting structurally symmetric subgraphs of the interconnection graph. The motivation for detecting such symmetries is, of course, the expectation that the amount of necessary computation can be reduced.

2) Declaration of the PMS Structure: The next type of input to the program is the labeled graph that represents the interconnection topology of the system. As noted in Section IV, all connections between components are modeled as duplex, i.e., information may flow in both directions along an arc in the graph. Thus the model uses nondirected graphs. The graph is described very simply to the program by means of an adjacency list. A section of a typical graph description input is shown in Table II.

In the table the component named P.IO.1 (input/output processor #1) is declared to be of type IO.Proc. This component type must have been declared during the first input phase, when component types were specified. K.IO.1 (input/output controller #1) and K.IO.2 (input/output controller #2) are seen to be declared identical components, both with Weibull reliability functions with a scale factor of 286.7/Mhr and a shape factor of 0.92 (refer to Table I).

3) Declaration of Reliability Requirements: We now come to the third kind of information necessary to calculate system reliability: a statement of what subset of system resources are required to be functional before the given task runs to completion on the system. This information can be supplied in a variety of ways and an example will help to make the subsequent discussion clearer.

Fig. 3 shows a dual-processor system. Each processor accesses memory and peripherals over a bus (S). The peripherals are dual-ported for access from both processors. Let us assume for a specific task that at least one of the processors, at least two of the memories, and at least one disk drive need to be functional for system success.

One way to convey these requirements is to explicitly enumerate the system states that are success states. For instance, in our example, $\{P.1, S.1, K.m.1, M.1, K.m.2, M.2, K.d.1, T.d.1\}$ is a full specification of one system success state. The program then has only to sum up the probability of occurrence of each state. This is objectionable for two reasons. First, the number of system success states can be large for systems of reasonable size. One can argue that only the minimal success states of the system need be considered since the probabilities of all the states subsumed by the minimal success states will cancel in the summation process. Even so, second, asking the user of the program to analyze and supply the set of minimal success states is objectionable. It is tantamount to asking him to do a major part of what the program should do to justify its use. Furthermore, if the user is relieved of the burden of analyzing the system states, he need not be experienced in the art of reliability computation. A larger base of users could then use the program as a design tool. Perhaps most important, the program could help to eliminate human error from the usually tedious PMS reliability calcultion task.

On the other hand, referring to our example again, a human being needs only the following brief statement for him to accomplish the task of reliability computation: "at least 1 P and at least 2 M's and at least 1 $T.d$ need to be functional." He then proceeds to use his knowledge of component behavior. He deduces that certain auxiliary components beyond those explicitly specified will need to succeed in order to create a system success state. For instance, if $P.1$, $M.1$, $M.2$, and $T.d.1$ must be functional to be part of a system success state, then $S.1$, $K.m.1$, $K.m.2$, and $K.d.1$ will additionally need to succeed.

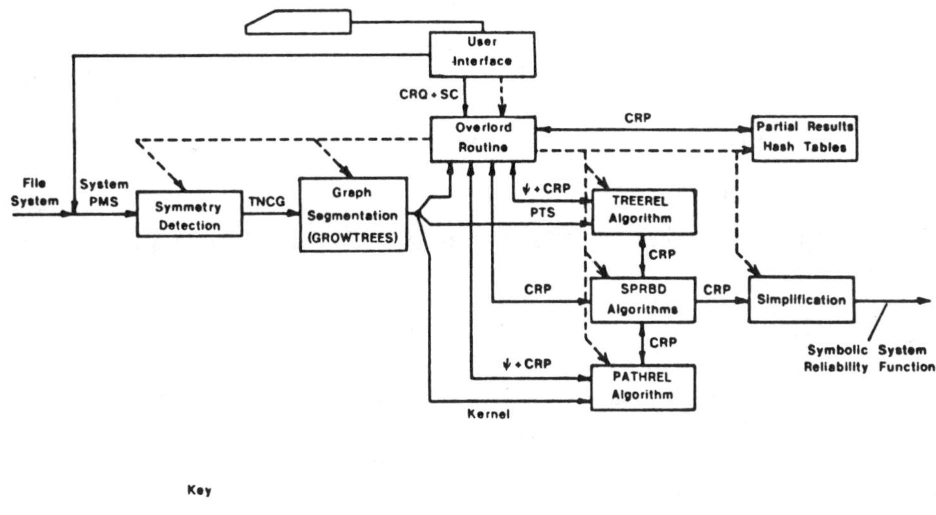

Fig. 2. The structure of the ADVISER program.

TABLE I
SAMPLE INPUT COMPONENT TYPE-DECLARATIONS

type #	typename	printname	rel fn	rel. fn. parameters
...	...	...	...	...
...	...	...		...
3	Cent.Proc	PC	Exponential	Lambda = 200 1/MHr
4	IO.Proc	PIO	Weibull	Scale = 385.3/MHr, Shape = 0.86
...	...	...		...
6	IO.Cont	KIO	Weibull	Scale = 286.7/MHr, Shape = 0.92

TABLE II
SAMPLE INPUTS DEFINING PMS INTERCONNECTIONS

component #	component name	component type	neighbor components
...	...	...	...
...	...		...
3	P.IO.1	IO.Proc	Unibus.1,K.IO.1,K.IO.2,...
4	K.IO.1	IO.Cont	P.IO.1,DISK.1,DISK.2,...
5	K.IO.2	IO.Cont	P.IO.1,TAPE.1,TAPE.2,...

This is an instance of application of the Communication Axiom.

Modeling the style of the requirements input after what would be expected by a human being, we chose a regular expression form. The primitives in the expression are operated upon by the standard logical AND and OR operators. The primitives are of the form "at least N of X" where N is integral and $N \geq 1$. "X" is the name of a previously declared component type. This is taken to mean "at least N components of type X should be functional." Two other possible forms for the primitives, namely "Exactly N of X" and "At most N of X," if allowed, lead to the conclusion that the system is noncoherent [34]. We therefore exclude these possibilities.

We refer to primitives such as "N of X" as *Atomic Requirements* and to Boolean combinations of them as *Compound Requirements.* Atomic requirements such as "N of X" are represented by the symbol $\psi(N, X)$. Within an atomic requirement $\psi(N, X)$, N is termed the *Integer Requirement* and X is termed the *Required Component Type* (or simply the *Required Type*). The simple grammar for compound requirements is shown below.

⟨requirements-expression⟩
    ::= ⟨conjunction⟩ | ⟨conjunction⟩ OR ⟨requirements-expression⟩
⟨conjunction⟩   ::= ⟨primitive⟩ | ⟨primitive⟩ AND ⟨conjunction⟩ | (⟨requirements-expression⟩)
⟨primitive⟩   ::= ⟨integer⟩ OF ⟨typename⟩

As alluded to in Section IV, three further forms of requirements input supplement the regular expression and provide further constraints on the evaluation, thus allowing a larger space of PMS structures to be handled. These ad hoc *side constraints* strive to include semantics of individual component behavior (since semantic behavior is not built into the program) as completely and as generally as possible in the context of PMS structures.

1) *Internal Port Connections:* In our model there is a one-to-one correspondence between arcs impinging on a vertex and I/O ports of the component represented by the vertex. Information flow *within* a component may be modeled by a directed graph, at least weakly connected, with the component's ports as its vertices. Information of this nature will allow the program to avoid considering erroneous paths, e.g., attempting to find a communication path between processor A and processor B via, say, a shared lineprinter controller C. A connectivity matrix is used to supply this information for any given PMS component. If the matrix is not supplied it is assumed that all pairs of ports communicate within the component.

2) *Intracomponent-type Communication:* In the blandest form of the model, since no component semantics are included, the Communication Axiom leads to finding K-edges between all pairs of critical components in any MCRS. However, in many cases information never passes between two components of the same type. For example, memories are passive components and usually never communicate with each other. When such passive behavior is to be taken into account, the use of the Communication Axiom must be modified if we are not to evaluate a pessimistic system reliability due to having erroneously assumed the necessity of some irrelevant intercomponent paths. The default assumption in this instance is that critical components of *like* type never actively communicate information, whereas critical components of *unlike* type will always need to communicate. The extra "constraint" being considered in this paragraph gives the user of ADVISER the ability to relax this default assumption in the case of selected critical component types. The choice of this default was not entirely arbitrary. Passive types of components (such as memories of various sorts, input/output transducers and buses) usually outnumber active types of components (such as processors and device controllers) in a typical PMS structure.

3) *Critical Component Clusters:* In order to have a functional system in certain PMS structures, it is not sufficient just to satisfy the lower bounds on the number of critical components of each critical component type. In addition, these functioning critical components need to satisfy criteria regarding how they are dispersed in the structure. The situation is best explained through an example. For instance, consider Fig. 4, which depicts a multiple processor system with an inter-processor bus link L. Let us assume that the processors do not share the same address space. Then for any processor to be useful when functional, at least some of its associated memory must be functional. Thus, if the minimal requirements for the PMS structure in the figure are

$$\psi(2, P) \text{ AND } \psi(4, M)$$

then the MCRS $(P_A, P_B, M_E, M_F, M_G, M_H)$ should not be part of a system success state. This kind of behavior is observable in multiprocessor systems such as Pluribus [30]. This situation can be viewed as an association or "clustering" of CCT's in substructures of the system. In other words, if the CCT's A and B are associated or clustered in this fashion, then in order for any functioning components of type A in a given substructure to play a useful role, components of type B must also be functional in that same substructure. We shall therefore refer to a *cluster* of critical component types related in this manner.

The notion of clustering CCT's is further refined. In the general case it is not sufficient just to cluster CCT's and satisfy minimal requirements for system success. Some lower bounds are usually in force on the number of components of each such clustered type that are to be functional in a specific substructure. The bounds effectively derive from sets of inequalities which relate the number of functioning components of various CCT's. Therefore, for this cluster of CCT's, we may have the following additional inequalities for each of the buses:

$$\begin{aligned} &\text{number of } P \geq 1 \\ &\text{number of } M \geq 2 * \text{number of } P. \end{aligned} \tag{1}$$

Thus, in Fig. 4, for instance, $(P_A, M_A, P_B, M_E, M_F, M_G)$ may *not* be a system success state, even though the clustering constraint is satisfied because it may be necessary to have at least two local M's functional per functioning P to achieve system success (if, for example, a processor needs a minimum of 8K of local memory for success and each M is a 4K board). Thus, the MCRS $(P_A, M_A, M_B, P_B, M_E, M_F)$ might be part of a system success state. This phenomenon of inequality relationships on the number of functioning components belonging to a set of clustered types in a substructure will be termed *bounded clustering* of critical component types. A cluster constraint to the program will consist of a set of CCT's and a set of inequalities that relate the number of functioning components of each CCT in the cluster, as in 1) above.

B. Program Algorithms

1) *Detection of Symmetries in the PMS Interconnection Graph:* Once the user has supplied the various inputs to the program, he may ask it to compute the reliability function. The program first attempts to detect symmetric substructures, if any, within the given PMS structure. The motivation for this,

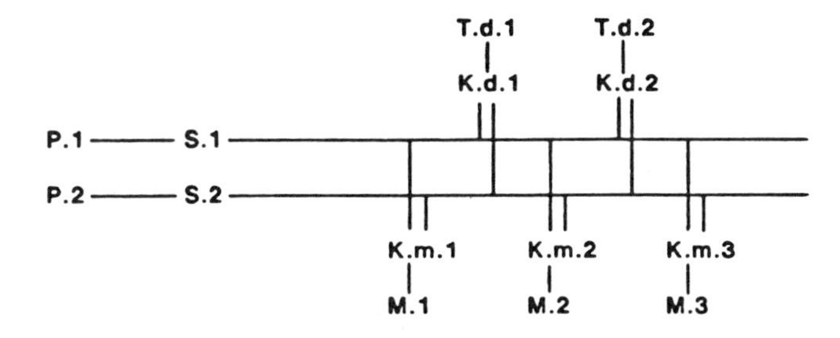

Key

Fig. 3. Example PMS structure for explanation of requirements input.

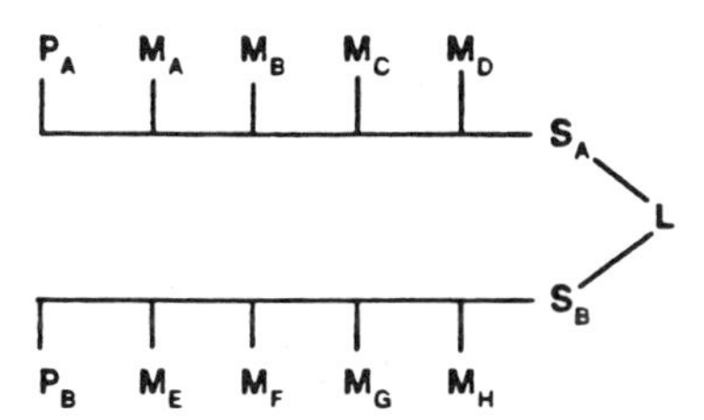

Fig. 4. Example of a PMS structure in which clustering of CCT's occurs.

as noted earlier, is to explore the possibility of avoiding duplication of effort.

Symmetry detection proceeds by assigning vertices of the interconnection graph into equivalence classes in three steps based on three equivalence relations, as follows.

Step 1: All vertices representing components of like type are assigned to the same equivalence class. Upon completion of this step there will be as many equivalence classes as there are distinct component types, say T.

Step 2: Each equivalence class generated in Step 1 is split into further equivalence classes based on the equal-degree relation, i.e., two vertices fall into the same class iff they have the same number of arcs impinging on them. At the end of this step, the maximum number of equivalence classes present will be at most $T*d_{\max}$, where $d_{\max}$ is the maximum degree of any vertex in the graph.

Step 3: The Neighbor Class Equivalence Relation (NCER) [35] is next applied to the classes resulting from Step 2 to finally detect symmetric subgraphs. The NCER is described further in Kini [33]. Roughly, two vertices will be equivalent under the NCER iff their neighboring vertices are equal in number and their corresponding neighbor vertices fall correspondingly into the same set of equivalence classes based on the NCER. At the end of this step there will be at most N NCER classes. Here N is the number of vertices in the PMS interconnection graph. This upper bound N on the number of classes generated by the NCER will occur in the extreme case that there is no structural symmetry in the graph and each component is of a distinct type.

The end result of this symmetry detection process is a set of equivalence classes into which the vertex set V of the PMS graph $G(V, E)$ is partitioned. Each class is related to other classes in a connectivity sense that derives from the symmetric connection of the vertices of that class to their neighbor vertices in their corresponding equivalence classes. The latter classes are therefore neighbors of the former class. Thus, these equivalence or neighbor classes and their connectivity relationships may be viewed as defining another graph called the *Typed Neighbor Class Graph* (*TNCG*) $G'(V', E')$. The members of the vertex set V' of the TNCG correspond uniquely to the equivalence classes on V under the NCER. The edges in the set E' map the connectivity of the vertices in V by the edges in E to the connectivity of the equivalence classes that those vertices comprise. G' may have vertices with self-loops, unlike the basic nondirected graph G, if the vertices in a given equivalence class are connected to *each other* in some symmetric fashion, thus making the equivalence class its own neighbor. Fig. 5 shows the effect of applying the symmetry detection algorithm to an example PMS structure.

2) Segmenting of the PMS Graph: Having detected symmetries in the PMS graph the program next investigates whether it is possible to segment the original PMS interconnection graph. If this is feasible then a divide-and-conquer approach may be applicable. We prefer the term "segment of G" rather than "partition of G," since the latter implies the subdivision of the vertex set induced by an equivalence relation. In the overall scheme, segments may be any form of PMS subgraph: 1) for which special reliability computation techniques are known, and 2) which can be recognized in G. By "special technique" for a kind of subgraph we mean an algorithm which, when given an instance of that kind of subgraph and an atomic requirement, returns a symbolic reliability expression for the subgraph. The expectation is, of course, that the algorithm is more efficient than simple path-finding.

At present a special algorithm exists for what are termed *Pendant Tree Subgraphs* (*PTS*), and the segmenting process in ADVISER proceeds by searching for these. PTS's are maximal rooted tree subgraphs of the PMS interconnection graph. Their roots are articulation vertices of the graph. Furthermore, the simple path between any pair of vertices in these tree subgraphs is the only path between those vertices in the overall interconnection graph G. It is common to find PTS's in most PMS structures. In particular, input/output subsystems typically assume this character, as in the examples of Fig. 6.

If arcs and vertices of such pendant trees, excluding their roots, are removed from the main PMS interconnection graph G, then the remaining vertices and arcs form a subgraph of G that is not tree-connected, i.e., contains cycles.[1] This will be referred to as the *Kernel*. The root vertices of the PTS's are termed interface vertices by virtue of their task of serving as communication "gateways" between components in the PTS's and components in the Kernel. The root of each PTS has dual status as member of the PTS as well as the Kernel. In view of this dual status, these interface vertices are accorded special treatment in the reliability calculation process.

The PTS's, along with the Kernel, form a natural set of

[1] This is not strictly true since the outcome depends on the criterion for maximality of the pendant tree when the entire PMS graph G is a tree and thus has no cycles [33, Sections 5.3 and 7.3.1].

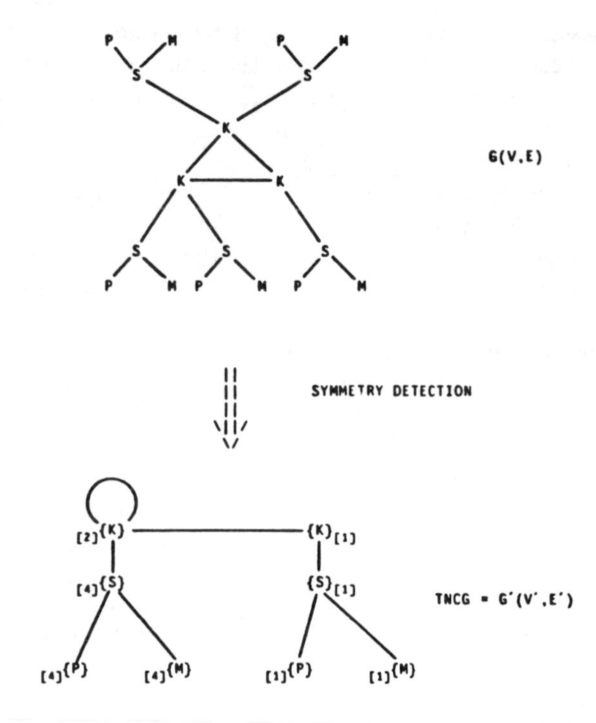

K,S --> Component Types
:] or [2]{K} --> Equivalence Class of components of type K with cardinality of two.

Fig. 5. Effect of applying symmetry detection algorithm to an example PMS structure. For details of this particular case, see Kini [33].

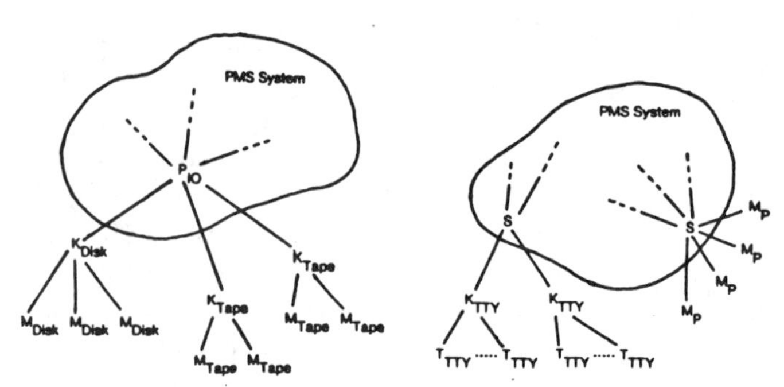

Fig. 6. Examples of Pendant Tree Subgraphs.

segments of G on the basis of which the reliability computation task may be divided. The choice of this segmenting scheme was motivated by the earlier development, during the course of this research, of an algorithm for computing the reliability functions for PTS's. *However, the scheme for making use of these segments in ADVISER does not depend on the segments being composed entirely of PTS's and the Kernel. The development of special techniques for subgraphs of G other than PTS's would allow an even finer segmenting of the graph without affecting the algorithms that make higher level decisions on how to use these segments.*

The program discovers the PTS's in a given PMS structure $G(V, E)$ by starting with those leaf vertices of G' which represent classes of leaf vertices of G. It is possible for leaf vertices of G' to represent classes of vertices of G tnat are *not* leaves of G (see [33, Section 4.3]). These "germinal trees" are then "grown" upward towards the root by adding on neighboring vertices of these leaves and at each step merging the germinal trees that overlap. This process continues (subject to termination conditions described in Kini [33]) until no more adding of vertices or merging of trees is possible. At this point a set of tree subgraphs of G' have been generated. Each of these trees in G' may represent one PTS of G or a set of PTS's. In the latter instance, all PTS's in the set will be symmetric.

C. The OVERLORD Routine

1) Generation of Feasible MCRS's: The OVERLORD routine in ADVISER is the heart of the program. In this routine critical components are "drawn" from the various segments in various ways to try to satisfy the various requirements. Each "draw" is then checked to see that requirements on G and other side-constraints (see Section V-A.3), as well as the Communication Axiom, are satisfied. The partial results of each successful draw are stored away in a special data structure. At the end of the drawing process the partial results are retrieved and merged to provide the system reliability function.

We clarify our explanation by evoking an analogy to drawing colored balls from urns. Balls are analogous to critical components and urns to the segments of G. The colors of the balls represent the various critical component types. Each urn contains a certain number (possibly zero) of balls of each color. The *minimal requirements* may then be rephrased as the desire to choose balls from urns in a way that satisfies a minimum on the total number of balls of *each* color that are chosen. This is further subject to side-constraints such as: 1) if some balls of color A are chosen from urn X, then some balls of color B must be chosen from X or else no balls may be chosen (clustering of component types), or 2) if colors A and B are to be simultaneously chosen from urn X, then a minimum of m balls of color A and n balls of color B must be chosen from the urn, with m and n being related via an inequality (bounded clustering of component types).

We now examine the process of making a "draw." Consider the simple case of a system where G has been segmented into five segments (urns) P_1 through P_5. Assume also that the only (atomic) requirement is $\psi(4, t)$. Furthermore, assume initially that each of the urns contains 4 or more balls of color t. Then the draw proceeds by generating the 5-compositions of the integer 4 as in Fig. 7(a).[2]

Each integer-part of each 5-composition represents the number of components of type t drawn from the corresponding segment of G. Since the preliminary assumption is that each segment contains at least four t's, all the 5-compositions in this

[2] A composition of the integer m into n parts, that is, an n-composition of m, is a representation of the form

$$m = k_1 + k_2 + \cdots + k_n, \qquad k_i \geq 0, \quad i = 1, 2, \cdots n$$

with regard to the particular order of the k_i's. Thus, there are exactly four 2-compositions of the integer 3, namely, $3 + 0$, $2 + 1$, $1 + 2$, and $0 + 3$. In general, there are $\binom{m+n-1}{n-1}$ n-compositions of the integer m (for a derivation, see Liu [36]). The n-compositions are not to be confused with *n-partitions* of the integer m. The latter take the same form as above except that $k_i > 0, i = 1, \cdots n$ *without* regard to the order of the k_i's. Thus, there are only two 2-partitions of the integer 3, namely, $3 + 0$, and $1 + 2$.

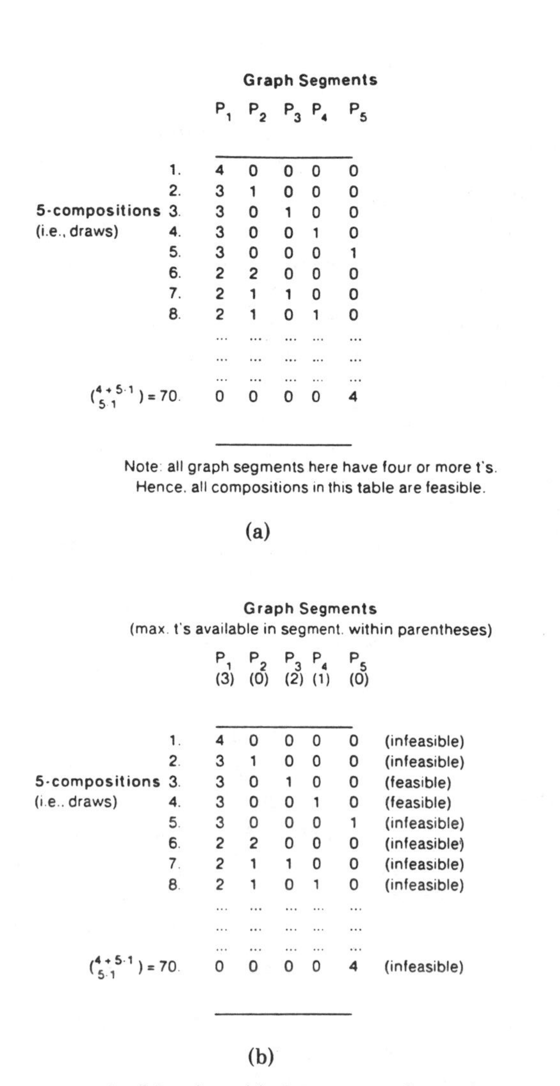

Graph Segments

		P_1	P_2	P_3	P_4	P_5
5-compositions (i.e., draws)	1.	4	0	0	0	0
	2.	3	1	0	0	0
	3.	3	0	1	0	0
	4.	3	0	0	1	0
	5.	3	0	0	0	1
	6.	2	2	0	0	0
	7.	2	1	1	0	0
	8.	2	1	0	1	0
		...	...	...	...	...
		...	...	...	...	...
		...	...	...	...	...
	$\binom{4+5-1}{5-1} = 70.$	0	0	0	0	4

Note: all graph segments here have four or more t's. Hence, all compositions in this table are feasible.

(a)

Graph Segments
(max. t's available in segment, within parentheses)

		P_1 (3)	P_2 (0)	P_3 (2)	P_4 (1)	P_5 (0)	
5-compositions (i.e., draws)	1.	4	0	0	0	0	(infeasible)
	2.	3	1	0	0	0	(infeasible)
	3.	3	0	1	0	0	(feasible)
	4.	3	0	0	1	0	(feasible)
	5.	3	0	0	0	1	(infeasible)
	6.	2	2	0	0	0	(infeasible)
	7.	2	1	1	0	0	(infeasible)
	8.	2	1	0	1	0	(infeasible)
		...	...	...	...	...	
		...	...	...	...	...	
		...	...	...	...	...	
	$\binom{4+5-1}{5-1} = 70.$	0	0	0	0	4	(infeasible)

(b)

Fig. 7. An example of drawing critical components from segments of G.

instance represent feasible draws. Once a draw is feasible it may be made, and it represents one possible alternative for satisfying the requirements. Of course, the side-constraints and the Communication Axiom must be satisfied before the MCRS, so drawn, constitutes part of a functioning system. Partial results are computed for an MCRS as follows. Consider composition 6, P_1 column, in Fig. 7(a). If P_1 is a PTS, then the special algorithm for PTS's is given the PTS P_1 and the atomic requirement $\psi(2, t)$ and will return a partial result reliability expression. Such partial results across the five columns will be combined in conjunction since the requirement $\psi(4, t)$ must be satisfied.

In general, not all of the segments of G will contain enough components of a given type to support a given atomic requirement on that type. Fig. 7(b) depicts an example of such a case. In this instance, against the requirement of four t's, none of the segments P_1 through P_5 are able to supply *all* four, and P_2 and P_5 contain no t's at all. These upper bounds on the number of components of type t that may be drawn from a particular segment can, in general, drastically curtail the number of feasible draws. The program actually generates all possibilities when the compositions of an integer are desired, testing each one against the upper bounds dictated by the contents of each segment for the particular case. The algorithm used in the ADVISER program to generate feasible compositions is an adaptation of the NEXTCOM algorithm in Nijenhuis and Wilf [37]. However, only those compositions which represent the feasible draws emerge from the generator function. The overhead for this scheme is very small compared to the computing requirements of other portions of the program.

Thus far in this section we have only considered atomic requirements of the type $\psi(m, t)$. In the more usual case the requirements will consist of a Boolean expression on such atoms. These expressions may be naturally divided into two classes: those that contain only conjunctions of atomic requirements (*Conjunctive Requirements*) and those that contain at least one disjunction in addition to possibly containing conjunctions (*Disjunctive Requirements*).

In the worst case all the segments of G contain enough components of each critical component type to satisfy each and every atom in the conjunctive requirement. In other words, all possible draws will be feasible. Then the total number of feasible draws over the entire conjunctive requirement

$$\bigwedge_{i=1}^{r} \psi(m_i, t_i) \text{ is given by } f_c = \prod_{i=1}^{r} \binom{m_i + n - 1}{n - 1}$$

where n is the number of segments of G and each segment contains at least m_i components of type t_i. In the purely disjunctive case (no conjunctions in the overall requirement) the requirement is

$\bigvee_{i=1}^{r} \psi(m_i, t_i)$ and the number of feasible draws is

$$f_d = \sum_{i=1}^{r} \binom{m_i + n - 1}{n - 1}.$$

The numbers f_c and f_d represent the upper bounds on the number of cases to be analyzed for purely conjunctive and purely disjunctive requirements respectively. However, in the more usual case of a mixture of conjunctions and disjunctions, the worst case bound can be much higher. A closed-form solution for the worst case bound for this intermediate variety of requirement expressions is unobtainable, since it depends on the values of the m_i's and n, as well as the positions and precedence order of the conjunctions and disjunctions in the expression.

The generation of all possible sets of compositions for a conjunctive requirement is a backtrack procedure which uses a stack discipline. This may be observed by considering the following example of a conjunctive requirement:

$$\psi(m_1, t_1) \wedge \psi(m_2, t_2) \wedge \psi(m_3, t_3). \tag{2}$$

Assume that there are n segments of G. Then for each n-compositions of m_2 have to be generated in sequence. In turn, for *each* of the latter, all n-compositions of m_3 have to be generated. This may be done in a systematic manner through the use of a stack as follows. Levels from bottom to top of the stack in the algorithm will correspond one-to-one to the atoms in the conjunctive requirement from left to right. Each level of the stack holds an n-composition of the corresponding requirement integer. In our example requirements expression (2), level j of the stack (j = 1, 2, 3) will hold the n-composition of m_j. The next n-composition in sequence is generated at any level k of the stack when all possible n-compositions have been exhausted at level $k + 1$. Subsequently, the composition generation cycle is restarted at level $k + 1$. The process terminates when all n-compositions have been exhausted at level 1 (bottom of the stack). For further details the reader is referred to Kini [33]. It is apparent that in the worst case at each level k of the stack, *all* the n-compositions m_k are generated repeatedly in sequence. The number of times this sequence is repeated at that level is given by

$$\prod_{r=1}^{k-1} \binom{m_r + n - 1}{n - 1}.$$

2) *Satisfying the Communication Axiom:* Each of the feasible draws generated constitutes an MCRS which may or may not be part of a functional system state. This depends, of course, on whether side-constraints have been met and the Communication Axiom has been satisfied. Checks are made with regard to these by the OVERLORD routine. For a given feasible draw the critical components chosen in that draw will be scattered in some fashion among the segments. In order to satisfy the Communication Axiom, the critical components in the pendant tree segments will have to communicate with each other and with critical components drawn from the Kernel, through paths in the Kernel. Moreover, information may flow in and out of the Kernel only through the root vertices of the pendant trees, since these are articulation vertices of G.

Thus, the question whether the Communication Axiom may possibly be satisfied by a given candidate draw may be separated into two concerns, namely

1) critical components in tree segments should be able to communicate with the component represented by the root vertex of that tree, and

2) the root vertices of the tree segments that contain the critical components of this candidate draw should be able to communicate with each other and with critical components drawn from the Kernel, via paths in the Kernel.

The former concern is addressed by the algorithm TREEREL developed for the PTS's. The latter concern is the domain of the OVERLORD routine (see [33, chs. 5 and 6]).

For each draw, therefore, the OVERLORD routine performs checks on the Kernel. For each such iteration, depending on which segments the critical components are drawn from, the set of relevant root vertices (and therefore their respective PTS's) and the set of critical components drawn from the Kernel are subject to change. The change is unpredictable and depends on the requirements expression, the scattering of the available critical components in various portions of the system, and the nature of the side-constraints. This unpredictability seems to imply that partial results may have to be rederived many times over during the process. However, a more efficient method is possible.

The reader will have noticed when the drawing process was described earlier that for any level r in the stack, the n-compositions of m_r were generated repeatedly except in the case of $r = 1$. The program is thus able to anticipate that certain partial results will be needed in several iterations. Such partial results are computed once initially and stored away in special hash tables. In general, in a compound requirement, many atoms in the expressions may refer to the same critical component type, say t. Let one such atom be $\psi(m_i, t)$. When compositions of the various m_i's are taken over the segments of G, the minimum number of critical components of type t that may be drawn from some segment, say p, is one. The maximum number of critical components of type t, say $m_{\max}$, that may be expected to be drawn from p is the lesser of: 1) the number of components of type t in p, and 2) the largest m_i in any atom of the form $\psi(m_i, t)$ in the compound requirement. Thus, the OVERLORD routine generates partial results for each such segment p for the set of atomic requirements $\{\psi,(j, t)\}$ where $j = 1, \cdots, m_{\max}$. This is done for each critical component type. These stored partial results are then later retrieved and used during the process of generating feasible compositions.

The Kernel is treated slightly differently, although even here such reusing of intermediate results is possible; the results are just of a different nature. The OVERLORD routine checks for the existence of K-edges in the Kernel that lead among the pendant trees (or what is equivalent, their root, or interface vertices) and other critical components drawn from the Kernel. Thus, in this case it generates and stores away partial results for such K-edges between *all possible pairs* of critical components and/or interface vertices in the Kernel.

In summary, all of the partial results that could possibly be used in the computation are generated once in the beginning. For each iteration, then, the OVERLORD routine retrieves and uses the appropriate partial results after ascertaining that the draw for that particular iteration will satisfy the various side-constraints and the Communication Axiom. The methods of representation and combination of the partial results are the subject of the next section.

3) *Representation of Reliability Expressions:* At all stages of the computation of the reliability function, the identity of each component in the structure is retained in the reliability expressions that are the partial results and the final reliability function. As a consequence, recalling that statistical independence of component failure behavior has been assumed, the structure of the partial results and the final reliability function will be very similar to that of a Boolean function in its minterm canonical form. Each partial result will be a function of the reliabilities of some subset of the system components. The expression that is the body of the function will consist of "minterms." Each minterm will consist of the algebraic product of the probabilities of success (reliabilities) of a subset of components. Each of these factors of a minterm will appear only once in the minterm and will be raised to the unit power. Each minterm will, in addition, be prefixed with a positive or negative sign. We shall term such an expression a *Canonical Reliability Polynomial.* An example of one is given below:

$$R_x = R_1 + R_2 + R_3 - R_1R_2 - R_2R_3 - R_1R_3 + R_1R_2R_3.$$

R_x is the system reliability for a 1-out-of-3 system. Such a system is functional only if at least one of its three components is functional. R_1, R_2, and R_3 are, respectively, the reliabilities of the three components. The above function could have been reduced to a noncanonical form if, say, $R_1 = R_2 = R$, whereupon

$$R_x = 2R + R_3 - 2R^2 - R_3^2 + R^2R_3, \quad R_1 = R_2 = R.$$

However, the canonical form is the most general and represents the reliability of a system wherein no two components have identical reliability functions. All noncanonical forms may be derived from the canonical form by appropriate algebraic substitution, although the reverse is not generally possible.

This, then, is one of two primary motives for retaining partial results in canonical form. In other words, the fact that two or more components in the system may have identical reliability functions does not change the canonical form, since its only proviso is that the components have statistically independent failure behavior.

The other equally important motive for retention of canonical form concerns the robustness and simplicity of the algorithm to combine the partial results in conjunction or disjunction. The algorithm and its data structures are the subject of [33, ch. 3]. It will suffice here to note two simple points.

First, factors in the minterms of a canonical reliability function are always raised to the unit power and are never replicated within the minterm. Hence, each minterm may be represented by a string of $(N + 1)$ bits (N system components + 1 sign bit) (this idea has been utilized before though not quite in the same fashion; see Gandhi *et al.* [38]), wherein each of the first N bits represents a unique factor (component). Furthermore, a *canonical reliability function is a set of minterms which may then be represented as an unordered list of bitstrings, each bitstring in the list representing one minterm.*

Second, operations on pairs of such lists will be composed of simple logical operations on pairs of bitstrings from the two lists. Thus, the resultant set of terms will contain bitstrings arising from the Cartesian product of the two input sets of bitstrings, using those logical operations.

A price is paid, however, for the use of the canonical form of the reliability function since the number of terms in the canonical form usually exceeds those in a simplified form for more complicated problems. Indeed, it is in the code that processes these lists of bitstrings where the ADVISER program spends much of its computation time. This problem is partially averted by assigning bits in the bitstrings for *statistically independent partial results*. This results in smaller lists in the canonical form. The partial results that are allotted bits in the bitstrings eventually become separate numeric calculations, the results of which are substituted into the main reliability function when it is numerically evaluated.

D. Program Output

1) Printing of Results: The final stage of the operation of the program consists of reducing the canonical form of the system reliability function that was generated, and then printing it out appropriately.

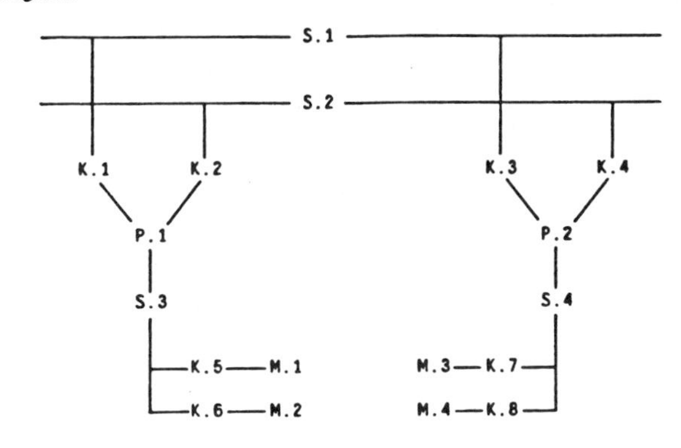

Requirements:

$$\psi(1,P) \wedge \psi(2,M)$$

Fig. 8. Example EXAMP.PMS—PMS diagram and requirements.

The reduction proceeds by noticing, from the component-type and interconnection graph declarations, which components are of the same type (i.e., have identical reliability functions). Appropriate substitutions and algebraic simplification are then performed to obtain the reduced function. The simplifications are rudimentary and limited to cancellation of like terms of opposite sign and the gathering of like terms of like sign. Complex symbolic manipulation such as factoring is not attempted, for example. The resulting noncanonical symbolic form is then a function of the symbolic reliabilities of the *component types*, and all individual component identities are lost as a result of the reduction. The program does, however, keep a copy of the canonical form should several printings be desired. The output of the program in its current state of development is a text file. This file will, upon the user's option, contain the text of either a Fortran function or a SAIL [39] program procedure, which computes the system reliability function $R_{sys}(t)$ for the input PMS structure under the input requirements. The procedure or function will have as a single parameter the time t at which the reliability is to be computed. The file may then be compiled with the appropriate compiler and used in numerical calculations that employ the generated reliability function. The advantage of having the system reliability output as a program is that different reliabilities may be used for the individual component types by editing the program rather than redoing the ADVISER computation. As a third option ADVISER will generate an output file containing the polynomials for the partial results and system reliability in a form suitable for input to the powerful MACSYMA symbolic manipulation program [29]. This makes it possible to carry out more advanced manipulations such as factoring and symbolic differentiation for sensitivity analysis on the system reliability function.

VI. An Example of the Use of ADVISER

We now present an example of the use of ADVISER in generating the symbolic reliability function for the PMS structure of Fig. 8. In the figure, two processors $P.1$ and $P.2$ can communicate through one or both of the two interprocessor buses (S.1 and S.2), and each has its own local bus (S.3 and S.4, respectively) with two disk memories apiece (M.1 through M.4). The requirements chosen for this structure were $\psi(1, P) \wedge \psi(2, M)$. The following series of listings show the input to ADVISER for the problem, a transcript of the teletype session with the program, and finally, the output from the program. Preceding each listing are comments to aid in its interpretation. ADVISER's output has been validated by comparison with several hand-generated reliability models.

LISTING 1—EXAMPLE EXAMP.PMS

The following is a listing of the command file EXAMP.PMS prepared for input to ADVISER in order to set up the example problem EXAMP. Listing 2 shows how it is used.

```
input title
EXAMP.PMS -- A dual bus, 2 processor-bus architecture

input types
dynabus         dbus        E      0.0001
k.dbus          ks          E      6.000
pdp11           p           W      8.00    0.89
unibus          ubus        E      0.0001
disk            ms          W      10.000  0.91
k.disk          km          W      6.000   0.86

input pms
s.1             dynabus     k.1    k.3
s.2             dynabus     k.2    k.4
k.1             k.dbus      s.1    p.1
k.2             k.dbus      s.2    p.1
k.3             k.dbus      s.1    p.2
k.4             k.dbus      s.2    p.2
p.1             pdp11       k.1    k.2     s.3
p.2             pdp11       k.3    k.4     s.4
s.3             unibus      p.1    k.5     k.6
s.4             unibus      p.2    k.7     k.8
k.5             k.disk      s.3    ms.1
k.6             k.disk      s.3    ms.2
k.7             k.disk      s.4    ms.3
k.8             k.disk      s.4    ms.4
ms.1            disk        k.5
ms.2            disk        k.6
ms.3            disk        k.7
ms.4            disk        k.8

input restriction selftalk
pdp11

set watch run
```

LISTING 2—EXAMPLE EXAMP.PMS

This listing shows the teletype session with ADVISER that solved the the problem and printed out the solution. User input is underlined. The "@ examp.pms" command causes the file shown in Listing 1 to be read and its lines executed as a series of commands. The reading of the command file causes the PMS structure and the component types to be defined. The command "input restriction selftalk" implements the Intra-Component-Type Communication side constraint discussed earlier. It states that members of the critical component class "pdp 11" communicate information among themselves. The commands set up in the command file may just as well be entered individually during the teletype session, but the command file option allows the user to think about and prepare an error-free input offline, with an editor. The print command prompts for the name of the SAIL procedure or Fortran function to be printed; the default is RSYS. The "print" command is used with the "macsyma" option to obtain the computed symbolic function in a form suitable for input to the MACSYMA symbolic manipulation system. Listing 3 below shows the symbolic reliability function printed out as a Fortran function that computes it numerically. Since it is similar, the equivalent SAIL procedure printed out with the second "print" command is not shown. Listing 4 shows the symbolic reliability function printed out in the form of a command file to the MACSYMA symbolic manipulation system. This may be used to load the reliability function into MACSYMA in order to perform various symbolic operations on it, e.g., differentiate it to study the sensitivity of the system reliability to various component class reliabilities or the reliability of a given component if that component was the member of a singleton component class in the description of the PMS structure.

```
 TOPS-20 Command processor 4(56047)-6
@adviser
ADVISER 2A(6)   Thursday  3 Sep 81   10:31:40
*@examp.pms
**input title
Title: EXAMP.PMS -- A dual bus, 2 processor-bus architecture
**
**input types
Input component types and associated print-names; end with blank line
Types    |        Print-names       | Rel.Fn.  |  Lambda   |  {Alpha}
-----------------------------------------------------
dynabus          dbus             E        0.0001
k.dbus           ks               E        6.000
pdp11            p                W        8.00     0.89
unibus           ubus             E        0.0001
disk             ms               W        10.000   0.91
k.disk           km               W        6.000    0.86

**input pms
Input graph in format (end with blank line):
Component name | Typename | Neighbour,Neighbour,....
------------------------------------------------------
s.1              dynabus          k.1      k.3
s.2              dynabus          k.2      k.4
k.1              k.dbus           s.1      p.1
k.2              k.dbus           s.2      p.1
k.3              k.dbus           s.1      p.2
k.4              k.dbus           s.2      p.2
p.1              pdp11            k.1      k.2      s.3
p.2              pdp11            k.3      k.4      s.4
s.3              unibus           p.1      k.5      k.6
s.4              unibus           p.2      k.7      k.8
k.5              k.disk           s.3      ms.1
k.6              k.disk           s.3      ms.2
k.7              k.disk           s.4      ms.3
k.8              k.disk           s.4      ms.4
ms.1             disk             k.5
ms.2             disk             k.6
ms.3             disk             k.7
ms.4             disk             k.8

**input restriction selftalk
Input list of Type names:
pdp11

**set watch run

[.00]

**

[.08]
*input requirements
Input boolean function (X of N AND/OR Y of M etc.):
1 of pdp11 and 2 of disk

[.16]
*get reliability
Generating symmetries.....
[.08]
Hashing core term lists.....
[.08]
Hashing tree term lists.....
[.07]

Setting up table space.....
Computing Reliability Function.....
CG: 1
######
        SC: 6
Collapsing term-list tree.....
######
```

```
[.25]
[.39]
Releasing table space.....
Reducing Reliability Function.....

Number of terms to be processed = 60. Here goes....
#################################################################
Doing Algebraic Simplification....
#############
Terms remaining = 8
[.74]
Done!

[1.45]
*openo examp.for

[.11]
*print reliability fortran
Name of output Fortran procedure [RSYS]: <carriage-return>

[.13]
*closo

[.10]
*openo examp.sai

[.14]
*print reliability sail
Name of output SAIL procedure [RSYS]: <carriage-return>

[.10]
*closo

[.09]
*openo examp.mcs

[.09]
*print reliability macsyma

[.03]
*closo

[.09]
*exit

[.00]

EXIT
@
```

LISTING 3—EXAMPLE EXAMP.PMS

This listing shows the reliability function computed by ADVISER for the PMS structure of Fig. 8, printed in the form of a Fortran function that computes it numerically. The names of ADVISER-generated temporary variables are prefixed with "*XXX*." The dollar sign "$" is used as a continuation character in the Fortran program.

```
C-----------------------------------------------------------------------
C ** FORTRAN Module for Reliability Function evaluation
C **      produced by ADVISER on Thursday,  3 Sep 81 at 10:33:43 for [4,1367]
C-----------------------------------------------------------------------
C ** Task Title: EXAMP.PMS -- A dual bus, 2 processor-bus architecture
C
C ** Requirements on the Structure were:
C
C       (1-OF-P AND 2-OF-MS)
C
C-----------------------------------------------------------------------
C
```

```
C ***  Begin Reliability Function evaluation code;

       REAL FUNCTION RSYS (T);
       IMPLICIT REAL (A-Z)

       WEIBUL(LAMBDA,ALPHA,TIME)=EXP(-(LAMBDA*1E-6*TIME)**ALPHA)

       DBUS = EXP(-0.000100 * 1E-6 * T)
       KS = EXP(-6.000000 * 1E-6 * T)
       P = WEIBUL( 8.000000 , 0.890000 , T )
       UBUS = EXP(-0.000100 * 1E-6 * T)
       MS = WEIBUL( 10.000000 , 0.910000 , T )
       KM = WEIBUL( 6.000000 , 0.860000 , T )
C ** End of expressions for calculating individual reliabilities;

       XXX1 = P * UBUS * MS**2 * KM**2

       XXX2 = 2.0 * P * UBUS * MS * KM  -  P * UBUS * MS**2 * KM**2

C ** End of template evaluating expressions;

       MODREL = 0

       MODREL = 2.0 * XXX1  -  XXX1**2  +  2.0 * DBUS * KS**2 * XXX2**
      $2  -  4.0 * DBUS * KS**2 * XXX1 * XXX2  +  2.0 * DBUS * KS**2 *
      $XXX1**2  -  DBUS**2 * KS**4 * XXX2**2  +  2.0 * DBUS**2 * KS**4
      $ * XXX1 * XXX2  -  DBUS**2 * KS**4 * XXX1**2
C **  End of System Reliability computation;

       RSYS = MODREL
       RETURN
       END
```

LISTING 4—EXAMPLE EXAMP.PMS

This listing shows the reliability function computed by ADVISER for the PMS structure of Fig. 8, printed in the form of a command file to MACSYMA. Compare to Listing 3.

```
/*
---------------------------------------------------------------------------
MACSYMA Module for Reliability Function manipulation
   produced by ADVISER on Thursday,  3 Sep 81 at 10:34:34 for [4,1367]
---------------------------------------------------------------------------
Task Title: EXAMP.PMS -- A dual bus, 2 processor-bus architecture



Requirements on the Structure were:

(1-OF-P AND 2-OF-MS)
---------------------------------------------------------------------------
*/


%%T1:
        P * UBUS * MS↑2 * KM↑2;
%%T2:
        2 * P * UBUS * MS * KM  -  P * UBUS * MS↑2 * KM↑2;
/* End of temporary variable initializations*/

System%Reliability: 0;
System%Reliability:
        2 * %%T1  -  %%T1↑2  +  2 * DBUS * KS↑2 * %%T2↑2  -  4 * DBUS
         * KS↑2 * %%T1 * %%T2  +  2 * DBUS * KS↑2 * %%T1↑2  -  DBUS↑2
         * KS↑4 * %%T2↑2  +  2 * DBUS↑2 * KS↑4 * %%T1 * %%T2  -  DBUS↑
        2 * KS↑4 * %%T1↑2
        ; /*End of System Reliability computation*/

FACTOR(%);
```

VII. Conclusion

This paper has outlined the methodology underlying the ADVISER program. ADVISER was intended as a first step towards the automation of reliability computation for Processor-Memory-Switch structures. Unlike earlier programs, ADVISER works directly from the interconnection structure of the PMS structure and a succinct statement of the operational requirements on the structure. It produces, subject to the limitations set forth in Section IV, the symbolic hard-failure reliability for the system, thereby allowing other reliability measures to be computed from it. The analysis strategy of segmenting the PMS interconnection structure into portions for which special reliability computation techniques are known, offers much promise. Since the burden of analysis of the structure is taken over by the program, more attention may be given to the issue of adequate problem specification. Furthermore, the program becomes accessible to a large class of users with less experience in reliability computation. Automation of reliability function computation also reduces the probability of human error. ADVISER, although currently modest in scope and capabilities, offers through its methodology the hope of more sophisticated and useful versions of the program that will address reliability issues not currently covered.

References

[1] C. G. Bell and A. Newell, *Computer Structures: Readings and Examples.* New York: McGraw-Hill, 1971.

[2] K. Greene and T. J. Cunningham, "Failure modes, effects and criticality analysis," in *Proc. 1968 IEEE Annu. Symp. on Rel.*, Boston, MA, 1968, p. 374.

[3] R. E. Barlow and H. E. Lambert, "Introduction to fault tree analysis," in *Reliability and Fault Tree Analysis: Theoretical and Applied Aspects of System Reliability and Safety Assessment*, R. E. Barlow, Ed. Philadelphia: Soc. Indust. and Appl. Math., 1975, pp. 7–35.

[4] U.S.N.R.C., "Reactor safety study—An assessment of accident risks in U.S. commercial nuclear power plants, WASH1400 (NUREG-75/014)," U.S. Nucl. Regulatory Commission, Tech. Rep., 1975, available from NTIS, Springfield, VA.

[5] S. Lapp and G. Powers, "Computer-aided synthesis of fault-trees," *IEEE Trans. Rel.*, vol. R-26, p. 2, Apr. 1977.

[6] R. Wilkov, "Analysis and design of reliable computer networks," *IEEE Trans. Commun.*, vol. COM-20, p. 660, Mar. 1972.

[7] E. Hansler, G. K. McAuliffe, and R. S. Wilkov, "Exact calculation of computer network reliability," *Networks*, vol. 4, pp. 95–112, 1974.

[8] A. Rosenthal, "Computing the reliability of complex networks," *SIAM J. Appl. Math.*, vol. 32, pp. 384–393, Mar. 1977.

[9] M. O. Ball, "Complexity of network reliability computations," *Networks*, vol. 10, pp. 153–165, 1980.

[10] A. C. Nelson, Jr., J. R. Batts, and R. L. Beadles, "A computer program for approximating system reliability," *IEEE Trans. Rel.*, vol. R-19, pp. 61–65, May 1970.

[11] J. L. Fleming, "RELCOMP: A computer program for calculating system reliability and MTBF," *IEEE Trans. Rel.*, vol. R-20, p. 102, Aug. 1971.

[12] P. O. Chelson and R. E. Eckstein, "Reliability computation from reliability block diagrams," Nat. Aeronaut. and Space Administration, Jet Propulsion Lab., Tech. Rep. 32-1543, Dec. 1971.

[13] Y. H. Kim, K. E. Case, and P. M. Ghare, "A method for computing complex system reliability," *IEEE Trans. Rel.*, vol. R-21, p. 215, May 1972.

[14] E. V. Krishnamurthy and G. Komissar, "Computer-aided reliability network analysis," *IEEE Trans. Rel.*, vol. R-21, p. 86, May 1972.

[15] A. Satyanarayana and A. Prabhakar, "New topological formula and rapid algorithm for reliability analysis of complex networks," *IEEE Trans. Rel.*, vol. R-27, pp. 82–100, June 1978.

[16] K.K. Aggarwal and S. Rai, "Symbolic reliability evaluation using logical signal relations," *IEEE Trans. Rel.*, vol. R-27, pp. 202–205, Aug. 1978.

[17] A. Satyanarayana and J. N. Hagstrom, "A new formula and an algorithm for the reliability analysis of multi-terminal networks," Univ. of California, Berkeley, Tech. Rep. ORC80-11, June 1980, available from Oper. Res. Cen., Univ. of California, Berkeley, CA.

[18] A. Avizienis, "Architecture of fault-tolerant computing systems," in *Proc. 5th Annu. Int. Symp. on Fault-Tolerant Comput.*, IEEE Comput. Soc., 1975, pp. 3–16.

[19] W. G. Bouricius, W. C. Carter, and P. R. Schneider, "Reliability modeling techniques for self-repairing computer systems," in *Proc. 24th Nat. Conf. ACM*, 1969, pp. 295–309.

[20] W. G. Bouricius, W. C. Carter, D. C. Jessep, P. R. Schneider, and A. B. Wadia, "Reliability modeling for fault-tolerant computers," *IEEE Trans. Comput.*, vol. C-20, pp. 1306–1311, 1971.

[21] F. P. Mathur, "Automation of reliability evaluation procedures through CARE—The computer-aided reliability estimation program," in *Proc. Fall Joint Comput. Conf.*, AFIPS, 1972, pp. 65–77.

[22] Y. W. Ng and A. Avizienis, "ARIES—An automated reliability estimation system for redundant digital structures," in *Proc. 1977 IEEE Annu. Rel. and Maintain. Symp.*, Jan. 1977, pp. 108–113.

[23] ——, "A unified reliability model for fault-tolerant computers," *IEEE Trans. Comput.*, vol. C-29, pp. 1002–1011, Nov. 1980.

[24] C. Landrault and J. C. Laprie, "SURF—A program for modeling and reliability prediction for fault-tolerant computing systems," in *Information Technology*, J. Moneta, Ed. Amsterdam, The Netherlands: North-Holland, 1978.

[25] R. E. Cox and H. D. Miller, *The Theory of Stochastic Processes.* London: Methuen, 1968.

[26] M. J. Knudsen, "PMSL, An interactive language for system-level description and analysis of computer structures," Ph.D. dissertation, Carnegie-Mellon Univ., Pittsburgh, PA, Apr. 1973.

[27] G. Powers and S. Lapp, "Computer-aided fault-tree synthesis," *Chem. Eng. Progress*, Apr. 1976.

[28] M. L. Shooman, *Probabilistic Reliability: An Engineering Approach.* New York: McGraw-Hill, 1968.

[29] The MathLab Group Laboratory for Computer Science, Massachusetts Inst. Technol., *MACSYMA Reference Manual*, version 9, 2nd printing. Cambridge, MA: Massachusetts Inst., of Technol., Dec. 1977.

[30] S. M. Ornstein *et al.*, "Pluribus—A reliable multiprocessor," in *AFIPS Conf. Proc.*, 1975, pp. 551–559.

[31] R. J. Swan, S. H. Fuller, and D. P. Siewiorek, "Cm*: A modular, multi-microprocessor," in *AFIPS Conf. Proc.*, vol. 46, 1977, pp. 637–644.

[32] J. A. Katzman, "A fault-tolerant computing system," Tandem Computers, Inc., Tech. Rep., 1977.

[33] V. Kini, "Automatic generation of reliability functions for processor-memory-switch structures," Ph.D. dissertation, Dep. Comput. Sci., Carnegie-Mellon Univ., Rep. CMU-CS-81-121, Apr. 1981.

[34] R. E. Barlow and F. Proschan, *Statistical Theory of Reliability and Life Testing.* New York: Holt, Rinehart, and Winston, 1975.

[35] J. Gaschnig, "A 'neighbors class' node partitioning algorithm for finding symmetry classes in graphs," to be published.

[36] C. L. Liu, *Introduction to Combinatorial Mathematics.* New York: McGraw-Hill, 1968.

[37] A. Nijenhuis and H. S. Wilf, *Combinatorial Algorithms for Computers and Calculators*, 2nd ed. New York: Academic, 1978.

[38] S. L. Gandhi, K. Inoue, and E. J. Henley, "Computer aided system reliability analysis and optimization," in *Computer-Aided Design: Proc. IFIP Working Conf. on Principles of Computer-Aided Design*, J. Vlietstra and R. F. Wielinga, Eds. Eindhoven, The Netherlands: IFIP, Oct. 1972, pp. 283–308.

[39] J. F. Reiser, Ed., Dep. Comput. Sci., Stanford Univ. SAIL, Stanford, CA, Rep. STAN-CS-76-574, Aug. 1976; also available from Nat. Tech. Inform. Service, Springfield, VA.

Vittal Kini (S'75–M'79) was born in New Delhi, India, on January 4, 1952. He received the B.Tech. degree from the Indian Institute of Technology, Bombay, India, in 1973, and the M.S. and Ph.D. degrees in 1975 and 1981, respectively, from Carnegie-Mellon University, Pittsburgh, PA, all in electrical engineering.

From 1974 through 1979 he held a Research Assistantship in the Department of Electrical Engineering at Carnegie-Mellon University, working in the areas of design-automation of digital hardware and fault-tolerant computing. In 1979 he joined the Information Sciences Institute of the University of Southern California, Marina del Rey, where he has carried out research into hardware descriptive languages, emulation of machine descriptions, and microcode verification. He is currently supervising a project at USC-ISI, investigating strategies for the testing and debugging of formal semantic definitions of programming languages (in particular, Ada) by transforming such definitions into executable program objects. His current research interests include fault-tolerant computing, reliability modeling, formal semantics of programming languages, design-automation of digital hardware, and the design of hardware descriptive languages.

Dr. Kini is a member of the IEEE Computer Society and Phi Kappa Phi.

Daniel P. Siewiorek (S'67–M'72–SM'79–F'81), for a photograph and biography, see page 671 of the July 1982 issue of this TRANSACTIONS.

Combinatorial Reliability Analysis of Multiprocessor Computers

Kai Hwang, Senior Member IEEE
Purdue University, West Lafayette
Tian-Pong Chang, Member IEEE
University of New Haven, West Haven

Key Words—**Multiprocessor system, Reliability evaluation, Combinatorial analysis.**

Reader Aids—
Purpose: Report on analysis
Special math needed for explanations: Combinatorics and probability
Special math needed to use results: Same
Results useful to: Computer system designers and reliability engineers

Abstract—**This paper proposes a combinatorial method to evaluate the reliability of multiprocessor computers. Multiprocessor structures are classified as crossbar switch, time-shared buses, and multiport memories. Closed-form reliability expressions are derived via combinatorial path enumeration on the probabilistic-graph representation of a multiprocessor system. The method can analyze the reliability performance of real systems like "C.mmp", "Tandem 16", and "Univac 1100/80". User-oriented performance levels are defined for measuring the performability of degradable multiprocessor systems. For a regularly structured multiprocessor system, it is fast and easy to use this technique for evaluating system reliability with statistically independent component reliabilities. System availability can be also evaluated by this reliability study.**

1. INTRODUCTION

Multiprocessor system (MPS) consists of several processors sharing multiple memory units and input/output channels via some interconnection structure. MPS provides higher system throughput, application flexibility, and better reliability/availability than conventional uniprocessor computers. Three categories of MPS interconnection structures: crossbar switches, time-shared buses, and multiport memories [1], are considered. We present a graph-theoretic model for evaluating MPS reliability through some combinatorial analysis [2].

Functionally, an MPS can assume one or more of the following operation modes: MIMD mode — multiple instruction streams and multiple data streams are executed by multiple interacting processors which update a shared memory space; M/SISD mode — multiple number of single-instruction and single-data (SISD) streams are independently executed by two or more processors; and SISD mode — only one of the processors survives to execute an SISD stream. M/SISD and SISD are degenerate cases of the MIMD mode. An MPS is *multiprocessing reliable,* i.f.f. it can be used for either MIMD or M/SISD operations.

We shall first present the graph reliability model. Various multiprocessor architectures are then evaluated by their corresponding probabilistic graphs. Two real-life multiprocessor computers, "C.mmp" and "Tandem/16" are analyzed using the proposed reliability evaluation method. Other commercial multiprocessors that can be evaluated by this method include the "Univac 1100/80", "IBM 370/168", "CDC Cyber-170", "Burrough B-7700", and "DEC System 10/KL-10" [6]. This combinatorial path-enumeration method can eliminate the calculation of unnecessary terms, a significant improvement over the methods by [3, 5]. This method can be also extended to study loosely-coupled multiple computers for distributed processing.

2. NOTATIONS AND DEFINITIONS

MPS	multiprocessor system or multiprocessor computer.
SISD	single-instruction stream and single data stream.
M/SISD	multiple SISD streams that are independent.
MIMD	multiple instruction streams and multiple data streams.
DFP	a data flow path in an MPS connects a processor and a memory unit via an interconnection path (bus, switch, or multiport).
n, m, b	the numbers of processors, memory units, and common buses in an MPS, respectively.
$E(\alpha, \beta)$	the event of an MPS with α-out-of-n:G processors updating β-out-of-m:G memory units in MIMD mode or M/SISD mode.
$G(n, i, p)$	$\binom{n}{i} p^i \bar{p}^{n-i}$ the probability of having exactly i good units out of n units, where p is the unit reliability and $\bar{p} = 1 - p$ is the unit unreliability.
x	$e^{-\lambda_x t}$, $\bar{x} \equiv 1 - x$, the reliability and unreliability of each processor, where λ_x is the failure rate. Similarly, y, $\bar{y}$ for memory units with λ_y; and z, $\bar{z}$ for crosspoint switches (with $\lambda_z = \lambda_c$), or for common buses (with $\lambda_z = \lambda_b$), or for multiport controllers (with $\lambda_z = \lambda_p$). All components of the same type are i.i.d.
w, u	The reliability of an interrupt bus between PDP11s in the "C.mmp", and of the DYNABUS (X and Y) in the "Tandem 16" system.
$R(t)$	Pr{$E(1, 1)$} the *system reliability* of using an MPS for either MIMD, SISD, or M/SISD operations.
$R_m(t)$	Pr{$E(2, 1)$} the *multiprocessing reliability* of using an MPS for either MIMD or M/SISD operations with at least two instruction streams.

$R_u(t)$ — $R(t) - R_m(t)$ the *uniprocessing reliability* of using a degraded MPS for exclusively SISD operations.

$R_{\alpha,\beta}(t)$ — $\Pr\{E(\alpha, \beta)\}$ the *threshold reliability* of using an MPS for MIMD operations for the event $E(\alpha, \beta)$. $R_{\alpha_1,\beta}(t) < R_{\alpha_2,\beta}(t)$, if $\alpha_1 > \alpha_2$; $R_{\alpha,\beta_1}(t) < R_{\alpha,\beta_2}(t)$, if $\beta_1 > \beta_2$.

S_k — $\{n_1, n_2, \ldots, n_j\}$ the partition of k good switches into j groups connected to j memory units such that $\sum_{l=1}^{j} n_l = k$, where n_l: the number of good switches connected to the l-th memory unit.

η — $\max\{n_l \mid n_l \varepsilon S_k\}$ the size of the largest switch group in the partition S_k.

ω — number of processors connected to k switches in set S_k.

$B(S_k, \gamma)$ — $\prod_{l=1}^{j} \binom{\gamma}{n_l} - \sum_{\phi=\eta}^{\gamma-1} \binom{\gamma}{\phi} B(S_k, \phi)$, for $\eta \leqslant \gamma < \alpha$ and $B(S_k, \eta) = 0$ the number of DFP patterns disqualified for the event $E(\alpha, \beta)$.

$D_{\alpha,\beta}(i, j, k)$ — number of distinct patterns of at most k DFPs qualified for the event $E(i, j)$, where $\alpha \leqslant i$ and $\max(\alpha, \beta) < k \leqslant ij$.

3. THE PROBABILISTIC GRAPH MODEL

A *probabilistic graph* is a finite set of weighted vertices connected by undirected edges. Each vertex, corresponding to a physical component in the MPS, is weighted by its reliability of functioning correctly. All component reliabilities are represented in vertices. The edges are used only to specify incidence relations between vertices. The vertex events are assumed statistically independent.

The hardware resources in an MPS are divided into three groups: *processors* (central processors, input/output processors, and attached processors); *memories* (main memories and peripheral storage devices); and *interconnection networks* (crossbar, buses, or multiport memory controllers). The simplest MPS structure is multiple time-shared buses as depicted in figure 1a. The graph representation of this *bus-structured* MPS is given in figure 1b. This is a *multi-partite graph,* whose vertices are divided into two or more disjoint "colored" subsets such that edges exist only between vertices of differently "colored" subsets. There are $n \times b \times m$ DFPs in the tri-partite graph of figure 1b.

When the number of buses increases to $b = n \times m$, there is a separate DFP between every processor-memory pair. Such a structure is known as the *crossbar-switched* MPS (figure 1c). Corresponding graph representation is shown in figure 1d. There are $n \times m$ DFPs in a crossbar MPS. The common-bus and crossbar switch are two extreme MPS structures. In between is the *multiport-memory* MPS depicted in figure 1e. This architecture is obtained by combining all crosspoint switches in the same crossbar column (figure 1c) into an *n-ported controller* attached to each memory unit. Topologically, the multiport MPS

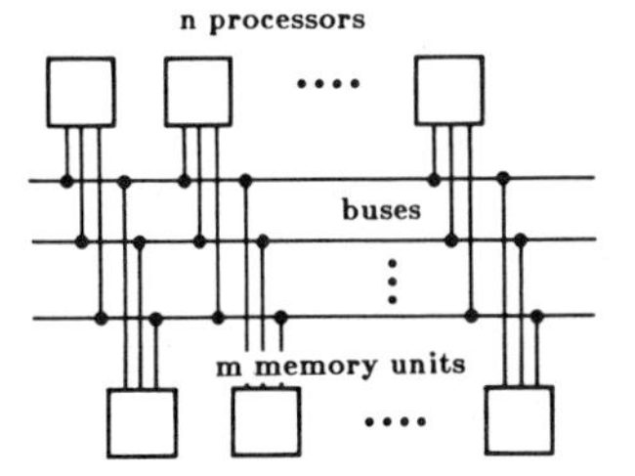

a. Bus-structured MPS.

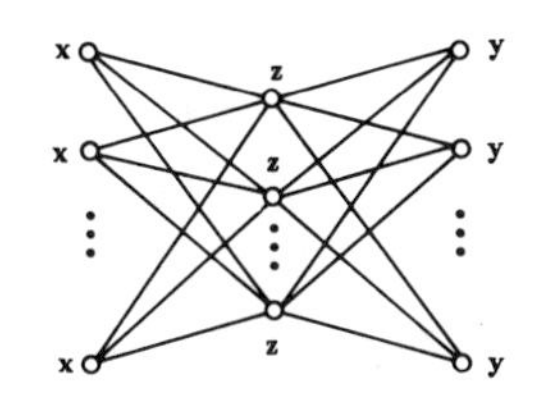

b. Graph representation of a.

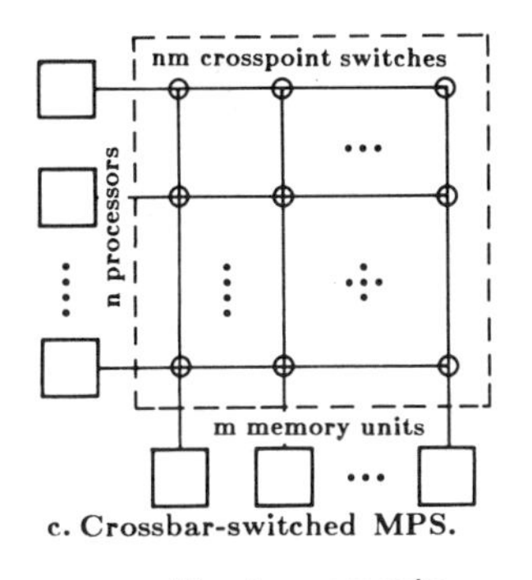

c. Crossbar-switched MPS.

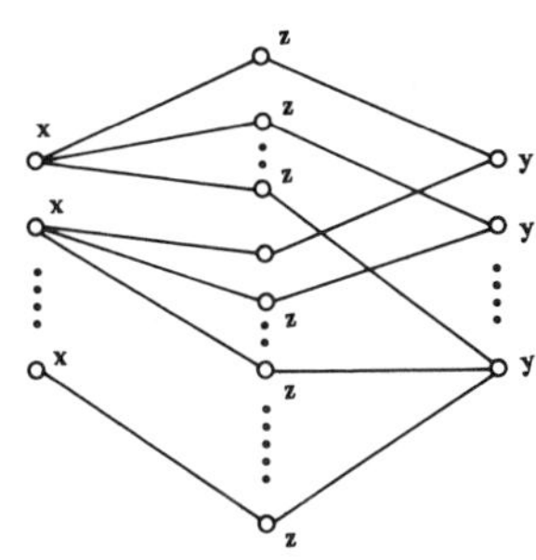

d. Graph representation of c.

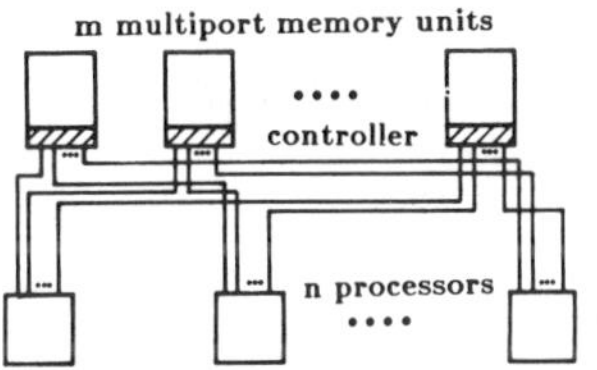

e. Multiport-Membor MPS.

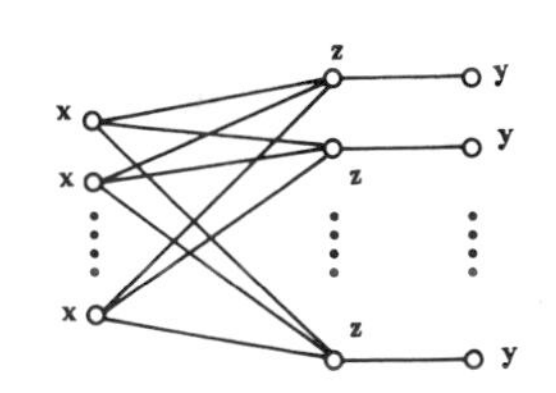

f. Graph representation of e.

Fig. 1. Multiprocessor architectures and their probabilistic-graph representations: a. Bus-structured MPS. b. Graph representation of bus-structured MPS. c. Crossbar-switched MPS. d. Graph representation of crossbar-switched MPS. e. Multiport-Member MPS. f. Graph representation of multiport-member MPS.

structure is represented by the graph shown in figure 1f. During each memory cycle, up-to-m DFPs can be established in this MPS.

4. CROSSBAR-SWITCHED MULTIPROCESSORS

For an MPS with an $n \times m$ crossbar switch, the multiprocessing reliabilities, $R_m(t)$ and $R_{\alpha,\beta}(t)$, are derived by path enumeration on the probabilistic graph in figure 1d. Instead of enumerating all $2^{nm} - 1$ DFPs, this graph-combinatorial approach eliminates many evaluation terms that can not contribute MIMD multiprocessing. Derived below is $R_m(t) = R(t) - R_u(t)$ after substituting the exponential component reliabilities. It is easy to verify that $R(t) - R_u(t) = \Pr\{E(2, 1)\} = R_{2,1}(t)$.

$$R_m(t) = \sum_{i=2}^{n} \sum_{j=1}^{m} G(n, i, x) \cdot G(m, j, y) \cdot \left[(1 - \bar{z}^{ij}) - \binom{i}{1} \sum_{k=1}^{j} \binom{j}{k} z^k \bar{z}^{ij-k}\right]$$

$$= \sum_{i=2}^{n} \sum_{j=1}^{m} \sum_{f=0}^{n-i} \sum_{g=0}^{m-j} \binom{n}{i} \binom{m}{j} \binom{n-i}{f} \binom{m-j}{g} \cdot [F_1(t) - F_2(t)] \quad (1)$$

where $F_1(t)$ contributes to $R(t)$ and—

$$F_1(t) \equiv \sum_{h=1}^{ij} (-1)^{f+g+h+1} \cdot \binom{ij}{h} \cdot \exp[-(i+f)\lambda_x - (j+g)\lambda_y - h\lambda_z]t$$

and $F_2(t)$ contributes to $R_u(t)$ and—

$$F_2(t) \equiv \sum_{k=1}^{j} \sum_{d=0}^{ij-k} (-1)^{f+g+d} \cdot \binom{i}{1}\binom{j}{k}\binom{ij-k}{d} \cdot \exp[-(i+f)\lambda_x - (j+g)\lambda_y - (k+d)\lambda_z]t.$$

For evaluating the threshold reliability, $R_{\alpha,\beta}(t)$, we have to enumerate on the number of legitimate DFP patterns, $D_{\alpha,\beta}(i, j, k)$ for i surviving processors updating j surviving memories via k surviving crosspoint switches with probabilities $G(n, i, x)$, $G(m, j, y)$, and $z^k \cdot \bar{z}^{(ij-k)}$, respectively. The following triple sums exhaust all the possibilities.

$$R_{\alpha,\beta}(t) = \sum_{i=\alpha}^{n} \sum_{j=\beta}^{m} \sum_{k=\max\{\alpha,\beta\}}^{ij} D_{\alpha,\beta}(i, j, k)\, G(n, i, x)\, G(m, j, y) z^k \bar{z}^{(ij-k)}$$

$$= \sum_{i=\alpha}^{n} \sum_{j=\beta}^{m} \sum_{k=\max\{\alpha,\beta\}}^{ij} \sum_{f=0}^{n-i} \sum_{g=0}^{m-j} \sum_{h=0}^{ij-k} D_{\alpha,\beta}(i, j, k) \cdot S(n, m, i, j, k, f, g, h, t) \quad (2)$$

$$D_{\alpha,\beta}(i,j,k) \equiv \begin{cases} \sum_{\text{All } S_k} \prod^{j} \binom{i}{n_l}, \text{ if } \alpha \leqslant \eta \\ \sum_{\text{All } S_k} \prod_{l=1}^{j} \binom{i}{n_l} - \sum_{\gamma=\eta}^{\alpha-1} \binom{i}{\gamma} B(S_k, \gamma), \text{ otherwise} \end{cases}$$

$$S(n, m, i, j, k, f, g, h, t) \equiv \binom{n}{i}\binom{m}{j}\binom{n-i}{f}\binom{m-j}{g}\binom{ij-k}{h} \cdot \exp[-(i+f)\lambda_x - (j+g)\lambda_y - (k+h)\lambda_z]t$$

5. MULTIPLE BUSES AND MULTIPORT MEMORIES

By enumerating DFPs on the graph of figure 1b, we can write below the threshold reliability of an MPS with α-out-of-n:G processors updating β-out-of-m:G memories via at most b common buses.

$$R_{\alpha,\beta}(t) = \sum_{i=\alpha}^{n} \sum_{j=\beta}^{m} \sum_{k=1}^{b} G(n, i, x)\, G(m, j, y)\, G(b, k, z)$$

$$= \sum_{i=\alpha}^{n} \sum_{j=\beta}^{m} \sum_{k=1}^{b} \sum_{f=0}^{n-i} \sum_{g=0}^{m-j} \sum_{h=0}^{b-k} \binom{n}{i}\binom{m}{j}\binom{b}{k}\binom{n-i}{f}\binom{m-j}{g}\binom{b-k}{h} Q(t) \quad (3)$$

$$Q(t) \equiv (-1)^{f+g+h} \cdot \exp[-(i+f)\lambda_x - (j+g)\lambda_y - (k+h)\lambda_z]t$$

Similarly for an MPS with m n-ported memories, we have—

$$R_{\alpha,\beta}(t) = \sum_{i=\alpha}^{n} \sum_{j=\beta}^{m} G(n, i, x) \cdot G(m, j, yz)$$

$$= \sum_{i=\alpha}^{n} \sum_{j=\beta}^{m} \sum_{f=0}^{n-i} \sum_{g=0}^{m-j} \binom{n}{i}\binom{m}{j}\binom{n-i}{f}\binom{m-j}{g} H(t)$$

$$H(t) \equiv (-1)^{f+g} \cdot \exp[-(i+f)\lambda_x - (j+g)(\lambda_y + \lambda_z)]t. \quad (4)$$

6. NUMERICAL RESULTS

Analytic results obtained in sections 4 and 5 are plotted in figure 2 under some parametric assumptions. In all three MPS architectures, we plot the results for an MPS having $n = 4$ processors, $m = 4$ memory units, and/or $b = 1$ common bus. Since common bus has the highest reliability, multiport memory is next, and crosspoint switches (in crossbar) have the lowest reliability, the numerical results are plotted under the assumptions $\lambda_p = \lambda_c/\sqrt{n}$ and $\lambda_b = \lambda_c/n$. All system components have comparable failure rates borrowed from the Carnegie-Mellon research group [3]. Plotted in figure 2 are the relative reliabilities of MPSs with 4 processors and 4 memory units interconnected by a 4 × 4 crossbar, or by 4-port memories, or by a single-bus structure in their resective architectural configurations as demonstrated in figure 1.

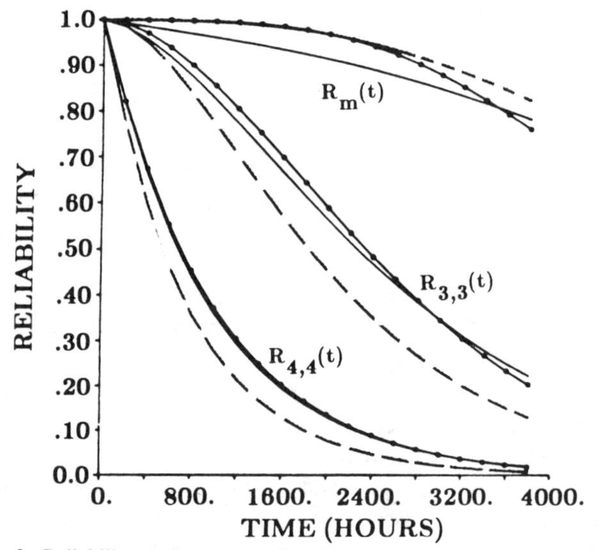

Fig. 2. Reliability performances of multi processor systems with crossbar (dotted lines), multiport-memory (dash lines), and common-bus (solid lines) interconnection structures.

7. EXAMPLE MULTIPROCESSOR SYSTEMS

Presented below are the reliability analyses of two example multiprocessor systems. The "C.mmp" has a crossbar structure developed at Carnegie-Mellon University for research purposes [3]. There are $n = 16$ processors (PDP-11s) and $m = 16$ memory units and 16 × 16 crosspoint switches in "C.mmp" (figure 3a). A probabilistic-graph representation of "C.mmp" is shown in figure 3b. Besides the processors (x), memory units (y), and switches (z), there is a separate bus (w) which connects all the UNIBUSs of PDP-11s for inter-processor interrupt and synchronization purposes.

The "Tandem/16" (figure 4a) has a combined structure of using two common buses (X and Y DYNABUS)

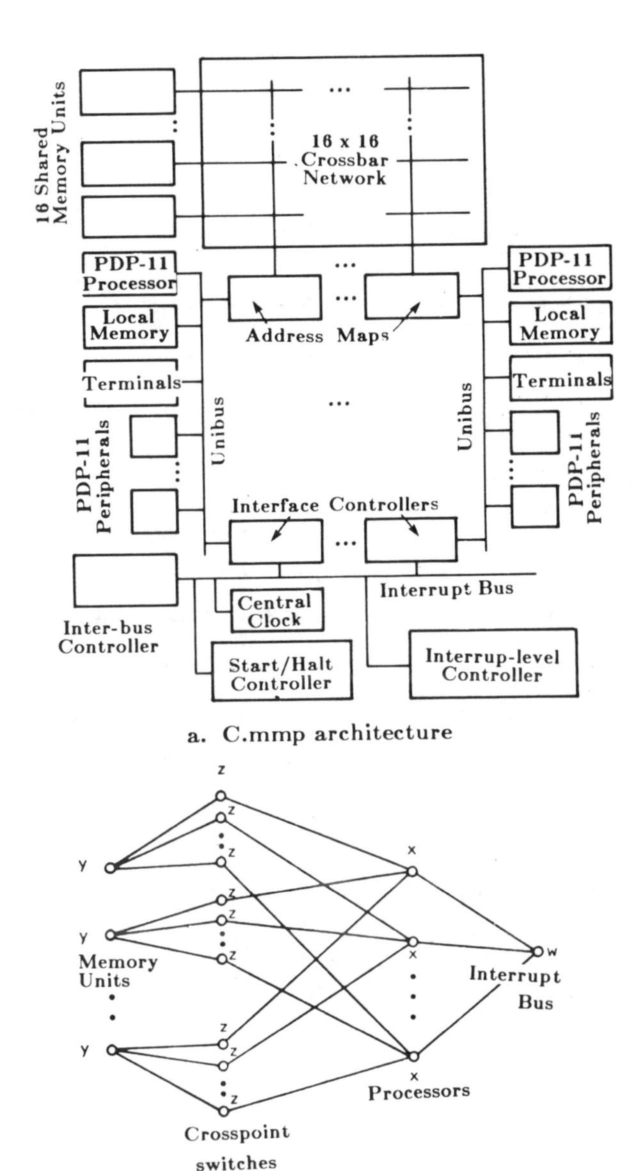

a. C.mmp architecture

b. Graph representation of C.mmp

Fig. 3. The "C.mmp" architecture and graph representation: a. C.mmp architecture; b. Graph representation of C.mmp.

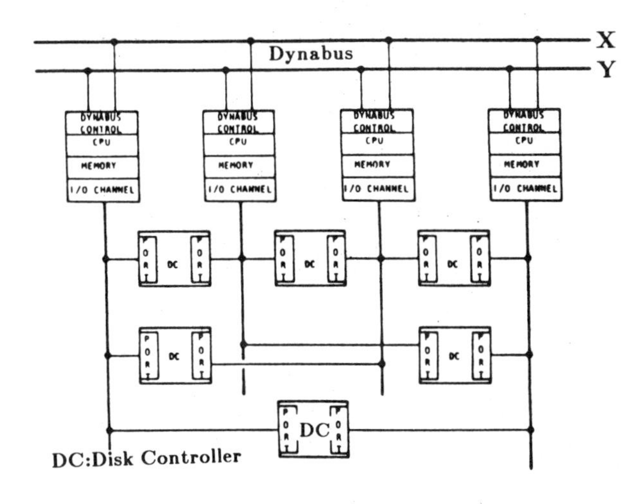

a. "Tandem" System with 4 processors

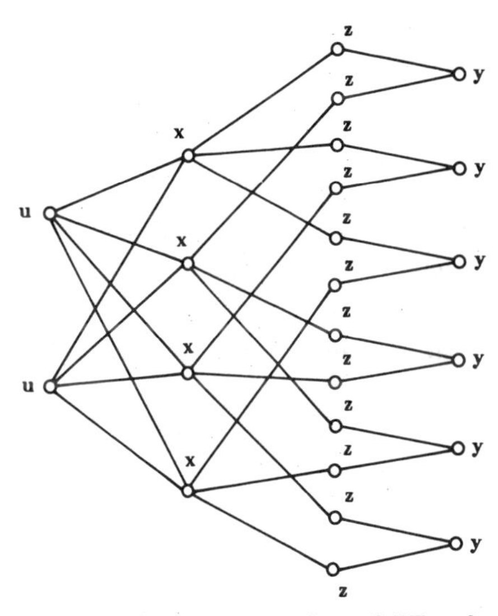

b. Graph representation of "Tandem"

Fig. 4. The "Tandem" architecture and graph representation: a. "Tandem" System with 4 processors; b. Graph representation of "Tandem".

and dual-port memories (shared disks), which are connected to the processors through multiple input/output channels [4]. This MPS structure can tolerate all types of single failures. The graph representation of the "Tandem/16" (figure 4a) is shown in figure 4b. The DYNABUS X and Y have reliabilities u. The dual-port disk controllers have reliabilities, z. The disk memory units have reliabilities, y.

Based on the graphs shown in figure 3b and figure 4b, we can derive various reliability expressions for "C.mmp" and "Tandem/16", respectively. Plotted in Figure 5 are the evaluation results, after substituting the component reliabilities by comparable numerical values as used by CMU [3]. For clarity, figure 5a shows the reliability measures of a "C.mmp-like" system with parameters $(n, m) = (8, 8)$, and figure 5b shows a "Tandem" system with $n = 4$ processors, $m = 6$ shared disk memories, 2 DYNABUSs, and 4 input/output channels. In both example systems, the threshold reliabilities decline more rapidly with the increase of performability levels, (α, β). This coincides with the prediction by Meyer [5] in a general degradable computing system.

ACKNOWLEDGMENTS

The authors wish to thank the referees and the Editor, whose comments and suggestions have greatly improved the quality of this paper. This research has been supported in part by Purdue Research Foundation under David Ross grant No. 0356, and in part by National Science Foundation under grant ECS-80-16580 with Purdue University.

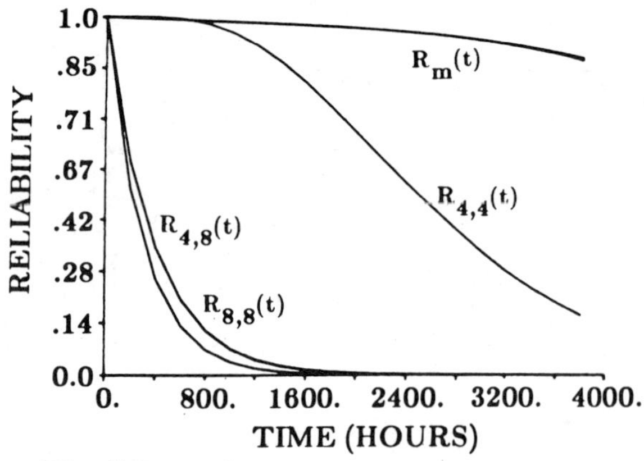

a. The "C.mmp" system with an 8x8 crossbar.

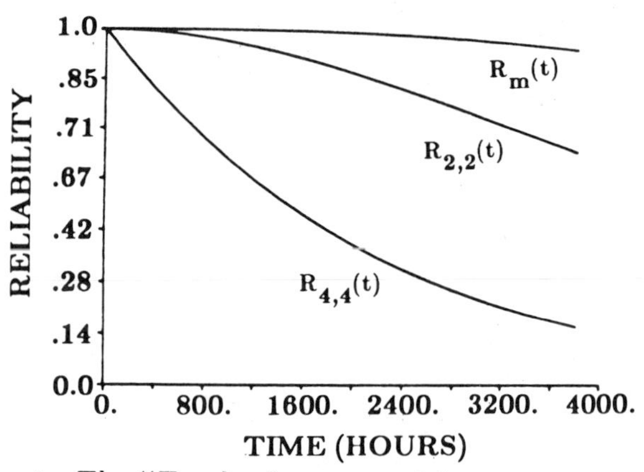

b. The "Tandem" system with 4 processors.

Fig. 5. Reliability performances of two real-life multiprocessor computers: a. The "C.mmp" system with an 8 × 8 crossbar; b. The "Tandem" system with 4 processors.

REFERENCES

[1] P.H. Enslow, Jr., "Multiprocessor organization — A survey," *ACM Computing Surveys,* vol 9, 1977 Mar, pp 103-129.

[2] K. Hwang, T.P. Chang, "Reliability and availability analysis of multiprocessor architectures using probabilistic graphs," Purdue University Technical Report TR-EE 80-44, 1980 Dec. Available from School of Electrical Engineering, Purdue University.

[3] R. Joobbiani, D.P. Siewiorek, "Reliability modeling of multiprocessor architectures," *Proc. First Int'l. Conf. Distributed Processing,* 1979 Oct, pp 384-398.

[4] J.A. Katzan, "A fault-tolerant computing system — Tandem/16," *Proc. IEEE Hawaii Int'l. Conf. System Sciences,* 1978 Jan, pp 1-18.

[5] J.F. Meyer, "On evaluating performability of degradable computing systems," *Proc. the Eighth Annual Int'l. Conf. Fault-Tolerant Computing,* (FTCS-8), 1978 Jun, pp 44-49.

[6] M. Satyanarayanan, "Commercial multiprocessing systems," *IEEE Computer,* vol 13, 1980 May, pp 75-96.

AUTHORS

Kai Hwang; School of Electrical Engineering; Purdue University; W. Lafayette, Indiana 47907 USA.

Kai Hwang (S'68, M'72, SM'81) is an Associate Professor of Computer Engineering at Purdue University. He received a PhD in electrical engineering and computer science from University of California at Berkeley. He is a Distinguished Visitor of the IEEE Computer Society. He is the author of two books on computer arithmetic and parallel processing computer architecture.

Tian-Pong Chang; Electrical Engineering Dept., University of New Haven, W. Haven, Connecticut 06516, USA.

Tian-Pong Chang (M'82) is an Assistant Professor of Electrical Engineering at University of New Haven. He received a PhD in computer engineering from Purdue University. He is a member of IEEE and ACM.

Manuscript TR81-138 received 1981 November 20; revised 1982 July 10.

★★★

Reliability and Fail-Softness Analysis of Multistage Interconnection Networks

Vladimir Cherkassky, Member IEEE
University of Minnesota, Minneapolis
Miroslaw Malek, Member IEEE
University of Texas, Austin

***Key Words*—Banyan, Crossbar switch, Fail-softness, Graph modeling, Multistage interconnection network, Reliability prediction.**

***Reader Aids*—**
Purpose: Report on analysis
Special math needed for explanations: None
Special math needed to use results: None
Results useful to: Computer-system designers and reliability engineers

***Abstract*—This paper presents a method for reliability analysis of multistage interconnection networks implemented with crossbar switching elements. Analytic estimates for the network reliability are derived, and the existence of a switching element with optimal fanout ensuring maximal network reliability is shown. A general quantitative measure for fail-softness evaluation of multistage interconnection networks is introduced. Under a single-fault assumption, the fail-softness improves with an increase in network size.**

1. INTRODUCTION

There is a growing interest in multiprocessor systems with increased performance and improved reliability. The reliability analysis of such systems is rare and usually inadequate for several currently developed multiprocessors. Most methods are based on graph-theoretical models of the multiprocessor systems [1-3]. The reliability of real-life systems like "C.mmp", "Tandem/16", and "Univac 1100/80" has been analyzed in [2] using combinatorial path enumeration on the probabilistic-graph representation of a multiprocessor system.

This paper is primarily concerned with multiprocessor systems based on multistage interconnection networks (MINs). We present reliability and fail-softness analysis for a large class of MINs which use $L \equiv \log_f N$ stages of $f \times f$ switching elements to connect N inputs to N outputs. A comprehensive survey of MINs is given in [4]. We first briefly describe a graph-theoretic model of MINs known as banyans. Then we present a model for reliability analysis and give a practical example of reliability prediction. Finally, we introduce a quantitative measure for fail-softness and analyze the fail-softness of MINs.

This research was supported in part by a grant from IBM Corporation.

2. SYSTEM MODEL AND NOTATION

The multiprocessor system model is composed of the *resources* (eg, processors, memory units, or computers) and the *interconnection network*.

We consider networks built of nominally identical $s \times f$ switches, ie, switches with s inputs (spreads) and f outputs (fanouts). A *network* is a directed graph where vertices (nodes) are of the following three types:

- Source nodes (inputs) which have indegree 0.
- Sink nodes (outputs) which have outdegree 0.
- Intermediate nodes which have positive indegree and outdegree.

A *banyan network* [5] is a network with a unique path from each source to each sink node. A *multistage interconnection network* (MIN) is a network in which vertices can be arranged in stages, with all the source vertices at stage (level) 0, and all the outputs at stage i connected to inputs at stage $i + 1$. An L-level banyan, or an L-stage MIN, is a network in which every path from any source to any sink has length L. An (f, L) banyan is an L-level banyan in which the indegree (spread-s) of every intermediate node equals its outdegree (fanout-f). An example of a (3, 2) banyan is shown in figure 1a. An (f, L) banyan graph can be viewed as a collection of $f \times f$ crossbar subgraphs (see figure 1b). Thus, (f, L) banyans represent graph-theoretic models of MINs built of nominally identical $f \times f$ switching elements. For example, an 8×8 MIN built of 3 stages of 2×2 switches can be denoted as a (2, 3) banyan, as shown in figure 2.

An example of a multicomputer system based on a banyan network is shown in figure 3; each computer (a processor with a local memory) is attached to both sides of a (3, 2) banyan. There is a single path between each ordered pair of computers (without traversing through an

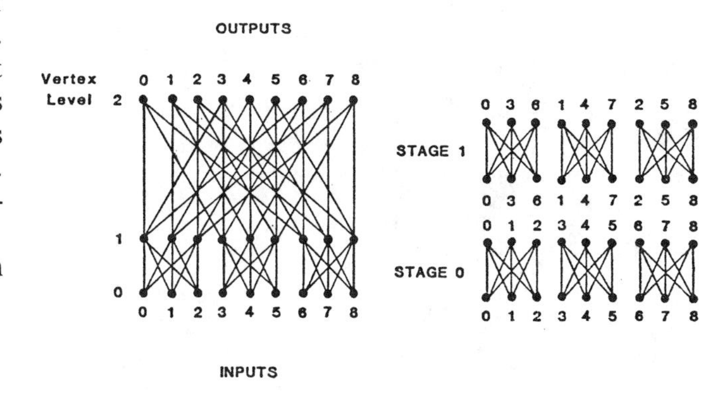

a (3, 2) Banyan b 3 × 3 Crossbar Subgraphs

Fig. 1. (3. 2) banyan graph and its crossbar subgraphs

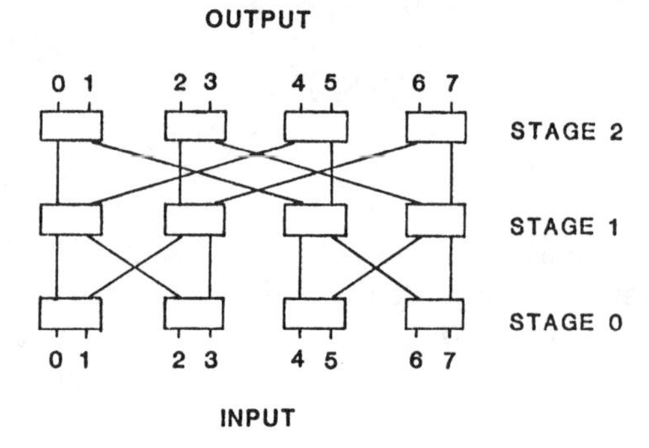

a Exchange box (traditional) representation

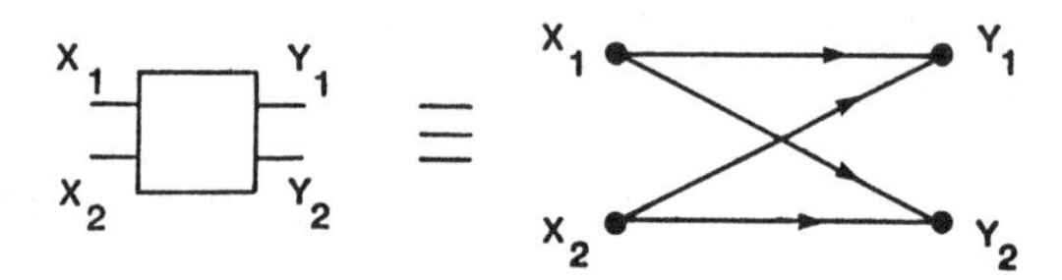

b 2 × 2 switch and its graph model

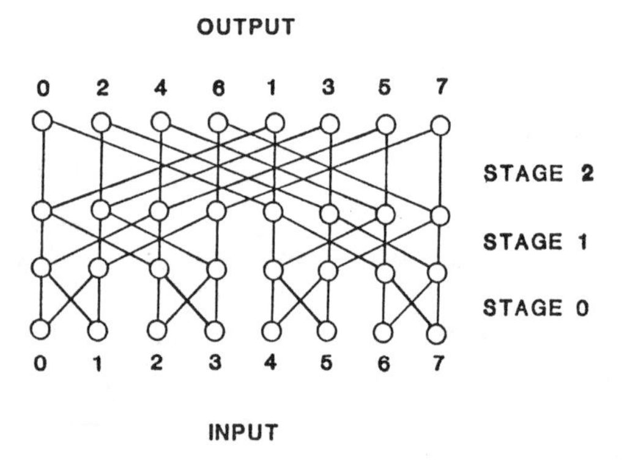

c Graph-theoretical representation of figure 2a as a (2, 3) banyan

Fig. 2. Two graphical representations of an 8 × 8 network

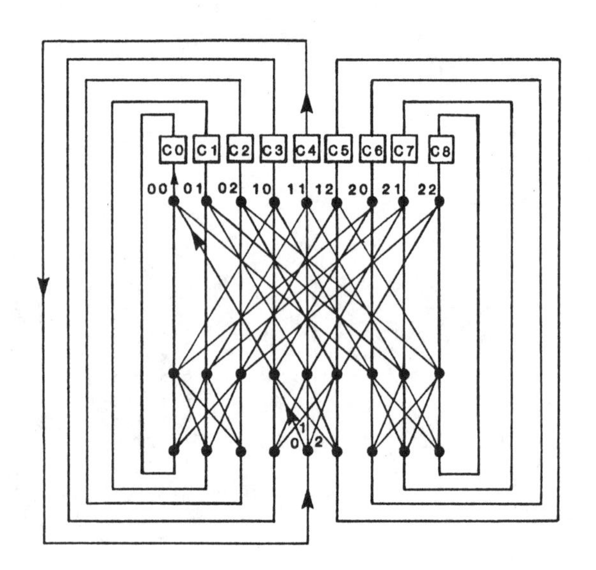

Fig. 3. A multicomputer system based on a (3, 2) *SW*-banyan with an example of routing from C4 to C0.

intermediate third computer), and the communication through the links is unidirectional. The control of the system is distributed and uses self-routing; that is, the routing information (routing tag) is embedded in the packet and travels along the path with the data. In such system, the routing tag is given by a packet destination address, ie, the special label assigned to each computer. The labeling and routing scheme is briefly described below. More detailed descriptions can be found in [6, 7]. A multicomputer system based on (f, L) banyan contains $N \equiv f^L$ computers; so each computer can be labeled by an L-digit number in base f. We assign the weights 0, 1, 2, ..., $f - 1$ to each outgoing link of every intermediate node (from left to right). The routing decisions are made in each stage by examining successive bits of the routing tag (see figure 3).

Notation & Nomenclature

(f, L) banyan — rectangular banyan with L stages and fanout f

N — network size, viz, the number of inputs/outputs in a rectangular (f, L) banyan $(N \equiv f^L)$

Bf — $(f \times f)$ crossbar switching element.

$\lambda(\text{Bf})$ — exponential failure rate of Bf.

Λ_{sys} — interconnection network failure rate

network failure — inability to perform an arbitrary input-output connection (in a single network pass)

3. RELIABILITY EVALUATION

Consider an (f, L) banyan with $N \equiv f^L$ resources attached to it (see figure 3). In the general case, an (f, L) banyan implementation requires $(N/f)*L$ switching elements Bf, since the number of Bf switches per stage equals N/f, and the number of stages is L. An example of such implementation for a (2, 3) banyan using B2 switches is shown in figure 2.

A banyan network has a single path between any input-output pair, and any single fault in the network disconnects some input-output resources. Thus, a banyan network is a non-redundant system, and its failure rate is the sum of all of its component (Bf switches) failure rates:

$$\Lambda_{sys} = (N/f) * L * \lambda(\text{Bf}) \tag{1}$$

A realistic model is derived for reliability of switching element Bf as a function of its fanout. In order to present an implementation-independent analysis, consider a graph model of a switch Bf (see figure 2b). In this model, link faults correspond to the faults on the internal lines of a switch, and node faults correspond to the faults on the input or output lines of a switch. The internal switch-complexity (the number of links) is proportional to f^2, and the interface switch-complexity (the number of nodes) and the number of buffers are proportional to f. The switch complexity associated with power supply and clocking is also proportional to f. The control-logic complexity for packet routing is associated with f multiplexers for distributed control or additional $f*\log_2 f$ control lines for centralized control. The priority logic (for conflict resolution in the case of packet contention) usually has complexity proportional to f (eg, round-robin or fixed priority scheme). Therefore, the switching-element failure-rate is:

$$\lambda(\mathrm{B}f) = \lambda_1 * f^2 + \lambda_2 * f + \lambda_3 * f * \log_2 f. \quad (2)$$

Here λ_1, λ_2, λ_3 are technology- and implementation-dependent positive constants. Eq. (2) describes a switch with centralized control. For distributed control the failure rate is given by (2) where $\lambda_3 = 0$. Constants λ_1, λ_2, λ_3 heavily depend on implementation factors, such as the bus width (the number of physical wires corresponding to each link in the banyan) and the buffer size. However, in this paper we are mostly interested in the network reliability as a function of its structural parameter (eg, network size N and fanout f).

Eq. (2) shows that the failure rate of a switching element grows with growth of its fanout, and, equivalently, its reliability decreases. Combining (1) and (2) produces the network failure rate:

$$\Lambda_{sys} = (N/f) * \log_f N * \lambda(\mathrm{B}f) = $$

$$= N * \log_2 N * \left(\frac{\lambda_1 * f}{\log_2 f} + \frac{\lambda_2}{\log_2 f} + \lambda_3 \right). \quad (3)$$

Eq. (3) implies an optimal fanout f_{opt} for a set of technology and design dependent constants λ_1, λ_2, λ_3, such that it corresponds to the minimal system failure rate. This optimal fanout can be found by minimizing:

$$\frac{\lambda_1 * f}{\log_2 f} + \frac{\lambda_2}{\log_2 f} + \lambda_3.$$

Eq. (3) shows that banyan network reliability depends on both network structural parameters (such as the network size and fanout) and on implementation factors (technology and switching element design).

4. RELIABILITY PREDICTION: A PRACTICAL EXAMPLE

This section evaluates the reliability of a 64 × 64 banyan network based on B2 and B4 switches, and predicts reliability of the network based on B8 switches. As mentioned earlier, reliability analysis is technology- and design-dependent. B2 and B4 switches have been designed and implemented using MSI and LSI technologies at the University of Texas at Austin. The circuit diagrams for B2 and B4 can be found in [8, 9]. Using circuit diagrams for B2 and B4, we predict their reliability using MIL-HDBK-217 method of stress analysis, failure rate and part count reliability prediction [9, 10]. Our calculations are based on the following assumptions:

1. The failure rate of a switching element is the sum of failure rates of IC chips, printed wiring board, and connectors.
2. The quality of every component meets the MIL B-class specifications [10].
3. The data width of the switching element is 16 bits.
4. Switching element can buffer one word (16 bits) on each input.
5. Operational temperature is 25C.

The failure rate for B2 is $0.336*10^{-6}$ failures/hour, and for B4 is $0.954*10^{-6}$ failures/hour.

Now we predict the reliability of a B8 switching element. Since the circuit diagram for B8 is not known (switch B8 has not been implemented), we cannot use the MIL-HDBK-217 method of reliability prediction. However, we can use (2) to predict the failure rate of B8. Since B2 and B4 switches use distributed-routing tag control, the constant λ_3 is 0. Now, the available information, viz, the failure rates for B2 and B4, is sufficient to determine $\lambda_1 = 0.033*10^{-6}$ failures/hour, and $\lambda_2 = 0.109*10^{-6}$ failures/hour. From (2) and (3), we can find the failure rate for B8 switching element and calculate the failure rate of a 64 × 64 network implemented with B2, B4, and B8 switches. See Table 1.

TABLE 1
Reliability prediction for 64 × 64 banyan network built of B2, B4 and B8 switches.

Switching element type	B2	B4	B8
Switching element failure rate (failures/hour)	$0.336*10^{-6}$	$0.954*10^{-6}$	$2.98*10^{-6}$
Number of switches Bf in the 64 × 64 network	192	48	16
Network failure rate (failures/hour)	$6.45*10^{-5}$	$4.58*10^{-5}$	$4.77*10^{-5}$

As anticipated, switching-element failure rate grows with its fanout. However, this is not the case with the system failure rate that is minimal for B4. Thus, B4 provides optimal reliability for a 64 × 64 banyan network and there is no reason, as far as reliability is concerned, to design a B8 switching element for this network. Of course, these conclusions are valid only if similar design methods and the same technology are used to implement B2, B4, and B8 switching elements.

5. FAIL-SOFTNESS ANALYSIS

An important and interesting issue of fail-softness (or

graceful degradation) of MINs is often addressed in qualitative form, as the system's ability to perform its function (maybe showing some degradation in performance) in case of system component failures. We define a fail-softness coefficient as a measure for fail-softness in distributed systems as —

Given a failure mode of a component C in distributed system S, *fail-softness coefficient* F with respect to a certain system performance criterion P is:

$$F(S, C) = \frac{P_1(S, C)}{P(S)} \tag{4}$$

$P(S) \equiv$ performance criterion in a fault-free system
$P_1(S, C) \equiv$ performance criterion after a failure of a component C.

We assume that the performance criterion decreases in the presence of faults in the system, so that $0 \leq F \leq 1$ and the bigger values of the fail-softness coefficient corresponds to better fail-softness. The above definition of fail-softness coefficient is quite general, and can be used to evaluate fail-softness of various distributed systems. For example, a similar approach has been used to evaluate fail-softness of tree-based local area networks [11].

In case of a MIN, a single fault in the network adversely affects only a portion of all possible input-output connections. Therefore, it is reasonable to define a performance criterion in a banyan network as the number of different input-output connections. For example, in a fault-free situation, an N-input, N-output banyan network can support N^2 different input-output connections, so that

$$P(S) = N^2 \tag{5}$$

Fail-softness analysis of an (f, L) banyan network graph with respect to single (vertex or link) faults is presented below.

1. *Vertex fault* (see figure 4-a). Since a faulty vertex at level i $(0 \leq i \leq L)$ can be reached by a path of length i from f^i inputs and by a path of length $L - i$ from f^{L-i} outputs [6], the number of disrupted input-output paths equals $f^i * f^{L-i} = f^L = N$. Thus, performance criterion in presence of a vertex fault is $N^2 - N$, and according to (4):

$$F = (N^2 - N)/N^2 = 1 - 1/N \tag{6}$$

2. *Link fault* (see figure 4-b). Since a faulty link at level k $(0 \leq k \leq L - 1)$ can be reached from f^k inputs and from f^{L-1-k} outputs [6], the total number of disrupted input-output connections is $f^k * f^{L-1-k} = f^{L-1} = N/f$. Thus, the performance criterion in presence of a link fault is $N^2 - N/f$, and —

$$F = (N^2 - N/f)/N^2 = 1 - 1/(f*N) \tag{7}$$

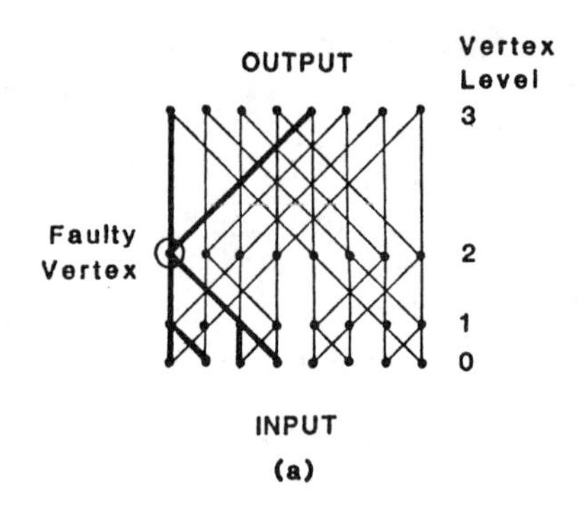

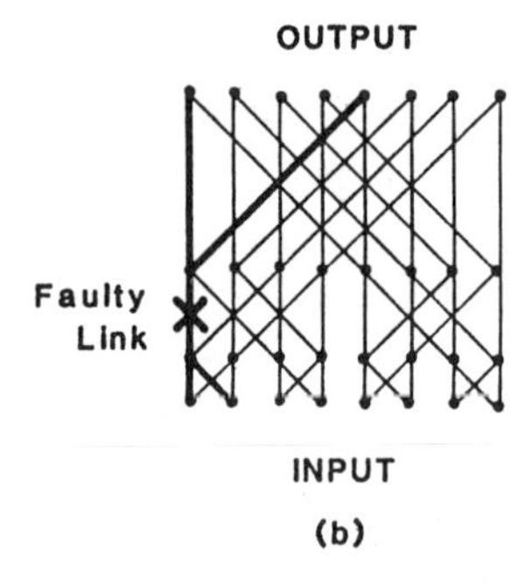

Fig. 4. Disrupted input-output paths in a (2, 3) banyan due to a vertex fault figure 4a, and link fault figure 4b.

For both types of faults, F improves with an increase in network size, N.

REFERENCES

[1] C. S. Raghavendra, A. Avizienis, M. Ercegovac, "Fault-tolerance in binary tree architectures," Digest of Papers 13th Fault-Tolerant Computing Symp. 1982, pp 360-364.

[2] K. Hwang, T. P. Chang, "Combinatorial reliability analysis of multiprocessor computers," *IEEE Trans. Reliability*, vol R-31, 1982 Dec, pp 469-473.

[3] R. Joobbiani, D. P. Siewiorek, "Reliability modeling of multiprocessor architectures," Proc. First Int'l Conf. Distributed Processing, 1979 Oct, pp 384-398.

[4] T.-Y. Feng, "A survey of interconnection networks," *Computer*, vol 14, 1981 Dec, pp 12-27.

[5] L. R. Goke, G. J. Lipovski, "Banyan networks for partitioning multiprocessor systems," Proc. First Symp. Computer Architecture, 1973, pp 21-28.

[6] U. V. Premkumar, "A theoretical basis for the analysis and partitioning of regular *SW*-banyans," PhD Dissertation, Univ. of Texas at Austin, 1981.

[7] A. Tripathi, G. J. Lipovski, "Packet switching in banyan networks," Proc. 6th Symp. on Computer Architecture, 1979, pp 160-167.

[8] A. Hung, M. Malek, "A 4 × 4 modular crossbar design for the multistage interconnection networks," Proc. Real-Time System Symp., Miami, 1981 Dec, pp 3-12.

[9] M. Malek, S. Cheemalavagu, M.-S. Juang, B. D. Rathi, "Design packaging, performance and self-diagnosis of banyan interconnection networks," Technical Report, Univ. of Texas at Austin, 1982.

[10] US Military Handbook, MIL-HDBK-217D, *Reliability Prediction of Electronic Equipment*, 1982.

[11] V. Cherkassky, M. Malek, G. J. Lipovski, "Fail-softness analysis of tree-based local area networks," Proc. 5th Int'l Conf. Distributed Computing Systems, 1985 May, pp 380-385.

AUTHORS

Vladimir Cherkassky; Department of Electrical Engineering; University of Minnesota; Minneapolis, Minnesota 55455 USA.

Vladimir Cherkassky (S'83, M'85) is an Assistant Professor of Electrical Engineering at the University of Minnesota. He received a PhD in computer engineering from the University of Texas at Austin in 1985. Prior to his present position, he worked as a development engineer and a consultant for Westinghouse Electric Corporation. His research interests include parallel architectures and fault-tolerant computing. He is a member of IEEE.

Miroslaw Malek; Department of Electrical and Computer Engineering; University of Texas; Austin, Texas 78712 USA.

Miroslaw Malek (M'78) is an Associate Professor of Electrical Engineering at the University of Texas at Austin, where he has served on the faculty since 1977 September. In 1977 he was a visiting scholar at the Department of Systems Design at the University of Waterloo in Canada. His research interests include parallel architectures, interconnection networks and fault-tolerant computing.

Malek received the MSc degree in Electrical Engineering in 1970 and his PhD in computer science in 1975, both from the Technical University of Wroclaw in Poland. He is a member of the IEEE and ACM. He served as General Chair'n of the 1984 Real-Time Systems Symposium. He is a consultant to industry in the areas of network performance, testability, and fault-tolerant computing.

Manuscript TR84-107 received 1984 August 30; revised 1985 September 18.

★ ★ ★

Design and Analysis of Dynamic Redundancy Networks

MENKAE JENG, MEMBER, IEEE, AND HOWARD JAY SIEGEL, SENIOR MEMBER, IEEE

***Abstract*—Most previous work in the fault-tolerant design of multistage interconnection networks (MIN's) has been based on improving the reliabilities of the networks themselves. For parallel systems containing a large number of processing elements (PE's), the capability to recover from a PE fault is also important. The dynamic redundancy (DR) network is investigated in this paper. It can tolerate faults in the network and support a system to tolerate PE faults without degradation by adding spare PE's, while retaining the full capability of a multistage cube network. The DR network can also be controlled by the same routing tags used for the multistage cube. Hence, with a recovery procedure added in the operating system, programs which can be executed in a system based on a multistage cube can be executed in a system based on the proposed network before and after a fault without any modification. A variation of the DR network, the reduced DR network, is also considered, which can be implemented more cost effectively than the DR while retaining most of the advantages of the DR. The reliabilities of DR-based systems with one spare PE and the reliabilities of systems with no spare PE's are estimated and compared, and the effect of adding multiple spare PE's is analyzed. It is shown that no matter how much redundancy is added into an MIN, the system reliability cannot exceed a certain bound; however, using the DR and spare PE's, this bound can be exceeded.**

***Index Terms*—Dynamic redundancy, fault tolerance, interconnection network, multistage cube, parallel processing, partitioning, PASM, reconfiguration, reliability analysis.**

I. Introduction

LARGE-SCALE parallel systems which employ a large number of processing elements (PE's) and an interconnection network for inter-PE communications have received increasing interest for applications which need fast computing power. These systems are vulnerable to failures because of the large number of components involved. One of the major design problems with such systems is to make these systems fault tolerant and reliable.

Fault-tolerant design for multistage interconnection network (MIN) based systems has been intensively studied. Most previous work has focused on designing fault-tolerant MIN's. An MIN is said to be fault-tolerant if under certain types of faults it can continue to provide a fault-free connection for any input–output pair. There are many methods to make an MIN fault tolerant. One example is to use error correcting codes to tolerate bit errors in the data and/or control paths [23]. Another example is to introduce redundant connection paths for any input–output pair such that at least one fault-free path is available in the presence of switch failures or link failures (e.g., [2]). Many fault-tolerant MIN's have been proposed recently (e.g., [3], [10], [14], [20], [22], [27], [30], [31], [34]–[36], [46], [48]. Surveys and comparisons of fault-tolerant MIN's are given in [1] and [4].

Adding redundancy into the network can increase the system reliability. However, the overall system reliability will be bounded by the reliabilities of other components such as the PE's. One possibility to further enhance system reliability is to add spare PE's. Most fault-tolerant MIN's do not have extra I/O ports for the spares. Some networks, such as the ANC [35] and ABN [20], contain more than N I/O ports. However, they are primarily designed for systems containing N processors (or PE's) where a single processor is connected to multiple I/O ports. The issue of adding spare PE's to these networks has not been addressed.

This paper investigates a fault-tolerant variation of the generalized cube (GC) which contains extra I/O ports for adding spare PE's, and analyzes how much reliability improvement can be obtained by using this approach. The GC network is chosen as a representative of the topologically equivalent class of multistage cube networks which include the baseline [47], the indirect binary n-cube [32], the omega [21], the Flip [7], the SW-banyan ($S = F = 2$) [18], and the multistage shuffle-exchange [45]. Multistage cube-type networks have been used or proposed for use in many systems, such as STARAN [8], BBN butterfly [11], IBM RP3 [33], PASM [43], Ultracomputer [19], the Ballistic Missile Defense Agency distributed processing test bed [26], [41], the Flow Model Processor of the Numerical Aerodynamic Simulator [6], and data flow machines [13].

In this paper, it is assumed that each PE is attached to an input port and an output port of an MIN. However, the results presented are also applicable if processors and memories are attached to different sides of an MIN. A PE which participates in the execution of tasks will be called a *functioning* PE, otherwise it is a *spare* PE. A spare PE will become functioning when a faulty functioning PE is detected and isolated. Fault

Manuscript received August 28, 1986; revised April 27, 1987. This work was supported by the Supercomputing Research Center under Contract 695 and the Rome Air Development Center under Contract F30602-83-K-0119. Preliminary versions of parts of this paper were presented at the Sixth International Conference on Distributed Computing Systems, May 1986, and at the 1986 Real-Time Systems Symposium, December 1986. A topology similar to the DR was described by Balasubramanian *et al.*, in *J. Parallel Distributed Comput.*, Aug. 1988.

M. Jeng is with the Department of Computer Science, University of Houston, Houston, TX 77204.

H. J. Siegel is with the Parallel Processing Laboratory, School of Electrical Engineering, Purdue University, West Lafayette, IN 47907.

IEEE Log Number 8718421.

Reprinted from *IEEE Transactions on Computers*, Volume 37, Number 9, September 1988, pages 1019–1029.

detection/location is not of direct concern here. There are procedures in the literature for detecting and locating faults in an MIN and an MIN-based system [12], [15], [24]. It is assumed that a faulty component can be successfully detected and located.

In Section II, the *dynamic redundancy (DR)* network and its properties will be discussed in detail. A variation of the DR network, the *reduced DR (RDR)* network, is considered in Section III. In Section IV, the reliabilities of systems based on DR or RDR are estimated and compared to an upper bound to which the system reliability can be improved by adding redundancy into MIN's only, and the sufficient condition to obtain reliability improvement by using the DR or RDR and adding spare PE's is given. Finally, in Section V, the complexity of the DR network is discussed.

II. The DR Network

A. Definition of the DR Network

The design of the DR network is based on the interconnection graph of the GC. A GC for $N = 8$ is shown in Fig. 1. In general, a GC with $N = 2^m$ I/O ports consists of m stages, where each stage consists of $N/2$ 2×2 interchange boxes. Fig. 2 shows an equivalent SW-banyan graph of GC [25], [40]. Let $p = g_{m-1} \cdots g_1 g_0$ be the binary representation of an arbitrary I/O port label. Then the m cube interconnection functions can be defined as

$$\text{cube}_i(g_{m-1} \cdots g_1 g_0) = g_{m-1} \cdots g_{i+1} \bar{g}_i g_{i-1} \cdots g_1 g_0 \quad (2.1)$$

where $0 \le i < m$, and $\bar{g}_i$ denotes the complement of g_i [38]. Stage i of a GC can implement the cube$_i$ function. Using the representation in Fig. 1, at stage i, link j and link cube$_i(j)$ can exchange data. Using the representation in Fig. 2, at stage i, switch j and switch cube$_i(j)$ can exchange data; i.e., switch j in stage i is connected to switch cube$_i(j)$ in stage $i-1, 0 \le j < N, 0 \le i < m$.

The DR network contains m stages, where $N = 2^m$. Stages are ordered from $m-1$ to 0 from the input side to the output side of the network. Each stage has $N + S$ switches followed by $3(N + S)$ links, as shown in Fig. 3 for $N = 8$ and $S = 2$. In addition, there are $N + S$ output switches. This allows for an initial set of N functioning PE's and S spares. PE's and switches of the network are physically numbered from 0 to $N + S - 1$. PE j of the system is connected to the input of switch j of stage $m-1$ and to the network output switch j. Each switch j at stage i of the network has three output links to stage $i-1$. The first link f_{-i} is connected to switch $(j - 2^i)$ mod $(N + S)$ of stage $i-1$, the second link f_0 is connected to switch j of stage $i-1$, and the third link f_{+i} to switch $(j + 2^i)$ mod $(N + S)$ of stage $i-1$. Switches at stage 0 are connected to the output switches.

A *row* of a DR network contains all the network switches having the same address, all links incident out of them, and the associated network input link. A row has the same address as its switches. Two rows of the DR network are said to be *adjacent* if their addresses are consecutive (modulo $N + S$).

B. Reconfiguring the DR Network

It is assumed that the system assigns PE 0 to PE $N-1$ as the functioning PE's at the beginning. When PE j or row j is detected faulty, physical PE p and row p of the network, $0 \le p < N + S$, will logically be renumbered $t(p)$:

$$t(p) = (p - k) \bmod (N + S), \quad (2.2)$$

where $k = j + S \bmod (N + S)$. PE's with logical addresses between 0 and $N-1$ will become new functioning PE's. Let Π denote the subnetwork which contains all rows of the network with logical addresses between 0 and $N-1$, then Π will be the subnetwork to provide the inter-PE communications for the new functioning PE's. It will be shown that no matter what j is, Π can logically perform all functions that a GC can perform.

Theorem 1: Stage i of $\Pi, 0 \le i \le m-1$, can perform cube$_i$ based on the logical addresses.

Proof: For any logical switch $t(p)$ at stage i of $\Pi, 0 \le i < m$, if $t(q) = \text{cube}_i(t(p))$, then it is obvious that $0 \le t(q) < N$. Hence, it is sufficient to show that there always exists a link for physical switch p to physical switch q. From (2.2),

$$t(p) = (p - k) \bmod (N + S) \quad (2.3)$$

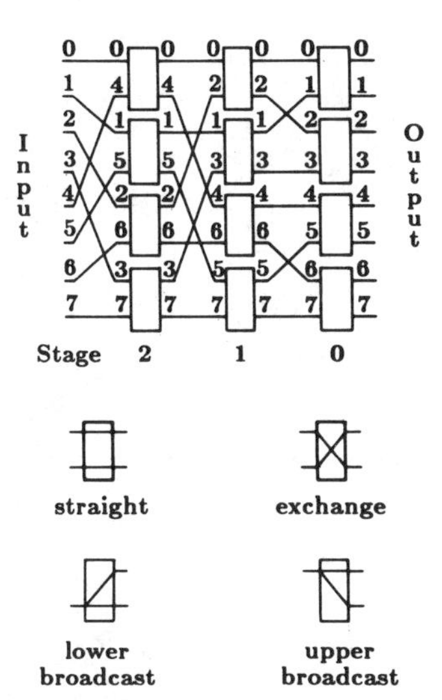

Fig. 1. Generalized cube network for $N = 8$. The four legitimate states of an interchange box are shown.

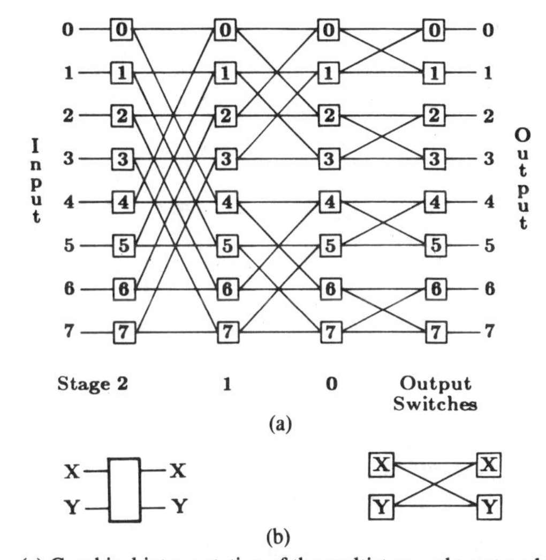

Fig. 2. (a) Graphical interpretation of the multistage cube network for $N = 8$. (b) Relationship between graphical interpretation and interchange box representation.

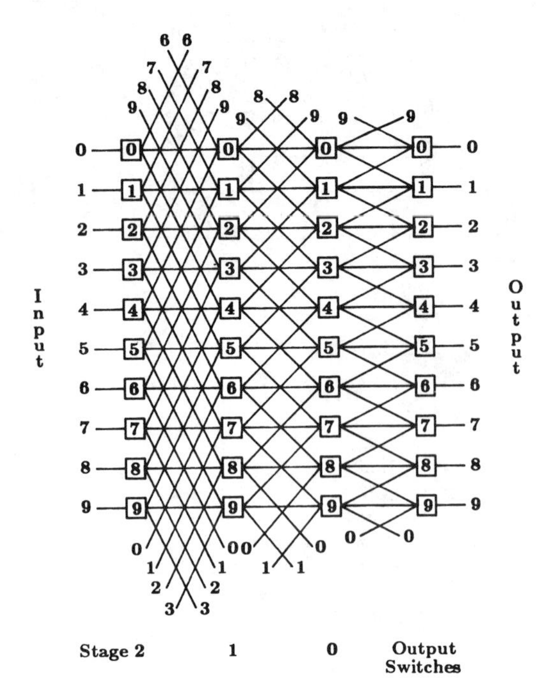

Fig. 3. A DR network for $N = 8$ and $S = 2$.

$$t(q) = (q - k) \bmod (N + S). \tag{2.4}$$

Case 1: Bit i of $t(p)$ is 0: Since bit i of $t(p)$ is 0, from (2.3),

$0 \le [t(p) + 2^i = (p - k) \bmod (N + S) + 2^i] < N + S$.

So, $(p - k) \bmod (N + S) + 2^i = (p - k + 2^i) \bmod (N + S)$.

Since $(q - k) \bmod (N + S) = t(q) = \text{cube}_i(t(p)) = t(p) + 2^i$,

$(q - k) \bmod (N + S) = (p - k + 2^i) \bmod (N + S)$.

So, $q = (p + 2^i) \bmod (N + S)$.

From the definition of the network, there is a link f_{+i} which connects switch p at stage i to switch $(p + 2^i) \bmod (N + S)$ at stage $i - 1$.

Case 2: Bit i of $t(p)$ is 1: The proof is similar to Case 1.

Corollary: In a DR network with no faults, stage i of Π can perform cube_i, $0 \le i \le m - 1$.

Proof: It follows from the Theorem 1 when $t(p)$ is the identity function ($k = 0$). □

Fig. 4 shows that Π contains a cube network for $N = 8$ and $S = 2$ with PE 7 faulty and $t(p) = (p - k) \bmod (N + S) = (p - 9) \bmod 10$. The solid lines show the cube subgraph of the DR (compare to Figs. 2 and 3). The DR network can tolerate any single PE or network fault. Multiple faults in S adjacent rows of the network and the associated S PE's can be tolerated by selecting Π to exclude those S rows and PE's.

C. Control of the DR Network

The DR network can be operated in either the circuit switched mode or packet switched mode. When a PE wants to transfer data, it will generate a *routing tag* as header of a message to establish a connection path. Each switch is set independently. The Exclusive-OR [41] and destination [21] routing tag schemes used for the GC can be used to control the DR network.

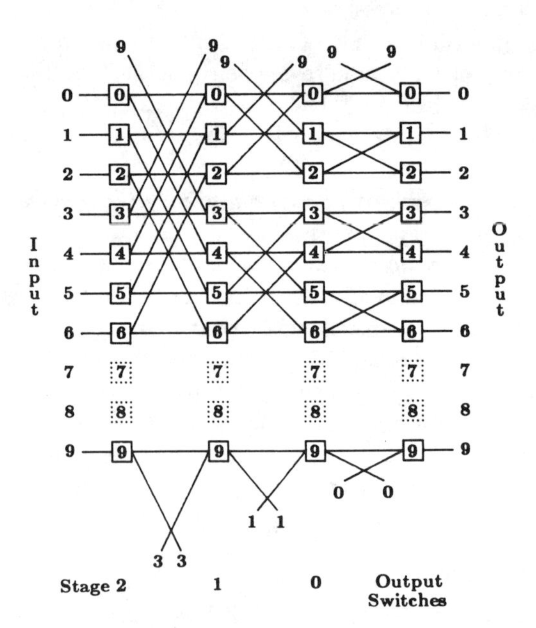

Fig. 4. Reconfiguration of a DR network with $N = 8$ and $S = 2$ when PE 7 is faulty. The solid lines show the cube subgraph of the DR network.

Control of the DR network is based on the logical addresses. Given a source PE with logical address X and a destination PE with logical address Y, the Exclusive-OR tag $E = e_{m-1} \cdots e_1 e_0$ can be derived by taking the bitwise Exclusive-OR of X and Y. Each switch in stage i of Π will examine e_i to determine which link to use. Let $W = w_{m-1} \cdots w_1 w_0$ be the logical address of the switch. If $e_i = 0$, then link f_0 is used. If $e_i = 1$ and $w_i = 0$, switch W will use link f_{+i}. Otherwise ($e_i = 1$ and $w_i = 1$) it will use link f_{-i}. Thus, for $e_i = 1$, W is connected to $\text{cube}_i(W)$.

This approach necessitates adding a one-bit flag in each switch to store the ith bit of the logical address of the switch, w_i. Observe that at stage i a connection path from a source logical PE X to a destination logical PE Y will use switch with logical address $W = y_{m-1} \cdots y_{i+1} x_i \cdots x_0$ (this is shown for the GC in [40]). Thus, $w_i = x_i$. Therefore, these flags can be set by each source PE sending its logical address during system initialization and after each reconfiguration due to a fault. Furthermore, as an alternative scheme that does not require a switch to store w_i, the source PE logical address X can be sent with each message, and x_i is used in place of w_i.

The DR network can also be controlled by using a destination tag which is the destination PE logical address Y. A switch W in stage i can examine y_i and w_i and use link f_{-i} when $y_i < w_i$, link f_0 when $y_i = w_i$, or link f_{+i} when $y_i > w_i$. As in the case for the Exclusive-OR scheme, x_i can be used in place of w_i.

For the GC, either routing scheme can include an m-bit *broadcast mask B*, where $b_i = 1$ means broadcast at stage i [41]. At a stage i switch in the DR network, when $b_i = 1$, if $w_i = 0$, links f_0 and f_{+i} will be used (upper broadcast), otherwise, links f_0 and f_{-i} will be used (lower broadcast). As before, x_i can be used in place of w_i.

D. Partitionability of the DR Network

The *partitionability* of an interconnection network is the ability to divide the network into independent subnetworks such that each subnetwork of size N' has all of the interconnection capabilities of a complete network of that same type with size N' [39]. When S is even, the DR network can be partitioned into two independent subnetworks π_0 and π_1 by setting all switches in stage 0 to f_0, where π_0 contains all even rows and π_1 contains all odd rows. Both π_0 and π_1 are of size $(N/2) + (S/2)$. The theory underlying this is similar to that for partitioning the ADM network, as discussed in [39] and [40]. An example of partitioning the DR network with $N = 8$ and $S = 2$ is shown in Fig. 5. Each PE in a partition has a *partition address* between 0 and $(N/2) + (S/2) - 1$. Let h_{π_i} be an address transformation which maps the physical addresses of switches in π_i to the partition addresses. Then $h_{\pi_0}(j) = j/2$ and $h_{\pi_1}(j) = (j-1)/2$.

Since π_0 and π_1 have all the interconnection capabilities of a complete DR network, both π_0 and π_1 can be partitioned again if both $N/2$ and $S/2$ are even. The network does not have to be partitioned into subnetworks of the same size. For example, consider a DR network with $S \geq 4$. It could first be partitioned into odd and even halves, and then just the even half is partitioned again, resulting in one subnetwork of size $(N/2) + (S/2)$ and two subnetworks of size $(N/4) + (S/4)$. In general, the physical addresses of all the switches in a subnetwork (partition) of size $(N + S)/2^v$ must agree in their low-order v bit positions, and each partition (subsystem) can tolerate faults independently.

The partitionability of the DR network provides the necessary capabilities for multiprocessor systems which can operate in multiple-SIMD mode and use spare PE's in each SIMD subsystem to enhance system reliability. Control and reconfiguration of each subnetwork for each SIMD subsystem will be the same as discussed in the previous section. The partitionability property of the DR network can also be exploited to establish virtual MIMD subsystems in an MIMD system. It can also be used to support partitionable SIMD/MIMD systems, i.e., systems capable of being partitioned into independent SIMD or MIMD machines of various sizes. PASM is such a system [43], [42]. The DR study was motivated by an investigation of incorporating fault tolerance into PASM.

E. Fault Recovery

In the recovery process, to ensure that the data for a task to be restarted are not polluted by faults, the task has to be rolled back to the beginning or to a point where a copy of clean data has been stored. The rollback distance depends on the fault detection techniques used and error latency [37] and will not be discussed here.

Unlike most other fault-tolerant MIN's, the proposed DR network does not provide multiple connection paths for every input–output pair. Fault tolerance of the DR network, as well as the system based on it, is achieved by reconfiguration. After reconfiguring the system, programs and data must be reloaded into the new functioning PE's to restart the task. Since the DR network eliminates network rows and PE's rather than finding a second connection path, there is no need to determine if a connection path is faulty and no need to modify the routing tags to reroute paths. In an SIMD environment, permuting data would need only one pass through the network after the recovery. Hence, with a recovery procedure added in the operating system, programs which can be executed in a system based on a GC can be executed in a DR-based system before and after a fault without any modification. The performance of the DR-based system will not be degraded after recovery.

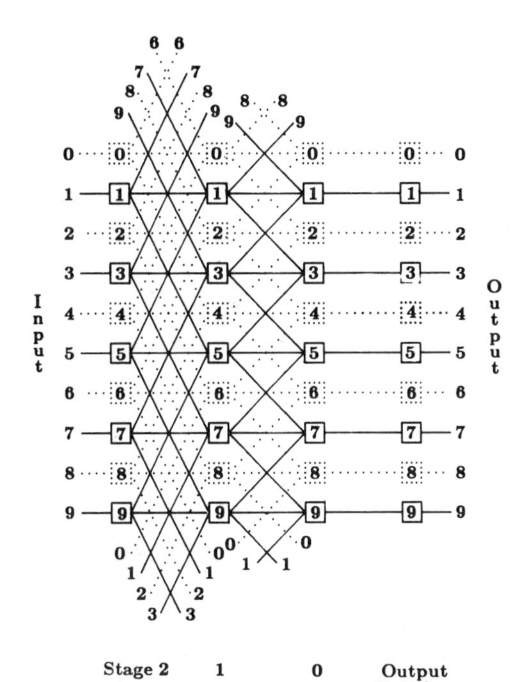

Fig. 5. Partitioning a DR network with $N = 8$ and $S = 2$ into two independent subnetworks. The solid lines show the subnetwork containing only odd number switches. The dotted lines show the subnetwork containing only even number switches.

III. The Reduced DR Network

Consider the proof of Theorem 1 in Section II-B. In stage i of the DR network, if bit i of $t(p)$, the logical address of a physical switch p, is 0, then link f_{+i} is used to perform cube$_i(t)(p)$); if bit i of $t(p)$ is 1, then link f_{-i} is used to perform cube$_i(t(p))$. Therefore, if the function $t(p)$ is modified such that bit i of $t(p)$ is always equal to bit i of p for some i, then switches in these stages will need only two links: f_0 and f_{+i}, or f_0 and f_{-i}. Thus, the complexity of the DR can be reduced without losing the capability to emulate the multistage cube network.

The graph of an RDR network of size $N + S$, $N = 2^m$ and $S = 2^s$, is a subgraph of a DR. The procedure to construct an RDR network is as follows.

1) At stages $m - 1$ to s, the RDR network has the same interstage connections as the DR network.

2) At stage $i, 0 \leq i < s$, a switch with physical address $p, 0 \leq p < N + S$, has two output links. If bit i of p is 0, then switch p has output links f_0 and f_{+i}; otherwise, it has output links f_0 and f_{-i}.

When $S = 1$ ($s = 0$), the RDR network is identical to the DR network. If $S > 1$ ($s > 0$), RDR has less links in stages $s - 1$ to 0. The interconnections f_{+i} and f_{-i} in stages $s - 1$ to 0 of the RDR network connect the switch physically numbered p to switch cube$_i(p)$. An RDR network with $N = 8$ and $S = 4$ is shown in Fig. 6.

Since the RDR network contains less links than DR, not

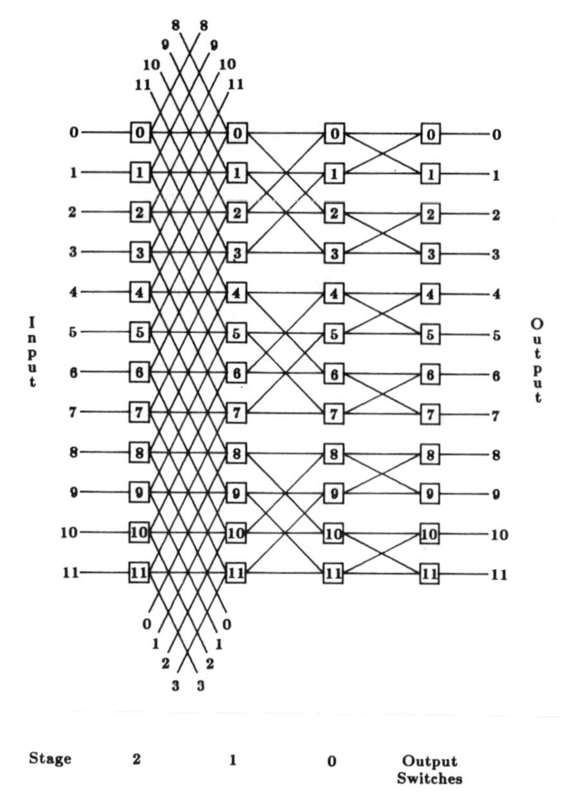

Fig. 6. An RDR network with $N = 8$ and $S = 4$.

any N adjacent rows of the RDR network can act as a GC. When a fault occurs, the set of new functioning PE's must be carefully selected so that the RDR network can still provide the full GC interconnection capabilities for the new functioning PE's. When a PE and/or a row of the RDR with physical address j fails, PE p and row $p, 0 \leq p < N + S$, will be logically renumbered $t'(p)$:

$$t'(p) = (p - k') \bmod (N + S), \tag{3.1}$$

where $k' = \lfloor (j + S)/S \rfloor \times S$. The notation $\lfloor x \rfloor$ represents the floor function, the greatest integer smaller than or equal to x. Consider Fig. 6 for example. If PE 1 is faulty, then $k' = 4$, PE's 0, 1, 2, and 3 will be isolated and the remaining PE's will become new functioning PE's.

The RDR network has the same interstage connections as the DR network in stages $m - 1$ to s. In stages $s - 1$ to 0 of the RDR, because k' is a multiple of S and $S = 2^s$, bit i of $t'(p)$ will be equal to bit i of p for $0 \leq i < s$. Therefore, Theorem 2 follows.

Theorem 2: Given the mapping function t', the RDR network can provide all the interconnection capabilities of a GC before and after a single fault occurs.

Proof: The proof is similar to that for Theorem 1 and hence is omitted here. □

Fault tolerance in the DR and RDR networks is different. In the DR network, multiple faults are tolerable if they are contained within S adjacent rows of the network and their associated PE's. In the RDR, multiple faults can be tolerated when each of them causes the same PE's to be isolated. Let p and q be two physical addresses of any two faulty components. If $\lfloor (p + S)/S \rfloor = \lfloor (q + S)/S \rfloor$, then these faults are tolerable. For example, in Fig. 6, multiple faults in both PE 0 and PE 3 are tolerable because both of them cause PE's 0, 1, 2, and 3 to be isolated, while multiple faults in both PE 3 and PE 4 are not tolerable because they cause different PE's to be isolated.

The RDR network has the same partitionability as the DR network. Even with the reduction of the number of links in stages $s - 1$ to 0, the RDR network can be partitioned into S independent subnetworks, each of which provides single fault tolerance for each subsystem.

One advantage of the RDR network is that it can be constructed from smaller DR and GC networks. First consider stages $m - 1$ to s of an RDR network. Recall that the RDR network can be partitioned into S independent subnetworks of size $(N/S) + 1$ by setting switches in stages 0 through $s - 1$ to straight; each subnetwork contains switches whose physical addresses agree in the low-order s bit positions. Based on reasoning similar to that for partitioning, stages $m - 1$ to s of an RDR can be decomposed into S DR subnetworks of size $(N/S) + 1$. Each DR subnetwork contains switches whose physical addresses agree in the low-order s bit positions. Second, in stages $s - 1$ to 0 of the RDR network, a switch with physical address p has links f_0 and f_{+i} if bit i of p is 0, or links f_0 and f_{-i} if bit i of p is 1. So the interconnection function f_{-i} or f_{+i} in stage $i, 0 \leq i < s$, is equivalent to cube$_i$. Therefore, stages $s - 1$ to 0 of the RDR network can be decomposed into $(N/S) + 1$ GC subnetworks of size S. Each GC subnetwork contains switches of RDR whose physical addresses agree in bit m to bit s. Therefore, an RDR network of size $N + S$ can be constructed by using S DR networks of size $(N/S) + 1$ and $(N/S) + 1$ GC networks of size S. Fig. 7 shows an example of constructing an RDR network of size 8 + 4 from four DR networks of size 2 + 1 and three GC networks of size 4.

Previously, the fault tolerance of the DR and RDR networks has been considered based on an implementation with $N + S$ switches per stage. Consider an RDR network implemented by DR and GC subnetworks as described above. If 2×2 interchange boxes are used in the GC subnetworks, then any single interchange box failure in the GC subnetworks can be tolerated. For example, consider the RDR network in Fig. 6. If the interchange box which corresponds to switches 1 and 3 in stages 1 and 0 (as well as the associated links) fails, the failure can be tolerated no matter how the RDR network is partitioned. If the RDR network is not partitioned, then failure of the interchange box can be considered as multiple faults which cause the same PE's (PE's 0, 1, 2, and 3) to be isolated and hence can be tolerated. If the RDR network is partitioned into two subnetworks, one containing even rows and the other containing odd rows, then the interchange box failure will cause PE's 1 and 3 in the odd partition to be isolated. So this failure is also tolerable. When the RDR network is partitioned into four subnetworks, the failure of the interchange box will affect two partitions, one containing rows 1, 5, and 9, and the other containing rows 3, 7, and 11. Each partition contains faults in one row of the subnetwork. Both partitions have to be reconfigured to tolerate this failure.

In general, since a 2×2 interchange box corresponds to switches in two rows of the RDR network, at most two partitions will be affected if an interchange box is faulty. If these

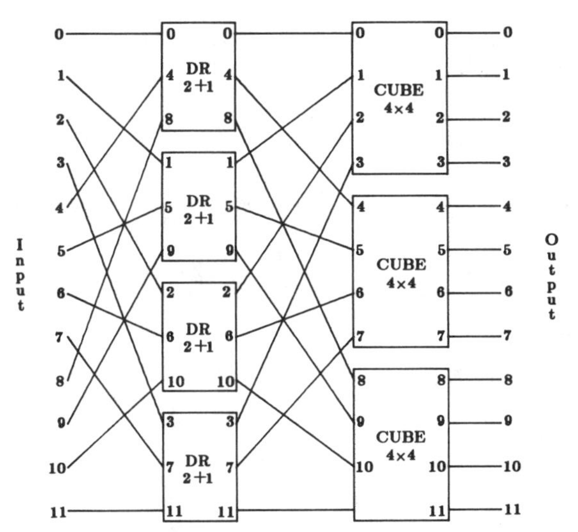

Fig. 7. An example of constructing an RDR network of size 8 + 4 from four DR networks of size 2 + 1 and three multistage cube networks of size 4.

two rows are in two different partitions, then each partition is assumed to contain one faulty row and can be reconfigured to isolate the faulty row. Assume that these two rows are in one partition of size $N' + S'$, where $N' = 2^{m'}$ and $S' = 2^{s'}$. Let the partition addresses of these two faulty rows be p' and q', where $q' = \text{cube}_i(p')$. The interchange boxes are used in stages 0 to $s' - 1$ only, so $0 \leq i < s'$. Since p' and q' differ in bit i only, $\lfloor (p' + S')/S' \rfloor = \lfloor (q' + S')/S' \rfloor$. From (3.1), the two faulty rows will cause the same PE's to be isolated and hence are tolerable.

The RDR network can be controlled by the same routing tags used for the DR network. In stages $m - 1$ to s, since the RDR has the same interstage connections, a switch of RDR will interpret a routing tag in the same way as in a DR network. In stage $s - 1$ to 0, since the subnetworks of the RDR are GC networks, a switch in these stages will simply interpret the Exclusive-OR tag or the destination tag in the same manner as in a GC. Control for the broadcast functions is similar.

IV. Reliability Prediction

A. Comparison of DR-Based Systems with S = 1 to Systems with No Spares

Most previous work in designing reliable MIN-based system has focused on adding redundancy in MIN's only. It will be shown that no matter how much redundancy is added into an MIN, the system reliability cannot exceed a certain bound; however, using the DR and one spare PE, the bound can be exceeded. The overall system reliabilities in both cases for a given mission time T will be estimated and compared. It is assumed that faults occur independently among different modules of a system and a reliable fault diagnosis and recovery process is used. Thus, the purpose of this subsection is to show that even though the DR is more complex than the GC, the DR with one spare provides better system reliability than a "perfect" GC variation with no spare PE.

First, consider a system containing N homogeneous PE's (no spares) and an $N \times N$ MIN. Let

- R_0 be the reliability of the system with no spares;
- R_p be the reliability of a single PE;
- R_n be the reliability of the multistage network.

The system reliability R_0 is

$$R_0 = C_1 \times R_p^N \times R_n \tag{4.1}$$

where C_1 is a factor which represents the reliability degradation due to other subsystems, such as the secondary memory storage and those modules which control and sequence the system.

Using a fault-tolerant MIN (not a DR) would increase R_n and therefore increase R_0. However, no matter how reliable an MIN is used, R_n is always less than one. Hence, R_0 can never exceed $C_1 R_p^N$. Let

$$R = C_1 R_p^N. \tag{4.2}$$

Then R can be considered as an upper bound of the system reliability that can be obtained by adding redundancy to the MIN only.

Now consider a DR-based system which contains $N + 1$ homogeneous PE's and a DR network of size $N + 1$. Let

- R_1 be the reliability of the overall system with $S = 1$;
- λ_p be the failure rate of an individual PE;
- λ_w be the failure rate of a network switch and all links incident out of it;
- ρ be the ratio between λ_w and λ_p, i.e., $\rho = \lambda_w/\lambda_p$;
- R_w be the reliability of a network switch and the associated links;
- R_r be the reliability of an individual row of the DR network.

The failure rate of a component depends on the complexity of the component, maturity of fabrication process, and other factors such as operating temperature. Both λ_p and λ_w are in terms of number of failures per million hours. Typical component failure rates are in the range of 0.1 to tens per million hours [44]. If both λ_p and λ_w are constant during the operational life of the system, then given a mission time T, the number of failures which may occur in a single PE during this period is $\lambda_p T$. For example, if $\lambda_p = 1$ failure per million hours and $T = 1000$ h, then $\lambda_p T = 10^{-3}$. The failure probabilities will follow the exponential distribution [44]. Thus, for a mission time T, $R_p = e^{-\lambda_p T}$ and $R_w = e^{-\lambda_w T}$. R_r is approximately equal to R_w^{m+1} because there are m switches plus one output switch in a row of the DR network. Replacing λ_w by $\rho\lambda_p$, then R_r can be expressed as

$$R_r = (e^{-\rho\lambda_p T})^{m+1} = R_p^{\rho(m+1)} = R_p^{\alpha} \tag{4.3}$$

where $\alpha = \rho(m + 1)$. Since a network switch is much smaller than a PE, λ_w is far less than λ_p. Hence, $\rho \ll 1$. For a PE or a row of DR to be usable in the system, both of them must be fault free, if they have the same physical address. The reliability R_1 of the overall system then is

$$\begin{aligned} R_1 &= C_2[(R_p R_r)^{N+1} + (N + 1)(R_p R_r)^N(1 - R_p R_r)] \\ &= C_2 e^{-(1+\alpha)\lambda_p N T}[1 + N(1 - e^{-(1+\alpha)\lambda_p T})] \end{aligned} \tag{4.4}$$

where C_2 is a factor similar to C_1.

To compare R_1 to the upper bound R, it is assumed that $C_1 = C_2 = 1$. Table I shows R_1 and R versus N for three different values of $\lambda_p T$ at $\rho = 0.1$. In Table I, when $\lambda_p T =$

TABLE I

SYSTEM RELIABILITIES: R IS THE UPPER BOUND RELIABILITY OF A SYSTEM CONTAINING NO SPARE PE. R_1 IS THE RELIABILITY OF A DR-BASED SYSTEM OF SIZE $N + 1$

N	$\lambda_p T = 10^{-4}$		$\lambda_p T = 5\times10^{-4}$		$\lambda_p T = 10^{-3}$	
	R	R_1	R	R_1	R	R_1
4	0.9996	1.0000	0.9980	0.9999	0.9960	0.9999
16	0.9984	0.9999	0.9920	0.9999	0.9841	0.9996
64	0.9936	0.9999	0.9684	0.9985	0.9379	0.9944
256	0.9747	0.9988	0.8798	0.9747	0.7740	0.9135
1024	0.9026	0.9799	0.5992	0.7078	0.3589	0.3662
4096	0.6639	0.7569	0.1289	0.0513	0.0166	0.0008

* $(C_1 = C_2 = 1, \rho = 0.1)$

10^{-3} or 5×10^{-4}, $R_1 > R$ for $N \leq 1024$; for $\lambda_p T = 10^{-4}$, $R_1 > R$ for $N \leq 4096$. In the following, it will be shown that R_1 can exceed the bound R for a wide range of N.

Theorem 3: If $N \leq [1 - (1 + \alpha)^2 \lambda_p T]/[\alpha(1 + \alpha/2)\lambda_p T]$, then $R_1 > R$.

Proof: Let $y = \lambda_p T$. Because $R = e^{-Ny}$ and $R_1 = e^{-(1+\alpha)Ny}[1 + N(1 - e^{-(1+\alpha)y})]$, proving $R_1 > R$ is equivalent to proving

$$e^{-\alpha Ny}[1 + N(1 - e^{-(1+\alpha)y})] > 1. \tag{4.5}$$

Since $e^{-\alpha Ny} = \sum_{k=0}^{\infty} (-\alpha Ny)^k/k!$ and $1 + N[1 - e^{-(1+\alpha)y}] = 1 - N\sum_{k=1}^{\infty} [-(1 + \alpha)y]^k/k!$, (4.5) becomes

$$\left(\sum_{k=0}^{\infty} \frac{(-\alpha Ny)^k}{k!}\right)\left(1 - N\sum_{k=1}^{\infty} \frac{[-(1 + \alpha)y]^k}{k!}\right) > 1. \tag{4.6}$$

Because $N \leq [1 - (1 + \alpha)^2 y]/[\alpha(1 + \alpha/2)y]$,

$$1 - \alpha Ny - \frac{\alpha^2 Ny}{2} - (1 + \alpha)^2 y \geq 0. \tag{4.7}$$

Since all the terms in the left side of (4.7) are positive, any nonconstant term must be less than one. Hence, $\alpha Ny < 1$, and $(1 + \alpha)y < (1 + \alpha)^2 y < 1$. Therefore,

$$\sum_{k=0}^{\infty} \frac{(-\alpha Ny)^k}{k!} > \sum_{k=0}^{3} \frac{(-\alpha Ny)^k}{k!}, \tag{4.8}$$

$$1 - N\sum_{k=1}^{\infty} \frac{[-(1 + \alpha)y]^k}{k!} > 1 - N\sum_{k=1}^{2} \frac{[-(1 + \alpha)y]^k}{k!}. \tag{4.9}$$

From (4.5), (4.6), (4.8), and (4.9), if

$$\left(\sum_{k=0}^{3} \frac{(-\alpha Ny)^k}{k!}\right)\left(1 - N\sum_{k=1}^{2} \frac{[-(1 + \alpha)y]^k}{k!}\right) > 1, \tag{4.10}$$

then $R_1 > R$. Hence, in the following, it will be proved that (4.10) is true. Since $1 - \alpha Ny - \alpha^2 Ny/2 - (1 + \alpha)^2 y \geq 0$ (see 4.7), so

$$1 - \alpha Ny - \frac{\alpha^2 Ny}{2} - \frac{(1 + \alpha)^2 y}{2} \geq 0,$$

$$Ny - \alpha(Ny)^2 - \frac{(\alpha Ny)^2}{2} - \frac{(1 + \alpha)^2 Ny^2}{2} \geq 0. \tag{4.11}$$

Because $\alpha Ny < 1$, so $1/6(1 + \alpha)(\alpha Ny)^3 < 1/6(1 + \alpha)(\alpha Ny)^2$. Hence,

$$\frac{1}{2}(1 + \alpha)(\alpha Ny)^2 - \frac{1}{6}\alpha(\alpha Ny)^2 - \frac{1}{6}(1 + \alpha)(\alpha Ny)^3$$
$$> \frac{1}{2}(1 + \alpha)(\alpha Ny)^2 - \frac{1}{6}\alpha(\alpha Ny)^2 - \frac{1}{6}(1 + \alpha)(\alpha Ny)^2$$
$$= (\alpha Ny)^2\left(\frac{1 + \alpha}{2} - \frac{\alpha}{6} - \frac{1 + \alpha}{6}\right) > 0.$$

Therefore,

$$\frac{(1 + \alpha)(\alpha Ny)^2}{2} - \frac{\alpha(\alpha Ny)^2}{6} - \frac{(1 + \alpha)(\alpha Ny)^3}{6} > 0. \tag{4.12}$$

Similarly,

$$\frac{1}{2}\alpha(1 + \alpha)^2 N^2 y^3 - \frac{1}{4}\alpha(1 + \alpha)^2 N^3 y^4 > 0. \tag{4.13}$$

$$\frac{1}{12}\alpha^3(1 + \alpha)^2 N^4 y^5 > 0. \tag{4.14}$$

By adding (4.11), (4.12), (4.13), and (4.14) together and incrementing both sides by 1, (4.10) is attained. Therefore, $R_1 > R$. □

When $N\lambda_p T$ is large, both R and R_1 become small. For example, if $\lambda_p T = 10^{-3}$ and $N = 1024$, then $N\lambda_p T \simeq 1$, $R \simeq 0.359$, and $R_1 \simeq 0.366$. In practice, for a system to be usable, the system reliability must be near one, i.e., $N\lambda_p T$ must be far less than one. Thus, in a practical system, $(1 + \alpha)^2 \lambda_p T \ll 1$. The condition in Theorem 3 can be approximated by the following inequality:

$$N \leq \frac{1}{\alpha\left(1 + \frac{\alpha}{2}\right)\lambda_p T}. \tag{4.15}$$

When N satisfies the condition above, then reliability improvement can be made by using a DR network of size $N + 1$ and one spare PE in a system. For example, if $\lambda_p T = 5 \times 10^{-4}$, and $\rho = 0.1$, then a reliability improvement can be made for any system with $N \leq 1024$. The reliability improvement in this case can be measured as follows:

$$\text{RIF} = \frac{1 - R}{1 - R_1} \tag{4.16}$$

where RIF is the reliability improvement factor [29]. For example, if $N = 256$ and $\lambda_p T = 10^{-4}$, RIF $= (1 - 0.9747)/(1 - 0.9988) \simeq 21$. If N is very large, R_1 may become smaller than R (see Table I). The reason is that when N is very large, the effect of adding one spare becomes small and the reliability enhancement is offset by the decrease in the network reliability. However, $R_1 < R$ does not imply $R_1 < R_0$ because the reliability of an $N \times N$ fault-tolerant MIN will also decrease as N increases and hence R_0 will become much less than R.

B. Effect of Multiple Spares in a Nonpartitionable System

In a DR-based or RDR-based system, using more spares $(S > 1)$ may tolerate more faults. However, it will be shown that if a system is not designed to operate in a partitionable environment, then little reliability improvement can be made by adding additional spares.

First, consider the reliability of a DR-based system. Let R_s denote the reliability of a nonpartitionable system containing $N + S$ homogeneous PE's and a DR network of size $N + S$ for inter-PE communication, and define $\mathrm{RIF}_{1,s}$ as follows:

$$\mathrm{RIF}_{1,s} = \frac{1 - R_1}{1 - R_s}. \tag{4.17}$$

The following theorem can be derived.

Theorem 4: For $2 \le S < N$, a) $R_s > R_1$; b) $\mathrm{RIF}_{1,s} \simeq 1$ if $(1 + \alpha)S\lambda_p T \ll 1$.

Proof: A DR-based system of size $N + S$ is available for applications which require N PE's if at least N adjacent PE's and their associated rows of the DR network are fault-free. Thus, if there are exactly $N + S, N + S - 1, \cdots,$ or N adjacent PE's and associated rows of the DR which are fault-free, then the system is usable. Let $R_c = R_p R_r = e^{-(1+\alpha)\lambda_p T}$. The possibility that all PE's and all rows of the DR are fault-free is R_c^{N+S}; the possibility that exactly $N + S - 1$ adjacent PE's and the associated rows of the DR are fault-free is $(N + S)R_c^{N+S-1}(1 - R_c)$. For a system to contain exactly i fault-free adjacent PE's and their associated rows of the DR, $N \le i \le N + S - 2$, there must be a faulty PE or component associated with each row on the "top and bottom (mod $N + S$)" of the i fault-free rows. For example, if $i = N + 1$ and PE's 2 to $N + 2$ and their associated rows of DR are fault-free, then PE 1 or row 1 of the DR must be faulty, and PE $N + 3$ or row $N + 3$ of the DR must be faulty, otherwise i will be $N + 2$ or $N + 3$ instead of $N + 1$. The PE's and rows numbered $N + 4$ to $N + S - 1$, and 0 may or may not be faulty. Thus, the possibility that exactly i adjacent PE's and their associated rows of the DR are fault-free is $(N + S)R_c^i(1 - R_c)^2$, where $N \le i \le N + S - 2$. Since all these conditions are mutually exclusive, the reliability R_s of the system is

$$\begin{aligned} R_s &= C(R_c^{N+S} + (N + S)[R_c^{N+S-1}(1 - R_c) + R_c^{N+S-2} \\ &\quad \cdot (1 - R_c)^2 + \cdots + R_c^{N+1}(1 - R_c)^2 + R_c^N(1 - R_c)^2]) \\ &= C\Bigg[R_c^{N+S} + (N + S)R_c^{N+S-1}(1 - R_c) \\ &\quad + (N + S)R_c^N(1 - R_c)^2 \sum_{i=0}^{S-2} R_c^i\Bigg] \end{aligned} \tag{4.18}$$

where C is a factor similar to C_1. Using $\sum_{i=0}^{S-2} R_c^i = (1 - R_c^{S-1})/(1 - R_c)$ and rearranging terms in (4.18) results in

$$R_s = CR_c^N[R_c^S + (N + S)(1 - R_c)]. \tag{4.19}$$

a) $R_s > R_1$ for $2 \le S \le N$

$$\frac{R_s}{R_1} = \frac{R_c^S + (N + S)(1 - R_c)}{R_c + (N + 1)(1 - R_c)}$$

$$= 1 + \frac{(1 - R_c)\left(S - \sum_{i=0}^{S-1} R_c^i\right)}{R_c + (N + 1)(1 - R_c)}$$

where $\sum_{i=0}^{S-1} R_c^i = (1 - R_c^S)/(1 - R_c)$. Since $R_c < 1$, so $S - \sum_{i=0}^{S-1} R_c^i > 0$ and $1 - R_c > 0$. Hence, $R_s/R_1 > 1$. Therefore, $R_s > R_1$ for $2 \le S < N$.

b) $\mathrm{RIF}_{1,s} \simeq 1$ if $(1 + \alpha)S\lambda_p T \ll 1$.

$$\mathrm{RIF}_{1,s} = \frac{1 - R_1}{1 - R_s} = \frac{1 - CR_c^N[R_c + (N + 1)(1 - R_c)]}{1 - CR_c^N[R_c^S + (N + S)(1 - R_c)]}.$$

For $(1 + \alpha)S\lambda_p T \ll 1, R_c^S = e^{-(1+\alpha)S\lambda_p T} \simeq 1 - (1 + \alpha)S\lambda_p T$ and $R_c \simeq 1 - (1 + \alpha)\lambda_p T$. Let $F_c = (1 + \alpha)\lambda_p T$. Then

$$\mathrm{RIF}_{1,s} \simeq \frac{1 - CR_c^N[1 - F_c + (N + 1)F_c]}{1 - CR_c^N[1 - SF_c + (N + S)F_c]}$$

$$= \frac{1 - CR_c^N(1 + NF_c)}{1 - CR_c^N(1 + NF_c)} = 1. \qquad \square$$

In general, $(1 + \alpha)S < N$, so $(1 + \alpha)S\lambda_p T < N\lambda_p T \ll 1$. Therefore, in practice, $\mathrm{RIF}_{1,s} \simeq 1$ holds for most systems. Table II shows the system reliabilities for different values of S and N. It shows that for a nonpartitionable system, a significant enhancement can be obtained by adding the first spare ($S = 1$), while only little gain can be made by further increasing S.

In a RDR-based system, not any N adjacent fault-free PE's and their associated rows of the RDR can be used. Therefore, R_s for an RDR-based system will be less than R_s for a DR-based system, for $S = 2^s$. So $\mathrm{RIF}_{1,s} \simeq 1$ holds for RDR also.

C. Effect of Multiple Spares in a Partitionable System

The partitionability of DR (or RDR) can provide the necessary capabilities for a partitionable system to incorporate spare PE's. In a system being designed to operate in multiple-SIMD mode, virtual multiple-MIMD mode, or partitionable SIMD/MIMD mode, according to the Theorem 4, only one spare is needed in each subsystem. If a system of size N is to be partitioned into $Q(= 2^q)$ subsystems of size $2^{m-q}, Q$ spare PE's are needed. The DR (or RDR) network to be used in the system will be of size $N + Q$. Because each partition can operate independently, the overall system reliability is the product of all the reliabilities of these partitions. Let R_q be the overall system reliability when the system is partitioned into Q subsystems. Then

$$R_q = C_3\left[R_c^{(N/Q)+1} + \left(\frac{N}{Q} + 1\right)R_c^{N/Q}(1 - R_c)\right]^Q \tag{4.19}$$

where C_3 is a factor similar to C_1 and C_2 and $R_c = R_p R_r$. Table III shows R_q and R_1 at three different values of $\lambda_p T$ for $C_3 = 1, q = m/2$, and $\rho = 0.1$. It shows that the system reliability can be greatly improved by adding multiple spares in a partitionable system.

V. Complexity and Comparison

An approach to adding spares is used in the GF11 SIMD machine project [9]. There a 576 input/output Benes network is used, where there are 512 "primary processors" and 64 spares. While the GF11 design is good for its intended applications, its overall architecture and centralized control Benes

TABLE II

RELIABILITIES OF NONPARTITIONABLE SYSTEMS AT DIFFERENT VALUES OF S. WHEN $S \geq 1$, A DR NETWORK OF SIZE $N + S$ IS USED

S	$\lambda_p T = 10^{-3}$		$\lambda_p T = 5\times10^{-4}$	
	N=64	N=256	N=64	N=256
0	0.9379	0.7740	0.9684	0.8798
1	0.9944	0.9135	0.9985	0.9747
2	0.9944	0.9135	0.9985	0.9747
3	0.9944	0.9135	0.9985	0.9747

* ($C_1 = C = 1$, $\rho = 0.1$)

TABLE III

SYSTEM RELIABILITIES: R_1 IS THE RELIABILITY OF A DR-BASED NONPARTITIONABLE SYSTEM OF SIZE $N + 1$. R_q IS THE RELIABILITY OF A DR-BASED OR RDR-BASED PARTITIONABLE SYSTEM OF SIZE $N + Q$

N	$\lambda_p T = 10^{-4}$		$\lambda_p T = 5\times10^{-4}$		$\lambda_p T = 10^{-3}$	
	R_1	R_q	R_1	R_q	R_1	R_q
4	1.0000	1.0000	0.9999	0.9999	0.9999	0.9999
16	0.9999	0.9999	0.9999	0.9999	0.9996	0.9999
64	0.9999	0.9999	0.9985	0.9997	0.9944	0.9991
256	0.9988	0.9999	0.9747	0.9980	0.9135	0.9923
1024	0.9799	0.9992	0.7078	0.9819	0.3662	0.9311
4096	0.7569	0.9930	0.0513	0.8454	0.0008	0.5261

* ($C_2 = C_3 = 1$, $\rho = 0.1$)

network were not designed for, and are therefore inappropriate for, emulating GC networks in multiple-SIMD, MIMD, and partitionable SIMD/MIMD systems.

The crossbar networks can readily accommodate spare processors and much more easily be reconfigured. However, they are seldom considered for use in large scale multiprocessor systems because of their high cost complexity, i.e., a cost of $O(N^2)$ switches and links for a size N network. A crossbar network can be a cost-effective solution when the entire network can be built in one chip [17]. However, current technology cannot put a large (e.g., $N = 2^7$) crossbar network into one chip or even a small set of chips. Hence, for large N, an MIN with a lower cost complexity of $O(N \log N)$ switches and links is much more favored. This is evidenced by the use of multistage cube-type networks in university projects such as PASM [43] and Ultracomputer [19], and industrial projects such as the Goodyear STARAN [8], the BBN butterfly [11], and the IBM RP3 [33].

Using a standard unique path MIN (in our case it is a GC) with size $2N$ for applications which require N PE's is another straightforward method of providing spare PE's. However, using a size $2N$ network requires adding N spare PE's to continue functioning as a multistage cube after a network fault occurs. The total hardware required in this approach will be more than double that for one network of size N and double the number of PE's. The extra N PE's will not be used until there is a fault (otherwise, this method will be a degraded recovery method for applications requiring $2N$ PE's). The cost overhead due to the N spare PE's will make this method become more expensive than the RDR approach.

The graph representation of a DR network is similar to that of an ADM network [27]. The ADM network was developed for applications which need network performance or permutation capabilities beyond that of the multistage cube. In order to increase network performance, ADM adopts a routing tag scheme other than the destination tag or the Exclusive-OR tag. The ADM employs a dynamic rerouting tag scheme with which, in some cases, faulty switches or links can be avoided by taking an alternative path. However, it is not a fault-tolerant network, e.g., there is only one disjoint path if both source address and destination address are even or both are odd. The ADM network was never intended to support the use of spare PE's. On the other hand, the DR network is designed for applications which need a fault-tolerant multistage cube type of network and spare PE's. With the DR network, any faulty PE, switch, or link can be tolerated by replacing the entire faulty row of the DR with a spare row, maintaining N PE's and N parallel paths through the network. Programs which can run in ADM-based systems cannot be executed in DR-based systems. The routing scheme used in the ADM network cannot be used to control the DR network, and many permutation patterns which can be done in ADM may not be realized in the DR network. Thus, while the ADM and the DR are topologically related, there are significant differences which include design goals, network control schemes, the way in which the networks are used, permutation capability, network fault tolerance and recovery procedures, and impact on system fault tolerance.

The complexity of the DR approximately equals that of the ADM network when S is much smaller than N. Compare the ADM and the GC based on their graph representations first. The link ratio between the ADM and the GC is 3/2. If links of their graphs are interpreted as interchange boxes and switching nodes are interpreted as links [Fig. 2(b)], then the ADM needs N more links than the GC, which interconnect interchange boxes within each stage of ADM [28]. Hence, the link ratio between the ADM and the GC is two. The link ratios between DR and GC are similar. Consider the reliability improvement obtained by using DR, from (4.16), the reliability improvement factor is about 21, compared to the upper bound R, for $\lambda_p T = 10^{-4}$, and about five for $\lambda_p T = 5 \times 10^{-4}$.

The link ratio between the RDR and the DR is $(3m - s)/3m$. However, the RDR network is more cost effective than the DR because it can be implemented more easily as discussed in Section III. When a system is operating in a partitionable environment, the reliability improvement factor obtained by using the RDR network is about 250, compared to R, for $\lambda_p T = 10^{-4}$, and about 60 for $\lambda_p T = 5 \times 10^{-4}$.

VI. CONCLUSION

The purposes of this paper are to investigate the possibility of adding redundancy to MIN's as well as to other subsystems to enhance the overall system reliability, and to analyze the improvement in reliability that can be obtained. While many other fault-tolerant multistage cube type of networks have been proposed [1], the network described here differs in that in addition to being fault tolerant, it also supports the inclusion of spare processors into the system. The DR network would be more expensive than the GC, but it can significantly improve system reliability over a large range of N.

The DR and RDR networks are designed to be partitionable into up to S independent subnetworks (subsystems), each of which is single-fault tolerant in terms of network faults or PE faults. Furthermore, the DR and RDR networks retain the same multistage cube capabilities for one-to-one, broadcast, and permutation connections before and after reconfiguration due to partitioning or a fault.

With a recovery procedure added in the operating system, the application programs can be executed before and after a fault without any modification (after being reloaded into the functioning PE's). The RDR network can be implemented more cost effectively than the DR while retaining most of the capabilities of the DR network. Fault-tolerant capabilities of DR and RDR are obtained by reconfiguration and no component of these networks is assumed to be fault-free.

References

[1] G. B. Adams III, D. P. Agrawal, and H. J. Siegel, "A survey and comparison of fault-tolerant multistage interconnection networks," *Computer,* pp. 14–27, June 1987.

[2] G. B. Adams III and H. J. Siegel, "The extra stage cube: A fault- tolerant interconnection network for supersystems," *IEEE Trans. Comput.,* vol. C-31, pp. 443–454, May 1982.

[3] —, "Modifications to improve the fault-tolerance of extra stage cube interconnection network," in *Proc. 1984 Int. Conf. Parallel Processing,* Aug. 1984, pp. 169–173.

[4] D. P. Agrawal and D. Kaur, "Fault tolerant capabilities of redundant multistage interconnection networks," in *Proc. 1983 Real-Time Syst. Symp.,* Dec. 1983, pp. 119–127.

[5] I. A. Baqai and T. Lang, "Reliability aspects of the Illiac IV computer," in *Proc. 1976 Int. Conf. Parallel Processing,* Aug. 1976, pp. 123–131.

[6] G. H. Barnes, "Design and validation of a connection network for many-processor multiprocessor systems," in *Proc. 1980 Int. Conf. Parallel Processing,* Aug. 1980, pp. 79–80.

[7] K. E. Batcher, "The flip network in STARAN," in *Proc. 1976 Int. Conf. Parallel Processing,* Aug. 1976, pp. 65–71.

[8] —, "STARAN series E," in *Proc. 1977 Int. Conf. Parallel Processing,* Aug. 1977, pp. 140–143.

[9] J. Beetem, M. Denneau, and D. Weingarten, "The GF11 supercomputer," in *Proc. 12th Symp. Comput. Architecture,* June 1985, pp. 108–115.

[10] L. Ciminiera and A. Serra, "A fault-tolerant connecting network for multiprocessor systems," in *Proc. 1982 Int. Conf. Parallel Processing,* Aug. 1982, pp. 113–122.

[11] W. Crowther, J. Goodhue, E. Starr, R. Thomas, W. Milliken, and T. Blackadar, "Performance measurements on a 128-node butterfly parallel processor," in *Proc. 1985 Int. Conf. Parallel Processing,* Aug. 1985, pp. 531–540.

[12] N. J. Davis IV, W. T-Y. Hsu, and H. J. Siegel, "Fault location techniques for distributed control interconnection networks," *IEEE Trans. Comput.,* vol. C-34, pp. 902–910, Oct. 1985.

[13] J. B. Dennis, G. A. Boughton, and C. K. L. Leung, "Building blocks for data flow prototypes," in *Proc. 7th Symp. Comput. Architecture,* May 1980, pp. 1–8.

[14] D. M. Dias and J. R. Jump, "Augmented and pruned $N \log N$ multistage networks: Topology and performance," in *Proc. 1982 Int. Conf. Parallel Processing,* Aug. 1982, pp. 10–11.

[15] T-Y. Feng and Q. Zhang, "Fault diagnosis of multistage interconnection networks with four valid states," in *Proc. 5th Int. Conf. Distributed Comput. Syst.,* May 1985, pp. 218–226.

[16] M. J. Flynn, "Very high-speed computing systems," *Proc. IEEE,* vol. 54, pp. 1901–1909, Dec. 1966.

[17] M. A. Franklin, "VLSI performance comparison of banyan and crossbar communication networks," *IEEE Trans. Comput.,* vol. C-30, pp. 283–291, Apr. 1981.

[18] L. R. Goke and G. J. Lipovski, "Banyan networks for partitioning multiprocessor systems," in *Proc. 1st Symp. Comput. Architecture,* Dec. 1973, pp. 175–189.

[19] A. Gottlieb, R. Grishman, C. P. Kruskal, K. P. McAuliffe, L. Rudolph, and M. Snir, "The NYU Ultracomputer—Designing an MIMD shared memory parallel computer," *IEEE Trans. Comput.,* vol. C-32, pp. 175–189, Feb. 1983.

[20] V. P. Kumar and S. M. Reddy, "Design and analysis of fault-tolerant multistage interconnection network with low link complexity," in *Proc. 12th Symp. Comput. Architecture,* June 1985, pp. 376–385.

[21] D. H. Lawrie, "Access and alignment of data in an array processor," *IEEE Trans. Comput.,* vol. C-24, pp. 1145–1155, Dec. 1975.

[22] C-T. Lea, "A load-sharing banyan network," in *Proc. 1985 Int. Conf. Parallel Processing,* Aug. 1985, pp. 317–324.

[23] J. E. Lilienkamp, D. H. Lawrie, and P-C. Yew, "A fault-tolerant interconnection network using error correcting codes," in *Proc. 1982 Int. Conf. Parallel Processing,* Aug. 1982, pp. 123–125.

[24] J. Maeng, "Self-diagnosis of multistage network-based computer systems," in *Proc. 1983 Int. Fault-Tolerant Comput. Symp.,* June 1983, pp. 324–331.

[25] M. Malek and W. W. Myre, "A description method of interconnection networks," *Distributed Processing Quarterly, IEEE Computer Society Tech. Comm. Distributed Processing Newsletter,* vol. 1, pp. 1–6, Feb. 1981.

[26] W. C. McDonald and J. M. Williams, "The advanced data processing testbed," *COMPSAC,* pp. 346–351, Mar. 1978.

[27] R. J. McMillen and H. J. Siegel, "Performance and fault-tolerance improvements in the inverse augmented data manipulator network," in *Proc. 9th Symp. Comput. Architecture,* Apr. 1982, pp. 63–72.

[28] —, "Evaluation of cube and data manipulator networks," *J. Parallel Distributed Comput.,* vol. 2, pp. 79–107, Feb. 1985.

[29] Y-W. Ng and A. Avizienis, "ARIES—An automated reliability estimation system for redundant digital structure," in *Proc. 1977 Reliability Maintainability Symp.,* Jan. 1977, pp. 108–113.

[30] K. Padmanabhan and D. H. Lawrie, "A class of redundant path multistage interconnection networks," *IEEE Trans. Comput.,* vol. C-32, pp. 1099–1108, Dec. 1983.

[31] —, "Fault tolerance schemes in shuffle-exchange type interconnection networks," in *Proc. 1983 Int. Conf. Parallel Processing,* Aug. 1983, pp. 71–75.

[32] M. C. Pease III, "The indirect binary n-cube microprocessor array," *IEEE Trans. Comput.,* vol. C-26, pp. 458–473, May 1977.

[33] G. F. Pfister, W. C. Brantley, D. A. George, S. L. Harvey, W. J. Kleinfelder, K. P. McAuliffe, E. A. Melton, V. A. Norton, and J. Weiss, "The IBM research parallel processor prototype (RP3): Introduction and architecture," in *Proc. 1985 Int. Conf. Parallel Processing,* Aug. 1985, pp. 764–771.

[34] C. S. Raghavendra and A. Varma, "INDRA: A class of interconnection networks with redundant paths," in *Proc. 1984 Real-Time Syst. Symp.,* Dec. 1984, pp. 153–164.

[35] S. M. Reddy and V. P. Kumar, "On fault-tolerant multistage interconnection networks," in *Proc. 1984 Int. Conf. Parallel Processing,* Aug. 1984, pp. 155–164.

[36] J. P. Shen and J. P. Hayes, "Fault tolerance of a class of connecting networks," in *Proc. 7th Symp. Comput. Architecture,* May 1980, pp. 61–71.

[37] K. G. Shin and Y. H. Lee, "Analysis of the impact of error detection on computer performance," in *Proc. 1983 Int. Fault-Tolerant Comput. Symp.,* June 1983, pp. 356–359.

[38] H. J. Siegel, "Analysis techniques for SIMD machine interconnection networks and the effects of processor address masks," *IEEE Trans. Comput.,* vol. C-26, pp. 153–161, Feb. 1977.

[39] —, "The theory underlying the partitioning of permutation networks," *IEEE Trans. Comput.,* vol. C-29, pp. 791–801, Sept. 1980.

[40] —, *Interconnection Networks for Large-Scale Parallel Processing: Theory and Case Studies.* Lexington, MA: Lexington Books, D. C. Health, 1985.

[41] H. J. Siegel and R. J. McMillen, "The multistage cube: A versatile interconnection network," *Computer,* vol. 14, pp. 65–76, Dec. 1981.

[42] H. J. Siegel, T. Schwederski, J. T. Kuehn, and N. J. Davis IV, "An overview of the PASM parallel processing system," in *Computer Architecture,* D. D. Gajski, V. M. Milutinovic, H. J. Siegel, and B. P. Furht, Eds. Washington, DC: IEEE Computer Society Press, 1987, pp. 387–407.

[43] H. J. Siegel, L. J. Siegel, F. C. Kemmerer, P. T. Mueller, Jr., H. E. Smalley, Jr., and S. D. Smith, "PASM: A partitionable SIMD/MIMD system for image processing and pattern recognition," *IEEE Trans. Comput.,* vol. C-30, pp. 934–947, Dec. 1981.

[44] D. P. Siewiorek and R. S. Swarz, *The Theory and Practice of Reliable System Design.* Bedford, MA: Digital, 1982, pp. 17–62.

[45] S. Thanawastien and V. P. Nelson, "Interference analysis of shuffle/exchange networks," *IEEE Trans. Comput.,* vol. C-30, pp. 545–556, Aug. 1981.

[46] N-F. Tzeng, P-C. Yew, and C-Q. Zhu, "A fault-tolerant scheme for multistage interconnection networks," in *Proc. 12th Symp. Comput. Architecture,* June 1985, pp. 368–375.

[47] C-L. Wu and T-Y. Feng, "On a class of multistage interconnection networks," *IEEE Trans. Comput.,* vol. C 29, pp. 694–702, Aug. 1980.

[48] C-L. Wu, T-Y. Feng, and M-C. Lin, "Star: A local network system for real-time management of imagery data," *IEEE Trans. Comput.,* vol. C-31, pp. 923–933, Oct. 1982.

Menkae Jeng (M'88) was born in Kauhsiung, Taiwan, on April 20, 1957. He received the B.S.E.E. degree from National Taiwan University, Taipei, Taiwan, in 1978, the M.S.E.E. degree from Mississippi State University, Starkville, in 1984, and the Ph.D. degree from the School of Electrical Engineering, Purdue University, West Lafayette, IN, in 1987.

While at Purdue, he was a Research Assistant on the PASM Parallel Processing Project. He is currently an Assistant Professor in the Department of Computer Science, University of Houston, Houston, TX. His research interest include computer architecture, parallel processing, fault-tolerant computing, design and analysis of algorithms, and operating systems for parallel computers.

Dr. Jeng is a member of the Tau Beta Pi and Eta Kappa Nu honorary societies.

Howard Jay Siegel (M'77–SM'82) was born in New Jersey on January 16, 1950. He received the S.B. degree in electrical engineering and the S.B. degree in management from the Massachusetts Institute of Technology, Cambridge, in 1972, the M.A. and M.S.E. degrees in 1974, and the Ph.D. degree in 1977, all in electrical engineering and computer science from Princeton University, Princeton, NJ.

In June 1976, he joined the School of Electrical Engineering, Purdue University, West Lafayette, IN, where he is a Professor and Director of the PASM Parallel Processing Project. He authored the book *Interconnection Networks for Large-Scale Parallel Processing,* and has consulted, given tutorials, coedited four books, and coauthored over 120 papers on parallel processing.

Dr. Siegel has been a guest editor of the IEEE TRANSACTIONS ON COMPUTERS (twice), an IEEE Computer Society Distinguished Visitor, a NATO Advanced Study Institute Lecturer, Chair of the IEEE Computer Society Technical Committee on Computer Architecture (TCCA), Chair of the ACM Special Interest Group on Computer Architecture (SIGARCH), Chair of the ACM/IEEE Workshop on Interconnection Networks for Parallel and Distributed Processing (1980), General Chair of the 3rd International Conference on Distributed Computing Systems (1982), Program Co-Chair of the 1983 International Conference on Parallel Processing, and General Chair of the 15th Annual International Symposium on Computer Architecture (1988). He is currently an associate editor of the *Journal of Parallel and Distributed Computing* and a member of the Eta Kappa Nu and Sigma Xi honorary societies.

Effect of Maintenance on the Dependability and Performance of Mulitprocessor Systems

Reprinted from *IEEE Transactions on Reliability*, Volume R-36, Number 2, June 1987, pages 208–215.

Chita R. Das, Member IEEE
Pennsylvania State University, University Park
Laxmi N. Bhuyan, Senior Member IEEE
University of Southwestern Lousiana, Lafayette
V. V. S. Sarma, Senior Member IEEE
Indian Institute of Science, Bangalore

Key Words—**Availability, Dependability, Maintenance, Multiprocessor System**

Reader Aids—
Purpose: Widen state of the art
Special math needed for explanations: Probability and statistics
Special math needed to use results: None
Results useful to: Multiprocessor designers, Reliability and system analysts.

Summary & Conclusions—**This paper presents analytic models for dependability and performance evaluation of multiprocessor systems with both on-line and off-line maintenance. Markov models are developed to compute the system reliability and performance availability incorporating the reliability of the maintenance processor.**

The maintenance processor failure is considered separately in order to emphasize its effect on system performance and dependability. The reliability of the maintenance processor can not be ignored for degradable multiprocessors. Probabilistic models are presented to compute the system downtime and the service cost for three off-line maintenance policies: scheduled maintenance (SM), unscheduled maintenance (UM), and scheduled & unscheduled maintenance (SUM). The SUM policy, that combines both SM and UM, can be used to give a compromise between cost and downtime.

1. INTRODUCTION

Multiprocessor systems are designed to provide high computing power with assured dependability. These systems can be categorized either as *loosely coupled* or *tightly coupled* systems [1]. A loosely coupled multiprocessor consists of a network of processors where a local memory is associated with each processor. A tightly coupled system, on the other hand, consists of a number of processors and globally addressable memory modules with an interconnection network between them. Both types of multiprocessors are quite dependable because of their ability to support graceful degradation in the event of failures [2, 3].

Graceful degradation in multiprocessors is accomplished through the automatic reconfiguration and recovery capabilities provided by an on-line maintenance processor [4]. The maintenance processor monitors the performance of each processor and diagnoses the faulty elements. If the fault is permanent the maintenance processor segregates the faulty module and reconfigures the system to a degraded mode. With the failure of the maintenance processor, the reconfiguration and recovery capabilities from permanent and certain transient faults are lost. The multiprocessor system remains operational until a subsequent fault occurs and then the system fails.

Prior studies on the dependability and performance issues of multiprocessors are based on the assumption that the reconfiguration process, provided by a maintenance processor, is taken into account by the coverage factor [2, 3, 5-11]. The effect of the maintenance processor reliability on system dependability has not been specifically modeled in these studies. Liu [4] has modeled steady-state availability of uniprocessor and 2-processor repairable systems with a maintenance processor. However, no general solutions for the transient reliability of N-processor loosely coupled or tightly coupled systems are addressed in that paper. Unlike a uniprocessor system, the complexity of a maintenance processor in a multiprocessing environment can be at least as complex as one of the host processors. Therefore, the effect of the maintenance processor reliability on system dependability can not be ignored and should be considered explicitly.

In addition, a second type of maintenance is provided by an off-line global maintenance facility that supplements the maintenance processor. This is based on several service policies such as scheduled maintenance (SM) and unscheduled maintenance (UM) [12, 13]. These service policies give different downtimes and costs over the system life-cycle. For example, SM at regular intervals is preferable because of its low cost. The UM is usually avoided for its high cost. The cost of UM is often difficult to quantify [14]. This paper considers a third type of maintenance policy in addition to SM and UM. This third policy is a combination of both SM and UM, denoted by scheduled and unscheduled maintenance (SUM). With this policy a system is repaired at regular intervals or upon system failure, which ever comes first. This is similar to the block replacement [12], except that instead of replacing all the components at the SM epoch only the faulty elements are repaired.

Most of the papers on repairable/nonrepairable computer systems are concerned with the reliability and availability issues [2-10, 15, 16]. Very little work has been done to quantify system downtime and service cost. Ingle & Siewiorek [7] have calculated the fractional improvement in reliability of multiprocessors due to SM assuming that the maintenance duration is negligible. Following them, Oda et al. [17] have presented reliability models for reconfigurable systems with SM. They have developed cost models assuming that a fixed number of units are repaired at each maintenance epoch; maintenance duration is neglected in this study.

But in practice each maintenance phase involves some regular maintenance checkup and the repair/replacement of the faulty elements. The system is unavailable to the user during this phase. Hence, frequent SM actions can lead to more system downtime. On the other hand, if the maintenance fequency is reduced, the system can fail much before the repair is due. This in turn also increases the downtime. Therefore, there exists an optimal interval when the system downtime is minimum. The same argument applies for the SM service cost when a penalty factor is associated with downtime. As the SUM uses both SM and UM principles, we anticipate that it will have a good cost and downtime behavior.

This paper is concerned with the effect of both on-line and off-line maintenance policies on the dependability and performance of multiprocessor systems. The system dependability is modeled based on task (job) requirement [2, 3, 7]. The task-based dependability assumes that the system remains operational as long as the minimum number of resources for the concurrent execution of a task are available on the system. Following Laprie [18] the term dependability in this paper is used to encompass both reliability and maintenability issues. Maintainability is characterized by downtime and service cost. Expressions for the reliability and computation availability of loosely coupled multiprocessors are first derived taking into account the maintenance processor failure. The computation availability of a system is directly proportional to the number of processors available for the execution of the task. Analytic models for computing the system downtime and cost over a finite life-cycle for SM, UM, and SUM are presented. The extension of the models to tightly coupled multiprocessors is briefly outlined.

2. NOTATION

M number of processors in the initial configuration of the multiprocessor.

N number of memories in the initial configuration of the multiprocessor.

I minimum number of processors required to keep the system operational.

J minimum number of memories required to keep the system operational.

K cost coefficient parameter for SM and UM.

λ_p, λ_p' constant failure rate of a host processor when the maintenance processor is UP, DOWN.

λ_{mp} constant failure rate of the maintenance processor.

λ_m, λ_m' constant failure rate of a memory when the maintenance processor is UP, DOWN.

$P_{i1}(t)$, $P_{i0}(t)$ Probability that at time t the system has i working processors and the maintenance processor is UP, DOWN.

$P_{i0}'(t)$ probability that at time t the system has failed with i processors after the maintenance processor is DOWN.

$P_{ij1}(t)$, $P_{ij0}(t)$ probability that at time t the system has i processors, j memories and the maintenance processor is UP, DOWN.

$P_{ij0}'(t)$ probability that at time t the system is failed, with i processors and j memories after the maintenance processor is DOWN.

C_p processor coverage parameter.

C_m memory coverage parameter.

C_u service cost rate.

C_r cost parameter representing the penalty for SM downtime.

$\overline{C}$ means service cost of the system over $T + T_r$ for SM.

C_T total maintenance cost over T_{total} of run.

$f_S(t)$, $F_S(t)$, $R_S(t)$ pdf, Cdf, reliability of the multiprocessor system at time t.

$CA_S(t)$ computation availability of the loosely coupled system at time t.

$BA_S(t)$ bandwidth availability of the tightly coupled system at time t.

$M(T)$ mean number of system failures in the interval $(0,T)$

t_g general repair time at maintenance epoch.

t_v travel time for UM.

t_p, t_p' repair time of a processor when the maintenance processor is UP, DOWN.

t_m, t_m' repair time of a memory when the maintenance processor is UP, DOWN.

t_{mp} repair time of the maintenance processor.

T maintenance interval.

T_r mean repair time at the SM epoch.

$T_{rd1}(t)$ mean repair time at time t if the system has failed due to exhaustion of components with the maintenance processor UP.

$T_{rd0}(t)$ mean repair time at time t if the system has failed due to failure of a component with the maintenance processor DOWN.

$T_{rd}(t)$ mean repair time of the system at time t if the system has failed.

T_{ru} mean repair time of the system if it has not failed before SM.

$\overline{T_{ds}}$ mean downtime of the system over $T + T_r$ hours for SM.

$\overline{T_{du}}$ mean downtime of the system for UM.

$\overline{T_d}$ mean downtime for SUM.

T_{SD} total system downtime in T_{total} of run.

$\Psi(z)$ $\lambda_{mp} \cdot \dfrac{1 - \exp(-zt)}{z}$, where z is any expression; this is a utility expression to simplify several of the equations. Ψ is dimensionless.

3. DEPENDABILITY WITH ON-LINE MAINTENANCE

Consider a loosely coupled multiprocessor system having M processors and a dedicated maintenance processor, as depicted in figure 1. The dependability analysis is based on the following assumptions.

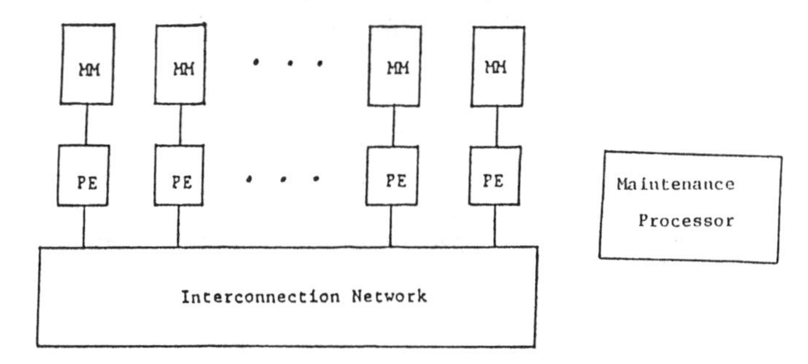

Fig. 1. A Loosely Coupled Multiprocessor.

1. The communication network is fault free and is not a bottleneck for system performance. Incorporation of network degradation makes the analysis intractable [2, 3].
2. A processor (including the maintenance processor) is either UP or DOWN.
3. There is no on-line repair facility for faulty elements. Only off-line maintenance is used for this purpose. All the faulty elements are perfectly repaired at the maintenance epoch.
4. The failure times of the processors are *s*-independent and follow an exponential distribution.

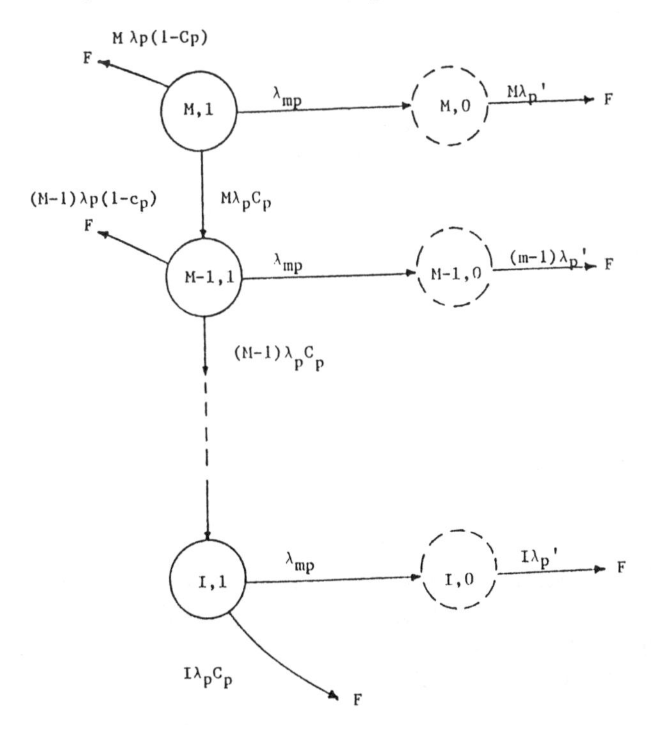

Fig. 2. Markov Model of a Loosely Coupled Multiprocesor with M Processors and One Maintenance Processor.

The Markov model for the multiprocessor is shown in figure 2. The system remains in a working state if there are at least I processors available for the execution of a task. All the dotted circles represent states with the maintenance processor down. As shown in the state transition diagram, the system can go to the failed mode F from any of the $(i, 1)$ working states with a rate $i\lambda_p(1 - C_p)$ if the fault is not covered by the maintenance processor. Here, $\lambda_p' \geqslant \lambda_p$ because the system may not be able to mask some of the transient faults in addition to the permanent faults when the maintenance processor is unavailable. The probability of state $(i, 1)$, for $I \leqslant i \leqslant M$, at time t is:

$$P_{i1}(t) = \binom{M}{i} C_p^{M-i} e^{-i\lambda_p t}(1 - e^{-\lambda_p t})^{M-i} e^{-\lambda_{mp} t}. \tag{3.1}$$

The probability of state $(i, 0)$ is:

$$P_{i0}(t) = \binom{M}{i} C_p^{M-i} e^{-i\lambda_p' t} \sum_{x=0}^{M-1} (-1)^x \binom{M-i}{x} \cdot$$

$$\Psi(i\lambda_p + x\lambda_p + \lambda_{mp} - i\lambda_p'). \tag{3.3}$$

If $\lambda_p = \lambda_p'$ (the maintenance processor status does not affect the processor failure rate) then (3.2) reduces to:

$$P_{i0}(t) = \binom{M}{i} C_p^{M-i} e^{-i\lambda_p t} \cdot$$

$$\sum_{x=0}^{M-i} (-1)^x \binom{M-i}{x} \Psi(x\lambda_p + \lambda_{mp}). \tag{3.3}$$

The system reliability is sum of probabilities all the working states having at least I processors:

$$R_S(t) = \sum_{i=I}^{M} (P_{i1}(t) + P_{i0}(t)). \tag{3.4}$$

The computation availability is proportional to the number of active processors:

$$CA_S(t) = \sum_{i=I}^{M} i(P_{i1}(t) + P_{i0}(t)). \tag{3.5}$$

Figure 3 shows the variation of system reliability with time for several task requirements and for several values of λ_{mp}. The initial configuration of the loosely coupled multiprocessor has 16 processors. The processor failure rate, λ_p is 0.2/1000 hours. The effect of the maintenance processor failure rate on $R_S(t)$ is quite distinct for $I = 12$. However, for $I = 16$, the maintenance processor reliability (for the chosen values of λ_{mp}) has no impact on $R_S(t)$. This is because the system fails mostly due to the failure of one of the host processors. Figure 4 depicts the effect of the maintenance processor reliability on $CA_S(t)$.

4. DEPENDABILITY WITH OFF-LINE MAINTENANCE

This section presents mathematical models to determine the average system downtime and cost associated with SM, UM, and SUM policies. The results obtained for these three maintenance policies are compared. The analysis is based on the following assumptions.

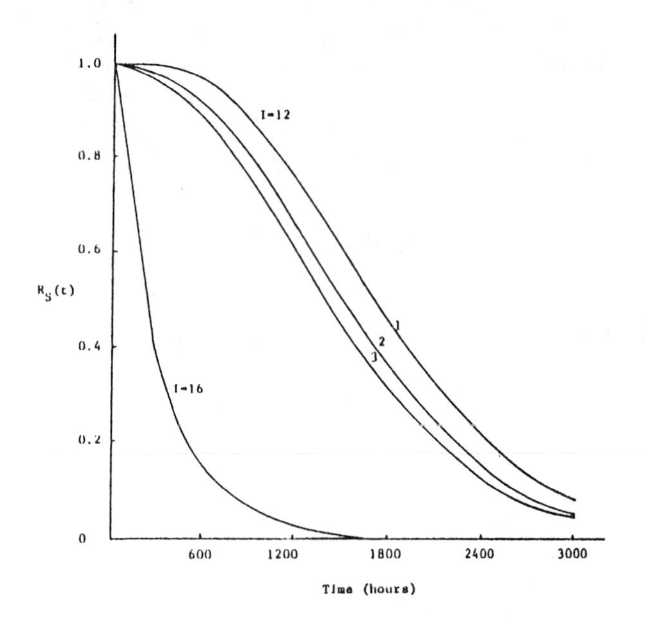

Fig. 3. Reliability Variation with Time, of a Loosely Coupled Multiprocessor for a Task Requiring I Processors.

$M = 16$, $\lambda_p = \lambda_p' = 0.2/1000$ hours, $C_p = 1.0$.
curve 1 → $\lambda_{mp} = 0$
curve 2 → $\lambda_{mp} = 0.2/1000$ hours
curve 3 → $\lambda_{mp} = 0.3/1000$ hours

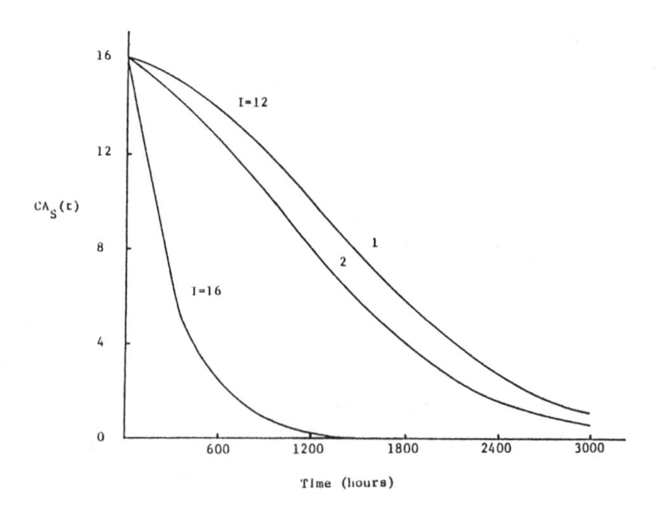

Fig. 4. Computation Availability Variation with Time of a Loosely Coupled Multiprocessor, for a Task Requiring I Processors.

$M = 16$, $\lambda_p = \lambda_p' = 0.2/1000$ hours, $C_p = 1.0$.
curve 1 → $\lambda_{mp} = 0$
curve 2 → $\lambda_{mp} = 0.3/1000$hours

1. All the faulty elements are perfectly repaired at a maintenance epoch and are returned to the system.
2. $t_p' \geqslant t_p$, because in the absence of the maintenance processor, the repairman has to detect a faulty processor before starting any acutal repair of the processor.
3. The service cost is proportional to the repair duration

4.1 *Scheduled Maintenance (SM)*

Let the system be repaired periodically at intervals of T hours. There can be two situations at the maintenance moment.

a. The system has failed at time t where $t \in (0, T)$.
b. The system has not failed in $(0, T)$.

The two situations are depicted in figures 5a and 5b respectively. The average repair time T_r at maintenance epoch is shown in figure 5c.

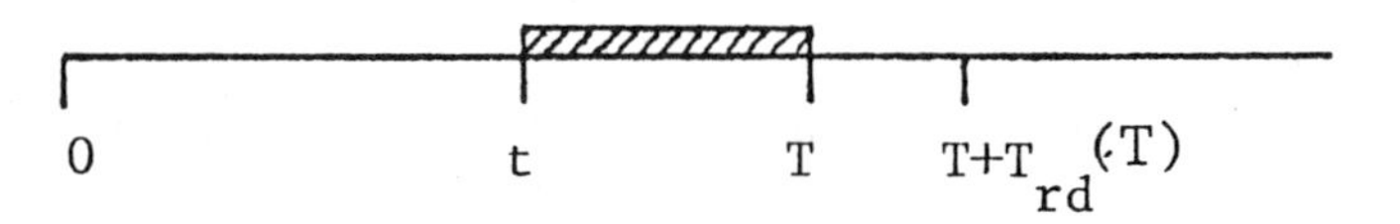

Fig. 5a. System Failed Between $(0, T)$ and SM at T.

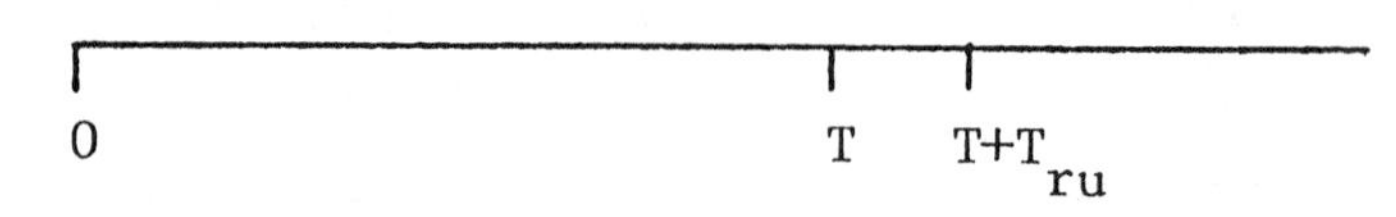

Fig. 5b. No System Failure, SM at T.

Fig. 5c. Average Repair Time at SM.

4.2 *System downtime model*

The mean downtime over $(T + T_r)$ hours is:

$$\overline{T_{ds}} = \left(\int_0^T (T - t) f_S(t) dt + F_S(T) T_{rd}(T)\right) + R_S(T)\, T_{ru} + t_g. \tag{4.1}$$

The first term in (4.1) represents the mean downtime if the system has failed during $(0, T)$. The integral part in (4.1) represents the mean amount of time the system was down, without any maintenance action for $t \leqslant T$. The second term in (4.1) represents the average repair time if the system has not failed in $(0, T)$.

Depending on whether or not the maintenance processor is available, the mean repair time $T_{rd}(T)$ in (4.1) mainly consists of two parts: $T_{rd1}(T)$ and $T_{rd0}(T)$ (assuming perfect coverage). The system could have failed at any time t, where $t \in (0, T)$, but the maintenance begins only at the maintenance epoch T. Therefore, the system state at time T is of interest.

$$T_{rd1}(T) = \frac{1}{F_S(T)} P_{(I-1)1}(T) \cdot (M - I + 1)t_p \quad (4.2)$$

where $P_{(I-1)1}(T)$ is the probability that the system is in the failed state $(I - 1)$ at time T. $(M - I + 1)$ is the number of processors to be repaired in this failed state.

$$T_{rd0}(T) = \frac{1}{F_S(T)} \sum_{i=I-1}^{M-1} P'_{i0}(T) \cdot ((M - i)t'_p + t_{mp}). \quad (4.3)$$

$P'_{i0}(T)$ in (4.3) represents the probability that the system has failed due to the failure of a processor after the maintenance processor has failed. Then, the mean repair time is:

$$T_{rd}(T) = T_{rd1}(T) + T_{rd0}(T). \quad (4.4)$$

If $\lambda_p = \lambda'_p$ then the required state probabilities for (4.2) and (4.3) can be written:

$$P_{(I-1)1}(t) = I \frac{\lambda_p}{\lambda_{mp}} \binom{M}{I} C_p^{M-I} \cdot \sum_{x=0}^{M-I} (-1)^x \binom{M-I}{x} \Psi((I + x)\lambda_p + \lambda_{mp}). \quad (4.5)$$

$$P'_{(i-1)0}(t) = i \binom{M}{i} C_p^{M-i} \cdot \sum_{x=0}^{M-i} (-1)^x \binom{M-i}{x} \frac{\lambda_p}{x\lambda_p + \lambda_{mp}} [\Psi(i\lambda_p) - \Psi((i + x)\lambda_p + \lambda_{mp})] \quad (4.6)$$

Similarly, the mean repair time if the system is in any one of the non-failed states at time $t = T$, is:

$$T_{ru} = \frac{1}{R_s(T)} \sum_{i=I}^{M} (P_{i1}(T) \cdot (M - i)t_p + P_{i0}(T) \cdot ((M - i)t'_p + t_{mp})). \quad (4.7)$$

Substitute (4.4) and (4.7) in (4.1); the mean down time over $(T + T_r)$ can be obtained. The mean repair time at the maintenance epoch is:

$$T_r = F_S(T)T_{rd}(T) + R_S(T)T_{ru} + t_g. \quad (4.8)$$

The mean total system downtime over T_{total} hours of operation is:

$$T_{SD} = \frac{\overline{T_{ds}} \cdot T_{total}}{(T + T_r)}. \quad (4.9)$$

4.3 *Cost model*

The service cost for SM can be modeled in two ways.

1. The cost is proportional only to the mean repair time. Then the cost decreases as the maintenance interval increases. This is appropriate for situations where the downtime has no impact on the uses. The mean repair cost in $(T + T_r)$ hours is:

$$\overline{C} = K \cdot C_u \cdot T_r \quad (4.10)$$

where K is a coefficient dependent on the type of the maintenance policy and on the type of the maintenance organization.

2. There is a penalty associated with the system downtime (excluding the actual repair time). Then the cost over $T + T_r$ hours is:

$$\overline{C} = (\int_0^T (T - t)\, C_r f_S(t)dt + F_S(T)\, KC_u T_{rd}(T)) + R_S(T)(KC_u T_{ru}) + t_g KC_u \quad (4.11)$$

where C_r is an appropriate penalty cost-rate associated with downtime. Now using (4.10) or (4.11) the total maintenance cost over T_{total} hours of operation is:

$$C_T = \frac{\overline{C} \cdot T_{total}}{T + T_r}. \quad (4.12)$$

4.4 *Unscheduled Maintenance (UM)*

UM refers to a policy where the system is repaired only when it fails. For any failure distribution the mean time to failure (MTTF) is:

$$\text{MTTF} = \int_0^\infty R_S(t)dt. \quad (4.13)$$

$\overline{T_{du}}$ is the mean repair time if the system has failed at MTTF:

$$\overline{T_{du}} = T_{rd}(\text{MTTF}) = T_{rd1}(\text{MTTF}) + T_{rd0}(\text{MTTF}) + t_g + t_v \quad (4.14)$$

where T_{rd1}(MTTF) and T_{rd0}(MTTF) are obtained from (4.2) and (4.3) respectively by substituting T = MTTF. A travel time parameter t_v is included in (4.14) for UM in addition to the general maintenance check-up time t_g.

Using (4.14) the total downtime in T_{total} of run, where $T_{total} \gg \text{MTTF} + \overline{T_{du}}$, is:

$$T_{SD} = \frac{\overline{T_{du}}\, T_{total}}{\text{MTTF} + \overline{T_{du}}} \quad (4.15)$$

$\frac{T_{total}}{\text{MTTF} + \overline{T_{du}}}$ represents the renewal function $M(T_{total})$

for the underlying failure distribution [19]. Similarly, the service cost over T_{total} hours is:

$$C_T = \frac{KC_u\overline{T_{du}}\, T_{total}}{\text{MTTF} + \overline{T_{du}}}. \tag{4.16}$$

4.5 *Scheduled & Unscheduled Maintenance (SUM)*

The system is repaired whenever it fails and at each SM epoch. Although SUM is similar to block maintenance [12], not all the components are replaced at the SM epochs; only the failed elements are repaired. With this policy the mean number of system failures (hence the UMs) in the interval $(0, T)$ is given by the renewal function [19]:

$$M(T) = F_S(T) + \int_0^T M(T - t)f_S(t)dt. \tag{4.17}$$

Hence, the mean downtime in the interval $T + T_r$ is:

$$\overline{T_d} = M(T) \cdot \overline{T_{du}} + T_r \tag{4.18}$$

where $\overline{T_{du}}$ is the mean repair time for the UM and is same as (4.14). T_r is the average repair time at the SM moment.

$$T_r \approx \frac{1}{T}\int_0^T \Bigg(\sum_{i=I}^{M} [P_{i1}(T - t - \overline{T_{du}})((M - i)tp)$$

$$+ P_{i0}(T - t - \overline{T_{du}})((M - i)t_p' + t_{mp})\,|\,(T > (t + \overline{T_{du}}))]$$

$$+ (t + \overline{T_{du}} - T)\,|\,(t + \overline{T_{du}}) \geqslant T\Bigg)dt + R_S(T)T_{ru} + t_g \tag{4.19}$$

where $(T - t + \overline{T_{du}})$ represents the residual time for which the system was up after the last failure at time t. The situation is depicted in figure 5d. The term $T - (t + \overline{T_{du}}) > 0$ indicates that the system was up at the SM and hence the corresponding repair time is computed for working states $(i, 1)$ and $(i, 0)$ respectively. If $t + \overline{T_{du}} \geqslant T$ (the system has failed at a time close to SM epoch and is still under repair at the SM) then the repair time extends beyond the start of the SM and is simply $(t + \overline{T_{du}} - T)$. The $R_S(T)\ T_{ru}$ in (4.19) represents the mean repair time if the system has not failed in $(0, T)$. T_{ru} is computed from (4.7).

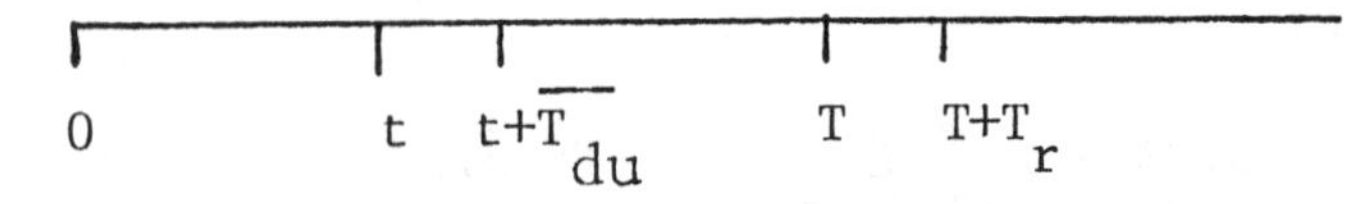

Fig. 5d. Residual Time, $(T - (t + \overline{T_{du}}))$, the System Was in Operation after Last Failure at t.

The system downtime and cost can now be computed from (4.9) and 4.12) respectively. However, the appropriate value of K should be used for the UM and SM repair times while computing C_T from (4.18).

The renewal function $M(T)$ reduces to $\dfrac{T}{\text{MTTF} + \overline{T_{du}}}$ when T is very large. In the event that $T < (\text{MTTF} + \overline{T_{du}})$ the computation of $M(T)$ from (4.17) becomes very tedious. Therefore, we use an approximate algorithm to compute $M(T)$. The algorithm assumes that when $T < (\text{MTTF} + \overline{T_{du}})$, more than one failure in the interval $(0, T)$ is neglected.

$$M(T) \approx \begin{cases} F_S(T), & \text{if } T < (MTTF + \overline{T_{du}}) \\ y + F_S(T - y \cdot (\text{MTTF} + \overline{T_{du}})), & \text{otherwise} \end{cases} \tag{4.20}$$

$$y \equiv \left| \frac{T}{\text{MTTF} + \overline{T_{du}}} \right|.$$

The downtime and the cost match within 10% of simulation results with this approximation.

Figure 6 depicts the variation of the system downtime with maintenance interval T for the SM. The t_p, t_p', and t_{mp} values are chosen arbitrarily. The downtime of the UM policy is also plotted in this graph. The downtime under SM can be minimized by choosing a proper value of T. The curves for the UM are plotted for travel time, $t_v = 0$. The inclusion of a finite value for t_v in (4.16) shifts the downtime curve for the UM upward.

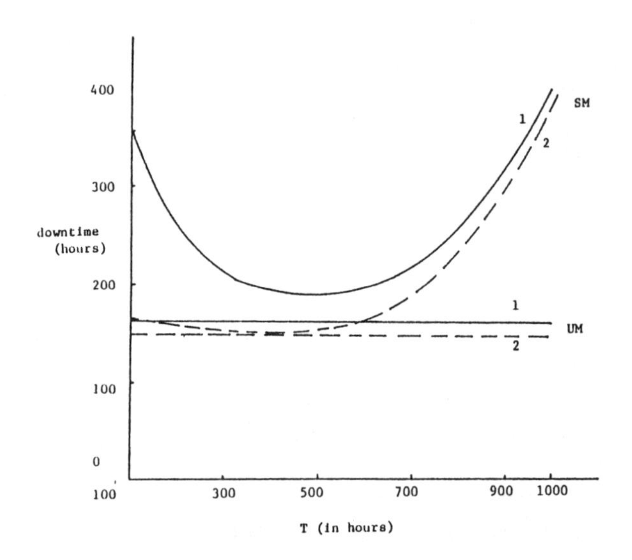

Fig. 6. Variation of Downtime with Interval T for SM for a Task Requiring $I = 12$ Processors.

———— $t_g = 2$ hours, - - - - - - - - $t_g = 0$ hours
$t_v = 0$ for UM
$t_p = t_{mp} = 5$ hours
$t_p' = 6$ hours

Figure 7 shows the variation of downtime and cost for the SUM for a task requiring 12 processors. The MTTF for this system configuration is 1484 hours. The downtime and cost decrease gradually as $T \rightarrow$ MTTF. When T exceeds MTTF + $\overline{T_{du}}$ the curves rise momentarily and then decrease. The rise of the curves at the MTTF is due to the fact that the

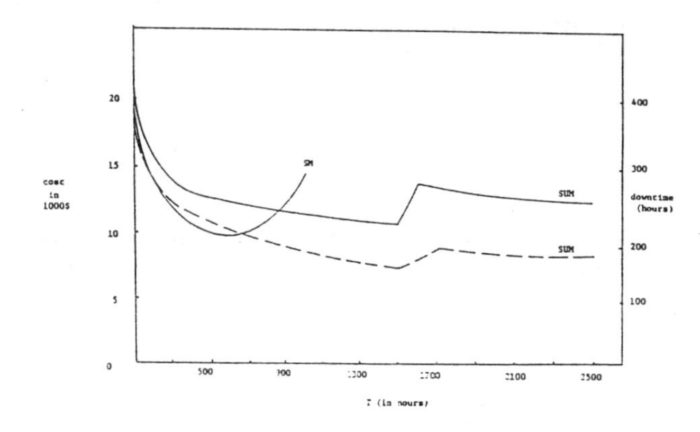

Fig. 7. Variation of Cost and Downtime with Interval T for SUM for a Task Requiring $I = 12$ Processors.

———— cost, $C_u = 55\$/\text{hour}$, $C_r = 30\$/\text{hour}$

- - - - - - - downtime, $K = 1$ for SM, $K = 1.5$ for UM

$t_p = t_{mp} = 5$ hours, $t'_p = 6$ hours

probability of a system failure before the SM interval T is very high which in turn adds the high cost and repair time of the UM to the system model. Figure 7 shows that the ideal interval T for SUM is just before the MTTF. However, unlike the SM, there is a minimum point for the cost/downtime curve just before $j \cdot$ MTTF, for $j = 1, 2, \ldots$ and $j \cdot \text{MTTF} < T_{total}$. Therefore, the SUM policy gives more freedom in selecting a maintenance interval, than SM.

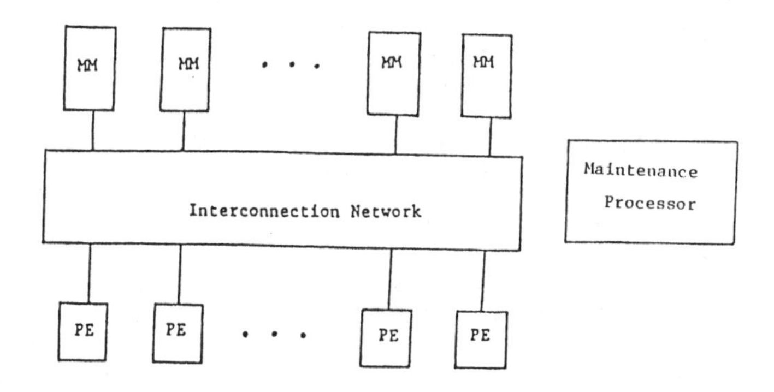

Fig. 8. A Tightly Coupled Multiprocessor.

5. EXTENSION TO TIGHTLY COUPLED SYSTEMS

This section extends the previous models to tightly coupled multiprocessors. The system organization, shown in figure 8, consists of M processors, N memory modules, and an interconnection network (IN) between the processors and memories. The IN may be a crossbar, a multistage interconnection network, or a multiple-bus organization [2]. Bandwidth availability (BA) is used as the performance-related reliability measure of these systems [9]. $BA_s(t)$ is defined as the mean amount of bandwith (BW) available on the system at time t. The BW in turn is defined as the mean number of memory modules remaining The BW of a tightly coupled multiprocessor very much depends on the type of the IN; thus our analysis is restricted to crossbar connections.

Assumptions

1. The IN is fault free, for simplicity.
2. Each processor generates a request to any of the memories with equal probability in each cycle.
3. The system is operational as long as I processors and J memories are available on it. □

The initial BW of a $M * N$ crossbar is [9]:

$$BW = N\left(1 - \left(1 - \frac{1}{N}\right)^M\right). \tag{5.1}$$

The state probabilities for this system with a dedicated maintenance processor are:

$$P_{ij1}(t) = \binom{M}{i} C_p^{M-i} e^{-i\lambda_p t} (1 - e^{-\lambda_p t})^{M-i} \cdot \binom{N}{j} C_m^{N-j} e^{-i\lambda_m t}(1 - e^{-\lambda_m t})^{N-j} e^{-\lambda_{mp} t} \tag{5.2}$$

$$P_{ij0}(t) = \binom{M}{i} C_p^{M-i} \binom{N}{j} C_m^{N-j} e^{-(i\lambda_p + j\lambda_m)t} \cdot \sum_{x=0}^{M-i} \sum_{y=0}^{N-j} (-1)^x \binom{m-i}{x} (-1)^y \binom{N-j}{y} \cdot \Psi(x\lambda_p + y\lambda_m + \lambda_{mp}) \tag{5.3}$$

$$BA_S(t) = \sum_{i=I}^{M} \sum_{j=J}^{N} \left(P_{ij1}(t) + P_{ij0}(t)\right) \cdot j\left(1 - \left(1 - \frac{1}{j}\right)^i\right). \tag{5.4}$$

The sum of the first bracketed terms in (5.4) represents the reliability of the system.

The basic analytic expressions for the three different off-line maintenance policies are also valid for the tightly coupled system. Only the mean repair times $T_{rd}(T)$ and T_{ru} in (4.1) are modified:

$$T_{rd1}(T) = \frac{1}{F_S(T)} \left(\sum_{j=J}^{M} P_{(I-1)j1}(T) \cdot [(M - I + 1)t_p + (N - j)t_m] + \sum_{i=I}^{M} P_{i(J-1)1}(T) \cdot [(M - i)t_p + (N - J + 1)t_m]\right). \tag{5.5}$$

The first term represents the situations where the system has failed due to the exhaustion of the processors, but the number of active memories is $\geq J$. The second term represents the situations when the system has failed due to exhaustion of memories and the number of active processors is $\geq I$.

Similarly, $T_{rd0}(T)$ in (4.3) is modified:

$$T_{rd0}(T) = \frac{1}{F_S(T)} \left(\sum_{i=I-1}^{M-1} \sum_{j=J}^{N} P'_{ij0}(T) \cdot [(M-i)t'_p + (N-J)t'_m + t_{mp}] + \sum_{i=I}^{M} \sum_{j=J-1}^{N-1} P'_{ij0}(T) \cdot [(M-i)t'_p + (N-j)T'_m + t_{mp}] \right). \quad (5.6)$$

The first term represents the situations where the system has failed due to the failure of a processor when the maintenance processor is unavailable. The second term represents the situations where the system has failed due to the failure of a memory when the maintenance processor is unavailable.

Similarly, the mean repair time for all the nonfailed states can be modified:

$$T_{ru} = \frac{1}{R_S(T)} \sum_{i=I}^{M} \sum_{j=J}^{N} [P_{ij1}(T) \cdot ((M-i)t_p + (N-j)t_m) + P_{ij0}(T) \cdot ((M-i)t'_p + (N-j)t'_m + t_{mp})]. \quad (5.7)$$

Using these equations the system downtime and cost for the tightly coupled system can be computed from (4.9), (4.12), (4.15), (4.16) for the three types of off-line maintenance policies.

ACKNOWLEDGMENT

This research was supported in part by USNSF Grant No. MCS-8513041. The facilities provided by The Center for Advanced Computer Studies, University of Southwestern Louisiana, Lafayette, are greatly appreciated.

REFERENCES

[1] K. Hwang, F. A. Briggs, *Computer Architecture and Parallel Processing,* McGraw-Hill, 1984.

[2] C. R. Das, L. N. Bhuyan, "Reliability simulation of multiprocessor systems", *Proc. Int. Conf. Parallel Processing,* 1985 Aug, pp 591-598.

[3] L. N. Bhuyan, C. R. Das, "Dependability evaluation of multicomputer networks", *Proc. Int. Conf. Parallel Processing,* 1986 Aug, pp 576-583.

[4] T. S. Liu, "The role of maintenance processor for a general purpose computer system", *IEEE Trans. Computers, Speical issue on Reliable and Fault-tolerant Computing,* vol C-33, 1984 Jun, 507-517.

[5] M. D. Beaudry, "Performance related reliability measures for computing systems", *IEEE Trans. Computers,* vol C-27, 1978 Jan, pp 540-547.

[6] T. C. K. Chou, J. A. Abraham, "Performance/Availability model of shared resource multiprocessors", *IEEE Trans. Reliability,* vol R-29, 1980 Apr, pp 70-74.

[7] A. D. Ingle, D. P. Siewiorek, "Reliability model for multiprocessor systems with and without periodic maintenance", *Proc. 7th FTCS,* 1977 Jun, pp 3-9.

[8] J. F. Meyer, "On evaluating the performability of degradable computing systems", *IEEE Trans. Computers,* vol C-29, 1980 Aug, pp 720-731.

[9] C. R. Das, L. N. Bhuyan, "Bandwidth availability of multiple-bus multiprocessors", *IEEE Trans. Computers, Special issue on Parallel Processing,* vol C-34, 1985 Oct, pp 918-926.

[10] A. Pedar, V. V. S. Sarma, "Architecture optimization of aerospace computing systems", *IEEE Trans. Computers,* vol C-32, 1983 Oct, pp 911-922.

[11] T. F. Arnold, "The concept of coverage and its effect on the reliability models of a repairable system", *IEEE Trans. Computers,* vol C-22, 1973 Mar, pp 251-254.

[12] R. E. Barlow, F. Proschan, *Mathematical Theory of Reliability,* John Wiley & Sons, 1965.

[13] W. P. Pierskalla, J. A. Voelker, "A survey of maintenance models: the control and survillance of deteriorating systems", *Naval Research Logistic Quarterly,* vol 23, 1976 Sep, pp 353-388.

[14] T. Y. Liang, "Optimum piggyback preventive maintenance policies", *IEEE Trans. Reliability,* vol R-34, 1985 Dec, pp 529-538.

[15] J. Losq, "Effect of failures on gracefully degradable systems", *Proc. FTCS-7,* 1977 Jun, pp 29-34.

[16] Y. W. Ng, A. Avizienis, "A reliability model for gracefully degrading and repairable fault-tolerant systems", *Proc. FTCS-7,* 1977 Jun, pp 22-28.

[17] Y. Oda, Y. Tohma, K. Furuya, "Reliability and performance evaluation of self-reconfigurable systems with periodic maintenance", *Proc. 11th FTCS,* 1981 Jun, pp 142-147.

[18] J. C. Laprie, "Dependable computing and fault-tolerance: concepts and terminology", *Proc. FTCS-15,* 1985 Jun, pp 2-11.

[19] K. Trivedi, *Probability and Statistics with Reliability, Queuing and Computer Science Applications,* Prentice-Hall, 1982.

AUTHORS

Dr. Chita R. Das; Department of Electrical Engineering; The Pennsylvania State University; University Park, Pennsylvania 16802 USA.

Chita R. Das (S'84, M'86) is an assistant professor in the Computer Engineering program at The Pennsylvania State University. He received his MSc degree in Electrical Engineering from Regional Engineering College, Rourkela, Sambalpur University, India in 1981, and PhD degree in Computer Science from the Center for Advanced Computer Studies, University of Southwestern Louisiana, Lafayette in 1986. His research interests include parallel/distributed computing, performance evaluation, and fault-tolerant computing.

Dr. Laxmi N. Bhuyan; The Center for Advanced Computer Studies; University of Southwestern Louisiana; Lafayette, Louisiana 70504-4330 USA.

Laxmi N. Bhuyan (S'81, M'82, SM'87) is an associate professor at the University of Southwestern Louisiana. He received the MS degree in Electrical Engineering from Regional Engineering College, Rourkela, Sambalpur University, India in 1979 and PhD degree in Computer Engineering from Wayne State University, Detroit in 1982. His research interests include parallel and distributed computer architecture, performance evaluation, fault-tolerant computing and local area networks. He has published extensively in the area of parallel and distributed processing. He is the Guest Editor for IEEE Computer Magazine, special issue on Interconnection Networks, scheduled for publication in 1987 June.

Dr. V. V. S. Sarma; Department of Computer Science and Automation; Indian Institute of Science; Bangalore - 560 012 INDIA.

V. V. S. Sarma (SM'82) is a professor at Indian Institute of Science, Bangalore. He received his ME and PhD degrees in Electrical Engineering from Indian Institute of Science, Bangalore in 1966 and 1970. His research interests are in fault-tolerant computing, pattern analysis, and artificial intelligence. He was a visiting professor at the University of Southwestern Louisiana, Lafayette from 1984 January to 1986 May.

Manuscript TR86-216 received 1986 May 15; revised 1987 January 2.

IEEE Log Number 15383

◄ TR ►

Realizing Fault-Tolerant Interconnection Networks via Chaining

NIAN-FENG TZENG, PEN-CHUNG YEW, AND CHUAN-QI ZHU

Abstract—**A scheme applicable to a wide class of multistage interconnection networks to enhance their fault-tolerant capability is proposed. Multiple paths between each input-output pair of a network are created by connecting together switching elements within the same stage. This scheme provides a network with alternative paths at every stage, requires a simple self-routing algorithm, and allows a network to become more robust as its size increases. An analysis is performed to obtain a quantitative measurement on the reliability improvement of the scheme.**

Index Terms—**Multiprocessors, multistage interconnection networks, network fault-tolerance, reliability analysis, routing schemes.**

I. Introduction

A multiprocessor system consists of many processors and memory modules interconnected by a network. To design a suitable interconnection network has long been recognized as one of the key issues in developing a multiprocessor system because overall system performance relies heavily upon the employed interconnection network. Many multistage interconnection networks (MIN's) have been proposed previously (see the references in [1]).

A MIN has several stages of small switching elements (SE's), which basically are crossbar switches. A path from an input to an output can be established using a destination-tag routing algorithm proposed by Lawrie [2]. However, most of the MIN's have only one unique path between each network input-output pair. Any single failure at a switching element or a link may render some outputs unreachable from certain inputs.

To overcome this problem, many schemes have been introduced to improve network fault-tolerant capability [4]-[11]. These schemes require some kind of redundancy to be built into a network, where redundancy could be in the form of hardware, time [4], or information [7]. Among these schemes, incorporating redundant hardware to provide multiple paths between every input-output pair is most widely used in designing fault-tolerant MIN's [5], [6], [8]-[11].

The interconnection networks proposed in [6], [8], and [9] have alternative paths only at some stages. Should a request in such a network encounter a fault on the way to its destination, it has to backtrack to a stage where alternative paths exist and use an alternative path. The time overhead in backtracking is significant. In addition, the control mechanism is very complicated because the network has to handle bidirectional signal flow. It is possible to eliminate the backtracking overhead by maintaining the status tables of paths. However, extra hardware for maintaining status tables may be substantial, and a new tag may be needed to use an alternative path.

Fault-tolerant MIN's proposed in [5], [10], and [11] provide alternative paths at every stage. A request in such a network can find an alternative path at the same stage where it encounters a fault. A network with this property has *strong reroutability* and is called a *strong reroutable* network. The overall control mechanism is simpler since no backtracking requests or status tables would be required. Although it may require some extra hardware, a strong reroutable network is perferable because the rerouting process incurs much less overhead.

Manuscript received May 26, 1987; revised November 20, 1987. This work was supported in part by National Science Foundation under Grants US NSF DCR84-06916 and US NSF DCR84-10110, the US Department of Energy under Grant US DOE-DE-FG02-85ER25001, and the IBM Donation.

N.-F. Tzeng is with the Center for Advanced Computer Studies, University of Southwestern Louisiana, Lafayette, LA 70504.

P.-C. Yew and C. -Q. Zhu are with Center for Supercomputing Research and Development, University of Illinois at Urbana-Champaign, Urbana, IL 61801.

IEEE Log Number 8719333.

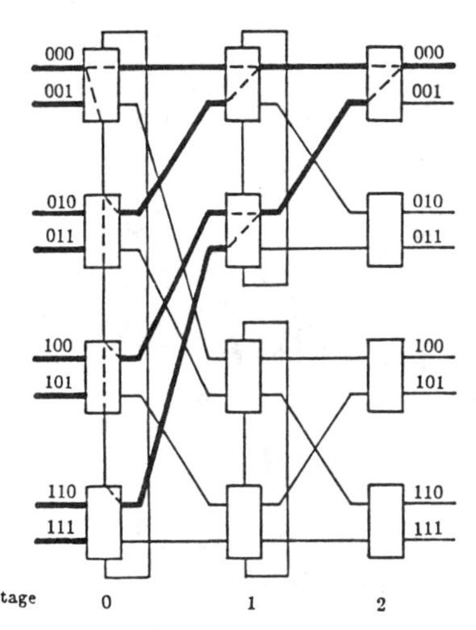

Fig. 1. One way to chain SE's in each stage to provide multiple paths.

In this correspondence, a fault-tolerance scheme which can be applied to a large class of MIN's is proposed. A network constructed with this scheme is a strong reroutable network. It requires a simple routing procedure. In Section II, a fault-tolerant network implemented with the proposed scheme is described. The reliability analysis of the network is presented in Section III. In Section IV, this fault-tolerance scheme is compared to other schemes.

II. Fault-Tolerant Multistage Interconnection Networks

A baseline network [3] built from 2×2 switching elements will be used as an example to illustrate our fault-tolerance scheme.

A. Network Configuration

A baseline network of size 8 consists of three stages each of which has four switching elements. A close inspection of a baseline network (see Fig. 1) reveals that a tree can be formed from each output (as the root of the tree) to all of the inputs of the network (as the leaves of the tree) with SE's as its nodes. All of the nodes with the same distance to the root form a level, and they happen to be the SE's in the same stage of the network. Our scheme is simply to chain together all of the nodes (i.e., SE's) in the *same level* of the tree by using extra links between the nodes and, hence, allows each node to have more than one path to the root. Redundant paths can be provided easily this way. Fig. 1 shows one way to connect the SE's in each stage to create redundant paths. The dotted lines in the figure illustrate four possible redundant paths between an input-output pair.

To permit SE's to be chained together, we provide each switching element with a chain-in link and a chain-out link (in addition to the original input links and output links). The "augmented" switching element functions like a 3×3 crossbar switch with a modified destination-tag routing algorithm which will be described later.

We use a naming mechanism similar to [3] to describe the configuration of the network. The stages are labeled in a sequence from 0 to $\log_2 N - 1$ with 0 for the leftmost stage. In each stage, an SE is named by the binary representation of its location in the stage, $p_0 p_1 \cdots p_{t-1} (t = \log_2 N - 1)$. Each input/output link of an SE is named by $t + 1$ bits, $p_0 p_1 \cdots p_{t-1} p_t$, in which the leftmost t bits are the same as the binary representation of the SE; the last bit p_t is 0 if the link is the upper link and p_t is 1 if it is the lower link.

Definition 1: A set of SE's in stage i, $0 < i < \log_2 N$, belongs to the same *partition* if $p_0 p_1 \cdots p_{i-1}$ in the binary representation of their names, $p_0 p_1 \cdots p_{t-1}$, has the same value. All of the SE's in stage 0 constitute a partition.

Reprinted from *IEEE Transactions on Computers*, Volume 37, Number 4, April 1988, pages 458–462.

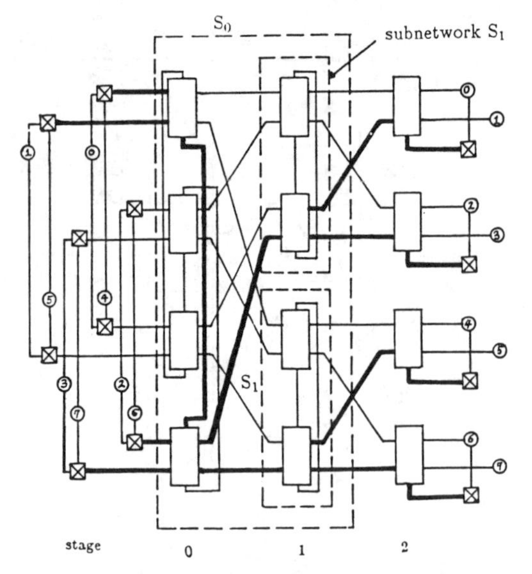

Fig. 2. The overall configuration of a completely chained network.

Definition 2: Any number of SE's in the same partition can form a *chain* by connecting the chain-in link of an SE to the chain-out link of itself or another SE within the same partition. A *complete chain* is a chain formed by connecting together all of the SE's within a partition. A network is called a *completely chained network* if all of its chains are complete chains.

The upper two SE's in stage 1 of Fig. 1 belong to a partition and so do the lower two SE's in the stage. Every chain in Fig. 1 is a complete chain because all SE's within a partition are connected together. According to Definition 2, SE's which constitute a chain can be connected in any order. A complete chain in stage 0 can also be connected as shown in Fig. 2. This connection avoids introducing unnecessary extra delay when an entire SE in stage 1 fails. A systematic approach to the construction of chains that can avoid extra delay due to an entire SE failure is given in [15].

Based on the recursive nature in the topology of a baseline network, the following theorem can be easily proved.

Theorem 1: At stage i, $0 \leq i \leq \log_2 N - 1$, there are 2^i partitions each with $N/2^{i+1}$ SE's.

From Theorem 1, we can obviously find out that 1) the number of complete chains in stage i is twice as many as those in stage $i - 1$, for $0 < i < \log_2 N$; and 2) each SE in the last stage is a partition by itself.

B. Routing Procedure

Assume a network input, labeled by $i_0 i_1 \cdots i_t$, is to be connected to a network output, labeled by $d_0 d_1 \cdots d_t$. The routing algorithm in the "chained" baseline network is essentially the same as that used in a regular baseline network (i.e., the one introduced by Lawrie in [2]).

At stage i, bit d_i of the destination label is used to route a request in an "augmented" SE. If $d_i = 0$, it is routed to the upper output link of the SE; if $d_i = 1$, it is switched to the lower output link. If the destined output link is "blocked" due to a conflict, a link failure, or a failure in the SE at the next stage to which the output link is connected, the request is routed through the chain-out link to another SE within the chain. The same d_i bit will be utilized by an identical routing algorithm in the new SE. If the destined output link in the new SE is again "blocked," this request will be routed to yet another new SE in the same chain. A request can be routed through as many SE's within a chain as needed. Eventually a "good" output link can be found and the request proceeds to the next stage. Because an identical routing algorithm is employed in each SE, the request will try the same output link of SE's within a chain when it traverses the chain. Thus, we have the following theorem.

Theorem 2: The routing algorithm proposed above can route a request from an input $i_0 i_1 \cdots i_t$ to its destination $d_0 d_1 \cdots d_t$ through as many SE's within a chain as needed.

C. Overall Network Consideration

A chained network has alternative paths at every stage, so it is a strong reroutable network. When a request encounters a failure or a conflict at an SE, it can traverse an alternative path through the chain-out link. Neither a status table nor backtracking requests is needed for the rerouting process. Better yet, it uses the same routing tag and the same routing procedure.

The Achilles heei of this scheme is its input links of the first stage (i.e., stage 0) and the output links of the last stage (i.e., stage $\log_2 N - 1$). If any of those links fails, no matter what fault-tolerance scheme is employed inside a network, we still cannot gain access to the network.

To remedy this problem, several approaches similar to the one shown in Fig. 2 can be employed. Each external component now is allowed to have access to two input links of the first stage through extra switches. It is known from Theorem 1 that each SE in the last stage is itself a partition. By taking the chain-out link of an SE back to its chain-in link, we can improve network performance because a chain-in/chain-out pair acts as a buffer to hold a blocked request [12]. However, forming a chain with one SE does not provide any fault-tolerant capability. If we use extra switches and connect an external component to an output as depicted in Fig. 2, a single fault in the last stage can be tolerated (the failure of an extra switch is regarded as the failure of a chain-out link).

In a high-performance network design, the width of a link is not uncommon to be 32 or more. Due to power consumption and speed consideration, several sets of pins are usually needed to provide clock, power, and control in an SE. Hence, it is highly unlikely that a failure in an SE will disable the entire SE, especially when efforts are made in the design to avoid such an undesirable situation. A failure in an SE is here assumed to affect only its *partial* functionality. Each link coming in or going out of an SE has a *module* associated with it [8]. A module contains all of the control mechanism needed to route a request through the connecting link. A module is called an *input* (or *output*) *module* if its connecting link is an input (or output) link of an SE. A *chain-in* (or *chain-out*) *module* can be similarly named.

An *element* is formed by an output module of an SE, an input module of an SE in the subsequent stage, and the link connecting them. The chain-out module of an SE, its connecting link, and the chain-in module of the connected SE also constitute an element. An element in the first stage contains an input module and its connecting link. Likewise, an output module (or a chain-out module and its associated switch) together with the connecting link in the last stage also form an element. An *input* (*output*) *element* in stage i, $0 \leq i < \log_2 N - 1$, is an element which contains an input (output) module of an SE in stage i. A *chain-in* (*chain-out*) *element* can be similarly defined. An element is considered to be *faulty* if any one of its components fails.

The fault model adopted in our subsequent analysis is that a fault disables an element instead of an SE or a link. An element is regarded as the *basic network unit* and a faulty element prevents a path involving that element from being realized. Also we consider only completely chained networks in our analysis since they have the maximum fault-tolerant capability.

III. Reliability Analysis

The following assumptions are made to facilitate our reliability analysis. 1) The event that an element becomes faulty is an independent event, and it occurs randomly. 2) A network is considered failed when the number and the locations of faulty elements prevent the connection of any path between an arbitrary input–output pair of the network.

An output element in stage i can be regarded as an input element in stage $i + 1$, so we consider only output elements and chain-out elements in each stage along with the input elements of stage 0 when the total number of elements is counted. Since a completely chained network has more than one path between each input–output pair, it can tolerate at least one faulty element. To calculate the upper bound on the number of tolerable faults in such a network is quite involved. First of all, let us consider the minimum number of fault-free elements required to provide full connectivity in the network.

A MIN comprises superimposed binary trees rooted at SE's in the first stage with SE's as their nodes and SE's in the last stage as their leaves. The binary tree with an SE in the first stage as its root is indicated by bold line segments in Fig. 2. A path from the root to every leaf exists if all elements along the tree are fault-free. Since an input link of the first stage is shared by two external components, it is sufficient for every external component to gain access to the network if one half of the input elements in stage 0 are fault-free. A path from every external component to the second stage is guaranteed, if the chaining elements needed to connect one half of SE's in the first stage as well as the two output elements of the rooted SE are fault-free (see Fig. 2). Therefore, $N/4 - 1$ chaining elements and two output elements in the first stage of a size N network must be functional. For middle stages, only two output elements in each chain are required and, from Theorem 1, there are 2^i chains at stage i, where $0 < i < n - 1$ ($n = \log_2 N$). An SE in the last stage can reach any one of the two attached external components if its chain-out elements are fault-free. From the above observation, we realize that the minimum number of fault-free elements required in a size N network is

$$N/2+(N/4-1+2)+\left(\sum_{m=1}^{n-2} 2\times 2^m\right)+N/2=9N/4-3. \quad (1)$$

The total number of elements in a network is $E = N + (3N\log_2 N)/2$. Hence, the maximum number of tolerable elements, denoted as B^u, is $N(3\log_2 N - 5/2)/2 + 3$. A network with size 8 has $B^u = 29$, as depicted in Fig. 2.

The number given by B^u is good for a completely chained network in which to backtrack requests is allowed or status tables are maintained, i.e., without strong reroutability. To a completely chained network with strong reroutability, let us obtain a conservative upper bound, denoted as B, based on the following assumption. Two faulty elements will cause a chained network to fail if 1) both faults are at the same SE and 2) one of them is an output element and the other is the chain-out element. The assumption is quite conservative because some SE's in a stage other than the last one can fail totally (i.e., the two output elements and the chain-out element simultaneously fail) without ruining the full connectivity of the network. For example, if an SE is not reachable due to failures in previous stages and in its chain-in element, then the SE is allowed to be totally faulty. The exact number of tolerable faults in a chain depends on the fault pattern existing in previous stages. Nonetheless, this assumption enables us to calculate the survival probability of chains, from which a conservative network survival probability can be derived. We have a substantial reliability improvement on a completely chained network even under this conservative assumption, as will be shown.

A complete chain with $M/2$ SE's has M output elements and $M/2$ chain-out elements. The worst case a chained network can tolerate is when all but two of the output elements in the complete chain are faulty, and all but one of the chain-out elements are good. Thus, a complete chain can tolerate at most $(M - 2) + 1 = M - 1$ faulty elements. From Theorem 1, stage i, $0 \le i \le \log_2 N - 1$, has 2^i complete chains each with $N/2^{i+1}$ SE's. For stages between 0 and $\log_2 N - 2$, the largest possible number of faults a network can tolerate is

$$\sum_{i=0}^{\log_2 N-2} 2^i\times(N/2^i-1)=N(\log_2 N-3/2)+1.$$

Each external component has access to two input elements in stage 0, so at most one half of the input elements in the stage can be tolerated. In the last stage, each SE can tolerate two faulty output elements, so up to N total faults are tolerable. The number of faulty elements the entire network can tolerate is $B = N(\log_2 N - 3/2) + 1 + N/2 + N = N\log_2 N + 1$.

A. Expected Number of Tolerable Faults

Faults in a network actually take place at random. In the following analysis, the number of faults which cause a network to fail is assumed to be a random variable. The previous conservative assumption is also used here. To simplify our analysis, we divide the elements of a network into three portions: 1) the input elements of the first stage, 2) the output elements and the chain-out elements in stages 0 through $\log_2 N - 2$, called intermediate stages, and 3) the output and the chain-out elements in the last stage.

First, let us consider the nth complete chain in an intermediate stage s, denoted by C_{sn}, where $0 \le s \le \log_2 N - 2$ and $1 \le n \le 2^s$. Assume $q_{sn}(i)$ is the probability that C_{sn} can still provide connections to the next stage after having i faulty elements. Then $q_{sn}(i)$ is expressed as a function of N, s, and i [10]. With $q_{sn}(i)$'s, we can derive the *survival probability* of intermediate stages after having k faulty elements by an iterative method. As shown in Fig. 2, a *subnetwork* S_j, $0 \le j \le \log_2 N - 3$, consists of three parts: a complete chain in stage j and two S_{j+1}'s. Note that an $S_{\log_2 N-2}$ contains a complete chain with two SE's in stage $\log_2 N - 2$. Assume i_{j1}, i_{j2}, and i_{j3} are the number of faulty elements in each of those three parts in S_j. Let I_j^k be a triplet $\langle i_{j1}, i_{j2}, i_{j3}\rangle$ such that $i_{j1} + i_{j2} + i_{j3} = k$, and $D_j(I_j^k)$ be the probability that a fault pattern I_j^k will occur in subnetwork S_j. Since each fault pattern is of equal probability, we have

$$D_j(I_j^k)=\frac{\binom{L_j^1}{i_{j1}}\binom{L_j^2}{i_{j2}}\binom{L_j^3}{i_{j3}}}{\binom{L_j}{k}}$$

where L_j is the total number of elements in subnetwork S_j, and L_j^1, L_j^2, L_j^3 are the total number of elements in each of those three parts in S_j, respectively, with $L_j^1 + L_j^2 + L_j^3 = L_j$. It is obvious that $L_j^1 = L_j/(\log_2 N - 1 - j)$ and $L_j^2 = L_j^3 = L_{j+1}$ with the boundary condition $L_{\log_2 N-2} = 6$. Let $Q_j(k)$ be the probability that subnetwork S_j can still function after having k faulty elements, then

$$Q_j(k)=\sum_{I_j^k\in U_j^k} D_j(I_j^k)q_{jn}(i_{j1})Q_{j+1}(i_{j2})Q_{j+1}(i_{j3}) \quad (2)$$

where $U_j^k = \{I_j^k \mid I_j^k = \langle i_{j1}, i_{j2}, i_{j3}\rangle, i_{j1} + i_{j2} + i_{j3} = k\}$, and $Q_{\log_2 N-2}(i) = q_{sn}(i)$ with $s = \log_2 N - 2$. We can compute (2) starting from stage $\log_2 N - 3$, and work all the way back to stage 0. After $\log_2 N - 2$ steps, we can obtain $Q_0(k)$, the survival probability of intermediate stages with k faulty elements.

Next, let us consider the survival probability for the input elements of the first stage and the survival probability of the last stage. Let $Q_f(k)$ be the probability that k input elements in stage 0 are faulty but they do not cause the network to fail; and $Q_l(k)$ be the survival probability of the last stage, i.e., stage $\log_2 N - 1$, after having k faulty elements. The expressions for $Q_f(k)$ and $Q_l(k)$ can be found in [10].

Having obtained $Q_f(k)$, $Q_0(k)$, and $Q_l(k)$, we can compute $Q(k)$, the survival probability of the entire network after k elements fail using similar steps in deriving (2).

With $Q(k)$'s, we can compute the probability $P(k)$ that k or fewer faults cause the network to lose full connectivity:

$$P(k)=1-Q(k). \quad (3)$$

Let $p(i)$ be the probability that the ith fault will cause the network to fail. We get

$$P(k)=\sum_{i=2}^{k} p(i). \quad (4)$$

TABLE I
$\bar{k}$, MTTF$_0$, MTTF, AND COST-EFFECTIVENESS RATIO OF NETWORKS

N	$\bar{k}$	$MTTF_0$	$MTTF$	$MTTF/MTTF_0$	μ/μ_0
4	4.5	$(8.0\lambda)^{-1}$	$(3.0\lambda)^{-1}$	2.7	1.2
16	11.5	$(53.3\lambda)^{-1}$	$(9.2\lambda)^{-1}$	5.8	2.6
64	27.2	$(298.7\lambda)^{-1}$	$(22.9\lambda)^{-1}$	13.0	5.8
256	61.8	$(1536.0\lambda)^{-1}$	$(53.2\lambda)^{-1}$	28.9	12.8
1024	137.1	$(7509.3\lambda)^{-1}$	$(118.9\lambda)^{-1}$	63.2	28.1

From (3) and (4), we can obtain all of $p(i)$'s recursively. Now, let $\bar{k}$ be the expected number of faulty elements that cause the network to lose full connectivity under the conservative assumption given above, then

$$\bar{k} = \sum_{i=2}^{B+1} ip(i)$$

where $B = N \log_2 N + 1$ is the conservative upper bound we obtained previously. Notice that $p(i) = 0$ for $i > B + 1$ because the network fails as $i = B + 1$. The k for a variety of network sizes N is given in Table I. It is interesting to see that as N increases, $\bar{k}$ also grows. The larger a network is, the better a completely chain network can survive from more faulty elements (because more redundant paths are provided). This is a sharp contrast to a regular network with only one unique path between each input–output pair. A strong reroutable fault-tolerant network [5], [10], [11] always possesses this property.

B. Mean Lifetime of a Network

The mean time to fail (MTTF) for completely chained networks is derived based on the assumptions that failures occur independently to elements with a constant rate λ. Let $R(t)$ be the reliability function of a chained network, then

$$R(t) = \sum_{i=0}^{E} Q(i) \binom{E}{i} (e^{-\lambda t})^{(E-i)} (1 - e^{-\lambda t})^i$$

where $E = N + (3N \log_2 N)/2$ is the total number of elements in the network. Under our conservative assumption, $Q(i)$ is 0 as $i > B$. The MTTF of the network can be obtained after a simple calculation as MTTF $= (1/\lambda) \sum_{i=0}^{E} (Q(i))/(E - i)$.

To understand the reliability improvement of a completely chained network over a regular network, we compare the two networks with the same size. Since a 2×2 "augmented" SE in a chained network is essentially a 3×3 crossbar switch, the hardware complexity of an element in a 2×2 "augmented" SE is roughly 3/2 times as complicated as that of an element in a regular 2×2 SE. Let us assume the failure rate to an element of a regular 2×2 SE be $(2/3)\lambda$ due to lower hardware complexity (comparing to λ for an element in a 2×2 "augmented" SE). A regular network fails after a single fault arises because there is only one unique path between each input–output pair. Its mean time to fail, denoted by MTTF$_0$, for various network sizes is given in Table I, where the corresponding MTTF is also listed. From the table, we can see that a large completely chained network performs far more reliably than a regular network. This is because larger completely chained networks have larger chains (i.e., chains with more SE's) and can thus provide more alternative paths. Their reliability is significantly increased even under the conservative assumption.

Now, a simple measure of cost effectiveness is given for comparison between chained networks and regular networks. An "augmented" SE has higher hardware complexity and, hence, higher cost than a regular SE. We assume the hardware cost of an SE is proportional to the number of *crosspoints* within an SE. The hardware cost of a 2×2 regular SE and an "augmented" SE is four units and nine units, respectively. Since a chained network and a regular network of the same size employ equal number of SE's, the hardware cost of a chained network is 9/4 times that of a regular network (ignoring the cost of extra switches in a chained network to connect external components). Let us define the cost-effectiveness measure μ of a network as the ratio of its MTTF to cost. In the last column of Table I, we list the ratio of μ, the measure for a chained network, to μ_0, the measure for a regular network. It can be seen that the ratio is always greater than 1 and grows rapidly as the network size increases. A chained network is more cost effective than a regular one.

IV. DISCUSSION AND COMPARISON TO OTHER SCHEMES

A chained network can provide fault tolerance as long as every chain in the network contains more than one SE. A fault-tolerant chained network is strong reroutable if every SE belongs to a chain (except SE's in the last stage). Even though we study only completely chained networks here because they provide the maximum fault-tolerant capability, other chaining configurations are also possible. For example, if pairs of SE's within a partition are chained together, it becomes Augmented Bidelta Network as proposed in [11].

A regular MIN exhibits unsatisfactory bandwidth performance because an internal link is shared by many paths. Request conflicts become increasingly serious as the network size grows. A scheme to pair two SE's in a stage to share load and to balance traffic is recently introduced in [13], where some implementation issues and network performance are detailed. Based on the same reason, a chained network enjoys the byproduct of bandwidth improvement as presented in [15].

An extra stage cube (ESC) network [6] is proposed to provide single-fault tolerance. Compared to a regular network built from $k \times k$ SE's, the hardware overhead of an ESC network with $k \times k$ SE's is $1/\log_k N + O(1/\log_k N)$; and it is $(2/k + 1/k^2) + O(1/\log_k N)$ for a baseline network with the proposed chaining scheme using $(k + 1) \times (k + 1)$ SE's, where N is the network size. The first term of these two expressions is attributed, respectively, to the extra stage in an ESC network and to the $(k + 1) \times (k + 1)$ SE's (i.e., "augmented" SE's) in a chained network. The second term is attributed to the extra multiplexers/demultiplexers. From the expressions, we realize that an ESC network has less hardware overhead only when k is small or N is very large. If we ignore the second term in the above expressions (i.e., the overhead due to multiplexers/demultiplexers), for $k = 8$, an ESC network has less hardware overhead only when N is larger than 2048; for $k = 16$, it has less overhead only when N is larger than 2^{31}.

In many previous fault-tolerance schemes, alternative paths are provided through the links *between* consecutive stages by adding extra stages [6], [9] or by using larger SE's in certain stages [8]. They provide a *fixed* number of alternative paths for any sized network. By contrast, the alternative paths in a chained network are provided through the links within the *same* stage. A chained network exploits all inherent paths in the tree structure embedded in the network. Hence, the number of alternative paths between each input–output pair grows exponentially as the network size increases. Although its alternative paths are not totally disjoint, a chained network is more reliable than a network with only a fixed number of alternative paths as the system size grows. This is illustrated below by the terminal reliability of various networks. The terminal reliability between a given network input–output pair is defined as the probability of existence of at least one path between them.

Recall that an element is considered to be the basic unit in our fault model. The reliability of an element is assumed to be a function of time $e^{-\lambda t}$ in deriving network MTTF previously. Now, suppose an element has constant reliability r instead, which takes the failure rate of both switching logics and the connecting link into account. To simplify our analysis, we assume the reliability of an input element at the first stage and an output element at the last stage is r as well. The network under consideration is of size $N = 2^n$. In an ESC network [6] or the Indra network [9] built from 2×2 SE's (i.e., $R = 2$), a path from S (a source node) to D (a destination node) consists of $n + 2$ elements, including the two elements for connecting S and D. The terminal reliability between S and D in an ESC network is $r^2(1 - (1$

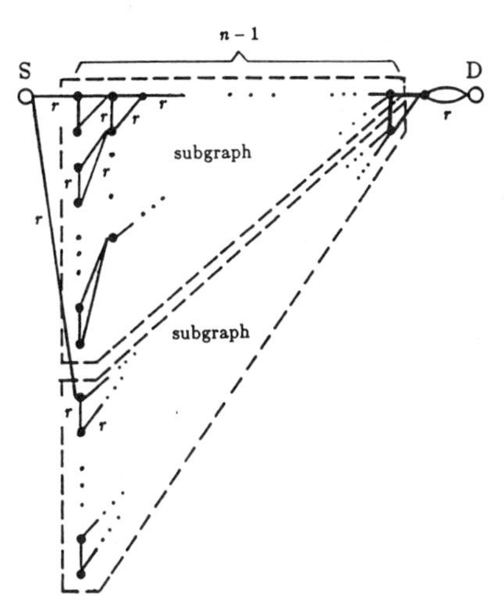

Fig. 3. The redundancy graph of a network constructed by chaining together two SE's in each partition.

TABLE II
TERMINAL RELIABILITY OF VARIOUS NETWORKS WITH r = 0.9

Network size	terminal reliability			
(N)	ESC	Gamma (max.)	Indra (R =2)	chained network (TR_m)
4	0.781	0.875	0.963	0.969
16	0.714	0.844	0.925	0.961
64	0.632	0.815	0.869	0.951
256	0.547	0.787	0.798	0.939
1024	0.466	0.759	0.717	0.926
4096	0.393	0.733	0.633	0.912

$- r^n)^2)$. The terminal reliability of the INDRA network is derived in [9], where an SE is treated as an entity with a given reliability value (which is different from our fault model) and a path from S to D is composed of $n + 1$ nodes. Thus, in citing the results, we have to be cautious about the slight difference caused by different fault models.

To calculate the terminal reliability of a completely chained network, denoted as TR, is very involved. Instead let us consider the terminal reliability of a network in which pairs of SE's within a partition are chained together. The redundancy graph [8] for a source-destination pair in such a chained network is depicted in Fig. 3, where a reliability value is associated with each link. Suppose its terminal reliability is represented by TR_m. It is obvious that TR_m = TR for $N = 4$ and $\mathrm{TR}_m <$ TR for $N > 4$. The reliability of subgraphs indicated in Fig. 3 is $g = r(1 - (1 - r)(1 - r^2))^{n-2}$. Now the terminal reliability between S and D can be derived by applying the bridge network result presented in [14]. We have

$$\mathrm{TR}_m = 2gr - (gr)^2 + 2(1-g)(1-r)gr^2(2r - r^2).$$

The terminal reliability of various fault-tolerant networks with r = 0.9 is summarized in Table II. The second column gives the terminal reliability of an ESC network. The maximum terminal reliability of a Gamma network [5] is taken from [9] with appropriate adjustment to reflect the difference of fault models; so is the terminal reliability of an Indra network with $R = 2$. The actual terminal reliability of a completely chained network (i.e., TR) is greater than or equal to figures shown in the last column. It can be found that a chained network always possesses a higher terminal reliability. The larger the network size is, the more a chained network outperforms other networks. Similar arguments can be applied to a modified omega network [8] as well.

V. CONCLUDING REMARKS

We propose a simple fault-tolerance enhancement scheme which can be applied to a wide class of multistage interconnection networks. Network fault-tolerant capability is greatly enhanced by chaining switching elements within the same stage together to provide multiple paths between each network input-output pair. It takes full advantage of the tree structure embedded in a network, so the number of alternative paths increases exponentially as the system size grows. A completely chained network possesses strong reroutability which permits a request to be rerouted by a simple routing procedure within the same stage when the request encounters a failure or a conflict. From the reliability analysis, we show that the network reliability is enhanced greatly as the network size becomes larger. It is also illustrated that a pair of source-destination nodes in a chained network has higher terminal reliability than in many other fault-tolerant networks.

The network bandwidth of a chained network is fundamentally improved due to the existence of buffering capability, as detailed in [15]. The behavior of a chained network with a single fault is also presented there. Although a fault can affect many paths, it is shown to have very little impact on bandwidth of a chained network, i.e., it performs almost equally well before and after the occurrence of the fault. For a network with buffers in its SE's, the chaining scheme is demonstrated [15] to be able to improve network delay effectively through dynamically regulating traffic among alternative paths to eliminate local congestion.

REFERENCES

[1] C.-L. Wu and T.-Y. Feng, *Tutorial: Interconnection Networks for Parallel and Distributed Processing*. IEEE Computer Science Press, 1984.

[2] D. H. Lawrie, "Access and alignment of data in an array processor," *IEEE Trans. Comput.*, vol. C-24, pp. 1145–1155, Dec. 1975.

[3] C.-L. Wu and T.-Y. Feng, "On a class of multistage interconnection networks," *IEEE Trans. Comput.*, vol. C-29, pp. 694–702, Aug. 1980.

[4] J. P. Shen and J. P. Hayes, "Fault tolerance of a class of connecting networks," in *Proc. 7th Annu. Symp. Comput. Architecture*, 1980, pp. 61–71.

[5] D. S. Parker and C. S. Raghavendra, "The gamma network: A multiprocessor interconnection network with redundant paths," in *Proc. 9th Annu. Symp. Comput. Architecture*, Apr. 1982, pp. 73–80.

[6] G. B. Adams III and H. J. Siegel, "The extra stage cube: A fault-tolerant interconnection network for supersystems," *IEEE Trans. Comput.*, vol. C-31, pp. 443–454, May 1982.

[7] J. E. Lilienkamp, D. H. Lawrie, and P.-C. Yew, "A fault tolerant interconnection network using error correcting codes," Dep. Comput. Sci. Rep. UIUCDCS-R-82-1094, Univ. Illinois Urbana-Champaign, June 1982.

[8] K. Padmanabhan, "Fault tolerance and performance improvement in multiprocessor interconnection networks," Ph.D. dissertation, Dep. Comput. Sci. Rep. UIUCDCS-R-84-1156, Univ. Illinois, Urbana-Champaign, May 1984.

[9] C. S. Raghavendra and A. Varma, "INDRA: A class of interconnection networks with redundant paths," in *Proc. Real-Time Syst. Symp.*, Dec. 1984, pp. 153–164.

[10] N.-F. Tzeng, P.-C. Yew, and C.-Q. Zhu, "A fault-tolerant scheme for multistage interconnection networks," in *Proc. 12th Int. Symp. Comput. Architecture*, June 1985, pp. 368–375.

[11] V. P. Kumar and S. M. Reddy, "Design and analysis of fault-tolerant multistage interconnection networks with low link complexity," in *Proc. 12th Int. Symp. Comput. Architecture*, June 1985, pp. 376–386.

[12] D. M. Dias and J. R. Jump, "Analysis and simulation of buffered delta networks," *IEEE Trans. Comput.*, vol. C-30, pp. 273–282, Apr. 1981.

[13] C.-T. A. Lea, "The load-sharing banyan network," *IEEE Trans. Comput.*, vol. C-35, pp. 1025–1034, Dec. 1986.

[14] P. M. Lin, B. J. Leon, and T.-C. Huang, "A new algorithm for symbolic system reliability analysis," *IEEE Trans. Reliability*, vol. R-25, pp. 2–15, Apr. 1976.

[15] N.-F. Tzeng, "Fault-tolerant multiprocessor interconnection networks and their fault-diagnoses," Ph.D. dissertation, Dep. Comput. Sci., Univ. Illinois, Urbana-Champaign, Aug. 1986.

Chapter 3: Multiterminal Reliability Evaluation

As mentioned in the earlier chapters, distributed system reliability is modeled as a collection of resources (processing elements, datafiles, programs, memory units, etc.) interconnected via an arbitrary communication network and controlled by a distributed operating system. Figure 1 illustrates a four-computer (node) distributed data processing system that can run programs across the system.

In this figure, FA denotes the set of files (FLs) available at a given computer site (node). FNs denote the files needed to execute a program (PM), and PG designates the set of programs to be executed at the node. Thus, each program can run on one or more computers and may frequently need files stored at other sites. Banking systems, travel agency systems, and power control systems are just a few examples of such a complex computing environment.

In a general distributed system, the following five reliability criteria are usually considered:

1. single source-to-terminal (SST) reliability—Probability that a computer called "source node" communicates successfully for datafiles with another computer called terminal node
2. multisource-to-terminal (MST) reliability—Probability that a number of source nodes simultaneously communicate successfully with a terminal node
3. source-to-multiterminal (SMT) reliability—Probability that a source node communicates simultaneously with a number of operative terminal nodes
4. multisource-to-multiterminal (MSMT) reliability—Probability that a number of source nodes simultaneously communicate successfully with a number of operative terminal nodes
5. network reliability—Probability that all operating nodes successfully communicate with each other

Criteria 2, 3, 4, and 5 are particularly important for a computer communication distributed system, and multiprocessor network. Further, for a distributed data processing system (DDPS), the reliability measure should capture the effects of redundant distribution of programs and datafiles (refer to Figure 1). In this case, we define DDPS reliability as the probability that all distributed programs can be executed successfully and are able to access all the required

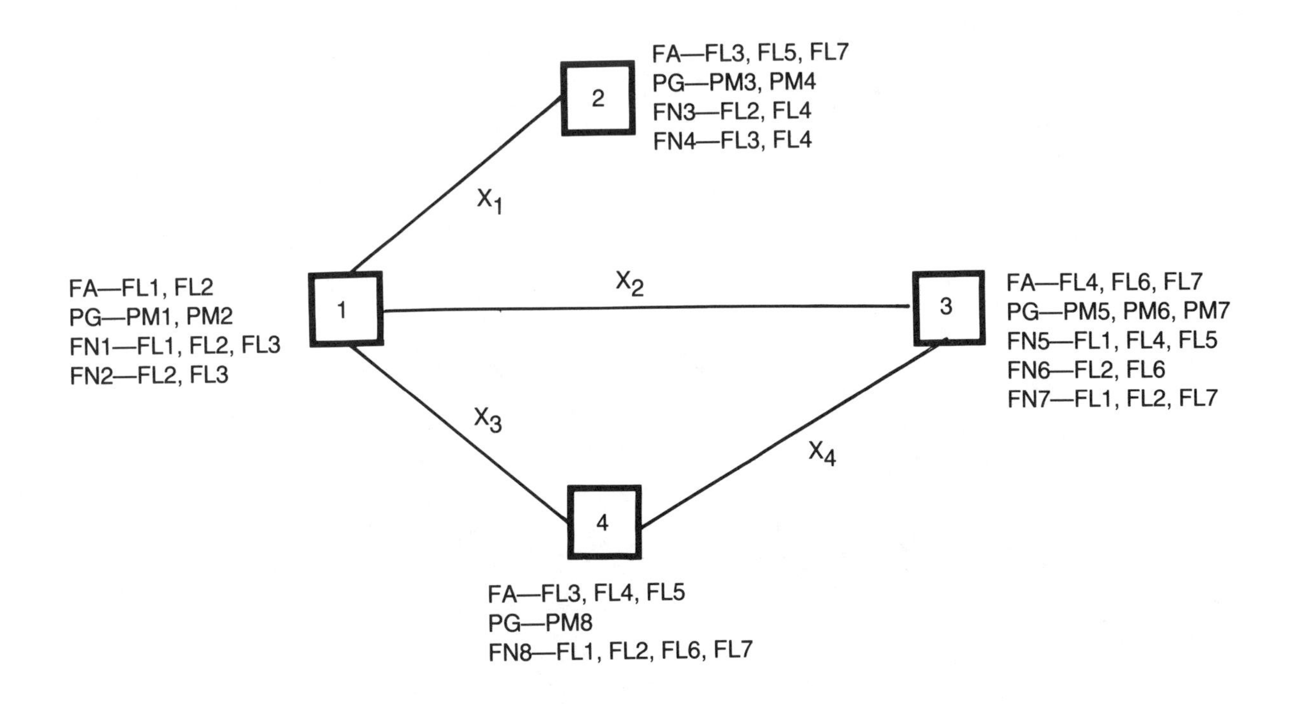

Figure 1: Example of a Distributed Data Processing System

remote files in spite of some faults present at either the computers or the connecting communication links.

Measures 2, 3, 4, 5, and DDPS reliability are best characterized by the term *multiterminal reliability*. Thus, multiterminal reliability is a generalized concept of reliability measures and is the probability of true value of a Boolean expression whose terms denote the existence of connections between subsets of resources. Depending on the problem, the multiterminal connection requirement falls into any one of the following three categories:

- OR-type
- AND-type
- AND-OR/OR-AND mixed type

If we know path set for all node pairs a priori, the logical relation considered above obtains the multiterminal connection expression; hence, the multiterminal reliability. Using this approach, multiterminal reliability is presented as an extended concept of terminal reliability. A major drawback of the algorithms based on this approach lies in the enumeration and manipulation of large numbers of events corresponding to multiterminal connection. An alternative technique using a breadth-first search approach has been suggested to obtain information on multiterminal connections. Another method utilizes the connection matrix of the system and applies extended conservative policy to derive multiterminal reliability. This technique represents a one-step approach as there is no need to compute multiterminal connections. Thus, this method avoids altogether the complexity involved in separately mapping the connections to their equivalent probability expression of reliability.

Reliability Evaluation in Computer-Communication Networks

K.K. Aggarwal
Regional Engineering College, Kurukshetra
Suresh Rai
Regional Engineering College, Kurukshetra

Key Words—**Computer-communication network, Reliability polynomial, Spanning tree.**

Reader Aids—
Purpose: Widen the state of art
Special math needed for explanations: Boolean algebra, Graph theory
Special math needed for results: Same
Results useful to: Reliability theoreticians

Abstract—**The network reliability for a computer-communication network is defined and a method based on spanning trees is proposed for its evaluation. The network reliability expression and *s-t* terminal reliability expression are compared assuming an equal probability of survival for each communication link. Examples illustrate the method.**

1. INTRODUCTION

The reliability analysis of a computer-communication network (CCN) using graph theoretic approach is based on modeling the network by a linear graph in which the nodes (vertices) correspond to computer centres (Hosts and Interface Message Processors) in the network, and edges correspond to the communication links. The terminal reliability, a commonly used measure of connectivity, is the probability of obtaining service between a pair of operative centres, called source and sink, in terms of reliability for each communication link/node in the network. The calculation of this probability is based not only on the minimum number of link successes (known as a path) needed for the service, but also on the total number of combinations of such successes. The calculation obviously does not take into account the communication between any other nodes but for the source and sink. In this paper, we find the probability of obtaining a situation in which each node in the network communicates with all other remaining communication centres (nodes). In the event that this probability, now onwards called 'Network Reliability' of a CCN, is to be calculated using the concepts of terminal reliability only, one can proceed by finding all possible paths between each of the $n(n-1)/2$ node pairs [1]. Since this is impractical for graphs with a large number of nodes, an alternative procedure is suggested using the concept of spanning trees.

2. ASSUMPTIONS

a. Links are half- or full-duplex. The network is modeled by a graph with nonoriented edges.

b. Link states are *s*-independent. Failure of one link does not affect the probability of failure of other links.

c. Each link and the network have only 2 states good (operating) and failed.

3. NOTATION & DEFINITIONS

G	reliability graph
S	system success function
n_i, x_i	node-*i* and edge-*i* in the graph G
C_i	vertex cut-set
$X_i(X_i')$	logical success (failure) of edge-*i*.
$p_i(q_i)$	reliability (unreliability) of edge-*i*
n, b	total numbers of nodes and edges in G
T_i	spanning tree-*i*
N	total number of spanning trees
d_{ij}	distance between T_i and T_j
d_m	max (d_{ij})
$\oplus$	ring sum operation
$\times$	cartesian products

Other, standard notation is given in "Information for Readers & Authors" at rear of each issue.

Some graph-theoretic definitions which are useful in developing the procedure are listed here.

Spanning tree: A tree T_i is said to be a spanning tree of graph G if T_i is a connected subgraph of G and contains all nodes of G. An edge in T_i is called a branch of T_i, while an edge of G that is not in T_i is called a chord. For a connected graph of n nodes and b edges, and spanning tree has $(n-1)$ branches and $(b-n+1)$ chords.

Distance between two spanning trees: The distance between two spanning trees T_i and T_j of a graph G is the number of edges of G present in one tree but not in the other. Let N_{ij} denote the number of edges in $T_i \oplus T_j$, then from [2, 7]—

$$d_{ij} = \frac{1}{2} N_{ij}; \tag{1a}$$

The maximum distance between any two spanning trees in G is

$$d_m = \frac{1}{2} \max\{N_{ij}\}$$

$$\leqslant (n-1), \text{ the rank of } G. \tag{1b}$$

No more than $(b-n+1)$ chords of a spanning tree T_i can be replaced to get another tree T_j. Hence

Reprinted from *IEEE Transactions on Reliability*, Volume R-30, Number 1, April 1981, pages 32–35.

$$d_m \leqslant (b - n + 1). \quad (1c)$$

Combine (1b) and (1c) to get:

$$d_m \leqslant \min\{n - 1, b - n + 1\}. \quad (1d)$$

4. ALGORITHM

From the definition of a spanning tree, any T_i will link all n nodes of G with $(n - 1)$ branches and hence represents the minimum interconnections required for providing a communication between all computer centres which are represented by nodes. Thus, the problem of studying the network reliability between any of the centres in the CCN is a problem of:

1. Enumerating all T_i's in the reliability graph corresponding to the network.
2. Interpreting Boolean algebraic statement of step #1 as probability expression.

For step #1, a simple approach is to use elementary tree—transformation which is based on the addition of a chord and deletion of an appropriate branch from a spanning tree T_i. Thus, starting from any spanning tree of G, one can generate all spanning trees by successive cyclic exchanges [3]. This approach, although straightforward, is difficult to computerize. Another method, described here, uses Cartesian products of $(n - 1)$ vertex cutsets C_i whose elements are the branches connected to any of the $(n - 1)$ nodes of G. Thus

$$C = C_1 \times C_2 \times \ldots \times C_{n-1}$$

$$= \mathop{\times}_{i=1}^{n-1} C_i \quad (2)$$

where C is a set of subgraphs of G with $(n - 1)$ branches. It has been proved [9] that any circuit of G with $(n - 1)$ branches will have an even number of identical appearances in C. If these terms are recognized, then deleted from C, the normalised Cartesian product C^* contains only those subgraphs which do not repeat an even number of times and are of cardinality $(n - 1)$. From the concept of spanning tree, C^* is, thus, the set of all T_i's of a connected graph G. A computer program for C^* based on (2) has been developed and is listed in the Supplement [12].

Example 1.

For a bridge network (Figure 1), the three vertex cutsets are:

$$C_1 = (x_1, x_2);\ C_2 = (x_4, x_5);\ C_3 = (x_1, x_3, x_4).$$

Using (2)—

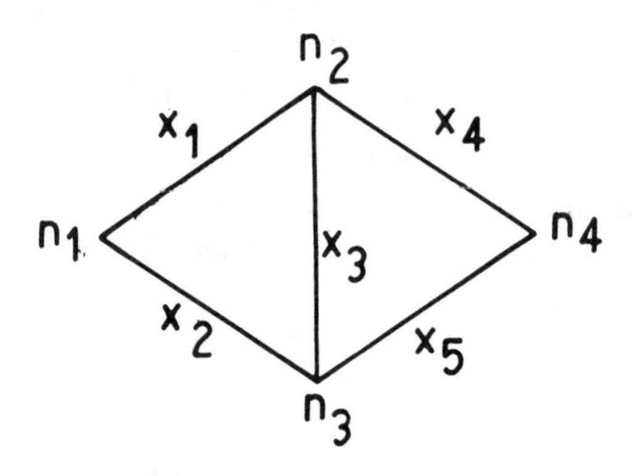

Fig. 1. A bridge network.

$$C = (x_1, x_2) \times (x_4, x_5) \times (x_1, x_3, x_4)$$

$$= (x_1x_4, x_1x_5, x_2x_4, x_2x_5) \times (x_1, x_3, x_4)$$

$$= (x_1x_3x_4, x_1x_3x_5, x_1x_4x_5, x_1x_2x_4, x_2x_3x_4, x_1x_2x_5,$$

$$x_2x_3x_5, x_2x_4x_5) \quad (3)$$

Since no term in (3) has an even number of identical appearances, C^* is the same as C. The 8 elements of set C^* thus represent 8 different spanning trees.

In step #2, a Boolean algebraic expression has a one-to-one correspondence with the probability expression if the Boolean terms are modified either canonically or conservatively until they represent a disjoint grouping. Obviously, the second approach is preferred since it results in a disjoint expression with many fewer terms as compared to canonical expansion. We present below an algorithm for finding the probability expression and hence the network reliability of CCN starting from a set of T_i's. Any of the existing methods of terminal reliability evaluation [1, 5, 6, 10, 11] could be used for the step. The proposed method is given here.

1. Consider a tree T_o amongst the list of T_i's. Arrange T_i's in ascending order of their distance from T_o; e.g., all spanning trees at distance-1 first, distance-2 next, and so on until all trees at distance-d_m have been considered. Choosing T_o as one of the central trees [3] helps in reducing the computer time and results in a simpler reliability expression.

For the purpose of network reliability, let system success S, be defined as the event of having at least one spanning tree with all its branches operative.

$$S = T_o \cup T_i \cup \ldots \cup T_{N-1} \quad (4)$$

2. Define F_i for each term T_i such that:

$$F_0 = T_0$$

$$F_i = T_0 \cup T_1 \cup \ldots \cup T_{i-1}\big|_{\text{Each literal of } T_i \to 1,\ \text{for } 1 \leqslant i \leqslant (N-1).}$$

The literals of T_i are assigned a value 1 (Boolean) which is substituted in any predecessor term in which they occur. F_i can be simplified by using elementary Boolean theorems [4].

3. Use exclusive-operator ξ [5] to get—

$$S(\text{disjoint}) = T_0 \bigcup_{i=1}^{N-1} T_i\, \xi(F_i). \quad (5)$$

Since, all terms in (5) are mutually exclusive, the network reliability expression R_s is obtained from (5) by changing X_i to p_i, and X'_1 to q_i, viz.,

$$R_s = S(\text{disjoint})\big|_{X_i(X'_i)\to p_i(q_i)} \quad (6)$$

For a d_m-distant spanning tree T_i from T_0, $\xi(F_i)$ is simply the intersection of complements of $(b - n + 1)$ chords with $(n - 1)$ tree branches of T_i; provided $d_m \equiv b - n + 1$. This observation is helpful in simplifying the computation involved in finding $\xi(F_i)$.

A computer programme in FORTRAN IV call REST (Reliability Evaluation using Spanning Tree) has been developed and tried successfully on TDC-316 computer. The source listing is given in the Supplement [12].

Example 2.

Take a simple bridge network as given in figure 1. The 8 spanning trees, obtained in (3), are rearranged as indicated in step 1. Take $T_0 = x_1x_3x_5$; then the 1- and 2-distant spanning trees are $(x_1x_3x_4,\ x_1x_4x_5,\ x_2x_3x_5,\ x_1x_2x_5)$ and $(x_2x_3x_4,\ x_1x_2x_4,\ x_2x_4x_5)$ respectively. Therefore, $S = X_1X_3X_5 \cup X_1X_3X_4 \cup X_1X_4X_5 \cup X_2X_3X_5 \cup X_1X_2X_5 \cup X_2X_3X_4 \cup X_1X_2X_4 \cup X_2X_4X_5$. The F_i's and $\xi(F_i)$'s for $i = 1, \ldots, 7$ are obtained as shown in Table 1.

TABLE 1

F_i	$\xi(F_i)$	F_i	$\xi(F_i)$
$F_1 = X_5$	X'_5	$F_5 = X_1 + X_5$	$X'_1X'_5$
$F_2 = X_3$	X'_3	$F_6 = X_3 + X_5$	$X'_3X'_5$
$F_3 = X_1$	X'_1	$F_7 = X_1 + X_3$	$X'_1X'_3$
$F_4 = X_3 + X_4$	$X'_3X'_4$		

From (6), the network reliability expression is—

$$R_s = p_1p_3p_5 + p_1p_3p_4q_5 + p_1p_4p_5q_3 + p_2p_3p_5q_1 + p_1p_2p_5q_3q_4 + p_2p_3p_4q_1q_5 + p_1p_2p_4q_3q_5 + p_2p_4p_5q_1q_3. \quad (7)$$

For the CCN having equal probabilities of survival p for each communication link, (7) simplifies to—

$$R_s = 8p^3 - 11p^4 + 4p^5. \quad (8)$$

In deriving (7) we have assumed perfect nodes. As computer outages account for as much as 90% of failures in most CCNs, we have to consider the reliability of nodes as less than 1 in such situations. In such a case, (7) is to be multiplied by a factor $(p_{n_1}p_{n_2}p_{n_3}p_{n_4})$, where p_{n_1} represents the reliability of node n_i [11].

5. COMPARISON

To compare the network reliability with terminal reliability, we consider the reliability polynomial generated by assuming a probability of survival p for each communication link. The properties of such a polynomial for terminal reliability are reported in [6]. For network reliability, the general polynomial is:

$$R_s(p) = p^{n-1} \sum_{i=0}^{b-n+1} c_i(1-p)^i;\ c_0 = 1$$

$$= \sum_{i=0}^{b-n+1} (-1)^i k_i\, p^{n-1+i};\ k_0 = N. \quad (9)$$

Some properties of $R_s(p)$ are—

1. The lowest degree for any term is $(n - 1)$ and its coefficient is the total number of spanning trees, N.
2. The sum of all the coefficients in the polynomial is unity i.e.,

$$\sum_{0}^{b-n+1} (-1)^i k_i = 1$$

3. The maximum possible degree for any term is the number of communication links in the network.
4. The terms are alternatively positive and negative.
5. The polynomial when plotted against p is S-shaped if there is no single communication link either in series or in parallel overall [8]. This means that there exists a value p_0, $0 < p_0 < 1$, such that $R_s(p_0) = p_0$, $R_s(p) \leqslant p$ for $0 \leqslant p \leqslant p_0$ and $R_s(p) \geqslant p$ for $p_0 \leqslant p \leqslant 1$. The practical implication of this result is that for a s-coherent system, only when all the components have attained sufficiently high reliability is the network reliability greater than the reliability of a single communication link. This property is illustrated in Table 2.

TABLE 2

p	0.5	0.6	0.7	0.8	0.9	1.0
R_s	0.2500	0.4423	0.6561	0.8441	0.9637	1.0000
R_{16}	0.3281	0.5635	0.7032	0.8641	0.9670	1.0000

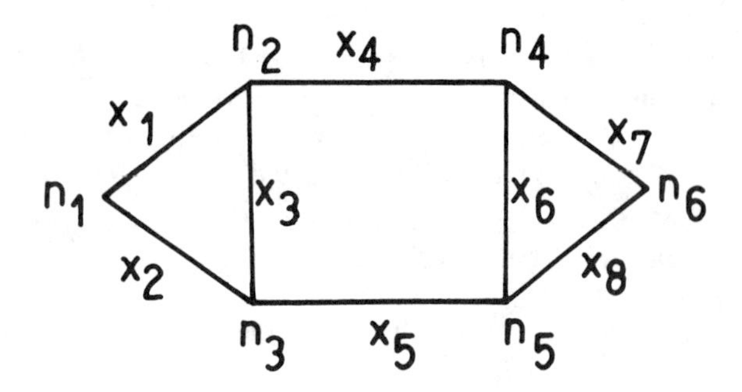

Fig. 2. 6-Node 8-link example.

For the 5-node, 7-edge graph [6, Fig. 7], the network reliability polynomial is—

$$R_s(p) = 21p^4 - 44p^5 + 32p^6 - 8p^7.$$

The coefficient of p^4 (viz. 21) specifies the total number of the spanning trees of the graph.

Example 3.

For figure 2, there are 30 spanning trees, each of cardinality 5. The network reliability polynomial, assuming each communication link has a reliability of p, is—

$$R_s(p) = 30p^5 - 65p^6 + 48p^7 - 12p^8. \quad (10)$$

If node 1 is source and node 6 is sink, then the terminal reliability polynomial is—

$$R_s(p) = 2p^3 + 4p^4 - 2p^5 - 13p^6 + 14p^7 - 4p^8. \quad (11)$$

For an easy comparison, the values of $R_s(p)$ and $R_{16}(p)$ for $0.5 \leqslant p \leqslant 1$ are listed in Table 2. It is evident that $R_s \leqslant R_{16}$ for any value of p. The difference $(R_{16} - R_s)$, however, is smaller in the high reliability region. There is a region of interest (from $p = 0.696$ to $p = 0.743$ for example 3) where terminal reliability is higher than the reliability of the individual link, yet the network reliability is lower than the reliability of individual link.

REFERENCES

[1] R.S. Wilkov, "Analysis and design of reliable computer networks", *IEEE Trans. Communications,* vol COM-20, 1972 Jun, pp 660-678.

[2] N. Christofides, *Graph Theory—An Algorithmic Approach,* Academic Press, New York 1975.

[3] N. Deo, *Graph Theory With Application To Engineering and Computer Science,* Prentice-Hall, Inc., Englewood Cliffs, NJ 1974.

[4] N.N. Biswas, *Introduction To Logic and Switching Theory,* Gordon and Breach Science Publishers, New York 1975.

[5] S. Rai, K.K. Aggarwal, "An efficient method for reliability evaluation of a general network", *IEEE Trans. Reliability,* vol R-27, 1978 Aug, pp 206-209.

[6] K.B. Misra, T.S.M. Rao, "Reliability analysis of redundant networks using flow graphs", *IEEE Trans. Reliability,* vol R-19, 1970 Feb, pp 19-24.

[7] W. Mayeda, *Graph Theory,* Wiley-InterScience, New York 1972.

[8] R.E. Barlow, F. Proschan, *Statistical Theory of Reliability and Life Testing: Probability Models,* Holt, Rinehart and Winston, Inc. New York 1975.

[9] M. Piekarski, "Listing of all possible trees of a linear graph", *IEEE Trans. Circuit Theory,* vol CT-12, 1965 Mar, pp 124-125.

[10] L. Fratta, U.G. Montanari, "A recursive method based on case analysis for computing network terminal reliability", *IEEE Trans. Communication,* vol COM-26, 1978 Aug, pp 1166-1177.

[11] K.K. Aggarwal, J.S. Gupta, K.B. Misra, "A simple method for reliability evaluation of a communication system", *IEEE Trans. Communication,* vol COM-23, 1975 May, pp 563-565.

[12] Supplement: NAPS document No. 03668-B, 5 pages in this Supplement. For current ordering information, see "Information for Readers & Authors" in a current issue. Order NAPS document No. 03668, 21 pages. ASIS-NAPS; Microfiche Publications; P.O. Box 3513, Grand Central Station; New York, NY 10017 USA.

AUTHORS

Dr. K.K. Aggarwal, Professor and Head; Electronics and Communication Engineering Department; Regional Engineering College; Kurukshetra 132 119 INDIA.

Dr. K.K. Aggarwal: For biography, see vol R-30; 1981 Apr, p 3.

Suresh Rai, Lecturer; Department of Electronics and Communication Engineering; Regional Engineering College; Kurukshetra 132 119 INDIA.

Suresh Rai: For biography, see vol R-27, 1978 Aug, p 211.

Manuscript TR79-78 received 1979 June 2; revised 1980 April 30, 1980 July 29.

★ ★ ★

Reprinted from *IEEE Transactions on Reliability*, Volume R-30, Number 4, October 1981, pages 325–334.

A New Algorithm for the Reliability Analysis of Multi-Terminal Networks

A. Satyanarayana
University of California, Berkeley
Jane N. Hagstrom
University of California, Berkeley

Key Words—**Topological formula, Exact reliability, Source to all-terminal reliability, Overall reliability, Network, Acyclic subgraph, Cylic subgraph, Reliability algorithm.**

Reader Aids—
Purpose: Advance the state of the art
Special math needed for explanations: Graph theory, probability theory
Special math need for results: Same
Results useful to: Reliability theoreticians & graph theorists

Abstract—**In a probabilistic network, source-to-multiple-terminal reliability (SMT reliability) is the probability that a specified vertex can reach every other vertex. This paper derives a new topological formula for the SMT reliability of probabilistic networks. The formula generates only non-cancelling terms. The non-cancelling terms in the reliability expression correspond one-to-one with the acyclic *t*-subgraphs of the network. An acyclic *t*-subgraph is an acyclic graph in which every link is in at least one spanning rooted tree of the graph. The sign to be associated with each term is easily computed by counting the vertices and links in the corresponding subgraph.**

Overall reliability is the probability that every vertex can reach every other vertex in the network. For an undirected network, it is shown the SMT reliability is equal to the overall reliability. The formula is general and applies to networks containing directed or undirected links. Furthermore link failures in the network can be *s*-dependent.

An algorithm is presented for generating all acyclic *t*-subgraphs and computing the reliability of the network. The reliability expression is obtained in symbolic factored form.

1. INTRODUCTION

A probabilistic network consists of a set of vertices and links that fail with some known joint probability distribution. In general failure events are not statistically (*s*-)independent. In a probabilistic network, any two vertices can communicate if they are both operative and if there is a path of operative vertices and links between them. Communication networks, electric power systems, water aqueducts, and transportation system are practical examples of such networks. Various performance measures can be defined for probabilistic networks [1, 2]. A commonly used performance measure is *terminal-pair reliability* which is the probability that a specified vertex can communicate with another specified vertex in the network. Perhaps the early interest in network reliability analysis grew from attempts to model complex systems as 2-terminal networks and thus most of the available literature concentrates on the terminal-pair reliability problem. For a review of the methods see [3, 4].

Recent developments in computer communication networks have led to an interest in computational techniques for more global reliability measures. One of these is the *source-to-multiple-terminal reliability* (SMT reliability) problem: "In a probabilistic network find the probability that a specified vertex (root vertex *s*) can reach every other vertex." A variety of practical systems could be modelled using this measure. As an example, consider a computer communication network. What is the probability that a particular computer can access every other computer in the network? In a variety of situations a performance measure of great interest is the *overall reliability,* viz., find the probability that every vertex can reach all other vertices in a probabilistic network. As an example of this measure, again in a computer network, one might require the probability that each computer can access all other computers. Recently, there has been considerable interest in computing the SMT and the overall reliability of lifeline networks, such as water pipelines, subjected to earthquakes [5]. In this paper we are concerned with the SMT and the overall reliability problems.

Wing & Demetriou [6] considered the problem of computing the overall reliability of a network by considering all possible vertex pairs of the network, but the approach quickly becomes infeasible even for networks of moderate size. Fu [7] used topological methods of electrical network analysis to obtain an approximate answer for the overall reliability problem.

Ball & Van Slyke [8] proposed a method of enumerating modified cutsets by backtracking to compute the overall reliability of an undirected graph. Satyanarayana [9] provided a formula and an algorithm for the same problem. It is difficult to compare the two because of differing features of the implementations. The algorithm in [8] produces numerical rather than symbolic output. This paper and [9] not only provide an algorithm, but an interesting topological observation: that is, the seemingly enormous size of the inclusion-exclusion expression for reliability can be considerably condensed. In [9] this observation is made and an algorithm is proposed only for the case of overall reliability of an undirected graph. For that case, [9] provides a more condensed expression than the present paper. However, the procedure for obtaining the coefficients of this highly condensed expression is much more cumbersome than the procedure in the present paper. Examples analyzed here and in [9] indicate that in their present forms the current algorithm is preferable for computing the overall reliability. Furthermore this paper also deals with the problem of SMT reliability of a directed graph, which [8] and [9] do not.

This paper presents a topological formula for SMT reliability of a probabilistic graph. The formula involves only non-cancelling terms of the reliability expression and these correspond one-to-one with certain acyclic subgraphs of the given network. The formula is general and applies to networks having directed as well as undirected links. Furthermore the failure events can be *s*-dependent. For an undirected network, we show that the overall reliability is identical with the SMT reliability and therefore the formula also applies for computing the overall reliability for undirected networks. Finally, we present an algorithm for computing the reliability of a given graph, and give the reference to a computer program for implementing the algorithm.

2. PRELIMINARIES

A *directed tree,* also called a *rooted tree,* is a connected directed graph with no cycles. In a rooted tree exactly one vertex, called the root, has indegree 0 while the rest of the vertices have indegree 1. A rooted tree has the property that there exists exactly one directed path from the root to every other vertex of the tree. Therefore, for a given probabilistic directed graph G, the SMT reliability $R_s(G)$ with respect to a specified root vertex *s,* is simply the probability that there exists at least one spanning rooted tree (rooted at *s*) in G such that all vertices and links in the tree are operative. This motivates the following new definition for a *t-graph.*

A *t-graph* is a connected directed graph with a particular vertex *s* called the root, having the property that every link of the graph lies on some rooted tree (rooted at *s*) of the graph. Clearly, any link that does not lie on a rooted tree of graph G makes no contribution to $R_s(G)$. An *acyclic t-graph* is a *t*-graph with no directed cycles while a *cyclic t-graph* must have at least one cycle. Henceforth by the term *rooted tree* or simply a *tree* of a *t*-graph G, we mean a spanning rooted tree of G rooted at *s*. Clearly, a rooted tree is an acyclic *t*-graph. A *t*-graph is termed *nontrival* if it is not a tree.

Let S be a subset of rooted trees of a *t*-graph G. Then S is termed a *formation* of G if every link of G lies on some rooted tree in S. Subset S is referred to as an *odd formation* if S consists of an odd number of trees and otherwise it is an *even formation.* The domination d(G) of a *t*-graph G is the number of odd minus the number of even formations of G. These definitions are illustrated in Fig. 1.

Consider a *t*-graph having m rooted trees. Using the Inclusion-Exclusion Principle of probability theory [10], $R_s(G)$ can be written as:

$$R_s(G) = \sum_i \varepsilon_i \Pr\{U_i\}. \tag{1}$$

U_i union of the *i*-th nonempty set of rooted trees of G and ε_i is +1 or −1 depending on whether the number of trees in U_i is odd or even respectively.

$\Pr\{U_i\}$ probability of the event that all trees in the union U_i are operative.

The sum is over all possible unions of m rooted trees of G and the number of terms in the explicit expression (1) will be $2^m - 1$. It is possible that $U_i = U_j$ for some $i \neq j$ and $\varepsilon_i = -\varepsilon_j$. Thus the terms $\Pr\{U_i\}$ and $\Pr\{U_j\}$ cancel. One of the most vexing problems in computing reliability using (1) is the appearance of pairs of identical terms with opposite signs which cancel. For practical complex networks, it can be easily shown that a majority of terms in (1) cancel. Indeed the cancellation problem here is identical to the one in the terminal-pair reliability problem [4]. In what follows, we derive a new topological formula which clearly characterizes the structure of both cancelling and noncancelling terms of (1). It will be shown that the noncancelling terms of (1) correspond one-to-one with the acyclic *t*-subgraphs of the given graph.

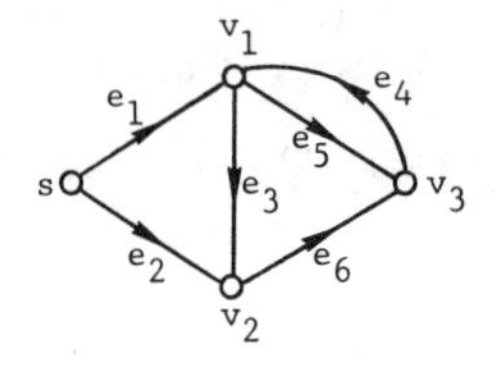

(a) A *t*-graph G

(b) Rooted Trees of G

Formation	
{t1,t2,t3,t4,t5}	odd
{t1,t2,t3,t4}	even
{t1,t2,t4,t5}	even
{t1,t3,t4,t5}	even
{t2,t3,t4,t5}	even
{t1,t2,t4}	odd
{t2,t3,t4}	odd
{t1,t4,t5}	odd
{t1,t3,t4}	odd
{t1,t4}	even

Domination of G

d(G) = 5 - 5 = 0

(c) Formations of G

Fig. 1. *t*-graph G, Rooted Trees of G, Formations and Domination of G.

3. THE FORMULA

SMT Reliability

Lemma 1: Any U_i in (1) is either an acyclic *t*-subgraph G_a or a cyclic *t*-subgraph G_c of the *t*-graph G. Conversely, each G_a and each G_c is identical to some U_i.

Proof: This follows from the definitions of U_i, G_a and G_c.

In view of Lemma 1, (1) can be rewritten as:

$$R_s(G) = \sum_i d(G_{a,i}) \Pr\{G_{a,i}\} + \sum_k d(G_{c,k}) \Pr\{G_{c,k}\}. \tag{2}$$

$G_{a,i}$ and $G_{c,k}$ — ith and kth acyclic and cyclic t-subgraphs of G respectively.

$d(G_{a,i})$ and $d(G_{c,k})$ — dominations of subgraphs $G_{a,i}$ and $G_{c,k}$.

$\Pr\{G_{a,i}\}$ and $\Pr\{G_{c,k}\}$ — probability that all vertices and links in $G_{a,i}$ and $G_{c,k}$ are operative.

The sum is over all possible acyclic and cyclic t-subgraphs of G.

We state the following two combinatoric results concering the domination of cyclic and acyclic t-graphs. For detailed exposition and proofs, see [11].

Theorem 1: The domination $d(G_c)$ of any cyclic t-graph G_c is zero.

Theorem 2: The domination $d(G_a)$ of any acyclic t-graph G_a is either +1 or −1 and furthermore $d(G_a) = (-1)^{b-n+1}$ where b and n are respectively the number of links and vertices of G_a.

Incorporating these two key results in (2), we have the topological formula (3) for SMT reliability of G.

$$R_s(G) = \sum_i (-1)^{b_i-n_i+1} \Pr\{G_{a,i}\} \tag{3}$$

where b_i and n_i are respectively the number of links and vertices of the ith acyclic t-subgraph $G_{a,i}$.

Formula (3) involves only non-cancelling terms and these correspond one-to-one with the acyclic t-subgraphs of the given graph. The sign to be associated with each term is simply obtained by a count of the number of links and vertices in the subgraph corresponding to the term. This can be compared with (1), where a count of the number of rooted trees is needed to determine ε_i. Indeed, formula (3) is not concerned with rooted trees at all. The formula will now be illustrated using the following examples.

Example 1: Consider the t-graph of Fig. 1. The graph has 5 rooted trees and therefore the total number of U_i's in (1) is $2^5 - 1 = 31$. However, the number of distinct U_i's (t-subgraphs) is 14. They are listed in Table 1 (a) and (b). For example, each of the 10 formations of the graph listed in Fig. 1 yields the same graph. Out of the 14 t-subgraphs, 11 are acyclic and 3 are cyclic. From Table 1b it can be seen that the cyclic t-subgraphs account for a total of 14 U_i's and by formula (3) these terms cancel out. Among the acyclic t-subgraphs, $G_{a,1}$ corresponds to 7 U_i's while the remaining correspond to unique U_i's. Table 1a shows that the dominations $d(G_{a,i})$ exactly coincide with the respective $\sigma_i = (-1)^{b_i-n_i+1}$

Example 2: Consider again the graph of Fig. 1. Let the failure of link e_1 increase the failure probability of link e_6, and vice versa. Similarly, say, for links e_3 and e_5. Thus we have, $\Pr\{e_1|e_6\} \neq \Pr\{e_1\}$, $\Pr\{e_3|e_5\} \neq \Pr\{e_3\}$. Let $p_i = \Pr\{e_i\}$, $p_{ij} = \Pr\{e_i|e_j\}$. By (3) we have:

$$\begin{aligned} R_s(G) &= p_1p_2p_3p_{53}p_{61} - p_1p_3p_{53}p_{61} - p_1p_2p_5p_{61} \\ &- p_1p_2p_3p_{61} - p_1p_2p_3p_{53} - p_1p_2p_4p_{61} + p_1p_3p_{53} \\ &+ p_1p_2p_5 + p_1p_3p_{61} + p_2p_4p_6 + p_1p_2p_{61}. \end{aligned} \tag{4}$$

The terms in (4) correspond one-to-one with the acyclic t-subgraphs of Table 1a.

TABLE 1(a)

Acyclic t−Subgraphs With Their Dominations for *Example* 1

i	$G_{a,i}$	$d(G_{a,i})$	*number of formations*		σ_i
			odd	*even*	
1	$e_1e_2e_3e_5e_6$	+1	4	3	+1
2	$e_1e_3e_5e_6$	−1	0	1	−1
3	$e_1e_2e_5e_6$	−1	0	1	−1
4	$e_1e_2e_3e_6$	−1	0	1	−1
5	$e_1e_2e_3e_5$	−1	0	1	−1
6	$e_1e_2e_4e_6$	−1	0	1	−1
7	$e_1e_3e_5$	+1	1	0	+1
8	$e_1e_2e_5$	+1	1	0	+1
9	$e_1e_3e_6$	+1	1	0	+1
10	$e_2e_4e_6$	+1	1	0	+1
11	$e_1e_2e_6$	+1	1	0	+1

TABLE 1(b)

Cyclic t−Subgraphs With Their Dominations for *Example* 1

k	$G_{c,k}$	$d(G_{c,k})$	*number of formations*	
			odd	*even*
1	$e_1e_2e_3e_4e_5e_6$	0	5	5
2	$e_1e_2e_4e_5e_6$	0	1	1
3	$e_1e_2e_3e_4e_6$	0	1	1

Overall Reliability of Undirected Graphs

A *symmetric digraph* is a directed graph in which for every link directed from i to j there is also a directed link from j to i. Consider the undirected graph G of Fig. 2a. The symmetric digraph G_d (Fig. 2b) is obtained by replacing every undirected link of G by two directed links in anti-parallel. In G_d the failure events of links are no longer *s*-independent and we have:

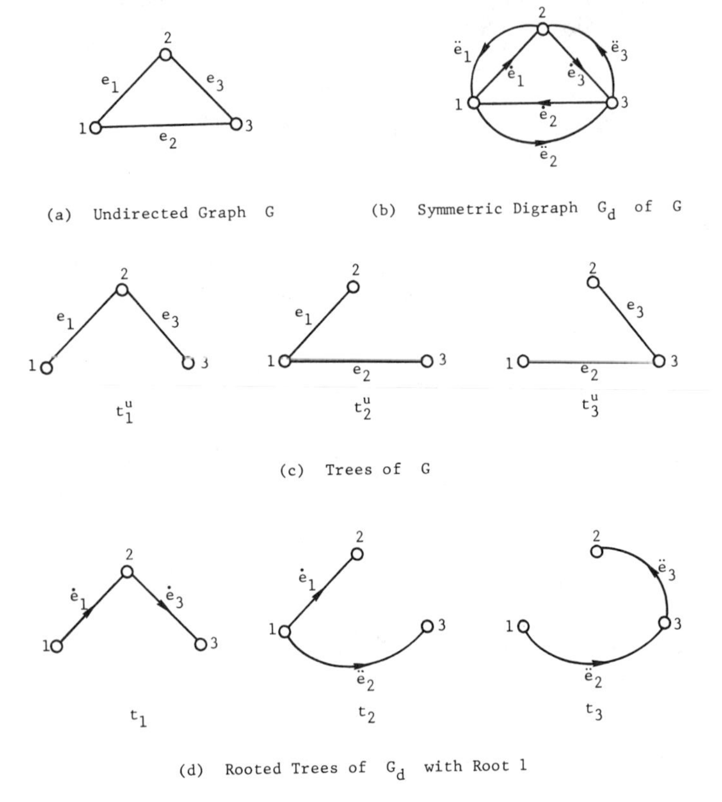

Fig. 2. Graph G, Symmetric Digraph G_d of G, Trees of G and Rooted Trees of G_d.

$$\Pr\{\dot{e}_i|\ddot{e}_i\} = \Pr\{\ddot{e}_i|\dot{e}_i\} = 1 \tag{5}$$

$$\Pr\{\dot{e}_i\} = \Pr\{\ddot{e}_i\} = \Pr\{e_i\} \tag{6}$$

for every link e_i of the undirected graph G. This situation implies that in G_d any $\dot{e}_i$ is operative i.f.f. $\ddot{e}_i$ is operative. The overall reliability of the undirected graph G is just the $\Pr\{t_1^u$ *or* t_2^u *or* t_3^u *is operating*$\}$. In view of (5), the event that t_1^u (see Fig. 2c) occurs is identical with the event that the rooted tree t_1 (see Fig. 2d) operates. Similarly the events that t_2^u and t_3^u are operating are identical respectively to the events that t_2 and t_3 are operating. Hence, the overall reliability of G is identical with the SMT reliability of G with any vertex designated as the source vertex *s*. Thus for any undirected graph G, the overall reliability is:

$$R_o(G) = R_s(G_d). \tag{7}$$

4. THE ALGORITHM

By (3) it is shown that the non-cancelling terms in $R_s(G_0)$ correspond one-to-one with the acyclic *t*-subgraphs of the given graph G_0. Thus the problem of computing $R_s(G_0)$ has become one of graph enumeration. Here we present an algorithm for generating the acyclic *t*-subgraphs of a given graph G_0 and evaluating the $R_s(G_0)$, or equivalently $R_o(G_0)$, if G_0 is undirected.

Neutral Link

For developing the algorithm, we need the concept of a neutral link in acyclic *t*-graphs. In [4] this concept has been introduced and studied in connection with acyclic *p*-graphs. The neutral link we define here is somewhat similar but not identical to the one described in [4]. A *neutral link* in an acyclic *t*-graph G is a link e whose deletion yields a subgraph G-e which is also an acyclic *t*-graph. As an example, consider the acyclic *t*-graph of Fig. 3. Links e_1, e_3, e_4, and e_6 are neutral since deleting any one of them results in a subgraph which also is *t* acyclic. However e_2 and e_5 are not neutral. We record the following four important properties of neutral links without proofs. For a detailed exposition of these results and other properties of acyclic *t*-graphs, see [12]. Property 4 forms the basis of the algorithm for generating all acyclic *t*-subgraphs of G_0.

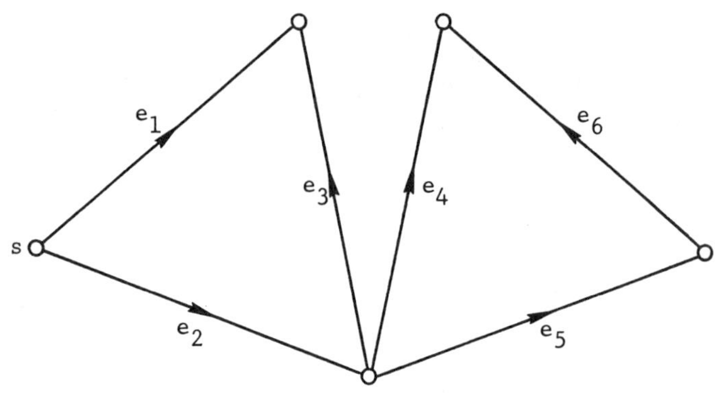

An Acyclic *t*-graph

Fig. 3.

Property 1: In an acyclic *t*-graph G_a, the neutral links are exactly those links directed into vertices of indegree > 1.

Property 2: Every non-trivial acyclic *t*-graph contains at least b-n+2 neutral links.

Property 3: Let $G_{a,i}$ be a non-trival *t*-subgraph of an acyclic *t*-graph G_a. Then every neutral link of $G_{a,i}$ is a neutral link of G_a.

Property 4: Every *t*-subgraph of an acyclic *t*-graph can be obtained by successive deletion of neutral links.

The Algorithm

The algorithm is similar to TRAP1 [4] and consists of the following stages.

1. If the given G_0 is undirected, obtain the corresponding symmetric digraph by replacing every link of G_0 with two directed links in anti-parallel. Choose a root vertex s arbitrarily and delete all links directed into s.
2. Successively decycle G_0 to obtain maximal acyclic t-subgraphs.
3. Obtain all t-subgraphs of these maximal t-subgraphs by deleting neutral links.

The algorithm grows a rooted directed tree with the following properties:

i. Vertices in the tree represent nonempty subgraphs of G_0, the root vertex being G_0. Any vertex, say k, corresponds one-to-one with a subgraph G_k which is (a) a cyclic or (b) an acyclic but not a t-graph or (c) an acyclic t-subgraph.
ii. A link directed from vertex i to j is weighted with the link deleted from the subgraph G_i to obtain G_j. For example if $G_j = G_i$-e, then in the rooted tree, the link directed from i to j is weighted with e.

In the rooted directed tree generated by the algorithm, a vertex i(j) is referred to as *father* (*child*) of j(i) when there exists a link directed from i to j. Vertex i is *ancestor* to j when i is contained in the path from the root vertex to j (i $\neq$ j). In the tree, vertices having the same father are termed *brothers*. A vertex i is *younger* (*elder*) brother of vertex j, if the algorithm generates the children of i later (earlier) than the children of j.

Starting from the root vertex, the algorithm grows a rooted directed tree by progressively generating children on all possible vertices of the tree. We use the following three rules for generating children on a vertex k of the tree.

Rule 1. If G_k is cyclic:

Let the set of links $\{e_1, e_2, ..., e_c\}$ constitute a simple directed cycle in G_k. Then $G_{ki} = G_k - e_i$, (i = 1, 2, ..., c), is a child of G_k, if $e_i \neq$ e, where e is the weight of the link incident into any elder brother or elder brother to an ancestor of k.

Rule 2. If G_k is an acyclic t-graph, but its father is not:

Obtain all neutral links $\{e_1, e_2, ..., e_m\}$ of G_k using Property 1. Then $G_{ki} = G_k - e_i$, (i = 1, 2, ..., m), is a child of G_k, if $e_i \neq$ e where e is the weight of the link incident into an elder brother or elder brother of an ancestor of k.

Rule 3. If G_k is an acyclic t-graph, and its father is an acyclic t-graph:

Let $\{e_1, e_2, ..., e_m\}$ be the weights of the links incident into the younger brothers (1, 2, ..., m) of k. Then $G_{ki} = G_k - e_i$, (i = 1, 2, ..., m), is a child of G_k, if e_i is neutral in G_k.

Rules 1 and 2 have a simple check to avoid duplicate generation of subgraphs. Indeed Rule 2 can be used in place of Rule 3 but Rule 3 has some advantages over Rule 2. Neutral links of G_k need not be obtained by testing for neutrality of every link of G_k while using Rule 3. By Property 3, every neutral link of G_k is a neutral link of its father and this knowledge of the father's neutral links is being used in Rule 3. Futhermore, Rule 3 does not need the check for avoiding duplicate subgraph generation. Since the number of vertices corresponding to Rule 3 is usually overwhelmingly large compared to those for Rule 2, a drastic improvement in speed of computation is achieved using Rule 3.

For a more rigorous presentation of the algorithm, see [12]. In what follows, we illustrate the algorithm by two simple hand-computed examples, one each for directed and undirected graphs.

Illustration for the Algorithm

Example 3: (A directed graph)

Consider the t-graph G_0 of Fig. 1a. Using the algorithm on the graph of Fig. 1a, we generate the rooted directed tree as shown in Fig. 4. Vertices of the tree of Fig. 4 represent subgraphs of G_0, while the root vertex '0' is G_0 itself. The darkened vertices of Fig. 4 represent the acyclic t-subgraphs of G_0.

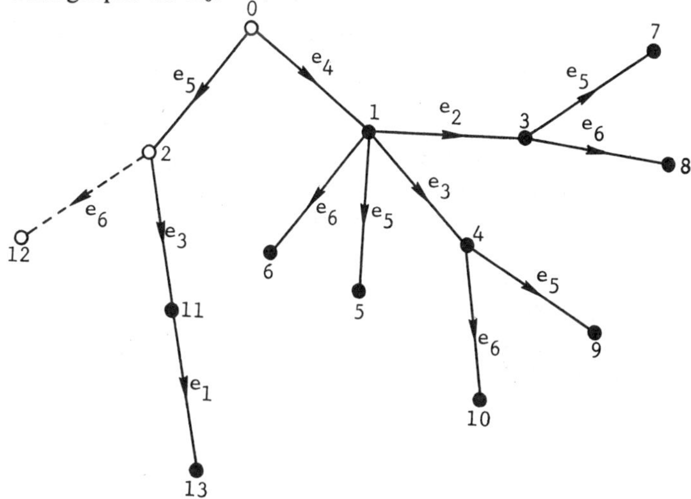

Rooted Directed Tree Obtained by the Algorithm for the Example G_0 of Figure 1a

Fig. 4.

G_0 is cyclic, hence consider a cycle (e_4, e_5). Applying Rule 1, the children of G_0 (viz. G_1, G_2) are obtained by deleting e_4 and e_5 respectively from G_0. The subgraphs G_1 and G_2 correspond one-to-one with vertices 1 and 2 of the rooted tree generated so far (see Fig. 4).

Consider the first child of G_0 (vertex 1). G_1 is an acyclic t-subgraph and since its father is not, we apply Rule 2 to generate children of G_1. Links e_2, e_3, e_5 and e_6 are neutral in G_1 and therefore G_3, G_4, G_5 and G_6 are obtained by deleting links e_2, e_3, e_5 and e_6 respectively from G_1. Now go to the first child of G_1 (vertex 3). Since the father of G_3 also is an acyclic t-graph we apply Rule 3 to obtain children of G_3. Consider the weights e_3, e_5 and e_6 of the links incident into the younger brothers of vertex 3 (namely vertices 4, 5 and 6). Among these, e_5 and e_6 are neutral in G_3 but e_3 is not neutral. Hence G_7 and G_8 are obtained by deleting neutral links e_5 and e_6 respectively from G_3. G_7 and G_8 are trivial graphs (i.e., rooted trees of G_0) and hence consider vertex 4 (younger brother of vertex 3). Applying Rule 3, the children G_9 and G_{10} are obtained by deleting neutral links e_5 and e_6 respectively from G_4. As G_9 and G_{10} are again trivial, we consider vertex 5, the younger brother of vertex 4. However, applying Rule 3, we find that G_5 has no children since e_6 is not neutral in G_5. Furthermore, G_6 also has no children since it has no younger brothers. Therefore we backtrack to vertex 1, the father of vertex 6.

Now consider vertex 2, younger brother of vertex 1. G_2 is cyclic, therefore consider a cycle (e_3, e_4, e_6) in G_2. Applying Rule 1, we obtain G_{11} and G_{12} by respectively deleting e_3 and e_6 from G_2. Notice that deletion of e_4 from G_2 yields a child but e_4 is the weight incident into the elder brother (vertex 1) of vertex 2. If deletion of e_4 were to be attempted, several subgraphs already generated as children of vertex 1 would get duplicated again as children of vertex 2. Here, the first child G_{11} of G_2 is an acyclic t-graph. Therefore, applying Rule 2, the only child G_{13} of G_2 is obtained by deleting neutral link e_1 from G_2. Indeed e_4 also is neutral in G_2 but Rule 2 does not consider e_4 since e_4 is the weight of the link incident into vertex 1 (an elder brother of an ancestor of vertex 11). Since G_{13} is trivial we proceed to vertex 12, the younger brother of the father of vertex 13. G_{12} is not a t-graph and hence we backtrack to vertex 0. Since we reached the root, the algorithm is terminated. The darkened vertices of Fig. 4 represent the acyclic t-subgraphs of the G_0 of Fig. 1a.

Each time a maximal acyclic t-subgraph G_M is obtained, the sign of the term corresponding to G_M is computed by $(-1)^{b_M-n_M+1}$ where b_M and n_M are respectively the links and vertices of G_M. For all other acyclic t-subgraphs, the sign is obtained by simply reversing the sign of its father. For example, in Fig. 4, G_1 is a maximal acyclic t-subgraph and its sign is $(-1)^{5-4+1} = +1$. The sign of the terms corresponding to G_3, G_4, G_5 and G_6 is -1 since their father has $+1$. Table 2 gives the terms of $R_s(G_0)$ along with their sign.

Example 4: (An undirected graph)

Consider the undirected graph G of Fig. 5a. The corresponding symmetric digraph is shown in Fig 5b. Let v_1 be the vertex choosen arbitrarily as the root of the digraph. Notice that in Fig. 5c all links directed into v_1 have been deleted to obtain G_0. All acyclic t-subgraphs of G_0 can be generated using the algorithm on similar lines as in Example 3. The rooted directed tree obtained using the algorithm is shown in Fig. 6. The darkened vertices represent acyclic t-subgraphs of G_0. By (3), (6) and (7), we have

$$\begin{aligned} R_s(G_0) = R_o(G) &= p_1p_2p_3p_4p_5 - p_2p_3p_4p_5 - p_1p_3p_4p_5 \\ &- p_1p_2p_3p_5 - p_1p_2p_3p_4 + p_2p_3p_5 + p_2p_3p_4 + p_1p_3p_5 \\ &+ p_1p_3p_4 + p_1p_2p_3p_4p_5 - p_1p_2p_4p_5 - p_1p_2p_3p_5 \\ &- p_1p_2p_3p_4 + p_1p_2p_5 + p_1p_2p_4 + p_1p_2p_3p_4p_5 \\ &- p_2p_3p_4p_5 - p_1p_3p_4p_5 - p_1p_2p_4p_5 + p_2p_4p_5 \\ &+ p_1p_4p_5 + p_1p_2p_3p_4p_5 - p_1p_2p_4p_5. \end{aligned}$$

where $p_i = \Pr\{e_i\}$ for i = 1, 2, ..., 5.

Features of the algorithm:

The algorithm generates only the non-cancelling terms in the reliability expression $R_s(G_0)$. No other general algorithm is available in the literature for SMT reliability

TABLE 2

Terms of $R_s(G_0)$ for Example 3

Term#	*G_i of G_0*	*Term*	*Sign*
1	G_1	$e_1e_2e_3e_5e_6$	+
2	G_3	$e_1e_3e_5e_6$	−
3	G_4	$e_1e_2e_5e_6$	−
4	G_5	$e_1e_2e_3e_6$	−
5	G_6	$e_1e_2e_3e_5$	−
6	G_7	$e_1e_3e_6$	+
7	G_8	$e_1e_3e_5$	+
8	G_9	$e_1e_2e_6$	+
9	G_{10}	$e_1e_2e_5$	+
10	G_{11}	$e_1e_2e_4e_6$	−
11	G_{13}	$e_2e_4e_6$	+

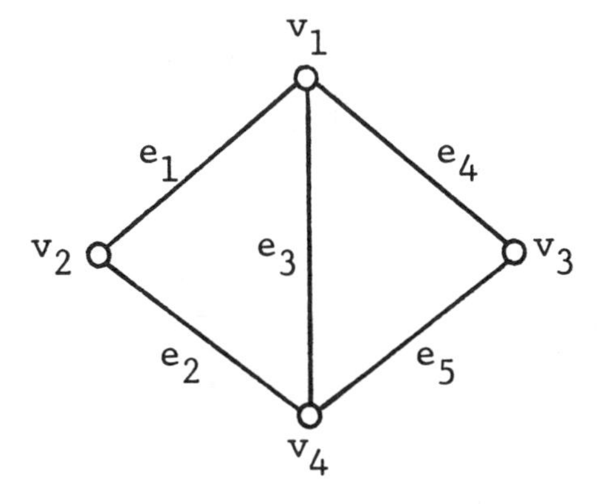

(a) An Undirected Graph G

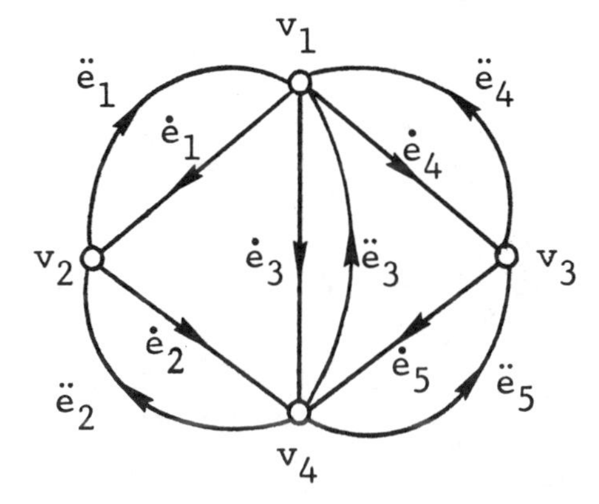

(b) Symmetric Digraph G_d of G

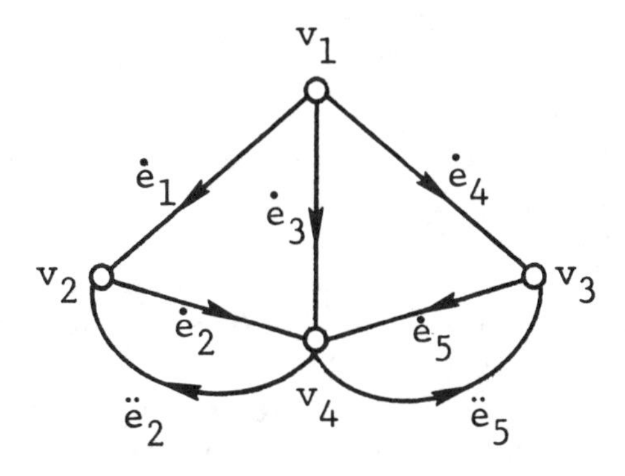

(c) Graph G_o Obtained from G_d with Root at v_1

Fig. 5. Graph G, Symmetric Digraph G_d of G and the G_o of G_d.

which involves only noncancelling terms.

The terms in $R_s(G_0)$ are generated in a tree structure and hence the expression obtained is in factored form. For the Example 3 (see Fig. 4),

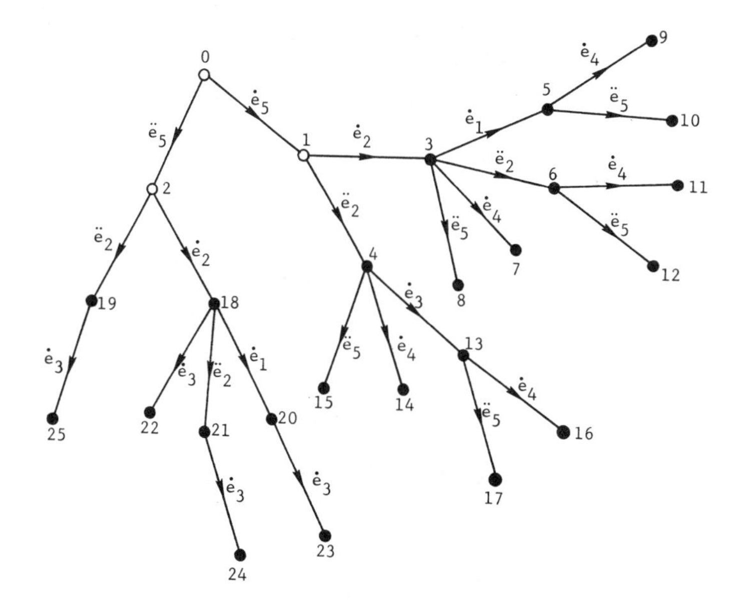

Rooted Directed Tree Obtained by the Algorithm for Example Graph of Figure 5

Fig. 6.

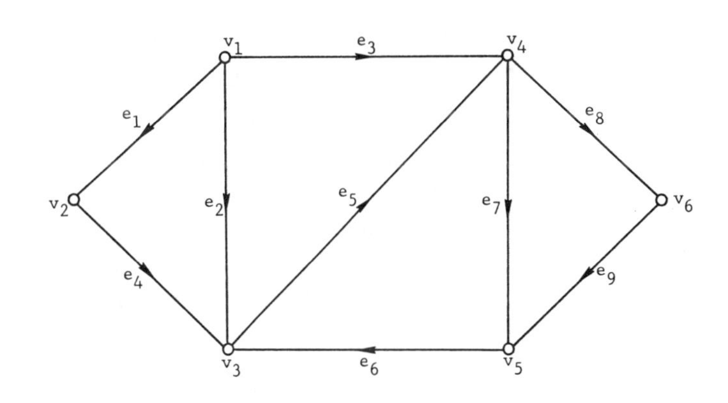

A Network for Example 5

Fig. 7.

$$R_s(G_0) = Y[X_4\{-1 + X_2(-1 + X_5 + X_6) + X_3(-1 + X_5 + X_6) - X_5 - X_6\} + X_5X_3(-1 + X_1)]$$

where $Y = \prod_{i=1}^{6} \Pr\{e_i\}$ and $X_i = 1/\Pr\{e_i\}$.

The symbolic expression once generated, can be used to compute the numerical reliability efficiently by mere substitution of element reliabilities. The algorithm has all other features of TRAP1 and hence is not discussed here any further.

5. COMPUTER IMPLEMENTATION

The algorithm has been implemented in Fortran and the program TRAP4 (*Topological Reliability Analysis*

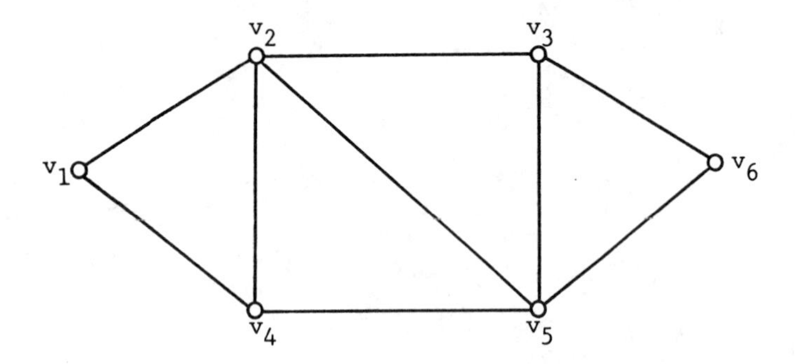

(a) An Undirected Graph G for Example 6

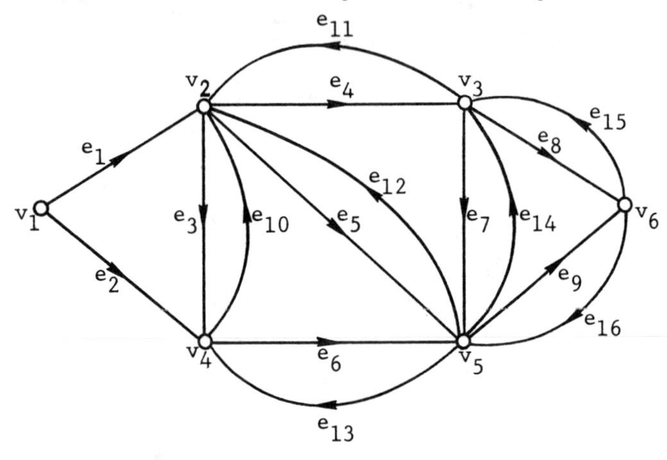

(b) Directed Graph G_o of G with Root Vertex v_1

Fig. 8. Graph G and the Digraph G_o of G.

TABLE 3

```
SAMPLE INPUT & OUTPUT OF PROGRAM TRAP4

 INPUT * * *

 4 6
 1 2
 1 3
 2 3
 4 2
 2 4
 3 4

 OUTPUT * * *

     THE INCIDENCE MATRIX OF GIVEN GRAPH

      -1 -1  0  0  0  0

       1  0 -1  1 -1  0

       0  1  1  0  0 -1

       0  0  0 -1  1  1

     INDIVIDUAL TERMS (ACYCLIC T- SUBGRAPHS)

      TERM #    SIGN        TERM
        1         1       1  3  6
        2         1       1  3  5
        3         1       1  2  6
        4         1       1  2  5
        5        -1       1  3  5  6
        6        -1       1  2  5  6
        7        -1       1  2  3  6
        8        -1       1  2  3  5
        9         1       1  2  3  5  6
       10         1       2  4  6
       11        -1       1  2  4  6

     SYMBOLIC RELIABILITY EXPRESSION OF GIVEN GRAPH
 RGO=Y*(X(4)*(1+X(2)*(-1+X(5)+X(6))+X(3)*(-1+X(5)+X(6))-X(5)-X(6))+
1X(3)*X(5)*(-1+X(1)))

     TOTAL NO. OF TERMS (ACYCLIC T- SUBGRAPHS)=   11
```

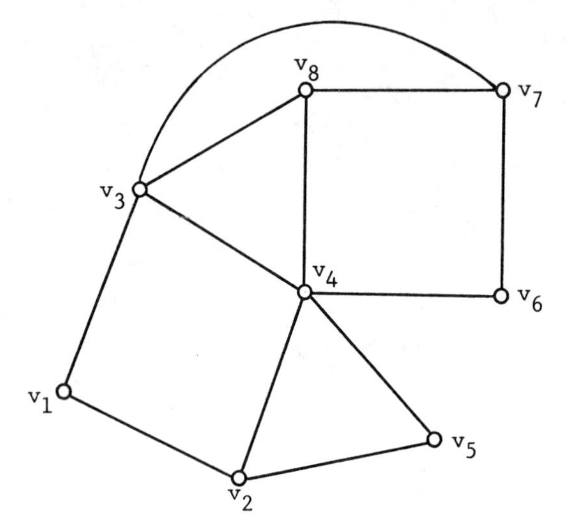

(a) An Undirected Graph G for Example 7

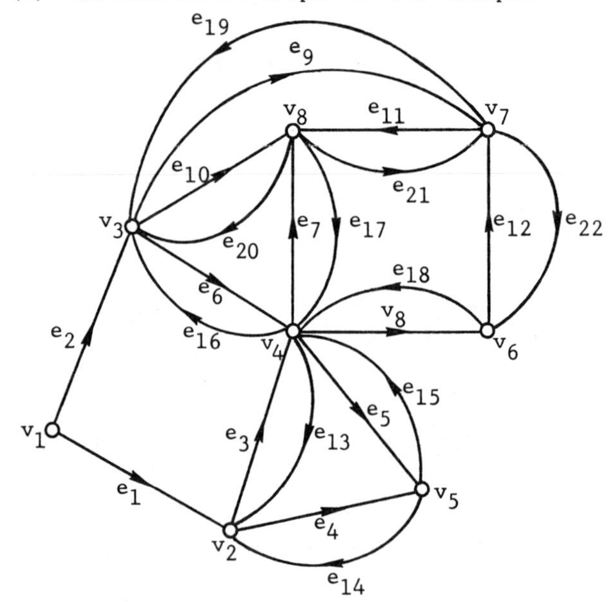

(b) Directed Graph G_o of G with Root Vertex v_1

Fig. 9. Graph G and the Digraph G_o of G.

Program - version 4) tested out on several examples. TRAP4 is similar to TRAP1 [4] but for the following changes. Rule 2 of TRAP1 is not needed for the present case since we are concerned only with spanning subgraphs. Furthermore, Rules 3 & 4 of TRAP1 are associated with neutral sequences, while the neutral link defined here is somewhat different. Therefore TRAP4 incorporates modified subroutines (namely Rules 2 & 3 using Property 1 and Property 3) in place of Rules 3 & 4 of TRAP1. For source listing of TRAP4, see [12].

For the example graph of Fig. 1a, a sample computer input and output of TRAP4 is shown in Table 3. We shall now show some typical results of TRAP4 run on a CDC 6400 computer.

Example 5:

The directed graph of Fig. 7 was analyzed using TRAP4. The symbolic expression for $R_s(G_o)$ is shown in

TABLE 4

COMPUTER SOLUTION FOR EXAMPLE 5

```
input * * *

6 9
1 2
1 3
1 4
2 3
3 4
5 3
4 5
4 6
6 5

output * * *

     the incidence matrix of given graph

      -1 -1 -1  0  0  0  0  0  0

       1  0  0 -1  0  0  0  0  0

       0  1  0  1 -1  1  0  0  0

       0  0  1  0  1  0 -1 -1  0

       0  0  0  0  0 -1  1  0  1

       0  0  0  0  0  0  0  1 -1

     individual terms (acyclic t- subgraphs)
```

term #	sign	term							
1	1	1	4	5	8	9			
2	1	1	4	5	7	8			
3	1	1	3	4	8	9			
4	1	1	3	4	7	8			
5	-1	1	4	5	7	8	9		
6	-1	1	3	4	7	8	9		
7	-1	1	3	4	5	8	9		
8	-1	1	3	4	5	7	8		
9	1	1	2	5	8	9			
10	1	1	2	5	7	8			
11	-1	1	2	5	7	8	9		
12	-1	1	2	4	5	8	9		
13	-1	1	2	4	5	7	8		
14	1	1	2	3	8	9			
15	1	1	2	3	7	8			
16	-1	1	2	3	7	8	9		
17	-1	1	2	3	5	8	9		
18	-1	1	2	3	5	7	8		
19	-1	1	2	3	4	8	9		
20	-1	1	2	3	4	7	8		
21	1	1	3	4	5	7	8	9	
22	1	1	2	4	5	7	8	9	
23	1	1	2	3	5	7	8	9	
24	1	1	2	3	4	7	8	9	
25	1	1	2	3	4	5	8	9	
26	1	1	2	3	4	5	7	8	
27	-1	1	2	3	4	5	7	8	9
28	1	1	3	6	8	9			
29	-1	1	3	4	6	8	9		
30	-1	1	2	3	6	8	9		
31	1	1	2	3	4	6	8	9	
32	1	1	3	6	7	8			
33	-1	1	3	6	7	8	9		
34	-1	1	3	4	6	7	8		
35	-1	1	2	3	6	7	8		
36	1	1	3	4	6	7	8	9	
37	1	1	2	3	6	7	8	9	
38	1	1	2	3	4	6	7	8	
39	-1	1	2	3	4	6	7	8	9

```
     symbolic reliability expression of given graph
 rg0=y*(x(6)*(-1+x(2)*(1+x(3)*(-1+x(7)+x(9))+x(5)*(-1+x(7)+x(9))-x(
17)-x(9))+x(3)*(1+x(4)*(-1+x(7)+x(9))-x(7)-x(9))+x(4)*(1+x(5)*(-1+x
1(7)+x(9))-x(7)-x(9))+x(5)*(1-x(7)-x(9))+x(7)+x(9))+x(7)*x(5)*(1+x(
12)*(-1+x(4))-x(4))+x(5)*(-1+x(2)*(1+x(4)*(-1+x(9))-x(9))+x(4)*(1-x
1(9))+x(9)))

     total no. of terms (acyclic t- subgraphs)=   39
```

Table 4. $R_s(G_0)$ has 39 non-cancelling terms. TRAP4 obtained the individual terms as well as the symbolic expression in 0.052 sec.

Expample 6:

Consider the undirected ARPA network of Fig. 8a. The directed graph G_0 of ARPA graph with v_1 as root vertex is shown in Fig. 8b. TRAP4 obtained the individual terms and the symbolic answer in 0.85 sec. $R_s(G_0)$ or equivalently $R_o(G)$ has 479 non-cancelling terms.

Example 7:

Finally, consider the example undirected graph of Fig. 9a. The G_0 of G with v_1 as root vertex is shown in Fig. 9b. TRAP4 took 8.0 secs. for the solution. G_0 has 3795 acyclic t-subgraphs.

ACKNOWLEDGMENTS

The authors are grateful to Professor Richard E. Barlow for creating an excellent environment for research. This work is supported by the US National Science Foundation under Grant PFR 7822265 and the US Department of Energy under contract EC-77-S-01-5105.

REFERENCES

[1] H. Frank, I.T. Frisch, "Analysis and design of survivable networks," *IEEE Trans. Commun. Technol.*, vol COM-18, 1970 Oct, pp 501-519.

[2] R.S. Wilkov, "Analysis and design of reliable computer networks," *IEEE Trans. Communications*, vol COM-20, No. 3, 1972 June, pp 660-678.

[3] P.M. Lin, B.J. Leon, T.C. Huang, "A new algorithm for symbolic system reliability analysis," *IEEE Trans. Reliability*, vol R-25, 1976 Apr, pp 2-15.

[4] A. Satyanarayana, A. Prabhakar, "New topological formula and rapid algorithm for reliability analysis of complex networks," *IEEE Trans. Reliability*, vol R-27, 1978 Jun, pp 82-100.

[5] R.E. Barlow, A.D. Kiureghian, A. Satyanarayana, "New methodologies for analyzing pipeline and other lifeline networks relative to seismic risk", *Proceedings of ASME at the Century 2 Pressure Vessels & Piping Conference*, San Francisco, California, Aug 12-15, 1980.

[6] O. Wing, P. Demetriou, "Analysis of probabilistic networks," *IEEE Trans. Commun. Technol.*, vol COM-12, 1964 Sep, pp 38-40.

[7] Y. Fu, "Application of topological methods to probabilistic communication networks," *IEEE Trans. Commun. Technol.*, vol COM-13, 1965 Sep, pp 301-307.

[8] M. Ball, R. Van Slyke, "Backtracking algorithms for reliability analysis," *Annals of Discrete Math.*, vol 1, 1977, pp 49-64.

[9] A. Satyanarayana, "Multiterminal network reliability", ORC 80-6, Operations Research Center, University of California, Berkeley, California, 1980 Mar. Available from Operations Research Center, University of California, Berkeley, CA 94720.

[10] K.L. Chung, *Elementary Probability Theory with Stochastic Processes*, Springer-Verlag, New York, 1979.

[11] A. Satyanarayana, J.N. Hagstrom, "Combinatorial properties of directed graphs useful in network reliability," *Networks*, to appear.

[12] A. Satyanarayana, J.N. Hagstron, "A new formula and an algorithm for the reliability analysis of multi-terminal networks," ORC 80-11, Operations Research Center, University of California, Berkeley, California, 1980 Jun. Available from Operations Research Center, University of California, Berkeley, CA 94720.

[13] Supplement: NAPS document No. 03751; 66 pages in this Supplement. For current ordering information, see "Information for Readers & Authors" in a current issue. Order NAPS document No. 03751, 66 pages. ASIS-NAPS; Microfiche Publications; P.O. Box 3513, Grand Central Station; New York, NY 10017 USA.

AUTHORS

A. Satyanarayana; Operation Research Center; University of California; Berkeley, CA 94720 USA.

A. Satyanarayana is on a visiting assignment and is on leave from Indian Telephone Industries Ltd., Bangalore, India. For biography, see vol R-27, 1978 Jun, p 100.

J.N. Hagstrom; Department of Quantitative Methods; University of Illinois at Chicago Circle; Box 4348; Chicago, IL 60680 USA.

Jane Nicholas Hagstrom is an Assistant Professor in the College of Business Administration at the University of Illinois at Chicago Circle. She received her BA in mathematics from Brown University. She received an MS and PhD in operations research from the University of California at Berkeley. She has been interested in topological issues of network reliability.

Manuscript TR80-58 received 1980 May 16; revised 1981 January 16, 1981 February 12.

★ ★ ★

Reprinted from *The Proceedings of the 1981 International Conference on Parallel Processing*, August 1981, pages 79–86. Copyright © 1981 by The Institute of Electrical and Electronics Engineers, Inc.

MULTITERMINAL RELIABILITY ANALYSIS OF DISTRIBUTED PROCESSING SYSTEMS*

Aksenti Grnarov and Mario Gerla
Computer Science Department
University of California
Los Angeles, California 90024

Abstract -- Distributed processing system reliability has been measured in the past in terms of point-to-point terminal reliability, or more recently, in terms of the 'survivability index' or 'team behavior.' While the first approach leads to oversimplified models, the latter approaches imply excessive computational effort. A novel, computationally more attractive measure based on multiterminal reliability is proposed. The measure is the probability of true value of a Boolean expression whose terms denote the existence of connections between subsets of resources. The expression is relatively straightforward to derive, and reflects fairly accurately the survivability of distributed systems with redundant processor, data base and communications resources. Moreover, the probability of such Boolean expression to be true can be computed using a very efficient algorithm. This paper describes the algorithm in some detail, and applies it to the reliability evaluation of a simple distributed file system.

1. Introduction

Distributed processing has become increasingly popular in recent years, mainly because of the advancement in computer network technology and the falling cost of hardware, particularly of microprocessors. Intrinsic advantages of distributed processing include high throughput due to parallel operation, modular growth, fault resilience and load leveling.

In a distributed processing system (DPS), computing facilities and communications subnetwork are interdependent of each other. Therefore, a failure of a particular DPS computer site will have a negative effect on the overall DP system. Similarly, failure of the communication subsystem will lead to overall performance degradation.

Recently, considerable attempts have been made to systematically investigate the survival attributes of distributed processing systems which are subject to failures or losses of processing or communication components. Two main approaches to DPS survivability evaluation have emerged:

a) In [MER 80] the term *survivability index* is used as a performance parameter of a DDP (distributed data processing) system. An objective function is defined to provide a measure of survivability in terms of node and link failure probabilities, data file distribution, and weighting factors for network nodes and computer programs. This objective function allows the comparison of alternative data file distributions and network architectures. Criteria can be included such as the addition or deletion of communication links, allocation of programs to nodes, duplication of data sets, etc. Constraints can be introduced which limit the number and size of files and programs that can be stored at a node. The main disadvantage of the survivability index is its computational complexity, which makes it practical only to DDP systems with, say, less than 20 nodes or links.

b) The second approach is a 'team' approach in which the overall system performance is related to both the operability and the communication connectivity of its 'member' components [HIL 80]. The performance index, defined axiomatically on the connectivity state space of the graph, captures the essentials of the 'team effect' and allows survivability cost/performance trade-offs of alternate network architectures. The basic advantage of the team approach is that performance degradation beyond the connected/disconnected state is measured. One disadvantage of the approach is that of being restricted to the homogeneous case and of ignoring other important details of real DPS's.

In this paper we propose a novel measure of DPS survivability, namely *multiterminal reliability*. We recall that in a communications network terminal reliability relative to node pair (i,j) is the probability that node i is connected to node j. We extend this notion to DPS's by defining the multiterminal reliability as follows:

Definition 1. The multiterminal reliability of a DPS consisting of a set of nodes (processors) $V=1,2,...,N$ is defined as

$$P_s = Prob\ C_{I_1,J_1} \oplus_1 C_{I_2,J_2} \oplus_2 \quad \oplus_{k-1} C_{I_k,J_k} \tag{1}$$

where:

$I_1,J_1,I_2,J_2,...,I_k,J_k$ are subsets of V

C_{I_j,J_j} denotes the existence of connections between all the nodes of the subset I_j and all the nodes of subset J_j

and

$\oplus_j$ has a meaning of OR or AND.

The choice of the subsets $I_1,J_1,...,I_k,J_k$ as well as the interpretation of the operator $\oplus_j$ $(j = 1, \cdots, K-1)$ depend on the event (task) whose survivability is being evaluated. Priority between operators is determined by parentheses in the same way as in standard logical expressions.

---

* This research was supported by the Office of Naval Research under contract N00014-79-C-0866. Aksenti Grnarov is currently on leave from the University of Skopje, Yugoslavia.

As an example, let us assume that the successful completion of a given task requires node A to communicate with node B *or* node C; and node D *and* E to communicate with node F *and* G. The multiterminal reliability of such task is given by

$$P_m = Prob\ (\ C_{I_1,J_1}\ OR\ C_{I_1,J_2}\ )\ AND\ C_{I_3,J_3}$$

where $I_1 = \{A\}$, $J_1 = \{B\}$, $J_2 = \{C\}$, $I_3 = \{D,E\}$ and $J_3 = \{F,G\}$.

The general definition of multiterminal reliability can be specialized to characterize the survivability of the following systems:

(A) Distributed Data Base System: For given link and computer center reliabilities, determine the reliability of a specific file allocation including redundant copies.

(B) Teamwork: Given link and processing node reliability, determine what distribution of the members will result in highest probability of a connection.

(C) Distributed Data Processing System: Given link and processing node reliability and (redundant) distribution of programs and data, determine the probability of successfully completing a specific application.

(D) Computer-Communication Network: Given link and node reliability, determine the probability of the network becoming partitioned.

Note that in all the above applications, system (or application) survivability is best characterized by some multiterminal reliability measure. In fact, terminal reliabilities alone could not be used to compute systems survivability because of the dependencies existing between the various events.

In this paper, an efficient algorithm for multiterminal reliability analysis is presented. The algorithm can be applied to oriented and non-oriented graph models of DPS's and can produce numerical results as well as symbolic reliability expressions.

The paper is organized in five sections. In Section 2, the application of Boolean algebra to multiterminal reliability is considered. Derivation of the algorithm is presented in Section 3. An example for determination of the multiterminal reliability is given in Section 4 . Some comments and concluding remarks are presented in the final section.

2. Boolean Algebra Approach

For reliability analysis a DPS is usually represented by a probabilistic graph $G(V,E)$ where $V = 1,2,...,N$ and $E = a_1 a_2,...,a_E$ are respectively the set of nodes (representing the processing nodes) and the set of directed or undirected arcs representing the communication links. To every DPS component i (processing node or link), a stochastic variable y_i can be associated. The weight assigned to the ith component represents the component reliability

$$p_i = Pr(y_i = 1)$$

i.e., the probability of the existence of the ith component. Variables are supposed to be statistically independent.

There are two basic approaches for computing terminal reliability [FRA 74]. The first approach considers elementary events and the terminal reliability of a connection from source s to termination t, by definition, is given by

$$P_{st} = \sum_{F(e)=1} P_e$$

where P_e is probability which corresponds to the event e and $F(e)=1$ means that the event is favorable, i.e., it includes a path from s to t.

The second approach considers larger events corresponding to the simple paths between terminal nodes. These events however are no longer disjoint and the terminal reliability is given by the probability of the union of the events corresponding to the existence of the paths.

The complexity of these approaches is caused in the first case by the large number of elementary events (of the order 2^n where n = the number of elements which can fail) and in the second case by the difficult computation of the sum of the probabilities of nondisjoint events (the number of joint probabilities to be computed is of the order 2^m where m = the number of paths between node pairs).

Fratta and Montanari [FRA 74] chose to represent the connection between nodes s and t by a Boolean function. This Boolean function is defined in such a way that a value of 0 or 1 is associated with each event according to whether or not it is favorable (i.e., the connection $C_{s,t}$ exists). Since the Boolean function corresponding to the connection $C_{s,t}$ is unique, this means that the connection $C_{s,t}$ can be completely defined by its Boolean function. Representing a connection by its Boolean function, the problem of terminal reliability can be stated as follows: Given a Boolean function F_{ST}, find a minimal covering consisting of nonoverlapping implicants. Once the desired Boolean form is obtained, the arithmetic expression giving the terminal reliability is computed by means of the following correspondences

$$x_i \rightarrow p_i$$

$$\bar{x}_i \rightarrow q_i = 1 - p_i$$

Boolean sum → arithmetic sum
Boolean product → arithmetic product

A drawback of the algorithms based on the manipulation of implicants is the iterative application of certain Boolean operations and the fact that the Boolean function changes at every step (and may be clumsy). The Boolean function may be simplified using one of the following techniques: absorption law, prime implicant form, irredundant form or minimal form. Any one of these procedures however requires a considerable computational effort. Therefore, it can be concluded that these algorithms are applicable only to networks of small size.

Recently, efficient algorithms based on the application of Boolean algebra to terminal reliability computation and symbolic reliability analysis were proposed in [GRN 79] and [GRN 80a] respectively. The algorithms are based on the representation of simple paths by 'cubes' (instead of prime implicants), on the definition of a new operation for manipulating the cubes, and on the interpretation of resulting cubes in such a way that Boolean and arithmetic reduction are combined.

The proposed algorithm for multiterminal reliability analysis is based on the derivation of a Boolean function for multiterminal connectivity and the extension of the algorithm presented in [GRN 80b] to handle both multiterminal reliability computation and symbolic multiterminal reliability analysis.

3. Derivation of the Algorithm

Before presenting the algorithm for multiterminal reliability analysis, it is useful to recall the definition of the path identifier from [GRN 79]:

Definition 2. The path identifier IP_k for the path π_k is defined as a string of n binary variables

$$IP_k = x_1 x_2 ... x_i ... x_n$$

where

$x_i = 1$ if the ith component of the DPS is included in the path π_k

$x_i = x$ otherwise

and n is the number of DPS components that can fail, i.e:

$n = N$ in the case of perfect links and imperfect nodes

$n = E$ in the case of perfect nodes and imperfect links

$n = N+E$ in the case of imperfect links and nodes.

As an example, let us consider a four node, five link DPS given in Figure 1, in which nodes are perfectly reliable and links are subject to failures. The sets of path identifiers for the connections $C_{S,A}$ and $C_{S,T}$ are given in Table 1 and Table 2 respectively.

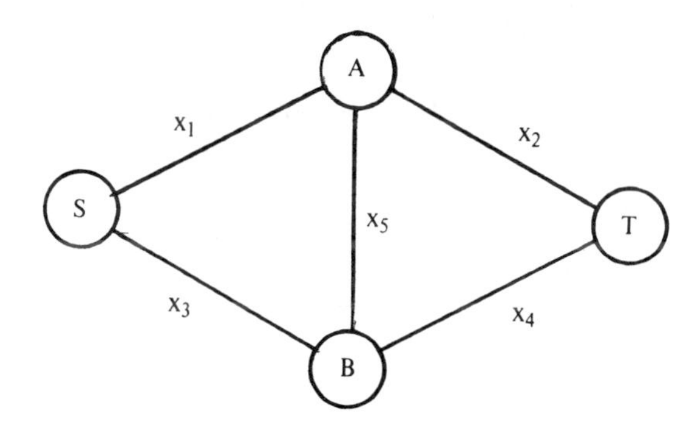

TABLE 1

PATH	IP
S x_1A	1xxxx
S x_3B x_5A	xx1x1
S x_3B x_4T x_2A	x111x

TABLE 2

PATH	IP
S x_1A x_2T	11xxx
S x_1A x_5B x_4T	1xx11
S x_3B x_4T	xx11x
S x_3B x_5A x_2T	x11x1

Figure 1. Example of DPS

Boolean functions corresponding to $C_{S,A}$ and $C_{S,T}$ given by their Karnaugh maps, are shown in Figure 2.

Instead of the cumbersome determination of elementary (or composite) events which correspond to a multiterminal connection, the multiterminal reliability can be determined from the Boolean function representing the connection. Moreover, the corresponding Boolean function can be obtained from path identifiers (Boolean

$F_{s,a}$, $x_5 = 0$

x_3x_4 \ x_1x_2	00	01	11	10
00			1	1
01			1	1
11		1	1	1
10			1	1

$F_{s,a}$, $x_5 = 1$

x_3x_4 \ x_1x_2	00	01	11	10
00			1	1
01			1	1
11	1	1	1	1
10	1	1	1	1

$F_{s,t}$, $x_5 = 0$

x_3x_4 \ x_1x_2	00	01	11	10
00			1	
01			1	
11	1	1	1	1
10			1	

$F_{s,t}$, $x_5 = 1$

x_3x_4 \ x_1x_2	00	01	11	10
00			1	
01			1	1
11	1	1	1	1
10		1	1	

Figure 2. Karnaugh Map Representation of the Connections $C_{S,A}$ and $C_{S,T}$

functions) representing terminal connections. For example, the Boolean function corresponding to the multiterminal connection

$$C_{mor} = C_{S,A} \ OR \ C_{S,T}$$

can be obtained as

$$F_{mor} = F_{S,A} \ U \ F_{S,T}$$

where U is the logical operation union. Karnaugh map of F_{mor} is shown in Figure 3.

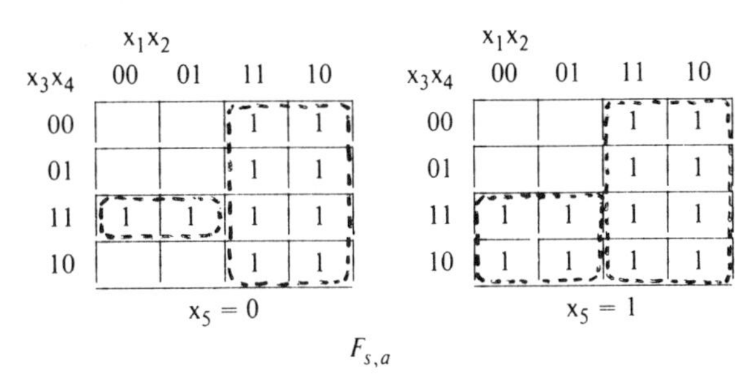

Figure 3. Karnaugh map representation of the connection $C_{mor} = C_{S,A}$ OR $C_{S,T}$

Covering the Karnaugh map with disjoint cubes, we can obtain F_{mor} as

$$F_{mor} = x_1 + \bar{x}_1 x_3 x_5 + \bar{x}_1 x_3 x_4 \bar{x}_5$$

i.e., multiterminal reliability is given by

$$P_{mor} = p_1 + q_1 p_3 p_5 + q_1 p_3 p_4 q_5$$

Analogously, the Boolean function corresponding to the multiterminal connection

$$C_{mand} = C_{S,A} \ AND \ C_{S,T}$$

can be obtained as

$$F_{mand} - F_{S,A} \wedge F_{S,T}$$

where Λ is the logical operation intersection.

According to the Karnaugh map representation (Figure 4), F_{mand} is given by

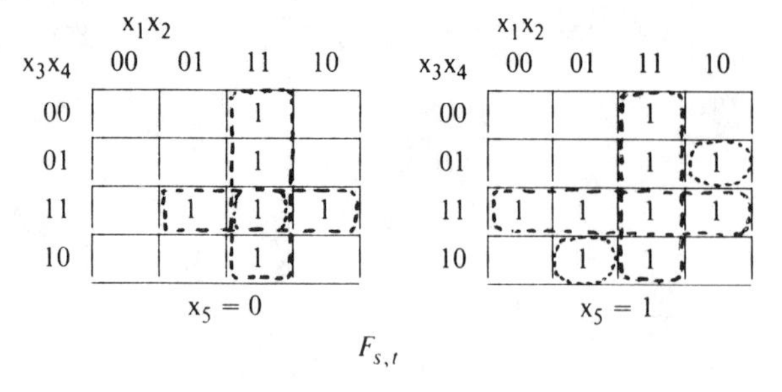

Figure 4. Karnaugh Map Representation of the connection $C_{mand} = C_{s,a}$ AND $C_{s,t}$

the following set of cubes

$$IP = (11xxx, xx111, 1xx11, x11x1, x111x, 1x11x)$$

Applying the algorithm REL [GRN 80b] we obtain that the multiterminal reliability is given by

$$P_{mand} = p_1p_2 + p_3p_4p_5(1-p_1p_2)$$

$$+ p_1p_4p_5q_2q_3 + q_1p_2p_3q_4p_5 + q_1p_2p_3p_4$$

Since the logical operations union and intersection satisfy the commutative and associative laws, previous results can be generalized as follows.

1) Multiterminal connection of OR type $C_{s,T}$ $(T = t_1, t_2, \cdots, t_k)$ is equal to

$$C_{s,T} = C_{s,t_1} \; OR \; C_{s,t_2} \; OR \; \cdots \; OR \; C_{s,t_k}$$

and the corresponding Boolean function $F_{s,T}$ can be obtained as

$$F_{s,T} = F_{s,t_1} \cup F_{s,t_2} \cup \cdots \cup F_{s,t_k}$$

2) Multiterminal connection of AND type $C_{s,T}$ $(T = \{t_1, t_2, \ldots, t_k\})$ is equal to

$$C_{s,T} = C_{s,t_1} \; AND \; C_{s,t_2} \; AND \cdots AND \; C_{s,t_k}$$

and the corresponding Boolean function $F_{s,T}$ can be obtained as

$$F_{s,T} = F_{s,t_1} \Lambda F_{s,t_2} \Lambda \ldots \Lambda F_{s,t_k}$$

In the case when all nodes from the set S have connections of the same type with all nodes from the set T, multiterminal connection can be written as $C_{S,T}$.

4. Determination of $F_{S,T}$

The determination of $F_{S,T}$ by Boolean expression manipulation or by determination of elementary events is a cumbersome and time consuming task. Hence, these methods are limited to DPS's that are very small in size.

However, since path identifiers can be interpreted as cubes, the Boolean function $F_{S,T}$ can be more efficiently obtained by manipulating path identifiers. In the sequel we present the OR-Algorithm and the AND-Algorithm for the determination of $F_{s,T}$ of type OR and AND respectively. Both algorithms are based on the application of the intersection operation [MIL 65]. Since the path identifiers have only symbols x and 1 as components, the intersection operation can be modified as follows:

Definition 3: The intersection operation between two cubes, say $c^r = a_1a_2 \cdots a_i \cdots a_n$ and $c^s = b_1b_2 \cdots b_i \cdots b_n$, is defined as

$$c^r \Lambda c^s = [(a_1 \Lambda b_1), (a_2 \Lambda b_2), \ldots, (a_i \Lambda b_i), \ldots, (a_n \Lambda b_n)]$$

where the coordinate Λ operation is given by

Λ	1	x
1	1	1
x	1	x

It can be seen that the intersection operation between two cubes c^r and c^s produce a cube which is common to both c^r and c^s. If $c^r \Lambda c^s = c^r$ this means that the cube c^r is completely included in the cube c^s. The modified intersection operation produces a cube which has only symbols x and 1 as coordinates, so the modified intersection operation can be applied again and again. Also, the previous fact allows us to apply the REL-Algorithm on the set of cubes obtained by the application of the modified intersection operation.

Let us suppose that the cubes corresponding to connections C_{s_1,T_1} and C_{s_2,T_2} are stored in lists L_1 and L_2 of length k_1 and k_2 respectively. Let c_i^j denote the jth element of the list L_i.

The OR-Algorithm for the computation of $F_{S,T}$ follows:

OR - Algorithm

STEP 1.

for i from 1 to k_1 do
    for j from 1 to k_2 do
        begin
            $c = c_1^i \Lambda c_2^j$; if $c = c_1^i$ then
                begin
                    delete c_1^i from list L_1 ;
                end
            else if $c = c_2^j$ then delete c_2^j from list L_2 ;
        end

STEP 2.
    Store undeleted elements from the lists L_1 and L_2 as new list L_1

END

As an example, the OR - algorithm is applied to the determination of $F_{s,T} = F_{s,a} \cup F_{s,t}$ for the DPS given in Figure 1. The lists L_1 and L_2 are

	L_1		L_2
c_1^1	1xxxx	c_2^1	11xxx
c_1^2	xx1x1	c_2^2	1xx11
c_1^3	x111x	c_2^3	xx11x
		c_2^4	x11x1

STEP 1:

STEP 1: $c_1^1 \wedge c_2^1 = c_2^1$ *delete* c_2^1

$c_1^1 \wedge c_2^2 = c_2^2$ *delete* c_2^2

$c_1^1 \wedge c_2^3 \neq c_1^1 \neq c_2^3$

$c_1^1 \wedge c_2^4 \neq c_1^1 \neq c_2^4$

$c_1^2 \wedge c_2^3 \neq c_1^2 \neq c_2^3$

$c_1^2 \wedge c_2^4 = c_2^4$ *delete* c_2^4

$c_1^3 \wedge c_2^3 = c_1^3$ *delete* c_1^3

STEP 2:

	L_1
c_1^1	1xxxx
c_1^2	xx1x1
c_1^3	xx11x

It can be seen that the OR - Algorithm produces a list with minimal number of elements which are cubes of the largest possible size. The same result could have been obtained from the identification of disjoint cubes directly in Fig. 3. Our method allows for the efficient generation of all disjoint cubes necessary for reliability analysis [GRN 79]. Next, we introduce the AND algorithm.

AND - Algorithm

```
STEP 1.
    for i from 1 to k1 do
        begin
            for j from 1 to k2 do

                c^j_{i+2} = c^i_1 Λ c^j_2 ;
                for k from 1 to k2-1 do
                    begin
                        m = k+1
                        while   c^k_{i+2} ≠ c^k_{i+2} Λ c^m_{i+2}   and
                        m ≤ k2 do
                            begin
                                c = c^k_{i+2} Λ c^m_{i+2} ;
                                if c = c^m_2 then delete c^m_i
                                from list L_i
                                m = m+1
                            end
                        if m ≤ k2 then delete c^k_{i+2} from
                        list L_{i+2}
                    end
            end
    STEP 2
        Store undeleted elements from lists L3, . . . , L_{k1+2}
        as a new list L1

END
```

As an example, the AND-Algorithm is applied to the determination of $F_{s,T} = F_{s,a} \wedge F_{s,t}$ for the DPS in Fig. 1.

STEP 1.

i = 1

Step 1.1

$$L_3 = \begin{cases} c_3^1 = c_1^1 \wedge c_2^1 = 11xxx \\ c_3^2 = c_1^1 \wedge c_2^2 = 1xx11 \\ c_3^3 = c_1^1 \wedge c_2^3 = 1x11x \\ c_3^4 = c_1^1 \wedge c_2^4 = 111x1 \end{cases}$$

Step 1.2

$c_3^1 \wedge c_3^2 \neq c_3^1 \neq c_3^2$

$c_3^1 \wedge c_3^3 \neq c_3^1 \neq c_3^3$

$c_3^1 \wedge c_3^4 = c_3^4$ *delete* c_3^4

$$L_3 = \begin{cases} 11xxx \\ 1xx11 \\ 1x11x \end{cases}$$

i = 2

Step 1.1

$$L_4 = \begin{cases} c_4^1 = 111x1 \\ c_4^2 = 1x111 \\ c_4^3 = xx111 \\ c_4^4 = x11x1 \end{cases}$$

Step 1.2

$$L_4 = \begin{cases} xx111 \\ x11x1 \end{cases}$$

i = 3

Step 1.1

$$L_5 = \begin{cases} 1111x \\ 11111 \\ x111x \\ x1111 \end{cases}$$

Step 1.2

$L_5 = x111x$

STEP 2.

	L_1
c_1^1	11xxx

c_1^2 1xx11
c_1^3 1x11x
c_1^4 xx111
c_1^5 x11x1
c_1^6 x111x

It can be seen that the AND - Algorithm also produces a list with minimal number of elements which are cubes of the largest possible size.

In the general case, the Boolean function corresponding to the connection $C_{S,T}$ where $S=\{s_1, s_2, ..., s_k\}$ and $T=\{t_1, t_2, ..., t_m\}$, can be obtained using the multiterminal Algorithm (m ⊕-Algorithm) described below:

m ⊕ - Algorithm

STEP 1:

Find the path identifiers for terminal connection s_1, t_1 and store them in the list L_1 ; $i \leftarrow 1$.

STEP 2:

Sort the path identifier in L_1 according to increasing number of symbols 1 (i.e. increasing path length);

STEP 3:

if $i \leqslant k$ continue. Otherwise go to step 5

STEP 4:

for $j = j_1, ..., m$ ($j_1 = 2$ if $i = 1$, otherwise $j_1 = 1$)

Step 4.1
Find the path identifiers for terminal connection s_i, t_j and store them in the list L_2

Step 4.2
Sort the path identifiers in L_2 according to the increasing number of symbols "1"

Step 4.3
Perform ⊕ -Algorithm on the lists L_1 and L_2

Step 4.4
$i \leqslant i+1$; go to step 2.

END

In the algorithm, ⊕ denotes OR or AND depending on the connection type. The sorting of the lists allows faster execution of the algorithm (starting with the largest cubes results in earlier deletion of covered cubes, i.e., faster reduction of the lists during the execution of Step 4.3).

Based on the previous results we can propose the following algorithm for multiterminal reliability analysis:

MUREL - Algorithm

STEP 1:

Derive the multiterminal connection expression corresponding to the event which has to be analyzed.

STEP 2:

Determine the Boolean function corresponding to the multiterminal connection by repetitive application of the m ⊕ - Algorithm.

STEP 3:

Apply the REL -Algorithm to obtain the multiterminal reliability expression or value.

Regarding the computational complexity of the MUREL-Algorithm, the following observations can be made:

i) The ⊕ -Algorithm can be implemented using only logical operations which generally belong to the class of the fastest instructions in a computer system.

ii) The m ⊕ -Algorithm produces a minimal set of maximal cubes (i.e., minimal irredundant form of the Boolean function).

iii) The REL-Algorithm is the fastest algorithm for the determination of the reliability expression or for the reliability computation from the set of cubes (path identifiers).

From the above considerations we conclude that the proposed algorithm can be applied to DPS of significantly larger size than was possible with other existing techniques.

In the following section, the algorithm is illustrated with an application to a small distributed system.

5. Example of Application of the Algorithm

As an example of the application of the algorithm we compute the survivability index for the simple DPS system shown in Figure 5 (the example is taken from [MER 80]). Assignment of files and programs to nodes is shown in figure 5.

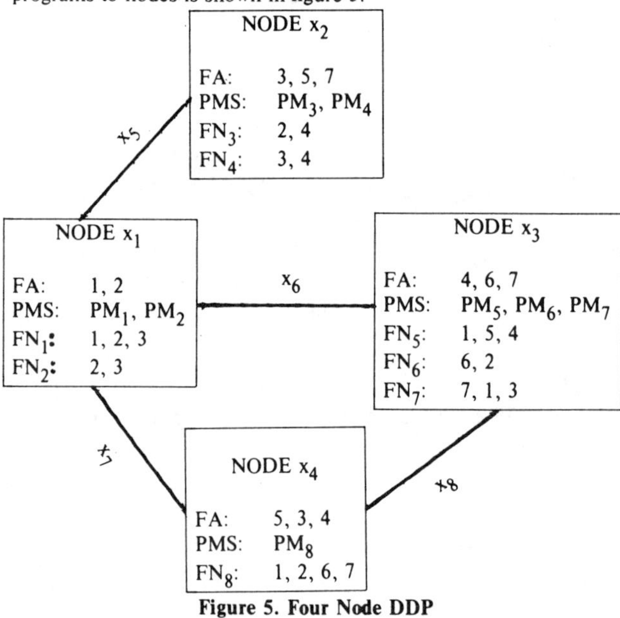

Figure 5. Four Node DDP

FA denotes the set of files available at a given node, FN_i denotes the files needed to execute program PM_i and PMS designates the set of programs to be executed at that node.

Let us assume that for a given application, we are interested in the survivability of program PM_3. Likewise, for another application, we need both programs PM_3 *and* PM_8 to be operational. We separately analyze these two cases using as a measure for survivability the multiterminal reliability (probability of program execution). The two problems can be stated as follows:

Given: node and link reliability, and file and program assignments to nodes.

Find: The survivability of:

1) Program PM_3

2) Both programs PM_3 and PM_8

Survivability of PM_3

The survivability PM_3 is equal to the multiterminal reliability of connection

$$C_{m3} = C_{2,I_1}\ OR\ C_{2,I_2}$$

where $I_1 = \{1,3\}$ and $I_2 = \{1,4\}$ The connections C_{2,I_1} and C_{2,I_2} are equal to

$$C_{2,I_1} = C_{2,1}\ AND\ C_{2,3}$$

$$C_{2,I_2} = C_{2,1}\ AND\ C_{2,4}$$

Paths and corresponding path identifiers for the connections $C_{2,1}$, $C_{2,3}$ and $C_{2,4}$ are shown in Figure 6.

$C_{2,1}$

paths	$F_{2,1}$
$x_1x_5x_2$	11xx1xxx

$C_{2,3}$

paths	$F_{2,3}$
$x_1x_5x_2x_6x_3$	111x11xx
$x_1x_5x_2x_7x_4x_8$	11x11x11

$C_{2,4}$

paths	$F_{2,4}$
$x_1x_5x_2x_7x_4$	11x11x1x
$x_1x_5x_2x_6x_3x_8x_4$	111111x1

Figure 6. Path and Path Identifiers Representing Connections $C_{2,1}$, $C_{2,3}$, and $C_{2,4}$

Applying the AND - Algorithm on $F_{2,1}$ and $F_{2,3}$, and $F_{2,1}$ and $F_{2,4}$ we obtain

F_{2,I_1}	F_{2,I_2}
111x11xx	11x11x1x
11x11x11	111111x1

Applying the OR - algorithm on F_{2,I_1} and F_{2,I_2} we obtain

F_{m3}

111x11xx
11x11x1x

Applying the REL - Algorithm on F_{m3} we obtain

$$P_{m3} = p_1p_2p_3p_4p_5p_6 + p_1p_2p_4p_5p_7(1 - p_3p_6)$$

Assuming $p_i = .95\ \forall\ i$, we have: $P_{m3} = .85$

5.2. Survivability of both PM_3 and PM_8

The survivability of PM_8 is equal to the multiterminal reliability of connection

$$C_{m8} = C_{4,I_3}$$

where $I_3 = \{1,3\}$. The connection C_{4,I_3} is equal to

$$C_{4,I_3} = C_{4,1}\ AND\ C_{4,3}$$

Paths and corresponding path identifiers for the connections $C_{4,1}$ and $C_{4,3}$ are shown in Figure 7.

$C_{4,1}$

paths	$F_{4,1}$
$x_4x_7x_1$	1xx1xx1x
$x_4x_8x_3x_6x_1$	1x11x1x1

$C_{4,3}$

paths	$F_{4,3}$
$x_4x_8x_3$	xx11xxx1
$x_4x_7x_1x_6x_3$	1x11x11x

Figure 7. Paths and Path identifiers for Connections $C_{4,1}$ and $C_{4,3}$

Applying the AND - Algorithm on $F_{4,1}$ and $F_{4,3}$ we obtain

F_{m8}

1x11xx11
1x11x11x
1x11x1x1

Applying the AND - Algorithm on F_{m3} and F_{m8} we obtain

F_m

1111111x
111111x1
11111x11

Applying the REL - Algorithm on F_m we obtain

$$P_m = p_1p_2p_3p_4p_5p_6p_7 + p_1p_2p_3p_4p_5p_6q_7p_8 + p_1p_2p_3p_4p_5q_6p_7p_8$$

Assuming p_i=0.95 ∀i, we have: $p_m = 0.778$

6. Conclusion

In the paper, the multiterminal reliability is introduced as a measure of DPS survivability and the MUREL-Algorithm for multiterminal reliability analysis of DPS is proposed. First, the event under study is expressed in terms of its multiterminal connection. Then the m $\oplus$ -Algorithm is used to translate the multiterminal connection into a Boolean function involving all the relevant system components. Finally, the multiterminal reliability is obtained from the Boolean function by application of the REL-Algorithm.

Preliminary computational complexity considerations show that the MUREL-Algorithm permits the survivability analysis of DPS of considerably larger size than using currently available techniques.

References

[GRN 79] A. Grnarov, L. Kleinrock, M. Gerla, "A New Algorithm for Network Reliability Computation", *Computer Networking Symposium,* Gaithersburg, Maryland, December 1979.

[GRN 80a] A. Grnarov, L. Kleinrock, M. Gerla, "A New Algorithm for Symbolic Reliability Analysis of Computer Communication Networks", *Pacific Telecommunications Conference,* Honolulu, Hawaii, January 1980.

[GRN 80b] A. Grnarov, L. Kleinrock, M. Gerla, "A New Algorithm for Reliability Analysis of Computer Communication Networks", *UCLA Computer Science Quarterly,* Spring 1980.

[HIL 80] G. Hilborn, "Measures for Distributed Processing Network Survivability, *Proceedings of the 1980 National Computer Conference,* May 1980.

[MER 80] R. E. Merwin, M. Mirhakak, "Derivation and use of a survivability criterion for DDP systems", *Proceedings of the 1980 National Computer Conference,* May 1980.

[MIL 65] R. Miller, Switching Theory, Volume I: Combinational Circuits, New York, Wiley, 1965.

A LINEAR-TIME ALGORITHM FOR COMPUTING K-TERMINAL RELIABILITY IN SERIES-PARALLEL NETWORKS*

A. SATYANARAYANA† AND R. KEVIN WOOD‡

Abstract. Let $G=(V, E)$ be a graph whose edges may fail with known probabilities and let $K \subseteq V$ be specified. The K-terminal reliability of G, denoted $R(G_K)$, is the probability that all vertices in K are connected. Computing $R(G_K)$ is, in general, NP-hard. For some series-parallel graphs, $R(G_K)$ can be computed in polynomial time by repeated application of well-known reliability-preserving reductions. However, for other series-parallel graphs, depending on the configuration of K, $R(G_K)$ cannot be computed in this way. Only exponential-time algorithms as used on general graphs were known for computing $R(G_K)$ for these "irreducible" series-parallel graphs. We prove that $R(G_K)$ is computable in polynomial time in the irreducible case, too. A new set of reliability-preserving "polygon-to-chain" reductions of general applicability is introduced which decreases the size of a graph, and conditions are given for a graph admitting such reductions. Combining all types of reductions, an $O(|E|)$ algorithm is presented for computing the reliability of any series-parallel graph irrespective of the vertices in K.

Key words. algorithms, complexity, network reliability, series-parallel graphs, reliability-preserving reductions

1. Introduction. Analysis of network reliability is of major importance in computer, communication and power networks. Even the simplest models lead to computational problems which are NP-hard for general networks [5], although polynomial-time algorithms do exist for certain network configurations such as "ladders" and "wheels" and for some series-parallel structures such as the well-known "two-terminal" series-parallel networks. In this paper, we show that a class of series-parallel networks, for which only exponentially complex algorithms were previously known [7], [8], can be analyzed in polynomial time. In doing this, we introduce a new set of reliability-preserving graph reduction of general applicability and produce a linear-time algorithm for computing the reliability of any graph with an underlying series-parallel structure.

The network model used in this paper is an undirected graph $G=(V, E)$ whose edges may fail independently of each other, with known probabilities. The reliability analysis problem is to determine the probability that a specified set of vertices $K \subseteq V$ remains connected, i.e., the K-terminal reliability of G. Computing K-terminal reliability was first shown to be NP-hard by Rosenthal [12], and it follows from Valiant [17] that the problem is #P-complete even when G is planar. Two special cases of this reliability problem are the most frequently encountered, the terminal-pair problem where $|K|=2$, and the all-terminal problem where $K=V$. These problems are also #P-complete [11], in general, although their complexities are unknown when G is planar.

In network reliability analysis, three reliability-preserving graph reductions are well-known: the series reduction, the degree-2 reduction (an extension of the series reduction for problems with $|K|>2$) and the parallel reduction. From the realiability viewpoint, we classify series-parallel graphs into two types, those which are reducible to a single edge using standard series, parallel and degree-2 reductions, and those

* Received by the editors February 23, 1982, and in final revised form June 15, 1984. This work was conducted at the Operations Research Center, University of California, Berkeley, California 94720, under Air Force Office of Scientific Research grant AFOSR-81-0122 and under U.S. Army Research Office contract DAAG29-81-K-0160.

† Department of Computer Science, Stevens Institute of Technology, Hoboken, New Jersey 07030.

‡ Department of Operations Research, Naval Postgraduate School, Monterey, California 93943.

which are not. The former type is "reducible" and the latter "irreducible." For example, the series-parallel graph of Fig. 1a is reducible if $K=\{v_1, v_2\}$, but irreducible for $K=\{v_1, v_6\}$. Thus, the reducibility of a series-parallel graph, for the purpose of reliability evaluation, depends on the nature of the vertices included in K. A more detailed exposition of this concept appears in § 2.

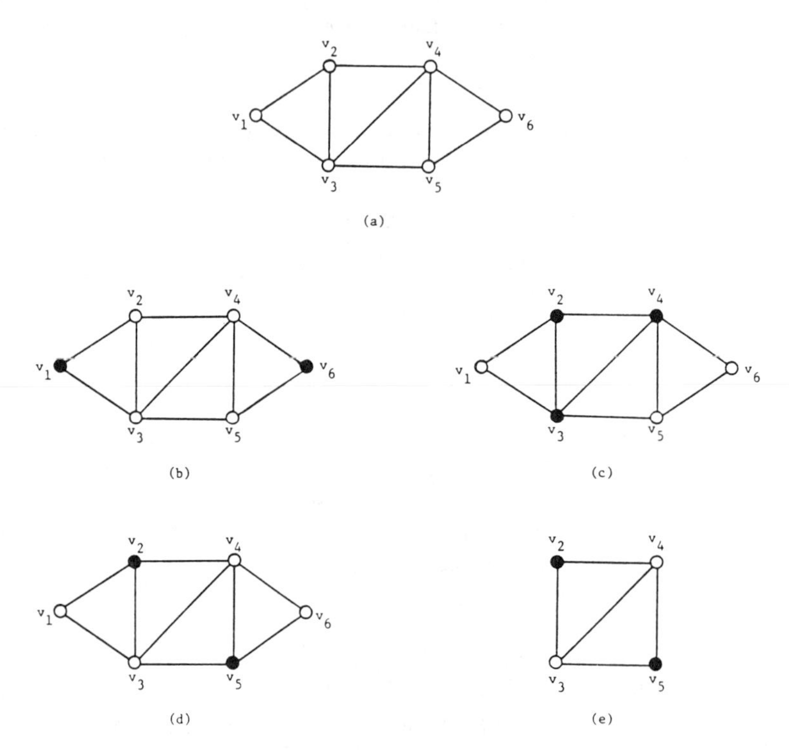

FIG. 1. *Reducible and irreducible series-parallel graphs. Note: Darkened vertices represent K-vertices.*

The K-terminal reliability of a reducible series-parallel graph can be computed in polynomial time. Several methods exist for the solution of the terminal-pair problem for such a graph, i.e., for a two-terminal series-parallel network [9], [15], and for $|K|>2$, direct extensions of the methods can be used. However, it has been believed that computing the reliability of irreducible series-parallel graphs is as hard as the general problem. (The use of series-parallel reductions with multi-state edges [13] is applicable to this problem although this has not been recognized. We do not follow this tack because of the simplicity and generality obtained by maintaining binary-state edges.) The purpose of this paper is threefold: (1) to introduce a new set of reliability-preserving graphs reductions called "polygon-to-chain reductions," (2) to show that by using these reductions, irreducible series-parallel graphs become reducible, and (3) to give a linear-time algorithm for computing the reliability of any graph with a series-parallel structure.

In a graph, a chain is an alternating sequence of vertices and edges, starting and ending with vertices such that end vertices have degree greater than 2 and all internal vertices have degree 2. Two chains with the same end vertices constitute a polygon. In § 3, we show that a polygon can be replaced by a chain and that this transformation will yield a reliability-preserving reduction. We discuss the relationship between irreducible series-parallel graphs and polygons in § 4. Using the polygon-to-chain reductions in conjunction with the three simple reductions mentioned earlier, a polynomial-time procedure is then outlined which will compute the reliability of any series-parallel graph. This procedure is very simple but not of linear-time complexity, so in § 5 we develop algorithm which is shown to operate in $O(|E|)$ time. This algorithm will compute the K-terminal reliability of any graph having an underlying series-parallel structure. Finally, in § 6, we briefly discuss an extension to the algorithm to reduce a nonseries-parallel graph as far as possible so that the algorithm could be used as a subroutine in a reliability analysis program for general networks.

2. Preliminaries. Consider a graph $G=(V, E)$ in which all vertices are perfectly reliable but any edge e_i may fail with probability q_i or work with probability $p_i=1-q_i$. All edge failures are assumed to occur independently of each other. Let K be a specified subset of V with $|K|\geqq 2$. When certain vertices of G are specified to be in K, we denote the graph G together with the set K by G_K. We will refer to the vertices of G belonging to K as the *K-vertices* of G_k. The *K-terminal reliability* of G, denoted by $R(G_K)$, is the probability that the K-vertices in G_K are connected. K-terminal reliability is a generalization of the common reliability measures, all-terminal reliability and terminal-pair reliability where $K = V$ and $|K|=2$, respectively.

Reliability of a separable graph. A *cutvertex* of a graph is a vertex whose removal disconnects the graph. A *nonseparable graph* is a connected graph with no cutvertices. A *block* of a graph is a maximal nonseparable subgraph.

Let $G=(V, E)$ be a separable graph and $v\in V$ be any cutvertex in G. G can be partitioned into two connected subgraphs $G^{(1)}=(V_1, E_1)$ and $G^{(2)}=(V_2, E_2)$ such that $V_1\cup V_2=V$, $V_1\cap V_2=v$, $E_1\cup E_2=E$ and $E_1\cap E_2=\emptyset$. Also, $E_1\neq\emptyset$ and $E_2\neq\emptyset$. Denote $K_1=K\cap V_1$ and $K_2=K\cap V_2$. If one of the K_i is null, say $K_1=\emptyset$, then $G^{(1)}$ is *irrelevant* and $R(G_K)=R(G^{(2)}_{K_2})$. Otherwise, assuming $K_1\neq\emptyset$ and $K_2\neq\emptyset$, it is well known that $R(G_K)=R(G^{(1)}_{K_1\cup v})R(G^{(2)}_{K_2\cup v})$. ($R(G_K)\equiv 1$ if $|K|=1$. Therefore, if $K_i=\{v\}$ then $R(G^{(i)}_{K_i\cup v})\equiv 1$ and the above statement is still true.) Thus the reliability of a separable graph can be computed by evaluating the reliabilities of its blocks separately. For this reason, we henceforth consider only nonseparable graphs.

Simple reductions. In order to reduce the size of graph G_K, i.e. reduce $|V|+|E|$, and therefore reduce the complexity of computing $R(G_K)$, *reliability-preserving reductions* are often applied: Certain edges and/or vertices in G are replaced to obtain G'; new edge reliabilities are defined; a new set K' is defined; and a multiplicative factor Ω is defined; all such that $R(G_K)=\Omega R(G'_{K'})$. The following three reliability-preserving reductions are well known and are called *simple reductions.*

A *parallel reduction* replaces a pair of edges $e_a=(u, v)$ and $e_b=(u, v)$ with a single edge $e_c=(u, v)$ and defines $p_c=1-q_aq_b$, $K'=K$, and $\Omega=1$.

Suppose $e_a=(u, v)$ and $e_b=(v, w)$ such that $u\neq w$, $\deg(v)=2$, and $v\notin K$. A *series reduction* replaces e_a and e_b with a single edge $e_c=(u, w)$, and defines $p_c=p_ap_b$, $K'=K$ and $\Omega=1$.

Suppose $e_a=(u, v)$ and $e_b=(v, w)$ such that $u\neq w$, $\deg(v)=2$, and $\{u, v, w\}\subseteq K$. A *degree-2 reduction* replaces e_a and e_b with a single edge $e_c=(u, w)$ and defines $p_c=p_ap_b/(1-q_aq_b)$, $K'=K-v$, and $\Omega=1-q_aq_b$.

Series-parallel graphs. The following definition should not be confused with the definition of a "two-terminal" series parallel network in which two vertices must remain fixed. No special vertices are distinguished here. In a graph, edges with the same end vertices are *parallel edges.* Two nonparallel edges are *adjacent* if they are incident on a common vertex. Two adjacent edges are *series edges* if their common vertex is of degree 2. Replacing a pair of series (parallel) edges by a single edge is called a series (parallel) *replacement.* A series-parallel graph is a graph that can be reduced to a tree by successive series and parallel replacements. Clearly, if a series-parallel graph is nonseparable, then the resulting tree, after making all series and parallel replacements, contains exactly one edge.

We wish to clarify the subtle difference between the term "replacement" used here and the term "reduction" used with respect to simple reductions. Replacement is a strictly graph-theoretic term indicating some edges or vertices from G are removed and then replaced by other edges or vertices to create a new graph G'. A reduction is defined, on the other hand, with respect to G, K, and edge reliabilities. A reduction includes the act of replacing edges or vertices in G to create G' along with defining edge reliabilities, K', and Ω, all such that $R(G_K) = \Omega R(G'_{K'})$, i.e. reliability is preserved. For example, in graph G as shown in Fig. 1a, series replacements are possible while no (reliability-preserving) simple reductions are possible in the corresponding G_K for $K = \{v_1, v_6\}$ (Fig. 1b). Motivated by the difference between graphs which allow replacements but, with K and edge reliabilities defined, do not allow reliability-preserving simple reductions, we distinguish between graphs which can and cannot be reduced by simple reductions.

Reducible and irreducible series-parallel graphs. Clearly, if G has no series or parallel edges, then for any K, G_K admits no simple reductions. If G is a series-parallel graph, then a simple reduction might or might not exist in G_K depending upon the vertices of G that are chosen to be in K. For example, consider the series-parallel graph G of Fig. 1a. The graph G_K, for $K = \{v_2, v_3, v_4\}$ as in Fig. 1c, can be reduced to a single edge by successive, simple reductions. On the other hand, for $K = \{v_1, v_6\}$, G_K admits no simple reductions (Fig. 1b). A series-parallel graph G_K is *reducible* if it can be reduced to a single edge by successive, simple reductions. If G_K is reduced to a single edge e_i using m reductions, then $R(G_K) = p_i \prod_{k=1}^{m} \Omega_k$ where Ω_k is the multiplicative factor defined by the kth reduction. Note that any series-parallel graph G is reducible for the all-terminal problem since any degree-2 vertex in G_V allows a degree-2 reduction.

It is possible for a (nonseparable) series-parallel graph to admit one or more simple reductions for a specified K and still not be completely reducible to a single edge. As an illustration, consider G_K of Fig. 1d. Two series reductions may be applied to this graph to obtain the graph of Fig. 1e, but no further simple reductions are possible. A graph G_K is an *irreducible* series-parallel graph if G_K cannot be completely reduced to a single edge using simple reductions.

Chains and polygons. In a graph, a *chain* χ is an alternating sequence of distinct vertices and edges, v_1, (v_1, v_2), v_2, (v_2, v_3), $v_3, \cdots, v_{k-1}$, (v_{k-1}, v_k), v_k, such that the internal vertices, $v_2, v_3, \cdots, v_{k-1}$, are all of degree 2 and the end vertices, v_1 and v_k, are of degree greater than 2. A chain need not contain any internal vertices, but it must contain at least one edge and two end vertices. The length of a chain is simply the number of edges it contains. A *subchain* is a connected subset of a chain beginning and ending with a vertex and containing at least one edge. Both the end vertices of a subchain may be of degree 2. The notation χ will also be used for a subchain with the usage differentiated by context.

Suppose χ_1 and χ_2 are two chains of lengths l_1 and l_2, respectively. If the two chains have common end vertices u and v, then $\Delta = \chi_1 \cup \chi_2$ is a *polygon* of length $l_1 + l_2$. In other words, a polygon is a cycle with the property that exactly two vertices of the cycle are of degree greater than 2. While this definition allows two parallel edges to constitute a polygon, we will initially require a polygon to be of length at least 3.

3. Polygon-to-chain reductions. In this section a new set of reliability-preserving reductions will be introduced which replace a polygon with a chain and always reduces $|V|+|E|$ by at least 1. Consider a graph G_K which does not admit any simple reductions but does contain some polygon Δ. In general, no such polygon need exist, but, if it does exist, then the number of possible configurations is limited.

Property 1. Let G_K be a graph which admits no simple reductions. If G_K contains a polygon, then it is one of the seven types given in the first column of Table 1.

Proof. This follows from the facts that (i) every degree-2 vertex of G_K is a K-vertex, (ii) there can be no more than two K-vertices in a chain, and (iii) the length of any chain in G_K is at most 3.

Polygon-to-chain transformations. Let Δ_j be a type j polygon in G_K, a graph which admits no simple reductions. Let u and v be the vertices in Δ_j such that $\deg(u)>2$ and $\deg(v)>2$. Then, $\Delta_j=\chi_j'\cup\chi_j''$, where χ_j' and χ_j'' are chains in G_K with common end vertices u and v. Replacing the pair χ_j' and χ_j'' by the corresponding chain χ_j, as in Table 1, is called a *polygon-to-chain transformation.*

In Theorem 1 we will prove that a polygon-to-chain transformation can be used to produce a reliability-preserving, polygon-to-chain reduction. It is useful here, however, to make the distinction between a polygon-to-chain reduction and a polygon-to-chain transformation, in the same manner that simple reductions and replacements are differentiated. A transformation is only a topological mapping of a graph G to a graph G' and ignores all considerations of reliability including K-vertices. A reduction includes the topological transformation as well as all reliability calculations and changes in K-vertices.

The proof technique of Theorem 1 requires that we first discuss the use of conditional probabilities for computing the reliability of a graph in a general context. Let $e_i=(u, v)$ be some edge of G_K and let F_i denote the event that e_i is working and $\bar{F}_i$ denote the complementary event that e_i has failed. Using rules of conditional probability, the reliability of G_K can be written as

$$R(G_K)=p_iR(G_K\,|\,F_i)+q_iR(G_K\,|\,\bar{F}_i)=p_iR(G'_{K'})+q_iR(G''_{K''}) \tag{1}$$

where

$$G'=(V-u-v+w,\ E-e_i),\qquad w=u\cup v,$$

$$K'=\begin{cases}K & \text{if } u, v\notin K,\\ K-u-v+w & \text{if } u\in K \text{ or } v\in K\end{cases}$$

and

$$G''=(V,\ E-e_i),$$

$$K''=K.$$

F_i and $\bar{F}_i$ are said to "induce" $G'_{K'}$ and $G''_{K''}$ from G_K, respectively. ("Induce" is not used in the standard graph-theoretic sense here.) $G'_{K'}$ is G_K with edge e_i contracted, and $G''_{K''}$ is G_K with edge e_i deleted.

Equation (1) can be applied recursively on the induced graphs and simple reductions made where applicable within the recursion. After repeated applications of the formula, the induced graphs are either reduced to single edges for which the reliability is simply the probability that the edge works, or some K-vertices become disconnected, in which case the reliability of the induced graph is zero. In this way, the reliability of any general graph may be computed. This method of computing the reliablity of a graph is known as "factoring" [10], [14] and is a special case of pivotal decomposition

TABLE 1
Polygon-to-chain reductions

Note: Darkened vertices represent K-vertices

Polygon Type	Chain Type	Reduction Formulas	New Edge Reliabilities
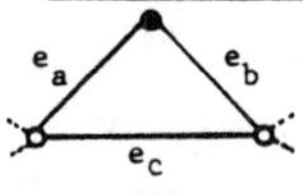 e_a, e_b, e_c (1)	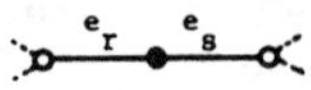 e_r, e_s	$\alpha = q_a p_b q_c$ $\beta = p_a q_b q_c$ $\delta = p_a p_b p_c\left(1 + \frac{q_a}{p_a} + \frac{q_b}{p_b} + \frac{q_c}{p_c}\right)$	$p_r = \frac{\delta}{\alpha + \delta}$
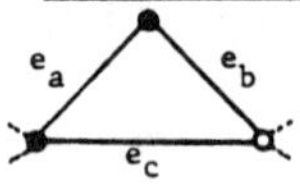 e_a, e_b, e_c (2)	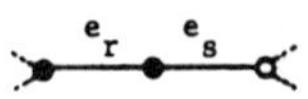 e_r, e_s	$\alpha = q_a p_b q_c$ $\beta = p_a q_b q_c$ $\delta = p_a p_b p_c\left(1 + \frac{q_a}{p_a} + \frac{q_b}{p_b} + \frac{q_c}{p_c}\right)$	$p_s = \frac{\delta}{\beta + \delta}$ $\Omega = \frac{(\alpha + \delta)(\beta + \delta)}{\delta}$
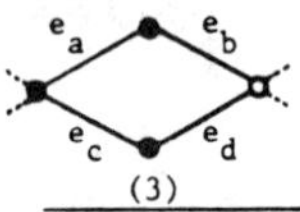 e_a, e_b, e_c, e_d (3)	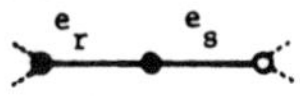 e_r, e_s	$\alpha = p_a q_b q_c p_d + q_a p_b p_c q_d + q_a p_b q_c p_d$ $\beta = p_a q_b p_c q_d$ $\delta = p_a p_b p_c p_d\left(1 + \frac{q_a}{p_a} + \frac{q_b}{p_b} + \frac{q_c}{p_c} + \frac{q_d}{p_d}\right)$	
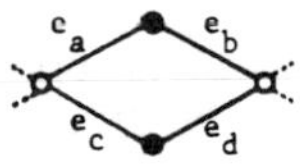 e_a, e_b, e_c, e_d (4)	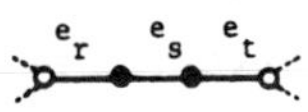 e_r, e_s, e_t	$\alpha = q_a p_b q_c p_d$ $\beta = p_a q_b q_c p_d + q_a p_b p_c q_d$ $\delta = p_a q_b p_c q_d$ $\gamma = p_a p_b p_c p_d\left(1 + \frac{q_a}{p_a} + \frac{q_b}{p_b} + \frac{q_c}{p_c} + \frac{q_d}{p_d}\right)$	
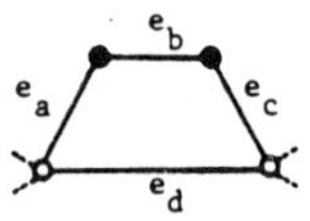 e_a, e_b, e_c, e_d (5)	$\lvert K\rvert > 2$ 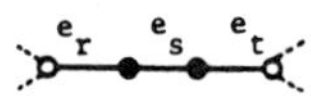 e_r, e_s, e_t See note	$\alpha = q_a p_b p_c q_d$ $\beta = p_a q_b p_c q_d$ $\delta = p_a p_b q_c q_d$ $\gamma = p_a p_b p_c p_d\left(1 + \frac{q_a}{p_a} + \frac{q_b}{p_b} + \frac{q_c}{p_c} + \frac{q_d}{p_d}\right)$	$p_r = \frac{\gamma}{\alpha + \gamma}$ $p_s = \frac{\gamma}{\beta + \gamma}$
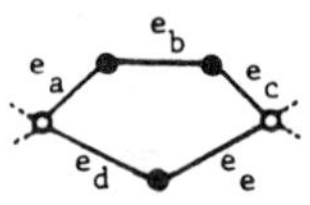 e_a, e_b, e_c, e_d, e_e (6)	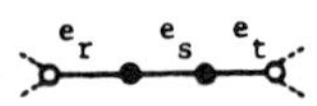 e_r, e_s, e_t	$\alpha = q_a p_b p_c q_d p_e$ $\beta = p_a q_b p_c (p_d q_e + q_d p_e)$ $+ p_b (q_a p_c p_d q_e + p_a q_c q_d p_e)$ $\delta = p_a p_b q_c p_d q_e$ $\gamma = p_a p_b p_c p_d p_e\left(1 + \frac{q_a}{p_a} + \frac{q_b}{p_b} + \frac{q_c}{p_c} + \frac{q_d}{p_d} + \frac{q_e}{p_e}\right)$	$p_t = \frac{\gamma}{\delta + \gamma}$ $\Omega = \frac{(\alpha + \gamma)(\beta + \gamma)(\delta + \gamma)}{\gamma^2}$
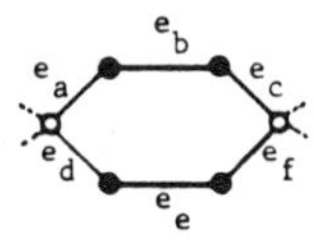 e_a, e_b, e_c, e_d, e_e, e_f (7)	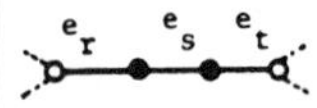 e_r, e_s, e_t	$\alpha = q_a p_b p_c q_d p_e p_f$ $\beta = p_a q_b p_c (q_d p_e p_f + p_d q_e p_f + p_d p_e q_f)$ $+ p_a p_b q_c p_f (p_d q_e + q_d p_e)$ $+ q_a p_b p_c p_d (q_e p_f + p_e q_f)$ $\delta = p_a p_b q_c p_d p_e q_f$ $\gamma = p_a p_b p_c p_d p_e p_f\left(1 + \frac{q_a}{p_a} + \frac{q_b}{p_b} + \frac{q_c}{p_c} + \frac{q_d}{p_d} + \frac{q_e}{p_e} + \frac{q_f}{p_f}\right)$	Note: For $\lvert K\rvert = 2$, new chain is 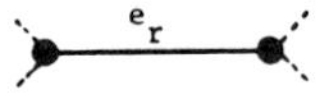 e_r $p_r = (p_b + p_a q_b p_c p_d)/\Omega$ $\Omega = p_b + p_a q_b p_c$

of a general binary coherent system [1]. For our purposes, factoring will only be applied to the edges of a single polygon or a chain.

Polygon-to-chain reductions.

THEOREM 1. *Suppose G_K contains a type j polygon. Let $G'_{K'}$ denote the graph obtained from G_K by replacing the polygon Δ_j with the chain χ_j having appropriately defined edge probabilities, and let Ω_j be the corresponding multiplication factor, all as in Table 1. Then, $R(G_K) = \Omega_j R(G'_{K'})$.*

We prove the exactness of reduction 7 only, since reductions 1-6 may be shown in a similar fashion. Figs. 2 and 3 illustrate the proof of the theorem. To improve readability in the proof, we drop the subscript "7" on α, β, δ, γ, and Ω even though, strictly speaking, these are functions of the type of reduction.

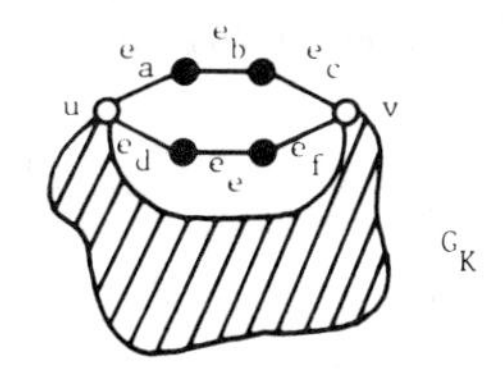

(a) *Schematic of a graph with a type 7 polygon.*

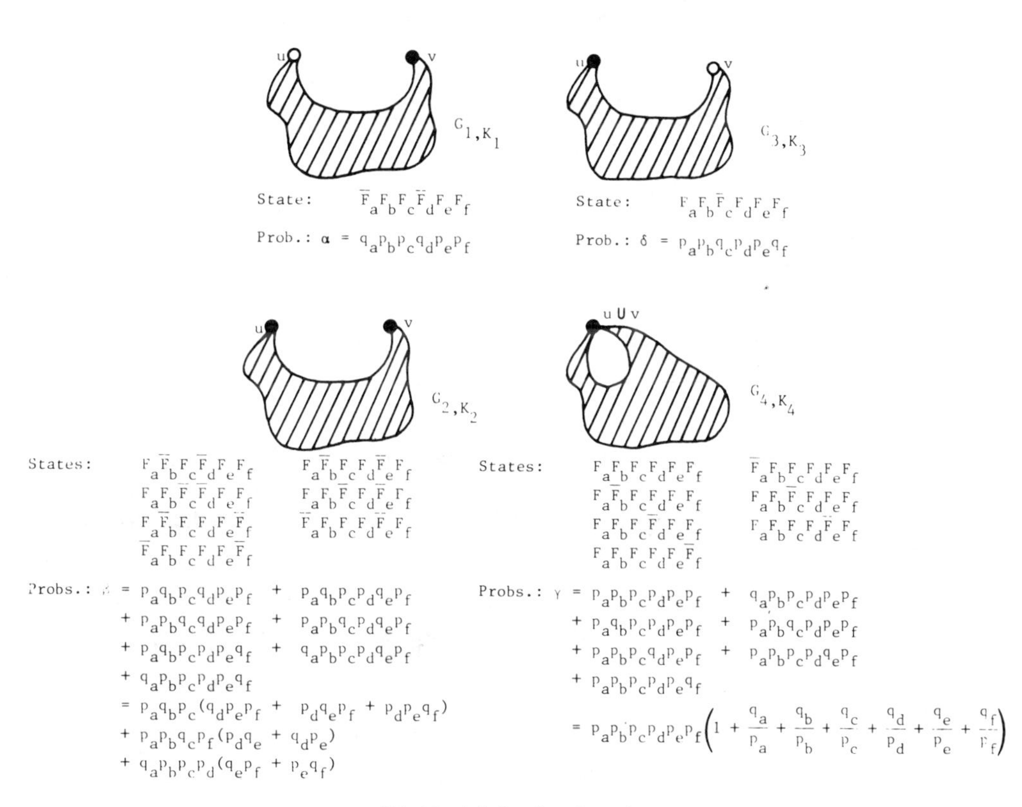

(b) *Nonfailed induced graphs.*

FIG. 2

Proof of Theorem 1. Let F_i be the event that edge e_i in the polygon is working and let $\bar{F}_i$ be the event that edge e_i has failed. $\underline{F}$ denotes a compound event or state such as $F_a F_b \bar{F}_c F_d \bar{F}_e F_f$, and $\mathbf{F}$ denotes the set of all 2^6 such states. Also, $z_i = 1$ if F_i occurs and $z_i = 0$ if $\bar{F}_i$ occurs. By conditional probability and extension of (1),

$$R(G_K) = \sum_{\underline{F} \in \mathbf{F}} p_a^{z_a} q_a^{1-z_a} \cdots p_f^{z_f} q_f^{1-z_f} R(G_K \mid \underline{F}). \tag{2}$$

Only sixteen of the possible sixty-four states are nonfailed states where $R(G_K \mid \underline{F}) \neq 0$. Each nonfailed state will induce a new graph with a corresponding set

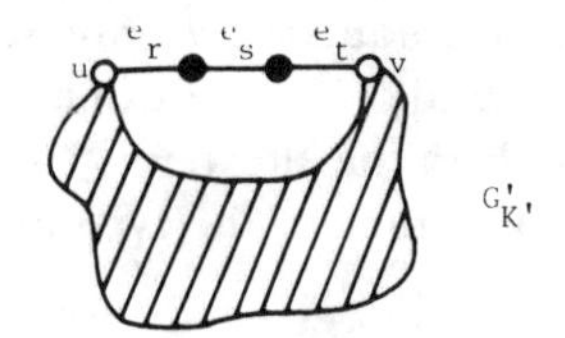

(a) *Graph of Fig.* 2 *with polygon replaced by chain.*

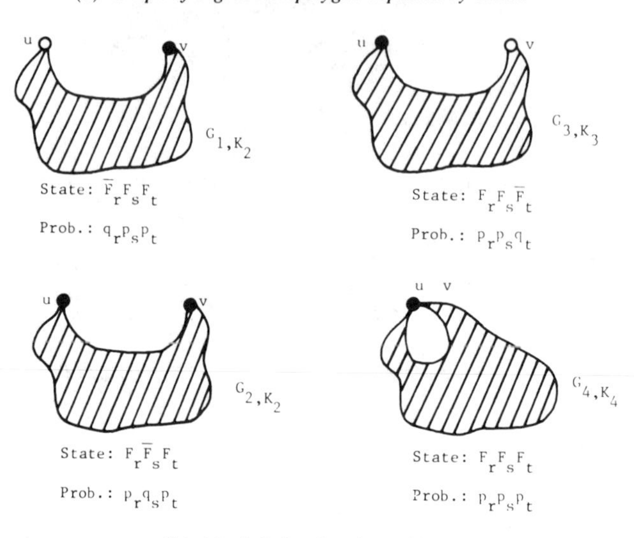

(b) *Nonfailed induced graphs*

FIG. 3

of K-vertices of which there only four different possibilities. Figure 2 gives these four graphs G_{i,K_i}, $i=1,2,3,4$, the states under which the graphs are induced, and the summed state probabilities in each case, α, β, δ, and γ. Thus, by grouping and eliminating terms, (2) is reduced to

$$R(G_K)=\alpha R(G_{1,K_1})+\beta R(G_{2,K_2})+\delta R(G_{3,K_3})+\gamma R(G_{4,K_4}). \tag{3}$$

Now $G'_{K'}$ is obtained from G_K by replacing the polygon with a chain u, e_r, v_1, e_s, v_2, e_t, w, and redefining K as shown in Fig. 3. Using conditional probabilities again,

$$\begin{aligned} R(G'_{K'})&=p_rq_sp_tR(G'_{K'}|(F_r\bar{F}_sF_t))+q_rp_sp_tR(G'_{K'}|(\bar{F}_rF_sF_t)) \\ &\quad+p_tp_sq_tR(G'_{K'}|(F_rF_s\bar{F}_t))+p_rp_sq_tR(G'_{K'}|(F_rF_sF_t)) \end{aligned} \tag{4}$$

where only the nonfailed states have been written.

The four nonfailed states of $G'_{K'}$ induce the same four graphs which the nonfailed states of G_K induce. Multiplying (4) by a factor Ω, we thus have

(5)

$$\Omega R(G'_{K'})=\Omega p_rq_sp_tR(G_{1,K_1})+\Omega q_rp_sp_tR(G_{2,K_2})+\Omega p_rp_sq_tR(G_{3,K_3})+\Omega p_rp_sp_tR(G_{4,K_4}).$$

Equating, term by term, the coefficients in (3) and (5) gives

$$\alpha=\Omega q_rp_sp_t=\Omega(1-p_r)p_sp_t, \qquad \delta=\Omega p_rp_sq_t=\Omega p_rp_s(1-p_t),$$

$$\beta=\Omega p_rq_sp_t=\Omega p_r(1-p_s)p_t, \qquad \gamma=\Omega p_rp_sp_t.$$

These four equations in the four unknowns Ω, p_r, p_s, and p_t may be easily solved to obtain

$$p_r = \frac{\gamma}{\alpha+\gamma}, \qquad p_s = \frac{\gamma}{\beta+\gamma},$$

$$p_t = \frac{\gamma}{\delta+\gamma}, \qquad \Omega = \frac{(\alpha+\gamma)(\beta+\gamma)(\delta+\gamma)}{\gamma^2},$$

which are the values given in Table 1 for a type 7 polygon. The reader may verify that when these values are substituted into (4), we obtain

$$\Omega R(G'_{K'}) = \alpha R(G_{1,K_1}) + \beta R(G_{2,K_2}) + \delta R(G_{3,K_3}) + \gamma R(G_{4,K_4}) = R(G_K). \qquad \square$$

It can be seen from Table 1 that polygon-to-chain reductions, like simple reductions, always reduce $|V|+|E|$ by at least 1.

Theorem 1 can be extended to give a result which can be useful for computing the reliability of a general graph. In a nonseparable graph, a *separating pair* is a pair of vertices whose deletion disconnects the graph. For example, vertices u and v in Fig. 2 are a separating pair. Using the same conditioning arguments as in the proof of Theorem 1, it can be shown that any subgraph between a separating pair can be replaced by a chain of 1, 2, or 3 edges to yield a reliability-preserving reduction. For two special cases, it has been shown that a subgraph between a separating pair can be replaced by a single edge [6]. The first case occurs when the subgraph including the separating pair has no K-vertices, and the second case occurs when the separating pair belongs to K. The fact that a chain can always be used to replace any subgraph, irrespective of the K-vertices, greatly increases the generality of any algorithm which uses this reduction.

4. Properties of series-parallel graphs. In this section we set down some properties of series-parallel graphs with respect to topology and reliability. We prove that a series-parallel graph must admit a polygon-to-chain reduction if all simple reductions have first been performed. Thus, every series-parallel graph is reducible irrespective of the vertices in K. Using this fact, we then outline a simple polynomial-time procedure for computing the reliability of such graphs.

The following property is a simple extension of the definition of a series-parallel graph.

Property 2. Let G' be the graph obtained from G by applying one or more of the following operations:

a series replacement;
a parallel replacement;
an inverse series replacement (replace an edge by two edges in series);
an inverse parallel replacement (replace an edge by two edges in parallel).

Then, G' is a series-parallel graph if and only if G is series-parallel.

Proof of Property 2 may be found in [3]. The next two properties show that the series-parallel structure of a graph is not altered by simple or polygon-to-chain reductions.

Property 3. Let G' be the graph obtained by a polygon-to-chain transformation on G. Then G' is a series-parallel graph if and only if G is series-parallel.

Proof. G' may be obtained from G by one or more series replacements, a parallel replacement, and one or more inverse series replacements, in that order. Thus, this property follows directly from Property 2. $\square$

Property 4. Let $G'_{K'}$ be the graph obtained from G_K by applying a simple reduction or a polygon-to-chain reduction on G_K. Then, G' is a series-parallel graph if and only if G is series-parallel.

Proof. A series or degree-2 reduction implements a series replacement, a parallel reduction implements a parallel replacement, and a polygon-to-chain reduction implements a polygon-to-chain transformation on G. Hence, by Properties 2 and 3, G' is a series-parallel graph if and only if G is a series-parallel. □

By next proving that every series-parallel graph G_K admits a simple reduction or a polygon-to-chain reduction, it will be possible to show that $R(G_K)$ can be computed in polynomial time for such graphs.

Property 5. Let G_K be a series-parallel graph. Then, G_K must admit either a simple reduction or one of the seven types of polygon-to-chain reductions given in Table 1.

Proof. If G_K admits a simple reduction, then we are done. If G_K has no simple reductions, then by Property 1, any polygon of G_K must be one of the seven types given in Table 1. Hence, we need only show that G contains a polygon. Let G' be the graph obtained by replacing all chains in G with single edges. If G' contains a pair of parallel edges, then the two chains in G corresponding to this pair of edges constitute a polygon. We argue that G' must contain a pair of parallel edges. If G' has no parallel edges, no simple reductions are possible in G' since all vertices in G' have degree greater than 2. Thus, G' and hence G are not series-parallel graphs, which is a contradiction. □

One simple procedure for computing $R(G_K)$ can now be outlined as follows: (1) Make all simple reductions; (2) find a polygon and make the corresponding reduction; and (3) repeat steps 1 and 2 until G_K is reduced to a single edge. If G_K is originally series-parallel, then Properties 4 and 5 guarantee that the above procedure eventually reduces G_K to a single edge. The reliability is calculated by initializing $M \leftarrow 1$, letting $M \leftarrow M\Omega_j$ whenever a polygon-to-chain reduction of type j is made, and letting $M \leftarrow M\Omega$, for $\Omega = 1 - q_a q_b$, whenever a degree-2 reduction is made on some edges e_a and e_b. At the end of the algorithm with a single remaining edge e_i, the reliability of the original graph is given by $R(G_K) = Mp_i$.

The total number of parallel and polygon-to-chain reductions executed by this procedure, before the graph is reduced to a single edge, is exactly $|E| - |V| + 1$. This is because the number of fundamental cycles in a connected graph is $|E| - |V| + 1$, and a parallel or polygon-to-chain reduction deletes exactly one such cycle [2]. The complexity of steps (1) and (2) above can be linear in the size of G, and thus, the running time of the whole procedure is at best quadratic in the size of G. In order to develop a linear-time algorithm, we have found it necessary to move the parallel reduction from the domain of simple reductions to the domain of polygon-to-chain reductions. Indeed, a parallel reduction is a trivial case of a polygon-to-chain reduction with a multiplier $\Omega = 1$. We will henceforth consider two parallel edges to be the type 8 polygon and the parallel reduction to be the type 8 polygon-to-chain reduction.

5. An $O(|E|)$ algorithm for computing the reliablity of any series-parallel graph. The objective here is to develop an efficient, linear-time algorithm for computing the reliability of any series-parallel graph. All results needed to present this algorithm have been established; however, some additional notation and definitions must be given.

If u and v are the end vertices of a chain χ, then u and v are said to be *chain-adjacent*. When it is necessary to distinguish these vertices, we will use the notation $\chi(u, v)$. A subchain with end vertices u and v will also be denoted $\chi(u, v)$ but in this case u and

v cannot be said to be chain-adjacent. The algorithm is presented next, followed by a proof of its validity and linear complexity. The algorithm reduces G_K to two edges in parallel and prints $R(G_K)$ if G is initially series-parallel (We stop at two edges in parallel instead of a single edge because these edges do not form a polygon by our definition; their end vertices do not have degrees greater than 2.), or prints a message that G is not series-parallel. Comments are enclosed in square brackets.

ALGORITHM.

Input: A nonseparable graph G with vertex set V, $|V| \geqq 2$, edge set E, $|E| \geqq 2$, and set $K \subseteq V$, $|K| \geqq 2$. Edge probabilities p_i for each edge $e_i \in E$.

Output: $R(G_K)$ if G is series-parallel or a message that G is not series-parallel.

Begin
$M \leftarrow 1$.
Perform all series reductions.
Perform all degree-2 reductions letting $M \leftarrow M\Omega$ for each such reduction.
Construct list, $T \leftarrow \{v \mid v \in V \text{ and } \deg(v) > 2\}$ marking all such v "onlist."
Mark all $v \notin T$ "offlist."
While $T \neq \emptyset$ and $|E| > 2$ **do**
  Begin
  Remove v from T.
  $i \leftarrow 1$. [Index of the next chain out of v to be searched]
  Until $i > 3$ or v is deleted or $\deg(v) = 2$ **do**
    Begin
    Search the ith chain out of v.
    $i \leftarrow i + 1$.
    If a polygon $\Delta(v, w)$ is found **then do**
      Begin
      Apply the appropriate type j polygon-to-chain reduction to $\Delta(v, w)$ to obtain $\chi(v, w)$, and let $M \leftarrow M\Omega_j$.
      $i \leftarrow i - 1$.
      If $\deg(v) = 2$ or $\deg(w) = 2$ **then do**
        Begin
        Apply all possible series and degree-2 reductions on the chain (or cycle) containing subchain $\chi(v, w)$ to obtain completely reduced chain $\chi(x, y)$ (or parallel edges (x, y) and (x, y)), letting $M \leftarrow M\Omega$ for each degree-2 reduction.
        If $y \neq v$ and y is "offlist" **then** mark y "onlist" and add y to T.
        If $x \neq v$ and x is "offlist" **then** mark x "onlist" and add x to T.
        End
      End
    End
  End
If $|E| = 2$ **then** print ("$R(G_K)$ is" $M(1 - q_a q_b)$) [for $E = \{e_a, e_b\}$]
  else print ("G is not series-parallel").
End.

The key to the algorithm is the way in which the "until" loop operates. This loop says: "Sequentially search chains incident to v reducing any polygons which are found and making any subsequent series and degree-2 reductions until either (a) v is shown to be chain-adjacent to three distinct vertices, or (b) v is completely deleted from G through the reductions, or (c) v becomes a degree-2 vertex through the reductions.

No chain is ever searched more than once each time this loop is entered. The correctness of the algorithm is not hard to show. Arguments similar to those presented here may be found in [16] where the problem is the recognition of two-terminal series-parallel directed graphs.

Suppose firstly that G consists of a single cycle. The initial series and degree-2 reductions will reduce G_K to two edges in parallel, T will be empty, and the algorithm therefore gives $R(G_K)$ correctly at the final step of the algorithm. Next, suppose that G does not consist of a single cycle, in which case T will not be empty and an initial search for a polygon will begin. Since all initial series and degree-2 reductions were performed, by Property 5, any polygon found must be one of the eight specified types. If a polygon is found and reduced, the resulting chain may, in fact, be a subchain. If this happens, some new series and degree-2 reductions may be admitted on the chain (or cycle) containing that subchain but nowhere else. All such reductions are made when applicable. Thus, every time the "until" loop of the algorithm is entered or iterated, the graph admits no series or degree-2 reductions, and only polygons of the eight given types can exist.

Vertices are continually removed from the stack T and replaced, at most two at a time, only when polygon-to-chain reductions are made. At most $|E|-|V|$ polygon-to-chain reductions can ever be made since each polygon-to-chain reduction removes exactly one of the $|E|-|V|+1$ fundamental cycles of G and the final reduced graph must retain at least one fundamental cycle. Therefore, at most $|V|+2(|E|-|V|)=2|E|-|V|$ vertices can ever pass through T before T becomes empty and the "while" loop must terminate. If $|E|=2$ at that point, then $R(G_K)$ is correctly given at the last step of the algorithm since only reliabilty-preserving series, degree-2, and polygon-to-chain reductions are ever performed. Property 4 proves that the original graph must have been series-parallel.

If $|E|>2$ when T becomes empty, then we must show that the reduced graph is not series-parallel and that the original graph was not series parallel. In this case, every vertex v with $\deg(v)>2$ is chain-adjacent to at least three distinct vertices. This is true since (i) every vertex v with $\deg(v)>2$ is initially put in the list T and its chain-adjacent vertices checked in the "until" loop and (ii) whenever the chain-adjacency of a vertex or vertices is altered (this can occur to at most two vertices at a time) after a polygon-to-chain reduction, then this vertex or vertices are returned to the list T if not already there. The following property proves that a graph with the given chain-adjacency structure is not series-parallel.

Property 6. Let G be a nonseparable graph such that all vertices v with $\deg(v)>2$ are chain-adjacent to at least three distinct vertices. Then, G is not a series-parallel graph.

Proof. Let G' be the graph obtained from G by first replacing all chains with single edges in a sequence of series replacements and then removing any parallel edges in a sequence of parallel replacements. By Property 2, G is a series-parallel if and only if G' is a series-parallel. Now, every vertex $v \in V'$ has $\deg(v)>2$ and there are no parallel edges in E'. Thus, G' admits no series or parallel replacements and cannot be series-parallel. Therefore, G cannot be series-parallel. □

This proves that if the algorithm terminates with $|E|>2$, the reduced graph is not series-parallel, and Property 4 proves that the original graph could not have been series-parallel either. This establishes the validity of the algorithm. We now turn our attention to its computational complexity.

In order to show that the algorithm is linear in the size of G, we use a multi-linked adjacency list to represent G. In this representation, for each vertex a doubly-linked list of adjacent vertices corresponding to incident edges is kept together with the associated edge probabilities. Every edge is represented twice since we are dealing with an undirected graph, and additional links are kept between both representations of each edge. Such an adjacency list can be initialized in $O(|V|+|E|)$ time for any graph. Using the above representation, any series, degree-2, or polygon-to-chain reduction can be carried out in constant time. Also, none of the reductions ever require the use of more vertices or edges after the reduction than before. This means that if any new edges or vertices must be defined, old ones can be reused and the size of the graph representation is never increased.

Now, initial series and degree-2 reductions are performed on $O(|V|)$ time only once and, consequently, may be ignored for purposes of complexity analysis. Consider the "until" loop of the algorithm. Each time chains emanating from the current vertex v are searched here, and l polygons are found and reduced, the maximum amount of work which can be performed is bounded by $C_1 + C_2 l$, where C_1 is a constant bounding the amount of work required to find three chains with distinct end vertices, and C_2 is a constant bounding the amount of work required to perform a polygon-to-chain reduction and any subsequent series and degree-2 reductions. That C_1 is, in fact, a constant is obvious. C_2 is a constant because there are only eight types of polygons to recognize and reduce, and because after reduction of $\Delta(v, w)$ to $\chi(v, w)$, any chain $\chi(x, y)$ containing $\chi(v, w)$ can have length at most 9. Thus $\chi(x, y)$ would require at most 8 series and degree-2 reductions to be completely reduced. This worst case could occur if $\deg(v) = \deg(w) = 2$ after the polygon-to-chain reduction and the subchains $\chi(x, v)$, $\chi(v, w)$, and $\chi(w, y)$, which were proper chains before the reduction, are at their maximum possible lengths of 3. (In the case that G is a cycle after a polygon-to-chain reduction, the maximum length of such a cycle is 6, and reduction of the cycle to two edges in parallel requires at most 4 series and degree-2 reductions.) Since at most $2|E|-|V|$ vertices ever pass through T, and since at most $|E|-|V|$ polygon-to-chain reductions will ever be performed, the work performed by the algorithm is bounded by $C_1(2|E|-|V|)+C_2(|E|+|V|)$. Under the connectivity assumptions $|E| \geqq |V|$, and we have therefore proven the following theorem:

THEOREM 2. *Let G be a nonseparable series-parallel graph. Then, for any K, $R(G_K)$ can be computed in $O(|E|)$ time.*

6. Extension to the algorithm. The algorithm of § 5 can be extended to make all possible simple and polygon-to-chain reductions in a nonseries-parallel graph. In this way, the extended algorithm can be used as a subroutine in a more general network reliability algorithm for computing $R(G_K)$ when G is not series-parallel. The complexity of computing $R(G_K)$ can often be reduced to some degree by this device.

Suppose the reduction algorithm of § 5 starts with a nonseries-parallel graph G. After termination of the algorithm, G_K may or may not have been partially reduced. From the proof of Property 6, the only possible remaining reductions are polygon-to-chain reductions. Each such polygon-to-chain reduction would correspond to a parallel edge replacement used to obtain the graph G' of that proof. Therefore, G_K can be totally reduced by first applying the algorithm and then finding and reducing any remaining polygons. This can easily be done by searching all chains emanating from all vertices v with $\deg(v) > 2$. In the worst case, each chain, and thus each edge, must be searched twice. Parallel chains can be recognized in constant time, and therefore, the added computation is $O(|E|)$ and the algorithm with the extension remains $O(|E|)$.

To illustrate the usefulness of the extended algorithm for a general graph, let us consider the ARPA computer network configuration as shown in Fig. 4a [4]. Suppose

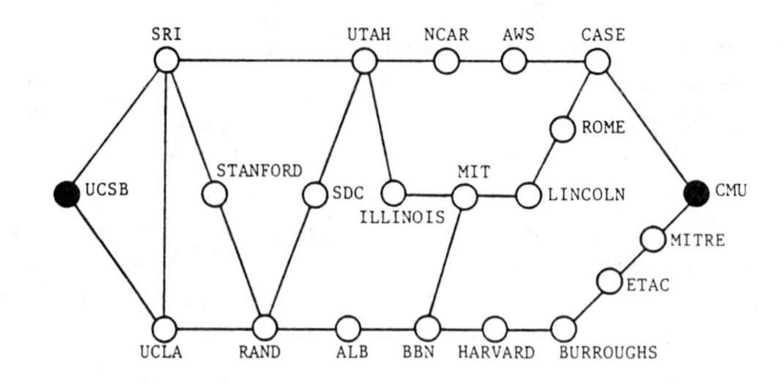

(a) ARPA *computer network.*

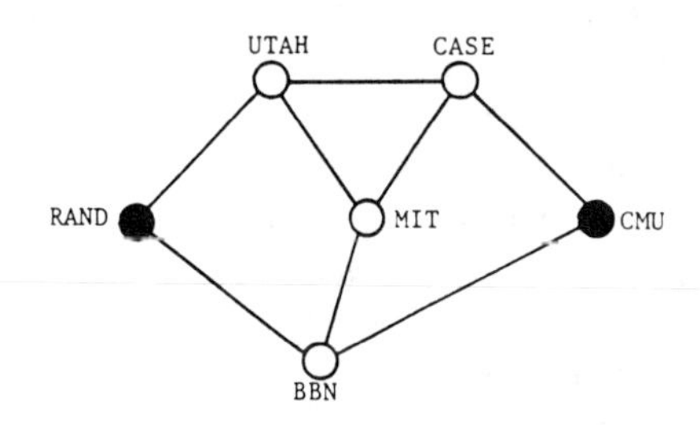

(b) *Reduced network.*

FIG. 4

we are interested in the terminal-pair reliability between UCSB and CMU. Application of the extended algorithm yields a reduced network as shown in Fig. 4b with redefined edge reliabilities and an associated multiplier. The original reliability problem is now equivalent to computing the terminal-pair reliability between RAND and CMU in the reduced network. In linear time the size of the network has been reduced considerably and, because computing the reliability of a general network is exponential in its size, a significant computational advantage should be gained.

REFERENCES

[1] R. E. BARLOW AND F. PROSCHAN, *Statistical Theory of Reliability*, Holt, Rinehart and Winston, New York, 1975.

[2] N. DEO, *Graph Theory with Applications to Engineering and Computer Science*, Prentice-Hall, Englewood Cliffs, NJ, 1974.

[3] R. J. DUFFIN, *Topology of series-parallel networks*, J. Math. Analysis Appl., 10 (1965), pp. 303–318.

[4] L. FRATTA AND U. MONTANARI, *A Boolean algebra method for computing the terminal reliability in a communication network*, IEEE Trans. Circuit Theory, CT-20 (1973), pp. 203–211.

[5] M. R. GAREY AND D. S. JOHNSON, *Computers and Intractability: A Guide to the Theory of* NP-*Completeness*, W. H. Freeman, San Francisco, 1979.

[6] J. N. HAGSTROM, *Combinatoric tools for computing network reliability*, Ph.D. Thesis, Dept. of IEOR, Univ. California, Berkeley, 1980.

[7] E. HANSLER, G. K. MCAULIFE, R. S. WILKOV, *Exact calculation of computer network reliability*, Networks, 4 (1974), pp. 95–112.

[8] P. M. LIN, B. J. LEON, T. C. HUANG, *A new algorithm for symbolic system reliability analysis*, IEEE Trans. Reliability, R-25 (1976), pp. 2–15.

[9] K. B. MISRA, *An algorithm for the reliability evaluation of redundant networks*, IEEE Trans. Reliability, R-19 (1970), pp. 146–151.

[10] F. MOSKOWITZ, *The analysis of redundancy networks*, AIEE Trans. (Commun. Electron.), 77 (1958), pp. 627–632.

[11] J. S. PROVAN AND M. O. BALL, *The complexity of counting cuts and of computing the probability that a graph is connected*, working Paper MS/S 81-002, College of Business and Management, Univ. Maryland, College Park, 1981.

[12] A. ROSENTHAL, *Computing reliability of complex systems*, Ph.D. Thesis, Dept. of EECS, Univ. California, Berkeley, 1974.

[13] ———, *Series and parallel reductions for complex measures of network reliability*, Networks, 11 (1981), pp. 323–334.

[14] A. SATYANARAYANA AND M. K. CHANG, *Network reliability and the factoring theorem*, Networks, to appear; Also, ORC 81-12, Operations Research Center, Univ. California, Berkeley, 1981.

[15] J. SHARMA, *Algorithm for reliability evaluation of a reducible network*, IEEE Trans. Reliability, R-25 (1976), pp. 337–339.

[16] J. VALDES, R. E. TARJAN AND E. L. LAWLER, *The recognition of series parallel digraphs*, this Journal, 11 (1982), pp. 297–313.

[17] L. G. VALIANT, *The complexity of enumeration and reliability problems*, this Journal, 8 (1979), pp. 410–421.

Reprinted from *IEEE Transactions on Software Engineering*, Volume SE-12, Number 1, January 1986, pages 42–50.

Distributed Program Reliability Analysis

V.K. PRASANNA KUMAR, MEMBER, IEEE, SALIM HARIRI, AND C.S. RAGHAVENDRA, MEMBER, IEEE

***Abstract*—The reliability of distributed processing systems can be expressed in terms of the reliability of the processing elements that run the programs, the reliability of the processing elements holding the required files, and the reliability of the communication links used in file transfers. We introduce two reliability measures, namely, *distributed program reliability* and *distributed system reliability* to accurately model the reliability of distributed systems. The first measure describes the probability of successful execution of a distributed program which runs on some processing elements and needs to communicate with other processing elements for remote files, while the second measure describes the probability that all the programs of a given set can run successfully. The notion of minimal file spanning trees is introduced to efficiently evaluate these reliability measures. Graph theory techniques are used to systematically generate file spanning trees that provide all the required connections. Our technique is general and can be used in a dynamic environment for efficient reliability evaluation.**

***Index Terms*—Distributed program, distributed system, graph theory, reliability, spanning tree.**

I. INTRODUCTION

THE development of computer networking and low-cost VLSI processing devices has led to an increased interest in Distributed Processing Systems (DPS) in which the computations are distributed among many processing elements (PE's). The DPS's provide potential increase in reliability, throughput, fault tolerance, resource sharing, and extensibility [4], [5], [14], [15]. While increase in performance is achieved because of running programs concurrently, the increase in reliability and fault tolerance is obtained through the redundancy in programs and data files, and the redundancy of paths in the communication network. DPS's provide a cost effective method for sharing hardware and software resources, and is easy to expand. Reliability analysis and evaluation algorithms for distributed programs are addressed in this paper.

In our model of distributed processing systems, programs and data files are distributed among many processing elements that exchange data and control information via communication links. Several processing elements cooperate in the execution of a program. A program running at a site may require files existing at some other sites. For the successful completion of a program, the local host, the processing elements holding the required files, and the interconnecting links must all be operational. With processing elements and communication links having a certain probability of being operational, there is a certain probability associated with the event that a program can be successfully executed.

Several reliability measures have been studied by researchers in the context of DPS, namely, *source-to-multiple-terminal reliability* (*SMT reliability*) [16], *computer network reliability* [3], *survivability index* [12], and *multiterminal reliability* [8]. The SMT reliability is defined as the probability that a specified processing node can reach every other processing element in the network. The computer network reliability is defined as the probability that each PE in the network can communicate with all other nodes [3]. The survivability index is a quantitative measure of the survivability of a DPS which is defined as the expected number of programs that remain operational after some combinations of nodes and links have failed [12]. The multiterminal reliability is the probability that paths exist between subsets of processing nodes which designate the source nodes and other subsets of nodes which designate the destination nodes [8].

The SMT and computer network reliability are good reliability measures for computer communication networks; however, for distributed processing systems, the reliability measures should capture the effects of redundant distribution of programs and data files. The survivability index is not applicable to large distributed systems because it finds the expected number of working programs for each state of the system. The multiterminal reliability algorithm discussed in [8] depends on decomposing the multiterminal connections between these two subsets (the source and destination subsets) into pairs between the PE that runs the program and all other PE's holding some files (Fs) required for execution of the program under consideration. The paths between pairs are combined to obtain the required multiterminal connections. If the number of paths is large, the path enumeration between all these pairs and their combinations will be computationally expensive.

In this paper, we develop an elegant approach based on graph theory to obtain the multiterminal connections required for executing a distributed program. It avoids applying the path enumeration among pairs of PE's as was done in multiterminal reliability algorithm; instead we traverse the graph to directly obtain all the required trees, called Minimal File Spanning Trees (MFST's). The probability that a distributed program is operational can then be evaluated by using these trees (MFST's) as input to an efficient terminal reliability algorithm [9]. We define another reliability measure, called distributed system relia-

Manuscript received February 25, 1985; revised July 1, 1985.

S. Hariri was supported by a Joint Fellowship from Damascus University and the Agency of International Development. C. S. Raghavendra was supported in part by the National Science Foundation under Grant ECS-8307077.

The authors are with the Department of Electrical Engineering—Systems, University of Southern California, Los Angeles, CA 90089.

IEEE Log Number 8405116.

bility, to describe the probability of having all the programs of a given set be operational in the DPS.

The organization of the paper is as follows. In the next section, notations and definitions that will be used throughout this paper are given. The derivation of the Distributed Program Reliability (DPR) algorithm and its evaluation using SYREL algorithm [9] are discussed in Section III. The DPR algorithm is generalized in Section IV to obtain the Distributed System Reliability (DSR) for a given redundant distribution of programs and files. In Section V, an illustrative example is used to explain these proposed algorithms. Finally, a brief summary and concluding remarks are presented.

II. Notations and Definitions

$G(v, e)$	An undirected graph in which the vertices represent the PE's and the edges represent the communication links.
n	Number of vertices in the network representing the DPS.
m	Number of programs that can run on the processing elements.
x_i	A node representing a processing element i.
$x_{i,j}$	A link between processing elements i and j.
$p_i(q_i)$	Probability that processing element x_i is up (down).
$p_{i,j}(q_{i,j})$	Probability that the edge $x_{i,j}$ is up (down).
t	A tree of a graph G which is defined as a sequence of vertices and edges.
$ED(t)$	Set of edges in t.
V_t	Set of vertices included in t.
E_{V_t}	Set of edges incident on V_t.
PA_i	Set of processing elements available that can run PRG_i (PRG_i denotes program i).
FN_i	Set of files (Fs) needed to execute PRG_i.
FA_i	Set of files available at processing element x_i.
$f_m(t)$	Set of missing files that are not available in V_t and are needed to execute the task under consideration.
FST	A spanning tree that connects the root node (the PE that runs the program under consideration) to some other nodes such that its vertices hold all the needed files (FN_i) for the program under consideration.
MFST	A minimal file spanning tree such that there exists no other file spanning tree which is subset of it.
$MFST(PRG_i)$	Set of all minimal file spanning trees associated with program i.

III. Derivation of DPR Algorithm

Consider the distributed processing system shown in Fig. 1 which consists of four processing elements that run three different programs distributed in redundant manner among the processing elements.

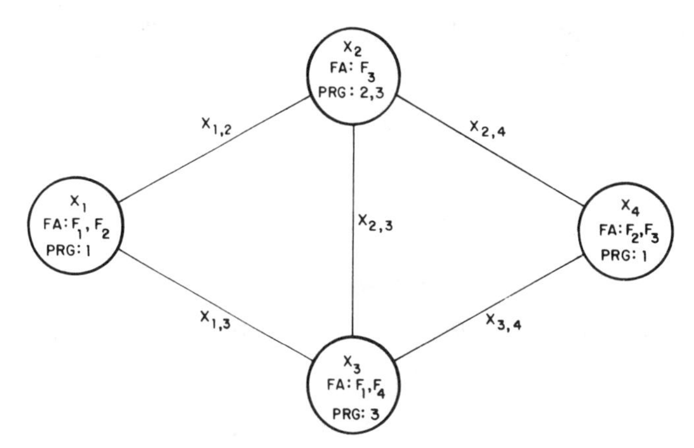

Fig. 1. A four-node distributed system.

For example, program PRG_1 can run successfully when either of x_1 or x_4 is working and it is possible to access the data files (F_1, F_2, F_3). If PRG_1 is running on x_1 which holds the files F_1 and F_2, it is required to access the file F_3 which is resident at x_2 or x_4. That is, additional PE's and links are needed to have access to that required file (F_3). In general, the set of nodes and links involved in running the given program and accessing its required files form a tree. We define a File Spanning Tree (FST) as a spanning tree that connects the root node (the PE that runs the program under consideration) to some other nodes such that its vertices hold all the required files for executing that program. For example, the following are some FST's that make PRG_1 run successfully on x_1.

$$x_1x_2x_{1,2}$$

$$x_1x_2x_3x_{1,2}x_{2,3}$$

$$x_1x_2x_4x_{1,2}x_{2,4}$$

$$x_1x_2x_3x_{1,3}x_{2,3}$$

$$x_1x_3x_4x_{1,3}x_{3,4}$$

$$x_1x_2x_3x_4x_{1,2}x_{2,3}x_{3,4}$$

$$x_1x_2x_3x_4x_{1,2}x_{2,4}x_{3,4}$$

$$x_1x_2x_3x_4x_{1,3}x_{2,3}x_{2,4}$$

$$x_1x_2x_3x_4x_{1,3}x_{3,4}x_{2,4}$$

Likewise, there will be several FST's when program PRG_1 runs on x_4. We introduce the notion of Minimal File Spanning Tree (MFST) to remove from the set of FST's all those that are supersets of some other FST's in that set. An $MFST_i$ is an FST_i such that there exists no other file spanning tree, say FST_j, which is subset of FST_i, i.e., the following is not true for all FST's:

$$\forall j \; FST_j \subseteq FST_i.$$

For example, the file spanning tree $x_1x_2x_4x_{1,2}x_{2,4}$ is not minimal because $x_1x_2x_{1,2}$ is also an FST. We are interested in finding all the MFST's to run a distributed program. For PRG_1 to run on either x_1 or x_4, the MFST's are

1) $x_1x_2x_{1,2}$

2) $x_1x_2x_3x_{1,3}x_{2,3}$

3) $x_3x_4x_{3,4}$

4) $x_2x_3x_4x_{2,4}x_{2,3}$.

Any one of these four MFST's can provide a successful execution of the program under consideration when all elements are working. The probability that a distributed program can run successfully be described in terms of the probability of having at least one of the MFST's operating. Hence, we define the Distributed Program Reliability (DPR) for a program as

DPR = Pr(at least one MFST of a given program is operational)

This can be written as

$$\mathrm{DPR} = \Pr\left(\bigcup_{j=1}^{n_{mfst}} \mathrm{MFST}_j\right) \tag{1}$$

where n_{mfst} is the number of MFST's that run the given program.

The evaluation of the reliability of executing a program on a distributed system can be determined by the following two steps.

1) Find all the Minimal File Spanning Trees (MFST's) for that program.

2) Apply any terminal reliability algorithm, such as SYREL to evaluate DPR.

A. MFST Algorithm

A DPS can be considered as an undirected graph $G(v, e)$ where the nodes represent the processing elements and the links represent the edges in that graph. This algorithm uses breadth-first search technique in traversing the graph representing a DPS to generate all the MFST's in which the roots are the processing elements that run a program, say PRG_i. The minimal file spanning trees are generated in nondecreasing order of their sizes, where the size is defined as the number of links in an MFST, and according to the following steps. First, all MFST's of size 0 are determined; this occurs when there exist some processing elements that run PRG_i and have all the needed files (which is denoted by the set FN_i) for its execution. Next, all MFST's of size 1 are determined; these trees have only one edge which connects the root node to some other node, such that the root node and the other node have all the files in FN_i. This procedure is repeated for all possible sizes of MFST's up to $n - 1$, where n is the number of nodes in the DPS.

The procedure used to construct the MFST's consists of a *checking* step and an *expanding* step. In the checking step, trees that have been generated so far, which are in a list called *TRY*, will be tested to determine whether or not the vertices V_t of each tree t in *TRY* has all the needed files (FN_i) for executing PRG_i. A tree t is an FST if its vertices have all the required files, i.e.,

$$\bigcup_{\forall i \subset V_t} FA_i \supseteq FN_i.$$

Let $f_m(t)$ be the set of all the missing files that are in the set FN_i but not available at the vertices of a tree t and *FOUND* be the list of all the minimal file spanning trees generated so far. According to this, for a tree to be FST, the set of files in FN_i and not in t is empty, which means that

$$f_m(t) = \Phi.$$

In addition to checking whether or not a tree is an FST, it is also necessary to check if it is covered by another FST that was constructed previously and stored in *FOUND*. This is done by checking that t is not superset of any MFST_i in *FOUND*, i.e.,

$$\forall\, \mathrm{MFST}_i \in FOUND,\ \mathrm{MFST}_i \cap t \neq \mathrm{MFST}_i$$

The tree $(x_1x_2x_3x_{1,2}x_{2,3})$ of Fig. 1 is an FST but it is not minimal because it is covered by another MFST $(x_1x_2x_{1,2})$.

Once the checking process is done, the list *TRY* will have all the trees that are not MFST's and therefore their $f_m(t)$'s are not empty. The expanding step is necessary to increase the size of each tree in *TRY* by connecting each vertex of the tree to a new adjacent vertex hoping that the new vertex will have the needed files. For example, the tree $(x_1x_3x_{1,3})$ is not an FST and therefore it is expanded by connecting to its vertices (x_1, x_3) one of the adjacent vertices to them such as (x_2, x_4). The added adjacent vertex might have all or some of the needed files. If that node has all the missing files, a new FST would be generated from adding the adjacent vertex. If it has some or none of the needed files, that node can be used as an intermediate node to access other set of adjacent nodes in the next expanding steps. The set of edges that can be added to any tree is called the Adjacent Edges of t, $AE(t)$. This set is obtained by first finding the set of edges incident on $V_t(E_{V_t})$ and then deleting from this set the edges included in t, i.e.,

$$AE(t) = E_{V_t} - ED(t).$$

For example, the set of adjacent edges of $x_1x_3x_{1,3}$ is computed be as

$$AE(x_1x_3x_{1,3}) = \{x_{1,3}, x_{1,2}, x_{2,3}, x_{3,4}\} - \{x_{1,3}\}$$
$$= \{x_{1,2}, x_{2,3}, x_{3,4}\}.$$

The expanded trees are stored in a temporary list called *NEW* which must also be checked to remove any replication among the expanded trees. This removal of the replicated trees is done in *CLEANUP* function. The formal description of the *MFST ALGORITHM* is given below.

MFST Algorithm

```
Step 1: Initialization
    TRY = PA_p                    * p denotes the program under consideration.
    FOUND = ∅                     * list TRY has initially the root nodes that will run program p.
      for all j ∈ TRY do          * the set of missing files is the difference between the set
        f_m(j) = FN_p - FA_j        of files that are needed for executing PRG_p and the set of
      od                            files available in the root node.
Step 2: Generating all MFST's
    repeat
      2.1 Checking step           * in this step each tree in list TRY is checked to determine
        for all t ∈ TRY do          whether or not it has all the required files for executing
          if CHECK(t) then          PRG_p.
            begin
              add t to FOUND
              remove t from TRY
            end
        od
      2.2. Expanding step         * in this step each tree in TRY is expanded by connecting
        NEW = ∅                     an adjacent edge to one of its vertices.
        for all t ∈ TRY do
          add EXPAND(t) to NEW
        od
        TRY = CLEANUP (NEW)
      until (TRY = ∅)
    function CHECK(t)             * this function returns a true value when t is an FST (f_m(t)
    begin                           = ∅) and is minimal, otherwise returns false value.
      CHECK = false
      if f_m(t) = ∅ then          * t is an FST and will be checked to see whether or not it
        begin                       is an MFST.
          if FOUND = ∅ or (∀ l ∈ FOUND, l ∩ t ≠ l)
            then
            CHECK = true
          else                    * t is removed because it is not an MFST.
            TRY = TRY - t
        end
    end (*CHECK*)
    function CLEANUP (NEW)        * this function removes from NEW all the replicated trees
    begin                           generated by the EXPAND function.
      for all t_i, t_j ∈ NEW do
        if t_i - t_j = ∅ then remove t_i from list NEW
      od
    end. (*CLEANUP*)
    functionEXPAND(t)             * this function constructs from t all trees that can be
    begin                           formed by adding to t one edge from its adjacent edge set
                                    AE(t).
      tmp = empty                 * list tmp is empty initially.
      AE(t) = E_{V_t} - ED(t)     * determine the set of adjacent edges.
      for all (x_{i,j} ∈ AE(t), i ∈ V_t ∧ j ∉ V_t) do
        begin                     * construct new tree newt by adding vertex j, which is an
          newt = t ∪ {x_{1,j}}      adjacent vertex to t.
          fm(newt) = fm(t) - FA_j * determine the set of missing files after adding vertex j
          add newt to tmp           to i.
        end
      EXPAND = tmp
    end (*EXPAND*)
```

B. Terminal Reliability Algorithms

Once all the minimal file spanning trees have been generated, the next step is to find the probability that at least one MFST is working which means that all the edges and vertices included in it are operational. Any terminal reliability evaluation algorithm based on path or cutset enumeration [1], [2], [6], [7], [9], [11], [13] can be used to obtain the distributed program reliability of the program under consideration.

In what follows, a brief adaptation of SYREL algorithm to the evaluation of DPR will be presented. Let us assume that there are nt MFST's $(T_1, T_2, \cdots, T_{nt})$ and $(E_1, E_2, \cdots, E_{nt})$ be the events in which these trees are working. The distributed program reliability is, by definition, given as

$$\text{DPR} = \Pr\left(\bigcup_{i=1}^{nt} E_i\right) \tag{2}$$

By using conditional probability, the events considered in (2) can be decomposed into mutually exclusive events as shown in (3).

$$\text{DPR} = \Pr(E_1) + \Pr(E_2) \cdot \Pr(\overline{E_1}|E_2) + \cdots \Pr(E_{nt}) \cdot \Pr(\overline{E_1} \wedge \overline{E_2} \wedge \cdots \overline{E_{nt-1}}|E_{nt}) \tag{3}$$

where $\Pr(\overline{E_1}|E_2)$ denotes the conditional probability that MFST_1 is in the failure state given that MFST_2 is in the operational state.

Hence, the DPR expression can be evaluated in terms of the probability of two distinct events. The first event indicates that a tree, say T_i, is in the operational state while the second event indicates that all the previous trees of T_i $(T_1 \cdots T_{i-1})$ are in the failure state given that T_i is in the operational state. The probability of the first event, $\Pr(E_i)$ is straightforward. It is equal to the product of the reliabilities of all the elements in T_i, i.e.,

$$\Pr(E_i) = \prod_{\forall k \in E_i} p_k.$$

The probability of the second event, $\Pr(\overline{E_1} \wedge \overline{E_2} \wedge \cdots \overline{E_{i-1}}|E_i)$, will be computed using conditional probability theory and standard Boolean operations according to the following steps.

1) Evaluating the minimal conditional sets of a given tree T_i. The elements of each of the minimal conditional sets are the minimal number of elements that must all be in the failure state which fails all the previous trees of T_i; for each tree, say T_j, in the set of previous trees PR_i, ($PR_i = \{T_1, T_2, \cdots T_{i-1}\}$), we find the set of elements, called the conditional sets, that are in T_j but not in T_i (since those are the only elements that can fail when T_i is assumed to be working). We then remove the dominated conditional sets to obtain the minimal conditional sets. For example, let us assume that we have three conditional sets $(x_{1,2}, x_{1,2}x_{3,4}, x_{1,2}x_{4,5}x_{6,7})$, the failure of element $x_{1,2}$ will lead to failure of the first tree and also the next two trees. Hence, the last two sets can be removed and therefore the minimal conditional set consists of only $x_{1,2}$.

2) Checking the minimal sets for common elements. Either one of the following two substeps will be performed based on whether or not the sets have common elements. a) If there are no common elements among these sets, then they can be considered as independent events. Hence, the probability of having each tree in PR_i not working will be the product of the probabilities that the minimal sets are not working. For example, let $(x_{1,2}, x_{3,4}, x_{5,6}x_{6,7})$ be the minimal conditional sets of a tree T_i. The failure of these sets will fail all the trees in PR_i. Since there is no common elements among these sets, the probability in which they fail is given as

$$\Pr(\overline{E_1} \wedge \overline{E_2} \wedge \cdots \overline{E_{i-1}}|E_i) = (1 - p_{1,2}) \cdot (1 - p_{3,4}) \cdot (1 - p_{5,6}p_{6,7}).$$

b) If they have some common elements, we first find the cut sets that will ensure that all the trees in PR_i are not working (when the elements of each cut has failed) and then make these cut sets mutually exclusive. For example, let $(x_{1,2}, x_{1,2}x_{5,6}, x_{1,2}x_{4,5})$ be three conditional sets that have $x_{1,2}$ as a common element. The two disjoint cut sets that ensure the failure of these sets are $\overline{x_{1,2}}$ and $\overline{x_{5,6}}\ \overline{x_{4,5}}\ x_{1,2}$. Hence, the probability that the trees in PR_i are not working is given by

$$\Pr(\overline{E_1} \wedge \overline{E_2} \wedge \cdots \overline{E_{i-1}}|E_i) = q_{1,2} + q_{5,6}q_{4,5}p_{1,2}.$$

The SYREL algorithm performs all these steps to efficiently evaluate the reliability for a given set of MFST's. The distributed program reliability evaluation can now be described as shown below.

DPR Algorithm

Step 1: Apply MFST Algorithm.

Step 2: Apply a terminal reliability algorithm, such as SYREL.

IV. Distributed System Reliability (DSR)

The distributed program reliability measures the reliability of one program in the system. However, for reliable distributed processing systems, it is important to obtain a global reliability measure that describes how reliable the system is for a given distribution of programs and files [10]. One way of measuring the reliability of the system is by determining the probability that all the distributed programs are working, i.e.,

$$\text{DSR} = \Pr\left(\bigcap_{i=1}^{m} \text{PRG}_i\right) \tag{4}$$

Let us denote the set of all minimal file spanning trees associated with a program PRG_i by $\text{MFST}(\text{PRG}_i)$. The distributed system reliability equation can be written in terms of the probability of the intersection of the set of all minimal file spanning trees of each program as shown in (5).

$$\text{DSR} = \Pr\left(\bigcap_{i=1}^{m} \text{MFST}(\text{PRG}_i)\right) \quad (5)$$

The intersection of the components of each $\text{MFST}(\text{PRG}_i)$ can be evaluated first by intersecting $\text{MFST}(\text{PRG}_1)$ and $\text{MFST}(\text{PRG}_2)$, then by intersecting this with the next MFST ($\text{MFST}(\text{PRG}_3)$), and so on until we intersect all the minimal file spanning trees. After each intersection, it is necessary to eliminate any dominated sets which are defined as follows. A set S_i dominates another set S_j if their intersection equals to S_i, i.e.,

$$S_i \cap S_j = S_i .$$

The steps of the DSR Algorithm are described next.

DSR Algorithm

step 1: For all i, generate $\text{MFST}(\text{PRG}_i)$.
  for $i=1$ to m **do**
  call *MFST ALGORITHM* for PRG_i
    $\text{MFST}(\text{PRG}_i) = FOUND$
  od
step 2: Intersecting all MFST's
  2.1 Initialization
    $list = \text{MFST}(\text{PRG}_1)$
  2.2 **for** $p=2$ **to** m **do**
    2.2.1 Pairwise addition
      $tmp = empty$
      for all $mf\,st \in \text{MFST}(\text{PRG}_p)$ **do**
          for all $l \in list$ **do**   * add to *tmp* the
          add $(mf\,st \cup l)$ to tmp union of the two
      od   sets l and $mf\,st$
    2.2.2 Eliminate all dominated sets
      for all $l_i, l_j \in tmp$ **do**
        if $l_i \cap l_j = l_i$ then remove l_j from tmp
        if $l_i \cap l_j = l_j$ then remove l_i from tmp
      od
    2.2.3 $list = tmp$
    od
  2.3 $\text{MFST}\left(\bigcap_{i=1}^{m} \text{PRG}_i\right) = list$
step 3: Determine DSR expression
  apply a terminal reliability algorithm, such as SYREL.

V. An Illustrative Example

The distributed processing system shown in Fig. 2 consists of six processing elements that can run four distributed programs that each one of them can run on two PE's. The allocations of these programs are described in the following sets.

$$PA_1 = \{x_1,x_6\} \quad PA_2 = \{x_3,x_4\}$$

$$PA_3 = \{x_3,x_4\} \quad PA_4 = \{x_2,x_5\}$$

The files needed for executing these programs are indicated in the following sets.

$$FN_1 = \{F_1,F_2,F_3\} \quad FN_2 = \{F_2,F_4,F_6\}$$

$$FN_3 = \{F_1,F_3,F_5\} \quad FN_4 = \{F_1,F_2,F_4,F_6\}$$

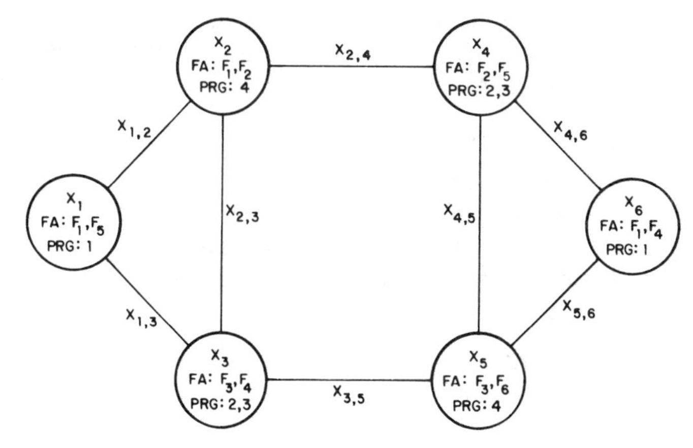

Fig. 2. An example of a distributed system.

The files available at each processing element and the programs that can run on each PE are shown in Fig. 2.

In what follows, we will briefly discuss the multiterminal reliability method [8] when it is used to obtain the multiconnections required for executing a program and compare it to our approach that will be explained also here. For example, if the reliability of program PRG_1 is to be evaluated, it is required first to determine the set of vertices that must be connected to ensure successful execution of PRG_1, i.e., the following set of vertices must be connected:

$$(x_1,x_2,x_3),\ (x_1,x_2,x_5)\quad (x_1,x_3,x_4),\ (x_1,x_4,x_5)$$

$$(x_6,x_2,x_3),\ (x_6,x_2,x_5)\quad (x_6,x_3,x_4),\ (x_6,x_4,x_5)$$

The set of all minimal file spanning trees of PRG_1 is derived by decomposing the elements of each set into pairs that are ANDed and ORed to obtain these trees as shown below.

$\text{MFST}(\text{PRG}_1) =$

$$[(x_1,x_2)\text{AND}(x_1,x_3)] \text{ OR } [(x_1,x_2)\text{AND}(x_1,x_5)] \text{ OR}$$

$$[(x_1,x_3)\text{AND}(x_1,x_4)] \text{ OR } [(x_1,x_4)\text{AND}(x_1,x_5)] \text{ OR}$$

$$[(x_6,x_2)\text{AND}(x_6,x_3)] \text{ OR } [(x_6,x_2)\text{AND}(x_6,x_5)] \text{ OR}$$

$$[(x_6,x_3)\text{AND}(x_6,x_4)] \text{ OR } [(x_6,x_4)\text{AND}(x_6,x_5)]$$

Hence, the path enumeration between the pairs must be applied 10 times. The AND algorithm to intersect the set of paths generated between the decomposed pairs will be applied eight times, and the OR algorithm seven times. It is clear from this example that for large distributed systems, where a large number of the processing elements cooperate on executing a distributed program, this approach is not efficient and computationally costly.

Our approach avoids the decomposition into pairs and the overhead operations involved in ANDing and ORing their combinations. Instead, we traverse the graph once to obtain directly all the required MFST's. We first find the DPR for the execution of programs PRG_1, PRG_2, PRG_3, and PRG_4 using DPR Algorithm developed in Section III, and then applying DSR Algorithm to obtain an expression of the system reliability for the given distribution of programs and files. The details of evaluation of DPR for PRG_1 only is given here.

A. Evaluating DPR for PRG_1

This program can run on either x_1 or x_6 and its minimal file spanning trees shown here are represented by only its edges since they imply the vertices involved in these trees, are obtained using the MFST Algorithm in the following order.

step 1: Initialization step yields

$TRY = PA_1 = \{x_1, x_6\}$

$FOUND = \emptyset$

$f_m(x_1) = FN_1 - FA_1 = \{F_1, F_2, F_3\} - \{F_1, F_5\}$
$= \{F_2, F_3\}$

$f_m(x_6) = FN_1 - FA_6 = \{F_1, F_2, F_3\} - \{F_1, F_4\}$
$= \{F_2, F_3\}$

step 2: generating all MFST'S.

2.1 Checking step for trees of size 0.

2.1.1 $CHECK(x_1) = false$ since $f_m(x_1) \neq \emptyset$

2.1.2 $CHECK(x_6) = false$ since $f_m(x_6) \neq \emptyset$

$TRY = \{x_1, x_6\}$

$FOUND = \emptyset$

2.2 Expanding step (generating all trees of size 1)

2.2.1 $EXPAND(x_1) = \{x_{1,2}, x_{1,3}\}$

$f_m(x_{1,2}) = f_m(x_1) - FA_2 = \{F_3\}$

$f_m(x_{1,3}) = f_m(x_1) - FA_3 = \{F_2\}$

$NEW = \{x_{1,2}, x_{1,3}\}$

2.2.2 $EXPAND(x_6) = \{x_{4,6}, x_{5,6}\}$

$f_m(x_{4,6}) = f_m(x_6) - FA_4 = \{F_3\}$

$f_m(x_{5,6}) = f_m(x_6) - FA_5 = \{F_2\}$

$NEW = \{x_{1,2}, x_{1,3}, x_{4,6}, x_{5,6}\}$

$TRY = CLEANUP(NEW) = \{x_{1,2}, x_{1,3}, x_{4,6}, x_{5,6}\}$

2.1 Checking step for trees of size 1.

2.1.1 $CHECK(x_{1,2}) = false$ since $f_m(x_{1,2}) \neq \emptyset$

2.1.2 $CHECK(x_{1,3}) = false$ since $f_m(x_{1,3}) \neq \emptyset$

2.1.3 $CHECK(x_{4,6}) = false$ since $f_m(x_{4,6}) \neq \emptyset$

2.1.4 $CHECK(x_{5,6}) = false$ since $f_m(x_{5,6}) \neq \emptyset$

$TRY = \{x_{1,2}, x_{1,3}, x_{4,6}, x_{5,6}\}$

$FOUND = \emptyset$

2.2 Expanding step (generating all trees of size 2)

2.2.1 $EXPAND(x_{1,2}) = \{x_{1,2}x_{1,3}, x_{1,2}x_{2,3}, x_{1,2}x_{2,4}\}$

$f_m(x_{1,2}x_{1,3}) = f_m(x_{1,2}) - FA_3 = \emptyset$

$f_m(x_{1,2}x_{2,3}) = f_m(x_{1,2}) - FA_3 = \emptyset$

$f_m(x_{1,2}x_{2,4}) = f_m(x_{1,2}) - FA_4 = \{F_3\}$

$NEW = \{x_{1,2}x_{1,3}, x_{1,2}x_{2,3}, x_{1,2}x_{2,4}\}$

2.2.2 $EXPAND(x_{1,3}) = \{x_{1,3}x_{2,3}, x_{1,3}x_{3,5}\}$

$f_m(x_{1,3}x_{2,3}) = f_m(x_{1,3}) - FA_2 = \emptyset$

$f_m(x_{1,3}x_{3,5}) = f_m(x_{1,3}) - FA_5 = \{F_2\}$

$NEW = \{x_{1,2}x_{1,3}, x_{1,2}x_{2,3}, x_{1,2}x_{2,4}, x_{1,3}x_{2,3}, x_{1,3}x_{3,5}\}$

2.2.3 $EXPAND(x_{4,6}) = \{x_{4,6}x_{5,6}, x_{4,6}x_{2,4}, x_{4,6}x_{4,5}\}$

$f_m(x_{4,6}x_{5,6}) = f_m(x_{4,6}) - FA_5 = \emptyset$

$f_m(x_{4,6}x_{2,4}) = f_m(x_{4,6}) - FA_2 = \{F_3\}$

$f_m(x_{4,6}x_{4,5}) = f_m(x_{4,6}) - FA_5 = \emptyset$

$NEW = \{x_{1,2}x_{1,3}, x_{1,2}x_{2,3}, x_{1,2}x_{2,4}, x_{1,3}x_{2,3},$
$x_{1,3}x_{3,5}, x_{4,6}x_{5,6}, x_{4,6}x_{2,4}, x_{4,6}x_{4,5}\}$

2.2.4 $EXPAND(x_{5,6}) = \{x_{5,6}x_{4,5}, x_{5,6}x_{3,5}\}$

$f_m(x_{5,6}x_{4,5}) = f_m(x_{5,6}) - FA_4 = \emptyset$

$f_m(x_{5,6}x_{3,5}) = f_m(x_{5,6}) - FA_3 = \{F_2\}$

$NEW = \{x_{1,2}x_{1,3}, x_{1,2}x_{2,3}, x_{1,2}x_{2,4}, x_{1,3}x_{2,3},$
$x_{1,3}x_{3,5},$
$x_{4,6}x_{5,6}, x_{4,6}x_{2,4}, x_{4,6}x_{4,5}, x_{5,6}x_{4,5},$
$x_{5,6}x_{3,5}\}$

$TRY = CLEANUP(NEW) = \{x_{1,2}x_{1,3}, x_{1,2}x_{2,3},$
$x_{1,2}x_{2,4}, x_{1,3}x_{2,3}, x_{1,3}x_{3,5}, x_{4,6}x_{5,6}, x_{4,6}x_{2,4},$
$x_{4,6}x_{4,5}, x_{5,6}x_{4,5}, x_{5,6}x_{3,5}\}$

2.1 Checking step for trees of size 2.

2.1.1 $CHECK(x_{1,2}x_{1,3}) = true$ since $f_m(x_{1,2}x_{1,3})$
$= \emptyset$

$TRY = TRY - \{x_{1,2}x_{1,3}\}$

$FOUND = \{x_{1,2}x_{1,3}\}$

2.12 $CHECK(x_{1,2}x_{2,3}) = true$ since $f_m(x_{1,2}x_{2,3}) = \emptyset$

$TRY = TRY - \{x_{1,2}x_{2,3}\}$

$FOUND = \{x_{1,2}x_{1,3}, x_{1,2}x_{2,3}\}$

2.1.3 $CHECK(x_{1,2}x_{2,4}) = false$ since $f_m(x_{1,2}x_{2,4})$
$\neq \emptyset$

2.1.4 $CHECK(x_{1,3}x_{2,3}) = true$ since $f_m(x_{1,3}x_{2,3})$
$= \emptyset$

$TRY = TRY - \{x_{1,3}x_{2,3}\}$

$FOUND = \{x_{1,2}x_{1,3}, x_{1,2}x_{2,3}, x_{1,3}x_{2,3}\}$

2.1.5 $CHECK(x_{1,3}x_{3,5}) = false$ since $f_m(x_{1,3}x_{3,5})$
$\neq \emptyset$

2.1.6 $CHECK(x_{4,6}x_{5,6}) = true$ since $f_m(x_{4,6}x_{5,6})$
$= \emptyset$

$TRY = TRY - \{x_{4,6}x_{5,6}\}$

$FOUND = \{x_{1,2}x_{1,3}, x_{1,2}x_{2,3}, x_{1,3}x_{2,3}, x_{4,6}x_{5,6}\}$

2.1.7 $CHECK(x_{4,6}x_{2,4}) = false$ since $f_m(x_{4,6}x_{2,4})$
$\neq \emptyset$

2.1.8 $CHECK(x_{4,6}x_{4,5}) = true$ since $f_m(x_{4,6}x_{4,5})$
$= \emptyset$

$TRY = TRY - \{x_{4,6}x_{4,5}\}$

$FOUND = \{x_{1,2}x_{1,3}, x_{1,2}x_{2,3}, x_{1,3}x_{2,3}, x_{4,6}x_{5,6},$
$x_{4,6}x_{4,5}\}$

2.1.9 $CHECK(x_{5,6}x_{4,5}) = true$ since $f_m(x_{5,6}x_{4,5})$
$= \emptyset$

$TRY = TRY - \{x_{5,6}x_{5,4}\}$

$FOUND = \{x_{1,2}x_{1,3}, x_{1,2}x_{2,3}, x_{1,3}x_{2,3}, x_{4,6}x_{5,6},$
$x_{4,6}x_{4,5}, x_{5,6}x_{5,4}\}$

2.1.10 $CHECK(x_{5,6}x_{3,5}) = false$ since $f_m(x_{5,6}x_{3,5})$
$\neq \Phi$

$FOUND = \{x_{1,2}x_{1,3}, x_{1,2}x_{2,3}, x_{1,3}x_{2,3}, x_{4,6}x_{5,6},$
$x_{4,6}x_{4,5}, x_{5,6}x_{5,4}\}$

$TRY = \{x_{1,2}x_{2,4}, x_{1,3}x_{3,5}, x_{4,6}x_{2,4}, x_{5,6}x_{3,5}\}$

A detailed description of the next iteration of applying *expanding* and *checking* steps is not shown here because of space limitation. However, the result of applying these two steps on the trees of length 2 in *TRY* yields the following.

$FOUND = \{x_{1,2}x_{1,3}, x_{1,2}\ x_{2,3}, x_{1,3}x_{2,3}, x_{4,6}x_{5,6}, x_{4,6}x_{4,5}, x_{5,6}x_{5,4},$

$x_{1,2}x_{2,4}x_{4,5}, x_{1,3}x_{3,5}x_{4,5}, x_{4,6}x_{2,4}x_{2,3}, x_{5,6}x_{3,5}x_{2,3}\}$

$TRY = \emptyset$

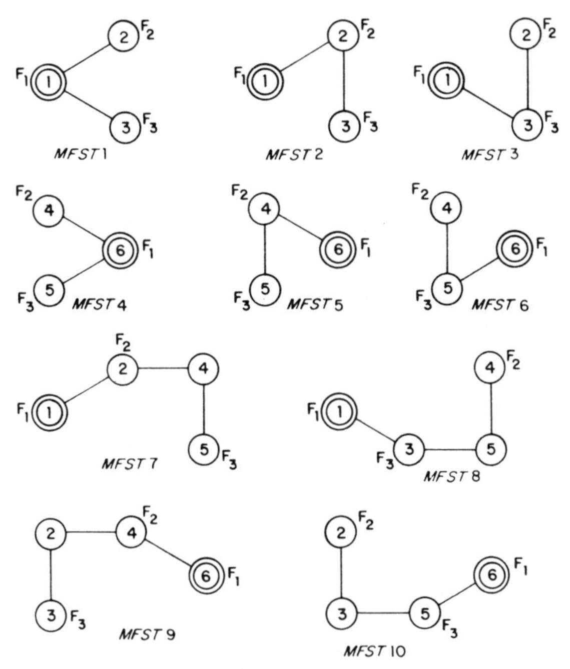

Fig. 3. MFST's for PRG_1.

The algorithm terminates because the list *TRY* becomes empty. Fig. 3 shows the MFST's obtained from applying the MFST Algorithm and also shows for each MFST the locations of the files needed for executing PRG_1. The double circles indicate the site on which the program can run.

Once MFST(PRG_1) has been found, applying the SYREL algorithm yields the following terms for the reliability expression of PRG_1.

$$T_1 = p_1p_2p_3p_{1,2}p_{1,3}$$

$$T_2 = p_1p_2p_3p_{1,2}p_{2,3}[q_{1,3}]$$

$$T_3 = p_1p_2p_3p_{1,3}p_{2,3}[q_{1,2}]$$

$$T_4 = p_4p_5p_6p_{4,6}p_{5,6}[(1 - p_1p_2p_3) + p_1p_2p_3(q_{1,2}q_{1,3} + q_{1,2}q_{2,3}p_{1,3} + q_{1,3}q_{2,3}p_{1,2})]$$

$$T_5 = p_4p_5p_6p_{4,5}p_{4,6}q_{5,6}[(1 - p_1p_2p_3) + p_1p_2p_3(q_{1,2}q_{1,3} + q_{1,2}q_{2,3}p_{1,3} + q_{1,3}q_{2,3}p_{1,2})]$$

$$T_6 = p_4p_5p_6p_{4,5}p_{5,6}q_{4,6}[(1 - p_1p_2p_3) + p_1p_2p_3(q_{1,2}q_{1,3} + q_{1,2}q_{2,3}p_{1,3} + q_{1,3}q_{2,3}p_{1,2})]$$

$$T_7 = p_1p_2p_4p_5p_{1,2}p_{2,4}p_{4,5}[q_3q_6 + q_3q_{4,6}q_{5,6}p_6 + q_6q_{1,3}q_{2,3}p_3 + q_{1,3}q_{2,3}q_{4,6}q_{5,6}p_3p_6]$$

$$T_8 = p_1p_3p_4p_5p_{1,3}p_{3,5}p_{4,5}[q_2q_6 + q_6q_{1,2}q_{2,3}p_2 + q_2q_{4,6}q_{5,6}p_6 + q_{1,2}q_{2,3}q_{4,6}q_{5,6}p_2p_6]$$

$$T_9 = p_2p_3p_4p_6p_{2,3}p_{2,4}p_{4,6}[q_1q_5 + q_1q_{4,5}q_{5,6}p_5 + q_5q_{1,2}q_{1,3}p_1 + q_{1,2}q_{1,3}q_{4,5}q_{5,6}p_1p_5]$$

$$T_{10} = p_2p_3p_5p_6p_{2,3}p_{3,5}p_{5,6}[q_1q_4 + q_1q_{4,5}q_{4,6}p_4 + q_4q_{1,2}q_{1,3}p_1 + q_{1,2}q_{1,3}q_{4,5}q_{4,6}p_1p_4]$$

$$\text{DPR} = \sum_{i=1}^{10} T_i$$

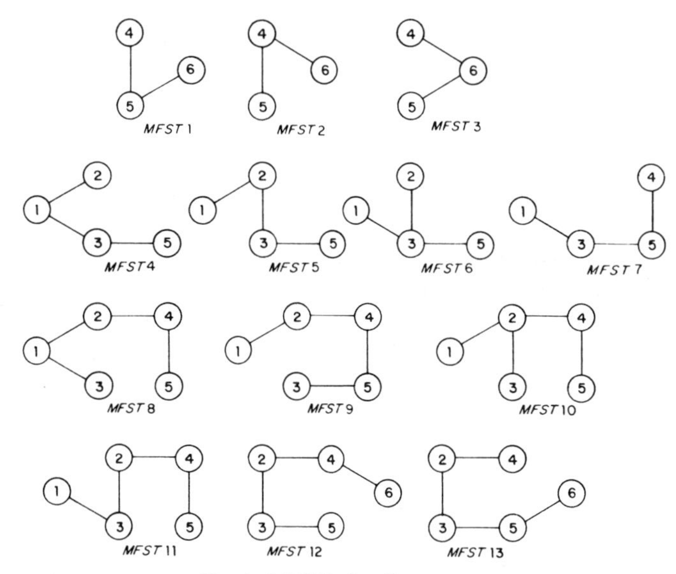

Fig. 4. MFST's for all programs.

B. Evaluating the DSR

The first step in the DSR Algorithm is to determine all the minimal file spanning trees for each program that can run on the system. The next step of this algorithm is to determine the set of all minimal spanning trees that guarantee successful execution of all the programs by recursively intersecting the MFST's of each program. These MFST's are shown in Fig. 4. The final step is applying SYREL to obtain the following terms for the DSR.

$$T_1 = p_4p_5p_6p_{4,5}p_{5,6}$$

$$T_2 = p_4p_5p_6p_{4,5}p_{4,6}[q_{5,6}]$$

$$T_3 = p_4p_5p_6p_{4,6}p_{5,6}[q_{4,5}]$$

$$T_4 = p_1p_2p_3p_5p_{1,2}p_{1,3}p_{3,5}[(1 - p_4p_6) + p_4p_6(q_{4,5}q_{4,6} + q_{4,5}q_{5,6}p_{4,6} + q_{4,6}q_{5,6}p_{4,5})]$$

$$T_5 = p_1p_2p_3p_5p_{1,2}p_{2,3}p_{3,5}q_{1,3}[(1 - p_4p_6) + p_4p_6(q_{4,5}q_{4,6} + q_{4,5}q_{5,6}p_{4,6} + q_{4,6}q_{5,6}p_{4,5})]$$

$$T_6 = p_1p_2p_3p_5p_{1,3}p_{2,3}p_{3,5}q_{1,2}[(1 - p_4p_6) + p_4p_6(q_{4,5}q_{4,6} + q_{4,5}q_{5,6}p_{4,6} + q_{4,6}q_{5,6}p_{4,5})]$$

$$T_7 = p_1p_3p_4p_5p_{1,3}p_{3,5}p_{4,5}[q_2q_6 + q_2q_{4,6}q_{5,6}p_6 + q_6q_{1,2}q_{2,3}p_2 + q_{1,2}q_{2,3}q_{4,6}q_{5,6}p_2p_6]$$

$$T_8 = p_1p_2p_3p_4p_5p_{1,2}p_{1,3}p_{2,4}p_{4,5}q_{3,5}[q_6 + q_{4,6}q_{5,6}p_6]$$

$$T_9 = p_1p_2p_3p_4p_5p_{1,2}p_{2,4}p_{3,5}p_{4,5}q_{1,3}q_{2,3}[q_6 + q_{4,6}q_{5,6}p_6]$$

$$T_{10} = p_1p_2p_3p_4p_5p_{1,2}p_{2,3}p_{2,4}p_{4,5}q_{1,3}q_{3,5}[q_6 + q_{4,6}q_{5,6}p_6]$$

$$T_{11} = p_1p_2p_3p_4p_5p_{1,3}p_{2,3}p_{2,4}p_{4,5}q_{1,2}q_{3,5}[q_6 + q_{4,6}q_{5,6}p_6]$$

$$T_{12} = p_2p_3p_4p_5p_6p_{2,3}p_{2,4}p_{3,5}p_{4,6}q_{4,5}q_{5,6}[q_1 + q_{1,2}q_{1,3}p_1]$$

$$T_{13} = p_2p_3p_4p_5p_6p_{2,3}p_{2,4}p_{3,5}p_{5,6}q_{4,5}q_{4,6}[q_1 + q_{1,2}q_{1,3}p_1]$$

$$\mathrm{DSR} = \sum_{i=1}^{13} T_i$$

If we assume that all the elements (processing elements and communication links), of the DPS shown in Fig. 2 have the same reliability and equal 0.9, then the reliability of executing PRG_1 is 0.9378 and the reliability of the distributed system for the given distribution of programs and files equals 0.842674.

VI. Summary and Conclusion

In this paper, distributed program reliability is addressed and an efficient algorithm based on graph theory is developed to obtain all the MFST's required for executing a distributed program directly by traversing the graph in breadth-first search manner. It avoids decomposing the multiconnections into pairs and combining the enumerated paths between the pairs. This is important because in large DPS's the number of PE's that cooperate on the execution of one program can be large enough to make the evaluation of its reliability by the known multiterminal reliability algorithm inefficient. We also developed a new reliability measure, distributed system reliability, that measures the probability of having all the distributed programs working. Our technique can be used to model the reliability of various distributed processing systems, such as distributed operating systems, and distributed database systems.

The DPR and DSR expressions can be used as objective functions in an optimization problem that is aimed at designing a distributed processing system with high reliability and minimum cost. These expressions can also be used for program sensitivity analysis, which quantitatively measures the improvement in the DPR or DSR when redundant copies of programs and/or data files are introduced to increase the reliability.

References

[1] J. A. Abraham, "An improved algorithm for network reliability," *IEEE Trans. Rel.*, vol. R-23, pp. 58-61, Apr. 1979.

[2] K. K. Aggarwal, K. B. Misra, and J. S. Gupta, "A fast algorithm for reliability evaluation," *IEEE Trans. Rel.*, vol. R-24, pp. 83-85, Apr. 1975.

[3] K. K. Aggarwal and S. Rai, "Reliability evaluation in computer-communication networks," *IEEE Trans. Rel.*, vol. R-30, pp. 32-35, Apr. 1981.

[4] D. W. Davies, E. Holler, E. D. Jensen, S. R. Kimbleton, B. W. Lampson, G. LeLann, K. J. Thurber, and R. W. Watson, *Distributed Systems—Architecture and Implementation* (Lecture Notes in Computer Science, vol. 105). Berlin, Germany: Springer-Verlag, 1981.

[5] P. Enslow, "What is a distributed data processing system," *Computer*, vol. 11, Jan. 1978.

[6] L. Fratta and U. G. Montanari, "A Boolean algebra method for computing the terminal reliability in a communication network," *IEEE Trans. Circuit Theory*, vol. CT-20, pp. 203-211, May 1973.

[7] A. Grnarov, L. Kleinrock, and M. Gerla, "A new algorithm for symbolic reliability analysis of computer communication networks," presented at the Pacific Telecommun. Conf., Jan. 1980.

[8] A. Grnarov and M. Gerla, "Multiterminal reliability analysis of distributed processing systems," in *Proc. 1981 Int. Conf. Parallel Processing*, Aug. 1981, pp. 79-86.

[9] S. Hariri and C. S. Raghavendra, "SYREL: A symbolic reliability algorithm based on path and cutset methods," USC Tech. Rep., 1984.

[10] S. Hariri, C. S. Raghavendra, and V. K. Prasanna Kumar, "Reliability measures for distributed processing systems," in *Proc. Int. Symp. New Directions in Comput.*, Trondheim, Norway, Aug. 1985.

[11] P. M. Lin, B. J. Leon, and T. C. Huang, "A new algorithm for symbolic system reliability analysis," *IEEE Trans. Rel.*, vol. R-25, pp. 2-15, Apr. 1976.

[12] R. E. Merwin and M. Mirhakak, "Derivation and use of a survivability criterion for DDP systems," in *Proc. 1980 Nat. Comput. Conf.*, May 1980, pp. 139-146.

[13] S. Rai and K. K. Aggarwal, "An efficient method for reliability evaluation," *IEEE Trans. Rel.*, vol. R-27, pp. 101-105, June 1978.

[14] D. A. Rennels, "Distributed fault-tolerant computer systems," *Computer*, vol. 13, pp. 55-56, Mar. 1980.

[15] J. A. Stankovic, "A perspective on distributed computer systems," *IEEE Trans. Comput.*, vol. C-33, pp. 1102-1115, Dec. 1984.

[16] A. Satyanarayana and J. N. Hagstrom, "A new algorithm for reliability analysis of multi-terminal networks," *IEEE Trans. Rel.*, vol. R-30, pp. 325-333, Oct. 1981.

V. K. Prasanna Kumar (M'84) was born in India. He received the B.S. degree in electronics engineering from Bangalore University, Bangalore, India, the M.S. degree from the School of Automation, Indian Institute of Science, Bangalore, and the Ph.D. degree in computer science from Pennsylvania State University, University Park, in 1983, where he was supported by an IBM fellowship.

Currently, he is an Assistant Professor with the Department of Electrical Engineering, University of Southern California, Los Angeles. His research interests include parallel processing, computer architecture, and distributed and VLSI computations.

Dr. Kumar is a member of the Association for Computing Machinery and the IEEE Computer Society.

Salim Hariri was born in Tartous, Syria, in 1954. He received the B.S.E.E. degree from Damascus University, Damascus, Syria, in 1977, and the M.S.E.E. degree from the Ohio State University, Columbus, in 1982.

He is currently working toward the Ph.D. degree in computer engineering at the University of Southern California, Los Angeles. His research interests include design of reliable distributed systems, fault-tolerant computing, fault diagnosis and detection, and design for testability.

C. S. Raghavendra (S'80-M'82) was born in India in March 1955. He received the B.Sc. (Hons.) degree in physics from Bangalore University, Bangalore, India, in 1973, the B.E. and M.E. degrees in electronics and communications from the Indian Institute of Science, Bangalore, in 1976 and 1978, respectively, and the Ph.D. degree in computer science from the University of California, Los Angeles, in 1982.

Since September 1982 he has been with the Department of Electrical Engineering—Systems of the University of Southern California, Los Angeles. His research interests are computer system architecture, fault-tolerant computing, and reliability analysis of networks and distributed systems. He is a consultant to the Hughes Aircraft Company.

Dr. Raghavendra is a recipient of the Presidendial Young Investigator Award for 1985.

Reliability Evaluation Algorithms For Distributed Systems*

Anup Kumar, Suresh Rai and D.P.Agrawal

Computer System Laboratory
Electrical & Computer Engineering
Box 7911, N.C.State University
Raleigh, NC 27695

ABSTRACT

This paper introduces two techniques for computing the reliability of a distributed computing system (DCS). The first scheme employs two steps and requires enumeration of multi-terminal connections which, in turn, leads to the reliability expression. The second technique, called FARE (Fast Algorithm for Reliability Evaluation), does not require an apriori knowledge of multi-terminal connections for computing the reliability expression. A performance parameter called "Communication Cost Index" (CCI), is also defined. We have compared our algorithms with an existing method in terms of computer time and memory requirement. To compare the performance of all three algorithms computer execution time is plotted against CCI.

Index Terms- communication cost index, distributed system, file distribution, reliability, 1-SUB and 0-SUB operation .

I. INTRODUCTION

Recently, Distributed Computing System(DCS) has become increasingly popular because it offers higher fault tolerance, potential for parallel processing, and better reliability in comparison with other processing environments [1]. A typical DCS consists of Processing Elements(PE), memory units, data files and programs as its resources. These resources are interconnected via a communication network that dictates how information could flow between PEs. Programs residing on some PEs can run using data files at other PEs as well. For successful execution of a program, it is essential that the PE containing the program and other PEs that have requisite data files, and communication links between them must be operational. Using this concept, Distributed System Reliability (DSR) is defined as the probability that all programs with distributed files can run successfuly in spite of some faults occurring in the PEs and/or in the communication links [2]. For a DCS the reliability measure should take into consideration the effect of redundant distributions of data files and programs. Based on these factors, many reliability indices are proposed: They include source-to-multiple-terminal reliability (SMT) [3], survivability index [4], computer network reliability [5], and multiterminal reliability [6]. SMT and computer network reliability are popular, and reasonable measures for DCS. The problem with survivability index is that the computation time for this index is prohibitively large even for relatively small networks [2]. The multi-terminal reliability algorithm works well as long as there are not too many alternate paths between source and destination pairs.

The algorithm proposed in [2] is a two step method. In this method, all multi-terminal connections are obtained by using graph traversal rather than applying path enumeration technique among the pairs of PEs. After finding out the multi-terminal connections, it requires any reliability evaluation algorithm as given in [7] to generate the reliability expression. Our first algorithm is also a two-step approach. The multi-terminal connection enumeration procedure is straightforward, and does not require apriori information regarding path set for all node pairs. To change the boolean expression of multi-terminal connections into an equivalent reliability expression, we have used a method proposed in [7]. The FARE utilizes a different concept and uses only one step to give the reliability expression from the connection matrix. It does not require generation of multi-terminal connections and thus, the complexity involved in converting multi-terminal connections into reliability expression is avoided.

*This work is part of the B-HIVE project and has been partially Supported by Army Research Office under contract no. DAAG 29-85-K-0236.

EH0300-4/90/0000/0121$01.00 © 1990 IEEE

The organization of the paper is as follows: Section II considers procedure development and describes terms useful for understanding the paper. Section III presents the two algorithms using pascal like structure, and a comparison of these methods is given in section IV. It also defines CCI, a parameter useful in monitoring the performance of algorithms. Section V concludes the paper.

II. PROCEDURE DEVELOPMENT

Grnarov and Gerla [6] have considered three possible types of multi-terminal connections (MC) required in a distributed computing environment. They are:

a) OR-type MC

b) AND-type MC, and

c) Mixed-type MC

In an OR-type MC, a program to be executed at a source node (SN) requires transaction of any one of the files residing on nodes (t1,t2, t3, . . . ,tm). It is assumed that the program is executable with the help of any one of these files. This situation is quite obvious when the nodes t1,t2, . . . ,tm have data files residing on them, and the main program is executed at the SN. Nodes, other than SN, are termed as terminal nodes (TN). Multi-terminal connection of AND-type exists when the transaction of files, residing at different node sites, is simultaneouly required for execution of a program Pi at the SN. Logically, it produces a list of paths with the minimum number of elements and thus, is a cube of the largest possible sizes [6]. In mixed-type connections, any logical combination of OR-type, and AND-type is possible. Besides confronting the problem of multi-terminal connections cited above, our algorithms may also solve various other subproblems defined later. Definitions 1 through 4 outline these problems. Such problems occur in computer networking, telecommunication system and electrical power distribution network. In all these cases, we are mainly concerned with the connectedness property of the topology. Algorithms make use of 0-SUB, and 1-SUB operations. They are explained in definition 5 and 6 respectively.

Definition 1- Single Pair Problem (SP) enumerates paths between a given source and a given terminal.

Definition 2- Single Terminal Problem (ST) enumerates paths between a given terminal and any number of sources.

Definition 3- Single Source Problems (SS) enumerates paths between a given source and any number of terminals.

Definition 4- All Pair Problem (AP) enumerates paths between a given number of source and terminal nodes.

Definition 5- A 0-SUB sustitute operation takes care of a 0-link where a 0 is replaced in the connection matrix (C). This is equivalent to removing a link from a graph [8].

Definition 6- A 1-SUB substitute operation is performed on 1-link using the following procedure:

```
Step 0. Input C
Step 1.
 for all link(s) = 1 do
   replace them by zeros
 od.
Step 2.
 for all column(s) corresponding to
 link(s) = 1 do
  find all row(s) r having entry = 1
    for each row r do
      row 1 <-- row 1 or row r
      delete row(s) r
    od.
  od.
Step 3.
 for all column(s) corresonding to
 link(s) = 1 do
   delete those columns
 od.
```

A 1-SUB operation is equivalent to contraction/fusion concept in graph theory [8].Note that 1-SUB operation reduces the size of C on which it is operated.

Definition 7:- A link xi is said to be 0(1)-link iff Xi=0(1). Otherwise, it is a variable link.

Decomposition Polices- They are used to obtain decomposition for the connection matrix or the 2-D array. By decomposing C we simply mean to perform 1-SUB and 0-SUB operation on to C to obtain a reduction in size of C. The events useful to decompose C are generated using conservative policy

(CP),exhaustive policy (EP) or extended conservative policy (ECP). Let us assume that (a1,a2,a3,... ,aN) are variables in first row of C. Three decomposition policies used in the paper are

1. Conservative Policy(CP): The N combinations of this scheme are shown below

a1 0 0 0 ... 0	1 0 0 0 ... 0
0 a2 0 0 ... 0	0 1 0 0 ... 0
0 0 a3 0 ... 0	0 0 1 0 ... 0
..	..
0 0 0 0 ... aN	0 0 0 0 ... 1

2. Exhaustive Policy(EP): In this scheme (2^N-1) combinations are defined and they are shown below

a1 0 0 0 ... 0	1 0 0 0 ... 0
0 a2 0 0 ... 0	0 1 0 0 ... 0
..	..
0 0 0 0 ... aN	0 0 0 0 ... 1
a1 a2 0 0	1 1 0 0 ... 0
a1 0 a3 0	1 0 1 0 ... 0
..	..
a1 a2 a3 ...aN	1 1 1 1 ... 1

3. Exhaustive Conservative Policy(ECP): Various N combinations in this scheme are shown below

a1	1
0 a2	0 1
0 0 a3	0 0 1
..	..
0 0 0 ... aN	0 0 0 ... 1

Note that ECP needs n combinations but has a scope of (2 -1) events of EP. A 0(1) in an event Ei refers 0(1)-SUB operation into C.

Node Vector(NV)- (NV)j of link xj is defined as the vector where 1(0) indicates the nodes that are connected(not connected) by link xj. In Table 2 (NV)2 is [010100]. In both the methods, presented here, we monitor continuously the node vector while decomposing 2D-array using an appropriate decomposition policy. Whenever a link variable is considered, its node vector updates the previous node vector. The updating is performed as a simple OR operation between two vectors, and is described as

(NV)new <- (NV)previous OR (NV)j

To illustrate the concept we consider an example from Figure 3. The 2D-array shown at the root has [000000] as node vector associated with it. After updating with respect to x1, the modified node vector is

$$(NV)new = [0\,0\,0\,0\,0\,0] \text{ OR } [1\,1\,0\,0\,0\,0] = [1\,1\,0\,0\,0\,0]$$

Terminal Node Vector(TNV)- It can be obtained form the knowledge of

i. type of muli-terminal connection sought
ii. set of source and terminal node

For example in Table 2 terminal node vectors are calculated for P1 to be executed on node 1. P1 requires files F1,F2 and F3 but F1 is already at node 1. Remaining files required are F2 and F3. These files are available on nodes (2,4) and (3,5) respectively. The type of connection can be (2 OR 4) AND (3 OR 5). Source node is 1. Hence terminal node vectors are [111ddd],[11dd1d],[1d11dd], and [1dd11d]; where 'd' denotes don't care condition.

Assumptions

1. Distributed System is modeled by a graph G where node represents processing elements and edges are communication links.
2. Failure of one link is s-independent of failures of other links.
3. Graph G does not have any self loop.
4. There are only two states of an link: working (good) or faulty (bad).

III. ALGORITHMS

Algorithm 1 uses CP whenever the node vector in question contains (tm-1) terminal nodes, otherwise it uses EP; here tm indicates the total number of terminal nodes. To help terminate the decomposition of 2D-array in subsequent steps two rules, as defined below, are used.

Rule 1- If the NV matches with TNV, then the node in question terminates. This means that the decomposition is not carried out any further in this branch of the tree.

Rule 2- If the node vector does not match with TNV and row 1 of 2D-array contains only 1's and 0's, then the decomposition of the node in question is terminated. Such nodes in the tree are called 'undesired leaves'.

Algorithm 1 [2-step method]
INPUT [Adjacency list, terminal node vector node vectors for links]

Step 1:Generating Connection matrix/2D array and vector indicating terminal nodes.
1a. generate connection matrix from given adjacency list
1b. if source node(i) = 1 then
begin
(i) Substitute link variable in column 1 by zero.
(ii) Add unity matrix to it
end;
else
begin
(i) row 1 <----> row i
(ii) column 1 <----> column i
(iii) substitute link variable in Column 1 by zero
(iv) add unity matrix to it.
end;
call it 2D-array

Step 2:Expansion Step
repeat
2a. Let there be N link variables in 1-row of 2-D array.
Matrix POSCOM=store various events Er using decomposition rule above.
2b. repeat
(i). for an Er use 1-sub operation with uncomplemented links in Er(use 0-sub for complemented links) to decompose 2D-array.
(ii). update the (NV) for Er.
until all events in POSCOM are exhausted.
until all decomposed 2D-arrays satisfy terminating rule.

Step3 :

After expansion, path from root to all the leaves stisfying only rule 1 of of terminating conditions, is traversed This gives multi-terminal connections.

Step 4:
From these multi-terminal connectiosn we can compute reliability expressiosn using an appropriate algorithm. Here, we have used E-operator [7] for obtaining the reliability expression.

Algorithm 2, does not require apriori information regarding multi-terminal connections and gives reliability expression directly. This method is called FARE (Fast Algorithm for Reliability Evaluation) as it avoids the complexity involved in enumerating multi-terminal connections, and then processing them into equivalent probability expressions of reliability. FARE utilizes ECP to decompose the 2D-array or connection matrix C. Here, the teminating rule is pretty simple and, involves matching of NV with TNV.

Algorithm 2 [FARE:Fast Algorithm for Relibility Evaluation]
INPUT:[Adjacency list, TNV, Node Vector for links]

Step 1: Generation of 2D-array is same as step 1 of algorithm 1

Step 2:Expansion Step
repeat
2a. Let there be N links in 1-row of 2D-array.
Matrix POSCOM = Store various events Er using ECP policy.
2b. Same as 2b of step 2 of Algorithm 1
until all decomposed 2D-arrays satisfy terminating rule.

Step 3:
(i) After expansion, paths from root to all the leaves are traversed to get logical expression of DSR.

(ii) To get expression of DSR we make the following susstitution:
pi <-- Xi
qi <-- Xi'

To illustrate the underlying concept we have solved an exmaple using both algorithms. Figure 1 illustrates a distributed system for which the reliability computation is carried out. Table 1 and 2 gives parameters like file distribution, TNVs, NVs for links, and also files required by the programs for their execution. The multiterminal connection obtained from algorithm 1 are shown in tree form in Figure 2. The final expression, after removing the redundant terms, is given as

x3x4+x1x3+x1x4+x1x2x6+x3x5x6+x3x5x7x8
+x1x2x7x8+x3x5x7x8.

A similar result is obtained using [2]. To generate the reliability expression from this multi-terminal connection we have used method [7]. DSR is given as

p3p4+p1p3q4+p1p4q3+p1p2p6q3q4+p3p5p6q1q4
+p3p5p7p8q1q4q6+p1p2p7p8q3q4q6.

Results, along with intermediate steps, from FARE are shown in Figure 3. Note that, FARE generates DSR directly, and is given as

p1p3+p1q3p4+q1p3p4+q1p3q4p5p6+p1q3q4p2p6
+q1p3q4p5q6p8p7+p1q3q4p2q6p7p8.

IV. COMPARISON

For comparing the algorithms with the one given in [2], we have studied how the execution time and average memory requirements for these algorithms vary according to the changes in

(i) toplogy of distributed system, and

(ii) file distribution on the PEs of the DCS.

In each case a parameter called the communication cost index (CCI) is evaluated. It takes into account the various relevant features of the problem. Precisely, it depends on: (i) file distribution matrix FD [f,n], (ii) connection matrix C[n,n], and (iii) the program matrix PN[p,f]. Entries in parentheses indicate the size of different matrices; p representing total number of programs, n the total number of nodes and f the total number of data files. File distribution matrix relates distribution of files on the nodes of the network. PN defines the file requirement for executing a program. Mathematically, they are considered as shown below

$$C_{jk} = \begin{cases} 1 & \text{if node j is connected to node k only when j <> k} \\ 0 & \text{otherwise} \end{cases}$$

$$PN_{lk} = \begin{cases} 1 & \text{if program l requires file k} \\ 0 & \text{otherwise} \end{cases}$$

$$FD_{ij} = \begin{cases} 1 & \text{if file i is on node j} \\ 0 & \text{otherwise} \end{cases}$$

It is obvious that FDij considers file distribution. When we want to execute a program Pv at node U such that at leat one file required for executing Pv is available at U, the FD is modfied. To generate the modified FD, hereinafter termed as FD*, all entries corresponding to 'available file' are made zero in FD. It means we are not intrested in bringing that file at node U if it is already present over it, and hence no cost is incurred for this situation. In the extreme case where node U does not have any one of the required files for Pv, FD* will be same as FD. Based on these parameter, we define CCI as follows-

$$CCI = \sum_{l=1}^{p} . \sum_{k=1}^{n} [PN_{lk} \ X \ \{FD_{ij} \ X \ C_{jk}\}^T]$$

1. Effect of Topology

In this study, we have run different set of programs and file distributions over various topologies starting from simple loop to a completely connected graph. These topologies are shown in Figure 4. Various sets of file distribution, program distribution and files needed for program execution are given in Table 3,4, and 5 respectively. The execution time is plotted against CCI. CCI does not depend on the type of algorithm used, hence it will be same for all three algorithms. Plots for 8-sets of data are shown in Figure 5. We observe a considerable increase in execution time of our algorithm 1, and the one given in [2] as the topology under consideration switches to a fully connected one. It is due to the fact that our algorithm 1 uses EP, where expansion of one node creates as many as 2^{e-1} nodes. e denotes maximum indegree of a node. Similarly in [2] the worst case can generate as many as $(n-1)^{e-1}$ intermediate trees, where n denotes the number of nodes in the graph of DCS. For FARE we have used ECP. Thus, it does not explode during the expansion of intremediate nodes as it does in the above two cases. In this case, the number of new nodes created from the expansion of a node is proportional to the maximum indegree of a node. It can be concluded from Figure 5 that the execution time for FARE does not increase as rapidly as the execution time for 2-step methods implying that FARE outperforms them. As regards average memory requirements Table 7 lists the values for all three methods. These values are generated using 'time' instruction on VAX-780 and a rigorous analysis need be done.

2. Effect of File Distribution

We have run 8-sets of file distribution, generated randomly, on various topologies for all three algorithms. The file distributions used are shown in Table 7. As in the earlier case, CCI for each set has been calculated. Execution time is plotted against CCI for all three algorithms. We have run FDs on three sets of topology and program. Three plots are shown in Figure 6. The minima in each curve of plot indicates minimum execution time on Y-axis and corresponding file allocation to be optimal for a given set of FDs. From the graphs shown in Figure 6, it is observed that 1-step method outperforms 2-step method. Table 8 lists the average memory requirements for different file distributions.

V CONCLUSION

In this paper we have described two reliability computation methods. First algorithm is 2-step method which requires the apriori knowledge of multiterminal connection for reliability evaluation. Second algorithm directly computes the reliability without enumerating multiterminal connections. It is obvious from the results that FARE outperforms 2-step methods. FARE permits the survivability analysis of DCS of considerably more modest size than using currently available techniques. For large system backtracking will be very complex, making it NP-complete [9]. The CCI defined above gives the information as to how these algorithms will behave for different sets of data. This index may be helpful in comparing various methods of reliability evaluations.

REFERENCES

1. D.P.Agrawal,"Advanced Computer Architecture", Tutorial Text, *Computer Society of the IEEE,* 376 pages.
2. V.K.Prasnnakumar, S.Hariri, and C.S.Raghavendra, "Distributed Program Reliability Analysis," *IEEE Trans. Software Engrg.,* Vol. SE-12, NO-1, pp. 42-50, Jan. 1986.
3. A.Satyanarayna, and J.N.Hagstrom, "New Algorithm for Reliability Analysis of Multiterminal Netowrks," *IEEE Trans. Reliability.,* Vol. R-30, pp. 325-333, Oct. 1981.
4. R.E.Merwin and M.Mirherkerk, "Derivation and Use of Survivability Criterion for DDP Systems," in *Proc. 1980 Nat'l Computer Conference,* May 1980, pp. 139-146.
5. K.K.Aggarwal and S.Rai, "Reliability Evaluation in Computer Communication Networks," *IEEE Trans. Reliability.,* Vol. R-30, pp. 32-35, April 1981.
6. A.Grnarov and M.Gerla, "Multiterminal Reliability Analysis of Distributed Processing Systems," in *Proc. 1981 Int'l Conf. on Parallel Processing,* Aug. 1981, pp. 79-86.
7. S.Rai and K.K.Aggarwal, "An Efficient Method for Reliability Evaluation," *IEEE Trans. Reliability,* Vol. R-27, June 1978, pp. 101-105.
8. M.N.S.Swamy and K.Tulsiraman, *Graphs, Networks and Algorithms,* John Wiley & Sons, NY, 1981.
9. A. V. Aho, J. E. Hopcroft and J. D. Ullman, *The Design and Analysis of Computer Algorithms,* Addison-Wesley Publishing Company, Massachusetts, 1974.

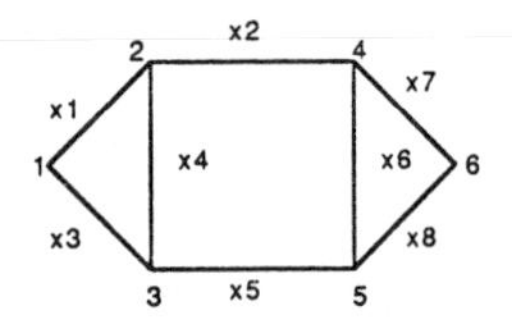

Figure 1. Topology of distributed network

node	files available	set of programs	files needed to execute a program
1	F1,F5	P1	
2	F1,F2	P4	FNP1 = {F1,F2,F3}
3	F3,F4	P2,P3	FNP2 = {F2,F4,F6}
4	F2,F5	P2,P3	FNP3 = {F1,F3,F5}
5	F3,F6	P4	FNP4 = {F1,F2,F4,F6}
6	F1,F4	P5	

Table 1. Distribution of programs, datafiles and files needed for program execution

link	node vector	terminal node vectror
1	110000	
2	010100	
3	101000	111ddd
4	011000	11dd1d
5	001010	
6	000110	1d11dd
7	000101	1dd11d
8	000011	

Table 2. Node vectors & terminal node vector for execution of program1 on node 1.

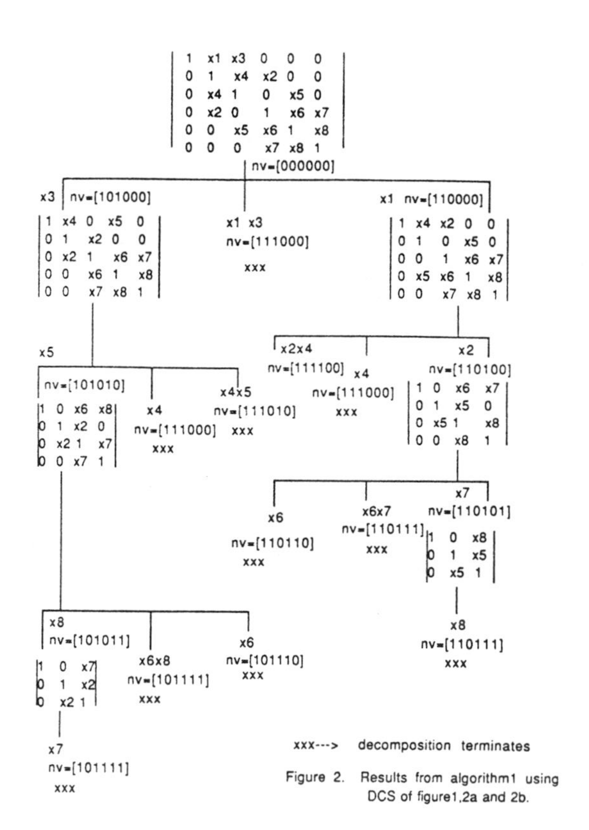

Figure 2. Results from algorithm1 using DCS of figure1,2a and 2b.

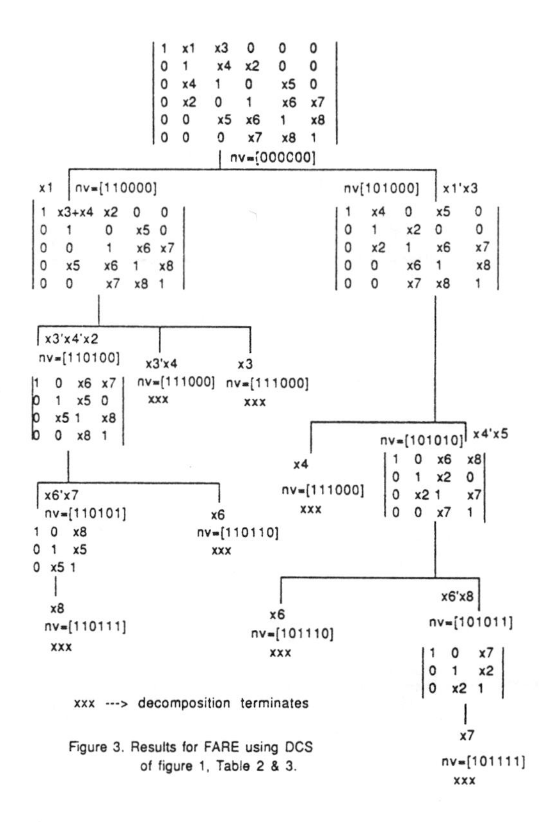

Figure 3. Results for FARE using DCS of figure 1, Table 2 & 3.

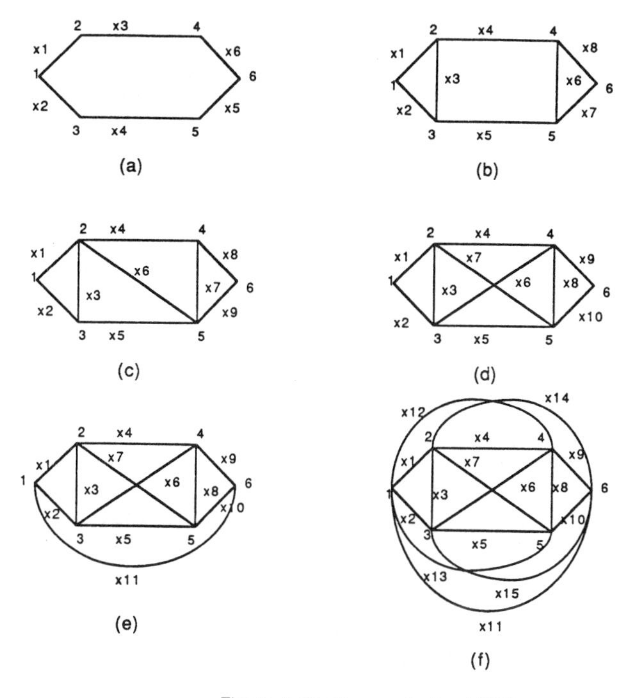

Figure 4. Topology variants of DCS

Figure 5. Plots of CCI and execution time for topology variations

□--> curve for FARE
o --> curve for algorithm in [2]
Δ --> curve for algorithm 1

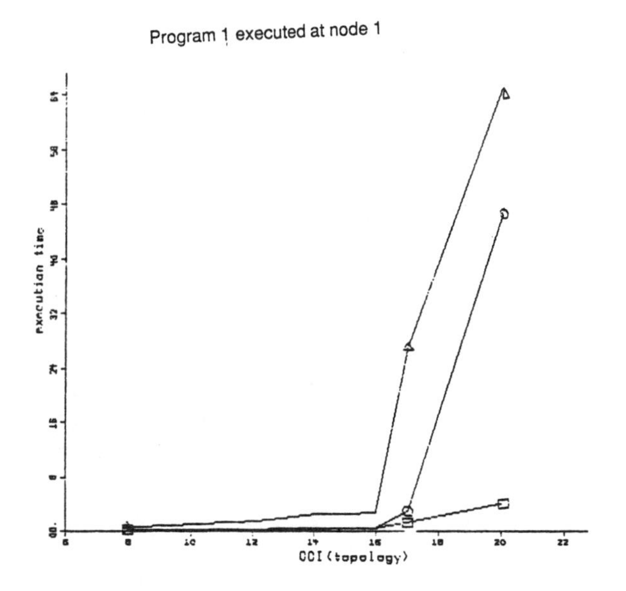

Figure 5. (contd.)

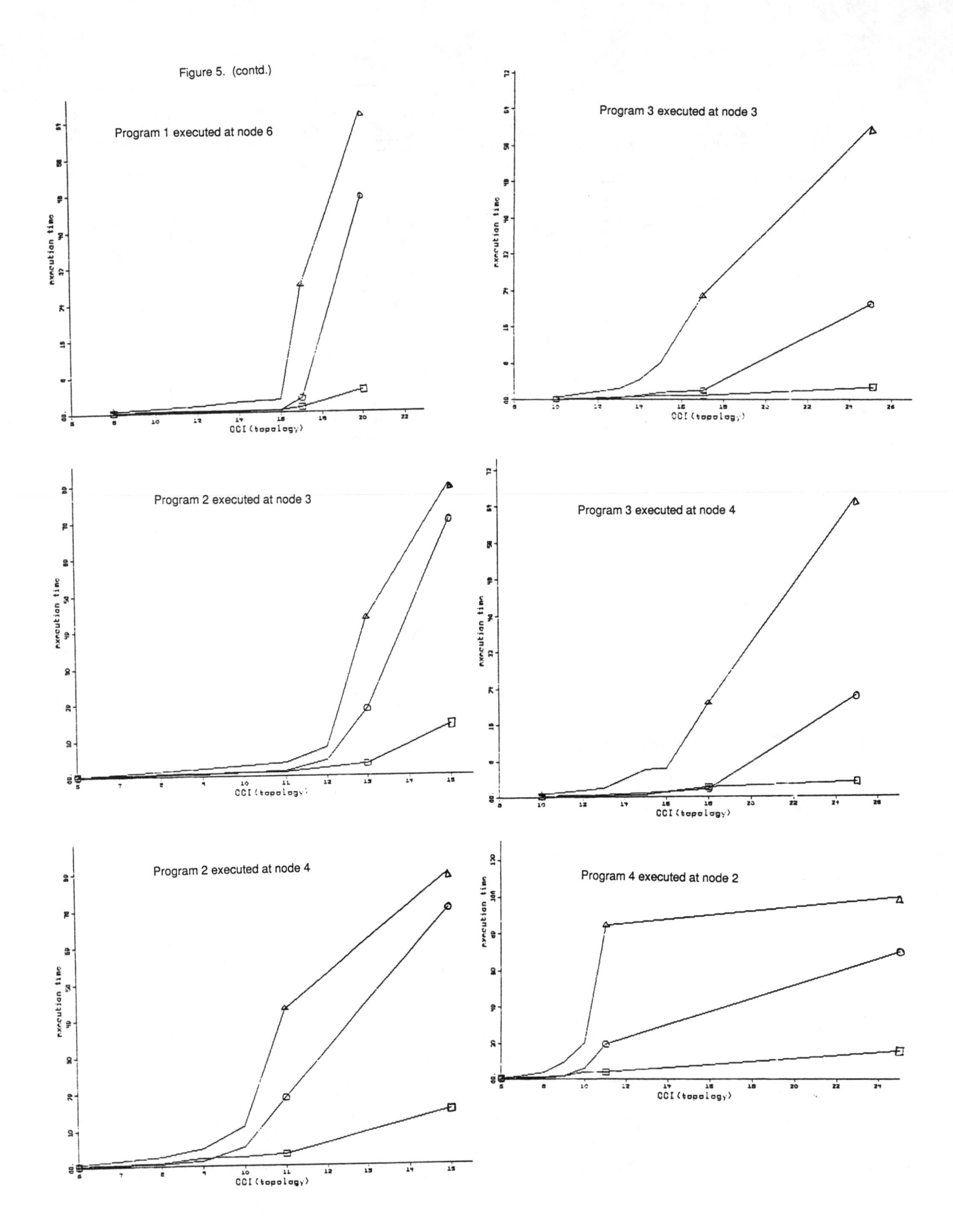

8C.4.8.

Figure 5. (contd.)

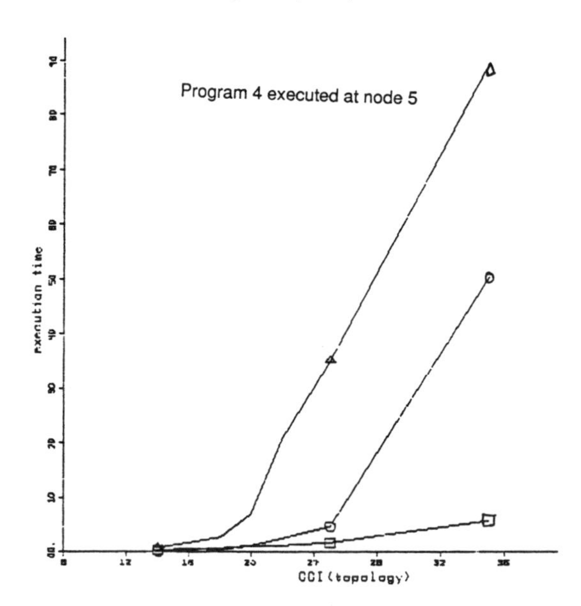

Program 3 executed at node 3

execution time

CCI (file distribution)

Figure 6. Plots of CCI and execution time for various file distribution

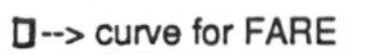
□--> curve for FARE
o --> curve for algorithm in [2]
▲ --> curve for algorithm 1

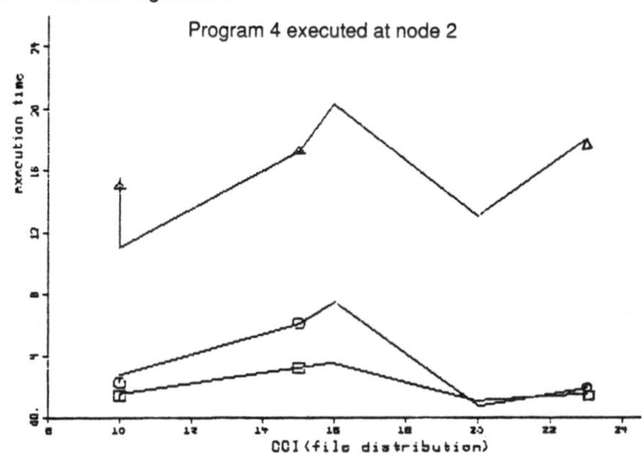

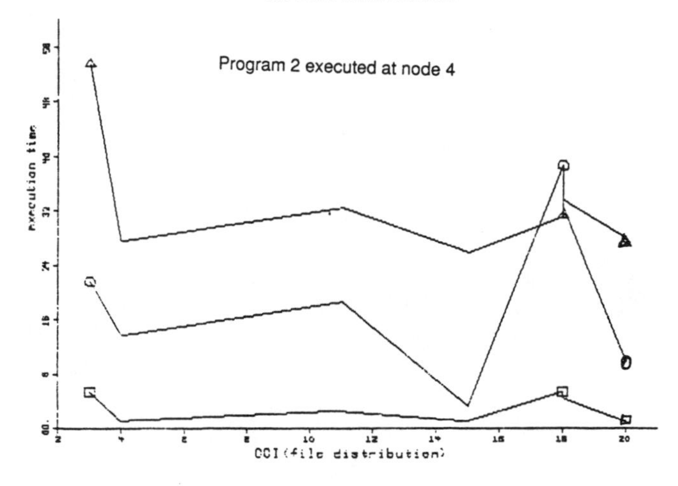

file	nodes
F1	1,2,3
F2	2,4
F3	3,5
F4	3,6
F5	1,4
F6	5

Table 3. File distribution

node	program
1	P1
2	P4
3	P2,P3
4	P2,P3
5	P4
6	P1

Table 4. Program distribution

program	files required
P1	F1,F2,F3
P2	F2,F4,F6
P3	F1,F3,F5
P4	F1,F2,F4,F6

Table 5. Files needed for execution of programs

files	set1 (nodes)	set2 (nodes)	set3 (nodes)	set4 (nodes)	set5 (nodes)	set6 (nodes)	set7 (nodes)	set8 (nodes)
F1	2,4,5	2,3,6	4,5,6	1,2,3	1,4,6	1,3,6	3,4,5	2,3,6
F2	4,5	3,5	2,3	4,5	2,5	3,6	1,2	3,5
F3	5,6	3,4	4,5	1,6	3,4	1,2	5,6	1,6
F4	3,4	2,3	1,3	2,4	2,5	4,5	5,6	2,6
F5	4,6	4,5	4,5	2,4	3,5	4,6	1,6	3,6
F6	6	3	6	3	5	5	5	4

Table 6. File distributions used for comparison

Topologies	algori- -thm 1	FARE	method [2]
a	43 k	28 k	26 k
b	47 k	37 k	47 k
c	48 k	29 k	47 k
d	54 k	30 k	34 k
e	51 k	32 k	39 k
f	57 k	36 k	37 k

Table 7. Average memory requirement for three algorithms
(For various topologies)
(This set is for program 1 executed on node 1)

file distribution	algori- -thm 1	FARE	method [2]
set1	55 k	39 k	40 k
set2	50 k	33 k	36 k
set3	50 k	36 k	38 k
set4	50 k	35 k	37 k
set5	51 k	35 k	37 k
set6	50 k	35 k	37 k
set7	50 k	35 k	34 k
set8	50 k	36 k	37 k

Table 8. Averege memory requirement for three algorithms
(For various file distribution)
(This set is for program 4 executed on node 2)

Chapter 4: Multimode and Dependent-Failure Analysis

Distributed system reliability has been extensively studied in the literature. Most methods use a two-state component model and statistically independent failures. Such simplifying assumptions can be explained by the analytical tractability of the problems being studied, which often precludes exact computations for networks of any size. This chapter presents the need for multimode components and a dependent-failure model and covers methods of computing system reliability using multimode components or dependent failures or both.

It is customary to use components with two states—which assumes only either up (good) or down (bad). In reality, components may operate in degraded states or nodes, lying between the fully operative state and the failed state. For example, a digital transmission link may experience a high bit error rate because of excessive noise. Data packets in error have to be retransmitted, and their effective capacity is reduced. Further, degraded communication capabilities between different locations are much more common than complete communication failures in commercial networks. All such examples rely on the multimode component assumption.

Several authors have studied multimode models to calculate the reliability of communication networks with statistically independent failure of components. To cope with a large number of network states, an approximation method based on generation of most probable states of the system has been developed. Recently, a faster algorithm has been proposed by Yang and Kubat (1987). Their technique generates the states by order of decreasing probability.

Algorithms for the computing reliability generally assume that component failures are statistically independent. Most of the time, this assumption is not realistic because (a) neighboring components are more likely to be affected by the same environmental condition (e.g., in communication networks, interference, jamming, and atmosphere disturbances can simultaneously affect several links) and (b)several components may depend on a common physical resource (power source, shared transmission line). Researchers have also mentioned the problem of correlated failures in command, communication, and control (C^3) networks although no significant analysis has been given. Spragins (1981) has shown that reliability analysis, based on the independent failure assumption may lead to dangerously optimistic results.

Researchers have addressed the issue of *s*-dependent failures under the assumption that components are either up or down. Listed references discuss a variety of approaches to the problem. An earlier work, called q-ψ model, tries to help solve the problem of failure dependencies. However, by using conditional probabilities in conjunction with the chain rule to find joint failures of related links, this theory implicitly considers a specific order of events in which the failures occur. Its major drawback is that the number of conditional probability terms grows exponentially with the size of the system. In addition, the terms must satisfy consistency requirements.

The E-model incorporates more general types of failure dependencies. This model still requires an exponential number of probability terms. The colored network model (CNM) determines the reliability of a communication network with specific kinds of dependent failures. The model's limiting aspect is that it unrealistically requires links incident to a communication center to fail in mutually exclusive groups. A recent model called EBRM (event-based reliability model) incorporates more general cases of independent component failures of communication networks without using conditional probabilities. This model satisfies the consistency requirement too. Heffes and Kumar (1986) use a dependence graph model for handling failure dependencies where the effects of the damage-causing events are utilized. Nodes that are separate in the dependence graph are independent—that is, they cannot be damaged by the same event.

Reprinted from *IEEE Transactions on Reliability*, Volume R-34, Number 3, August 1985, pages 236–240.

Multistate Block Diagrams and Fault Trees

Alan P. Wood, Member IEEE
ESL, Inc., Sunnyvale

Key Words—**Multistate system, Multistate component, Block diagram, Fault tree.**

Reader Aids—
Purpose: Widen state of the art
Special math needed for explanations: Probability theory
Special math needed for results: Same
Results useful to: Reliability engineers and analysts

Abstract—**This paper shows how to model a multistate system with multistate components using binary variables. This modeling technique allows current binary algorithms for block diagrams and fault trees to be applied to multistate systems. Several multistate examples are presented, and some cases in which computational efficiency can be enhanced are discussed.**

1. INTRODUCTION

Most reliability calculations are performed assuming that components and systems are either functioning or failed. This dichotomy is often a reasonable assumption, but the assumption is sometimes made simply because there are no applicable results dealing with more complicated state spaces. There are many situations for which the ability to consider multiple states would be useful in a reliability context. A component or system might have a useful partially-operating mode. For example, if one of two turbines in a power plant is undergoing repair, the plant may be able to generate 50% of its rated electric capacity which is appreciably better than being completely shut down. It may also be useful to differentiate among different failure modes. A valve might fail to open, or fail closed, or fail ruptured, and these different failure modes might have very different effects on system operation. The enlarged state space can be used for actual quantities rather than just qualitative measures. For example, states 0, 1, ..., 100 might be water temperature in degrees Celsius.

Block diagrams and fault trees are used by reliability engineers to find the probability distribution for the state of the system. In the binary case, there are many computer programs with block diagrams or fault trees as input and system reliability as output. Some work has also been done on algorithms for ternary components and systems [2]. This paper shows that multistate components and systems can be analyzed using existing binary algorithms.

Notation for binary systems [1]

- X_i — state of component i; 0 ≡ failed, 1 ≡ functioning.
- P_i — $\Pr\{X_i = 1\}$.
- $\mathbf{X}$ — vector of component states, $(X_1, ..., X_n)$.
- $\phi(\mathbf{X})$ — state of the system; 0 ≡ failed, 1 ≡ functioning; called the system structure function.

Notation for multistate systems [5]

- N_i — best state of component i.
- X_i — state of component i; $X_i \in \{0, 1, ..., N_i\}$.
- $\mathbf{X}$ — vector of component states, $(X_1, ..., X_n)$.
- X_{ij} — indicator; $X_i = 0$, if $X_i < j$; $X_i = 1$, otherwise.
- P_{ij} — $\Pr\{X_{ij} = 1\}$.
- M — best state of the system.
- $\phi(\mathbf{X})$ — state of the system; $\phi(\mathbf{X}) \in \{0, 1, ..., M\}$; called the system structure function.

$\phi^k(\mathbf{X}) = 0$ if $\phi(\mathbf{X}) < k$, 1 if $\phi(\mathbf{X}) \geq k$.

Nomenclature for binary systems [1]

Monotone: a system for which $\phi(\mathbf{0}) = 0$, $\phi(\mathbf{1}) = 1$, and $\phi(\mathbf{X})$ is nondecreasing in $\mathbf{X}$.

Relevant: component i is relevant if there exist component states $(X_i, ..., X_{i-1}, X_{i+1}, ..., X_n)$ such that $\phi(X_1, ..., X_{i-1}, 0, X_{i+1}, ..., X_n) = 0$, and $\phi(X_1, ..., X_{i-1}, 1, X_{i+1}, ..., X_n) = 1$.

Series: a system for which $\phi(\mathbf{X}) = \Pi_{i=1}^{n} X_i = \min(X_i)$ = 1 if all components function, 0 otherwise.

Parallel: a system for which $\phi(\mathbf{X}) = 1 - \Pi_{i=1}^{n} (1 - X_i) = \max(X_i)$ = 0 if all components fail, 1 otherwise.

k-out-of-n:G: a system for which $\phi(\mathbf{X}) = 1$ if at least k components function, 0 otherwise.

Reliability: probability that the system functions (no repair or replacement) = $\Pr\{\phi(\mathbf{X}) = 1\} = \mathrm{E}\{\phi(\mathbf{X})\}$.

Minimal cut set: a minimal set of components such that if all components in the set fail, the system fails.

Minimal path set: a minimal set of components such that if all components in the set function, the system functions.

Nomenclature for multistate systems [5]

Monotone: a system for which $\phi(\mathbf{0}) = 0$, $\phi(N_1, ..., N_n) = 1$, and $\phi(\mathbf{X})$ is nondecreasing in $\mathbf{X}$.

Relevant: component i is relevant if there exist component states $(X_1, ..., X_{i-1}, X_{i+1}, ..., X_n)$ such that $\phi(X_1, ..., X_{i-1}, 0, X_{i+1}, ..., X_n) < \phi(X_1, ..., X_{i-1}, N_i, X_{i+1}, ..., X_n)$.

Series: a system for which $\phi(\mathbf{X}) = \min(X_i)$.

Parallel: a system for which $\phi(\mathbf{X}) = \max(X_i)$.

k-out-of-n:G: a system for which $\phi(\mathbf{X}) = \max\{j$: at least k components are at or above level $j\}$.
Series at level j: a system for which $\phi(\mathbf{X}) = j$ iff $\min(X_i) = j$.
Parallel at level j a system for which $\phi(\mathbf{X}) = j$ iff $\max(X_i) = j$.
k-out-of-n:G at level j: a system for which $\phi(\mathbf{X}) = j$ iff at least k components are at or above level j.
Reliability at level k: probability that the system state is k or higher $= \Pr\{\phi(\mathbf{X}) \geq k\} = \Pr\{\phi^k(\mathbf{X}) = 1\} = \mathrm{E}\{\phi^k(\mathbf{X})\}$.

2. MULTISTATE STRUCTURE FUNCTIONS

The reliability literature of the past 10 years contains many extensions from binary structure functions [1] to multistate structure functions; [3] and [5] contain bibliographies. Most of those extensions pertain to s-coherent structure functions. A binary s-coherent structure function is a binary monotone structure function for which all components are relevant. There are at least 14 ways to define relevant components in multistate systems [3]. The multistate definition of relevant component used in this paper is the least restrictive definition contained in [3]. The formulation in section 3 of this paper only requires multistate monotone structure functions, but the concept of relevance is useful in applying the results.

Extending the concepts of series, parallel, and k-out-of-n:G to multistate systems is straightforward. In addition, these concepts are defined for a particular system state as shown in the nomenclature list. The following relationships exist among these concepts.

1. A series system is n-out-of-n:G; a parallel system is 1-out-of-n:G.
2. $M = \min\{N_i\}$ for a series system; $M = \max\{N_i\}$ for a parallel system.
3. A series system is a series system at level j for all j. This is also true for parallel and k-out-of-n:G systems.

3. REPRESENTATION BY BINARY VARIABLES

This derivation assumes that $\phi(\mathbf{X})$ is a monotone multistate structure function, meaning that the system state never increases when a component deteriorates. To relate multistate systems to binary systems, binary variables X_{ij} and $\phi^k(\mathbf{X})$ are introduced to represent the state of the components and system. Block diagrams and fault trees can be formulated in terms of these binary variables to calculate the reliability of the system at a given level. By definition, the state of component i is the sum of the X_{ij}, and the system state $\phi(\mathbf{X})$ is the sum of the $\phi^k(\mathbf{X})$.

$$X_i = \sum_{j=1}^{N_i} X_{ij}, \qquad \phi(\mathbf{X}) = \sum_{k=1}^{M} \phi^k(\mathbf{X})$$

Each $\phi^k(\mathbf{X})$ is a binary monotone structure function which only depends on binary components.

$$\phi^k(\mathbf{X}) \equiv \phi^k(X_{11}, X_{12}, \ldots, X_{1,N_1}, X_{21}, \ldots, X_{2,N_2}, \ldots, X_{n,N_n})$$

The advantages of this formulation are:

1. Multistate components and multistate monotone structure functions are represented in terms of binary components and binary monotone structure functions.
2. Computer programs for the binary case can be applied to the multistate case with adjustments in the model rather than the program.
3. It is easy to calculate reliability at system level k as $\mathrm{E}\{\phi^k(\mathbf{X})\}$. It is also easy to calculate other measures of system effectiveness such as $\mathrm{E}\{\phi(\mathbf{X})\} = \Sigma_{k=1}^{M} \mathrm{E}\{\phi^k(\mathbf{X})\}$ or the mean system utility, $\mathrm{E}\{U(\phi(\mathbf{X}))\} = \Sigma_{k=1}^{M}(U(k) - U(k-1))\,\mathrm{E}\{\phi^k(\mathbf{X})\}$, where $U(k)$ is the utility of state k, and $U(0) = 0$.

The disadvantages of this formulation are:

1. There are many more components, $(N_1 + 1) \times (N_2 + 1) \times \ldots \times (N_n + 1)$, as opposed to the original n.
2. Many of the binary variables could be irrelevant. If $\phi(\mathbf{X})$ represents a system that is series at level k, then the only relevant components for $\phi^k(\mathbf{X})$ are $X_{1k}, X_{2k}, \ldots, X_{nk}$. It is possible, in principle, to eliminate these irrelevant components leaving M binary coherent structure functions of n components each, but this may be more work than solving the original problem.
3. Mutual s-independence among the components is lost. Even if the original multistate components are s-independent, the binary components X_{ij} are correlated in general since $X_{ij} = 1$ implies $X_{ik} = 1$ for all $k \leq j$. The binary structure functions are also correlated since $\phi^j(\mathbf{X}) = 1$ implies $\phi^k(\mathbf{X}) = 1$ for all $k \leq j$.

These disadvantages are inherent to reliability calculations for multistate systems since the state space is complicated, some states of some components are often irrelevant for determining the probability of a particular system state, and all states of a single component are correlated.

4. BLOCK DIAGRAMS

A block diagram is a logic diagram composed of series and parallel operators. However, a block diagram is not limited to systems which have only series and parallel combinations of components since it is well known that a system can be represented in terms of its minimum path sets or minimum cut sets. Each block in a block diagram represents a binary random variable. The series operator replaces n blocks in series with $\prod_{i=1}^{n} X_i$ while the parallel operator replaces n blocks in parallel with $1 - \prod_{i=1}^{n} (1 - X_i)$.

If the components in a block diagram are s-independent, the calculation of system reliability is simple—just replace X_i by P_i. If the components are s-dependent or if the same component appears in multiple places in the system block diagram (example 1), conditional probability expansions are used, eg,

$$E\{\phi(\mathbf{X})\} = P_1 E\{\phi(\mathbf{X})|X_1 = 1\} + (1 - P_1) E\{\phi(\mathbf{X})|X_1 = 0\}.$$

Block diagrams are used to calculate either the probability that the system functions or the probability that the system fails, depending on how the model is formulated.

Example 1: Consider the system shown in figure 1. If component 1 functions, then the system functions. If not, then one of components 2 and 3 must function, and 2-out-of-3 of components 4-6 must function in order for the system to function. Since components 4-6 appear in two places in the block diagram, a conditional probability expansion must be used to calculate system reliability.

$$E\{\phi(\mathbf{X})\} = P_4 E\{\phi(\mathbf{X})|P_4 = 1\} + (1 - P_4) E\{\phi(\mathbf{X})|P_4 = 0\}$$

$$= 1 - (1 - P_1)\{1 - [P_2 + P_3 - P_2P_3]$$

$$\cdot [(P_5 + P_6 - P_5P_6)P_4 + P_5P_6(1 - P_4)]\}$$

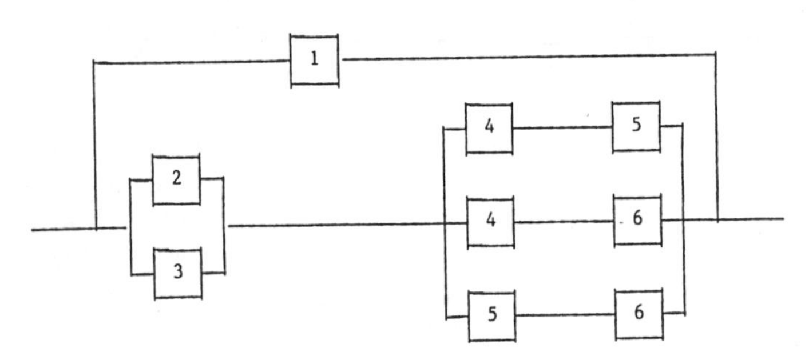

Fig. 1. Success Block Diagram.

For multistate systems, reliability at level k is determined by constructing the block diagram corresponding to $\phi^k(\mathbf{X})$ and then calculating $\Pr\{\phi^k(\mathbf{X}) = 1\}$. If the components are s-independent and if, for each component i, at most one binary variable X_{ij} appears at most once in the block diagram, then reliability can be calculated as though the binary components were s-independent. Otherwise, conditional probability expansions are necessary.

Example 2: Consider a twin-engine jet which can land normally if one engine is at full power and the other engine is at half power. It can land on a foamed runway if one engine is at full power or if both engines are at half power. It will crash if one engine is at half power and the other engine is failed. Let the component state space be {0 = failed, 1 = half power, 2 = full power}, and let the system state space be {0 = crash, 1 = land on foamed runway, 2 = land normally}.

$$\phi(2, 2) = \phi(2, 1) = \phi(1, 2) = 2$$

$$\phi(0, 2) = \phi(2, 0) = \phi(1, 1) = 1$$

$$\phi(0, 1) = \phi(1, 0) = \phi(0, 0) = 0$$

$$\phi^1(\mathbf{X}) = \max(X_{12}, X_{22}, X_{11}X_{21})$$

$$\phi^2(\mathbf{X}) = \max(X_{11}X_{22}, X_{12}X_{21})$$

Block diagrams for ϕ^1 and ϕ^2 are shown in figure 2. Both the component number and the minimum required component state appear inside the blocks in the block diagram. A conditional probability expansion yields:

$$E\{\phi^1(\mathbf{X})\} = P_{12} E\{\phi^1(\mathbf{X})|X_{12} = 1\}$$

$$+ (1 - P_{12}) E\{\phi^1(\mathbf{X})|X_{12} = 0\}$$

$$= P_{12} + (1 - P_{12})P_{22} + (1 - P_{12})(1 - P_{22}) \times$$

$$E\{\phi^1(\mathbf{X})|X_{12} = 0, X_{22} = 0\}$$

$$= P_{12} + (1 - P_{12})P_{22} + (1 - P_{12})(1 - P_{22}) \times$$

$$(P_{11} - P_{12})(P_{21} - P_{22})$$

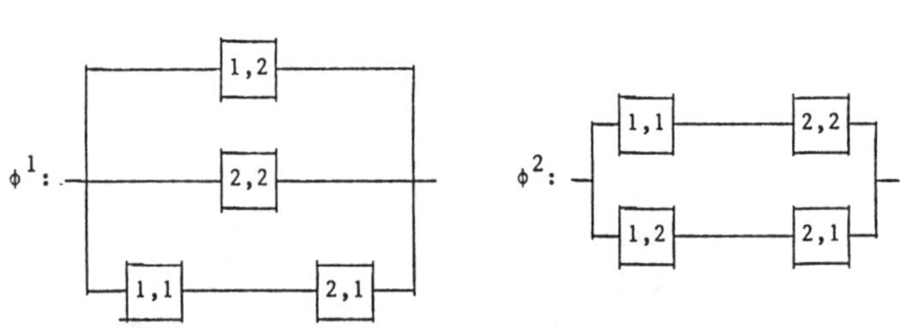

Fig. 2. Multistate Block Diagrams.

where $E\{X_{11}|X_{12} = 0\} = \Pr\{X_1 = 1\} = \Pr\{X_1 \geq 1\} - \Pr\{X_1 \geq 2\} = P_{11} - P_{12}$. Some computer programs might need to be modified to handle this type of s-dependence.

The block diagrams can be simplified in some cases. Consider a multistate system which is k-out-of-n:G at each system level m where k can depend on m. Then, for each level k,

$$\phi^k(\mathbf{X}) = \phi^k(X_{1k}, X_{2k}, \ldots, X_{nk}).$$

Component states other than k are unimportant in determining the value of ϕ^k. This means that if the original multistate components are s-independent, then the binary variables X_{ik} are s-independent in each ϕ^k. No conditional probability expansions are necessary since the X_{ik}'s are s-independent for fixed k, and for every i, only one X_{ik} appears in the block diagram for ϕ^k. This situation occurs in example 3.

Example 3: Let $\phi(X_1, X_2, X_3)$ be parallel at level 1, 2-out-of-3:G at level 2, and series at level 3, and let X_1, X_2, X_3 be mutually s-independent.

$$\phi^1(\mathbf{X}) = 1 - (1 - X_{11})(1 - X_{21})(1 - X_{31})$$

$$\phi^3(\mathbf{X}) = X_{13}X_{23}X_{33}$$

$$\phi^2(\mathbf{X}) = 1 - (1 - X_{12}X_{22})(1 - X_{12}X_{32})(1 - X_{22}X_{32})$$

The block diagram for each ϕ^k is shown in figure 3. Reliability is calculated using s-independence for ϕ^1 and ϕ^3, but conditional probabilities are necessary to calculate $\Pr\{\phi^2(\mathbf{X}) = 1\}$ correctly.

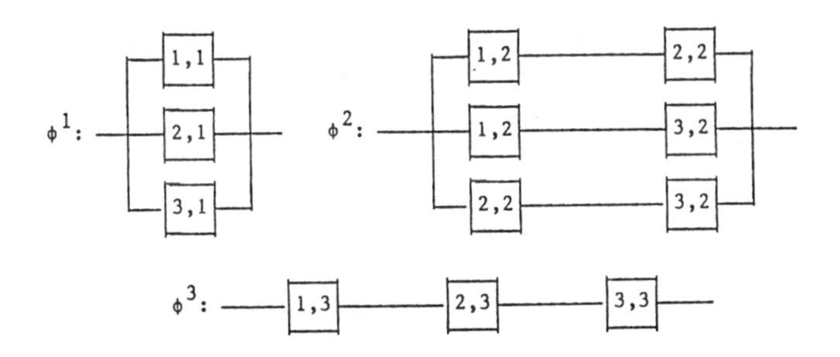

Fig. 3. Multistate Block Diagrams.

An even more specialized case occurs when all the ϕ^k have identical block diagrams. Then a system reliability analysis requires only a single block diagram with different input probabilities for each system level. This occurs for a system which is series, parallel, or k-out-of-n:G at all levels. It also occurs in example 4.

Example 4: A power plant generates 0%, 25%, 50%, 75%, or 100% (corresponding to states 0, 1, 2, 3, and 4) of rated electric capacity depending on the condition of the turbine and the amount of steam flow reaching the turbine. Three turbines are available, and the one which can maximize power output is always used. Components 1-3 are the turbines, and component 4 represents the rate of flow at the turbine. The block diagram for all levels of this system is shown in figure 4. Assuming s-independent components,

$$\Pr\{\phi^k(\mathbf{X}) = 1\} = P_{4k}[1-(1 - P_{1k})(1 - P_{2k})(1 - P_{3k})].$$

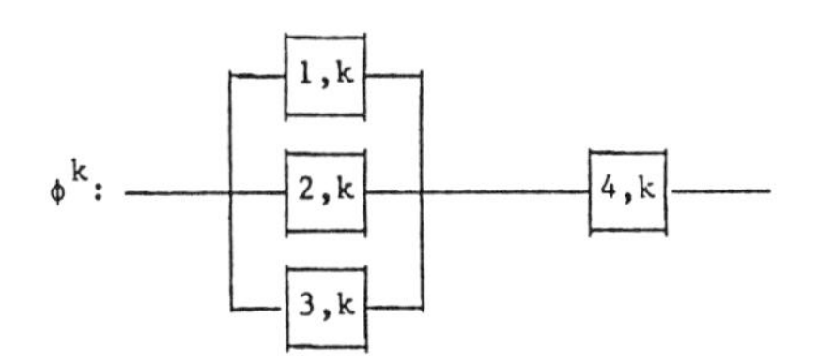

Fig. 4. Multilevel Block Diagram.

5. FAULT TREES

Fault trees and event trees are logic diagrams consisting of a top event and a structure delineating the ways in which the top event can occur. The term fault tree is used when the top event is system failure while the term event tree is used for system success. The tree structure consists of *and gates* and *or gates* which perform the same functions in the fault tree as the series and parallel operators in the block diagram.

Computer programs based on fault trees generally proceed by determining the minimal cut sets of a system. Minimal cut sets are useful to reliability engineers since they provide a qualitative measure of the most important components in the system. Let K_j be the minimal cut set j; $j = 1, 2, \ldots, t$. The value of $\phi(\mathbf{X})$ is calculated from:

$$\phi(\mathbf{X}) = \prod_{j=1}^{t} \left[1 - \prod_{i \in K_j} (1 - X_i)\right]. \quad (1)$$

If the components are s-independent and each component appears in at most one minimal cut set, then taking s-expectations in (1) yields:

$$\mathrm{E}\{\phi(\mathbf{X})\} = \prod_{j=1}^{t} \left[1 - \prod_{i \in K_j} (1 - P_i)\right]. \quad (2)$$

Conditional probabilities are used if a component appears in more than one minimal cut set. Current fault-tree computer-programs might need to be slightly modified to properly account for s-dependence.

The extension of fault-tree analysis to the multistate case is very similar to the extension of block diagrams. A fault tree or event tree is constructed for each system level. Basic events are X_{ik} for event trees and $1 - X_{ik}$ for fault trees. A binary fault-tree computer-program is used to determine $\mathrm{E}\{\phi^k(\mathbf{X})\}$ for each system level k. More information on fault trees and a different multistate formulation can be found in [3].

Example 5: The event trees for the structure function contained in example 3 are shown in figure 5.

The special cases pertaining to multistate block diagrams have straightforward analogies to fault trees and event trees. In particular if the ϕ^k have identical block diagrams, the fault trees or event trees for each ϕ^k are also identical. This is very useful computationally since the minimum cut sets need to be found only once.

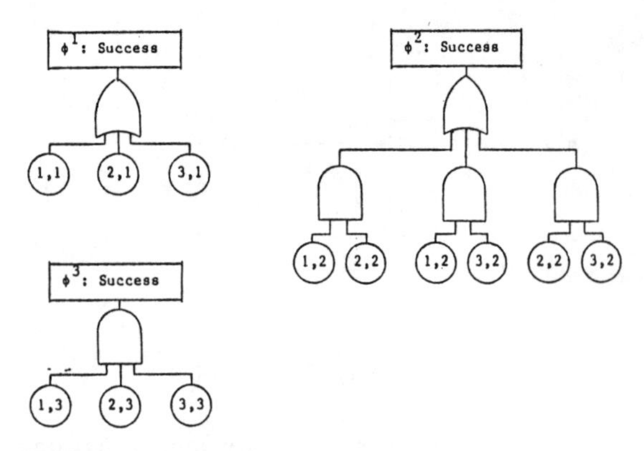

Fig. 5. Multistate Event Trees.

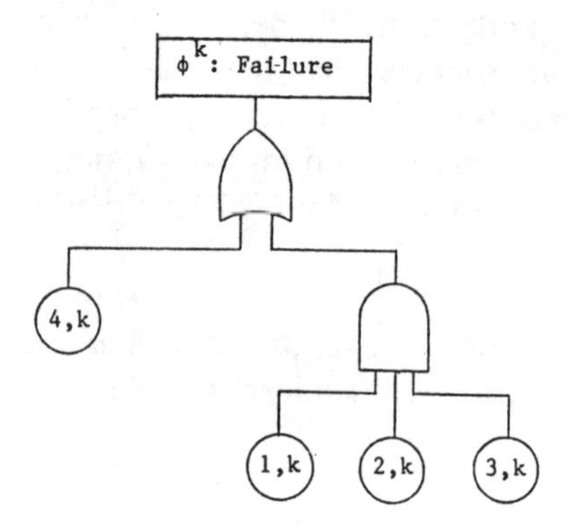

Fig. 6. Multilevel Fault Tree.

Example 6: The fault tree for the system contained in example 4 is shown in figure 6. The minimal cut sets for ϕ^k are $\{(4, k)\}$ and $\{(1, k), (2, k), (3, k)\}$. Using (2) and assuming *s*-independent components, $\mathrm{E}\{\phi^k(\mathbf{X})\} = P_{4k}[1 - (1 - P_{1k})(1 - P_{2k})(1 - P_{3k})]$, the same result derived in example 4.

REFERENCES

[1] R. E. Barlow, F. Proschan, *Statistical Theory of Reliability and Life Testing*, Holt, Rinehart, & Winston, Inc., 1975.

[2] L. Cardarola, "Fault-tree analysis with multistate components," *Synthesis and Analysis Methods for Safety and Reliability Studies*, G. Apostolakis, S. Garribba, and G. Volta eds., Plenum Press, 1980.

[3] Y. Hatoyama, "Reliability analysis of 3-state systems," *IEEE Trans. Reliability*, vol R-28, 1975 Dec, pp 386-393.

[4] F. Ohi, T. Nishida, "On multistate coherent systems," *IEEE Trans. Reliability*, vol R-33, 1984 Oct, pp 284-287.

[5] A. Wood, "Multistate reliability", PhD thesis, Stanford University, Dept. of Oper. Res. and Dept. of Stat., Stanford, California. 1983.

AUTHOR

Dr. Alan P. Wood; ESL, Inc.; POBox 3510; Sunnyvale, California 94088-3510 USA

Alan Wood (M'85, born 1955) received a BA and MS in mathematics from the University of Colorado in 1976 and 1978. He received a MS in statistics and a PhD in operations research from Stanford University in 1983. From 1978-1983, he analyzed nuclear power plant reliability at General Electric and Los Alamos Technical Associates. His current work involves the development of mathematical models to analyze electronic circuit reliability at ESL, Inc.

Manuscript TR83-157 received 1983 October 14; revised 1985 May 28.

★★★

Reprinted from *IEEE Transactions on Reliability*, Volume R-34, Number 5, December 1985, pages 473–479.

Reliability Evaluation of Multistate Systems with Multistate Components

Terje Aven
University of Oslo, Oslo

Key Words—**Multistate system, Multistate component, s-Coherent system, Reliability bound, Path set, Cut set.**

Reader Aids—
Purpose: Present two algorithms
Special math needed for explanations: Elementary probability theory
Special math needed to use results: None
Results useful to: Reliability analysts

Abstract—**This paper presents two efficient algorithms for reliability evaluation of monotone multistate systems with s-independent multistate components. The algorithms are based on the Doulliez & Jamoulle decomposition method. Algorithm 1 requires the minimal paths to be known; Algorithm 2 requires the minimal cuts to be known (the state of the system need not be specified for each vector of component states). Computer programs for implementing the algorithms are given. Computational-times are presented, and compared with the "Inclusion-Exclusion Method" and the "State Enumeration Method". The results demonstrate clearly the superiority of the algorithms to the two other methods.**

1. INTRODUCTION

In reliability theory a key problem is to find how the reliability of a complex system can be determined from knowledge of the reliabilities of its components. One inherent weakness of traditional theory is that the system and the components are described just as functioning or failed. This approach oversimplifies many real situations where the systems and their components are capable of assuming a whole range of levels of performance, varying from perfect functioning to complete failure.

Fortunately, some recent papers [3-4, 6-11] have made important contributions to a theory of multistate systems of multistate components. However, little attention has been devoted to finding efficient algorithms for (exact) reliability computation of multistate systems with multistate components. Some formulae, based on the inclusion-exclusion principle are established [6, 8, 10], but they are practical only if the number of minimal paths or cuts is very small. An algorithm, which extends Abraham's network algorithm [1] to multistate systems, has recently been reported [9].

Finding efficient algorithms is very important since reliability computations of multistate systems with multistate components can be extremely time-consuming.

2. NOTATION, NOMENCLATURE, ASSUMPTIONS

Notation

n	number of components
$M_i + 1$	number of states of component i $(0 < M_i < \infty)$, $i = 1, 2, \ldots, n$
x_{ij}	state j of component i; $j = 0, 1, 2, \ldots, M_i$, $x_{i0} < x_{i1} < \ldots < x_{iM_i}$, $i = 1, 2, \ldots, n$
S_i	$\{x_{i0}, x_{i1}, \ldots, x_{iM_i}\}$, $i = 1, 2, \ldots, n$
α	$(\alpha_1, \alpha_2, \ldots, \alpha_n)$, $\alpha_i \in S_i$, $i = 1, 2, \ldots, n$, α any name
S	state space; $S = \times_{i=1}^{n} S_i = \{\mathbf{x}; x_i \in S_i, i = 1, 2, \ldots, n\}$
$\psi(\mathbf{x})$	state of the system when the vector of component states is $\mathbf{x}$
$M + 1$	number of states of the system $(0 < M < \infty)$
ψ_j	state j of the system, $j = 0, 1, 2, \ldots, M$, $\psi_0 < \psi_1 < \ldots < \psi_M$ (ψ is a function from $S - \{\psi_0, \psi_1, \ldots, \psi_M\}$)
ψ_j'	a given state of the system
n_p	number of min-paths to level ψ_j'
$\mathbf{y}^l$	min-path l to level ψ_j', $l = 1, 2, \ldots, n_p$
n_c	number of min-cuts to level ψ_j'
$\mathbf{z}^l$	min-cut l to level ψ_j', $l = 1, 2, \ldots, n_c$
X_i	state of component i $(X_i \in S_i)$; r.v.
$R_i(j)$	$\Pr\{X_i \geq x_{ij}\}$, $i = 1, 2, \ldots, n$, $j = 0, 1, \ldots, M_i$
$R(j)$	$\Pr\{\psi(\mathbf{X}) \geq \psi_j\}$, $j = 0, 1, \ldots, M$
$Q(j)$	$\Pr\{\psi(\mathbf{X}) < \psi_j\}$, $j = 0, 1, \ldots, M$
$\mathbf{x} = \mathbf{y}$	iff $x_i = y_i$, for all i
$\mathbf{x} \leq \mathbf{y}$	iff $x_i \leq y_i$, for all i
$\mathbf{x} < \mathbf{y}$	iff $x_i \leq y_i$, for all i and there exists at least one i such that $x_i < y_i$.
$(i_1, i_2, \ldots, i_n)$	$(x_{1i_1}, x_{2i_2}, \ldots, x_{ni_n})$
$\prod_{j \in \phi}$	1

Nomenclature

Path vector: A vector $\mathbf{y}$ is a path vector to level ψ_j iff $\psi(\mathbf{y}) \geq \psi_j$. A path vector to level ψ_j, $\mathbf{y}$, is minimal iff $\psi(\mathbf{x}) < \psi_j$ for all $\mathbf{x} < \mathbf{y}$.

Min-path: Minimal path vector to level ψ_j'.

Cut vector: A vector $\mathbf{z}$ is a cut vector to level ψ_j iff $\psi(\mathbf{z}) < \psi_j$. A cut vector to level ψ_j, $\mathbf{z}$, is minimal iff $\psi(\mathbf{x}) \geq \psi_j$ for all $\mathbf{x} > \mathbf{z}$.

Min-cut: Minimal cut vector to level ψ_j'.

Assumptions

1. The system is a *multistate monotone system*:
   A. $\psi(\mathbf{x})$ is non-decreasing in each argument
   B. $\psi(\mathbf{0}) = \psi_0$, and $\psi(\mathbf{M}) = \psi_M$
2. The X_i's are mutually s-independent.

3. TWO ALGORITHMS FOR EVALUATING $R(j')$ ($Q(j')$)

The two algorithms are based upon the Doulliez & Jamoulle [5] decomposition method, in which the state space is decomposed into certain sets: a set of *acceptable* states, a set of *non-acceptable* states, and sets of *unspecified* states. For algorithm 1, a state $\mathbf{x}$ is acceptable iff $\psi(\mathbf{x}) \geq \psi_{j'}$ and non-acceptable iff $\psi(\mathbf{x}) < \psi_{j'}$. For algorithm 2, a state $\mathbf{x}$ is non-acceptable iff $\psi(\mathbf{x}) \geq \psi_{j'}$ and acceptable iff $\psi(\mathbf{x}) < \psi_{j'}$. Each set of unspecified states is again decomposed, and so forth until there is no set of unspecified states left. Associated with each set of unspecified states C there are two *limiting state points* $\mathbf{b}^0$ and $\mathbf{b}$ such that:

$C = \{\mathbf{x} \in S, \mathbf{b} \leq \mathbf{x} \leq \mathbf{b}^0\}$, for alg. 1

$C = \{\mathbf{x} \in S; \mathbf{b}^0 \leq \mathbf{x} \leq \mathbf{b}\}$, for alg. 2.

Initially, $\mathbf{b}^0 = \mathbf{M}$ and $\mathbf{b} = \mathbf{0}$ for alg. 1,

whereas $\mathbf{b}^0 = \mathbf{0}$ and $\mathbf{b} = \mathbf{M}$ for alg. 2.

Associated with each set of acceptable states A there is a critical value $\mathbf{v}^0 \in C$ (C is the set of unspecified states which is decomposed) such that:

$A = \{\mathbf{x} \in C; \mathbf{v}^0 \leq \mathbf{x}\}$, for alg. 1

$A = \{\mathbf{x} \in C; \mathbf{x} \leq \mathbf{v}^0\}$, for alg. 2.

To each set of non-acceptable states B there is associated a critical value $\mathbf{v}$ such that:

$B = \{\mathbf{x} \in C; x_i < v_i$ for at least one $i\}$, for alg. 1

$B = \{\mathbf{x} \in C; x_i > v_i$ for at least one $i\}$, for alg. 2.

The $\Pr\{\mathbf{X} \in A\}$ are easy to compute; and because the generated A's are non-overlapping and include all acceptable states, the $R(j')$ ($Q(j')$) is the sum of the $\Pr\{\mathbf{X} \in A\}$ terms. To make the number of sets of unspecified states as small as possible, a large $\mathbf{v}$ and a small $\mathbf{v}^0$ are sought for algorithm 1, whereas a small $\mathbf{v}$ and a large $\mathbf{v}^0$ are sought for algorithm 2.

Algorithm 1.

1. Set $R = 0, k = 1, b_i^0 = M_i, b_i = 0; i = 1, 2, \ldots, n$.
2. Choose a min-path $\mathbf{y}^{\ell_0}$ such that:

$\mathbf{y}^{\ell_0}$ maximizes $H(\mathbf{y}^\ell) \equiv \sum_{i=1}^{n} (b_i^0 - \max\{y_i^\ell, b_i\}), \mathbf{y}^\ell \leq \mathbf{b}^0$.

Set $\bar{y}_i = \min\{y_i^\ell, \mathbf{y}^\ell \leq \mathbf{b}^0\}, i = 1, 2, \ldots, n$.

3. Set $v_i^0 = \max\{y_i^{\ell_0}, b_i\}$

$v_i = \max\{\bar{y}_i, b_i\}, i = 1, 2, \ldots, n$.

4. Set $p_i = R_i(v_i^0) - R_i(b_i^0 + 1), i = 1, 2, \ldots, n$,

$R \leftarrow R + \prod_{i=1}^{n} p_i$.

5. Let $a_d, d = 1, 2, \ldots, s$ be the i's, $i \in \{1, 2, \ldots, n\}$, satisfying $v_i < v_i^0$. If no such i exists, then set $s = 0$. If $s \geq 1$, set for $d = 1, 2, \ldots, s, i = 1, 2, \ldots, n$

$$b_i^0(d + k - 1) = \begin{cases} v_i^0 - 1, & \text{for } i = a_d \\ b_i, & \text{otherwise} \end{cases}$$

$$b_i(d + k - 1) = \begin{cases} v_i^0, & \text{for } i < a_d \\ v_i, & \text{otherwise.} \end{cases}$$

Set $k \leftarrow k - 1 + s$.

6. If $k \neq 0$, then set $b_i^0 = b_i^0(k)$

$b_i = b_i(k), i = 1, 2, \ldots, n$,

Go To 2

Else Stop; R equals the sought value $R(j')$.
End of Algorithm 1.

Algorithm 2.

1. Set $Q = 0, k = 1, b_i^0 = 0, b_i = M_i; i = 1, 2, \ldots, n$.
2. Choose a min-cut $\mathbf{z}^{\ell_0}$ such that

$\mathbf{z}^{\ell_0}$ maximizes $H(\mathbf{z}^\ell) \equiv \sum_{i=1}^{n} (\min\{z_i^\ell, b_i\} - b_i^0), \mathbf{z}^\ell \geq \mathbf{b}^0$.

Set $\tilde{z}_i = \max\{z_i^\ell, \mathbf{z}^\ell \geq \mathbf{b}^0\}, i = 1, 2, \ldots, n$.

3. Set $v_i^0 = \min\{z_i^{\ell_0}, b_i\}$

$v_i = \min\{\tilde{z}_i, b_i\}, i = 1, 2, \ldots, n$.

4. Set $p_i = R_i(b_i^0) - R_i(v_i^0 + 1), i = 1, 2, \ldots, n$,

$Q \leftarrow Q + \prod_{i=1}^{n} p_i$.

5. Let $a_d, d = 1, 2, \ldots, s$ be the i's, $i \in \{1, 2, \ldots, n\}$, satisfying $v_i > v_i^0$. If no such i exists, then set $s = 0$. If $s \geq 1$, set for $d = 1, 2, \ldots, s, i = 1, 2, \ldots, n$

$$b_i^0(d + k - 1) = \begin{cases} v_i^0 + 1, & \text{for } i = a_d \\ b_i^0, & \text{otherwise} \end{cases}$$

$$b_i(d + k - 1) = \begin{cases} v_i^0, & \text{for } i < a_d \\ v_i, & \text{otherwise.} \end{cases}$$

Set $k \leftarrow k - 1 + s$.

6. If $k \neq 0$, then set $b_i^0 = b_i^0(k)$

$b_i = b_i(k)$, $i = 1, 2, \ldots, n$,

Go To 2
Else Stop; Q equals the sought value $Q(j')$.

End of Algorithm 2.

Remarks (Remarks 3 and 4 are proved in the appendix)

1. k represents the number of stored sets of unspecified states, which is the number of stored limiting state points.
2. $\mathbf{b}^0(\cdot)$ and $\mathbf{b}(\cdot)$ are the limiting state points for the stored sets of unspecified states.
3. There always exists an ℓ such that $\mathbf{y}^\ell \leq \mathbf{b}^0$ ($\mathbf{z}^\ell \geq \mathbf{b}^0$) (this statement is equivalent to "$\mathbf{b}^0$ is acceptable").
4. A state $\mathbf{x} \in A$ (B) is acceptable (non-acceptable).
5. In step 2, any $\mathbf{y}^{\ell_0} \leq \mathbf{b}^0$ ($\mathbf{z}^{\ell_0} \geq \mathbf{b}^0$) can be chosen. To find a small $\mathbf{v}^0$ for algorithm 1 and a large $\mathbf{v}^0$ for algorithm 2, $\mathbf{y}^{\ell_0}$ ($\mathbf{z}^{\ell_0}$) has been chosen to maximize $H(\mathbf{y}^\ell)$, $\mathbf{y}^\ell \leq \mathbf{b}^0$ ($H(\mathbf{z}^\ell)$, $\mathbf{z}^\ell \geq \mathbf{b}^0$). The rule for choosing ℓ_0 is very efficient concerning the number of sets of unspecified states and required computer times.
6. The values of $R(Q)$ are non-decreasing and give lower bounds on $R(j')$ ($Q(j')$). Non-increasing upper bounds on $R(j')$ ($Q(j')$) are given by $1 - \sum_B \Pr\{B\}$. The bounds converge to $R(j')$ ($Q(j')$). To incorporate the upper bounds into algorithms 1 and 2, the following statement is added to step 4 [5]:

Algorithm 1. Add to step 4: Set —

$$p_i^{(1)} = R_i(v_i) - R_i(b_i^0 + 1),$$

$$p_i^{(2)} = R_i(b_i) - R_i(v_i),$$

$$p_i^{(3)} = p_i^{(1)} + p_i^{(2)}; \text{ for } i = 1, 2, \ldots, n$$

$$Q \leftarrow Q + \sum_{i=1}^{n} p_i^{(2)} \left(\prod_{m=1}^{i-1} p_m^{(1)}\right) \prod_{m=i+1}^{n} p_m^{(3)}.$$

Algorithm 2. Add to step 4: Set —

$$p_i^{(1)} = R_i(b_i^0) - R_i(v_i + 1),$$

$$p_i^{(2)} = R_i(v_i + 1) - R_i(b_i + 1),$$

$$p_i^{(3)} = p_i^{(1)} + p_i^{(2)}; \text{ for } i = 1, 2, \ldots, n$$

$$R \leftarrow R + \sum_{i=1}^{n} p_i^{(2)} \left(\prod_{m=1}^{i-1} p_m^{(1)}\right) \prod_{m=i+1}^{n} p_m^{(3)}.$$

For algorithms 1 & 2, add to step 1:

$Q = R = 0$.

The Q's & R's satisfy —

$R \leq R(j') \leq 1 - Q$.

For algorithms 1 & 2, instead of the stopping rule: "if $k = 0$, then Stop", a rule for example of the form "if $|1 - Q - R| < \epsilon$, then Stop" can be used.

By repeated applications of the algorithm we can find $R(j')$ for different values of j'; the min-paths (min-cuts) for each level must be known.

Only examples of application of alg. 1 are given. The first, very simple, example illustrates the use of the algorithm step by step.

4. EXAMPLE 1

Let $n = 3$, $M_i = 2$, $R_i(1) = 0.9$, $R_i(2) = 0.7$, for $i = 1, 2, 3$,

$n_p = 3$,

$\mathbf{y}^1 = (2, 0, 2)$, $\mathbf{y}^2 = (2, 1, 1)$, $\mathbf{y}^3 = (1, 2, 2)$.

The algorithm:

1. $R = 0$, $k = 1$, $\mathbf{b}^0 = (2, 2, 2)$, $\mathbf{b} = (0, 0, 0)$
2. $\ell_0 = 1$, $\mathbf{y}^{\ell_0} = (2, 0, 2)$ (for example) $\tilde{\mathbf{y}} = (1, 0, 1)$.
3. $\mathbf{v}^0 = \mathbf{y}^{\ell_0} = (2, 0, 2)$, $\mathbf{v} = \tilde{\mathbf{y}} = (1, 0, 1)$.
4. $\mathbf{p} = (0.7, 1.0, 0.7)$, $R = 0.490$.
5. $a_1 = 1$, $a_2 = 3$, $s = 2$

   $\mathbf{b}^0(1) = (1, 2, 2)$, $\mathbf{b}(1) = (1, 0, 1)$

   $\mathbf{b}^0(2) = (2, 2, 1)$, $\mathbf{b}(2) = (2, 0, 1)$, $k = 2$.

6. $\mathbf{b}^0 = (2, 2, 1)$, $\mathbf{b} = (2, 0, 1)$; goto 2.

2. $\ell_0 = 2$, $\mathbf{y}^{\ell_0} = (2, 1, 1)$, $\tilde{\mathbf{y}} = \mathbf{y}^{\ell_0} = (2, 1, 1)$
3. $\mathbf{v}^0 = \mathbf{v} = \mathbf{y}^{\ell_0} = (2, 1, 1)$.
4. $\mathbf{p} = (0.7, 0.9, 0.2)$, $R = 0.490 + 0.126 = 0.616$.
5. $s = 0$, $k = 1$;
6. $\mathbf{b}^0 = \mathbf{b}^0(1) = (1, 2, 2)$, $\mathbf{b} = \mathbf{b}(1) = (1, 0, 1)$; goto 2.

2. $\ell_0 = 3$, $\mathbf{y}^{\ell_0} = (1, 2, 2)$, $\tilde{\mathbf{y}} = \mathbf{y}^{\ell_0} = (1, 2, 2)$.
3. $\mathbf{v}^0 = \mathbf{v} = \mathbf{y}^{\ell_0} = (1, 2, 2)$.

4. $\mathbf{p} = (0.2, 0.7, 0.7)$, $R = 0.616 + 0.098 = 0.714$.

5. $s = 0$, $k = 0$.

6. $R(j') = 0.714$; Stop.

5. EXAMPLE 2

Let $n = 8, 14, 15$, or 16; $M_i = 4$ for $i = 1, 2$; $M_i = 3$ for $i = 3, 4, ..., 8$; $M_i = 2$ for $i = 9, ..., 12$; $M_i = 1$ for $i \geq 13$;

$R_i(1) = 0.9$ for $i \geq 1$; $R_i(2) = 0.8$ for $i = 1, ..., 12$; $R_i(3) = 0.7$ for $i = 1, ..., 8$; $R_i(4) = 0.6$ for $i = 1, 2$;

$n_p = 8, 12, 13$, or 14.

When $n = 16$ and $n_p = 14$ the min-paths are given in table 1. The min-paths when $n = 16$ and $n_p = 13$ are obtained from table 1 by removing the last row; the min-paths when $n = 15$ and $n_p = 14$ are obtained by removing the last column, and so on. The probabilities $Q(j')$ for different values of (n_p, n) were computed by means of the computer program 1 in the appendix and the two "Inclusion-Exclusion Method" (IEM) programs in the Supplement [12]; in addition $Q(j')$ was computed for $n = 8$, $n_p = 8, 12, 13, 14$ by means of the "State Enumeration Method" (SEM) program in the Supplement [12]. For the IEM, $Q(j')$ is computed by using the equality

$$Q(j') = \sum_{h=1}^{n_p} (-1)^{h+1} S_h, \tag{5.1}$$

$$S_h \equiv \sum_{1 \leq i_1 < i_2 < \ldots < i_h \leq n_p} \Pr\left\{ \bigcap_{r=1}^{h} [\mathbf{X} \geq \mathbf{y}^{i_r}] \right\} \tag{5.2}$$

$$= \sum_{1 \leq i_1 < \ldots \leq n_p} \prod_{m=1}^{n} R_m \left(\max_{1 \leq r \leq h} y_m^{i_r} \right).$$

TABLE 1
Min-paths

(3	3	0	0	1	1	1	2	1	1	0	2	1	1	0	0)
(0	1	3	1	1	3	2	2	2	1	1	1	0	1	1	1)
(0	4	0	2	3	1	2	0	0	2	2	0	1	1	0	0)
(2	0	0	2	2	0	2	3	1	1	1	0	1	0	1	0)
(4	0	1	3	1	0	2	0	2	2	0	1	1	0	1	1)
(2	2	2	1	3	0	0	1	2	0	1	1	0	0	1	1)
(1	1	1	1	0	2	3	2	0	2	2	1	1	1	0	1)
(1	0	1	3	2	0	1	2	1	1	2	0	0	0	1	1)
(1	1	2	0	1	3	3	2	2	2	1	0	0	1	0	1)
(3	4	2	3	0	1	0	0	0	1	2	1	1	1	0	0)
(4	0	2	2	2	0	1	0	1	1	1	2	0	0	0	1)
(0	1	2	2	1	1	3	0	2	1	1	2	1	1	1	0)
(1	1	3	3	0	2	2	1	1	1	0	0	0	1	0	1)
(2	1	3	1	1	1	2	0	0	0	2	2	1	1	1	1)

For the SEM, all states $\mathbf{x}$ are considered and if x is acceptable, $\Pr\{\mathbf{X} = \mathbf{x}\}$ is computed. All programs were run on VAX 11/750 under UNIX at the University of California, Berkeley. The $Q(j')$'s are given in table 2. Table 3 gives the required computer times (CPU-times) using algorithm 1 and the IEM (for the IEM the smallest CPU time for the two programs is stated). The CPU-times using algorithm 1 are much smaller than the corresponding CPU-times using the IEM for large n_p. On the other hand, the CPU-times using the IEM are smaller than the corresponding CPU-times using algorithm 1 for small n_p and large n. There exist very few real systems with small n_p and large n. The CPU-times for IEM are approximately doubled as n_p increases by 1; this is anticipated since the number of addends in (5.1) is $2^{n_p} - 1$.

TABLE 2
$Q(j')$

		n			
		8	14	15	16
n_p	8	0.700	0.504	0.482	0.465
	12	0.750	0.548	0.533	0.517
	13	0.753	0.566	0.552	0.534
	14	0.758	0.579	0.564	0.545

TABLE 3
CPU-times in seconds using "Alg. 1"
and the "Inclusion-Exclusion Method" ("IEM")

			n			
			8	14	15	16
n_p	8	"Alg. 1"	1.1	2.5	2.9	3.3
		"IEM"	1.4	2.1	2.1	2.1
	12	"Alg. 1"	1.4	4.9	6.5	7.2
		"IEM"	12.0	16.5	18.0	18.8
	13	"Alg. 1"	1.7	5.9	8.2	9.5
		"IEM"	22.7	33.9	33.9	36.5
	14	"Alg. 1"	1.8	6.7	10.0	10.9
		"IEM"	50.1	74.1	73.9	78.5

TABLE 4
CPU-times in seconds using the
"State Enumeration Method" ("SEM")

		$n = 8$
n_p	8	63.5
	12	90.8
	13	93.7
	14	99.9

Table 4 gives the CPU-times using the SEM. The reason why $Q(j')$ has not been computed for $n > 8$ with this method is that the CPU-times are so large. For example, if $n = 14$ and $n_p = 8$, the CPU-time is approximately $63.5 \cdot 81$ second ≈ 86 minutes. Table 5 gives the number of sets of unspecified states (#C) (= #A = #B = the number of visits to step 2); the maximum value of k, km = max number of stored sets of unspecified states; and the total number of states. Because km is so low ($km \leq 20$), the equality $\mathbf{v}^0 = \mathbf{v}$ occurs very often.

TABLE 5
Number of C sets (#C), maximum value of k (km) and, number of states

			n			
			8	14	15	16
n_p	8	#C	71	245	299	297
		km	10	17	17	16
	12	#C	125	545	712	771
		km	12	16	20	18
	13	#C	136	628	835	1009
		km	12	17	20	19
	14	#C	143	775	835	1233
		km	11	17	20	19
number of states			102400	33177600	66355200	132710400

TABLE 6
Input data for the Example 2, section 5, when n = 8 and n_p = 12.

8	12							
4								
4								
3								
3								(M_i)
3								
3								
3								
3								
0.9								
0.8								
0.7								
0.6								
0.9								
0.8								
0.7								
0.6								
0.9								
0.8								
0.7								
0.9								
0.8								
0.7								($R_i(j)$)
0.9								
0.8								
0.7								
0.9								
0.8								
0.7								
0.9								
0.8								
0.7								
0.9								
0.8								
0.7								
3	3	0	0	1	1	1	2	
0	1	3	1	1	3	2	2	
0	4	0	2	3	1	2	0	
2	0	0	2	2	0	2	3	
4	0	1	3	1	0	2	0	
2	2	2	1	3	0	0	1	
1	1	1	1	0	2	3	2	($\mathbf{y}^i$)
1	0	1	3	2	0	1	2	
1	1	2	0	1	3	3	2	
3	4	2	3	0	1	0	0	
4	0	2	2	2	0	1	0	
0	1	2	2	1	1	3	0	

APPENDIX

Proof of remarks 3

Proof "$\mathbf{y}^\ell \le \mathbf{b}^0$": Let $k \le u \le k - 1 + s\,(s \ge 1)$, and let i_u be such that $b_i^0(u) = b_i^0$, $i \ne i_u$, and $b_{i_u}^0(u) = v_{i_u}^0 - 1$. It is sufficient to show that there exists an ℓ_u such that $\mathbf{y}^{\ell_u} \le \mathbf{b}^0(u)$ under the assumption that there exists an ℓ such that $\mathbf{y}^\ell \le \mathbf{b}^0$; initially, $\mathbf{y}^\ell \le \mathbf{b}^0$ for some ℓ since $\psi(\mathbf{b}^0) = \psi_M \ge \psi_{j'}$. We have $\tilde{y}_{i_u} \le v_{i_u} \le v_{i_u}^0 - 1 = b_{i_u}^0(u)$. But $\tilde{y}_{i_u} = y_{i_u}^{\ell_u}$ for some ℓ_u where $\mathbf{y}^{\ell_u} \le \mathbf{b}^0$. It follows that $\mathbf{y}^{\ell_u} \le \mathbf{b}^0(u)$; remember that $b_i^0(u) = b_i^0$, $i \ne i_u$. *Q.E.D.*

The proof for algorithm 2 is similar.

Proof of remark 4

Proof "A, Algorithm 1": It suffices to show that $\mathbf{v}^0$ is acceptable. From the definition of v_i^0 we have $\mathbf{v}^0 \ge \mathbf{y}^{\ell_0}$. Hence $\psi(\mathbf{v}^0) \ge \psi(\mathbf{y}^{\ell_0}) \ge \psi_{j'}$, ie, $\mathbf{v}^0$ is acceptable.

Proof "B, Algorithm 1": Assume $b_r < v_r$. Let $b_r \le x_r < v_r$, $\mathbf{b} \le \mathbf{x} \le \mathbf{b}^0$. We must show that $\mathbf{x}$ is non-acceptable. Since $x_r < \tilde{y}_r(= v_r)$, it follows that there does not exist any $\mathbf{y}^\ell$ such that $\mathbf{y}^\ell \le \mathbf{x}$. Hence $\mathbf{x}$ must be non-acceptable. *Q.E.D.*

The proofs for algorithm 2 are similar.

Computer programs, written in Fortran 77, for Algorithm 1 and 2.

Below are some comments concerning Program 1 (similar comments apply to Program 2). The program assumes $n \le n1$, $n_p \le np1$, max $k \le kk$, $M_i \le m1$, where $n1 = 30$, $np1 = 30$, $kk = 1000$, $m1 = 9$. The values of $n1$, $np1$, kk, $m1$ can be increased. An example of input data for the program is given in table 6; the example is the input data for "Example 2, n = 8, n_p = 12". The output data of the program are the input data plus the sought probability, $R(j')$.

```
c       program 1

        parameter(n1=30,np1=30,kk=1000,m1=9)
        integer bo(n1),b(n1),bbo(n1,kk)
        integer bb(n1,kk),vo(n1),v(n1),m(n1)
        integer y(n1,np1),ymin(n1),j,lo,l,i
        integer np,s,k,h1,h2,c1,c2,k1,g
        real rr(n1,0:m1+1),p(n1),r,r1

        read(5,*) n,np
        write(6,10) n,np
10      format(2i4)
        do 20 i=1,n
        read(5,*) m(i)
        write(6,25) m(i)
20      continue
```

```
   25 format(i2)
      do 50 i=1,n
      rr(i,0)=1
      do 40 j=1,m(i)
      read(5,*) rr(i,j)
      write(6,60) rr(i,j)
   40 continue
      write(6,55)
   50 continue
   55 format(//)
   60 format(f6.4)
      do 90 l=1,nc
      read(5,*) (y(i,l),i=1,n)
      write(6,100) (y(i,l),i=1,n)
   90 continue
  100 format(30i2/)

      k=1
      do 102 i=1,n
      bo(i)=m(i)
      ymin(i)=bo(i)
  102 continue

  105 s=0
      h1=-1
      r1=1
      lo=1
      do 150 l=1,np
      do 130 i=1,n
      if (y(i,l).gt.bo(i)) go to 150
  130 continue
      h2=0
      do 145 i=1,n
      if (y(i,l).lt.ymin(i)) ymin(i)=y(i,l)
      g=max(y(i,l),b(i))
      h2=h2+bo(i)-g
  145 continue
      if (h2.gt.h1) then
      h1=h2
      lo=l
      end if
  150 continue
      do 160 i=1,n
      v(i)=max(ymin(i),b(i))
      vo(i)=max(y(i,lo),b(i))
      c1=vo(i)
      c2=bo(i)+1
      p(i)=rr(i,c1)-rr(i,c2)
      r1=r1*p(i)
  160 continue
      r=r+r1

      k=k-1
      k1=k
      do 200 i=1,n
      if (v(i).ne.vo(i)) then
      s=s+1
      k=k1+s
      do 190 j=1,n
      if (j.ne.i) bbo(j,k)=bo(j)
      if (j.eq.i) bbo(j,k)=vo(j)-1
      if (j.lt.i) bb(j,k)=vo(j)
      if (j.ge.i) bb(j,k)=v(j)
  190 continue
      end if
  200 continue
      if (k.eq.0) then
      write(6,210) r
      stop
      end if
  210 format(//,f8.6)
      do 220 i=1,n
      bo(i)=bbo(i,k)
      b(i)=bb(i,k)
      ymin(i)=bo(i)
  220 continue
      go to 105
      end


c     program 2

      parameter(n1=30,nc1=30,kk=1000,m1=9)
      integer bo(n1),b(n1),bbo(n1,kk)
      integer bb(n1,kk),vo(n1),v(n1),m(n1)
      integer z(n1,nc1),zmax(n1),j,lo,l,i
      integer nc,s,k,h1,h2,c1,c2,k1,g
      real rr(n1,0:m1+1),p(n1),r,q,q1

      read(5,*) n,nc
      write(6,10) n,nc
   10 format(2i4)
      do 20 i=1,n
      read(5,*) m(i)
      write(6,25) m(i)
   20 continue
   25 format(i2)
      do 50 i=1,n
      rr(i,0)=1
      do 40 j=1,m(i)
      read(5,*) rr(i,j)
      write(6,60) rr(i,j)
   40 continue
      write(6,55)
   50 continue
   55 format(//)
   60 format(f6.4)
      do 90 l=1,nc
      read(5,*) (z(i,l),i=1,n)
      write(6,100) (z(i,l),i=1,n)
   90 continue
  100 format(30i2/)
```

```
      k=1
      do 102 i=1,n
      b(i)=m(i)
102   continue

105   s=0
      h1=-1
      q1=1
      lo=1
      do 150 l=1,nc
      do 130 i=1,n
      if (z(i,l).lt.bo(i)) go to 150
130   continue
      h2=0
      do 145 i=1,n
      if (z(i,l).gt.zmax(i)) zmax(i)=z(i,l)
      g=min(z(i,l),b(i))
      h2=h2+g-bo(i)
145   continue
      if (h2.gt.h1) then
      h1=h2
      lo=l
      end if
150   continue
      do 160 i=1,n
      v(i)=min(zmax(i),b(i))
      vo(i)=min(z(i,lo),b(i))
      c1=bo(i)
      c2=vo(i)+1
      p(i)=rr(i,c1)-rr(i,c2)
      q1=q1*p(i)
160   continue
      q=q+q1

      k=k-1
      k1=k
      do 200 i=1,n
      if (v(i).ne.vo(i)) then
      s=s+1
      k=k1+s
      do 190 j=1,n
      if (j.ne.i) bbo(j,k)=bo(j)
      if (j.eq.i) bbo(j,k)=vo(j)+1
      if (j.lt.i) bb(j,k)=vo(j)
      if (j.ge.i) bb(j,k)=v(j)
190   continue
      end if
200   continue
      if (k.eq.0) then
      r=1-q
      write(6,210) r
      stop
      end if
210   format(//,f8.6)
      do 220 i=1,n
      bo(i)=bbo(i,k)
      b(i)=bb(i,k)
      zmax(i)=bo(i)
220   continue
      go to 105
      end
```


ACKNOWLEDGMENT

This work was carried out while I was visiting the Operations Research Center, University of California, in 1984 Spring. I thank Dr. Richard E. Barlow at the Operations Research Center for making the visit possible. The work has been supported by the Norweigian Counsil for Scientific and Industrial Research and the Johan and Mimi Wesmanns Foundation. The support is gratefully acknowledged.

REFERENCES

[1] J. A. Abraham, "An improved algorithm for network reliability," *IEEE Trans. Reliability*, vol R-28, 1979 Apr, pp 58-61.

[2] R. E. Barlow, F. Proschan, *Statistical Theory of Reliability and Life Testing*, Holt, Rinehart and Winston, 1975.

[3] R. E. Barlow, A. S. Wu, "Coherent systems with multistate components", *Mathematics of Operations Research*, vol 4, 1978, pp 275-281.

[4] D. A. Butler, "Bounding the reliability of multistate systems", *Operations Research*, vol 30, 1982, nr 3, pp 530-544.

[5] P. Doulliez, J. Jamoulle, "Transportation networks with random arc capacities", *RAIRO, Recherche Operationnelle Operations Research*, vol 3, 1972, pp 45-60.

[6] E. El-Neweihi, F. Proschan, J. Sethuraman, "Multistate coherent systems", *J. Appl. Prob.*, vol 15, 1978, pp 675-688.

[7] W. S. Griffith, "Multistate reliability models", *J. Appl. Prob.* vol 17, 1980, pp 735-744.

[8] J. C. Hudson, K. C. Kapur, "Reliability analysis for multistate systems with multistate components", *AIIE Trans.*, vol 15, 1983, pp 127-135.

[9] J. C. Hudson, K. C. Kapur, "Reliability bounds for multistate systems with multistate components. *Operations Research*, vol 33, 1985, pp 153-160.

[10] B. Natvig, "Two suggestions of how to define a multistate coherent system", *Adv. Appl. Prob.*, vol 14, 1982, pp 434-455.

[11] S. M. Ross, "Multivalued state component systems," *Annals of Prob.*, vol 7, 1979, pp 379-383.

[12] Supplement, NAPS document No. 04299-B, 7 pages in this Supplement. For current ordering information, see "Information for Readers & Authors" in a current issue. Order NAPS document No. 04299, 00 pages ASIS-NAPS; Microfiche Publications; POBox 3513, Grand Central Station; New York, NY 10163 USA.

AUTHOR

Dr. Terje Aven; Rogaland College; 4000 Stavanger, NORWAY.

Terje Aven was born in Stavanger, Norway, on 1956 June 6. He received his MS and PhD degrees in mathematical statistics from University of Oslo in 1981 and 1984. From 1980 to 1984, he was an assistant professor in statistics at University of Oslo. Since 1984 he has been an associate professor at Rogaland College, Rogaland Research Institute, Petroleum Research Center.

Manuscript TR84-057 received 1984 June 21; revised 1985 August 31.

★★★

Reprinted from *IEEE Transactions on Communications*, Volume COM-29, Number 10, October 1981, pages 1482–1491.

Communication Network Availability Predictions Based on Measurement Data

JOHN D. SPRAGINS, MEMBER, IEEE, JAMES D. MARKOV, M. W. DOSS, STEPHANIE A. MITCHELL, AND DAVID C.SQUIRE

Abstract—**Published data describing the probability distributions of data communication system reliability parameters such as availability are very sparse. This paper describes a set of data collected in order to model communication line availability by monitoring operational communication systems. Probability distribution functions for single line availabilities and for the durations of failures are given. A heuristic model for estimating availabilities for more complex systems is presented along with computer simulation data validating the accuracy of the model. The results presented make possible more realistic availability predictions for common types of networks than have previously been computable.**

INTRODUCTION

A VARIETY of designs for distributed function computer communication systems have been developed since the late 1960's, with a wide variety of such systems currently being developed and installed in countries around the world. One of the key concerns in the design of such systems is the reliability of telephone systems for transmitting data, since the great majority of the data communication links currently in use are telephone links.

Reliability of telephone systems is not a new concern, but the new systems being designed, with their emphasis on real-time operation, have forced reliability to be considered in a different way. Prior to this time, a great deal of effort was spent on understanding bit error characteristics of telephone circuits. This information has allowed various types of error detecting and correcting methods, and corresponding data transmission protocols, to be developed to ensure almost error-free transmission of data. As these error-handling techniques become more successful, long-term communication line outages are becoming the dominant factor limiting data communication reliability. In contrast to short-term bit errors, the new concern is over failures of telephone lines that last minutes or hours.

A considerable amount of very fine work has been done to understand bit error rates. Examples of this work can be found in [1]-[4]. This paper addresses the concern over failures lasting longer than 1 min to augment the information about bit error rates and provide an overall reliability model for a transmission line. Longer term outages such as those motivating this study have received far less attention than bit errors, so no references discussing long-term outages in a manner comparable to [1]-[4] have been published. This paper contains more detailed measurements for long-term outages than have previously been published.

The primary reliability measure studied here is system availability, which can be defined as the probability the system is operational at a randomly chosen instant of time (normally interpreted as the instant when a user wishes to use the system). An equivalent measure is system unavailability, one minus availability, which gives the probability that the system is not operational at a randomly chosen instant of time. Most of the plots given here are distribution functions for unavailability.

The measurement data included here were obtained for systems of the form indicated in Fig. 1. Two logical machines are interconnected by a communication line provided by a PTT or communications common carrier in the manner shown. The distance between the two machines is variable from a mile or so up to thousands of miles. Furthermore, the routing of the line between the two machines is normally unknown to the customer. (The actual routing distance may be several times the airline distance between the end points [5].) The two machines may be in any of a variety of countries or even in different countries.

MEASUREMENT DATA

Sources of Data

The next few sections of this paper summarize a set of data which can be used to characterize communication lines. To better appreciate these results, it is important to understand the sources of the data. Despite several obvious weaknesses in the sampling process, described below, the data obtained have proven to be remarkably consistent. Important parameters such as the mean and the 90th percentile of availability distributions (the two primary parameters used in the theoretical analysis which follows) tended to differ for different systems observed by only a small fraction of a percent.

Most of the data presented here come from teleprocessing (TP) system logs, which have been compiled by users of TP systems for their own purposes. In some cases, copies of the actual logs were obtained and detailed data were available.

Paper approved by the Editor for Computer Communication of the IEEE Communications Society for publication without oral presentation. Manuscript received February 26, 1980; revised January 30, 1981. This work was based in part upon work supported by the National Science Foundation under Grant ENG 78-03384 and Tektronix, Inc. through its graduate education program.

J. D. Spragins is with the Department of Electrical Engineering, Clemson University, Clemson, SC 29631.

J. D. Markov and M. W. Doss are with the IBM Corporation, Research Triangle Park, NC 27709.

S. A. Mitchell is with the IBM Corporation, Winston-Salem, NC 27105.

D. C. Squire is with Tektronix, Inc., Beaverton, OR 97077.

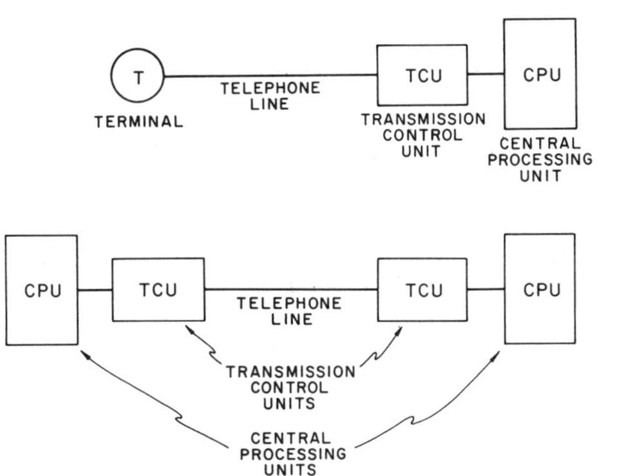

Fig. 1. System description.

In others, only reduced summaries were obtained. The data were compiled for a total of 118 multidrop lines, with 415 branches or drops, in North America and Europe. Some lines were configured with multiple branches or drops, while others were point to point. For clarity, the information here is presented on a per-drop basis, with each drop called a branch. The observation times for lines ranged from three months to three years, with most systems (about 90 percent) monitored for at least six months of continuous operation. (While a considerable amount of information with shorter observation periods was obtained, this was only used for reasonableness checks on the results.) A total of 1233 line-months and 3799 branch-months of observations are included in the data. (This excludes a small sample of 11 high-speed, 50 kbit/s lines, used for a curve examining speed variations given below, but otherwise excluded from the summary data.) The earliest data reported date back to 1965, with additional data collected over a ten year period since then.

A list of the major categories of lines considered, and the numbers of branches (drops) in each category, is given in Table I.

The data presented here should not be interpreted as being from an ideal statistical sampling process. The amount of data for some classes of lines is not really adequate, and the sources of data have not been properly distributed over the various classes of TP systems or over various geographical regions. Nevertheless, it has practical value and is more comprehensive than any similar data published previously. In addition, the TP system logs used were not all recorded under the same conditions. In particular, the definition of a failure was not clearly established; a system or subsystem was defined to have failed when the persons maintaining the log for that system recorded a failure, with various operators displaying different degrees of tolerance for the level of system degradation which they considered to merit recording. In addition, it was not always possible to fully separate out the effects of line or branch failures from those of failures of other equipment such as modems, terminal gear, etc. Although the data given are felt to be dominated by telephone line failures, other failures doubtless enter in also. Unfortunately, the data do not allow the effects of those other failures to be quantified.

TABLE I
LINE CLASSES STUDIED; SEE SECTIONS ON SPEED VARIATIONS AND COUNTRY AND DISTANCE VARIATIONS FOR DEFINITIONS OF CLASSES

Class	Number of Branches
Short	14
Long (national)	381
International	20
North America	271
Europe	144
Low Speed	87
Medium Speed	328
High Speed	11
Total (all save high speed)	415

The procedure used to record the data was, in general, consistent. The TP system operators maintained logs on the operation of each of the branches on their systems. Failures and durations of failures were recorded. For most of the lines, a threshold of 1 min of unsatisfactory operation was used to declare a branch failed. For some of the lines, the threshold used was 10 min.

Although the limited amount of published data on telephone line failures, such as those in [6], are not fully adequate for computing availability statistics, they have also been used for reasonableness checks on the data given here. The statistics published by common carriers tend to give higher average availabilities than those given here. The primary reason for this appears to be almost unavoidable differences in the manner in which line failure statistics are obtained. One important difference in failure statistics is the fact that branches were considered to have failed, for the purposes of this study, as soon as the system operators maintaining logs recorded failures, while the common carriers normally cannot record a failure until it is reported to them. This normally means, among other things, that short-duration self-clearing failures are not included in their database. The precise definitions of repair times used here and by the common carriers also differ in a less quantifiable manner. Hence, statistics should not be expected to agree precisely.

Availability Function

An overall probability distribution function for all 415 branches was obtained from the data. First, an availability figure was obtained for each branch by calculating the probability that the branch was operational (i.e., performing satisfactorily) during the observation period.

Each individual branch availability A_i was calculated by

$$A_i = \frac{\sum_{j=1}^{n} u_j}{\sum_{j=1}^{n} u_j + \sum_{k=1}^{m} d_k} \tag{1}$$

where

u_j = branch up time in minutes

d_k = branch down time in minutes

and summations are over all the time intervals during which the branch was observed. Another way to express this is simply

$$A_i = \frac{\sum_{j=1}^{n} u_j}{\text{total observation time}} . \tag{2}$$

After calculating each branch availability, an overall availability distribution function was determined by

$$F(A) = \Pr [A \leqslant a] . \tag{3}$$

An overall summary curve, summarizing results for all branches observed, is shown in Fig. 2 where the probabilities are expressed in percentages for all the branches observed. The curve plotted is actually a distribution function for unavailability, one minus availability, since this gives a plot which is easier to interpret. The two curves are related by

$$F(A_C) = 1 - F(A) \tag{4}$$

with A_C representing unavailability.

It should be noted that some of the branches were unavailable less than 0.1 percent of the time, while others were unavailable over 10 percent of the time. The total range of observed values, over three orders of magnitude base ten, indicates a very large variation in the behavior of the branches. The amount of variation is important since it reflects the uncertainty associated with the random selection of a branch. It should also be noted that 69 percent of all the branches were available 99 percent of the time or better. That is, about two thirds of the branches showed an unavailability of less than one percent.

In order to better understand the branch availability figures, the effects of several different parameters were examined. The results are discussed below.

Speed Variations

The data were separated into three speed classifications (see Fig. 3). Most (328) of the branches observed were transmitting data at 2000 or 2400 bits/s which is classified as medium speed in Fig. 3. Some of the branches (87) transmitted data at lower speeds, such as 1200 bits/s, 600 bits/s and lower; these lower speed branches are grouped and classified accordingly. In addition, a curve obtained from a small sample of 11 high-speed branches (otherwise excluded from the summary data) is included for comparison purposes.

Some improvement in availability at lower speeds is indicated, but the variation within each class is more significant than the variations between classes. It is possible that a substantial percentage of the difference between availabilities observed at different speeds is due to degraded performance failures, which appear to occur more frequently at higher speeds. It is also possible that some of the improvement at lower speeds is due to less careful recording of failures in the logs of system users whose lines operated at these speeds, since these systems tended to have less critical performance requirements.

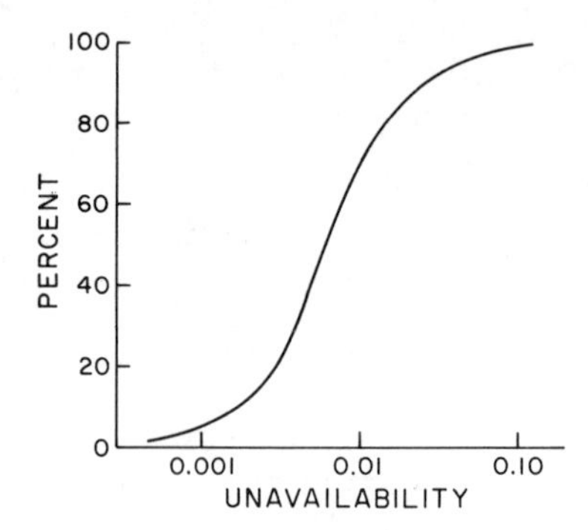

Fig. 2. Overall availability.

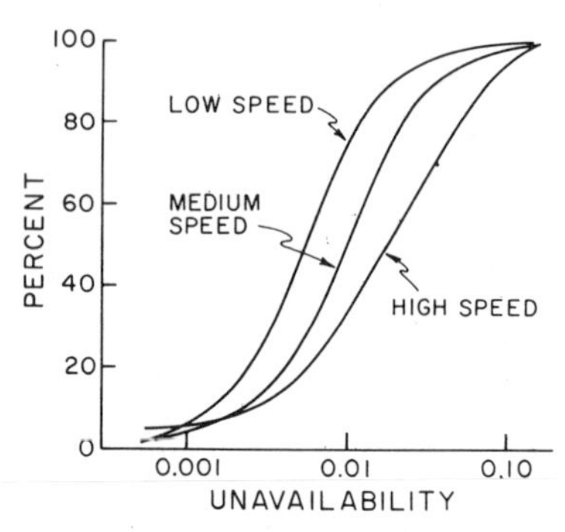

Fig. 3. Speed variations.

Country and Distance Variations

Some of the branches observed passed through more than one country. In general, they were long branches on the order of 100 mi or more. Three classifications of branches are shown in Fig. 4. Long branches are arbitrarily classified as those longer than 50 airline mi. These long branches are separated into those within one nation, called national branches, and branches that passed through more than one country, called international branches. For contrast, branches shorter than 50 mi that existed within the boundaries of one country are also shown, even though the sample size (see Table I) is small.

Shorter branches are seen to be more reliable than longer branches. A physical interpretation of this may be that shorter branches are made up of fewer pieces of equipment, have less circuit miles of transmission, fewer weather disturbances, and involve fewer people who may inadvertently disrupt the operation. Longer branches may be subject to more of these effects. Similarly, longer branches may be thought to contain within them two or more short branch segments whose failures would be additive, thus decreasing the availability of longer branches. There is also a difference between longer branches that are international and those contained within one country. In an explanation of this, a number of factors can be cited. There are differences between the equipment in different countries, so mismatches can occur; when repair of a failure is called for, national responsibility for the repair work may be in question, which can delay the time required to make the repair, etc.

Another set of comparison figures is shown in Fig. 5. Here two countries are compared for differences in the overall reliability of the branches. While these differences are certainly discernible, the variation within either of these countries is more significant than the variation between the countries.

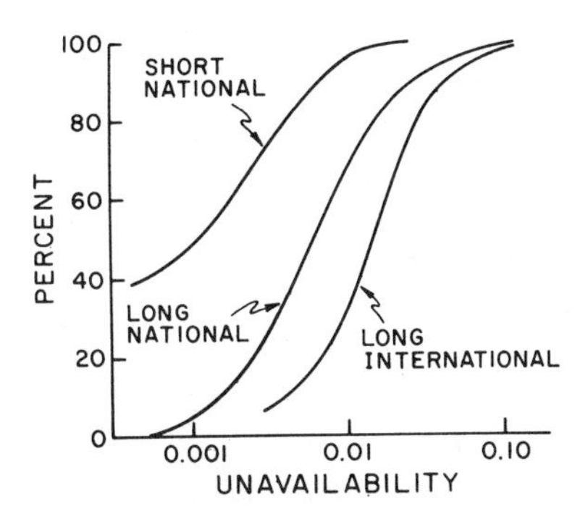

Fig. 4. National and international branches.

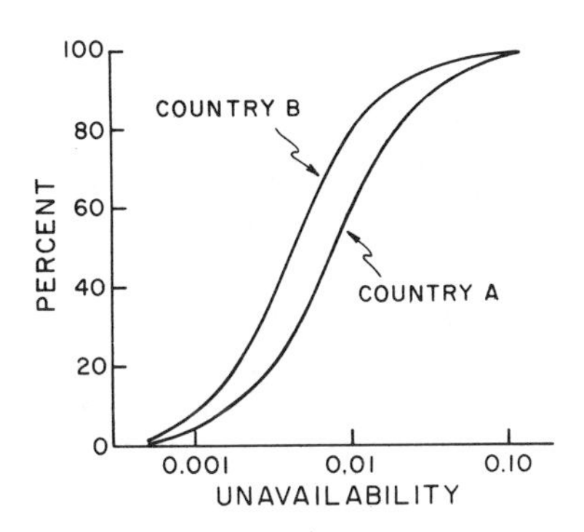

Fig. 5. Country variations.

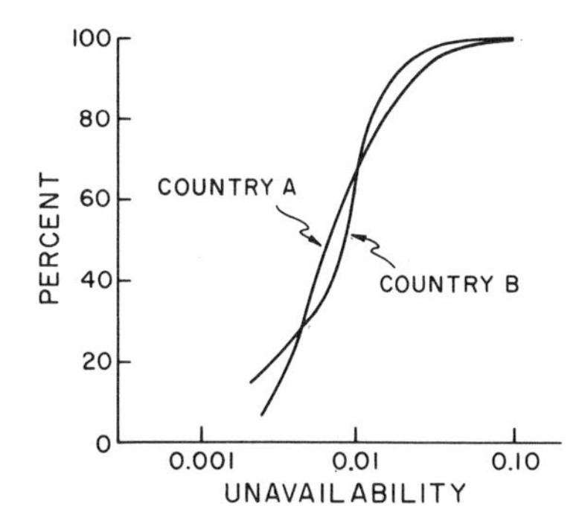

Fig. 6. Similar industry variations in two different countries.

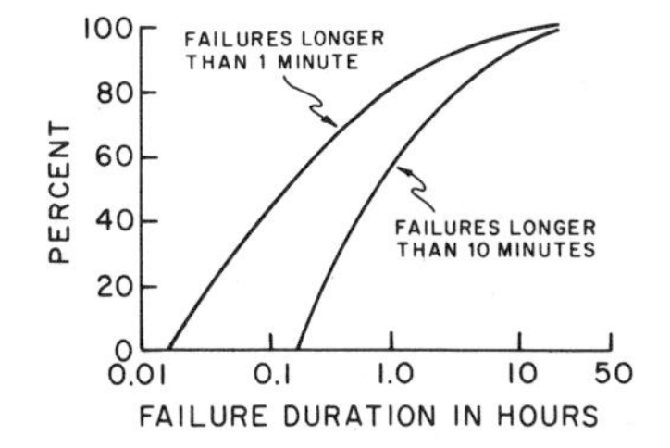

Fig. 7. Distribution of failure durations.

There are, of course, errors in the recording of the data presented here. In addition, errors may also be made in interpreting the results. To help appreciate the subtle factors involved, a supplementary set of data is presented in Fig. 6. Here the same two countries are compared as in Fig. 5, but for these data, a common industry that used similar data communication systems was selected in both countries. These two curves are much more closely spaced. A way to view this is to visualize that different thresholds of acceptable system performance are imposed in different systems. Some systems are monitored more closely than others; accordingly, failures are recorded with respect to different acceptable performance levels. The reasons for these variations are too complex to elaborate on here, but they seem to exist, and they are reflected in the data. Once again, however, the variations within one country are considered more significant than the differences between the countries.

Down Time Distribution

In addition to the availability of a branch, it is important to understand the length of time a failure lasts. Given that the branch is down, a probability distribution is determined for the duration of the down time. Note, from the data, that this is a conditional probability function based on the condition that the failure has already lasted 1 min. While failures occurred that lasted less than 1 min, they were not included. The distribution for duration of the down time is expressed by

$$F(D) = \Pr[D \leqslant t \mid D \geqslant 1 \text{ min}]. \tag{5}$$

This distribution function is shown in Fig. 7. Once again these probabilities are expressed as percentages. From the figure we see that some of the failures lasted longer than 10 h and 16 percent of the failures lasted longer than 1 h. It can also be observed that 50 percent of the failures lasted less than 6 min. In addition to data for failures lasting longer than 1 min, a curve is drawn in a similar fashion for failures lasting longer than 10 min. This was done by changing the condition and counting only failures that lasted longer than 10 min.

One other related plot of interest is given in Fig. 8 which (very roughly) plots duration of failures versus mean time between failures (more precisely, mean time between failures of at least this duration), on a log-log scale, for a 2400 bit/s line. Very short duration failures (or bit errors) as well as longer duration failures were considered for this plot. The plot indicates that errors affecting at least one bit (42 ms duration at 2400 bits/s) occur roughly every 40 s. (This corresponds to a burst error rate [3] of approximately 10^{-5}.) Errors of at least 1 min duration occur about once per day per branch. A 1 h disturbance occurs approximately once each month, and disturbances lasting one day should (by extrapolating the curve slightly beyond the range for which good data are available) occur approximately once each year per branch.

The values plotted in Fig. 8 have been obtained from a variety of different sources, including [1]-[4], which reflect measurements obtained under varying conditions. This is a very general and very approximate set of data and reflects an attempt to simply summarize, and provide a rule of thumb, for understanding the overall performance of a branch selected at random and to bring within a common perspective short bit errors and longer failures which are often treated separately.

Further details of this measurement study are contained in [7]. An earlier paper by Provetero [8], based on a subset of the data considered here, revealed strong correlations between failures of different communication lines serving a common node, i.e., the standard assumption of independent failures for such lines was shown to be false. Theoretical analysis of failure dependencies are given in [9], [18], [19]. The more extensive data considered here have further confirmed that failures of

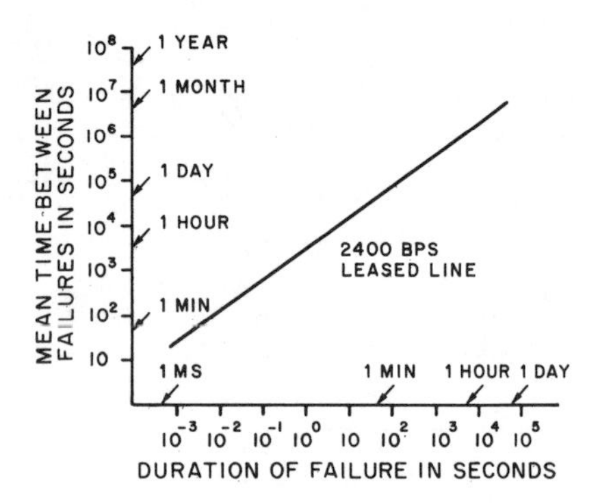

Fig. 8. Mean time between failures versus outage duration.

such lines are dependent, but better measurement data on failure dependencies could not readily be extracted from the measurements obtained.

CAUSES OF VARIABILITY IN RELIABILITY PARAMETERS

The primary causes of variability in line failure rates or similar reliability parameters appear to be the tremendous variety of equipment types and ages in the telephone plant (by far the most common source of communication links) and the correspondingly tremendous variety in operating environments for this equipment. A detailed description of the main factors causing reliability problems in the common carrier telephone plant has been given in another paper [5].

The telephone plants in the U.S., and in many other countries, are among the most complex systems ever designed and installed by mankind. With design lifetimes for equipment of 20–40 years, and the rapid pace of technological change that has occurred within this time period, a large variety of different types and ages of equipment have been interconnected to form the installed plant. Further, operating environments for equipment range from almost ideal to cases where there are so many undesirable environmental factors (temperature variation, humidity problems, corrosive atmospheres, strong electromagnetic fields, and other factors) that it is remarkable that the equipment operates at all. Three or four orders of magnitude variation in percentage down time or failure rates for different lines or branches do not appear to be at all surprising when these factors are considered.

NETWORKS CONSISTING OF MULTIPLE LINES

Although the measurement data given here give reasonable estimates for the distribution functions for some important reliability measures for one line (recall the configuration for measurements in Fig. 1), they do not readily yield comparable data for more complex configurations consisting of multiple branches interconnected in any of a variety of ways. A heuristic model for approximating the corresponding distributions for at least the most common types of networks is given here, along with the results of a computer simulation study to verify the accuracy of the heuristic model. Additional details on the heuristic model and its validation are given in [10], [11].

Heuristic Model Introduction

The model discussed here is a simple heuristic model for approximately computing probability distribution functions for communication system availabilities. The model assumes that failures of different branches are independent (or at least that their availabilities are independent) despite the comments given earlier about this assumption not being valid. Even the independent failure case has not been treated adequately so far, however, and extensions to handle dependent failure parameters will have to come later. All computations given here will be based on use of the overall probability distribution function in Fig. 2, since this curve is the most realistic one to assume for a general analysis.

A general formula for the availability A of a system composed of a number of independently failing subsystems (with independent repairs also) can be written in the form of a sum of products of availabilities A_i of the individual subsystems. General techniques for computing such availabilities are given in [12], [13].

The main sources of difficulty in finding percentile values for overall system availability, for reasonably complex systems, stem from the facts that percentiles are defined in terms of integrals of the appropriate probability densities and that probability laws for sums of products of random variables (even independent random variables) are normally very difficult to evaluate. (A quick verification of this can be obtained by consulting any good probability theory text, e.g., [14], for the formula giving the probability density for the product of even two independent random variables.) Some limit theorems, such as the central limit theorem, might be applicable if there is a fairly large number of terms summed; alternatively, the central limit theorem would imply that the limiting distribution for the product of a large number of independent random variables is log normal. An important constraint for the case of interest, though, is that all the random variables of interest are between zero and one (since availabilities are probabilities), and this constraint is difficult to incorporate in the limit theorems, which can be expected to converge slowly without the constraint being used. Hence, the alternative heuristic approach described below has been developed.

The analysis technique is based on first fitting a reasonable analytic curve to availability data for one branch, then approximating a corresponding analytic curve for the overall system. A reasonable analytic curve (found by trial and error) for availability of one branch is a beta density [15]:

$$F_{A_i}(a_i) = \begin{cases} \dfrac{\Gamma(r+s)}{\Gamma(r)\Gamma(s)} a_i^{r-1}(1-a_i)^{s-1} & 0 \leqslant a_i \leqslant 1 \\ 0 & \text{elsewhere} \end{cases} \tag{6}$$

with parameters $r = 50$ and $s = 0.5$. This density yields a mean availability of 0.990 and a 90th percentile availability of 0.973. These values agree well with the mean and 90th percentile values of 0.990 and 0.975, respectively, which can be computed from the data given in the previous sections of this paper or with the corresponding values of 0.988 and 0.973, respectively, computed from [7].

The beta distribution curve [cumulative distribution curve for $1 - a_i$ obtained from the density in (1)] is compared with the overall summary curve from Fig. 2 in Fig. 9. Although the agreement between the two curves is excellent for unavailabil-

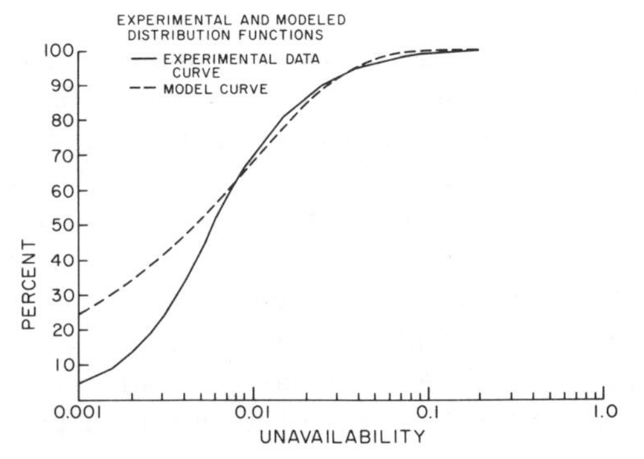

Fig. 9. Comparison of experimental and analytical unavailability distribution functions.

ities approximately above the 60th percentile, the two curves differ appreciably toward the left ends of the two plots. (The difference is greatly exaggerated by plotting the curves on a semilog scale, however; the two curves would be virtually indistinguishable on a linear scale plot.) Fortunately, the two curves agree closely for the cases of primary interest to the designer of commercial communications based systems, although different parameters chosen to improve the fit in other ranges might be advisable when working with systems, such as some military systems, with extremely high availability requirements. Extremely reliable communication lines are of little concern to many system designers, since they cause no problems for system users. The theoretical curve was chosen to give good estimates of the percentage of users of the more common commercial systems likely to obtain unsatisfactory reliability rather than to estimate how many customers would have reliability much better than they require. (It should also be noted that the measurement data are least likely to be accurate in the regions where the fit is poor, since the measurement period needed to obtain statistically significant estimates of unavailability for extremely reliable branches considerably exceeds the observation periods actually used, and the nonrigorous "statistical sampling plan" used here should not be expected to give trustworthy results for the tails of the distributions.)

The rest of this paper uses the fitted curve as if it were the true distribution function for unavailability since having an analytical formula greatly simplifies many of the computations. Later comparisons between analytical and simulation results compare the computations made with the heuristic model with simulation results obtained under the assumption the fitted curve is valid. The resulting comparisons are reasonably valid for the percentages of branches or systems likely to cause reliability problems, but they may underestimate the percentage of systems with extremely high availability.

The first and second moments of the density (6), $\overline{A_i}$ and $\overline{A_i^2}$, are needed for the analysis below. They can be expressed in terms of the parameters r and s as

$$\overline{A_i} = \frac{r}{r+s} \tag{7}$$

$$\overline{A_i^2} = \frac{r(r+1)}{(r+s)(r+s+1)}. \tag{8}$$

Alternatively, r and s can be expressed in terms of $\overline{A_i}$ and $\overline{A_i^2}$ as

$$r = \frac{\overline{A_i}(\overline{A_i} - \overline{A_i^2})}{\overline{A_i^2} - (\overline{A_i})^2} \tag{9}$$

$$s = \frac{(1 - \overline{A_i})(\overline{A_i} - \overline{A_i^2})}{\overline{A_i^2} - (\overline{A_i})^2}. \tag{10}$$

Approach to Handling Complex Systems

As has been stated earlier, techniques given in [12], [13] can be used to compute the availabilities of complex systems in terms of sums of products of availabilities of individual subsystems, but the formulas for computing probability distribution functions for functions of random variables are too complex for direct computation of the distribution functions for availabilities, even if availabilities are assumed to be independent. Reasons for eliminating the use of limit theorems from serious consideration have also been discussed. Instead, the approach adopted here is to use techniques given in [12], [13] to compute the first two moments of system availability from the first two moments for individual branches, assuming that branch availabilities are independent. Another beta distribution is then fitted to these first two moments, and this distribution is used for subsequent calculations. One rationale for this approach is that assuming a beta density for the characterization of the complex system is as logical as assuming a beta density for characterizing one branch. Simulation results confirming the validity of this approach are given later.

Computation of appropriate first and second moments for the complex system is fairly simple for cases with independent A_i's. The means can be computed by simply substituting mean values for each of the A_i's in the formula for A, while second moments can be readily computed (at least for cases where the formula for A is reasonably simple) by utilizing the facts that second moments for products of independent random variables multiply and variances for sums of independent random variables add. These first two moments can then be used to find a new beta density to characterize the complex system by using (9) and (10) to find the parameters of this new density.

Since the availabilities of series (cascade) and parallel combinations of N independent branches are given by

$$A_{\text{series}} = \prod_{i=1}^{N} A_i \tag{11}$$

$$A_{\text{parallel}} = 1 - \prod_{i=1}^{N} (1 - A_i) \tag{12}$$

the first two moments for these two simple classes of complex systems are given, for independent a_i's, by

$$\overline{A}_{\text{series}} = \prod_{i=1}^{N} \overline{A_i} \tag{13}$$

$$\overline{A}_{\text{series}}^{\,2} = \prod_{i=1}^{N} \overline{A_i^2} \tag{14}$$

$$\bar{A}_{\text{parallel}} = 1 - \prod_{i=1}^{N} (1 - \bar{A}_i) \qquad (15)$$

$$\bar{A}_{\text{parallel}}^{2} = 1 - 2\prod_{i=1}^{N} (1 - \bar{A}_i) + \prod_{i=1}^{N} (1 - 2\bar{A}_i + \overline{A_i^2}) \quad (16)$$

with all terms in each product equal when the a_i's are identically distributed.

As [12], [13] show, most availability block diagrams for complex systems of real interest can be reduced by iterative application of the series and parallel reduction formulas, so the equations given suffice for most cases. A few cases where these formulas do not suffice also exist, but techniques are given in [12], [13] for handling these cases also. Although the resulting equations may, in some cases, be horribly messy, they can always be expanded to express total availability in terms of sums of products of availabilities of individual subsystems, so simple modifications of the formulas given here suffice for handling these more complex cases. The formulas given above are sufficient to handle all examples discussed in this paper.

Simulation Studies

Since the heuristic model developed here is far from rigorous, computer simulation has been used to examine the accuracy of the approach. The simulation runs were based on an assumption that the analytical curve in (1) was the true density characterizing one branch. Random variables obeying this distribution were generated to simulate availabilities of individual branches in a complex system and the rules in [12], [13] used for determining the corresponding total system availabilities.

The beta-distributed branches were simulated using what Tocher calls the "top hat" method [16]. This involves computing the inverse of the cumulative distribution function so that the availability becomes the dependent variable and the probability becomes the independent variable. If the probabilities are then selected from a uniformly distributed function with values between zero and one, the corresponding availabilities follow the correct density function.

The simulation was run using some modified statistical application programs. The statistical application package used has a program for use with beta densities that computes the availability given the right tail probability. This program was modified so that it would generate a table of values for probabilities from 0.001 to 0.999 in steps of 0.001. Uniformly distributed random numbers were generated using a standard random number generation function. All branches were assumed to have the beta distribution defined by $r = 50, s = 0.5$.

Systems were defined to the program as appropriate combinations of series and parallel branches. Once the system was defined, a random number was generated for each branch and the corresponding availability was selected from the tabulated data. The system availability was then calculated, for this sample, using the relationships in (10) and (11) for series and parallel combinations. (Only networks consisting of iterative combinations of series and parallel blocks have been simulated.) The process was repeated for a number of samples specified at program run time. The cumulative distribution function was then calculated by converting the availabilities to unavailabilities and accumulating their occurrences into bins specified at run time by a "step size" declaration. The resulting distribution of availabilities is a histogram approximation to the probability density function for unavailability. To compute the cumulative distribution function, the values in the bins were simply accumulated. The resulting data were then plotted.

The statistical application package also has a program for computing right tail probabilities given the beta distribution parameters. This program was modified to generate theoretical curves shown on the same figures. The parameters of the beta distribution to be computed, r and s, were calculated using the heuristic technique previously described.

Results

Typical comparisons between theoretical and simulation curves for six different systems are shown in Figs. 10-15. In addition, a plot of percentage error is shown on each figure. Derived values of r and s, maximum error between the two curves, and theoretical values for the means and 90th percentiles of availability are shown in Table II. Maximum error for any system was +6 percent. The match between simulation and heuristic model appears to be excellent, indicating that the approach gives realistic results under the assumptions that the fitted beta distribution is the true distribution characterizing one branch and that different branches fail independently. Since the agreement between the two curves in Fig. 9 is excellent for higher percentile values, the agreement with measured results for complex systems should also be good for higher percentile values (say 80th or higher percentiles) if these complex systems were composed of branches which failed independently.

The results plotted in Figs. 10-15 are directly applicable to the analysis of particular communication paths in some standard forms of networks. For example, a particular communication path in a multidrop system consists of one or more branches in cascade so Figs. 10-13 are applicable. Similar computations give path availabilities in star and loop configurations. More general networks with alternate routing may require separate computations, but the techniques are applicable.

IMPLICATIONS OF RESULTS

Since different communication lines display extreme variability in their reliability parameters (as the measurement data in Figs. 2-6 indicate), when systems are designed entirely on the basis of mean availability values, an appreciable percentage of unhappy system users is inevitable. The techniques developed here at least give approaches to estimating the percentage of unhappy users to be expected. The design can then be altered to appropriately change this percentage if this is desirable.

Both the dependent failure problem and the variable parameter problem indicate that alternative approaches should be carefully considered in lieu of relying heavily on real-time data communications in systems which require extreme reliability.

TABLE II
TABLE OF RESULTS

System	r	s	Maximum Difference	Availability Mean	Availability 90th Percentile
1 branch	50.00	0.50	+2 percent	99.0 percent	97.3 percent
2 branches cascaded	49.75	1.00	+4 percent	98.0 percent	93.9 percent
5 branches cascaded	49.02	2.50	+6 percent	95.1 percent	91.0 percent
10 branches cascaded	47.83	5.00	−4 percent	90.5 percent	85.1 percent
2 10's in parallel	282.19	2.55	−3 percent	99.1 percent	98.3 percent
2 20's in parallel	171.20	5.76	−3 percent	96.7 percent	94.9 percent

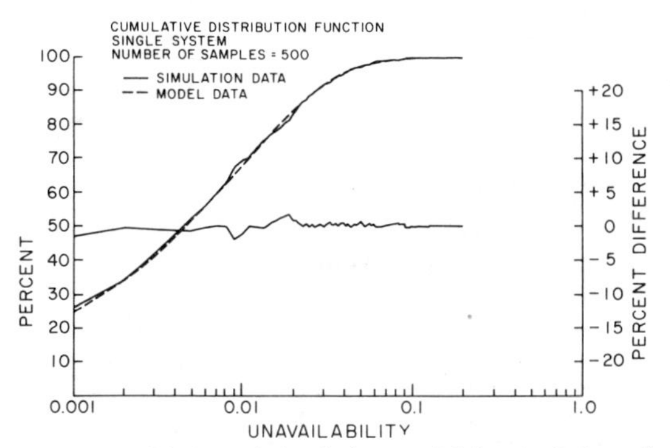

Fig. 10. Validation of simulation model for single branch.

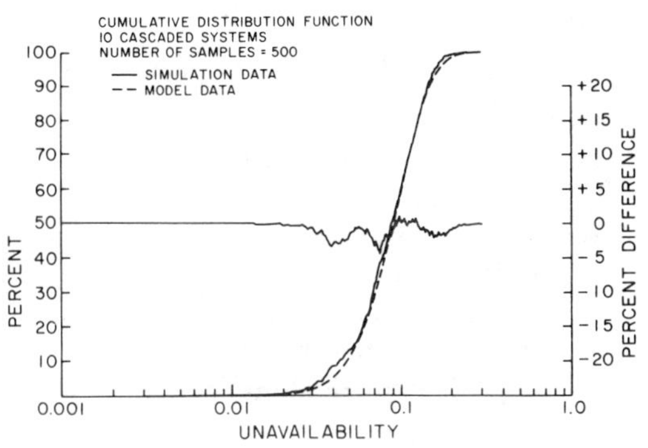

Fig. 13. Comparison between heuristic and simulation models for ten cascaded branches.

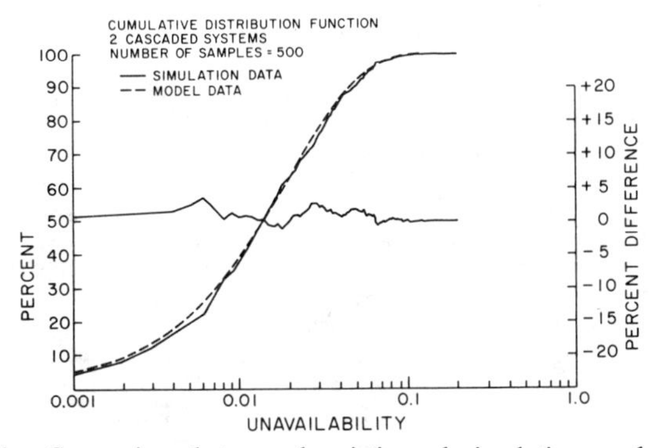

Fig. 11. Comparison between heuristic and simulation models for two cascaded branches.

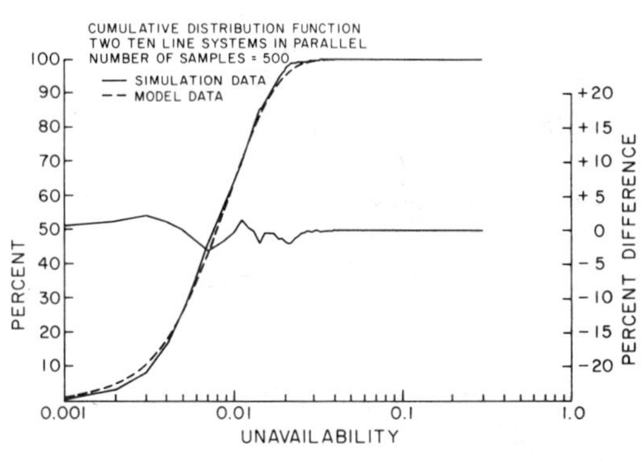

Fig. 14. Comparison between heuristic and simulation models for two ten branch systems in parallel.

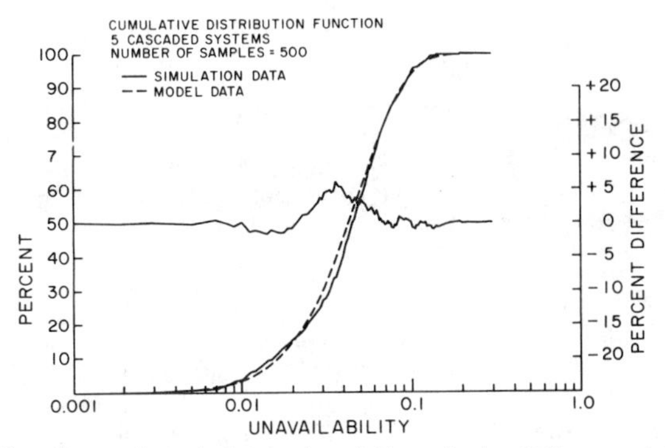

Fig. 12. Comparison between heuristic and simulation models for five cascaded branches.

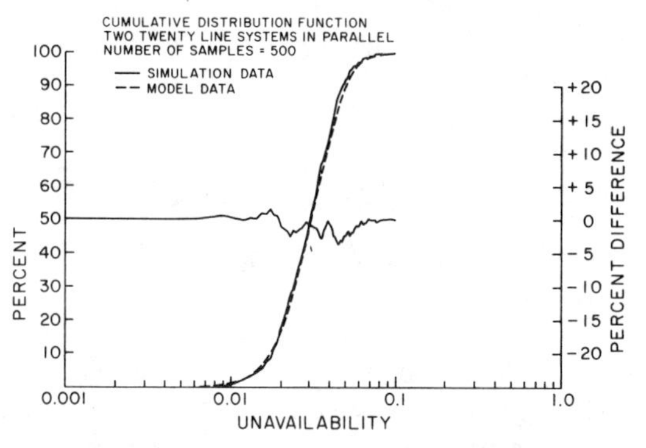

Fig. 15. Comparison between heuristic and simulation models for two twenty branch systems in parallel.

In some cases, it may be possible to reduce such systems' dependence on real-time data communications by putting all the equipment and files needed during normal operation at one location or by providing a stand-alone mode of operation which can be used to survive communications line outages. (Both approaches, applied to operational systems, are described in [17].) Other design procedures will be better in different situations, but there are too many possibilities to attempt enumeration in this paper. One other possibility is that some systems which are desired may not be feasible.

The measurement data included here and the heuristic model for extending results to handle more complex networks consisting of multiple branches allow computation of more realistic reliability predictions than have heretofore been computable. They now make it possible to find reasonable estimates for the probability of system availability falling within certain ranges, rather than simply estimates of mean availability.

EXTENSIONS TO RESEARCH

Two primary types of extensions to this work are currently under consideration. The first is modification of the simulation runs to generate random variables obeying the measured distribution function for one branch rather than the fitted distribution function. The resulting simulation data should then be slightly more accurate, although little change in higher level percentile values is anticipated. The other modification, which is felt to be considerably more important, is to modify the techniques to at least approximately reflect the dependencies between failures of different branches. Alternative approaches to the problems, including some based on the statistical theory of extreme values, are also being considered.

Further research on developing better techniques for computing the availability of systems with dependent failures (but ignoring the variability of parameters problem) is also being pursued. New results from this research, which extend the techniques in [9] to handle more general forms of distributed networks, are given in [18], [19].

Possibly the most obvious area where the need for further work is suggested by this paper is in obtaining better measurement data. A measurement program comparable in magnitude to the Bell System 1969-1970 connection survey [2]-[4] is needed in order to obtain truly trustworthy reliability statistics. Until such a measurement program is attempted, there will always be "a problem with the reliability of the reliability statistics offered" (quoting one of the reviewers of this paper). Hopefully, this paper may help motivate a more systematic measurement program.

ACKNOWLEDGMENT

The authors wish to acknowledge the efforts of the many people who conscientiously collected the data used in this paper. In particular, thanks go to M. A. Berk, J. Provetero, and A. Jama of IBM Corporation who assembled and organized various parts of this information into a useful description. Also, thanks go to B. O. Evans and E. H. Sussenguth of IBM for their support in doing this work.

REFERENCES

[1] A. A. Alexander, R. M. Gryb, and D. W. Nast, "Capabilities of the telephone network for data transmission," *Bell Syst. Tech. J.* vol. 39, pp. 431–476, May 1960.

[2] F. P. Duffy and T. W. Thatcher, Sr., "1969–70 connection survey: Analog transmission performance on the switched telecommunications network," *Bell Syst. Tech. J.*, vol. 50, pp. 1311–1347, Apr. 1971.

[3] M. D. Balkovic, H. W. Klancer, S. W. Klare, and W. G. McGruther, "1969–70 connection survey: High-speed voiceband data transmission performance on the switched telecommunications network," *Bell Syst. Tech. J.*, vol. 50, pp. 1349–1384, Apr. 1971.

[4] H. C. Fleming and R. M. Hutchinson, Jr., " 1969–70 connection survey: Low speed data transmission performance on the switched telecommunications network," *Bell Syst. Tech. J.*, vol. 50, pp. 1385–1405, Apr. 1971.

[5] J. Spragins, "Data transmission over the common carrier telephone plant: Factors affecting its reliability," in *Conf. Rec. 1978 Int. Conf. Commun.*, Toronto, Ont., Canada, June 1978, pp. 3.1.1–3.1.5.

[6] Bell Syst. Tech. Ref., "Data communications using voiceband private line channels," PUB41004, Oct. 1973.

[7] J. D. Markov, M. W. Doss, and S. A. Mitchell, " A reliability model for data communications," in *Conf. Rec. 1978 Int. Conf. Commun.*, Toronto, Ont., Canada, June 1978, pp. 3.4.1–3.4.5.

[8] J. Provetero, "Availability of voice grade private wire telephone lines," in *Proc. IEEE Fall Electron. Conf.*, Chicago, IL, Oct. 1971, pp. 392–397.

[9] J. Spragins, "Dependent failures in data communication systems," *IEEE Trans. Commun.*, vol. COM-25, pp. 1494–1498, Dec. 1977.

[10] J. Spragins and D. Squire, "Data communication network reliability calculations with real world distributions for reliability parameters," in *Proc. 1979 Comput. Networking Symp.*, Gaithersburg, MD, Dec. 1979, pp. 110–116.

[11] D. Squire, "A simulation study of a heuristic technique for approximating availability percentiles for cascaded independent systems," M.S. thesis, Oregon State Univ., Corvallis, Dec. 1978.

[12] J. A. Buzacott, "Finding the MTBF of repairable systems by reduction of the reliability block diagram," *Microelectron. Rel.*, vol. 6, pp. 105–112, 1967.

[13] ——, "Network approaches to finding the reliability of repairable systems," *IEEE Trans. Rel.*, vol. R-19, pp. 140–146, Nov. 1970.

[14] A. Papoulis, *Probability, Random Variables, and Stochastic Processes*. New York: McGraw-Hill, 1965.

[15] S. S. Wilks, *Mathematical Statistics*. New York: Wiley, 1962.

[16] K. D. Tocher, *The Art of Simulation*. London, England: English Univ. Press, 1963.

[17] R. O. Hippert, L. R. Palouneck, J. Provetero, and R. O. Skatrud, "Reliability, availability and serviceability design considerations for the supermarket and retail store systems," *IBM Syst. J.*, vol. 14, no. 1, pp. 81–95, 1975.

[18] J. Spragins and J. Assiri, "Communication network reliability calculations with dependent failures," in *Conf. Rec. 1980 Nat. Telecommun. Conf.*, Houston, TX, Dec. 1980, pp. 25.2.1–25.2.5.

[19] J. Assiri, "Development of dependent failure reliability models for distributed communication networks," Ph.D. dissertation, Oregon State Univ., Corvallis, June 1980.

John D. Spragins (S'56–M'59) received the B.S. degree in electrical engineering from Oklahoma State University, Stillwater, in 1956 and the M.S. and Ph.D. degrees in electrical engineering from Stanford University, Stanford, CA, in 1958 and 1964, respectively.

Since 1980 he has been a Professor of Electrical and Computer Engineering at Clemson University, Clemson, SC. Prior to this he was an Associate Professor at Oregon State University, Corvallis, Research Staff Member and Advisory Engineer at IBM Corporation, Research Triangle Park, NC, Principal Engineer at General Electric Computer Equipment Division, Phoenix, AZ, and Assistant Professor at Arizona State University, Tempe. His current research interests include developing improved reliability and performance models for computer communications networks, new network configurations, and networking protocols. He has published over 40 reports and papers in these and allied fields.

★

James D. Markov received the B.S. degree in electrical engineering from the University of Akron, Akron, OH, in 1961, and the M.S. degree in engineering sciences from the University of Alabama, Tuscaloosa, in 1970.

He is a Senior Engineering Manager with IBM's Systems Communication Division, Research Triangle Park, NC. He is currently the Manager of System Projects for Local Networks. Prior to his current position, he was Manager of Networking Architecture, a position he held from 1975 to 1978. Since 1970 he has been working in the field of communications system architecture with a major emphasis on the development of SNA. During the 1960's he worked on the development of the guidance and control system for the Saturn vehicle as part of the Apollo program.

Mr. Markov is a member of Sigma Tau.

M. W. Doss, photograph and biography not available at the time of publication.

Stephanie A. Mitchell was born in Greensboro, NC. She received the B.S. degree in mathematics (magna cum laude) from Wake Forest University, Winston-Salem, NC, in 1972.

She joined IBM in its Systems Communications Division in 1972. In 1976 she attended IBM's Systems Research Institute and transferred to IBM's Data Processing Division, Winston-Salem, in 1977.

David C. Squire received the B.S.E.E. and M.S.E.E. degrees from Oregon State University, Corvallis, in 1967 and 1979, respectively, through a joint program offered by Tektronix, Inc. and Oregon State University.

He worked at NASA's Ames Research Center as a Research Engineer from 1967 to 1969 and joined Tektronix, Inc., Beaverton, OR, in 1969 where he is now Manager of the Raster Scan Terminal Development Group.

Reprinted from *IEEE Transactions on Communications*, Volume COM-34, Number 1, January 1986, pages 82–84.

Reliability Modeling and Analysis of Communication Networks with Dependent Failures

Y. F. LAM AND VICTOR O. K. LI

Abstract—**This paper presents a new model to study the reliability of communication networks in which link failures are statistically dependent. The approach tries to identify and model explicitly the events that cause communication link failures. No conditional probabilities are needed, and so two major difficulties inherent to them, namely, an exponential number of conditional probabilities to deal with and a consistency requirement to satisfy, are avoided. For reliability computations, some existing algorithms for finding network reliability can be used with minor modifications and no significant increase in computational complexity.**

I. Introduction

One important performance measure of a communication network is reliability. Reliability analysis of networks or other complex systems has been studied for many years, and numerous algorithms and evaluation techniques have been proposed. (See [3] for a general review.) However, almost all of them make the assumption that component failures are statistically independent. For most real-world situations this assumption of independence does not hold. There have been very few known attempts to study the reliability of communication networks with interdependent components. One approach is to specify statistical dependencies between network components by conditional probabilities of failure, so that the joint probability of failures of two (or more) dependent components can be evaluated using chain rule expansion. A major problem with this approach is that the number of conditional probabilities required is exponential in the number of fail-prone components. Furthermore, the set of conditional probabilities has to satisfy a consistency requirement. (See [4] for a discussion of these problems.)

A q-ψ model was developed in [10] as an attempt to simplify certain types of failure dependencies between the communication links of a network. Unfortunately, the model does not satisfy the consistency requirement [4]. More recently, a new ϵ-model was developed in [6] to incorporate more general types of failure dependencies. It still employed conditional probabilities to specify dependencies, but the authors made use of standard rules of probability and the consistency constraint to reduce the total number of parameters (conditional probabilities) that have to be initially specified for the model. However, the minimum number of parameters required is still exponential in the number of fail-prone communication links.

Since using conditional probabilities to specify failure dependencies presents such inherent problems, a totally different approach was taken in [4] to avoid them. A simple colored network model (CNM) was used to model a specific kind of failure dependency between communication links. The CNM can be easily transformed to a network whose links are perfectly reliable and whose nodes fail independently, so that its reliability can be evaluated using numerous existing techniques. The restriction of the model is that links incident to a communication center have to fail in mutually exclusive groups. This assumption is not always valid, and so more flexible models have to be sought.

This paper presents a new model called the event-based reliability model (EBRM). It incorporates more general cases of interdependent component failures of communication networks without using conditional probabilities. The model is simple and flexible, and network reliability can be computed using some known and efficient algorithms with minor modifications. The EBRM is described in Section II. Section III shows how a known algorithm can be easily adapted for reliability computations in the EBRM. Section IV discusses some advantages of the EBRM, and conclusions are given in Section V.

II. The Event-Based Reliability Model

One major reason why the links of a communication network do not fail independently is that there exist events which may cause the simultaneous failures of several links. For example, in a communication center, several outgoing links may share a significant amount of common equipment, in which case the assumption of independent failures obviously does not hold. Also, links within the same geographic vicinity are likely to be affected simultaneously by the same environmental impacts. Such situations can be explicitly modeled in the following manner. Communication centers and links are represented as usual by vertices and edges of a graph, respectively, while failure-causing events are modeled by "event elements" which are added to the affected edges (links). An event element is said to be in the "down" mode when the corresponding failure-causing event occurs, and is said to be in the "up" mode otherwise. All failure-causing events are assumed to be independent and occur with known probabilities. The resulting model is called the event-based reliability model (EBRM).

A simple example is shown in Fig. 1. The network consists of four stations (A, B, C, and D) connected in a bridge configuration, and there are nine event elements scattered on the links. The relationship between link failures and the modes of event elements is not difficult to visualize. Consider link A-B. It is governed by three event elements (1, 3, and 9), and so the link operates if and only if all these three event elements are in the "up" mode. If an event element is in the "down" mode, then all links affected by that event element will fail.

It seems that the EBRM only considers simultaneous failures. There may be situations wherein if a link fails, then some other links will fail with higher probabilities. There may also be situations wherein a sequence of failures occur as a chain reaction to affect a number of links. However, in network reliability computations, most of the time one is asked to find the probabilities of certain states of the given network, not the probabilities of the occurrences of certain failure scenarios. This important concept can be made clear by considering the use of conditional probabilities to model dependent failures. Let $P(A)$ denote the probability that link A of a given network is up. Using chain rule expansion, $P(AB) = P(A)P(B|A) = P(B)P(A|B)$. One might say that $P(A)P(B|A)$ and $P(B)P(A|B)$ can represent two different failure scenarios and, therefore, may not have the same value. This contradiction is due to a wrong conception of $P(AB)$. Note that $P(AB)$ is the probability of a state of two links, not the probability of occurrence of a particular failure scenario affecting the two links. If there are different scenarios

Paper approved by the Editor for Computer Communication of the IEEE Communications Society. Manuscript received September 6, 1984; revised May 8, 1985. This paper was presented at IEEE INFOCOM, Washington, DC, March 1985. This work was supported in part by the Air Force Office of Scientific Research, Air Force Systems Command, under Grant AFOSR 84-0269, and by the Joint Services Electronics Program under Contract F49620-85-C-0071.

The authors are with the Department of Electrical Engineering, University of Southern California, Los Angeles, CA 90089.

IEEE Log Number 8406423.

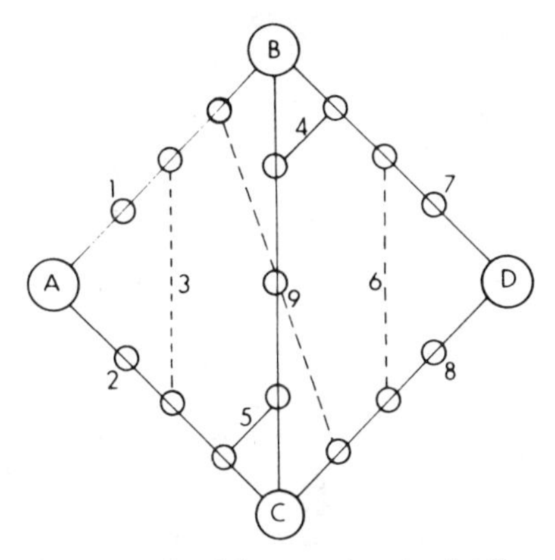

Fig. 1. An example of the event-based reliability model.

associated with these two links, all of them should contribute to $P(AB)$. Conditional probabilities represent how the links of a network are correlated, and do not explicitly represent how a particular failure occurs. In the EBRM, each failure scenario can be represented by an event element covering the appropriate links. If two failure scenarios affect the same links, they can be represented by a single event element. An event element can also represent a chain of failures affecting multiple links. Although a chain of failures may not occur all at once, it is reasonable to assume that they occur within a very short time period compared to the "up" and "down" time intervals of network links, especially in a well-maintained system. Thus, the steady-state probability of observing the network in the middle of a chain of failures is negligible, so that the chain of failures can be assumed to occur simultaneously. Therefore, such nonsimultaneous failures can also be handled by the EBRM.

The identification and measurement (or estimation) of event element parameters is a practical problem which depends on the particular situations of a given network. One possible source of information is the history of past network failures, which should include causes of failures in addition to the "up" and "down" times of network components. Details of network topology and equipment used should also be known in order for a reliability analyst to try to model and analyze the network. It is not unreasonable to assume that such information can be obtained. The following simple example illustrates how event elements can be identified from a knowledge of the equipment used in a network.

In Fig. 2(a), Station A and Station B are linked through a radio channel, while Station C is wired to Stations A and B. In Station A, a single data processing unit is used to process the data to and from the two communication links, while different transmitter/receiver units are needed since the communication links are of different types. The same is true for Station B. In Station C, a single unit does everything since the two incident links are of the same type. Fig. 2(b) shows how this situation can be modeled by the EBRM. Event element 1 represents the data processing unit at Station A. If this event element is in the "down" mode, which means that the data processing unit at Station A is down, then the two links incident at Station A will all be down. Event element 2 represents the radio link between Stations A and B, and includes the radio transmitter/receiver units at both stations. If only this event element is in the "down" mode, the radio link cannot function, but both stations can still communicate with Station C. The other event elements can be similarly interpreted.

III. Reliability Computations in the EBRM

Computing network reliability under the EBRM is not more difficult than that under an ordinary network model which assumes independent link failures. This is due primarily to the fact that independent event elements have been used to model failure dependencies. Numerous existing network reliability

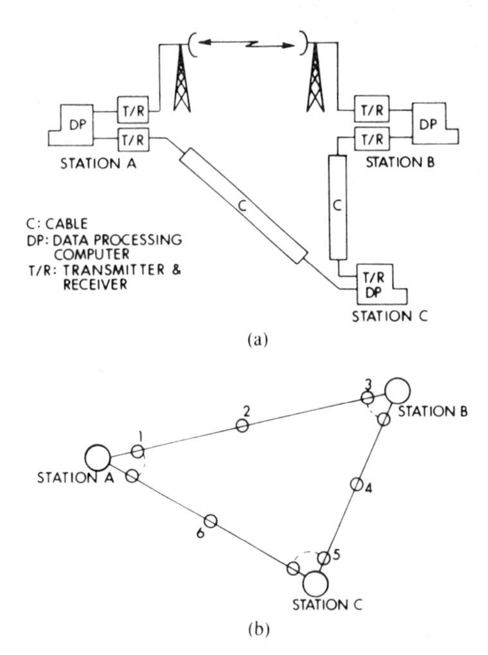

Fig. 2. Identifying event elements.

algorithms may be used. In the following example, an efficient algorithm developed in [7] and [8] is used. The algorithm is capable of finding the probability that a source node can reach a specified set of nodes in a given network. This is a very general reliability measure, and it contains the source-to-single-terminal reliability and the source-to-all-terminal reliability as special cases. The algorithm outputs a list of subsets of network components corresponding to acyclic subgraphs of the network. For each subset, the probability that all its components are in the operating mode is computed. The network reliability is then calculated from these probabilities. (Readers are referred to the original papers for details of this algorithm.) In the EBRM, the probability that a group of network components are in the operating mode is simply the probability that all the event elements involved are in the "up" mode. The latter can be easily computed since all event elements are independent. Identifying subsets of event elements from subsets of links is a simple task. Given a subset of k links, the corresponding subset of event elements can be obtained by checking off the event elements associated with each of the k links. In the worst case it takes $O(mk)$ steps for each subset of k links, where m is the maximum number of event elements on a link. This amount of additional work does not make the algorithm significantly worse in terms of computational complexity.

An example given in [7] will be used here as an illustration. The network is shown in Fig. 3, and it is assumed that all nodes are perfectly reliable. If links c and d are considered as a single bidirectional link, then this network is the same as the one shown in Fig. 1. Let $P(A \rightarrow D)$ denote the probability that Station A can reach Station D, and $P(abc)$ denote the probability that links a, b, and c are operating. Then, from [7], the probability that node A can reach node D is given by

$$\begin{aligned} P(A \rightarrow D) = {} & P(ae) + P(bf) + P(acf) \\ & + P(bde) - P(abde) - P(abcf) \\ & - P(acef) - P(bdef) - P(abef) \\ & + P(abcef) + P(abdef). \end{aligned} \tag{1}$$

Now assume that failures of the links in Fig. 3 are due to the

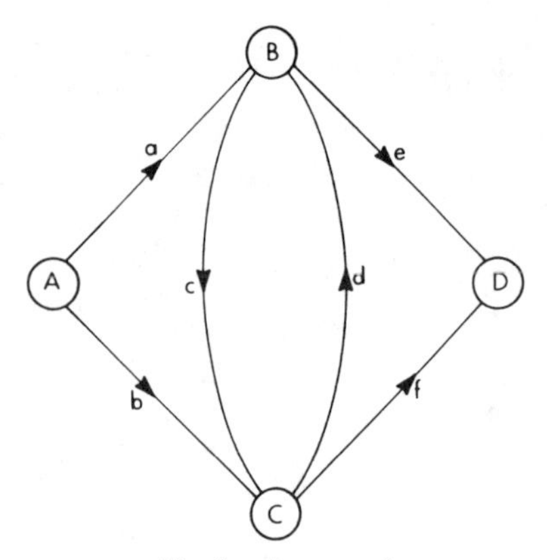

Fig. 3. An example.

event elements shown in Fig. 1. As has been mentioned above, the probability that a group of links are all operating is just the probability that all event elements involved are in the "up" mode. Let $P(1\ 2\ 3)$ denote the probability that event elements 1, 2, and 3 (as shown in Fig. 1) are in the "up" mode. After simple manipulation and simplification of (1), the probability that node A can reach node D will become

$$\begin{aligned} P(A \rightarrow D) = {} & P(1\ 3\ 4\ 6\ 7\ 9) + P(2\ 3\ 5\ 6\ 8\ 9) \\ & + P(1\ 3\ 4\ 5\ 6\ 8\ 9) + P(2\ 3\ 4\ 5\ 6\ 7\ 9) \\ & - P(1\ 2\ 3\ 4\ 5\ 6\ 7\ 9) - P(1\ 2\ 3\ 4\ 5\ 6\ 8\ 9) \\ & - P(1\ 3\ 4\ 5\ 6\ 7\ 8\ 9) - P(2\ 3\ 4\ 5\ 6\ 7\ 8\ 9) \\ & + P(1\ 2\ 3\ 4\ 5\ 6\ 7\ 8\ 9). \end{aligned} \tag{2}$$

Since event elements are statistically independent, $P(A \rightarrow D)$ can be easily calculated from the above equation. Some other reliability evaluation algorithms can also be used in the EBRM. Examples include the algorithm given in [1] and [2] to find the probability (or its upper and lower bounds, if the algorithm is not run to completion) that two nodes are connected in a network, the iterative algorithms given in [9], and the technique reported in [5] to approximate network reliability by efficient enumeration. Their modification is similar to the above example, and will not be detailed here.

IV. Advantages of the EBRM

The use of EBRM in reliability modeling and analysis of networks with dependent failures has the following advantages.

1) Since event elements of the EBRM are statistically independent, the model is always consistent.

2) Theoretically, there can be as many as $2^N - 1$ event elements in a network of N communication links. But in real-world situations most of them do not exist. For example, it is unlikely that there is an event which affects two links that are very far apart. It is also unlikely that there is an event which affects several links that are scattered around a large network. Failure dependencies in a network are usually found among neighboring links, so that a large number of event elements "cannot" exist. The important thing to note here is that in the EBRM, if an event element is not present, then it never shows up in any reliability computations. In the case of using conditional probabilities, the number of parameters to be handled cannot be reduced, even though some components are in fact independent or conditionally independent from others. Therefore, the EBRM will always have fewer parameters to deal with than the use of conditional probabilities.

3) Event elements and their probabilities of occurrence are physically more meaningful than conditional probabilities of failure. The EBRM gives a better understanding of component correlations and causes of component failures, which are of crucial importance to the maintenance and improvement of network performance. The EBRM can also explicitly model failures caused by jamming, atmospheric conditions, and other similar natural or man-made factors. These usually affect more than one communication link simultaneously, and are often the major causes of dependent failures.

4) The EBRM covers the traditional assumption of independent failures as a special case. This corresponds to every event element falling on only one link. As new information is gained or observed, the model can be updated by adding new event elements, with changes to only a small number of existing parameters. In the case of using conditional probabilities, when new failure dependencies are observed, a much larger number of existing parameters have to be modified. This means that the EBRM is more adaptive to changes in network operating conditions.

5) Although the EBRM is a more general model, its reliability computation is not more difficult than that of the traditional model which assumes independent failures. It has been shown in the previous section how a known and very efficient algorithm for computing network reliability can be applied to the EBRM with a simple modification and no significant increase in computational complexity.

V. Conclusions

A new event-based reliability model has been proposed for studying the reliability of communication networks in which link failures are statistically dependent. The EBRM uses independent event elements to model dependent link failures, and so avoids the inherent complexity of using conditional probabilities to model statistical dependencies. This approach also helps users better understand the causes and effects of network failures, and the model is flexible and adaptive to network growth and changes. This paper has also shown how a known and efficient algorithm for computing network reliability can be easily adapted for reliability computations in the EBRM.

References

[1] W. P. Dotson, Jr., "An analysis and optimization technique for probabilistic graphs," Ph.D. dissertation, Air Force Inst. Technol., Wright-Patterson AFB, OH, Aug. 1976.

[2] W. P. Dotson and J. O. Gobien, "A new analysis technique for probabilistic graphs," *IEEE Trans. Circuits Syst.*, vol. CAS-26, pp. 855–865, Oct. 1979.

[3] C. L. Hwang, F. A. Tillman, and M. H. Lee, "System-reliability evaluation techniques for complex/large systems—A review," *IEEE Trans. Reliability*, vol. R-30, pp. 416–423, Dec. 1981.

[4] Y. F. Lam and V. O. K. Li, "On reliability calculations of network with dependent failures," in *Proc. IEEE GLOBECOM*, San Diego, CA, Dec. 1983, pp. 1499–1503.

[5] V. O. K. Li and J. A. Silvester, "Performance analysis of networks with unreliable components," *IEEE Trans. Commun.*, vol. COM-32, pp. 1105–1110, Oct. 1984.

[6] S. N. Pan and J. Spragins, "Dependent failure reliability models for tactical communications networks," in *Proc. Int. Conf. Commun.*, 1983, pp. 765–771.

[7] A. Satyanarayana and A. Prabhakar, "New topological formula and rapid algorithm for reliability analysis of complex networks," *IEEE Trans. Reliability*, vol. R-27, pp. 82–100, June 1978.

[8] A. Satyanarayana, "A unified formula for analysis of some network reliability problems," *IEEE Trans. Reliability*, vol. R-31, pp. 23–32, Apr. 1982.

[9] D. R. Shier, "Iterative algorithms for calculating network reliability," Dep. Math. Sci., Clemson Univ., Clemson, SC, Tech. Rep. 457, 1984.

[10] J. Spragins and J. Assiri, "Communication network reliability calculations with dependent failures," in *Proc. Nat. Telecommun. Conf.*, 1980, pp. 25.2.1–25.2.5.

RELIABILITY BOUNDS FOR NETWORKS WITH STATISTICAL DEPENDENCE

Eddy H. Carrasco and Charles J. Colbourn

Computer Communications Networks Group
Department of Computer Science
University of Waterloo
Waterloo, Ontario, N2L 3G1
CANADA

ABSTRACT

Many bounds for the all-terminal reliability of a network have been proposed, but most assume that link failures are statistically independent. This paper develops a lower bound for the all-terminal reliability of a network when statistical dependence of link failures occurs; in particular, a bound is produced when information about failure of links and failure of pairs of links is given. The value produced is an absolute lower bound, which holds under the most pessimistic assumptions about unspecified statistical dependencies. Moreover, the bound can be computed in polynomial time; this distinguishes it from many available bounds.

1. INTRODUCTION

Reliability assessment in computer networks is a complex task; part of the complexity arises from the statistical dependencies which hold between link failures. We investigate an approach here to obtain efficiently computable lower bounds on reliability when limited information about statistical dependencies is available. A computer network is typically modelled as a *probabilistic graph;* the undirected edges of the graph represent bidirectional communication links, and the nodes represent sites of the network. Failure probabilities are associated with each edge of the network. In this setting, the *all-terminal reliability* of the network is the probability that the operational edges provide communication paths between all pairs of nodes. Most studies have assumed that link failures are statistically independent; the resulting measure is a unique probability that the network is operational. Even with the assumption of statistical independence, all-terminal reliability is hard to compute [7], which has led to investigations of efficiently computable bounds [2,3,6].

When information about statistical dependence is limited, little is known. Hailperin [4] developed a linear programming model, which Zemel [10] used to obtain the best possible bounds when the only information given concerns failures of individual links and the dependencies between link failures are unknown. The most pessimistic assumptions are made about the unknown statistical dependencies in this model, as must be the case when an absolute bound on reliability is desired. These are called the *first-order* bounds. Improvements on these bounds can only be obtained by exploiting information about the failure probabilities of pairs of links; the result of employing this information in Hailperin's model gives the *second-order* bounds. Assous [1] developed second-order bounds for the two-terminal reliability problem using Hailperin's model; to obtain the lower bound, however, one must solve a quadratic programming problem, and no computationally efficient technique is known for this. Hence, Assous produces a simple heuristic technique for producing a lower bound using graph-theoretic techniques and the second-order information. In section 3, we develop a similar heuristic lower bound in the case of all-terminal reliability.

It is important to explore the relation with other reliability measures which incorporate statistical dependencies. Lam and Li [5] develop a model which incorporates very general information about statistical dependence; however, their evaluation procedure is a small modification of [8], an exponential time algorithm. Shier and Spragins [9] also develop a general model, and give an exponential time exact algorithm and a sequence of easier approximation problems. However, even their approximations examine the collection of all minimal pathsets, an exponentially large set. In addition, the approximations in the Shier-Spragins technique are obtained by truncating a summation with alternating signs; hence they obtain an approximation, but not a bound. As stated earlier, we are concerned with obtaining efficient (=

Reprinted from *The Proceedings of INFOCOM '86*, 1986, pages 290–292.

EH0300-4/90/0000/0157$01.00 © 1986 IEEE

polynomial time) absolute bounds; this is quite different from existing work in the literature.

2. FIRST-ORDER BOUNDS

In the weakest model, we assume that each edge e_i has a success probability p_i satisfying $a_i \leq p_i \leq b_i$. No other information about failures is known, and no assumption of statistical independence is made. The network is *coherent,* however, in that the failure of an edge cannot make a failed network operational. In this context, Hailperin [4] showed that the tightest lower bound on all-terminal reliability is obtained by solving the linear program

$$minimize \quad \sum_{S \epsilon F} Y_S$$

subject to

$$\sum_{S | i \epsilon S} Y_S \leq b_i, \quad 1 \leq i \leq e$$

$$\sum_{S | i \epsilon S} Y_S \geq a_i, \quad 1 \leq i \leq e$$

$$\sum_{S \subseteq \{1,\ldots,e\}} Y_S \leq 1$$

and nonnegativity constraints. In this linear program, F is the set of operational configurations of the network, and S varies over all configurations, both failed and operational.

The direct application of Hailperin's model is computationally intractable, since there are an exponential number of variables and constraints. Zemel [10] showed how to solve an equivalent problem efficiently, and later Assous [1] developed a simple computational method which we describe next. Assous showed that the minimum value L achieved by Hailperin's linear program satisfies

$$L = max(0, 1 - \min_{S \epsilon F^*} \sum_{j \epsilon S} (1 - a_j))$$

where F^* is the set of all spanning trees of the network. Thus L is simply one minus the weight of a minimum weight spanning tree of the network obtained using $1 - a_j$ as edge weights.

Although computationally appealing, the first-order bounds are very poor indeed for any practical purposes. For example, when p=0.9 for each link and the network has more than ten nodes, the lower bound is zero. One cannot fault the bounds for this, as there is no hope of obtaining a better bound unless additional information is provided. Nevertheless, the need for better bounds is clear.

3. SECOND-ORDER BOUNDS

In an effort to improve the first-order bounds, we assume that in addition to the previous model, for every pair i,j of edges, we have bounds on q_{ij}, the probability that edges i and j fail simultaneously. In particular, we suppose that $\underline{b}_{ij} \leq q_{ij} \leq \bar{a}_{ij}$. We continue to use first-order constraints, $\bar{b}_i \leq q_i \leq \bar{a}_i$, where $q_i = 1-p_i$.

Once again, Hailperin's model can be used to set up a linear program; here, however, there is no easy way to circumvent the exponential size of the linear program. We therefore resort to heuristic techniques. Given the network N we first find the most reliable spanning tree T of N with respect to the first-order information, as before. For each edge $e=(x,y)$ of T, let L_e be an upper bound on the probability that x and y have no operational path between them. Then a lower bound on the reliability is given by

$$1 - \sum_{e \epsilon T} L_e \qquad (*)$$

To obtain the first-order bound, we simply observe that $L_e \leq \bar{a}_e$. However, second-order information can be used to improve this.

For each $e \epsilon T$, $e=(x,y)$, construct a network N_e by deleting each edge of T from N and setting the weight of each edge f to $\bar{a}_{ef}$. Find a minimum weight path P from x to y, if one exists. Let W_e be the weight of the path P if one is found, ∞ otherwise. Now observe that $L_e \leq W_e$. In fact, W_e is an upper bound on the probability that edge e and path P fail simultaneously. One might consider including further $x-y$ paths, but then third- and higher-order information would be required to bound L_e; since only one path can be chosen, we select the most reliable. Combining the two constraints, we have $L_e \leq min(\bar{a}_e, W_e)$. Substituting the values of L_e obtained into (*) above, we obtain a lower bound on the all-terminal reliability. This bound can, of course, be no worse than the first-order bound, and is typically much better.

We have tested the new lower bound on a number of networks; to do so, we assume that pairs of failures are statistically independent, but make no assumptions about triples, quadruples, and so on.

The results of these tests are not surprising. Once statistical independence cannot be assumed, the bounds are significantly weaker than bounds which assume statistical independence.

4. CONCLUDING REMARKS

Our computational experience suggests that the second-order bounds fare very poorly against bounds which assume statistical independence of failures. Nevertheless, in certain contexts no information about high-order correlations is available, and the assumptions about statistical correlations are dangerous. In these contexts, the second-order bound developed here proves to be a definite asset; the improvement in the accuracy over the first-order bound makes the collection of second-order information worthwhile.

Acknowledgments

The contributions of Tim Brecht, Aparna Ramesh, and Nancy Ross are gratefully acknowledged. Research of the first author is supported by the Universidad Central de Venezuela, and of the second author by NSERC Canada under grant number A0579.

References

[1] J.Y. Assous, "Bounds for terminal reliability", preprint, Temple University, 1984.

[2] M.O. Ball and J.S. Provan, "Bounds on the reliability polynomial for shellable independence systems", *SIAM J. Alg. Disc. Meth.* 3 (1982) 166-181.

[3] C.J. Colbourn and D.D. Harms, "Bounding all-terminal reliability in computer networks", CCNG Report E-123, University of Waterloo, 1985.

[4] T. Hailperin, "Best possible inequalities for the probability of a logical function of events", *Amer. Math. Monthly* 72 (1965) 343-359.

[5] Y.F. Lam and V.O.K. Li, "Reliability modelling and analysis of communication networks with dependent failures", *Proceedings of INFOCOM85,* 1985, pp. 196-199.

[6] M.V. Lomonosov and V.P. Polesskii, "Lower bound of network reliability", *Prob. Inf. Trans.* 8 (1972) 118-123.

[7] J.S. Provan and M.O. Ball, "The complexity of counting cuts and of computing the probability that a graph is connected", *SIAM J. Comput.* 12 (1983) 777-788.

[8] A. Satyanarayana, "A unified formula for the analysis of some network reliability problems", *IEEE Trans. Rel.* R-31 (1982) 23-32

[9] D.R. Shier and J.D. Spragins, "Exact and approximate dependent failure models for telecommunications networks", *Proceedings of INFOCOM85,* 1985, pp. 200-205.

[10] E. Zemel, "Polynomial algorithms for estimating network reliability", *Networks* 12 (1982) 439-452.

Incorporating Dependent Node Damage in Deterministic Connectivity Analysis and Synthesis of Networks

H. Heffes and A. Kumar

AT&T Bell Laboratories, Holmdel, New Jersey 07733

Survivability of a node vulnerable network is often assessed in terms of the (node) connectivity of the graph that represents the logical topology of the network. When the damage causing events have widespread impact then, owing to the physical layout of the network facilities, each event can destroy several nodes. As a survivability measure, therefore, we define the generalized connectivity as the *minimum number of events* (rather than the minimum number of node removals) required to disconnect the network. To model the possible effects of damage causing events, we introduce the notion of a *dependence graph* on the nodes of the network and a set of *admissible cliques* in this graph. Nodes that are nonadjacent in the dependence graph are independent, i.e., they cannot be damaged by the same event, and each event destroys an admissible clique of nodes. We present techniques for calculating or bounding the generalized connectivity of given network graphs, and for synthesizing minimum link networks with prescribed generalized connectivity.

1. INTRODUCTION

In analyzing and synthesizing survivable communication networks, one is sometimes concerned with *dependencies* (i.e., simultaneous failure of sets of network resources) *introduced by events with widespread impact* (e.g., bombs, earthquakes (cf. Wood [15]), power failures). Standard network analysis and synthesis techniques, however, often include the assumption that failures of network components are independent. For example, the connectivity analysis question as generally stated is: "What is the minimum number of nodes that must be removed from the network so that the remaining network is disconnected (i.e. the node connectivity)?" [2, 5, 6]. In computing connectivity, no attention is usually paid to the number of underlying damage causing events that can disconnect the network. For example, if the nodes whose removal can disconnect the network are in close proximity then it is possible for *one* event, which can affect a large geographical area, to disconnect the network. A typical synthesis

NETWORKS, Vol. 16 (1986) 51–65

CCC 0028-3045/86/010051-15$04.00

problem is to find a network, with the minimum number of links, that has a prescribed node connectivity (in the usual sense), again without regard to the number of underlying events that can disconnect the network [2, 3, 6]. In this paper we introduce a *generalized definition of node connectivity*, namely, the number of damage causing *events* needed to disconnect the network, and present analysis and synthesis results for this generalized notion of connectivity.

To model the ways in which damage causing events affect network nodes, we introduce the notion of a dependence graph on these nodes and a set of admissible cliques in this graph. The interpretation of this model is that non-adjacent nodes in the dependence graph cannot be destroyed by a single event and consequently, the nodes destroyed by each event must be fully connected in the dependence graph (i.e., must form a clique). Further, a clique is admissible if and only if there exists an event that can destroy all the nodes in this clique. In this paper we study two subclasses of such dependence models, namely, weak dependence models and strong dependence models. In a weak (resp., strong) dependence model, nodes that are adjacent in the dependence graph may (resp., must) be destroyed together. A somewhat similar approach for modeling failure dependence in the stochastic reliability framework was proposed recently by Lam and Li [12].

In Section 2 we present two examples which demonstrate that the usual connectivity analysis and synthesis procedures are unsuitable in the presence of dependencies. In Section 3 we define our dependence model and the generalized notion of connectivity. Section 4 deals with the synthesis and analysis problems for networks in which the dependencies can be modeled with a strong dependence model. We show that networks having minimum number of links for prescribed generalized connectivity are obtained by using standard techniques to synthesize a reduced network, on coalesced nodes, without dependencies, and using the result of this synthesis to define an appropriate network on the original set of nodes. For the connectivity analysis of networks and strong dependence models in which the restrictions of the network graph to the components of the dependence graph are connected, we show that we can work with a collapsed network without dependencies and use standard analysis algorithms. For networks not satisfying this property we suggest a minimum set covering formulation and show how some computational savings can be realized. Section 5 deals with analysis and synthesis under a weak dependence model. The problem is shown to be related to a clique partition problem which is known to be NP-complete. Owing to this complexity we only discuss methods for obtaining upper and lower bounds on the generalized connectivity. For the synthesis problem we provide lower bounds on the minimum number of edges required to achieve a given generalized connectivity. For the case where each component of the dependence graph has two nodes, we present a synthesis algorithm which achieves a lower bound on the number of edges.

2. TWO EXAMPLES

These examples help motivate our generalization of graph connectivity. The following is a model of 12 geographically distributed nodes (indexed $a,b,c, \ldots, l$) of a network. Dotted lines between pairs of nodes indicate that a single event will destroy both the nodes simultaneously. Assume that the cost of a network is the number of

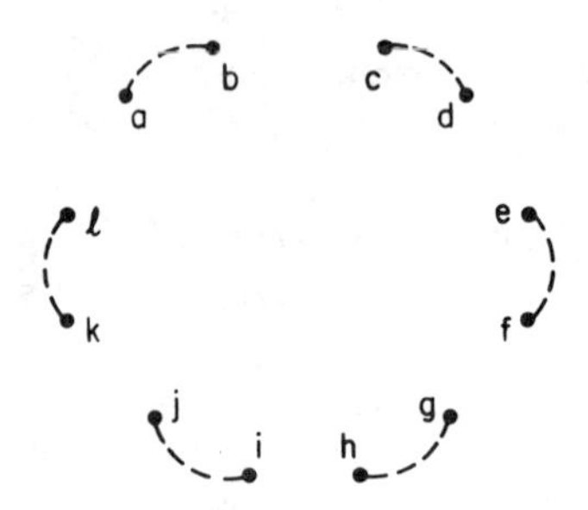

links, and consider the problem of finding a minimum cost link topology under the constraint that at least four events are required to disconnect the network.

We show in Section 4 that at least 18 links are needed and that G_1^* is an optimal topology (cf. Proposition 4.2 and Theorem 4.2). Consider, on the other hand, the graph G_1, with 12 nodes and 18 edges, obtained by using a standard synthesis technique (cf. [6, p. 333]), with the maximum achievable connectivity of three. It can be shown

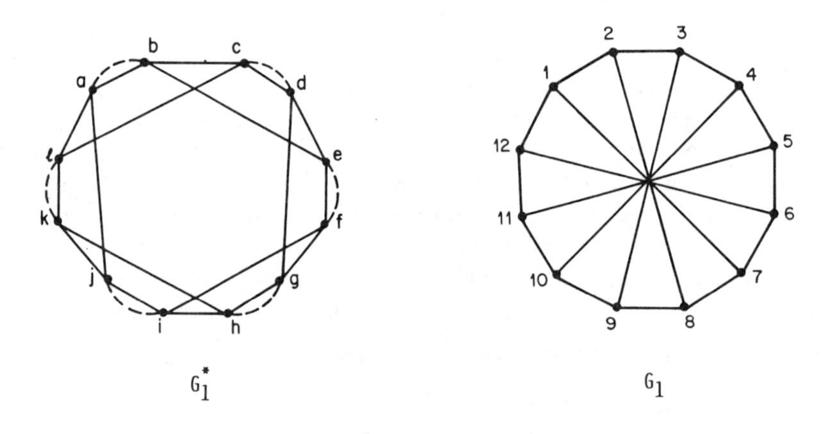

that there is no (one-to-one) mapping from $\{1,2, \ldots, 12\}$ onto $\{a,b, \ldots, l\}$ such that the corresponding mapping of graph G_1 requires at least four events to disconnect it.

Observe that G_1^* has a node connectivity (or *logical* connectivity) of three in the usual sense. Thus, in the presence of dependence between nodes, this is not a correct measure of the *physical* connectivity. Secondly, even though a graph with a node connectivity of three does the job, any such graph will not do. G_1 was obtained using a standard synthesis procedure, and no matter in what way one tries to map it onto the physical nodes, it can be disconnected with three events.

In the second example we consider the 12 geographically distributed nodes displayed at the beginning of this section where now a dotted line between a pair of nodes indicates that it is possible (but not necessary) to destroy both nodes with a single event. The problem again is to find a minimum cost topology such that at least four events are required to disconnect it. Note that G_1^* is no longer adequate since, in the presence of this weaker dependence, three events can disconnect it. Since now it is

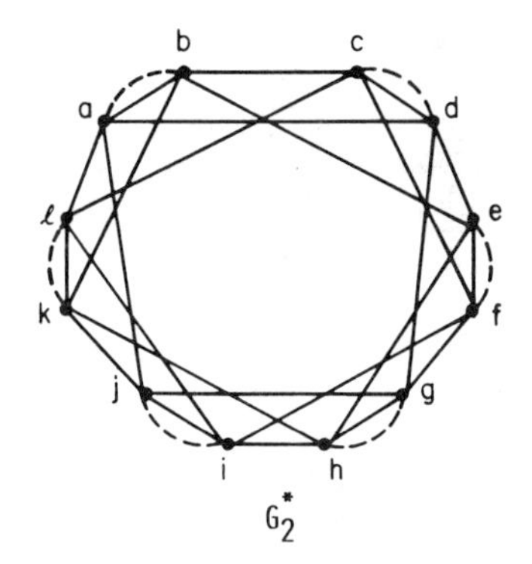

G_2^*

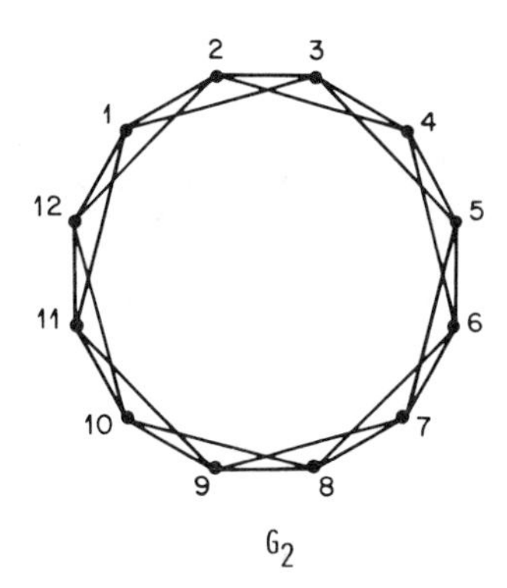

G_2

possible to destroy individual nodes, at least 24 links are required. The graph G_2^* has 24 links and requires at least four events to disconnect it. Hence it must be optimal. As before, the graph G_2 is obtained by using a standard synthesis procedure [6] for obtaining a graph of connectivity four with 12 nodes and 24 links. Again there is no way of mapping G_2 onto $\{a,b, \ldots, l\}$ so that it cannot be disconnected by three or fewer events. In any case, we contend that it is futile to start with a graph of connectivity four and attempt to map it onto the node set. A synthesis procedure must, *from the very beginning,* take account of the dependencies between the nodes.

In the following sections, we formalize the two notions of dependence, generalize the definition of node connectivity, and present analysis and synthesis results.

3. DETERMINISTIC MODELS FOR DEPENDENT NODE DAMAGE

Let $G = (V,E)$ denote an undirected graph G with node set $V = \{v_1,v_2, \ldots, v_n\}$ and edge set E. An *event* is any single occurrence that can cause damage to network elements. We consider only node damage in this paper and assume that all nodes are vulnerable.

Definition. A graph $D = (V,\delta)$ is said to be a *dependence graph* on V (or, equivalently, δ is said to be a *dependence relation* on V) when $\{v_i,v_j\} \notin \delta$ iff there does not exist an event that can simultaneously destroy v_i and v_j.

From this definition, it follows that the nodes destroyed by a single event must be fully connected in the dependence graph (i.e., they must form a clique). We note that the idea of representing dependencies between a finite number of entities via a "neighbor graph" is prevalent in the theory of Markov Random Fields (e.g., see [9]). In practice, a dependence graph can be thought of as a way to model the fact that certain pairs of nodes are too far apart to possibly be influenced by the same damage causing event.

We draw the edges of network graphs with solid lines and the edges of dependence graphs with dashed lines.

Example:

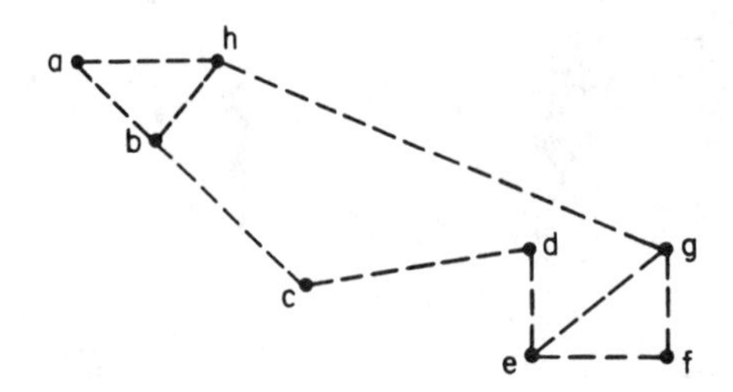

This is a dependence graph D on the set of nodes $V = \{a,b,c, \ldots, h\}$. Each of the sets of nodes $\{b,d\}$, $\{b,c,d\}$, $\{d,e,f,g\}$, etc., cannot be destroyed by a single event since each contains at least one pair of nodes that is not adjacent in D. This representation, however, does not fully answer the question as to which sets of nodes *can* be destroyed by single events. Consider the sets $\{a,b,h\}$ and $\{e,f,g\}$. Nothing that we have said so far excludes the possibility that these sets can be destroyed by single events. If we want to, say, include the possibility that $\{a,b,h\}$ can be destroyed by one event but $\{e,f,g\}$ cannot then we will have to do so explicitly. This motivates the definition of admissible cliques that follows this example. ■

Definition. (i) Given a dependence graph $D = (V,\delta)$, a set of fully connected nodes in D, i.e., a clique, is said to be an *admissible clique* iff there exists an event that can simultaneously destroy *exactly* the nodes in the clique. We denote the set of admissible cliques by C. A set of cliques in C that partitions D is an *admissible clique partition.*

(ii) Denoting cliques in D by C_i, $i = 1, 2, \ldots$, if C is such that $C_1 \in \mathsf{C}$, $C_2 \subset C_1 \Rightarrow C_2 \in \mathsf{C}$ then we say that (D,C) is a *weak dependence model.*

(iii) Suppose D has fully connected components. When a clique is admissible iff it is a component of D then we say that (D,C) is a *strong dependence model.*

The set of admissible cliques C models all the possible sets of nodes that can be destroyed by single events. If C models weak dependence and if a set of nodes can be destroyed by an event then for each subset of this set there is an event that can destroy exactly this subset. If C models strong dependence then each event destroys exactly the nodes in one component of D, i.e., all the nodes in a component of D are up or down together. Note further that, given a set of admissible cliques C, in general, there may not exist an admissible clique partition of D. If (D,C) is a weak dependence model, however, then, by the usual way of getting a partition from a covering, it can be shown that there is always an admissible clique partition.

Letting $\Delta = (D,\mathsf{C})$ we denote by (G,Δ) a network graph on the node set V along with a dependence graph D on V and a set of admissible cliques in D. For a strong dependence model, since the notation (D,C) is redundant, we simply denote a strong dependence model (or strong dependence graph) by $\mathbf{D}$; along with the network graph we have the notation $(G,\mathbf{D})$. Φ denotes the dependence graph with no edges. Hence (G,Φ) denotes the graph G with no dependence between its nodes.

Definition. The *(generalized) connectivity* of (G,Δ), denoted by $\omega(G,\Delta)$, is the minimum number of events required to disconnect G (i.e., after the occurrence of these events, there is no network path between at least one pair of surviving nodes), or to destroy all the nodes in G.

Thus, $\omega(G,\Phi)$ is the usual connectivity of G, except that we differ from the literature in taking the usual node connectivity of a fully connected graph on n nodes to be n. It will become clear from what follows why we need to differ in this way. In previous work the connectivity of a fully connected graph is either left undefined [5] or is $n - 1$ [2].

Since each admissible clique partition of D corresponds to a way of destroying all the nodes in V, it follows, from the above definitions, that $\omega(G,\Delta)$ is no more than the cardinality of the minimum admissible clique partition of D. Further observe that if G is fully connected then, depending on Δ, $\omega(G,\Delta)$ can have any value between 1 and n.

4. ANALYSIS AND SYNTHESIS UNDER STRONG DEPENDENCE

Let G be a connected graph on the set of nodes V, and $\mathbf{D}$ be a strong dependence model on V. In this section we show how $(G,\mathbf{D})$ can be analyzed to obtain $\omega(G,\mathbf{D})$, and how a minimum link graph G^* can be synthesized so that $\omega(G^*,\mathbf{D}) = \omega_o$ for a given ω_o.

Let $\mathbf{D}_1, \mathbf{D}_2, \ldots, \mathbf{D}_N$ denote the N (fully connected) components of $\mathbf{D}$ and let $V_1, V_2, \ldots, V_N$ be the node sets of these components. For every pair of nodes $s, t \in V$, such that there is no $s - t$ path all of whose nodes belong to the component(s) of $\mathbf{D}$ to which s and t belong, define $\omega_{st}(G,\mathbf{D})$ to be the minimum number of events required to disconnect the pair s, t without destroying s or t. The following Lemma is immediate:

Lemma 4.1. If it is possible to disconnect $(G,\mathbf{D})$ without destroying all the nodes, then $\omega(G,\mathbf{D}) = \min\{\omega_{st}(G,\mathbf{D}): s, t \in V, \omega_{st}(G,\mathbf{D}) \text{ defined}\}$, otherwise $\omega(G,\mathbf{D}) = N$.

Property P. $(G,\mathbf{D})$ has the property P if each of the subgraphs obtained by restricting G to the node sets $V_1, V_2, \ldots$ and V_N is connected.

If $(G,\mathbf{D})$ has property P, then obtain $(G'(\mathbf{D}),\Phi)$ on the node set $(d_1,d_2, \ldots, d_N)$ by coalescing the components of $\mathbf{D}$ into single nodes. The component $\mathbf{D}_i$ yields the node d_i, and there is a link in $G'(\mathbf{D})$ between d_i and d_j if there is a link in G between some node in V_i and some node in V_j.

Proposition 4.1. If $(G,\mathbf{D})$ has property P, then for all $s, t \in V$, $s \in V_i$ and $t \in V_j$, $i \neq j$, for which $\omega_{st}(G,\mathbf{D})$ is defined, $\omega_{st}(G,\mathbf{D}) = \omega_{d_i d_j}(G'(\mathbf{D}),\Phi)$.

Proof. See Appendix I. The basic idea is that, owing to property P, every path between a pair of nodes in $G'(\mathbf{D})$ corresponds to at least one path in G. ■

Theorem 4.1. If $(G,\mathbf{D})$ has property P, then $\omega(G,\mathbf{D}) = \omega(G'(\mathbf{D}),\Phi)$.

Proof. Follows easily from Lemma 4.1 and Proposition 4.1. ■

INCORPORATING DEPENDENT NODE DAMAGE

Theorem 4.1 shows that the calculation of $\omega(G,\mathbf{D})$, in the presence of the property P, reduces to the usual connectivity calculation on a graph with no dependence between its nodes. Thus once the reduction from $(G,\mathbf{D})$ to $G'(\mathbf{D})$ has been done, standard connectivity analysis can be used. This result also implies that if $(G,\mathbf{D})$ has property P then, owing to our definition of the connectivity of a fully connected graph, $\omega(G,\mathbf{D})$ can have any (integer) value between 0 and N except $N - 1$. The reason for our difference from the standard definition should now be clear. If $(G,\mathbf{D})$ has property P and if $N - 1$ components of $\mathbf{D}$ are destroyed, the nodes in the remaining component are still connected.

Before proceeding to the analysis problem when $(G,\mathbf{D})$ does not have the property P, we solve the synthesis problem for strong dependence. Given a strong dependence model $\mathbf{D}$ on a set of nodes $V = \{v_1,v_2, \ldots, v_n\}$, let $\{V_1,V_2, \ldots, V_N\}$, $V_i \subset V$, be as before. Given $1 \leq \omega_o \leq N$, we need a graph $G^* = (V,E^*)$, such that $\omega(G^*,\mathbf{D}) = \omega_o$ and $|E^*| = \min\{|E|:G = (V,E),\omega(G,\mathbf{D}) = \omega_o\}$ where $|E|$ denotes the cardinality of the set E.

Proposition 4.2. Let $G = (V,E)$ be such that $\omega(G,\mathbf{D}) = \omega_o$. Then, letting $|V_i| = n_i$,

(a) if $2 \leq \omega_o \leq N - 1$, then $|E| \geq \Sigma_{i=1}^{N} (n_i - 1) + \lceil N\omega_o/2 \rceil$,
(b) if $\omega_o = N$, then $|E| \geq \Sigma_{i=1}^{N} (n_i - 1) + (N(N - 1))/2$,

where $\lceil x \rceil$ denotes "the smallest integer greater than or equal to x." (Observe that the lower bounds for $\omega_o = N - 1$ and $\omega_o = N$ are the same.)

Proof. Consider the subgraphs obtained by restricting G to V_i, $1 \leq i \leq N$. Let, for every i, $1 \leq i \leq N$, V_{ij}, $1 \leq j \leq k_i$, partition the set V_i into node sets corresponding to the components of the subgraph obtained by restricting G to V_i; i.e., $V_i = V_{i1} \cup V_{i2} U \cdots \cup V_{ik_i}$, $V_{ij_1} \cap V_{ij_2} = \phi$, $1 \leq j_1 < j_2 \leq k_i$, and if $v_1 \in V_{ij_1}$ and $v_2 \in V_{ij_2}$ then v_1 and v_2 are not adjacent in G. (*cf.* the following picture and also observe that if $(G,\mathbf{D})$ has property P then $k_i = 1$, $1 \leq i \leq N$). Let $|V_{ij}| = n_{ij}$, $1 \leq i \leq N$, $1 \leq j \leq k_i$. The

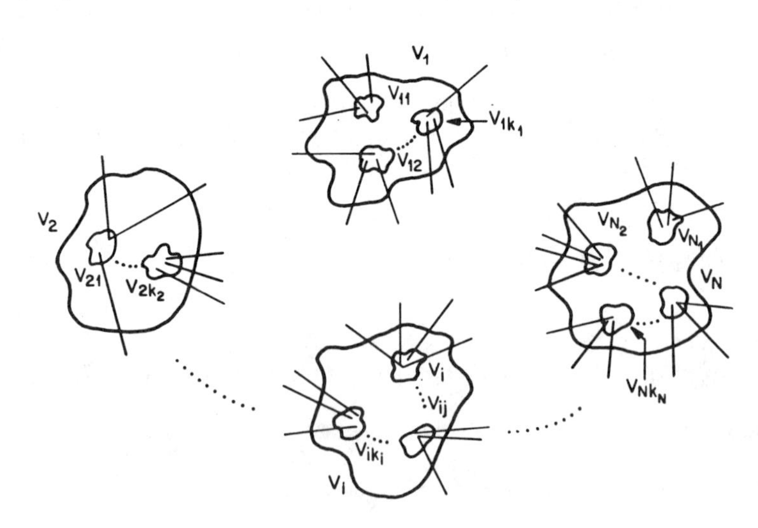

proof (see Appendix I) relies on the observation that the "degree" of each set V_{ij} must be at least ω_o (for $2 \leq \omega_o \leq N - 1$). The number of links needed to satisfy this requirement is the least when $k_i = 1$, $1 \leq i \leq N$. ■

Algorithm A1. Given V, $\mathbf{D}$, as in Proposition 4.2 and $1 \leq \omega_o \leq N$, $\omega_o \neq N - 1$. Synthesize a graph G^* on V as follows.

a. For every i, $1 \leq i \leq N$ coalesce the nodes in V_i into a single node. Thus we now have N nodes $d_1, d_2, \ldots, d_N$.
b. Synthesize a graph G' on these nodes such that $\omega(G',\Phi) = \omega_o$. This will require $N(N - 1)/2$ edges if $\omega_o = N$, and $\lceil N\omega_o/2 \rceil$ edges if $2 \leq \omega_o < N - 1$ (and, of course, $N - 1$ edges if $\omega_o = 1$).
c. Return now to the node sets V_i, $1 \leq i \leq N$. Put a spanning tree on each of the V_i. This requires $\Sigma_{i=1}^{N} (n_i - 1)$ edges. Put a link between any node in V_i and any node in V_j if d_i and d_j are adjacent in G'. This yields the graph G^*. ■

Theorem 4.2. A graph G^* obtained via Algorithm A1 is such that $\omega(G^*,\mathbf{D}) = \omega_o$ and $|E^*| = \min \{|E|:G = (V,E),\omega(G,\mathbf{D}) = \omega_o\}$

Proof. Immediate from Proposition 4.2 and Theorem 4.1. ■

We have left out the case $\omega_o = N - 1$ in Algorithm A1. Since the lower bounds for $\omega_o = N - 1$ and $\omega_o = N$ in Proposition 4.2 are the same, and synthesis for N achieves this lower bound, one is no worse off by synthesising for $\omega_o = N$.

In practice, V and $\mathbf{D}$ will correspond to some geographical distribution of nodes. In step (c) of the algorithm, therefore, it may be preferable to choose minimum length spanning trees, and if a link is needed between the node sets V_i and V_j, the link may be put between the two nodes that are the shortest distance apart. Better still, if in step (b) we define the cost of placing a link between d_i and d_j to be the minimum distance between a node in V_i and a node in V_j, then we could ask for the minimum cost graph with connectivity ω_o. A heuristic for this problem is presented in [14].

Our results for $(G,\mathbf{D})$ with property P become intuitively clear if one observes that each component of $\mathbf{D}$ is equivalent to a single node in an ordinary graph in the following senses: (i) Either all the nodes in a component are dysfunctional or all are functional, and (ii) if two or more edges in G are each incident on a component, then it is possible to "enter" the component by any one of these edges and "leave" by any of the others.

We turn now to the connectivity analysis of graphs with a strong dependence model on their nodes, but possibly without the property P. We return to our notation in the proof of Proposition 4.2 and obtain a graph $G'(\mathbf{D})$ on the set of nodes $\{d_{ij}, 1 \leq i \leq N, 1 \leq j \leq k_i\}$, where each d_{ij} is obtained by coalescing the nodes in V_{ij}. There is an edge in $G'(\mathbf{D})$ between $d_{i_1j_1}$ and $d_{i_2j_2}$, $i_i \neq i_2$, if there is an edge in G between a node in $V_{i_1j_1}$ and a node in $V_{i_2j_2}$. In the strong dependence graph on the node set of $G'(\mathbf{D})$, the node sets $\{d_{ij}, 1 \leq j \leq k_i\}$, for all i, $1 \leq i \leq N$, are fully connected. We call this dependence graph $\mathbf{D}'(G)$. Our discussion so far and Lemma 4.1 in particular should make the following fact clear.

Lemma 4.2. (i) If, in $(G,\mathbf{D})$, the pair of nodes $s, t \in V$ can be disconnected (without destroying either s or t) then, letting $s \in V_{i_1j_1}$ and $t \in V_{i_2j_2}$, $\omega_{st}(G,\mathbf{D}) = \omega_{d_{i_1j_1}d_{i_2j_2}}(G'(\mathbf{D}),\mathbf{D}'(G))$
(ii) $\omega(G,\mathbf{D}) = \omega(G'(\mathbf{D}),\mathbf{D}'(G))$

Lemma 4.2 shows that, in general, the connectivity analysis problem reduces, via the above described reduction procedure, to the connectivity analysis of a graph $(G,\mathbf{D})$ in which the restrictions of G to the components of $\mathbf{D}$ are graphs with empty edge sets. Consider a pair of nodes s, t of such a graph. If $\omega_{st}(G,\mathbf{D})$ exists then it can be calculated by formulating the problem as a minimum set covering problem (cf. [8, p. 301]). One considers all the $s - t$ paths in G. A component of $\mathbf{D}$ "covers" a path if removal of the nodes of the component breaks the path. The problem is to find the minimum number of components of $\mathbf{D}$ that cover all the $s - t$ paths in G (or, equivalently, that include an $s - t$ node cut). This is clearly a minimum set covering problem with unit costs. Explicit enumeration of all $s - t$ paths is impractical, however. In [1], Bellmore et al. report computational experience with the minimum set covering formulation of the connectivity problem for ordinary graphs. They use an iterative algorithm in which all $s - t$ paths are not generated at the outset, but only as they constrain the optimal solution.

It is not necessary to calculate $\omega_{st}(G,\mathbf{D})$ for every s, t pair to show that $\omega(G,\mathbf{D}) = \omega_o$. For calculating the connectivity of a graph there is a result, due to Kleitman [11], which produces a considerable reduction in the number of s, t pairs for which connectivity needs to be calculated. This result extends to graphs with strong dependence as follows.

Proposition 4.3. Choose a component $\mathbf{D}_i$ of $\mathbf{D}$ and let $\alpha_i = \min \{\{\omega_{st}(G,\mathbf{D}): s \in V_i, t \neq s, \omega_{st}(G,\mathbf{D}) \text{ defined}\}, N\}$. If $\alpha_i \geq \omega_o$ then $\omega(G,\mathbf{D}) = \omega_o$ if $\omega(\overline{G}_i,\overline{\mathbf{D}}_i) = \omega_o - 1$, where $(\overline{G}_i,\overline{\mathbf{D}}_i)$ is obtained by removing all the nodes in V_i and all the edges incident to them.

Proof. Essentially the same argument as in [11]. ■

In order to verify that $\omega(G,\mathbf{D}) \geq \omega_o$, this proposition can be applied iteratively. Suppose the sequence of components of $\mathbf{D}$ that are chosen in this process is $\mathbf{D}_{i_1}, \mathbf{D}_{i_2}, \ldots, \mathbf{D}_{i_{\omega_o}}$. Then the number of $\omega_{st}(G,\mathbf{D})$ calculations that are required is equal to

$$\sum_{j=1}^{\omega_o} \frac{n_{i_j}(n_{i_j} - 1)}{2} + n_{i_j}\left(n - \sum_{l=1}^{j} n_{i_l}\right).$$

where $n = \sum_{i=1}^{N} n_i$. If $\omega_o < N$ then it can be shown that in order to minimize the above function, the ω_o smallest components should be chosen.

5. ANALYSIS AND SYNTHESIS UNDER WEAK DEPENDENCE

Let G be a connected graph on the set of nodes V and let D be a dependence graph on V and C a set of admissible cliques in D such that $\Delta = (D,C)$ is a weak dependence model. We show that the problem of calculating $\omega_{st}(G,\Delta)$ is NP-hard. We also present upper and lower bounds on $\omega(G,\Delta)$, and a synthesis procedure for the case where every component of D has two nodes. The following proposition and its proof are due to D. M. Topkis.

Proposition 5.1. The problem of showing whether $\omega_{st}(G,\Delta) \leq k$ is NP-complete.

Proof. There are instances of the problem that are equivalent to the clique partition problem which is known to be NP-complete [7]. For suppose that both s and t are adjacent to every other node in G (but not to each other). Then to destroy all $s - t$ paths, every other node must be destroyed. Consider the subgraph D' of D on the nodes $V - \{s,t\}$. If all cliques of D are admissible then clearly $\omega_{st}(G,\Delta)$ is the number of elements in the minimum clique partition of D'. ■

We can, however, draw upon the strong dependence results in Section 4 to establish bounds on $\omega(G,\Delta)$. It is clear that each admissible clique partition of D corresponds to a set of events that can destroy all the nodes in V. Thus if we consider only this set of events, and the subgraph of D obtained by deleting all edges that are not between the nodes of a clique in the chosen clique partition, then we obtain a strong dependence model, say $\mathbf{D}$. Henceforth we denote an admissible clique partition of D by $\mathbf{D}$, i.e., the corresponding strong dependence graph. Now for every such $\mathbf{D}$, if only the corresponding events are allowed, at least $\omega(G,\mathbf{D})$ events are required to disconnect G. Further, given $\omega(G,\Delta)$ events that disconnect G, and using the "closure-under-inclusion" property of C, we can obtain a clique partition $\mathbf{D}^*$ of D such that $\omega(G,\mathbf{D}^*) = \omega(G,\Delta)$. So we obtain

Theorem 5.1. (i) For each admissible clique partition $\mathbf{D}$ of D, $\omega(G,\Delta) \leq \omega(G,\mathbf{D})$.
(ii) $\omega(G,\Delta) = \min \{\omega(G,\mathbf{D})$: $\mathbf{D}$ is an admissible clique partition of $D\}$.

For each admissible partition of D into cliques, $\omega(G,\mathbf{D})$ can be calculated as discussed in Section 4. Proposition 5.1 suggests that it is difficult to devise an efficient algorithm for the minimization in Theorem 5.1. Upper and lower bounds, however, can be obtained.

Each admissible clique partition $\mathbf{D}$ of D yields an *upper bound* $\omega(G,\mathbf{D})$. For each such $\mathbf{D}$, a weaker but more easily computable upper bound can be obtained by *adding edges* to G to obtain a G' such that $(G',\mathbf{D})$ has property P. A *lower bound* can be obtained by *adding edges* to D to make its components fully connected and letting all cliques be admissible; this increases the possible ways in which nodes in V can be destroyed. Further, with the components of the augmented D fully connected the clique partition problem becomes trivial. Letting D' denote the augmented D and C' the set of all cliques of D', to calculate $\omega_{st}(G,\Delta')$, one need only consider the restriction of D' to $V - \{s,t\}$ which is also fully connected. Let $\mathbf{D}'$ be the union of this restriction and the two single node "graphs" s and t; then $\omega_{st}(G,\Delta') = \omega_{st}(G,\mathbf{D}')$. If D has only one component then this will yield the trivial lower bound of one. The bound will be better when (G,Δ) is a model for a network spread over a large geographical area and D has several components.

We turn now to the synthesis problem for weak dependence. Given a weak dependence model Δ on a set of nodes $V = \{0,1, \ldots, n-1\}$ and ω_o, $1 \leq \omega_o \leq n$, the problem is to find a graph $G^* = (V,E^*)$, such that $\omega(G^*,\Delta) = \omega_o$ and $|E^*| = \min \{|E|: G = (V,E), \omega(G,\Delta) = \omega_o\}$.

Proposition 5.2. Let $G = (V,E)$ be such that $\omega(G,\Delta) = \omega_o$.

(i) If $\mathbf{D}$ is an admissible partition of D into cliques and $N(\mathbf{D})$ is the number of components of $\mathbf{D}$, then

$$|E| \geq n - N(\mathbf{D}) + \left\lceil \frac{N(\mathbf{D})\omega_o}{2} \right\rceil, \qquad 2 \leq \omega_o \leq N(\mathbf{D}) - 1$$

and

$$|E| \geq n - N(\mathbf{D}) + \frac{N(\mathbf{D})(N(\mathbf{D}) - 1)}{2}, \qquad \omega_o = N(\mathbf{D})$$

(ii)

$$|E| \geq \begin{cases} \left\lceil \dfrac{n\omega_o}{2} \right\rceil, & 2 \leq \omega_o \leq n - 1 \\ \dfrac{n(n-1)}{2}, & \omega_o = n \end{cases}$$

Proof. (i) is immediate from Theorem 5.1 and Proposition 4.2.

(ii) Follows from (i) after observing that individual nodes can be damaged by each event and hence, for this partition of D, $N(\mathbf{D}) = n$. ∎

We do not as yet have a general synthesis procedure. In general the number of edges in the lower bounds in Proposition 5.2 will not suffice. For the case where n is even, each component of D has two nodes (as in the examples in Section 2), all cliques are admissible, and $\omega_o \geq 2$ is even, the following procedure yields an optimal graph. It is optimal because it uses $n\omega_o/2$ edges. We point out here that in order to understand why the procedure works it is necessary to understand the standard synthesis procedure in [6, p. 332].

Algorithm A2. $V = \{0,1,2, \ldots, n-1\}$, n even; each component of D has two nodes, i.e., D has $n/2$ components; ω_o is even and $2 \leq \omega_o \leq n/2$.

(i) Let each component of D be a clique. Consider the resulting strong dependence relation $\mathbf{D}$, and use Algorithm A1 to synthesize a graph G_1^* such that $\omega(G_1^*,\mathbf{D}) = \omega_o$. This will require $n/2(\omega_o/2 + 1)$ edges. Let the pairs of dependent nodes be $\{0,1\}$, $\{2,3\}$, $\{4,5\}, \ldots, \{n-2, n-1\}$. Choose graph G_1^* such that there is an edge between each node in a pair, and an edge between node $2i - 1$, $1 \leq i \leq n/2$, and nodes $(2i - 1 + 2j - 1) \bmod n$, $1 \leq j \leq \omega_o/2$. The following figure illustrates this for $n = 16$, $\omega_o = 6$, with solid lines denoting the edges just inserted. Not all the edges have been shown. The graph with solid edges can be shown to have an ordinary connectivity of $\omega_o/2 + 1$.

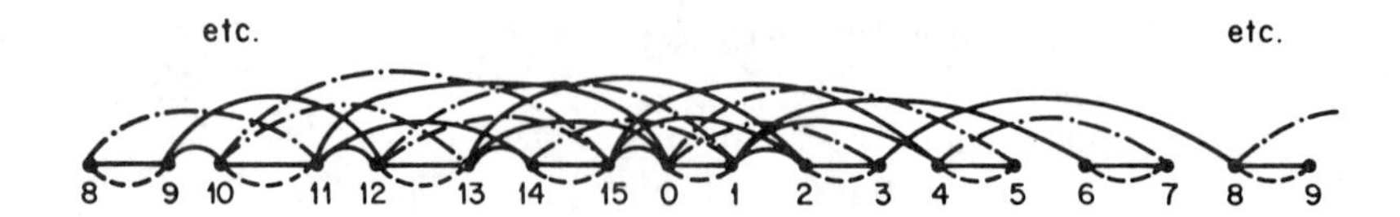

(ii) Add $n/2(\omega_o/2 - 1)$ more edges as follows. Place an edge between node $(2i - 1)$, $1 \leq i \leq n/2$, and nodes $(2i - 1 - (2j - 1)) \bmod n$, $1 < j \leq \omega_o/2$. In the example above the dotted-dashed lines indicate these edges. Call this graph G^*. It can be shown to have an ordinary connectivity of ω_o, and it can also be checked that $\omega(G^*,\Delta) = \omega_o$. This follows from the fact that between any pair of nonadjacent nodes in G^* there are ω_o paths that are pairwise "component" disjoint. Note also that $G^{*\prime}(\mathbf{D})$ is the graph that will be obtained if the standard synthesis procedure is used to construct a graph with connectivity ω_o on $n/2$ nodes. Thus whether individual nodes or entire components of D are damaged by events, at least ω_o events are required to disconnect G^*. Since only $n\omega_o/2$ edges have been used, G^* is optimal. ∎

6. DISCUSSION

We have generalized the definition of node connectivity to incorporate dependent failures and have presented analysis and synthesis results based on this generalized notion of connectivity. To represent damage dependence between nodes, we have introduced the concept of a dependence graph and a set of admissible cliques of this graph; together these constitute a dependence model. We have presented analysis and synthesis results for two classes of dependence models, namely weak dependence and strong dependence. Intermediate classes of models can also be studied; our results Theorem 5.1(i) and Proposition 5.2(i) will apply if there exist admissible clique partitions in these models.

One way of assessing the degree of dependence between node failures, in a given application, is to use the results of concomitant research in which we construct a damage model, in a stochastic setting, and study the degree of correlation between node failures [10]. The results there are in terms of the intensity of the damage scenario (i.e., number of events per unit area), the damage radius of an event and the spacing of the resources.

Our work has concentrated on connectivity which is only one measure of network robustness, useful under low traffic conditions. Connectivity, as a measure of network robustness, has another limitation that must be recognized. It is concerned only with whether a network is disconnected following the occurrence of a certain number of events, not with how "badly" it is disconnected. Two networks could have the same connectivity but after that many events have occurred (in the worst possible way for either network), one network could have many more of its surviving node pairs able to communicate than the other. An alternative method [13] for assessing the survivability of a network is to calculate the fraction of surviving node pairs that are connected as a function of the number of damage causing events, assuming that the events occur in the worst possible way. If the network is connected to begin with then this function will be 1 up to $\omega(G,\Delta) - 1$ events and then will fall below 1. For given $\omega(G,\Delta)$, a network may be considered to be good if the function does not decrease too rapidly as the number of events increases beyond $\omega(G,\Delta) - 1$.

We have considered only dependent node damages in this paper. Clearly the extension of these results to incorporate dependent link and link/node failures is an important area for future investigation. Finally we note that more work is needed on both the analysis and synthesis problems under weak dependence.

In the development of the results in this paper we have been motivated to obtain algorithms which use standard graph connectivity algorithms, where appropriate, for which experience and software packages exist. This has been accomplished as evidenced by the network reduction techniques in the analysis and synthesis results for strong dependence.

We are grateful to R. Johnson, P. R. Sokkappa, G. S. Subramanian, and D. M. Topkis for making themselves available for discussions, and to an anonymous referee for helpful comments.

APPENDIX I

Proof of Proposition 4.1

(a) $\omega_{st}(G,\mathbf{D}) \geq$ no. of paths in any set of pairwise node disjoint $d_i - d_j$ paths in $G'(\mathbf{D})$. For, from the construction of $(G'(\mathbf{D}),\Phi)$ and the fact that $(G,\mathbf{D})$ has property P, it is clear that any $d_i - d_j$ path in $G'(\mathbf{D})$ corresponds to at least one $s - t$ path in G. Further if two $d_i - d_j$ paths in $G'(\mathbf{D})$ are node disjoint (except, of course, for their terminal nodes), then there exist two corresponding $s - t$ paths in G such that, except for nodes in V_i and V_j, they do not both use nodes in the same component of $\mathbf{D}$. So these two paths cannot be simultaneously damaged by an event that does not destroy s or t. The inequality (a) then follows from the definition of $\omega_{st}(G,\mathbf{D})$.

(b) $\omega_{st}(G,\mathbf{D}) \leq$ number of nodes in any d_i, d_j node-cut in $G'(\mathbf{D})$. For suppose after destroying all the components of $\mathbf{D}$ corresponding to a d_i, d_j node-cut in $G'(\mathbf{D})$, there still exists an $s - t$ path in G. This path corresponds to a surviving $d_i - d_j$ path $G'(\mathbf{D})$ which is contradictory to our assumption that a d_i, d_j node-cut was removed. Inequality (b) is thus established.

From (a) and (b),

$$\left\{\begin{matrix}\text{min number of nodes in a}\\ d_i,\ d_j \text{ node-cut in } G'(\mathbf{D})\end{matrix}\right\}$$

$$\geq \omega_{st}(G,\mathbf{D}) \geq \left\{\begin{matrix}\text{max number of paths in a set of}\\ \text{pairwise node disjoint } d_i - d_j \text{ paths in } G'(\mathbf{D})\end{matrix}\right\}$$

By Menger's Theorem [4], the left- and right-hand terms are equal and it follows that $\omega_{st}(G,\mathbf{D}) =$ min number of nodes in a d_i, d_j node-cut in $G'(\mathbf{D}) = \omega_{d_i d_j}(G'(\mathbf{D}),\Phi)$. ■

Proof of Proposition 4.2

Refer to the notation and the figure in the sketch of this proof in the main text.

(a) For $2 \leq \omega_o \leq N - 1$, consider the restriction of G to V_{ij}. This is a connected graph. In order that $\omega(G,\mathbf{D}) = \omega_o$, the number of edges between nodes in V_{ij} and the node sets $V_1, V_2, \ldots, V_{i-1}, V_{i+1}, \ldots, V_N$ must be at least ω_o; otherwise after removing the ($\omega_o - 1$ or fewer) nodes to which these edges connect, there are at least two node sets left and V_{ij} is disconnected from the other surviving nodes. Hence, noting that a link between (say) $V_{i_1 j_1}$, and $V_{i_2 j_2}$ can serve the purpose for both these sets,

$$|E| \geq \frac{\omega_o}{2} \sum_{i=1}^{N} k_i + \sum_{i=1}^{N} \sum_{j=1}^{k_i} (n_{ij} - 1)$$

where the second term arises from the fact that each V_{ij} is connected. Therefore,

$$|E| \geq \sum_{i=1}^{N} \left(n_i + k_i \left(\frac{\omega_o}{2} - 1 \right) \right)$$

$$\geq \sum_{i=1}^{N} \left(n_i + \left(\frac{\omega_o}{2} - 1 \right) \right) \qquad (\text{since } k_i \geq 1 \text{ and } \omega_o \geq 2)$$

$$= \sum_{i=1}^{N} (n_i - 1) + \frac{N\omega_o}{2}.$$

And since $|E|$ is an integer, $|E| \geq \sum_{i=1}^{N} (n_i - 1) + \lceil N\omega_o/2 \rceil$.

(b) If $\omega_o = N$ then clearly, for every i, k_i must be one, otherwise if some k_i is more than 1 then $N - 1$ events can destroy all the node sets $V_1, V_2, \ldots, V_{i-1}, V_{i+1}, \ldots, V_N$ leaving a pair of nodes in V_i disconnected. We now have N node sets, each of which is connected within itself, and must have a link to every other node set. Hence

$$|E| \geq \sum_{i=1}^{N} (n_i - 1) + \frac{N(N-1)}{2}. \qquad \blacksquare$$

References

[1] M. Bellmore, M. J. Greenberg, and J. J. Jarvis, Multi-commodity disconnecting sets. *Management Sci.* **16** (1970) B427–B433.

[2] F. T. Boesch, Graph theoretic models for network reliability studies. Electrical Eng. and Computer Sc. Dept., Stevens Inst. of Tech., Technical Report #8010, December, 1980.

[3] F. T. Boesch and A. P. Felzer, A general class of invulnerable graphs. *Networks* **2** (1972) 261–283.

[4] L. R. Ford, Jr., and D. R. Fulkerson, *Flows in Networks*. Princeton University Press, Princeton, NJ, 1962.

[5] H. Frank and I. T. Frisch, Analysis and design of survivable networks. *IEEE Trans. Comm. Tech.* **COM-18** (1970) 501–519.

[6] H. Frank and I. T. Frisch, *Communication, Transmission and Transportation Networks*. Addison-Wesley, Reading, MA, 1971.

[7] M. R. Garey and D. S. Johnson, *Computers and Intractability: A Guide to the Theory of NP-Completeness*. Freeman, San Francisco, 1979.

[8] R. S. Garfinkel and S. L. Nemhauser, *Integer Programming*. John Wiley & Sons, New York, 1972.

[9] G. R. Grimmett, A theorem about random fields. *Bull. London Math. Soc.* **5** (1973) 81–84.

[10] H. Heffes and A. Kumar, Stochastic damage models and dependence effects in the survivability analysis of communication networks. *IEEE Journal on Selected Areas in Communications*, to appear.

[11] D. J. Kleitman, Methods for investigating connectivity of large graphs. *IEEE Trans. Circuit Theory*, **CT-16** (1969) 232–233.

[12] Y. F. Lam and V. O. K. Li, Reliability modeling and analysis of communication networks with dependent failures. *Proceedings IEEE INFOCOM*, 1985, pp. 196–199.

[13] P. R. Sokappa, Private Communication.

[14] K. Steiglitz, P. Weiner, and D. J. Kleitman, The design of minimum-cost survivable networks. *IEEE Trans. Circuit Theory* **CT-16** (1969) 455–460.

[15] R. K. Wood, "Efficient calculation of the reliability of lifeline networks subject to seismic risk." Operations Research Center, Univ. of Cal., Berkeley, Research Report, No. ORC 80-13, June 1980.

Received February, 1985
Accepted August, 1985

Chapter 5: Performability

Most existing network reliability models simply provide measures of network connectivity. Connectivity measures remain important in analyses in which component availabilities, rather than component failures, are considered. Thus, in telecommunication applications, it is not only physical failure of a component but also availability of a component for message processing that matters. Unavailability of a component is caused either by its failure or by a network control mechanism that effectively blocks any incoming traffic to a saturated or overloaded unit of the network.

In this chapter, we consider papers describing method(s) for performability evaluation in computing systems, gracefully degrading systems, and telecommunication or computer networks.

Experience with current telecommunication networks suggests that connectivity failures are relatively rare and affect, at most, a small percentage of network users at any time. The failure of one or more network components, however, may lead to increases in network congestion and delay and to decreases in the available throughput. These phenomena can be experienced simultaneously by all users; they warrant developing a unified framework for studying performance and reliability, popularly called performability.

Performability evaluation is also mandatory for distributed and parallel real-time systems. Such systems typically exhibit properties of concurrency, timeliness, fault tolerance, and degradable performance. They require unified and general modeling techniques employed in strict evaluations of performance and reliability.

Trivedi (1982) has given a simple example to illustrate the notion of performability. Let $X_1, X_2, \ldots, X_n$ be the times to failure of *n* processors. Then after a period of time $Y_1 = \min\{X_1, X_2, \ldots, X_n\}$, only (n-1) processors will be active and the computing capacity of the system will drop to (n-1). The cumulative computing capacity that the system supplies until all processors have failed is then given by the random variable: $C_n = nY_1 + (n-1)(Y_2 - Y_1) + \cdots + (Y_n - Y_{n-1})$. Some authors have called C_n a *computation before failure* while others term it *performability*. The concept of performability is thus quite general and, depending on the choice of the base model and the performance variable, can be specified by usual notions of (strict) performance, reliability, and reliability-related concepts such as availability and maintainability.

But, there exists many open questions, such as high-level system performance models for multicomputers, and much more research needs to be done. Similarly, modeling of MINs has not been well studied although papers dealing with some generic techniques have been included in this chapter. For example, Kubat (1986) provides a brief survey of the network reliability and performance models in the current literature and a theoretical basis for a joint assessment of reliability and performance for highly reliable networks. The underlying model assumes that, at most, a small number of components can be down at a time, and that the average repair/replacement time of a failed component is small when compared with the average up times of network components. The behavior of the network or distributed real-time systems is, then, described by a stochastic process. Generally, knowledge of the probability distribution function of performance variable(s) is enough to determine performability values for accomplishment sets of interest to the user.

Reprinted from *IEEE Transactions on Computers*, Volume C-27, Number 6, June 1978, pages 540–547.

Performance-Related Reliability Measures for Computing Systems

M. DANIELLE BEAUDRY, STUDENT MEMBER, IEEE

***Abstract*—We have developed measures which reflect the interaction between the reliability and the performance characteristics of computing systems. These measures can be used to evaluate traditional computer architectures, such as uniprocessors and standby redundant systems; gracefully degrading systems, such as multiprocessors, which can react to a detected failure by reconfiguring to a state with a decreased level of performance; and distributed systems. This analysis method, which provides quantitative information about the tradeoffs between reliability and performance, is demonstrated in several examples.**

***Index Terms*—Computer performance, computer reliability, graceful degradation.**

I. Introduction

IN GENERAL, systems with multiple copies of a resource (for example, computer systems with several processors) are configured in four ways to achieve reliability.

1) *Massive redundant systems*, which use techniques such as triple-modular redundancy (TMR) [1], N-modular redundancy (NMR) [2], and self-purging redundancy [3], execute the same task on each equivalent module and vote on the output.

2) *Standby redundant systems* execute tasks on their active modules. Upon detection of the failure of an active module, these systems attempt to replace the faulty unit with a spare unit [4].

3) *Hybrid redundant systems* are composed of a massive redundant core with spares to replace failed modules [3].

4) *Gracefully degrading systems* may use all modules to execute tasks, i.e., all failure-free modules are active. When a module failure is detected, these systems attempt to reconfigure to a system with one fewer module [5].

The reliability measures applied to these various architectures are typically:

1) the cumulative distribution of the random variable t_F the time at which the system first fails, given that the system was in some initial state at time 0, or equivalently, the *system reliability*, $R(t) = \Pr\,[t_F > t \mid \text{initial system state}]$,

2) the *mean time to failure* (MTTF) which is the expected value of t_F,

$$\text{MTTF} = E[t_F] = \int_0^\infty R(t)\,dt,$$

3) the *mission time*, i.e., the time at which the system reliability reaches a specific value, and

4) the *system availability*, the steady-state probability that the system is operational.

These measures are appropriate for ultra-reliable systems, i.e., massive redundant, standby redundant, and hybrid redundant systems; however, they may not be sufficient for evaluating gracefully degrading systems. In contrast to the ultra-reliable architectures, gracefully degrading systems may use redundancy in computing resources to gain performance as well as to increase system reliability. Gracefully degrading systems can be represented as falling somewhere between the extremes of ultra-reliable systems and high performance parallel systems, e.g., Illiac IV [6], in terms of the tradeoffs between performance and reliability gained by the use of redundancy. Gracefully degrading systems react to a detected failure by reconfiguring to a state which may have a decreased level of performance. For example, if a single processor of a multiprocessor system fails, the system may continue to operate without the faulty processor, but has a lower level of performance until the processor can be repaired and then reconfigured into the system again. This difference in performance levels leads us to define the *computation capacity* of a system state as the amount of useful computation per unit time available on the system in that state. We can thus characterize gracefully degrading computing systems by the flexible use of redundancy of computing resources, recovery and/or reconfiguration capability, and states with different computation capacities. Because gracefully degrading systems have active states that differ in capacity to execute tasks, and because system failure occurs when the system capacity falls below some value, we define the following performance-related reliability measures.

1) The *computation reliability* $R^*(t, T)$ is the probability that, at time t, the system is in an unfailed state and correctly, i.e., without entering a failed state, executes a task of length T started at time t, given an initial system state.

2) The *mean computation before failure* (MCBF) is the expected amount of computation available on the system before its first failure, given an initial system state.

Manuscript received August 12, 1977; revised February 23, 1978. This work was initiated while the author held an IBM Graduate Fellowship and was supported in part by the National Science Foundation under Grant MCS-05327 and by the Air Force Office of Scientific Research under Grant 77-3325.

The author is with the Center for Reliable Computing, Digital Systems Laboratory, Departments of Electrical Engineering and Computer Science, Stanford University, Stanford, CA 94305.

3) The *computation thresholds* are t_T, the time at which the computation reliability reaches a specific value for a task of length T; and T_t, the maximum task length which the system can correctly execute starting at time t, with a specific computation reliability.

4) The *computation availability* a_c is the expected value of the computation capacity of the system at time t or in steady-state operation.

5) The *capacity threshold* t_c is the time at which the computation availability reaches a specific value.

These performance-related reliability measures take into account the different levels of system performance and the system's varying capability to execute a computation task or mission. In the following sections, we will use some Markov models to derive expressions for these measures and demonstrate their applicability to various systems.

II. Performance-Related Reliability Measures

In order to demonstrate these reliability measures, we use a Markov model to describe a gracefully degrading system. This model is used as a framework in which to develop the performance-related reliability measures defined in the previous section. We assume that the system is composed of modules which have exponentially distributed failures, i.e., the probability that the module has not failed before time t, given that it was initially functional, is $e^{-\lambda t}$, where λ is the failure rate of the module and $1/\lambda$ is its MTTF. If these modules have recovery or repair capability, then we also assume that the recovery or repair times are exponentially distributed, although for practical situations this is not a necessary assumption [7]. In a Markov model, each state represents a different system configuration and the transition probabilities between system states are characterized by the number of modules, the state (correct, undetected failure, detected failure, etc.) of the modules, their failure rates, their recovery or repair rates, and the *coverage* of the state, the probability that the system reconfigures correctly, given that it has detected a failure in a module [8], [9]. We note, however, that the derivation of these measures is not restricted to Markov models, which are primarily used in this section to determine the probabilities of being in particular system states.

If F is the set of system failure states and the initial state of the system is I, then the system is in an unfailed state with probability:

$$\sum_{i \notin F} P_i(t)$$

where $P_i(t)$ is the probability that the system is in state i at time t, given that it was initially in state I. Standard techniques [10] can be used to determine the values of $P_i(t)$ for Markov processes. Briefly, the transition probabilities are used to set up a system of differential equations of the form:

$$\frac{dP_i(t)}{dt} = \sum_{j \neq i} (p_{ji} P_j(t) - p_{ij} P_i(t)),$$

where $p_{ij}\, dt$ is the probability that a transition occurs from state i to state j in an infinitesimal time interval dt. These equations can be solved for $P_i(t)$ given I, the initial state of the system, by setting the initial conditions $P_I(0) = 1$ and $P_i(0) = 0$ for $i \neq I$. We can also use this method to determine the reliability and the MTTF for the system being modeled by making the states in F absorbing. This is accomplished by effectively removing the transitions out of the failed states which corresponds to setting:

$$p_{fj} = 0 \qquad \text{for all } f \in F \text{ and for all states } j.$$

If the computing system can recover from a failed state by repair of its component modules, then we can examine the steady-state behavior of the system to obtain $P_i(\infty)$, the steady-state or limiting probabilities. Assuming that these limits exist (a good assumption for any system of practical interest) this is accomplished by setting:

$$\frac{dP_i(t)}{dt} = 0$$

in the system of differential equations shown above and then solving the resulting linear system of equations for $P_i(\infty)$. The availability of the repairable system is then:

$$a = \sum_{i \notin F} P_i(\infty).$$

This summarizes the use of Markov models to determine the traditional reliability measures.

When we consider the amount of computation available on the system, we must recognize that the various states have different computation capacities. We consider the computation available in state i:

$$T = \alpha_i t \text{ or equivalently } t = T/\alpha_i$$

where $\alpha_i > 0$ is the *computation capacity*, the amount of useful computation per unit time, of the system in state i. T is expressed in computation units, e.g., instructions or CPU hours, and the computation capacity is expressed in the same computation units over time units, e.g., instructions/second or CPU hours/hour. Many factors can affect the computation capacity of a state. For example, the system overhead due to dynamic testing, error detection, recovery, or operating system functions decreases the computation capacity of the system. Performance measures such as throughput and execution speed can be used to determine a reasonable value for the computation capacity. Also, α_i can reflect the expected demand for the computation available in the different system states.

If we now examine the Markov chain in terms of the computation variable T instead of the time variable t, we can make the substitution:

$$dt = dT/\alpha_i$$

in the state transition probabilities as follows:

$$p_{ij}\, dt = p_{ij}\, dT/\alpha_i.$$

This technique, which transforms a time domain representation of the system into a computation domain representation, is used to determine the computation reliability and the MCBF of the system. The probability that the system is in state j after an amount T of computation is then $P_j^*(T)$; this is the computation domain analog to $P_j(t)$ in the time domain representation. Since we want to consider how much computation the system provides before it first reaches a failed state, we make all system failure states absorbing. The transformed Markov chain can now be used to determine the *capacity function* of the various system states:

$$C_i(T) = 0 \qquad \text{for all } i \in F$$

$$C_i(T) = \Pr[\text{system executes a task of length } T \mid \text{state was } i \text{ at the beginning of the computation}] \qquad \text{for all } i \notin F.$$

The capacity function is one minus the cumulative distribution function of T_F, the amount of computation available on the system before it first reaches a failed state. This function is the analog of the time domain reliability. The capacity function is computed by setting up the differential equations for $P_j^*(T)$ using the computation domain representation of the Markov chain with absorbing failure states, solving this system of equations assuming that the initial state of the system was i, i.e., $P_i^*(0) = 1$ and $P_j^*(0) = 0$ for $j \neq i$, and then setting:

$$C_i(T) = \sum_{j \notin F} P_j^*(T).$$

The capacity distribution is used to determine the system's mean computation before failure as follows:

$$\text{MCBF} = \int_0^\infty C_I(T)\, dT$$

where I is the state of the system at $t = 0$. The MCBF is thus a measure of the system's ability to execute computing tasks. We also use the capacity function to define the *computation reliability* of the system:

$$\begin{aligned} R^*(t, T) &= \sum_{i \notin F} C_i(T) P_i(t) \\ &= \Pr[\text{system executes a task of length } T \mid \text{state} = i \text{ at } t] \cdot \Pr[\text{State} = i \text{ at } t \mid \text{State} = I \text{ at } t = 0]. \end{aligned}$$

If we set $T = 0$ in the computation reliability, we see that

$$R^*(t, 0) = R(t)$$

and if we set $t = 0$:

$$R^*(0, T) = C_I(T).$$

Analogous to the mission time concept, the computation thresholds, t_T and T_t, are the solutions to

$$R^*(t_T, T) = \xi \quad \text{and} \quad R^*(t, T_t) = \xi.$$

The computation availability, $a_c(t)$, of a system is thus:

$$a_c(t) = \sum_{i \notin F} \alpha_i P_i(t).$$

We note that

$$\int_{t_1}^{t_2} a_c(t)\, dt$$

is the expected amount of computation available on the system during the time interval (t_1, t_2). If the Markov chain converges, we can define the steady-state computation availability a_c:

$$a_c = \sum_{i \notin F} \alpha_i P_i(\infty)$$

The capacity threshold of the system t_c is the time at which the expected system capacity becomes equal to some value and thus t_c is just the solution to

$$a_c(t_c) = n.$$

For systems which have a constant computation capacity α in all their unfailed states, the computation reliability is just:

$$R^*(t, T) = R(t + T/\alpha)$$

since the system must remain active for the additional time T/α to do T amount of computation. Because the computation capacity is constant, the total amount of computation available on such a system is just αt_F, where t_F is the time at which the system first fails. Hence, the system's mean computation before failure is

$$\text{MCBF} = \alpha \text{ MTTF}$$

and similarly, the computation availability for a system with constant computation capacity is

$$a_c(t) = \alpha \sum_{i \notin F} P_i(t)$$

where $P_i(t)$ are the state probabilities of the time domain Markov model. If the system is in steady state, then

$$a_c = \alpha \sum_{i \notin F} P_i(\infty) = \alpha a$$

where a is the traditional availability.

We have thus used a Markov model to determine the performance-related reliability measures defined in Section I for systems which have states of differing computation capacity, as well as for systems with constant computation capacity. We want to note here that these models can become complex, especially for systems with various types of component modules, e.g., processing elements, storage modules, interconnection networks. Also, determining a suitable value of the computation capacity for a particular system state may be difficult; however, a significant body of research has addressed the problem of evaluating system performance [11], [12].

III. Examples of Performance-Related Measures in System Evaluation

In this section, we use Markov models to represent several systems, some of which distribute their tasks to a number of processing subsystems. We also develop the performance-related reliability measures described in the previous sections and use them to evaluate various systems.

Example 1: Systems Without Repair

The first example studies a gracefully degrading system composed initially of N identical processing modules. The Markov chain for this system is shown in Fig. 1, where state i

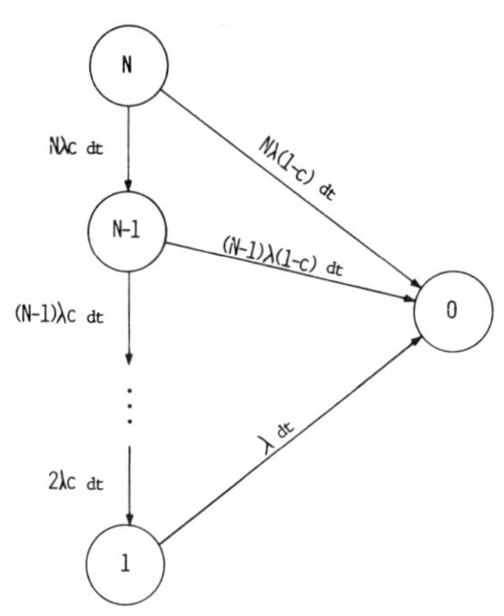

Fig. 1. Time domain representation of the Markov model for a gracefully degrading system without recovery or repair. $c_i = c$, for $i = 2, 3, \cdots, N$.

has i operational units. The failure rate in state i is just $i\lambda$ because the failure of any one of the i active modules will cause a transition. The transition results in state $i - 1$ if the module failure is covered; otherwise, the transition leads to state 0, the system failure state. The coverage c_i is the probability that the system reconfigures given that a failure has been detected while in state i, for $i = 2, 3, \cdots, N$. The differential equations, where $P_i(t)$ is written as P_i, describing this system are

$$\frac{dP_N}{dt} = -N\lambda P_N$$

$$\frac{dP_i}{dt} = -i\lambda P_i + (i+1)\lambda c P_{i+1} \qquad \text{for } i = 1, 2, \cdots, N-1$$

$$\frac{dP_0}{dt} = \lambda P_1 + \lambda(1-c) \sum_{j=2}^{N} jP_j$$

As shown in [13], if each state has the same coverage, $c_i = c$ for $i = 2, 3, \cdots, N$, and the system is initially in state N then

$$P_i(t) = c^{N-i} e^{-i\lambda t} \binom{N}{i} (1 - e^{-\lambda t})^{N-i} \qquad \text{for } i = 1, 2, \cdots, N.$$

The system reliability is

$$R(t) = \sum_{i=1}^{N} P_i(t)$$

and its mean time before failure is

$$\frac{1}{\lambda} \sum_{i=1}^{N} \frac{c^{N-i}}{i}.$$

If each processing module of the system has the same ability to execute computing tasks, and if we can arbitrarily distribute the tasks over all the available processors, then state i has computation capacity $i\alpha$ where α is the computation capacity of the system with a single processor. The assumption that the computation capacity of a state is a linear function of the number of active processors is quite unrealistic; however, it simplifies the example used here to demonstrate the analysis technique necessary to determine both traditional and performance-related reliability measures. The computation domain representation of the Markov model for this system is shown in Fig. 2, where the transitions from state i, $i = 1, 2, \cdots, N$, are transformed by the substitution:

$$dt = dT/\alpha_i = dT/i\alpha.$$

For example,

$$p_{N,N-1} = N\lambda c\, dt = N\lambda c\, dT/\alpha_N = N\lambda c\, dT/N\alpha = \lambda c/\alpha\, dT.$$

The differential equations for this system are

$$\frac{dP_N^*}{dT} = -\frac{\lambda}{\alpha} P_N^*$$

$$\frac{dP_i^*}{dT} = -\frac{\lambda}{\alpha} P_i^* + \frac{\lambda c}{\alpha} P_{i+1}^* \qquad \text{for } i = 1, 2, \cdots, N-1$$

$$\frac{dP_0^*}{dT} = \frac{\lambda}{\alpha} P_1^* + \frac{\lambda(1-c)}{\alpha} \sum_{j=2}^{N} P_j^*.$$

As shown in [13], analysis of this transformed Markov chain yields the capacity function of each state:

$$C_i(T) = e^{-\lambda T/\alpha} \sum_{j=0}^{i-1} \frac{1}{j!} \left(\frac{\lambda c T}{\alpha}\right)^j \qquad \text{for } i = 1, 2, \cdots, N.$$

The mean computation before failure is

$$\text{MCBF} = \int_0^\infty C_I(T)\, dT = \frac{\alpha}{\lambda} \sum_{j=0}^{N-1} c^j.$$

The computation availability of the system is

$$a_c(t) = \sum_{i=1}^{N} \alpha_i P_i(t) = \sum_{i=1}^{N} i\alpha P_i(t).$$

We can now use these results to analyze a two-processor standby redundant system and a two-processor gracefully degrading system. Table I summarizes the appropriate traditional and performance-related reliability measures for these two systems. For the standby redundant system, we assume that the failure rate of the spare module is the same as that of the active module. The computation capacity of the standby redundant system is constant and for the purposes of this paper is considered to be unity, i.e., $\alpha = 1$, when this system is operational. The computation capacity of the gracefully degrading system is 2α when both processors are functioning correctly and α when only one is operating. We assume that $\alpha < 1$, i.e., that there is some loss of performance due to the parallel operation of both processors. Both systems have the same coverage c.

If we consider the traditional reliability measures, then the reliability and MTTF of the standby redundant system are higher than those of the gracefully degrading system. If we examine the computation reliability, we see that the gracefully degrading system has a higher probability of executing a long computing mission. Fig. 3 shows a graph of the various values of time λt and computation λT such that the computation reliability $R^*(t, T) = 0.99$. We see that during initial system operation, the gracefully degrading system has more expected computation available than the standby system; but that later in the mission, the standby

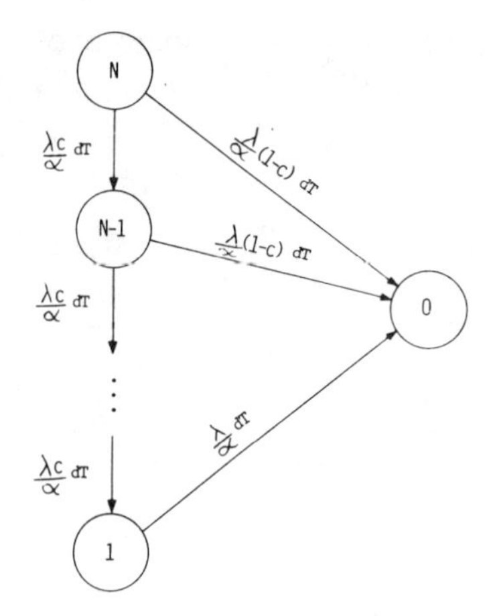

Fig. 2. Computation domain representation of the Markov model of Fig. 1.

TABLE I
RELIABILITY MEASURES FOR A TWO-PROCESSOR STANDBY REDUNDANT SYSTEM AND FOR A TWO-PROCESSOR GRACEFULLY DEGRADING SYSTEM

Measure	Standby Redundant	Gracefully Degrading
Reliability	$e^{-\lambda t}(1+c) - ce^{-2\lambda t}$	$2ce^{-\lambda t} + (1-2c)e^{-2\lambda t}$
MTTF	$\dfrac{c+2}{2\lambda}$	$\dfrac{2c+1}{2\lambda}$
Computation Reliability	$e^{-\lambda(t+T)}(1+c) - ce^{-2\lambda(t+T)}$	$e^{-\lambda(t+T/\alpha)}\left[2c + e^{-\lambda t}\left(1 - 2c + \dfrac{c\lambda T}{\alpha}\right)\right]$
MCBF	$\dfrac{c+2}{2\lambda}$	$\dfrac{c+1}{\lambda}\alpha$
Computation Availability	$e^{-\lambda t}(1+c) - ce^{-2\lambda t}$	$2\alpha e^{-\lambda t}[c + (1-c)e^{-\lambda t}]$

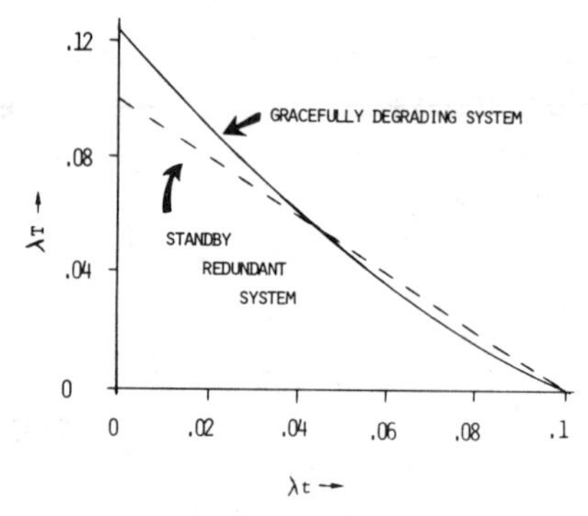

Fig. 3. Graph of values of time λt and computation λT such that the computation reliability of the gracefully degrading system is constant. $\alpha = 0.9$ and $c = 0.99$.

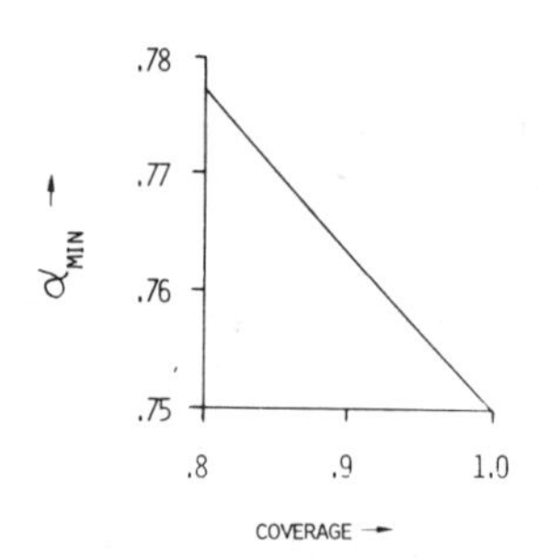

Fig. 4. Minimum capacity level of processors in the gracefully degrading system needed to provide the same MCBF as the standby redundant system, as a function of coverage.

system has greater expected computation. If we examine the mean computation before failure for the two systems, we see that the gracefully degrading system has greater computation power, i.e., its MCBF is greater than that of the standby redundant system, if

$$\alpha\frac{1+c}{\lambda} > \frac{2+c}{2\lambda}.$$

This implies that the computation capacity of the multiprocessor modules must be greater than $\alpha_{\min}$, which is plotted in Fig. 4 as a function of coverage.

In order to execute a task of length T with reliability ξ, we must start it at some time before λt_T, one of the computation thresholds. As shown in Table II, these computation thresholds are lower for the gracefully degrading system; this means that a task must be introduced to this system earlier if $R^*(t, T) \geq \xi$. The computation availability is used in Fig. 5 to plot $\lambda t_{\max}$, the time at which the expected capacity of the gracefully degrading system falls below that of the standby redundant system as a function of α, the capacity of a single processor of the gracefully degrading system. As expected, the gracefully degrading system has considerable added capacity especially for high values of α. The capacity threshold is the time at which the system's total expected capacity reaches some value. The gracefully degrading

TABLE II
COMPUTATION THRESHOLDS λt_T FOR TASKS OF LENGTH λT AND COMPUTATION RELIABILITY WHERE $\alpha = 0.9$ AND COVERAGE = 0.99

ξ	λt_T for Standby Redundant	λt_T for Gracefully Degrading	Task Length
0.9999	0.006	0.004	
0.999	0.027	0.024	$\lambda T = 0.0001$
0.99	0.100	0.096	
0.999	0.018	0.015	
0.99	0.090	0.085	$\lambda T = 0.01$

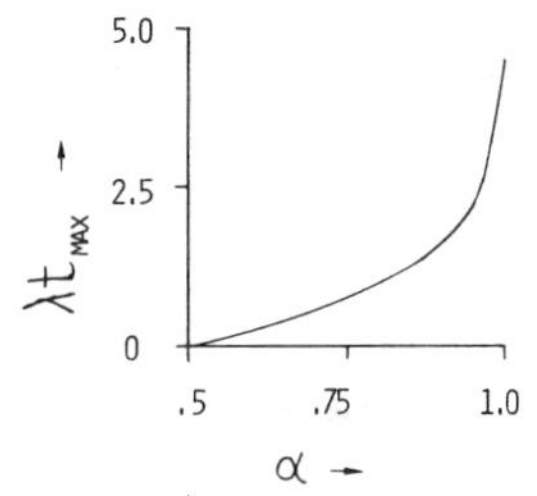

Fig. 5. Time λt_{max} at which the computation availability of the gracefully degrading system becomes lower than that of the standby redundant system as a function of the computation capacity α for coverage = 0.99.

system's expected capacity falls to a value of 1, the computation capacity of the standby redundant system, at time λt_c as shown in Fig. 6. Thus the performance-related reliability measures have enabled us to compare these two systems in terms of their ability to execute computing tasks.

Example 2: Systems With Repair

Our next comparison is between a uniprocessor and a multiprocessor which has two processors. All the processors are assumed to be repairable. Fig. 7 shows the Markov chain modeling this system in the time domain. We assume that the uniprocessor has computation capacity 1 and failure rate μ; and that the processors of the multiprocessor have capacity α and failure rate λ. The mean time to repair for the processors, $1/\rho$, is assumed to be approximately equal in both systems. The differential equations for the time domain representation are

$$\frac{dP_2}{dt} = -2\lambda P_2 + \rho P_1 = 0$$

$$\frac{dP_1}{dt} = 2\lambda c P_2 - (\rho + \lambda)P_1 + \rho P_0 = 0$$

$$\frac{dP_0}{dt} = 2\lambda(1 - c)P_2 + \lambda P_1 - \rho P_0 = 0.$$

This system of equations can be solved [10] for the steady-state probabilities $P_i(\infty)$, $i = 0, 1, 2$. The availability of the multiprocessor is thus

$$a = P_1(\infty) + P_2(\infty) = \frac{\rho(\rho + 2\lambda)}{\rho^2 + 2\lambda\rho(2 - c) + 2\lambda^2}$$

and its computation availability is

$$a_c = \alpha P_1(\infty) + 2\alpha P_2(\infty) = \frac{2\rho(\rho + \lambda)}{\rho^2 + 2\lambda\rho(2 - c) + 2\lambda^2}\alpha$$

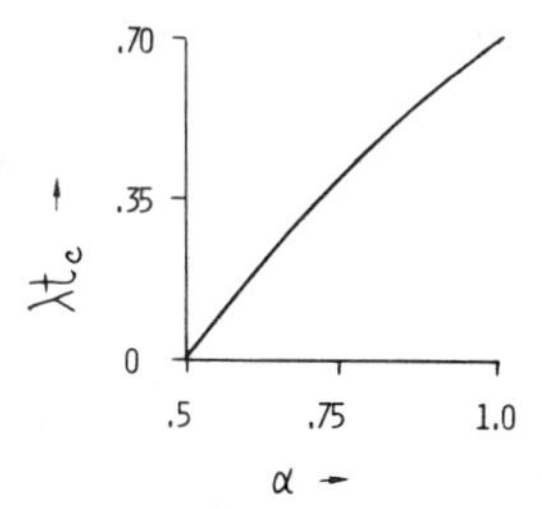

Fig. 6. Capacity threshold λt_c for the gracefully degrading system as a function of the computation capacity α where $a_c(t_c) = 1$ for coverage = 0.99.

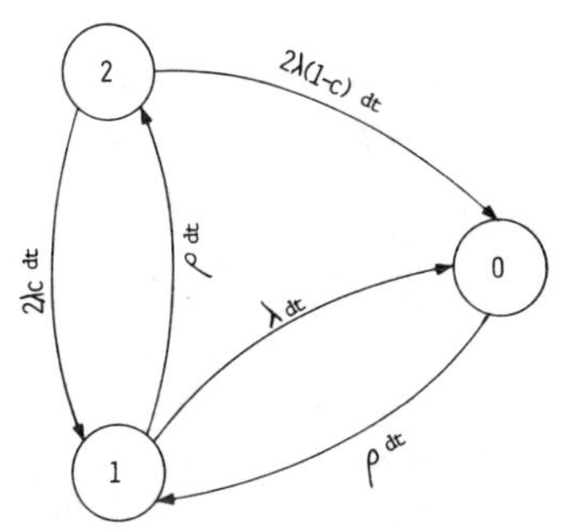

Fig. 7. Time domain representation of the Markov model for a multiprocessor with repair.

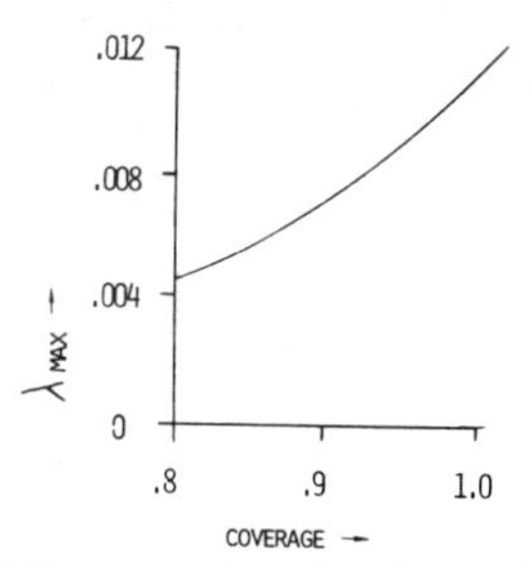

Fig. 8. Maximum failure rate λ_{max} for the processors of the multiprocessor to have the same availability as the uniprocessor, as a function of coverage for $\mu = 0.002$ hr^{-1} and $\rho = 0.11$ hr^{-1}.

where we assume that the computation capacity of state 2 is 2α and that of state 1 is α. The availability of the uniprocessor is

$$a = \rho/(\rho + \mu)$$

as is its computation availability, since we have assumed its computation capacity to be unity. If we want both systems to be equally available, then the failure rate of the multiprocessor modules must be bounded. Fig. 8 plots the bound, λ_{max}, as a function of the coverage of the multiprocessor system. The values of μ and ρ are typical values for a communications processor such as the Interface Message Processor (IMP) of the ARPA network [13]. This system is currently being implemented as a multiprocessor, the PLURIBUS [14]. If we examine the computation availabilities, we see that for a system with $\rho \gg \lambda$, the multiprocessor is as computationally powerful on the average as the uniprocessor for $\alpha > 0.5$. Any added computation capacity can thus be used to execute additional computing tasks.

Example 3: Computation Center Multiprocessor

In this example, we use some performance-related reliability measures to evaluate the effect of a possible change in the configuration of the Stanford Linear Accelerator Center (SLAC) multiprocessor. The SLAC multiprocessor uses three separate central processing units (CPU's), two IBM System/370 Model 168s, and an IBM System/360 Model 91, hence its name, the Triplex [16]. These three processors function as one logical computing resource under thc control of asymmetric multiprocessing system (ASP). ASP resides on one of the Model 168s and its primary functions are to support the interactive computation load and to provide the multi-CPU interface. It is also responsible for spooling and scheduling functions, and for all communications to and from the operator(s). Each processor has a resident operating system, OS/VS2 on the Model 168s and OS/MVT on the Model 91, which provides batch service. Fig. 9 shows how the four major control programs, ASP, two OS/VS2's, and OS/MVT are configured on the three CPU's.

The Triplex system must provide enough performance to meet a high level of computation demand, and it must be highly available as well. The staff of the computation center carefully monitors the system and as a result, data are gathered on both the performance and the reliability characteristics of the Triplex. Because most jobs or tasks are input to the system by means of the interactive interface between the user at a terminal and the system, a system failure in the Triplex, called a *service interruption*, is defined to be an interruption in interactive service. Either of the Model 168 CPU's can support the ASP program, so we consider only three system states.

State 2: Both Model 168s are functioning correctly and both batch and interactive service are available.

State 1: A CPU-only failure has occurred causing the loss of a single CPU or one of the Model 168s is scheduled for preventive maintenance.

State 0: A failure has occurred which has caused a service interruption.

From a reliability point of view, the Model 91 CPU is primarily a part of the system's batch computing resource and is not included in this example of the analysis technique.

Statistical data [17], which include software, hardware, and operator induced system service interruptions, were used to determine the steady-state probabilities of the system states:

$$P_2 = 0.80$$

$$P_1 = 0.18$$

$$P_0 = 0.02.$$

The availability of this system is thus approximately 0.98, i.e., 98 percent of the time the system is functioning. Since ASP is the most sensitive control program in the Triplex, it has been suggested that its functions be distributed between both Model 168 CPU's, thus making the system more resistant to failures in a single CPU, as shown in the configuration of Fig. 10.

Since performance is also an important issue we must examine the computation availability of the two possible system configurations. We assume that the computation capacities of the system states have the values shown in Table III, where k is the computation capacity due to the ASP-resident Model 168 CPU, while the capacity due to the other Model 168 CPU, which executes batch functions is unity. Since the ASP functions use a significant amount of the computation resource of the ASP-resident CPU, $k < 1$. α is the hypothesized computation capacity due to each Model 168 CPU of the suggested distributed ASP system. Again, since both processors handle interactive and other system computation demands, α, the computation capacity of each CPU, is less than one. The *overhead*, i.e., the computing power which is needed to support the interactive and multiprocessor functions, is thus $(1 - k)$ for the current system and $2(1 - \alpha)$ for the suggested system.

Examining the computation availabilities of the two system configurations yields information about the tradeoffs between reliability and performance. For example, even if the suggested system's availability is significantly improved, e.g., P_2 remains 0.8 but P_1 is 0.2 which implies that $P_2 + P_1 = 1$, its performance must nevertheless be at a level suggested by the graph of Fig. 11. This graph plots the

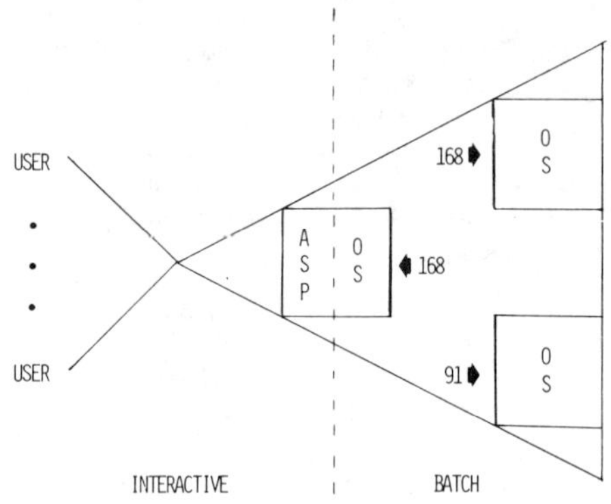

Fig. 9. Current configuration of the SLAC Triplex.

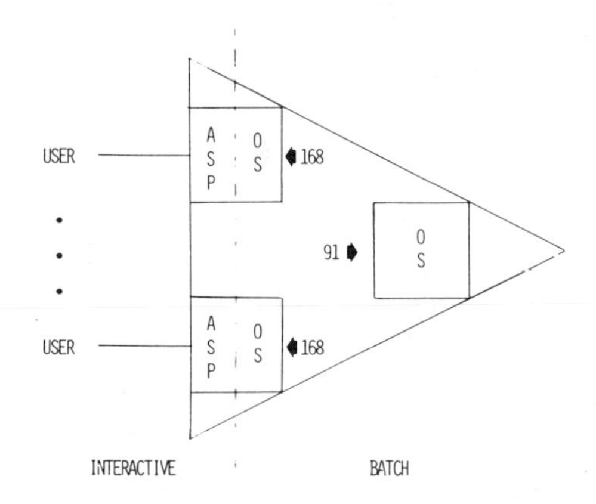

Fig. 10. Suggested configuration of the SLAC Triplex.

TABLE III
COMPUTATION CAPACITIES FOR SLAC TRIPLEX SYSTEM STATES

State	Computation Capacity Current System	Suggested System
2	$1 + k$	2α
1	k	α
0	0	0

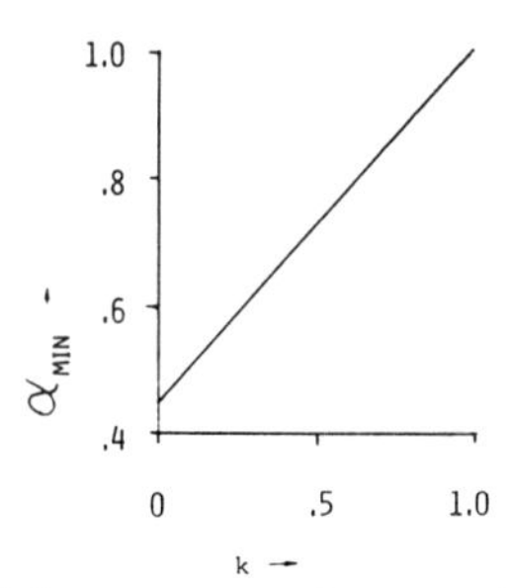

Fig. 11. Graph of minimum computation capacity α_{min} due to a single CPU in the suggested configuration as a function of the computation capacity k of the ASP-resident CPU of the current system.

minimum value of α needed to provide the same computation availability as the current system, thus imposing performance bounds on the suggested system design. Alternatively, the suggested system must confine the overhead due to distributing the interactive and multiprocessor functions.

IV. Conclusions

The Markov models presented allowed us to develop measures to analyze the effects of distributing the computation load in a multiprocessor or gracefully degrading system. Clearly the models are not detailed enough and refining them will lead to considerable added complexity. Simulation will probably be necessary to examine systems with different types of redundant modules. Work also needs to be done to determine good estimates for the computation capacity of various configurations of these redundant components.

The reliability measures presented in this paper are especially important in the evaluation and comparison of computing systems which can distribute their tasks. These measures give us insight into the expected response of the system to a computational demand, taking into account various system characteristics, e.g., hardware and software redundancy and efficiency, and time of task execution. The need for these measures is especially clear when processing systems are used in applications where both reliability and performance are important.

Acknowledgment

The author would like to thank Prof. E. J. McCluskey and the Digital Systems Laboratory Reliability and Testing Seminar (RATS), as well as the referees, for their helpful comments and suggestions; and also acknowledge the computing support of the SLAC computation center.

References

[1] J. von Neumann, "Probabilistic logics and the synthesis of reliable organisms from unreliable components," in *Automata Studies*, C. E. Shannon and J. McCarthy, Eds. Princeton, NJ: Princeton University Press, 1956, pp. 43–98.

[2] F. P. Mathur and A. Avizienis, "Reliability analysis and architecture of a highly redundant digital system: Generalized triple modular redundancy with self-repair," in *Proc. 1970 SJCC, AFIPS Conf. Proc.*, vol. 36, pp. 375–383, May 1970.

[3] J. Losq, "A highly efficient redundancy scheme: Self-purging redundancy," *IEEE Trans. Comput.*, vol. C-25, pp. 569–578, June 1976.

[4] W. G. Bouricius, W. C. Carter, D. C. Jessep, and P. R. Schneider, "Reliability modeling techniques for self-repairing computer systems," in *Proc. 24th ACM Nat. Conf.*, pp. 295–305, Aug. 1969.

[5] B. R. Borgerson and R. F. Freitas, "A reliability model for gracefully degrading and standby-sparing systems," *IEEE Trans. Comput.*, vol. C-24, pp. 517–525, May 1975.

[6] G. H. Barnes, R. M. Brown, M. Kato, D. J. Kuck, D. L. Slotnick, and R. A. Stokes, "The ILLIAC IV computer," *IEEE Trans. Comput.*, vol. C-17, pp. 746–757, Aug. 1968.

[7] J. Laprie, "Reliability and availability of repairable structures," in *Proc. 1975 Int. Symp. Fault-Tolerant Comput.*, pp. 87–92, June 1975.

[8] W. G. Bouricius, W. C. Carter, D. C. Jessep, P. R. Schneider, and A. B. Wadia, "Reliability modeling for fault-tolerant computers," *IEEE Trans. Comput.*, vol. C-20, pp. 1306–1311, Nov. 1971.

[9] T. F. Arnold, "The concept of coverage and its effect on the reliability model of a repairable system," *IEEE Trans. Comput.*, vol. C-22, pp. 251–254, Mar. 1973.

[10] W. Feller, *An Introduction to Probability Theory and Its Applications, Volume I*. New York: Wiley, 1968, Ch. XVII.

[11] H. C. Lucas, "Performance evaluation and monitoring," *ACM Computing Surveys*, vol. 3, pp. 79–91, Sept. 1971.

[12] L. Svobodova, *Computer Performance Measurement and Evaluation Methods: Analysis and Applications*. New York: American Elsevier, 1976.

[13] M. D. Beaudry, "Performance considerations for the reliability analysis of computing systems," Ph.D. dissertation, Digital Syst. Lab., Stanford Univ., Stanford, CA, Apr. 1978.

[14] F. E. Heart, R. E. Kahn, S. M. Ornstein, W. R. Crowther, and D. C. Wadia, "The interface message processor for the ARPA network," in *Proc. 1972 SJCC, AFIPS Conf. Proc.*, vol. 40, pp. 551–567, 1972.

[15] F. E. Heart, S. M. Ornstein, W. R. Crowther, and W. B. Barker, "A new minicomputer/multiprocessor for the ARPA Network," in *1973 NCC, AFIPS Conf. Proc.*, pp. 529–537, 1973.

[16] D. W. Lynch, "The Triplex guide and visitor's handbook," *SCIP User Note*, 1975.

[17] M. D. Beaudry, "A statistical analysis of service interruptions in the SLAC Triplex multiprocessor," Digital Syst. Lab., Stanford Univ., Stanford, CA, Tech. Rep. 141, Apr. 1978.

M. Danielle Beaudry (S'74) was born in Washington, DC, on October 18, 1947. She received the B.S. degree in physics from Massachusetts Institute of Technology, Cambridge, in 1969, and the M.S. degrees in electrical engineering and statistics, and the Ph.D. degree in electrical engineering, all from Stanford University, Stanford, CA, in 1974, 1977, and 1978, respectively.

From 1969 through 1974, she held positions with Fairchild Semiconductor and Hewlett-Packard as a programmer. While completing her doctoral studies, she held an IBM Graduate Fellowship and a Research Assistantship in the Digital Systems Laboratory, Stanford University, where she is currently a Research Associate working on fault-tolerant computing. Her research interests include reliability and performance modeling of computing systems.

Ms. Beaudry is a member of the IEEE Computer Society, the Association for Computing Machinery, and Women in Science and Engineering.

Reprinted from *IEEE Transactions on Computers*, Volume C-29, Number 8, August 1980, pages 720–731.

On Evaluating the Performability of Degradable Computing Systems

JOHN F. MEYER, SENIOR MEMBER, IEEE

***Abstract*—If the performance of a computing system is "degradable," performance and reliability issues must be dealt with simultaneously in the process of evaluating system effectiveness. For this purpose, a unified measure, called "performability," is introduced and the foundations of performability modeling and evaluation are established. A critical step in the modeling process is the introduction of a "capability function" which relates low-level system behavior to user-oriented performance levels. A hierarchical modeling scheme is used to formulate the capability function and capability is used, in turn, to evaluate performability. These techniques are then illustrated for a specific application: the performability evaluation of an aircraft computer in the environment of an air transport mission.**

***Index Terms*—Degradable computing systems, fault-tolerant computing, hierarchical modeling, performability evaluation, performance evaluation, reliability evaluation.**

I. INTRODUCTION

DURING the past decade, performance evaluation and reliability evaluation have emerged as important technical disciplines within computer science and engineering. Evaluations of computer performance and computer reliability are each concerned, in part, with the important question of computer system "effectiveness," that is, the extent to which the user can expect to benefit from the tasks accomplished by a computer in its use environment. With regard to effectiveness issues, modeling of computer performance (see [1]–[3], for example) has stressed the need to represent the probabilistic nature of user demands (workload) and internal state behavior, under the assumption that the computer's structure is fixed (that is, there are no permanent changes in structure due to faults). On the other hand, modeling of computer reliability (beginning with the pioneering work of Bouricius *et al.* [4]) has stressed representation of the probabilistic nature of structural changes caused by transient and permanent faults of the computer.

In the face of these traditional modeling distinctions, we consider an important class of computing systems wherein system performance is "degradable," that is, depending on the history of the computer's structure, internal state, and environment during some specified "utilization period" T, the system can exhibit one of several worthwhile levels of performance (as viewed by the user throughout T). In this case we find that performance evaluations (of the fault-free system) will generally not suffice since structural changes, due to faults, may be the cause of degraded performance. By the same token, traditional views of reliability (probability of success, mean time to failure, etc.) no longer suffice since "success" can take on various meanings and, in particular, it need not be identified with "absence of system failure."

Modeling needs for (gracefully) degradable systems were first investigated by Borgerson and Frietas [5] in connection with their analysis of the PRIME system [6]. Although they recognized the need to formulate the probability of each possible level of performance, that is, the probability of k "crashes" during T for $k = 0,1,2,\cdots$, their evaluation effort dealt mainly with the question of reliability (the probability of no crashes during T). Other studies employing Markov models have likewise emphasized the evaluation of reliability oriented measures (see [7]–[9], for example).

Some recent investigations, on the other hand, have dealt with measures aimed at quantifying performance as well as reliability. In particular, Beaudry [10] has introduced a number of computation related measures for degradable computing systems and has shown how to formulate these measures in terms of a transformed Markov model. In examining reconfiguration strategies for degradable systems, Troy [11] has distinguished levels of performance according to "workpower" and has formulated system effectiveness (referred to as "operational efficiency") as expected workpower. In another recent study, Losq [12] has investigated degradable systems in terms of degradable resources, where each resource is modeled by an irreducible, recurrent, finite-state Markov process.

In the discussion that follows, we describe a general modeling framework that permits the definition, formulation, and evaluation of a unified performance-reliability measure referred to as "performability." It is shown that performability relates directly to system effectiveness and is a proper generalization of both performance and reliability. A critical step in performability modeling is the introduction of the concept of a "capability function" which relates low-level system behavior to user-oriented levels of performance. A hierarchical modeling scheme is used to formulate the capability function, and capability is used, in turn, to evaluate performability.

II. SYSTEM MODELS

A computing system, as it operates in its use environment, may be viewed at several levels. At a low level, there is a de-

Manuscript received June 19, 1978; revised February 22, 1980. This research was supported by the NASA Langley Research Center under Grant NSG 1306.

The author is with the Department of Electrical and Computer Engineering and the Department of Computer and Communication Sciences, University of Michigan, Ann Arbor, MI 48109.

tailed view of how various components of the computer's hardware and software behave throughout the utilization period. At this level, there is also a detailed view of the behavior of the computer's "environment," where by this term we mean both man-made components (user input, peripheral subsystems, etc.) and natural components (radiation, weather, etc.) which can influence the computer's effectiveness. The computer, together with its environment, will be referred to as the "total system." A second view of the total system is the user's view of how the system behaves during utilization, that is, what the system accomplishes for the user during the utilization period. A third, even less detailed view, is the economic benefit derived from using the system, that is, the computing system's worth (as measured, say in dollars) when operated in its use environment.

To formalize these views, we postulate the existence of a *probability space* $(\Omega,\mathcal{E},P)$ that underlies the total system, where Ω is the *sample space,* $\mathcal{E}$ is a set of *events* (measurable subsets of Ω), and $P: \mathcal{E} \rightarrow [0,1]$ is the *probability measure* (see [13], for example). This probability space represents all that needs to be known about the total system in order to describe the probabilistic nature of its behavior at the various levels described above. It thus provides a hypothetical basis for defining higher level (i.e., less detailed) models. In general, however, it will neither be possible nor desirable to completely specify Ω, $\mathcal{E}$, and P.

In the discussion that follows, let S denote the *total system,* where S is comprised of a *computing system* C and its *environment* E. At the most detailed level, the behavior of S is formally viewed as a stochastic process [14], [15]

$$X_S = \{X_t \mid t \in T\} \tag{1}$$

where T is a set of real numbers (observation times) called the *utilization period* and, for all $t \in T$, X_t is a random variable

$$X_t{:}\Omega \rightarrow Q$$

defined on the underlying description space and taking values in the *state space* Q of the total system. Depending on the application, the utilization period T may be discrete (countable) or continuous and, in cases where one is interested in the long-run behavior, it may be unbounded [e.g., $T = \boldsymbol{R}_+ = [0,\infty)$]. The state space Q embodies the *state sets* of both the computer and its environment, i.e.,

$$Q = Q_C \times Q_E$$

where Q_C and Q_E can, in turn, be decomposed to represent the local state sets of computer and environmental subsystems. For our purposes, it suffices to assume that Q is discrete and, hence, for all $t \in T$ and $q \in Q$, "$X_t = q$" has a probability (i.e., $\{\omega \mid X_t(\omega) = q\} \in \mathcal{E}$). The stochastic process X_S is referred to as the *base model* of S. An instance of the base model's behavior for a fixed $\omega \in \Omega$ is a *state trajectory* $u_\omega{:}T \rightarrow Q$ where

$$u_\omega(t) = X_t(\omega), \qquad \forall t \in T. \tag{2}$$

(The term "state trajectory" derives from modern usage in the theory of modeling [16]; synonyms in the more specific context of stochastic processes are "sample function," "sample path," and "realization" [14], [15].) Thus, corresponding to an underlying outcome $\omega \in \Omega$, u_ω describes how the state of the total system changes as a function of time throughout the utilization period T. Accordingly, the "description space" for the base model is the set

$$U = \{u_\omega \mid \omega \in \Omega\} \tag{3}$$

which is referred to as the (state) *trajectory space* of S.

As generally defined, the concept of a base model thus includes the type of queueing models used in computer performance evaluation [2], [3] and the kind of Markov or semi-Markov models employed in reliability evaluation [7]–[9]. The intent of the definition, however, is the inclusion of less restricted base models which can represent simultaneous variations in the computer's structure and internal state (via the state set Q_C) and environment (via the state set Q_E). In other words, the emphasis here is on the modeling of degradable computing systems where changes in structure, internal state, and environment can all have an influence on the system's ability to perform. Accordingly, these base models may be regarded as generalized performance models, where structure is allowed to vary, or equivalently as generalized reliability models where variations in internal state and/or the computational environment are taken into account.

In formal terms, the user-oriented view of system behavior is likewise defined in terms of the underlying probability space $(\Omega,\mathcal{E},P)$. Here we assume that the user is interested in distinguishing a number of different "levels of accomplishment" when judging how well the system has performed throughout the utilization period (one such level may be total system failure). The user's "description space" is thus identified with an *accomplishment set* A whose elements are referred to alternatively as *accomplishment levels* or (user-visible) *performance levels.* A may be finite, countably infinite, or uncountable (in the last case, A is assumed to be an interval of real numbers). Thus, for example, the accomplishment set associated with a nondegradable system is $A = \{a_0, a_1\}$ where a_0 = "system success" and a_1 = "system failure." In their modeling of the PRIME system, Borgerson and Freitas [5] viewed accomplishment as the set $A = \{a_0,a_1,a_2, \cdots\}$ where a_k = "k crashes during the utilization period T." If the user is primarily concerned with system "throughput," a continuous accomplishment set might be appropriate, i.e., $A = \boldsymbol{R}_+ = [0,\infty)$, where a number $a \in A$ is the "throughput averaged over the utilization period T."

In terms of the accomplishment set, *system performance* is formally viewed as a random variable

$$Y_S{:}\ \Omega \rightarrow A \tag{4}$$

where $Y_S(\omega)$ is the accomplishment level corresponding to outcome ω in the underlying description space. Similarly, assuming that the economic gain (or loss) derived from using the system is represented by a real number r (interpreted, say, as r dollars), *system worth* is a random variable defined as

$$W_S{:}\ \Omega \rightarrow \boldsymbol{R} \text{ (the real numbers)} \tag{5}$$

where $W_S(\omega)$ is the worth associated with outcome ω.

The terminology and notation defined previously is summarized in Table I. Note that, at this point in the development, there are no implied relationships between these three views; their only common bond so far is that they are representations

TABLE I
TERMINOLOGY AND NOTATION FOR SYSTEM MODELS

Model	Description Space
Base model X_S	Trajectory space U
System performance Y_S	Accomplishment set A
System worth W_S	The real numbers $\mathbb{R}$

of the same system or, more formally, that they are defined on the same underlying probability space. To be useful, however, the base model X_S should support the performance variable Y_S in an appropriate manner (indeed, the term "base" is suggestive of this need) and, in turn, Y_S should support the worth variable W_S. The precise nature of these connections, as they relate to the system's effectiveness, is developed in the section that follows.

III. EFFECTIVENESS, PERFORMABILITY, AND CAPABILITY

When applied to computing systems, "system effectiveness" (see [1], for example) is a measure of the extent to which the user can expect to benefit from the tasks accomplished by a computer in its use environment. More precisely, if benefit is identified with the worth W_S of the system then *system effectiveness* is expected worth, i.e., the expectation (expected value) of the random variable W_S; in short

$$\text{eff}(S) = E[W_S]. \tag{6}$$

(An implicit assumption here is that W_S is defined such that $E[W_S]$ exists; see [13], for example.) Because a direct evaluation of eff(S), using the definition of W_S, is generally not feasible (cf., our earlier remarks concerning the hypothetical nature of the underlying probability space), we wish to establish connections among the base model X_S, system performance Y_S, and system worth W_S which can be used in the process of evaluating eff(S).

To express system effectiveness in terms of system performance, the user's view of system worth must be compatible with system performance to the extent that W_S can be formulated as a function of Y_S. More precisely, we assume there exists a *worth function* $w: A \to \mathbf{R}$ such that, for all $\omega \in \Omega$,

$$W_S(\omega) = w(Y_S(\omega)). \tag{7}$$

If $a \in A$, $w(a)$ is interpreted as the "worth of performance level a." As for the performance variable Y_S, a natural measure that quantifies both system performance and reliability (i.e., the ability to perform) is the probability measure induced by Y_S. We refer to this unified performance-reliability measure as the "performability of S" which, in terms of our modeling framework, can be generally defined as follows.

Definition 1: If S is a total system with performance Y_S taking values in accomplishment set A, then the *performability of* S is the function p_S where, for each measurable set B of accomplishment levels ($B \subseteq A$),

$$p_S(B) = P(\{\omega \mid Y_S(\omega) \in B\}).$$

Since P is the probability measure of the underlying probability space, the interpretation of performability is straightforward, that is, for a designated set B of accomplishment levels, $p_S(B)$ is the probability that S performs at a level in B. The requirement that B be "measurable" says simply that the corresponding event $\{\omega \mid Y_S(\omega) \in B\}$ must lie in the underlying event space, insuring that the right-hand probability is defined.

If the performance variable Y_S is continuous then A must be continuous and, hence (by an earlier assumption), A is some interval of real numbers, including the possibility that $A = \mathbf{R} = (-\infty,\infty)$. In this case (or if Y_S happens to be discrete and yet real-valued), we know from probability theory that the induced measure p_S is uniquely determined by the *probability distribution function* of Y_S (see [13], for example), i.e., by the function F_{Y_S} (which we write simply as F_S) where, for all $b \in A$,

$$F_S(b) = P(\{\omega \mid Y_S(\omega) \leq b\}). \tag{8}$$

Moreover, p_S can then be expressed as the Lebesgue–Stieltjes measure induced by F_S (cf., [13, sec. 4.5]), that is, for any (measurable) set B of accomplishment levels, the performability value of B is given by

$$p_S(B) = \int_B dF_S(b). \tag{9}$$

In particular, if B is a single interval $B = (b_0,b_1]$ where $b_0 < b_1$, then

$$p_S(B) = F_S(b_1) - F_S(b_0).$$

This special case has practical significance since it quantifies the ability of S to perform within the specified limits b_0 and b_1.

If, on the other hand, Y_S is a discrete random variable then each singleton set $B = \{a\}(a \in A)$ is measurable and p_S is uniquely determined by the *probability distribution* of Y_S, i.e., by the set of values

$$\{p_S(a) \mid a \in A\} \tag{10}$$

where $p_S(a)$ denotes $p_S(\{a\})$. Given this distribution, if B is a set of accomplishment levels then $p_S(B)$ can be written as the sum

$$p_S(B) = \sum_{b \in B} p_S(b). \tag{11}$$

Hence, the probability distribution of Y_S or, equivalently, the restriction of p_S to single accomplishment levels, suffices to determine the performability. For this reason, when Y_S is discrete the performability of S can be alternatively defined as follows.

Definition 1a: If S is a total system with performance Y_S taking values in accomplishment set A and, moreover, Y_S is a discrete random variable, then the *performability of* S is the function p_S where, for each accomplishment level $a(a \in A)$,

$$p_S(a) = P(\{\omega \mid Y_S(\omega) = a\}).$$

Note that Definition 1a is essentially the restriction of Definition 1 to single accomplishment levels (which have probabilities defined when Y_S is discrete). Conversely, given Definition 1a, its extension to Definition 1 can be obtained (in principle) by application of (11) to each subset B of A.

To justify the notion of performability in the context of system effectiveness, if we assume the existence of a worth function w [see (7)], then the real-valued random variable W_S is a function w of the random variable Y_S. Moreover, we know that w is a "measurable" function (e.g., see [13, sec. 3.8]) since, prior to (7), we assumed that W_S was a random variable. (Indeed, condition (7) is actually stronger than needed; its advantages, however, are its simplicity and the fact that it serves the purpose of the present discussion.) Hence, we are able to appeal to the well-developed theory of functions of a random variable and, particularly, expectations where, again, it is convenient to distinguish two cases. If the performance variable Y_S is continuous (and thus real-valued) with probability distribution function F_S (8) then

$$E[w(Y_S)] = \int_A w(a)\, dF_S(a) \qquad (12)$$

where the integral is a Lebesgue–Stieltjes integral. In case Y_S is discrete, then (12) still applies provided the levels in A are represented by real numbers. However, independent of whether Y_S is real-valued, a simpler and more familiar formulation holds in this case where, if $p_S(a)$ is as defined in (10), then

$$E[w(Y_S)] = \sum_{a \in A} w(a) p_S(a). \qquad (13)$$

By the definition of system effectiveness (6) and the fact that $W_S = w(Y_S)$ (7), we have

$$\mathrm{eff}(S) = E[W_S] = E[w(Y_S)]$$

and, accordingly, (12) and (13) are formulations of the effectiveness of S. Moreover, we see that each formula involves the worth function w and the performability p_S [although p_S does not occur explicitly in (12), recall that F_S characterizes p_S (9)]. In other words, relative to the system-user interface delineated by the accomplishment set A, effectiveness evaluation may be decomposed into worth evaluation (on the user side of the interface) and performability evaluation (on the system side). Consequently, looking in toward the system, performability emerges as a key measure with regard to evaluations of system effectiveness.

To further justify the concept of performability, we note that traditional evaluations of computer performance and computer reliability are concerned with special types of performability. Performance evaluation is concerned with evaluating p_S under the assumption that the computer part of S is fixed (i.e., its structure does not change as the consequence of internal faults). Reliability evaluation is concerned with evaluating $p_S(B)$ where B is a designated subset of accomplishment levels associated with system "success." If A is finite, a performability evaluation can alternatively be regarded as $|A|$ reliability evaluations, one for each singleton "success set" $B = \{a\}$, and the evaluation may actually be carried out in this manner. As this process is generally more complex than a typical reliability evaluation procedure (in particular, it involves distinguishing all the performance levels as well as determining their probabilities), we reserve the term "reliability evaluation" to mean the evaluation of "probability of success" for some specified success criterion B. Moreover, even when $|A| = 2$, we find (as discussed later in this section) that performability models represent a proper extension of models typically employed in reliability evaluation.

As a final remark regarding justification, we have found that when system performance is not degradable (as in the case, for example, with fault-tolerant architectures which employ standby sparing [4], N modular redundancy [17], or combinations thereof), it is possible to treat performance and reliability as separate issues in the process of evaluating system effectiveness. On the other hand, if performance is degradable, it can be shown (see [18]) that the more general concept of performability must typically be invoked (as in (12) and (13), for example) when evaluating system effectiveness.

With performability established as the object of the evaluation process, we are now in a position to specify how the base model process X_S (1) must relate to the performance variable Y_S (4) if it is to support an evaluation of the performability p_S. To precisely state this relationship, we suppose that X_S is specified by its finite-dimensional probability distributions (or by information that determines these distributions) and we let Pr denote the probability measure (defined on a σ algebra of subsets of U) which is uniquely determined by these finite-dimensional distributions (see [14], for example). If Pr is defined for a trajectory set $V (V \subseteq U)$ then, relative to the underlying measure P,

$$\Pr(V) = P(\{\omega | u_\omega \in V\}), \qquad (14)$$

i.e., $\Pr(V)$ is the probability that an observed state trajectory u_ω [see (2)] lies in the set V. In practice, however, $\Pr(V)$ will be calculated directly from the finite-dimensional distributions that determine Pr. The measure Pr thus serves to formally describe the probabilistic nature of the base model X_S.

For X_S to support Y_S, we now impose the following restrictions. We assume first that the base model is refined enough to distinguish the levels of accomplishment perceived by the user, that is, for all $\omega, \omega' \in \Omega$,

$$Y_S(\omega) \neq Y_S(\omega') \text{ implies } u_\omega \neq u_{\omega'} \qquad (15)$$

where u_ω and $u_{\omega'}$ are the state trajectories associated with outcomes ω and ω'. This implies that each trajectory $u \in U$ is related to a unique accomplishment level $a \in A$. In addition, we assume that the probabilistic nature of Y_S is determinable from that of X_S. More precisely, if B is a measurable set of accomplishment levels, i.e., the set $\{\omega | Y_S(\omega) \in B\}$ is in the domain of the underlying measure P, then we require that the corresponding trajectory set

$$U_B = \{u_\omega | Y_S(\omega) \in B\}$$

lie in the domain of the base model measure Pr; in short

$$\text{If } B \text{ is measurable then Pr is defined for } U_B. \qquad (16)$$

Given that conditions (15) and (16) are satisfied, we can establish a link between X_S and Y_S which, in the context of effectiveness modeling [18], is generally referred to as the "capability" of S. Adopting this terminology, we have

Definition 2: If S is a system with trajectory space U and accomplishment set A then the *capability function of* S is the function $\gamma_S: U \rightarrow A$ where $\gamma_S(u)$ is the level of accomplishment resulting from state trajectory u, that is,

$$\gamma_S(u) = a \qquad \text{if for some } \omega \in \Omega,\ u_\omega = u \text{ and } Y_S(\omega) = a.$$

Condition (15) insures that the capability function γ_S is well-defined (i.e., it deserves the name "function"), for if $u_\omega = u_{\omega'}$ then $\gamma_S(u_\omega) = \gamma_S(u_{\omega'})$. Condition (16) guarantees that the inverse γ_S^{-1} of the capability function (γ_S^{-1} is a relation between A and U but generally not a function) will carry sets that are measurable with respect to Y_S into sets that are measurable with respect to X_S. To substantiate this fact, suppose that B is a measurable set of accomplishment levels. Then the inverse image of B is the trajectory set

$$\gamma_S^{-1}(B) = \{u \mid \gamma_S(u) \in B\}$$

or, equivalently, by the definition of γ_S (Definition 2),

$$\gamma_S^{-1}(B) = \{u_\omega \mid Y_S(\omega) \in B\} = U_B.$$

But condition (16) insures that Pr is defined for U_B and, hence, $\gamma_S^{-1}(B)$ is measurable where

$$\Pr(\gamma_S^{-1}(B)) = \Pr(U_B). \tag{17}$$

In effect, therefore, conditions (15) and (16) say that γ_S can be viewed as a random variable defined on the probability space (with measure Pr) induced by the base model X_S. Of more practical significance, however, is the fact that, under these conditions, X_S and γ_S suffice to determine the performability of S. To argue the latter, if B is measurable then, by (14),

$$\Pr(U_B) = P(\{\omega \mid u_\omega \in U_B\}) = P(\{\omega \mid Y_S(\omega) \in B\})$$

which, by Definition 1, is just the performability of S for accomplishment levels B, i.e.,

$$\Pr(U_B) = p_S(B). \tag{18}$$

Combining (17) and (18), we conclude that

$$p_S(B) = \Pr(\gamma^{-1}(B)), \tag{19}$$

substantiating the fact that X_S and γ_S (which determine Pr and γ_S^{-1}, respectively) suffice to support an evaluation of the performability p_S.

In view of what has just been observed, if X_S and Y_S admit to the definition of a capability function γ_S (in which case we presume that conditions (15) and (16) are satisfied), the pair (X_S, γ_S) is said to constitute a *performability model of* S. If B is a (measurable) set of accomplishment levels, the inverse image $\gamma_S^{-1}(B) = U_B$ is referred to as the *trajectory set of* B, where its determination requires an analysis of how levels in B relate back down via γ_S^{-1} to trajectories of the base model. $p_S(B)$ is then determined by a probability analysis of $\gamma_S^{-1}(B)$. In case Y_S is discrete (Definition 1a), it suffices to consider inverse images of the form $\gamma_S^{-1}(a)$ where $a \in A$. Methods of implementing this process in the discrete case are discussed in Section IV.

The role of a capability function in performability evaluation is similar to that of a "structure function" [19] in reliability evaluation. However, even when performability is restricted to reliability, the concept of a capability function is more general. The special class which corresponds to the use of structure functions in "phased mission" analysis (see [20], for example) may be characterized as follows. Let S be a system where Q is the state space of the base model and $A = \{0,1\}$ is the accomplishment set (here, 1 denotes "success" and 0 denotes "failure"). Then a capability function γ_S is *structure-based* if there exists a decomposition of T into k consecutive time periods $T_1, T_2, \cdots, T_k$ and there exist functions $\varphi_1, \varphi_2, \cdots, \varphi_k$ with $\varphi_i: Q \rightarrow \{0,1\}$ such that, for all $u \in U$,

$$\gamma_S(u) = 1 \quad \text{iff } \varphi_i(u(t)) = 1, \tag{20}$$

for all $i \in \{1, 2, \cdots, k\}$ and for all $t \in T_i$. In the context of "phased mission" analysis, T_i is referred to as the ith *phase* (of the mission) and φ_i is the *structure function* of the ith phase. For each function φ_i, the inverse image $\varphi_i^{-1}(1)$ can be interpreted as the set of "success states" of the ith phase and, accordingly, (20) says that S performs successfully ($\gamma_S(u) = 1$) if and only if $u(t)$ is a success state throughout each phase. Thus, the advantage of a structure-based formulation is that each phase may be treated independently when determining the set $\gamma_S^{-1}(1)$ of all successful state trajectories.

If system success is viewed in structural terms, as is the case in most reliability studies, a structure-based capability function will usually suffice. On the other hand, when success relates to system performance we find that capability may no longer be expressible in terms of locally defined success criteria as specified by the structure functions φ_i. The following example serves to demonstrate this fact.

Let $S = (C,E)$ where C represents a distributed computer comprised of n subsystems, and E represents the computer's workload. Suppose further that system "throughput" (i.e., the user-visible work rate of C in E) varies as a function of the number of faulty subsystems. For our purposes here, it suffices to assume that the workload E is constant and, hence, the operational states of S can be represented by the state space $Q = \{q_0, q_1, \cdots, q_n\}$ where state q_i corresponds to "i faulty subsystems." The variation in throughput is described by a function $\tau: Q \rightarrow \boldsymbol{R}_+$ where $\tau(i)$ = the throughput of S in state q_i. Assuming S is used continuously throughout a utilization period $T = [0,h]$ of duration $h > 0$, the base model of S is a stochastic process $X_S = \{X_t \mid t \in [0,h]\}$ where each X_t is a random variable taking values in Q. (The probabilistic nature of X_S is not an issue here.) As for performance, suppose that the user is interested in the average throughput of the system, where the average is taken over the utilization period T. Suppose further that system "success" is identified with a minimum average throughput $\bar{\tau}$. Then the capability function of S is the function $\gamma_S: U \rightarrow \{0,1\}$ where

$$\gamma_S(u) = \begin{cases} 1 & \text{if } \dfrac{1}{h}\displaystyle\int_0^h \tau(u(t))dt \geq \bar{\tau} \\ 0 & \text{otherwise.} \end{cases} \tag{21}$$

Due to the inherent memory of the integration operation,

we find that γ_S does not admit to a structure-based formulation. To verify this fact with a simple 2-state example, suppose $Q = \{q_0,q_1\}$, $\tau(q_0) > \tau(q_1)$, and $\bar{\tau} = (\tau(q_0) - \tau(q_1))/2$. Then, according to (21), $\gamma_S(u) = 1$ if the total time for which $u(t) = q_0$ is at least $h/2$. In particular, this says that more than one trajectory results in success, i.e., $|\gamma_S^{-1}(1)| > 1$. To prove that γ_S is not structure-based let us suppose to the contrary, that is, there exist phases $T_1,T_2,\cdots,T_k$ and structure functions $\varphi_1,\varphi_2,\cdots,\varphi_k$ such that (20) is satisfied. If we let R_i denote the success states of phase i, i.e., $R_i = \varphi_i^{-1}(1)$, then $R_i \neq \phi$, for all i, or otherwise no trajectory results in success. It must also be the case that $R_i \neq \{q_0,q_1\}$ for all i, for if $R_i = \{q_0,q_1\}$ (all states are success states during phase i) then the condition $\varphi_i(u(t)) = 1$, $\forall\ t \in T_i$ [see (20)] is always satisfied, that is, phase i can be ignored when determining whether u spends at least half its time in state 0. This is clearly impossible if the duration of T_i is at least $h/2$. If the duration of T_i is less than $h/2$, trajectories u and v can be found such that $u(t) = v(t)$, $\forall\ t \in (T - T_i)$, and yet $\gamma_S(u) \neq \gamma_S(v)$ contradicting the ability to ignore phase i. The only remaining alternative is that $|R_i| = 1$, for all i, that is, each phase has exactly one success state which, in turn, implies that there is exactly one success trajectory u. This contradicts our initial observation that $|\gamma_S^{-1}(1)| > 1$ and proves that γ_S is not structure-based.

We can conclude, therefore, that even in the case of two accomplishment levels, the concept of a capability function (Definition 2) represents a proper extension of relations between state behavior and system performance that are typically assumed in the theory of reliability. Moreover, we have found that this extension permits the phases of a utilization period to be "functionally dependent" in a precisely defined sense, whereas the phases associated with a structure-based capability function must be functionally independent. The reader is referred to [21] for a more complete discussion of functional dependence and its implications.

IV. Performability Evaluation

As established in the previous section [see (19)], if (X_S,γ_S) is a performability model then the performability of S for a set of accomplishment levels B may be expressed as $p_S(B) = \Pr(\gamma_S^{-1}(B))$. Accordingly, one method of evaluating a particular $p_S(B)$ is to 1) determine $\gamma_S^{-1}(B)$ and then 2) evaluate $\Pr(\gamma_S^{-1}(B))$. Since the "distance" between the base model X_S and the accomplishment set A may be considerable, step 1) can be facilitated by introducing additional models between X_S and A.

In general, each intermediate model is defined in a manner similar to that of the base model. More precisely, if there are $m + 1$ levels in the hierarchy, the *level-i model* ($i = 0,1,\cdots,m$, where level-0 is the least detailed model at the "top" of the hierarchy) is a stochastic process

$$X^i = \{X^i_t \mid t \in T^i\}, \qquad T^i \subseteq T$$

where, for a fixed $t \in T^i$, X^i_t is a random variable taking values in a set Q^i, the *state space* of X^i. The state space Q^i is generally composed of two components, i.e.,

$$Q^i = Q^i_c \times Q^i_b$$

where Q^i_c is the *composite state set* and Q^i_b is the *basic state set* (at level-i). States in the composite part Q^i_c represent a less detailed view of the operational status of the system than do states in Q^{i+1}, such that the state behavior at level-$(i + 1)$ uniquely determines the composite state behavior at level-i (this will be made more precise in a moment). States in Q^i_b, on the other hand, represent basic information not conveyed by states in Q^{i+1}, i.e., Q^i_b is a coordinate set of the base model state space Q. In case there is no composite (alternatively, basic) part at level-i, $Q^i_c(Q^i_b)$ is simply deleted, that is, $Q^i = Q^i_b$ $(Q^i = Q^i_c)$. In particular, the above definition precludes a composite state set at level-m (the "bottom" level of the hierarchy) and, hence, $Q^m = Q^m_b$.

In specifying the model hierarchy, it is convenient to view X^i as a pair of processes which determine the projections on Q^i_c and Q^i_b, respectively. (If one of Q^i_c or Q^i_b does not exist, this pair reduces to a single process.) More precisely, given Q^i_c, the *composite process* (at level-i) is the stochastic process

$$X^i_c = \{X^i_{c,t} \mid t \in T^i_c\}, \qquad T^i_c \subseteq T^i$$

where the random variables $X^i_{c,t}$ take values in Q^i_c. For a fixed outcome ω in the underlying sample space Ω, a *composite state trajectory* is a function $u_{c,\omega}\colon T^i_c \rightarrow Q^i_c$ where $u_{c,\omega}(t) = X^i_{c,t}(\omega)$; the *composite trajectory space* is the set $U^i_c = \{u_{c,\omega} \mid \omega \in \Omega\}$. Similar definitions, terminology, and notation apply to the *basic process* X^i_b. To permit extension of either X^i_c or X^i_b to larger time bases, a fictitious state ¢ is adjoined to each of Q^i_c and Q^i_b so that if $t \notin T^i_c$ (similar remarks apply to the basic part) then $X^i_{c,t}$ is defined to be a degenerate random variable that always assumes the value ¢, i.e.,

$$X^i_{c,t}(\omega) = ¢, \qquad \text{for all } \omega \in \Omega. \tag{22}$$

If X^i_c and X^i_b are so extended to T^i, and we take X^i to be the process whose projections on Q^i_c and Q^i_b are X^i_c and X^i_b, respectively, then X^i is uniquely determined by X^i_c and X^i_b. (Note that the processes X^i_c and X^i_b may be statistically dependent.)

By the previous observation, we can alternatively regard the level-i model as the pair of processes

$$X^i = (X^i_c,X^i_b)$$

which is a convenient view for the purpose of specifying interlevel relationships. With this identification, a state trajectory of X^i is viewed as a pair of trajectories, i.e., the *trajectory space* U^i (at level-i) is taken to be the set $U^i_c \otimes U^i_b$ where

$$U^i_c \otimes U^i_b = \{(u_{c,\omega}, u_{b,\omega}) \mid \omega \in \Omega\}.$$

In case there is no composite (alternatively, basic) state set at level-i, the above representations of X^i and U^i are understood to be their appropriate single component versions.

The required relationship of these models to the base model, the accomplishment set, and the capability function is prescribed by the following definition.

Definition 3: If S is a total system with base model X_S and capability function γ_S, the collection $\{X^0,X^1,\cdots,X^m\}$ of level-0 to level-m models is a *model hierarchy for S* if the following conditions are satisfied.

a) $X^m = X^m_b$, that is, the bottom model is comprised only of a basic process.

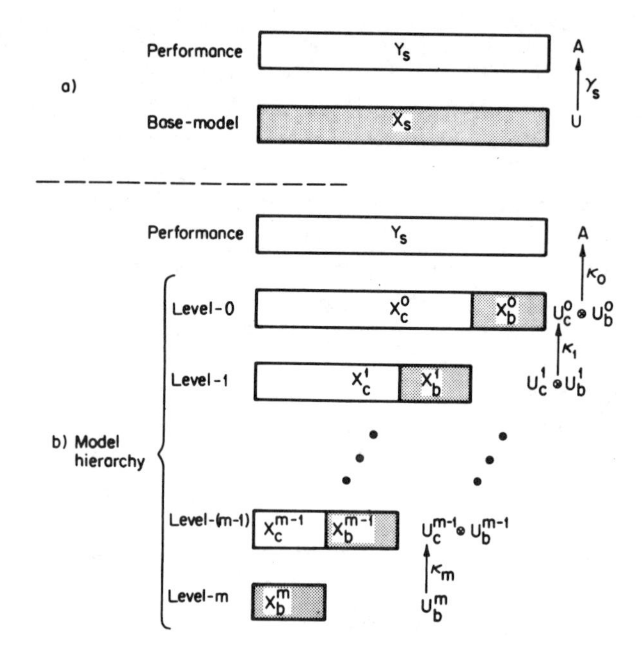

Fig. 1. Total system S with (a) base model X_S and (b) model hierarchy for S.

b) If each model X^i is extended to the utilization period T, the base model X_S is the stochastic process $X_S = \{X_t | t \in T\}$ where $X_t = (X_{b,t}^m, X_{b,t}^{m-1}, \cdots, X_{b,t}^0)$. Accordingly, the state space of X_S is $Q = Q_b^m \times Q_b^{m-1} \times \cdots \times Q_b^0$ and the trajectory space U is represented by the set $U_b^m \otimes U_b^{m-1} \otimes \cdots \otimes U_b^0$.

c) For each level i, there exists an *interlevel translation* κ_i where

$$\kappa_0: U_c^0 \otimes U_b^0 \to A$$

$$\kappa_i: U_c^i \otimes U_b^i \to U_c^{i-1} \qquad (1 < i < m)$$

$$\kappa_m: U_b^m \to U_c^{m-1}$$

such that the capability function γ_S can be decomposed as follows. If $u \in U$ where $u = (u_m, u_{m-1}, \cdots, u_0)$ with $u_i \in U_b^i$, then

$$\gamma_S(u) = \kappa_0(\kappa_1(\cdots \kappa_{m-1}(\kappa_m(u_m), u_{m-1}), \cdots), u_0). \qquad (23)$$

The terminology and notation of Definition 3 is summarized in Fig. 1 where Fig. 1(a) is the original model and Fig. 1(b) is the hierarchical model.

A model hierarchy thus provides a step-by-step formulation of the capability function in terms of interlevel translations of state trajectories, beginning with a translation of the bottom model. It also permits the expression of capability relative to higher level (less detailed) views of total system behavior. More precisely, let $\overline{U}^i$ denote the Level-i trajectory space, along with all the basic trajectory spaces of higher level models, i.e.,

$$\overline{U}^i = U^i \otimes U_b^{i-1} \otimes \cdots \otimes U_b^0.$$

(Note that, at the extremes, $\overline{U}^0 = U^0$ and $\overline{U}^m = U$.) Then the *level-i based capability function* is the function

$$\gamma_i: \overline{U}^i \to A$$

defined inductively as follows. If $i = 0$ and $u \in \overline{U}^0$, then

$$\gamma_0(u) = \kappa_0(u). \qquad (24)$$

If $i > 0$ and $(u,u') \in \overline{U}^i$ where $u \in U^i$ and $u' \in U_b^{i-1} \otimes \cdots \otimes U_b^0$, then

$$\gamma_i(u,u') = \gamma_{i-1}(\kappa_i(u),u'). \qquad (25)$$

It is easily shown that γ_i has its intended interpretation, i.e., if u and u' correspond to a base model trajectory v then $\gamma_i(u,u') = \gamma_S(v)$. In particular, if $i = m$ then $\gamma_m = \gamma_S$.

The practical significance of the model hierarchy, however, is the ability to formulate the inverse of γ_S via the inverses of the γ_i, thereby providing a step-by-step, top-down method of elaborating a set of accomplishment levels B. Beginning with level-0-based capability, by (24) we have

$$\gamma_0^{-1}(B) = \kappa_0^{-1}(B). \qquad (26)$$

Assuming that $\gamma_{i-1}^{-1}(B)$ has been determined, by (25) it follows that

$$\gamma_i^{-1}(B) = \bigcup_{(u,u') \in \gamma_{i-1}^{-1}(B)} (\kappa_i^{-1}(u),u'), \qquad (27)$$

where $(\kappa_i^{-1}(u),u') = \{(v,u') | \kappa_i(v) = u\}$. This process is iterated until $i = m$, yielding $\gamma_m^{-1}(B) = \gamma_S^{-1}(B)$. Actual implementations of this procedure can exploit simplified characterizations and decompositions of trajectory sets so as to avoid the manipulation of individual trajectories (see [18] and [22]). Thus a model hierarchy for S is useful not only in the process of model construction but also in the process of model solution.

Although the purpose of this investigation has been to establish the foundations of performability modeling and evaluation, it is helpful to illustrate these ideas in the context of a specific application. The example we consider is an aircraft computing system of the type being developed for next-generation commercial aircraft [23], [24]. Such systems have provided an impetus for the work described herein and are representative of degradable computing systems that operate in a real-time control environment. Although space limitations necessitate a "scaled-down" example, it should suffice to illustrate the basic concepts and constructs. More extensive applications of this type have been investigated subsequent to the work described in this paper; in particular, see [25]–[27] which describe a performability evaluation of the SIFT computer [24].

Beginning at the highest level of description, the total system $S = (C,E)$ is a fault-tolerant computer C which operates in the environment E of a portal-to-portal flight of a commercial aircraft. The user is interested in fuel efficiency, timeliness, and safety, and accordingly, five levels of accomplishment are distinguished:

- a_0: low-fuel consumption, no diversion to an alternate landing site, and safe
- a_1: high-fuel consumption, no diversion, and safe
- a_2: low-fuel consumption, diversion, and safe
- a_3: high-fuel consumption, diversion, and safe
- a_4: unsafe (crash).

The model hierarchy consists of four levels, beginning with a high mission-oriented model and proceeding through aircraft and computational task levels to the bottom level model of the computer. More precisely, employing the notation developed above and assuming $T = [0,h]$, these models are as follows.

Level-0: The model at this level is a simple, user-oriented model which relates directly to the accomplishment levels distinguished above. At this high level, we assume that the model will be fully elaborated at the next lower level (i.e., no part of the state set is basic) and hence $Q^0 = Q_c^0$. The composite state set is taken to be the set

$$Q_c^0 = \{(q_1,q_2,q_3) | q_i \in \{0,1\}\}$$

where

$$q_1 = \begin{cases} 0 & \text{if the flight is fuel efficient,} \\ 1 & \text{otherwise,} \end{cases}$$

$$q_2 = \begin{cases} 0 & \text{if the flight is not diverted,} \\ 1 & \text{otherwise,} \end{cases}$$

$$q_3 = \begin{cases} 0 & \text{if the flight is safe,} \\ 1 & \text{otherwise.} \end{cases}$$

Thus, for example, the state

$$q = (0,1,0)$$

says that the flight is fuel efficient and safe but has to be diverted to an alternate landing site. By the interpretation of these states, we are modeling the status of the system on the completion of a flight and thus the time base consists of a single observation time at the end of utilization. More precisely (and to illustrate the notation developed earlier),

$$T_c^0 = \{h\}$$

and, accordingly, the composite process is a single random variable

$$X_c^0 = \{X_{c,h}^0\}$$

with trajectory space

$$U_c^0 = Q_c^0.$$

Since there is no basic process at this level, $X^0 = X_c^0$, thereby completing the description of the level-0 model. The interlevel translation κ_0: $U_c^0 \rightarrow A$ (see Definition 3c); note that $U_c^0 = Q_c^0$ represents U^0 due to the lack of U_b^0) follows immediately from the preceding definitions of Q_c^0 and A, and is given by Table II.

TABLE II
FUNCTION TABLE OF TRANSLATION κ_0

$u=(q_1,q_2,q_3)$	$\kappa_0(u)$
0 0 0	a_0
0 0 1	a_4
0 1 0	a_2
0 1 1	a_4
1 0 0	a_1
1 0 1	a_4
1 1 0	a_3
1 1 1	a_4

Level-1: At this level of the hierarchy, the model describes the extent to which various aircraft related tasks can be accomplished (the composite part) and the weather condition at the destination airport (the basic part). Although computing systems for advanced commercial aircraft will be called on to realize a variety of control functions, it suffices (for the purpose of illustration) to consider two functional tasks: fuel control (FC) and automatic landing (AL). The FC task encompasses functions such as engine control and navigation which, in a more refined model, might be treated as individual tasks. The AL task is comprised of functions required to automatically land the aircraft in zero-visibility weather; prior to landing, AL is interpreted as (computer-implemented) checkout of the AL system.

To satisfy the requirements of a model hierarchy, construction of the level-1 model must rely on knowledge of how these tasks and the condition of the weather relate to states of the level-0 model. In this regard, we presume the following knowledge, where T/C denotes the takeoff/cruise phase of the flight and L denotes the landing phase; "loss" of a functional task during a specified period of time (e.g., a phase) means failure to accomplish that task at some time during the specified period.

a) The flight is fuel efficient iff FC is accomplished throughout T/C and L.

b) The flight is diverted iff there is zero-visibility weather at the landing site (just prior to L) and either FC is lost during T/C or AL is lost during the last half of T/C.

c) The flight is unsafe iff AL is lost during an AL (attempted iff there is zero-visibility weather and the flight is not diverted) or the fuel consumption is "excessive"; excessive fuel consumption results iff FC is lost during both halves of T/C or during both T/C and L.

Translation between the level-1 and level-0 models (and thus the construction at level-1) will also rely on knowledge of the computational demands (workload) imposed by the aircraft, where we can presume the following.

d) FC is in demand throughout T/C and L (and hence is accomplished whenever it can be).

e) AL is not demanded during the first half of T/C, is in demand during the last half of T/C, and is in demand during L iff an AL is attempted [see condition c)].

Given this knowledge, we find that the set

$$Q_c^1 = \{0,1, \cdots ,7\}$$

suffices as the composite state set, where the states $q \in Q_c^1$ are interpreted as follows (here, for task A and time period B, "A during B" means A can be accomplished throughout B; "no A during B" means the opposite, i.e., if A is demanded then A is lost during B):

$$q = \begin{cases} 0 & \text{if FC during } T/C \text{ and AL during the last half of } T/C, \\ 1 & \text{if FC during } T/C \text{ and no AL during the last half of } T/C, \\ 2 & \text{if FC during one half of } T/C \text{ and no FC during the other half,} \\ 3 & \text{if no FC during both halves of } T/C, \\ 4 & \text{if FC and AL during } L, \\ 5 & \text{if FC and no AL during } L, \\ 6 & \text{if no FC and AL during } L, \\ 7 & \text{if no FC and no AL during } L. \end{cases}$$

Note that our interpretation of these states is representative of "supply" as opposed to "demand," i.e., the computer's ability to supply the computations demanded by the aircraft; considerations of demand [conditions d) and e)] will be incorporated in the translation between Level-1 and Level-0. With this interpretation, it suffices to observe the system at the end of each phase, i.e.,

$$T_c^1 = \{t_2,t_3\}$$

where t_2 is the end of T/C ($t_2 = h - \frac{1}{2}$ hour) and $t_3 = h$. (Time t_1 is introduced later at level-2 of the hierarchy.) Accordingly, the composite part of the level-1 model is the process

$$X_c^1 = \{X_{c,t_2}^1, X_{c,t_3}^1\}, \tag{28}$$

where, by the definition of Q_c^1, X_{c,t_2}^1 takes values in $\{0,1,2,3\}$ and X_{c,t_3}^1 takes values in $\{4,5,6,7\}$; the composite trajectory space is therefore the Cartesian product

$$U_c^1 = \{0,1,2,3\} \times \{4,5,6,7\}. \tag{29}$$

The basic part of the level-1 model involves only the weather and is much easier to specify. Here

$$Q_b^1 = \{0,1\}$$

can serve as the state set where, if $q \in Q_b^1$, then

$$q = \begin{cases} 1 & \text{if there is zero-visibility weather at the landing site just prior to } L, \\ 0 & \text{otherwise.} \end{cases}$$

Moreover, with this interpretation, a single observation at time t_2 (the end of T/C) suffices; hence, the basic part is the (single variable) process

$$X_b^1 = \{X_{b,t_2}^1\}.$$

To complete the construction according to the general procedure described in conjunction with Definition 3, X_c^1 and X_b^1 are then extended to the time base $T^1 = T_c^1 \cup T_b^1 = \{t_2,t_3\}$ so as to establish a common time base for both processes. Here, since $T_c^1 = T^1$, only X_b^1 requires extension, resulting in the process

$$X_b^1 = \{X_{b,t_2}, X_{b,t_3}\} \tag{30}$$

[where $X_{b,t_3} = ¢$; see (22)] with trajectory space

$$U_b^1 = \{0,1\} \times \{¢\}. \tag{31}$$

Combining the composite process X_c^1 (28) with the basic process X_b^1 (30), the level-1 model is described by the pair of processes

$$X^1 = (X_c^1, X_b^1)$$

with the (combined) trajectory space [see (29) and (31)]

$$\begin{aligned} U^1 &= U_c^1 \otimes U_b^1 \\ &= \{0,1,2,3\} \times \{4,5,6,7\} \times \{0,1\} \times \{¢\}. \end{aligned}$$

Finally, to establish that this model can indeed support the composite part (and hence all) of the level-0 model, we must be able to specify an interlevel translation $\kappa_1 : U^1 \to U_c^0$ [see Definition 3a)] which is consistent with the aircraft's in-flight behavior [conditions a)–c)] and the aircraft's computational demands [conditions d) and e)]. The required translation is obtained by applying just these conditions, for example, if $u \in U^1$ where $u = (2,5,1,¢)$ (which says FC can be accomplished during one-half but not all of T/C; FC can be accomplished during L but AL cannot be; there is zero-visibility weather at the landing site just prior to L) then, by conditions a)–e), it follows that $\kappa_1(u) = (1,1,0)$ (the flight is fuel inefficient, diverted, and safe). The remaining values of κ_1 are determined in a like manner.

Level-2: At this level, the computational capacity of the computer C is represented in terms of its ability to execute FC and AL computations during specified phases of the utilization period. We assume that this ability will be fully determined by the computer model at level-3 and, hence, the state set Q^2 coincides with the composite state set Q_c^2. The latter is taken to be the set

$$Q_c^2 = \{(q_1,q_2) \mid q_i \in \{0,1\}\}$$

where, for a given phase of T, the coordinates q_1 and q_2 are interpreted as follows:

$$q_1 = \begin{cases} 0 & \text{if FC computations can be executed throughout the phase,} \\ 1 & \text{otherwise} \end{cases}$$

$$q_2 = \begin{cases} 0 & \text{if AL computations can be executed throughout the phase,} \\ 1 & \text{otherwise.} \end{cases}$$

To support the composite part of the level-1 model (28), it suffices to distinguish three such phases, obtained from the level-1 phases by adding an observation time t_1 midway through T/C. More precisely, if t_2 and t_3 are as defined for level-1 then

$$T^2 = \{t_1,t_2,t_3\}$$

where $t_1 = t_2/2$, thereby distinguishing phase $[0,t_1]$ (first half of T/C), phase $[t_1,t_2]$ (second half of T/C), and phase $[t_2,t_3]$ (L). Accordingly, the level-2 model (comprised of a composite part only) is the process

$$X^2 = X_c^2 = \{X_{c,t_1}^2, X_{c,t_2}^2, X_{c,t_3}^2\}$$

where variable X_{c,t_i}^2 ($i = 1,2,3$) takes values in $Q_c^2 = \{0,1\}^2$, e.g., if $X_{c,t_2}^2 = (1,0)$ then the computer is unable to execute FC computations at some time during $[t_1,t_2]$ but has the resources to execute AL computations during this period. The level-2 trajectory space is therefore the set

$$U^2 = U_c^2 = \{0,1\}^2 \times \{0,1\}^2 \times \{0,1\}^2 \tag{32}$$

which is refined enough to admit an interlevel translation κ_2 from U^2 into U_c^1 (29). Specification of this translation follows immediately from the definitions of U^2 and U_c^1; for example, if $u \in U^2$ where $u = [(0,1),(1,1),(0,1)]$ (FC computations can be executed throughout the first and third phases, but not the second; AL computations cannot be executed throughout any

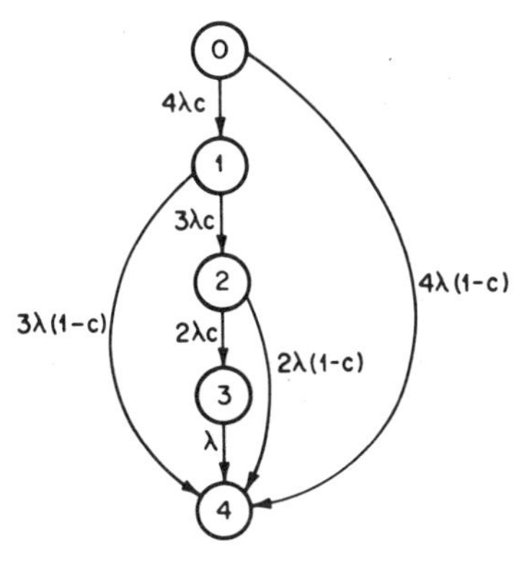

Fig. 2. Transition graph of the Level-3 Markov model.

TABLE III
COMPUTATIONAL CAPABILITY OF C

Level-3 state	Level-2 phase and computational task: $[0,t_1]$ FC	$[0,t_1]$ AL	$[t_1,t_2]$ FC	$[t_1,t_2]$ AL	$[t_2,t_3]$ FC	$[t_2,t_3]$ AL
0	X	x	X	X	X	X
1	X	X	X	X	X	X
2	X		X			X
3		X		X	X	
4						

phase), then $\kappa_2(u) = (2,5)$ (FC can be accomplished during one-half but not all of T/C; FC can be accomplished during L but AL cannot be).

Level-3: This model, at the bottom level of the hierarchy, describes variations in the computer's structure caused by faults which occur during utilization. Here we find that conventional stochastic models employed in reliability evaluation (e.g., continuous-time Markov models [7]–[9]) can adequately support higher level models of the type described previously. In particular, for this example, we suppose that C is a reconfigurable, fault-tolerant computer whose resources consist of four (essentially identical) processing subsystems. The integrity of these subsystems is represented by the state set

$$Q^3 = Q_b^3 = \{0,1,2,3,4\}$$

where state $q = i$ means that exactly i subsystems are faulty. Assuming that subsystems fail permanently at a constant rate λ (failures/hour) and letting c denote the "coverage" [4], the bottom model $X^3 = X_b^3$ is taken to be the continuous-time time-homogeneous Markov process described by the graph of Fig. 2. The level-3 trajectory space is the set $U^3 = U_b^3$ of all functions of the form

$$u: [0,h] \to Q^3$$

which can be realized by a Markov process of this type (see, for example, [15, ch. 8.3]).

To establish the interlevel translation $\kappa_3: U^3 \to U_c^2$, it is necessary to know the extent to which available computer resources (the fault-free subsystems) can support the execution of FC and AL computations during each of the level-2 phases. Generally, such knowledge will depend on a number of factors including the user's computational priorities during each phase, the computational demands of the computer's operating system, and the processing capacity of each (fault-free) subsystem. To avoid further elaboration (that would really not serve our purpose here), let us suppose these factors have already been examined, yielding the information summarized in Table III. An "X" entry in the table indicates that the resources of level-3 state i (designated by the row) are configured during the level-2 phase (designated by the column) so as to permit execution of the computational task (designated by the column); absence of an "X" signifies the contrary situation. Note that, as at level-1 and level-2 of the hierarchy, this information is representative of supply as opposed to demand, e.g., in states 0, 1, and 3 during phase $[0,t_1]$, the computer is able to execute AL computations even though it will not be called on to do so.

This information, along with certain properties of the base model X^3 (Fig. 2), suffices to determine the interlevel translation κ_3. For example, suppose that $u \in U^3$ (where u: $[0,h] \to Q^3$) where $u(0) = 0$, $u(t_1) = 2$, $u(t_2) = 3$, and $u(t_3) = 3$. Then, since resources fail permanently, it follows that $u(t) \in \{0,1,2\}$, for all $t \in [0,t_1]$; $u(t) \in \{2,3\}$, for all $t \in [t_1,t_2]$; and $u(t) = 3$, for all $t \in [t_2,t_3]$. This says, in turn (see Table III), that the computer's ability to execute AL computations is lost at some time during $[0,t_1]$, AL and FC are lost during $[t_1,t_2]$ (even though AL capability is recovered by the end of the phase), and there is no ability to execute AL computations throughout $[t_2,t_3]$ (due to a reassignment of computational priorities at the beginning phase 3). In other words, for the trajectory u in question, the only computations that are executed successfully are FC computation's throughout phases $[0,t_1]$ and $[t_2,t_3]$; hence $\kappa_3(u) = ((0,1),(1,1),(0,1))$, resulting in a trajectory that was illustrated earlier at level-2.

By similar arguments, each trajectory $u \in U_b^3$, when sampled at times $0,t_1,t_2$, and t_3, can be assigned a (unique) state trajectory $\kappa_3(u) \in U^3 = U_c^3$, thereby determining the interlevel translation κ_3. This, then, completes the specification of the four-level model hierarchy for S and, therefore, the performability model (X_S,γ_S) where $X_S = (X_b^3,X_b^1)$ and γ_S is determined by the interlevel translations κ_0-κ_3 according to (23).

To "solve" the model (i.e., evaluate the performability p_S), we apply the two step procedure stated at the outset of this section. Implementation of step 1), i.e., calculation of the trajectory sets $\gamma_S^{-1}(a)$ for each $a \in A$ (such sets suffice since A is finite, see Definition 1a, relies on the notion of a level-i based capability function [see (24) and (25)] and on successive formulations of $\gamma_i^{-1}(a)$ for $i = 0,1,2,3$ [see (26) and (27)]. Step 2) of the procedure, i.e., calculation of the probabilities $\Pr(\gamma_S^{-1}(a))$ for each $a \in A$, can be implemented via introduction of an equivalent "phased" base model and the subsequent application of specially developed formulas involving "interphase" and "intraphase" matrices. (See [28] for a detailed discussion of this technique.)

The solution procedure we have just outlined is a subject in itself and, without further elaboration, it would be difficult to illustrate its application. We can, however, illustrate the kind

TABLE IV
PERFORMABILITIES OF SYSTEMS S_1 AND S_2

a	$P_{S_1}(a)$	$P_{S_2}(a)$
a_0	0.999994	0.999821
a_1	3.4×10^{-6}	3.4×10^{-5}
a_2	0	0
a_3	2.6×10^{-6}	1.4×10^{-4}
a_4	2.2×10^{-9}	5.1×10^{-8}

of results that have been obtained for specific instances of the preceding example. In the first instance (S_1), the environment is a short flight from Washington, DC to New York City where the duration h is 1 h and the probability of zero-visibility (Category III) weather at JFK is 0.011. In the second instance (S_2), the environment is a longer flight from Washington, DC to Los Angeles where h is 5.5 h and the probability of Category III weather is 0.019. Both instances assume a computer C where the failure rate of each processing subsystem is $\lambda = 10^{-3}$, the coverage is ideal ($c = 1$), and the initial state X_0^3 is 0 with probability 1. The resulting performabilities are given in Table IV. Note that, in both instances, Level a_3 (low-fuel consumption, diversion, and safe) is impossible to accomplish, due to the probabilistic nature of the base model (Fig. 2) and the way computational tasks are allocated to available computer resources (Table III).

V. SUMMARY

Since traditional distinctions between performance and reliability become blurred when performance is degradable, we have proposed that the two be dealt with simultaneously via a unified measure called performability. After formalizing this measure in probability-theoretic terms, it was shown that performability is a natural component of system effectiveness and is a proper generalization of both performance and reliability. Performability modeling needs were then characterized via the concept of a capability function and it was demonstrated that capability is a proper extension of the kind of structure-behavior relationships that are typically assumed in reliability models. Finally, a hierarchical modeling scheme was introduced to facilitate both model construction and model solution and, in particular, to permit formulation of the capability function via interlevel translations. These concepts were then illustrated for an aircraft computing system.

It is hoped that the results of this paper can serve as a foundation for future work on unified performance-reliability models and (model-based) solution methods. Our experience to date with further developments [21], [28] and applications [25]-[27] of this methodology suggests that performability evaluation is indeed feasible, thereby providing a means of assessing the kind of performance-reliability interaction that is characteristic of degradable systems.

ACKNOWLEDGMENT

This research was conducted at the Systems Engineering Laboratory of the University of Michigan with the able cooperation of Research Assistants R. Ballance, D. Furchtgott, and L. Wu. The work has also benefited from valuable discussions with S. Bavuso at the NASA Langley Research Center, as well as with many other members of the NASA staff.

REFERENCES

[1] L. Svobodova, *Computer Performance Measurement and Evaluation Methods: Analysis and Applications.* New York: Elsevier, 1976.
[2] D. Ferrari, *Computer Systems Performance Evaluation.* Englewood Cliffs, NJ: Prentice-Hall, 1978.
[3] H. Kobayashi, *Modeling and Analysis: An Introduction to System Performance Evaluation Methodology.* Reading, MA: Addison-Wesley, 1978.
[4] W. G. Bouricius, W. C. Carter, and P. R. Schneider, "Reliability modeling techniques for self-repairing computer systems," in *Proc. ACM 1969 Nat. Conf.*, Aug. 1969, pp. 295-305.
[5] B. R. Borgerson and R. F. Freitas, "A reliability model for gracefully degrading and standby-sparing systems," *IEEE Trans. Comput.*, vol. C-24, pp. 517-525, May 1975.
[6] H. B. Baskin, B. R. Borgerson, and R. Roberts, "PRIME-A modular architecture for terminal-oriented systems," in *1972 Spring Joint Computer Conf., AFIPS Conf. Proc.*, vol. 40. Washington, DC: Spartan, 1972, pp. 431-437.
[7] J. C. Laprie, "Reliability and availability of repairable structures," in *Dig. 1975 Int. Symp. Fault-Tolerant Computing,* Paris, France, June 1975, pp. 87-92.
[8] Y.-W. Ng and A. Avižienis, "A reliability model for gracefully degrading and repairable fault-tolerant systems," in *1977 Proc. Int. Symp. on Fault-Tolerant Computing,* Los Angeles, CA, June 1977, pp. 22-28.
[9] A. Costes, C. Landrault, and J.-C. Laprie, "Reliability and availability models for maintained systems featuring hardware failures and design faults," *IEEE Trans. Comput.*, vol. C-27, pp. 548-560, June 1978.
[10] M. D. Beaudry, "Performance-related reliability measures for computing systems," *IEEE Trans. Comput.*, vol. C-27, pp. 540-547, June 1978.
[11] R. Troy, "Dynamic reconfiguration: An algorithm and its efficiency evaluation," in *Proc. 1977 Int. Symp. Fault-Tolerant Computing,* Los Angeles, CA, June 1977, pp. 44-49.
[12] J. Losq, "Effects of failures on gracefully degradable systems," in *Proc. 1977 Int. Symp. Fault-Tolerant Computing,* Los Angeles, CA, June 1977, pp. 29-34.
[13] P. E. Pfeiffer, *Concepts of Probability Theory.* New York: McGraw-Hill, 1965.
[14] E. Wong, *Stochastic Processes in Information and Dynamical Systems.* New York: McGraw-Hill, 1971.
[15] E. Çinlar, *Introduction to Stochastic Processes.* Englewood Cliffs, NJ: Prentice-Hall, 1975.
[16] B. P. Zeigler, *Theory of Modelling and Simulation.* New York: Wiley, 1976.
[17] F. P. Mathur and A. Avižienis, "Reliability analysis and architecture of a hybrid-redundant digital system: Generalized triple modular redundancy with self-repair," in *Proc. 1970 Spring Joint Computer Conf., AFIPS Conf.*, vol. 36. Washington, DC: Spartan, 1970, pp. 375-383.
[18] J. F. Meyer, "Models and techniques for evaluating the effectiveness of aircraft comput. systems," Semiannu. Status Rep. 3, NASA Rep. CR158992 (NTIS Rep. N79-17564/2GA), Jan. 1978.
[19] Z. W. Birnbaum, J. D. Esary, and S. C. Saunders, "Multicomponent systems and structures and their reliability," *Technometrics*, vol. 3, pp. 55-77, Feb. 1961.
[20] J. D. Esary and H. Ziehms, "Reliability of phased missions," in *Reliability and Fault Tree Analysis,* SIAM, Philadelphia, PA, 1975, pp. 213-236.
[21] R. A. Ballance and J. F. Meyer, "Functional dependence and its application to system evaluation," in *1978 Proc. Johns Hopkins Conf. Inform. Sciences and Systems,* Johns Hopkins, Univ., Baltimore, MD, Mar. 1978, pp. 280-285.
[22] J. F. Meyer, "Models and techniques for evaluating the effectiveness of aircraft computing systems," Semiannu. Status Rep. 4, NASA Rep. CR158993 (NTIS, Rep. N79-17563/4GA), July 1978.
[23] A. L. Hopkins, Jr., T. B. Smith, III, and J. H. LaLa, "FTMP—A highly reliable fault-tolerant multiprocessor for aircraft," *Proc. IEEE*, vol. 66, pp. 1221-1239, Oct. 1978.

[24] J. H. Wensley, L. Lamport, J. Goldberg, M. W. Green, K. N. Levitt, P. M. Melliar-Smith, R. E. Shostak, and C. B. Weinstock, "SIFT: Design and analysis of a fault-tolerant computer for aircraft control," *Proc. IEEE,* vol. 66, pp. 1240-1255, Oct. 1978.

[25] D. G. Furchtgott and J. F. Meyer, "Performability evaluation of fault-tolerant multiprocessors," in *1978 Dig. Government Microcircuit Applications Conf.,* Monterey, CA, Nov. 1978, pp. 362-369.

[26] J. F. Meyer, D. G. Furchtgott, and L. T. Wu, "Performability evaluation of the SIFT computer," in *Proc. 1979 Int. Symp. Fault-Tolerant Computing,* Madison, WI, June 1979, pp. 43-50.

[27] J. F. Meyer, D. G. Furchtgott, and L. T. Wu, "Performability evaluation of the SIFT computer," *IEEE Trans. Comput.,* vol. C-29, pp. 501-509, June 1980.

[28] L. T. Wu and J. F. Meyer, "Phased models for evaluating the performability of computing systems," in *1979 Proc. Johns Hopkins Conf. Information Sciences and Systems,* Baltimore, MD, Mar. 1979, pp. 426-431.

John F. Meyer (M'60-SM'71) received the B.S. degree from the University of Michigan, Ann Arbor, the M.S. degree from Stanford University, Stanford, CA, and the Ph.D. degree in communication sciences, also from the University of Michigan, in 1957, 1958, and 1967, respectively.

He is currently a Professor in the Department of Electrical and Computer Engineering and the Department of Computer and Communication Sciences, University of Michigan. He is also associated with their graduate program in computer, information, and control engineering and is a member of the Systems Engineering Laboratory. In addition to his university affiliations, he is a consultant to several firms. During the past 20 years, he has been active in computer research and has published widely in the areas of system modeling and fault-tolerant computing. In the summer of 1977 he was a Visiting Researcher at the Laboratoire d'Automatique et de d'Analyse des Systèmes, Toulouse, France, and in 1975, during an academic leave, he was affiliated with the Direction de l'Informatique de la Société Thomson-CSF, Paris. Prior to joining the Michigan faculty in 1967, he was a Research Engineer at the California Institute of Technology Jet Propulsion Laboratory where his contributions included the first patent issued to the National Aeronautics and Space Administration.

Dr. Meyer is a member of Sigma Xi, Tau Beta Pi, Eta Kappa Nu, the Association for Computing Machinery, and the American Association for the Advancement of Science. In the IEEE Computer Society, he served as Chairman of the Technical Committee on Fault-Tolerant Computing from 1976-1979. He is also a member of the Publications Committee and has served as a Guest Editor of the IEEE TRANSACTIONS ON COMPUTERS.

Reprinted from *IEEE Transactions on Computers*, Volume C-34, Number 2, February 1985, pages 101–1109.

Evaluating Response Time in a Faulty Distributed Computing System

HECTOR GARCIA-MOLINA AND JACK KENT

Abstract —**This paper presents an evaluation technique which is useful for studying both the performance and the reliability of a distributed computing system. The distributed system is evaluated from the point of view of a user who submits a request for service. Our technique computes the average time to successful completion of this request, taking into account the system failures or repairs which may occur before the request is completed. Given a model of the system and its failures, the performance–reliability measures are computed in an automatic numerical fashion. The technique is computationally intensive, so it is limited to relatively small systems. However, it can produce results for many interesting cases without an inordinate amount of computation.**

Index Terms —**Distributed computing system, failure models, graceful degradation, performance evaluation, reliability evaluation, response time.**

I. Introduction

In this paper we present a technique which evaluates, in a combined fashion, the performance and reliability of a distributed computing system. Given a task with known computational requirements, this technique computes the expected time to successfully complete it. The evaluation takes into account crashes, where work is lost, and reconfigurations, where work is suspended.

Classical reliability evaluation techniques study *system* measures such as expected throughput and mean time to failure [6], [9], [15]. Our approach is orthogonal: we study performance and reliability *from the point of view of the user* submitting a particular task to the system. To a user, the response time of his task is just as important as the system measures. This is especially true for lengthy tasks such as long lived database transactions [11] (e.g., to process an insurance claim) or engineering design tasks (e.g., to design and lay out a VLSI chip). Such tasks can take hours or even days to complete, and the impact of failures on their response time can be very significant.

Unlike classical techniques, ours is very sensitive to software crash recovery strategies, and hence is especially well suited for their evaluation. For example, it can be used to study commit protocols, data backup mechanisms, or the frequency of transaction recovery points. Classical measures are not as useful for such comparisons, but of course, they are important for other types of evaluations. Thus, our technique does not replace them, it only complements them.

Manuscript received June 18, 1982; revised August 23, 1983 and July 14, 1984. This work is based upon work supported in part by the National Science Foundation under Grant ECS-8019393.

The authors are with the Department of Electrical Engineering and Computer Science, Princeton University, Princeton, NJ 08544.

Our work is based upon that of Beaudry [1] and Castillo and Siewiorek [4]. We have combined and generalized their failure models, making them more appropriate for distributed computing systems. However, the mathematical analysis of our model is considerably more complex, so we must use numerical methods. These methods are computationally intensive, so a central issue in the numerical analysis we present is the efficient (and accurate) computation of the performance measures. Specifically, we use dynamic programming and careful error bounding to control the computation costs. Even so, our technique is limited to relatively small systems, but we believe it is still useful for many cases of interest.

In the next section, we present our model through a detailed example. In Sections III and IV we outline the analysis, and in Section V we return to the example and give some results. Finally, in Section VI we discuss the complexity of our numerical procedures and make closing comments.

II. The Model

To explain our system model, let us consider the following example. A system has two computers C1 and C2, connected via a very reliable network. Each user communicates directly with one or the other computer. Users must access a database, which, for purposes of reliability, is replicated. Each computer manages one of the copies. Fig. 1 illustrates the system.

Updates to the duplicated database must be synchronized; otherwise the database may become inconsistent or we may end up with "copies" which are different. Notice that the algorithm which performs the updates must function in the face of failures. Many such update algorithms have been suggested in the literature [2], [13], [14].

We assume that only the hardware (i.e., computers C1 and C2) may fail. We do not consider software failures. (These assumptions only apply to the example.)

The first step in our evaluation strategy is to model the system (both hardware and software) by a Markov process [12], [15], to which we add extra information. Fig. 2 shows the model for the system of Fig. 1 running an update algorithm that performs reconfigurations after failures or repairs. State 1 represents the system when both computers are in normal operation. The arcs out of state 1 represent the "rates" at which the system can leave this state. (Transition times are exponentially distributed.) Notice that the rates from state 1 to states 6 and 7 are equal because we are assuming that both

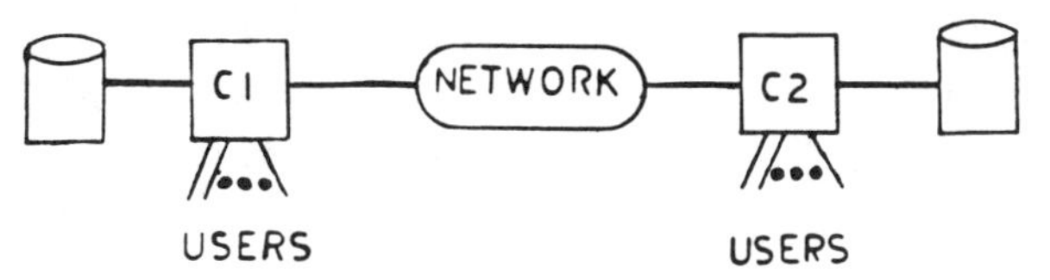

Fig. 1. A replicated database system.

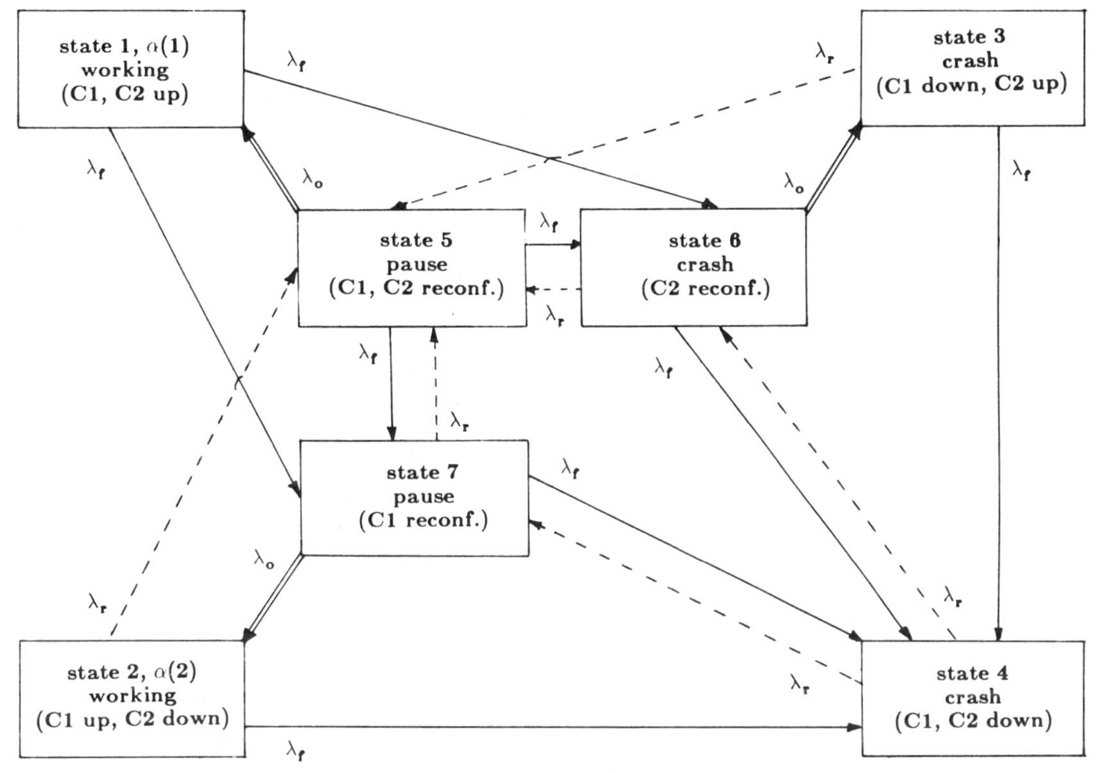

Fig. 2. The model for the system of Fig. 1.

computers fail at the same rate λ_f.

When computer C2 fails, the system does not go into state 2 (C1 = up, C2 = down) but into a reconfiguration state 7 where no updates are processed. During the reconfiguration, C1 may fail or C2 may be repaired, so in state 7 there not only are transitions into state 2 (C1 = up, C2 = down), but also into state 4 (C1 = down, C2 = down) and state 5 (C1 and C2 reconfiguring together). The rate from state 7 to state 2 is the rate at which reconfigurations of this type complete. Keep in mind that to keep the example simple, we have assumed that the network does not fail.

Our technique evaluates the system from the point of view of a user submitting a service request. Thus, in what follows, we must model the system from that same perspective. For example, suppose that the user in question is connected to computer C1 and is submitting a request to update the database (say, a request to accumulate monthly interest in the saving accounts of a bank). From his point of view, we can classify the states of Fig. 2 into pause, crash, and working states. A *pause state* is one in which processing of the user request is temporarily suspended. In our example, we assume that reconfiguration states 5 and 7 are pause states. In them, C1 halts updates, but work performed up to that point is not lost.[1]

[1]For this example, we assume that C1 goes into reconfiguration mode as soon as C2 fails.

States 3, 4, and 6 are similar to the pause states in that computer C1 does not perform user work in these cases. However (in this example), we choose to classify them as *crash states* because we assume that user requests must be restarted from scratch after computer C1 goes down. (When C1 comes up after a crash, it does not know what occurred while it was down. Updates which conflict with its unfinished updates could have been performed during the down time. So, C1 is playing it safe and aborting all unfinished updates. Of course, this is not the only alternative. Different alternatives will yield different performance–reliability models.)

Finally, we classify states 1 and 2 as working states. In a *working state,* computer C1 makes progress on the user requests. Depending on the hardware, the load, the communication delays, and the algorithms used, work on user requests will proceed at different rates in each of the working states. We model this by associating with each state i a work rate or power $\alpha(i)$. This power can be interpreted as follows. Say the user request we are considering takes X units of time to execute on a reference system with no failures, and with no interference from other requests. Then, if the same request is run in our system and no failures occur, it will complete in $X\alpha(i)$ units of time. (Note we are assuming that execution times are deterministic.)

In our example, to arrive at the values for the power of states 1 and 2 we must study the performance of the system in a no failure period. This is not a trivial task, but standard performance evaluation techniques [7] can be used. Notice that the system may have higher power (i.e., smaller α) in a state where some components have failed. For example, in state 2, updates may be performed faster than in state 1. This occurs because in state 1 updates must use a two phase commit protocol [10] which involves at least two rounds of messages between C1 and C2. On the other hand, in state 2 C1 is processing updates by itself, so the messages are not needed.

Given a model like the one of Fig. 2, we can now compute the expected time to complete an update request submitted at C1. Let us suppose that the system is in state 1 and the request takes X s to complete on the reference system. With a certain probability, the system will stay in state 1 more than $X\alpha(1)$ s, and the user request will be completed in $X\alpha(1)$ s. However, several other sequences of events might occur. For example, computer C1 might crash before the user request is completed. Eventually, the computer is repaired and the system returns to state 1. Here, the user request is restarted and manages to complete before any other failure occurs. The sequence of events described will occur with a certain probability. For that sequence, we can also compute the expected time to the completion of the user request, including the wasted time before the crash and the duration of the crash itself. By considering all possible sequences of events, we can compute the expected duration of the user request. (Since there is an infinite number of possible sequences, we will stop our computation when we have considered all sequences which contribute a "significant" amount to the expected value.)

In a similar fashion, we may compute the expected time to successfully execute the request of X s given that the system is initially in state 2. We can also study the expected duration of the same request given that it encountered a failure. This will give us an idea of how much a user will be inconvenienced when a failure does occur. We may also study the performance and reliability of the system from the point of view of users at computer C2. But before we can do this, we must build a new model.

In summary, we believe that the model we have presented summarizes in a convenient way the characteristics of both the system hardware and the software crash recovery strategy. The performance–reliability measures are easy to understand, and as we will see when we return to our example in Section V, they can provide useful insight into the operation of the system.

In closing this section, note that many assumptions are built into our model. Some deal with the particular system being studied (e.g., state 6 is a crash state), but these can easily be changed by altering the model instance used. Other assumptions are inherent to our modeling technique. Of these, many are used in reliability studies (e.g., exponential distributions) and have been discussed extensively in the literature [6], [15]. Of the remaining inherent assumptions, we comment on three important ones.

1) We assume that the time to complete the task in state i is deterministic and equal to $X\alpha(i)$. While there are other ways to model the tasks, ours captures in a very simple way the notion that the task needs to perform a definite amount of work before it can complete. However, if necessary, this approach can be extended as described by [4]: Assume that X can take on values from a finite set $\{x_1, x_2, \cdots, x_m\}$. For each x_j, $f_x(x_j)$ is the probability that X takes on that value. Given that X took on the value x_j, then the user request will take $x_j\alpha(i)$ s to execute in state i. To compute the expected duration of the task, we compute the expected duration of each x_j (using our technique), multiply it by $f_x(x_j)$, and add the results.

2) We assume that nonworking states are either pause or crash states, although in some only a fraction of the outstanding tasks may be aborted. This situation can be modeled by splitting such states into two: one crash and one pause state. The rate into the crash state will be proportional to the fraction of tasks that are aborted.

3) We assume that the power of a state $\alpha(i)$ is independent of events in the past. However, in some cases this may not be valid. For example, in Fig. 2, the power of state 1 may be different after a crash because backlogged tasks increase the system load. This effect can be approximated by adding a new state 1′ to model this high load period. Arcs coming into state 1 would be rerouted to 1′, and a new arc would lead from 1′ to 1.

III. Reduction to a Work Graph

To compute the expected duration of a task, we proceed in two steps. In the first step, we reduce the system model to a *work graph*. This graph summarizes all the information that is necessary to compute our measure. In the second step, the work graph is analyzed and the final results are obtained. This section describes the first step, while Section IV covers the second one.

The work graph contains only the working states of the original model. Each arc represents a possible transition between work states. In a transition, the system may go through pause and crash states, but not through other work states. Each arc in the work graph from node i to j is labeled with the following four items.

1) $r_p(i,j)$ = the probability of going from state i to state j in the original model without visiting a crash state.

2) $t_p(i,j)$ = the expected time to reach state j after having left state i, given that no crash states were visited in the transition.

3) $r_c(i,j)$ = the probability of going from state i to state j in the original model and visiting at least one crash state.

4) $t_c(i,j)$ = the expected time to reach state j after having left state i, given that a crash state was visited in the transition.

Fig. 3 shows the work graph corresponding to the model of Fig. 2 for the indicated sample values. For example, if the system is in state 1 and there is a failure, there is a 0.489 probability that the next working state is 2, and that it is reached without a crash. (For instance, the system could go from 1 to 7 to 2, or 1, 7, 5, 7, 2, or many other possibilities.) The expected time for this type of transition is 0.01 time units. Similarly, with probability 0.013, the failure could take the system to state 2 via a crash state. The expected time for this is 0.736 units.

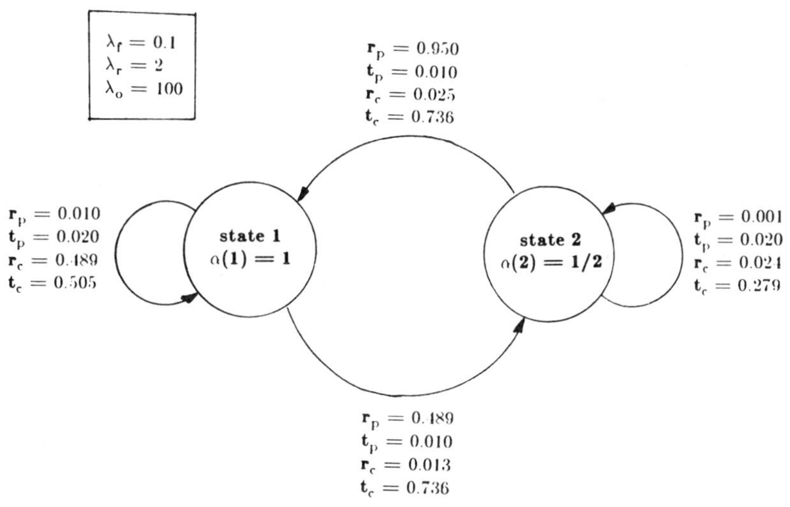

Fig. 3. The work graph for the model of Fig. 2.

The values on the labels can be computed with standard Markov chain techniques. The key idea is to create an intermediate system model that contains two copies of the original model. If the system is in a state in the first copy, no crashes have occurred since the last working state, but if it is in the second copy, a crash has taken place. Arcs coming into crash state j in the first copy are changed so that they lead to state j in the second copy (thus recording that a crash occurred).

To compute the work graph arcs out of node k, we proceed as follows. Working states in the intermediate model (both copies) are made absorbing. A new start state k_s is added, with appropriate arcs to the other states. Now, $r_p(k,j)$ is the absorption probability of state j in the first copy (i.e., the probability of going from k_s to j without crashes). Similarly, $r_c(k,j)$ is the absorption probability of state j in the second copy; $t_p(k,j)$ is the mean absorption time (or first passage time) to j in the first copy; and $t_c(k,j)$ is the mean absorption time to j in the second copy. All these values can easily be computed (e.g., [12], [15]).

To illustrate, consider the simple model of Fig. 4(a). (We use this model and not the one of Fig. 2 because it illustrates the computation of the work graph more clearly.) Fig. 4(b) shows the corresponding intermediate model. For each state k in the original model, its first copy is labeled k_p, and its second k_c. The new start states for both working states are included.

IV. The Analysis

To compute the expected duration of a task (requiring X computing units) we consider the sequences of events that may lead to its completion. For each sequence, we compute the probability that it occurs and its expected duration. These values are multiplied and added for all sequences, giving the final result.

Due to space limitations, we cannot give a full description of this analysis here. Instead, we only sketch the key ideas and refer the reader to [8] for the details and proofs.

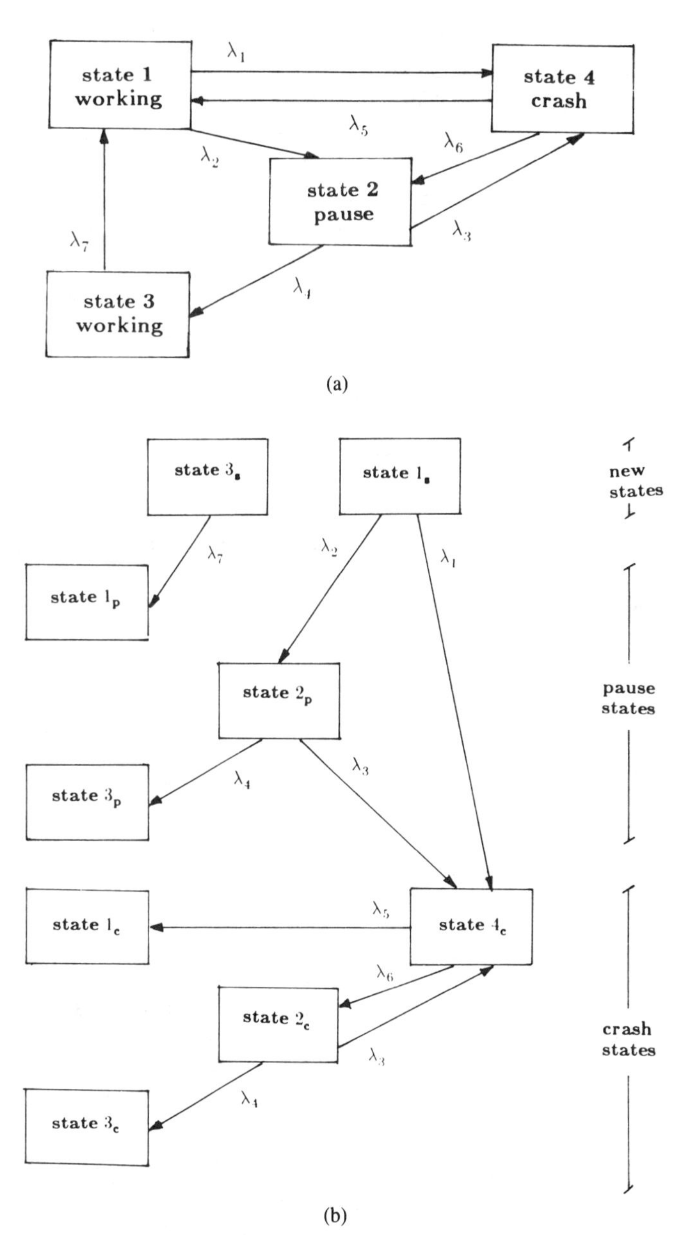

Fig. 4. (a) A simple model. (b) Its intermediate model for computing work graph.

A. The Elementary Subsequences

Fig. 5(a) shows a typical sequence of events we want to consider. In this diagram, we represent the execution history of a user request. On the left-hand side the beginning time of the request is indicated by an "S" (start). On the right-hand side the request is successfully completed at "F" (finish). In between, a number of interruptions may occur. These interruptions are indicated by a "P" if a pause occurred, and by a "C" if a crash occurred. Over a period where work was performed, we indicate the state the system was in.

When a crash occurs, the user request must be restarted from scratch. This means that the events which follow a crash are independent of the previous history of the request. Thus, the analysis of the sequence can be broken up into the analysis of the subsequences between crashes. For example, the probability that the entire sequence takes place is the probability that the events leading up to the first crash occur, times the probability that the events between the first and second crash occur, and so on.

There are only two types of subsequences from which all other sequences are built. These subsequences are shown in Fig. 5(b) and (c). In the first of these subsequences, the user request is started (or restarted after a crash) in state s_1, a number of pauses occur, and finally a crash occurs before the user request can complete. The second subsequence is similar, except that the user request is completed before a crash interferes.

In the rest of this section, we describe how the probability and the duration of the sequence of Fig. 5(b) can be obtained. The analysis of the one in Fig. 5(c) is slightly more complex but similar, and is not covered here.

Let $s_1, s_2, \cdots, s_m$ be the states in the sequence, and let Y_i be the time that the system spent in state s_i. The Y_i are exponentially distributed with mean $1/\lambda(s_i)$ where $\lambda(s_i)$ is the total rate out of state s_i in the original system model. If $\alpha(s_i)$ is the power of the system in state s_i, then $Y_i/\alpha(s_i)$ was the amount of work performed on the user request in state s_i. Since the user request needing X s on a reference system did not finish in this subsequence, then

$$\frac{Y_1}{\alpha(s_1)} + \frac{Y_2}{\alpha(s_2)} + \cdots \frac{Y_m}{\alpha(s_m)} < X. \tag{1}$$

Let us call the probability that this occurs F_m. That is,

$$F_m = \Pr\left[\frac{Y_1}{\alpha(s_1)} + \frac{Y_2}{\alpha(s_2)} + \cdots + \frac{Y_m}{\alpha(s_m)} < X\right]. \tag{2}$$

Notice that F_m is really a function of $s_1, s_2, \cdots, s_m$, and not just of m. The probability that the subsequence occurs is F_m times the probability that the system goes to states $s_1, s_2 \cdots s_m$ when the state transitions occur. (Recall that when the system leaves state s_j, state s_{j+1} is just one of the possible working states the system may go to.) So,

$$\Pr\left[\begin{array}{c}\text{subsequence of}\\ \text{Fig. 5(b) occurs}\end{array}\right] = F_m \prod_{j=1}^{m} r_p(s_{j-1}, s_j). \tag{3}$$

Recall that $r_p(s_{j-1}, s_j)$ is given in the work graph. We can

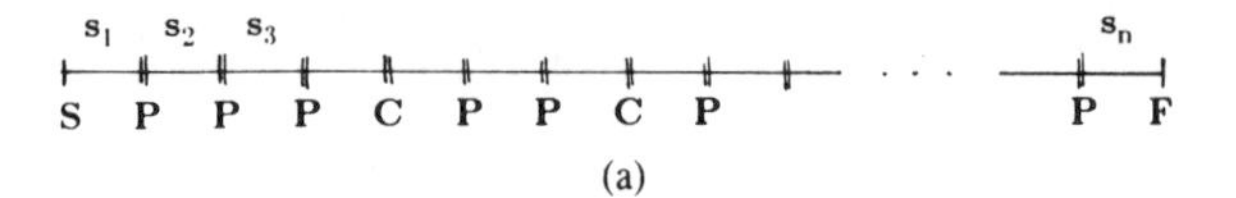

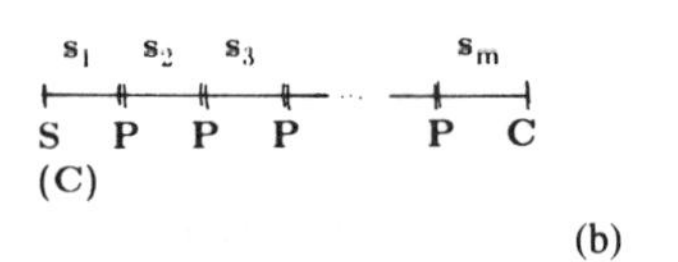

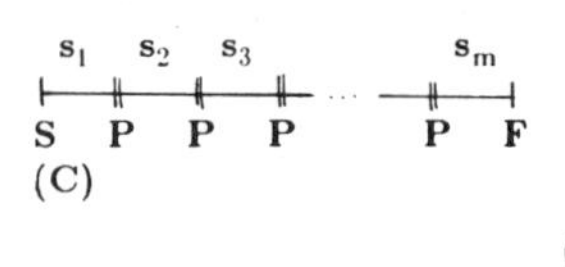

Fig. 5. (a) A sequence of events. (b) First elementary subsequence. (c) Second elementary subsequence.

interpret $r_p(s_0, s_1)$ as the probability that the system is in state s_1 when the subsequence begins.

The value of F_m can be computed as

$$F_m = \int_0^{\theta_1}\int_0^{\theta_2} \cdots \int_0^{\theta_m} f_{Y_i}(y_i) \cdots f_{Y_m}(y_m)\, dy_m \cdots dy_1 \tag{4}$$

where $f_{Y_j}(y_j)$ is the probability density function of Y_j, and

$$\theta_j = \alpha(s_j)\left(X - \frac{y_1}{\alpha(s_1)} - \frac{y_2}{\alpha(s_2)} - \cdots - \frac{y_{j-1}}{\alpha(s_{j-1})}\right). \tag{5}$$

(The value θ_j is the maximum value Y_j can take without the user request being completed, given that Y_k took on a value y_k, $1 \le k < j$.)

It is possible to obtain a closed form expression for the integrals of (4). However, the closed form expression contains differences of similar exponential terms, and is hence susceptible to numerical errors in evaluation. Furthermore, some of the expressions which follow will be evaluated numerically, so we will use numerical evaluations exclusively. By substituting the Taylor series expansion for $f_{Y_j}(y_j)$ in (4) and integrating, we find that

$$F_m = \sum_{i_1=0}^{\infty} \cdots \sum_{i_m=0}^{\infty} \frac{(-1)^m}{(i_1 + \cdots + i_m + m)!} \prod_{j=1}^{m} [-\phi_j]^{i_j+1} \tag{6}$$

where

$$\phi_j = \lambda(s_j)\alpha(s_j)X. \tag{7}$$

(See [8] for details.)

The expected duration of the subsequence of Fig. 5(b), given that the subsequence occurred, is given by

$$\sum_{j=1}^{m-1} t_p(s_j, s_{j+1}) + E[Y_1 + Y_2 + \cdots + Y_m \mid \text{sequence 5(b) occurred}]. \tag{8}$$

Again, $t_p(s_j, s_{j+1})$ is obtained from the work graph.

To compute the expected value in (8), we need the joint probability density function of $Y_1, Y_2, \cdots, Y_m$, conditioned by the fact that the subsequence occurred. This is simply $f_{Y_1}(y_1) \cdots f_{Y_m}(y_m)$ divided by the probability that (1) holds. Therefore, we can compute the expectation as

$$E[Y_1 + \cdots + Y_m \mid \text{sequence 5(b) occurred}] = \frac{C_m}{F_m} \quad (9)$$

where

$$C_m = \int_0^{\theta_1} \cdots \int_0^{\theta_m} (y_1 + \cdots + y_m) f_{Y_1}(y_1) \cdots f_{Y_m}(y_m)\, dy_m \cdots dy_1 . \quad (10)$$

(θ_j is defined by (5). Again, C_m is really a function of the states $s_1, s_2, \cdots, s_m$ and not just of m.) It can be shown [8] that C_m is given by

$$C_m = \sum_{i_1=0}^{\infty} \cdots \sum_{i_m=0}^{\infty} \frac{(-1)^m X[\alpha(s_1)(i_1 + 1) + \cdots + \alpha(s_m)(i_m + 1)]}{(i_1 + i_2 + \cdots + i_m + m + 1)!} \prod_{j=1}^{m} [-\phi_j]^{i+1} \quad (11)$$

[where ϕ_j is defined by (7)].

Thus, we have seen that the analysis of the subsequence of Fig. 5(b) reduces to the computation of two numbers: F_m and C_m. Similarly, the analysis of the second subsequence involves two numbers, which for reference we call G_m and D_m. These numbers must be computed for many sequences, so it is crucial to evaluate them efficiently.

B. Efficient Evaluation

The key observation is that when we need F_m, G_m, C_m, and D_m, we will have already computed similar values for the subsequence of $m - 1$ states $s_1, s_2, \cdots, s_{m-1}$. Thus, we can use any previously obtained results to speed up our computations. (Using previous values is called dynamic programming.)

For example, to compute F_m, we note that it can be written as

$$F_m = \phi_1\phi_2 \cdots \phi_m \sum_{j=0}^{\infty} (-1)^{j-m} f(m,j) \quad (12)$$

where $f(m,j)$ are the terms of (6) with divisor $j!$. These terms can be computed recursively as follows:

For $m = 1$: $f(1,0) = 0$

$f(1,1) = 1$

$f(1,j) = \phi_1 f(1,j-1)/j, \quad \text{for } j \geq 2;$

For $m \geq 2$: $f(m,j) = 0, \quad \text{for } j < m$

$$f(m,j) = \frac{\phi_m f(m,j-1) + f(m-1,j-1)}{j}, \quad \text{for } j \geq m .$$

Thus, if the terms for F_{m-1} are known, the ones for F_m are easy to find, and computing F_m reduces to the evaluation of the single sum in (12). The only remaining problems are how to truncate the series, and how to evaluate the resulting error term.

Since the term $f(m,j)$ has a divisor $j!$, the terms in the summation of (12) will eventually start decreasing in magnitude. Furthermore, the terms are alternating in sign. This means that once the terms start decreasing, we can truncate the series, and the error will be less than the magnitude of the first dropped term [5].

The tricky point is deciding where the terms start decreasing in magnitude. If all the ϕ's are less than 1, then it is possible to show that the terms always decrease in magnitude as j increases. However, if some of the ϕ's are greater than 1, the terms may grow and shrink several times as j grows before reaching the point where they start decreasing monotonically. However, this last point can be identified without too much effort and the series can be truncated. This is discussed in [8].

The evaluation of C_m, G_m, and D_m is similar and is not discussed here.

C. Computing the Performance–Reliability Measures

Let E_j be the expected time for successful completion of the user request, given that the request was started when the system was in working state j. Let us assume that we need E_j for all working states j. (This simplifies the description of the computations.)

Our computation strategy centers on a procedure which we call VISIT. It takes as input a start state i, and a sequence of events S like the one in Fig. 5(a). The procedure computes the total contribution to E_i of S, and of all other sequences that start with S but go on to additional pauses or crashes.

For example, consider the work graph from our example (Fig. 3). To compute E_1, we simple call VISIT with initial state 1 and a null sequence. It computes the probability that the task completes in state 1 (using G_1 and D_1), the expected time for this, and accumulates the product in a global variable. Next, VISIT considers longer sequences that start in state 1. The key observation is that these can be computed with recursive calls to VISIT. For instance, to evaluate sequences of the form "1 → pause → 2 → · · ·" or "1 → pause → 1 → · · ·", we simply call VISIT with that sequence of events.

Sequences of the form "1 → crash → j → · · ·" could also be evaluated recursively, but this is not necessary. The contribution of these sequences is

$$P(\Delta + E_j) \quad (13)$$

where P is the probability of "1 → crash → j" (computable from F_1), and Δ is the expected duration of "1 → crash" (computable from C_1). Of course, E_1 and E_2 are not known yet, so procedure VISIT, instead of directly giving the desired values, will produce a system of linear equations that must be solved later. In return for this, we cut the number of recursive calls to VISIT roughly in half.

Procedure VISIT cannot go on calling itself indefinitely, so at some point the calls must stop. As the sequences of events get longer and longer, their probability decreases and their contribution to the E_j's becomes negligible. Thus, each time VISIT is called, it computes a bound for its contributions. If the bound is small enough, VISIT terminates immediately. The derivation of this bound is not trivial, and is discussed in [8].

Also notice that VISIT must track the errors introduced in the computations, and must guarantee that they do not exceed some predetermined limit. The errors arise when the series for F_m, G_m, C_m, and D_m are truncated, and when VISIT decides not to evaluate a sequence of events. The error handling details are also discussed in [8].

In addition to the expected durations E_j, other measures of interest may be computed. For example, the expected duration given that a failure occurred can be easily computed as follows. The probability that the user request completes in the initial state $s_1 = j$ without interruptions is $e^{-\phi_1}$ (where $\phi_1 = \lambda(s_1)\alpha(s_1)X$) and the contribution of this case to E_j is $\alpha(s_1)X$. Therefore, the expected duration of the user request, given that the initial state was $s_1 = j$ and that a failure (or repair) occurred, is given by

$$\frac{E_j - \alpha(s_1)Xe^{-\phi_1}}{1 - e^{-\phi_1}}. \tag{14}$$

In many cases this measure will be more sensitive than E_j to the performance of the crash recovery algorithms because it measures the impact of failures on the users which are directly affected by the failures.

Another possibility is to study the individual sequences of events that may take place. For example, the computational procedures may also report each high probability sequence that is considered, its probability of occurring, its expected duration, and its contribution to the total expected duration of the user request. These data can be viewed as a "distribution" for the duration of the user request, and may provide the algorithm and system designer with valuable insight. For example, with these data the most common sequences of failures and events can be identified, and it may then be possible to optimize the crash recovery algorithms based on this.

V. The Example Revisited

In this section, we give a brief example to illustrate the use of our evaluation technique and the results it can produce. Suppose that we wish to compare two alternatives for managing two copies of a database in a distributed computing system. The first alternative is the one discussed in Section II, and is described by the model of Fig. 2. The second alternative is not to have reconfigurations after crashes. In this case, updates to the database will only be processed when both computers are up. A crash of either computer causes a suspension of all updates, but no work is lost. (In contrast with the first alternative, a recovering computer now knows that no conflicting updates were performed while it was down, so it does not have to cancel its own pending updates.) Fig. 6(a) shows the model for this alternative. In this case, regardless of the location of the users, all states except 1 are pause states.

Let us assume that after studying the hardware and the algorithms, we come up with the following values.

1) The average time between computer failures is 10 h ($= 1/\lambda_f$).

2) The average time to repair a computer is 0.5 h ($= 1/\lambda_r$).

3) The average reconfiguration time is 0.01 h ($= 1/\lambda_o$).

4) In both alternatives, the power of state 1 is $\alpha(1) = 1$.

5) In the reconfiguration alternative, $\alpha(2) = 0.5$.

Fig. 6(b) shows the work graph for the second alternative, assuming that the user requests are to be run on computer C1. Fig. 3, presented earlier, shows the work graph for the reconfiguration alternative for these same values.

Table I presents some of the results that can be obtained by

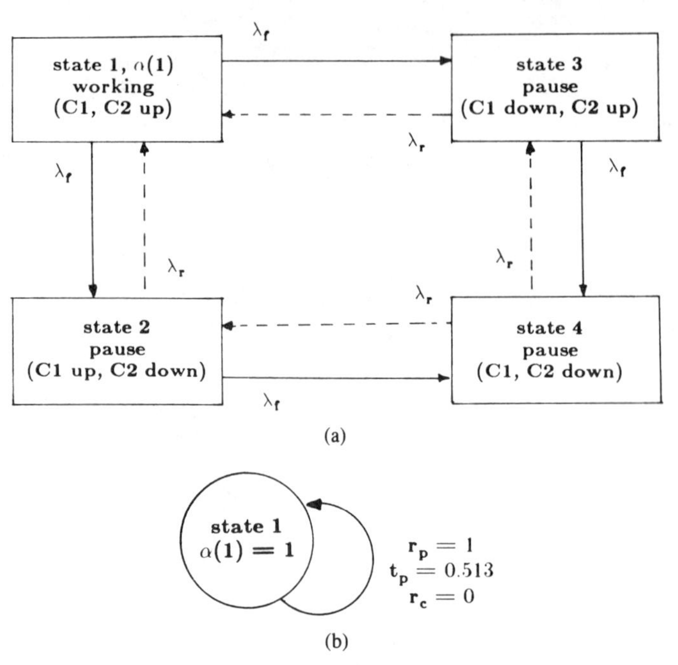

Fig. 6. (a) The model for the no reconfigurations alternative. (b) The work graph for the model of (a).

analyzing the two work graphs. We assume that the task in question is long, so we look at values of X in the range of a few hours. (Recall that X is the time it takes to execute the user request on the reference system.) As expected, as X decreases, the expected time to complete the request approaches $\alpha(i)X$ where i is the starting node. As X grows and the probability of encountering a failure grows, the expected duration increases. (With the reconfiguration alternative, the requests started when the system is in state 2 take less time to execute because of the reduced overhead for processing update requests; see Section II.) Notice that for small X, the reconfiguration alternative is better because work can be performed while computer C2 is down. However, as X grows, the no reconfiguration strategy is best. In this case, the benefits of performing work when C2 is down are out weighed by the risk of losing work when C1 crashes.

Table I also gives the expected duration of a user request given that a failure occurs. As expected, this measure is more sensitive to variations in crash recovery strategies. It is especially interesting to note the case for $X = 0.25$ (i.e., 15 min). Even with this relatively short task, the difference in expected execution times (given a failure) is significant: users submitting such tasks will be much less annoyed by a failure in a reconfiguration system. This means that our evalution technique can yield interesting results even for short tasks.

The results of Table I are only a small fraction of the interesting measures that can be computed for comparing the two crash recovery strategies. For example, we can also compute the expected duration of requests started when the system is in a state different than 1. We should also study the response time of other types of requests that will be processed (e.g., read-only data accesses). Finally, as discussed in the introduction, to make a fair and complete comparison of the alternatives, we should also investigate other measures like the throughput of the system, the steady state probabilities, and the simplicity of the algorithms.

TABLE I
EXPECTED DURATION OF USER REQUESTS (ALL ERROR BOUNDS LESS THAN 1 PERCENT)

Computing Requirements X	Duration With Reconfigurations			Duration Without Reconfigurations	
	starting in state 1	starting in state 2	starting in state 1 given failure occurred	starting in state 1	starting in state 1 given failure occurred
0.25	0.265	0.151	0.553	0.276	0.775
0.50	0.534	0.331	0.852	0.551	1.04
0.75	0.807	0.536	1.16	0.827	1.30
1.00	1.08	0.761	1.47	1.10	1.57
3.00	3.54	3.06	4.19	3.31	3.68
5.00	6.49	5.99	7.36	5.51	5.81
8.00	12.1	11.6	13.2	8.82	9.03

VI. CONCLUSIONS

There are several potential problems with the technique we have described. First, we do not know whether the original Markov model is an accurate representation of the distributed system. Unfortunately, this question pertains to all models, and only an implementation will provide an answer.

A second problem is that the system may have very many states, making the Markov model difficult to construct. In our example there were only two components that could fail and the model had seven states. But as the number of components increases, the number of states can grow very fast. This is a common problem in reliability modeling, and several strategies like lumping states together have been proposed [3], [15]. In our specific case, there are two factors which ameliorate this problem.

1) We do not wish to model an entire distributed system (which could have hundreds of processors and communication channels), but only to model the system from the point of view of a user request. Most requests will only interact with a small number of system components, and it is these components we wish to model. Thus, we believe that for many user requests, the "system" model will be of manageable size.

2) We are mainly interested in comparing crash recovery strategies. Even if we are limited to comparing these strategies in "small" systems, we will still be able to draw interesting conclusions.

A third potential problem is the computational complexity of the evaluation technique. The procedure is computationally intensive, again limiting us to relatively "small" models. But as we have just argued, we believe that this is an acceptable limitation.

To see just how "small" the model must be, we can study the complexity of our procedure. It is not hard to see that the conversion to the work graph will take on the order of

$$|A|\,|NA|^3$$

operations where $|A|$ is the number of absorbing states in the modified model (equal to 2 times the number of working states in the original model) and $|NA|$ is the number of nonabsorbing states (equal to 2 times the number of nonworking states).

Unfortunately, the computational requirements of the last step of our technique are greater and also harder to evaluate. This step is a depth-first search of the work graph, so the number of steps (i.e., calls to procedure VISIT) will depend on the branching factor of the work graph and the depth of the search. If N is the number of *working* states, the branching factor can be as high as N. (Keep in mind that the number of working states is considerably smaller than the number of states in the model.) However, in most cases the branching factor will effectively be smaller than N because the probabilities associated with some arcs will be very small. For example, in a system with five components, the probability of going from the state where all components are in operation to the state where only one component works will typically be very small. Thus, from the point of view of the search, the node for all components working will effectively have less than five outgoing arcs. The depth of the search depends on the desired error bound, as well as on the value of the ϕ's ($\phi_i = \lambda(i)\alpha(i)X$). Within each step, we must compute F_n, G_n, C_n, and D_n. The number of iterations required for these computations again depends on the desired error bound and the ϕ values.

If all the ϕ's are less than 1, then the series converge rapidly, and relatively large work graphs can be analyzed. (This occurs when for all working states, the time needed to service the user request in that state is less than the average time between failure or repair events in that state. This will occur, for example, in transaction processing systems where requests are short.) As the time required by a request grows, some of the ϕ's will become greater than 1, making it more time consuming to compute F_n, G_n, C_n, D_n and increasing the depth.

To illustrate the computational requirements of the search step, we briefly present execution time statistics for some sample work graphs. We looked at the problem of managing N copies of a database under the same assumptions and with a similar strategy to the one discussed in Section II. For each value of N, the model has $4N - 1$ states, of which N are working states. Working state i, $1 \leq i \leq N$, represents the system when $i - 1$ computers (other than C1) have failed. We assume that working state i has $\alpha(i) = 0.5(N - i + 1)$, to be compatible with Fig. 2. Using the values of λ_f, λ_r, and λ_o of Section V, we produced a family of work graphs with N nodes each.

Fig. 7 shows the running time of the search step (as a

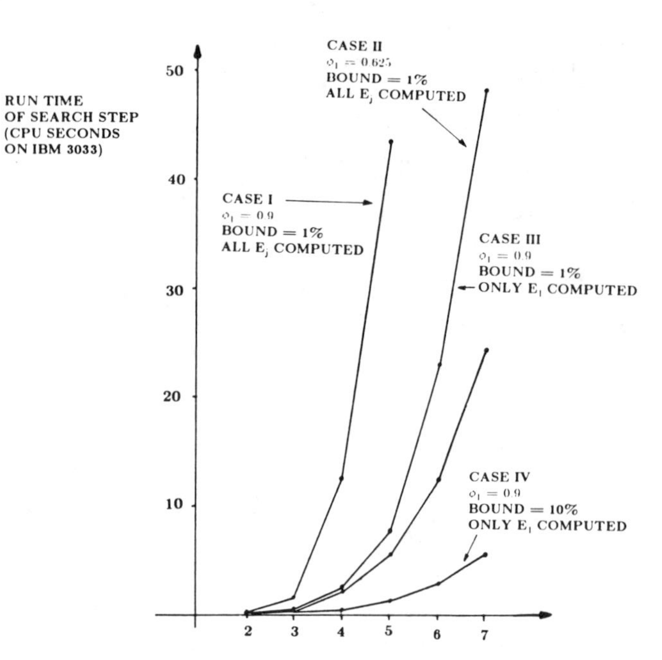

Fig. 7. Running time of search.

function of N) for four cases. These times are on an IBM 3033 computer. Case I is for value of $\phi_1 = 0.9$ and an error bound of one percent. (Since the work graphs have different values of $\lambda(1)$ and $\alpha(1)$, we show our results for constant ϕ_1 and not for constant X.) When $\phi_1 = 0.90$, the user request takes almost as long to complete in state 1 (with no failures) as the average time between failures. Thus, we do not expect to be interested in values of ϕ_1 much beyond 0.90. (Notice that all other ϕ's in this case have values larger than one.)

As the value of ϕ_1 (and the other ϕ's) drops, the execution times drop rapidly. Fig. 7 case II shows the times when $\phi_1 = 0.625$ and the error bound is one percent (but all other ϕ's are still greater than 1).

One of the reasons why execution times increase with N is that in each case we are computing N measures, E_1–E_N. If we are only interested in one E_j, we can reduce the computation by roughly a factor of N. This is because the measures for the other working states can be computed with less accuracy. Fig. 7 case III illustrates this for the case where ϕ_1 is 0.90. Another way to reduce the execution time is to reduce the desired error bound. Case IV of Fig. 7 shows the running times for the previous case where the error bound has been dropped to 10 percent.

These examples show that our technique has some clear limitations, but that it can nevertheless be used to study some interesting "small" systems. We believe that the results obtained can be helpful in the evaluation and comparison of crash recovery strategies, and we are currently using this technique to study distributed database algorithms.

Acknowledgment

Several useful suggestions and comments were provided by R. Cordon, S. Davidson, D. Skeen, and the referees.

References

[1] M. D. Beaudry, "Performance-related reliability measures for computing systems," *IEEE Trans. Comput.*, vol. C-27, pp. 540–547, June 1978.

[2] P. A. Bernstein and N. Goodman, "Concurrency control in distributed database systems," *ACM Comput. Surveys*, vol. 13, no. 2, pp. 185–221, June 1981.

[3] J. A. Buzacott, "Markov approach finding failure times of repairable systems," *IEEE Trans. Reliability*, vol. R-19, no. 4, pp. 128–134, Nov. 1970.

[4] X. Castillo and D. P. Siewiorek, "A performance-reliability model for computing systems," in *Proc. 10th Int. Symp. Fault-Tolerant Computing*, Kyoto, Japan, Oct. 1980, pp. 187–192.

[5] G. Dahlquist and A. Bjorck, *Numerical Methods.* Englewood Cliffs, NJ: Prentice-Hall, 1974.

[6] B. S. Dhillon and C. Singh, *Engineering Reliability, New Techniques and Applications.* New York: Wiley, 1981.

[7] D. Ferrari, *Computer Systems Performance Evaluation.* Englewood Cliffs, NJ: Prentice-Hall, 1978.

[8] H. Garcia-Molina, "A technique for evaluating the performance and reliability of distributed computing systems," Dep. Elec. Eng. Comput. Sci., Princeton Univ., Princeton, NJ, Tech. Rep. 291, Dec. 1981.

[9] F. A. Gay and M. L. Ketelsen, "Performance evaluation for gracefully degrading systems," in *Proc. 9th Int. Symp. Fault-Tolerant Computing*, Madison, WI, June 1979, pp. 51–58.

[10] J. N. Gray, "Notes on database operating systems," in *Operating Systems: An Advanced Course*, R. Bayer *et al.*, Eds. Berlin: Springer-Verlag, 1979, pp. 393–481.

[11] ——, "The transaction concept: Virtues and limitations," in *Proc. 7th VLDB Conf.*, Sept. 1981, pp. 144–154.

[12] P. G. Hoel, S. C. Port, and C. J. Stone, *Introduction to Stochastic Processes.* Boston, MA: Houghton Mifflin, 1972.

[13] W. H. Kohler, "A survey of techniques for synchronization and recovery in decentralized computer systems," *ACM Comput. Surveys*, vol. 13, no. 2, pp. 149–183, June 1981.

[14] J. B. Rothnie and N. Goodman, "A survey of research and development in distributed database management," in *Proc. 3rd VLDB Conf.*, Tokyo, Japan, 1977, pp. 48–62.

[15] C. Singh and R. Billinton, *System Reliability Modeling and Evaluation.* London: Hutchinson, 1977.

Hector Garcia-Molina received the B.S. degree in electrical engineering from the Instituto Tecnologico de Monterrey, Mexico, in 1974, the M.S. degree in electrical engineering from Stanford University, Stanford, CA, in 1975, and the Ph.D. degree in computer science, also from Stanford University, in 1979.

He is an Assistant Professor in the Department of Electrical Engineering and Computer Science, Princeton University, Princeton, NJ. His research interests include distributed computing systems and database systems.

Dr. Garcia-Molina is a member of ACM and the IEEE Computer Society.

Jack Kent received the B.S. degree in computer studies from Northwestern University, Evanston, IL, in 1980, and the M.S. degree in computer science from Princeton University, Princeton, NJ, in 1982, where he is currently a graduate student.

His research interests include database systems and operating systems.

Reprinted from *IEEE Transactions on Communications*, Volume COM-34, Number 6, June 1986, pages 564–568.

Reliability Analysis for Integrated Networks with Application to Burst Switching

PETER KUBAT

Abstract—**System performance and reliability are jointly assessed for highly reliable telecommunication or computer networks. The underlying model assumes that at most a small number of components can be "down" at a time, and that the average repair/replacement time of a failed component is small when compared to the average uptimes of network components. At steady state the system is assumed to follow a regenerative stochastic process.**

This methodology is used to evaluate highly distributed voice/data integrated networks. The performance measure selected for this application is the traffic loss rate.

INTRODUCTION

TRADITIONALLY, the reliability of communication networks has been measured in terms of network connectivity. Connectivity measures such as the probability that the network is connected, the fraction of connected nodes, and similar measures were considered (for more details see Ball [1] and references therein). The calculation of such measures is very complex [2], yet for networks with many highly reliable components (nodes, links) it can be expected that the resulting value will be extremely close to a given limit. For instance, the probability of a network being connected converges to 1 as the number of components grows, provided the reliability of an individual component is not too low and the network is not of some regular structure, e.g., ring, star, or similar [3]–[6]. This confirms the empirical observation that for a network with a large number of components, node separation caused by component failure is rare. The network may still fail, however, because of other causes.

A more general definition of network failure may be as follows: we say that the network fails if its performance reaches a point outside the acceptable performance limits. For networks which operate close to their capacity, even the failure of a single component may be crucial. Although the failed component may not cause connectivity failure, it may cause serious delays, blocking of calls/messages, buffer overflow, loss of traffic, or other unacceptable performance, thus causing the network to fail.

Research efforts to assess reliability using performance measures other than connectivity are rare. For example, Bonaventura *et al.* [7] and Meyer [8], [9] enumerated all the possible states of a computer system, calculated the steady-state probabilities and system performance in each state, and then obtained the overall performance as defined by the average system performance in the steady state. The large (exponential) size of the state space, its intractability, and its inherent technical complications in calculating the steady-state probabilities sometimes make this approach difficult to use. In a recent paper, Li and Silvester [10] avoided this pitfall by considering the average performance across the most probable states only.

In this paper, we develop a methodology to assess the reliability and performance of communication networks. The networks are assumed to have highly reliable components, so multiple failures are unlikely to occur; after repair, the network is returned to its fully operational state and will be "as good as new"—i.e., all components will return to perfect condition.

This methodology is applied to burst switching—a new technology developed for switching digitized voice and data traffic. Burst-switched networks are characterized by highly dispersed small switches, dispersed control, and improved bandwidth efficiencies (for details see Amstutz [11]).

The measure of performance considered in this application is the traffic loss rate. The advantage of this measure is its ability to accommodate both voice and data traffic and also its ability to effectively evaluate traffic loss rate due to a node or link failure.

THE MODEL—ASSUMPTIONS AND ANALYSIS

Consider a network with C components (i.e., N nodes and L links, $C = N + L$). The components are highly reliable. From time to time, however, they fail. Sometimes even a group of components may fail simultaneously. When a component fails, a repair/replacement procedure is initiated. After repair, the component returns to full function and is assumed to be "as good as new." For the sake of a tractable analysis, we will make the following assumptions.

- The network components are highly reliable; component uptimes are large (order of magnitude: years or months, say).
- The repair times are much shorter than the uptimes (order of magnitude: days or hours).
- The call/packet/message interarrival times and processing times are much shorter than the repair times (order of magnitude: seconds or milliseconds).
- The network is in a steady state; when a failure occurs, the traffic flow in the remaining working components reaches equilibrium quickly (order of magnitude for the time to reach stationarity: seconds or minutes).

In practice, these assumptions are not very restrictive and apply to a large number of real networks.

Since the downtimes are assumed to be relatively short when compared to the uptimes, we may therefore assume, for simplicity, that the probability of another component(s) failure during the downtime is zero. More precisely, we assume that a component failure (or simultaneous failure of a group of components) prevents any future failures until the completion of the repair. There are M distinct failure types (states), each state having a nonzero probability of occurrence.

Under these assumptions, the behavior of such a network can be described by a stochastic process $X(t)$. In this process, the process state will oscillate between the fully operational state (state 0) and a "repair" or "down" state i, in which the system is "down" because a failure of type i has occurred. After repair, the process always returns to the state 0 (Fig. 1).

Furthermore, let us assume that U_i, the time to the i-type

Paper approved by the Editor for Voice/Data Networks of the IEEE Communications Society. Manuscript received December 14, 1984; revised July 29, 1985.

The author is with the Telecommunications Research Laboratory, GTE Laboratories, Inc., Waltham, MA 02254.

IEEE Log Number 8608519.

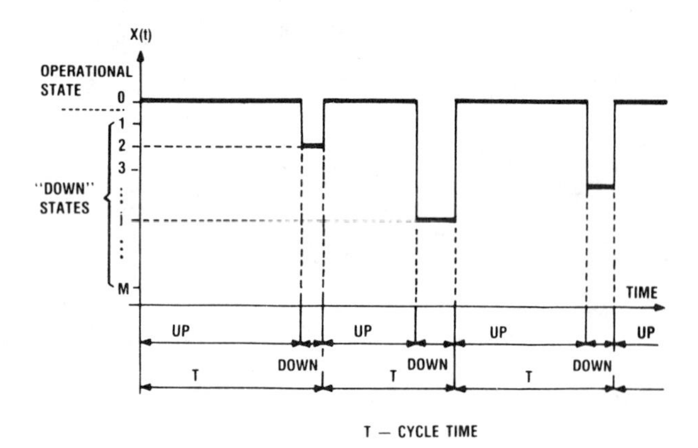

Fig. 1. A realization of the process $X(t)$.

failure (uptime), is exponentially distributed with mean $1/\lambda_i$; $i = 1, \cdots, M$, and all the uptimes are independent. The repair time (downtime) D_i for the ith type failure may have a general distribution with mean $E[D_i]$.

Consequently, the times at which the system becomes fully operational, i.e., $X(t)$ changes from state i to state 0, can be regarded as regeneration points of the process. At these points, the process (probabilistically) restarts itself. An up–down cycle or cycle time T is defined as the time between successive regeneration points. Let T_0 be the time spent in state 0 during a cycle. Since $T_0 = \min U_i$, the mean uptime is

$$E[T_0]=E[\min_{1\leq i\leq M} U_i]=\int_0^{\infty}\left(\prod_{i=1}^{M} e^{-\lambda_i t}\right) dt=\Lambda^{-1} \quad (1)$$

where

$$\Lambda=\sum_{i=1}^{M} \lambda_i. \quad (2)$$

The probability that the process $X(t)$ will jump from state 0 to state i during the cycle is λ_i/Λ. The expected cycle time is then

$$E[T]=\Lambda^{-1}+(\Lambda^{-1})\sum_{j=1}^{M} \rho_j \quad (3)$$

where

$$\rho_j=\lambda_j E[D_j], \qquad j=1, \cdots, M. \quad (4)$$

Let T_j be the time spent in state j during a cycle; then

$$E[T_j]=\frac{\lambda_j}{\Lambda} E[D_j]=\rho_j/\Lambda, \qquad j=1, \cdots, M. \quad (5)$$

Following Ross [12, p. 84] we get the steady-state probabilities for the process:

$$p_j=\lim_{t\to\infty} P\{X(t)=j\}$$
$$=E[T_j]/E[T]=\rho_j p_0, \qquad j=1, \cdots, M \quad (6)$$

where

$$p_0=1/\left(1+\sum_{j=1}^{M} \rho_j\right). \quad (7)$$

When the process $X(t)$ is in state i, its performance (reward) is r_i, measured as a rate of units/time. Thus, $X(t)$ is an alternating renewal process with rewards.

It follows directly from Theorem 3.6.1 in Ross [12, p. 78] that the average performance (reward) per unit time, denoted by AR, is

$$\text{AR}=\frac{E\text{ [reward during a cycle]}}{E\text{ [cycle time]}}$$
$$=\frac{\frac{r_0}{\Lambda}+\sum_{i=1}^{M}\frac{\lambda_i}{\Lambda} r_i E[D_i]}{\frac{1}{\Lambda}+\sum_{i=1}^{M}\frac{\lambda_i}{\Lambda} E[D_i]}=\frac{r_0+\sum_{i=1}^{M}\rho_i r_i}{1+\sum_{i=1}^{M}\rho_i}. \quad (8)$$

It may be of interest to note that in a rather different setting, a formula similar to (8) was obtained by Keilson [13].

We will use AR, the average reward per unit time, as a joint measure of performance and reliability for our system. AR can also be written as

$$\text{AR}=\sum_{i=0}^{M} r_i p_i. \quad (9)$$

Application to Burst Switching

Burst switching [11], [14], [15] is a recently developed method for switching digitized voice and data characters in an integrated way. It is characterized by highly dispersed small switches, dispersed control, and improved bandwidth efficiencies. A burst switch consists of many small "link switches" (each with up to 16 voice/data ports) and 0, 1, or more high-capacity "hub switches." The interconnections between switches are called links. Each link group may handle up to 16 link switches, and the hub switch may interface with up to 256 link groups. The data and voice are switched through the same circuits in the same way. However, while data are switched in the same way as voice, burst switching does differentiate between data and voice. Burst switching switches voice samples at higher priority than data, so that a voice burst may have the first chance at resources in the case of contention. This minimizes the loss of voice samples (clipping). Data characters will be buffered, and thus no data samples are lost, although their delivery will be delayed. A typical burst switch with one hub and a number of link groups is shown in Fig. 2.

In a good network design it is advantageous to interconnect the switches with at least two links so that if a link fails, the traffic can be rerouted. Since we assume that only one component can fail at a time, we will suppose, for simplicity, that the connecting links never fail. If, however, for a given network configuration we cannot assume this, the model may be adjusted to accommodate unreliable links (for instance, treat links in a manner similar to nodes). The switches can fail, and we assume that at most one switch fails at a time. The reliability of ports and voice/data terminals will not be considered in assessing the overall reliability of the network.

Traffic rate, measured either in bits/unit time or bursts/unit time, is used here to illustrate how one can jointly assess network performance and reliability. Other measures are, of course, possible and can be tailored to a particular network scenario.

A. Burst Switch with One Hub Only

Consider a burst switch configuration similar to the one in Fig. 2 with N link switches and one hub. Define:

State 0	fully operational state, all switches are "up"
State i	link switch i is "down," $i=1, \cdots, N$
State $N+1$	hub switch ($N+1$) is "down"
λ_i	failure rate of switch i

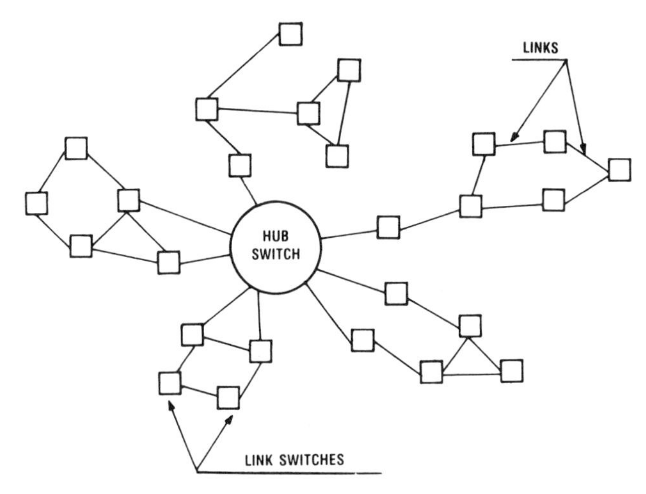

Fig. 2. A typical burst switch layout.

$E[D_i]$	mean downtime of switch i, $i=1, \cdots, N+1$
V_i	input traffic rate for voice (kbits/s) in node i, $i=1, \cdots, N$
W_i	input traffic rate for data (kbits/s) in node i, $i=1, \cdots, N$
f_{ij}	fraction of voice traffic originated in i which will terminate in j
g_{ij}	fraction of data traffic originated in i which will terminate in j
I_{ij} =	$\begin{cases} 1 & \text{if the traffic originated in } i \text{ must go to link switch } j \text{ via hub} \\ 0 & \text{otherwise} \end{cases}$
A_i	indirect loss of voice traffic rate due to failure of switch i $(i=1, \cdots, N+1)$
B_i	indirect loss of data traffic rate due to failure of switch i $(i=1, \cdots, N+1)$
A_0	loss of voice traffic rate in a fully operational network
B_0	loss of data traffic rate in a fully operational network.

By the direct traffic loss due to failure of switch i, we will understand the traffic which should have been originated in i and the loss of traffic originated in node j $(j \neq i)$ which has been designated for i. By the indirect traffic loss due to failure of switch i, we mean all other traffic loss other than direct loss, such as transit loss, loss of traffic (in other nodes) due to congestion, buffer overflow, clipping, etc. In the fully operational network, all traffic losses are due to congestion, buffer overflow, or clipping, etc. With this notation, the direct traffic loss rate due to link switch failure is

$$V_i+W_i+\sum_{\substack{k=1\\k\neq i}}^{N}(V_kf_{ki}+W_kg_{ki}) \tag{10}$$

and indirect loss is $A_i + B_i$. Thus, we define the loss r_i associated with the failure of link switch i as the total traffic loss associated with ith link switch failure, i.e.,

$$r_i=A_i+B_i+V_i+W_i+\sum_{\substack{k=1\\k\neq i}}^{N}(V_kf_{ki}+W_kg_{ki}). \tag{11}$$

When the hub switch fails, the associated loss is

$$r_{N+1}=A_{N+1}+B_{N+1}+\sum_{i=1}^{N}\sum_{j\neq i}^{N}(V_if_{ij}+W_ig_{ij})I_{ij}. \tag{12}$$

Note that, since no terminals are directly connected to the hub, the traffic loss r_{N+1} represents, in fact, the indirect traffic (in this case transit) loss rate. It follows from (9) that the overall traffic loss rate at equilibrium is

$$\begin{aligned}\mathrm{AR}&=\sum_{i=0}^{N+1}r_ip_i\\&=\sum_{i=1}^{N}\left[A_i+B_i+V_i+W_i+\sum_{\substack{k=l\\k\neq i}}^{N}(V_kf_{ki}+W_kg_{ki})\right]p_i\\&\quad+\left[A_{N+1}+B_{N+1}+\sum_i\sum_{j\neq i}(V_if_{ij}+W_ig_{ij})I_{ij}\right]p_{N+1}\\&\quad+(A_0+B_0)p_0.\end{aligned} \tag{13}$$

A performance measure closely related to AR but dimensionless is a traffic efficiency (TE) defined as

$$\mathrm{TE}=\frac{\text{total traffic}-\text{traffic loss}}{\text{total traffic}}=1-\frac{\mathrm{AR}}{\text{total traffic}}. \tag{14}$$

B. Burst Switch with One Hub and One (or More) Outside Line(s)

Consider a layout with $N - 1$ link switches, as shown in Fig. 3. There is a certain fraction of traffic coming from outside and going out via the hub. The outside line may (or may not) fail. For modeling purposes, the outside line can be terminated by a "dummy" link switch (denoted by N), and the analysis can proceed as in the previous section. If there is more than one outside line, each line can be replaced by a "dummy" link switch.

Sometimes we will be interested in the "effective outgoing traffic rate." Assume, for simplicity, that we can ignore the indirect losses A_i and B_i of traffic due to the node i failure. In this case, the average traffic loss for the outside line (node N) is

$$\sum_{i=1}^{N-1}(V_if_{iN}+W_ig_{iN})p_i \tag{15}$$

and thus, the effective outgoing traffic rate is

$$\begin{aligned}\mathrm{EOT}&=\sum_{i=1}^{N-1}(V_if_{iN}+W_ig_{iN})-\sum_{i=1}^{N-1}(V_if_{iN}+W_ig_{iN})p_i\\&=\sum_{i=1}^{N-1}(V_if_{iN}+W_ig_{iN})(1-p_i).\end{aligned} \tag{16}$$

C. Burst Switch with More than One Hub

Consider a burst switch layout, as shown in Fig. 4. When the number of hub and link switches is relatively small, we can adapt the method discussed in detail in previous subsections to this network. However, when the total number of link and hub switches is large, the possibility of simultaneous multiple failures must be taken into consideration. One way to accommodate this is via decomposition. More specifically, cut each line connecting two hubs and add a dummy link switch to the end of each half-line as shown in Fig. 5. Now, the network has been broken down into smaller one-hub subnets. Each one-hub subnet can be solved as we have shown above, provided we supply outside traffic rates. These outside rates can be approximated at first; however, in steady state the rate IN and OUT must be equal. Thus, a number of iterations are required to satisfy this requirement. The total traffic loss rate for the entire network is then the sum of the subnetworks' traffic loss rates.

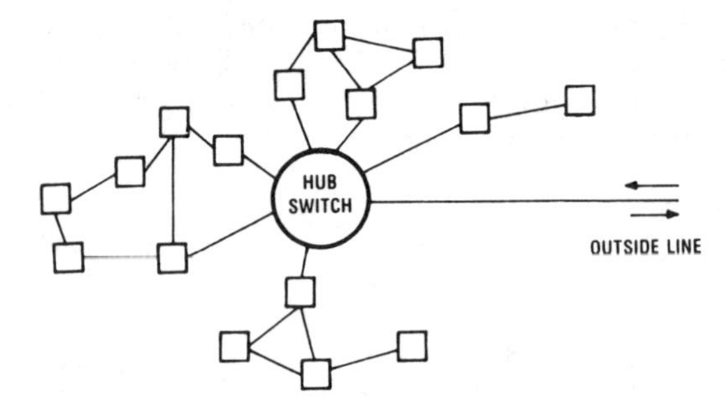

Fig. 3. A burst switch with one hub and one outside line.

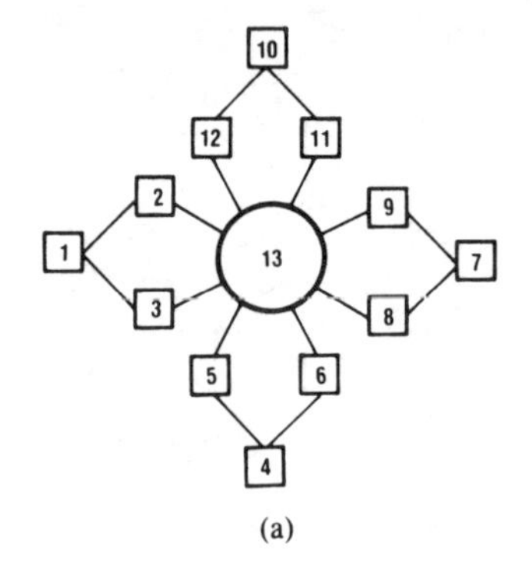

(a)

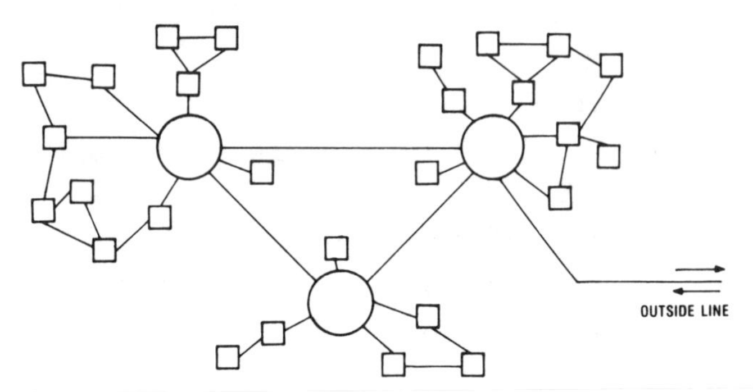

Fig. 4. A burst switch with three hubs and one outside line.

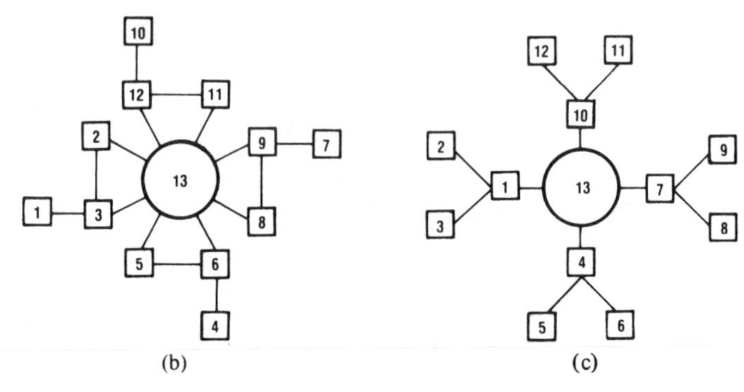

(b) (c)

Fig. 6. Network layout for Examples 1, 2, and 3.

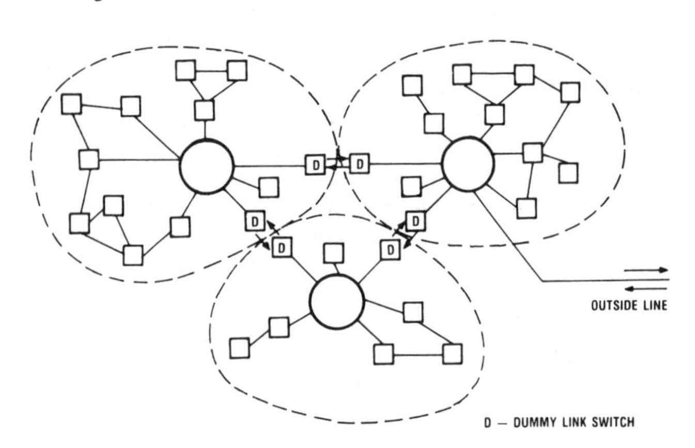

Fig. 5. Decomposition of a burst switched network.

EXAMPLES

Consider a simple burst-switched network with one hub and 12 link switches, as shown in Fig. 6. For simplicity, assume only one type of traffic, say data.

Example 1: Assume network architecture as in Fig. 6(a) with the relevant parameters given in Table I. The routing matrix is given in Table II.

In this example, nodes i = 1, 4, 7, 10 may represent nodes where host computers are located. The other nodes have only terminals. The hub performs switching functions only.

We have that

$$\Lambda = \sum_{i=1}^{13} \lambda_i = 33$$

$$p_0 = 1/\left(1 + \sum_{i=1}^{13} \rho_i\right)$$

$$= 1/[1 + 4(0.0109) + 8(0.0246) + 0.0027] = 0.8044$$

$$p_i = p_0\rho_i = \begin{cases} 0.0088 & \text{for } i = 1,\ 4,\ 7,\ 10 \\ 0.0198 & \text{for } i = 2,\ 3,\ 5,\ 6,\ 8,\ 9,\ 11,\ 12 \\ 0.0022 & \text{for } i = 13. \end{cases}$$

TABLE I
INPUT DATA FOR EXAMPLES 1, 2, AND 3

Switch i	λ_i (1/year)	$E[D_i]$ years	W_i (kb/s)	ϱ_i
1,4,7,10	2	2/365	300	0.0109
2,3,5,6,8,9, 11,12	3	3/365	10	0.0246
13	1	1/365	—	0.0027

TABLE II
ROUTING MATRIX FOR EXAMPLES 1, 2, AND 3

DESTINATIONS (columns j); SOURCES (rows i)

i \ j	1	2	3	4	5	6	7	8	9	10	11	12
1	5	3	3	1	1	1	1	1	1	1	1	1
2	11	0	0	1	1	1	1	1	1	1	1	1
3	11	0	0	1	1	1	1	1	1	1	1	1
4	1	1	1	5	3	3	1	1	1	1	1	1
5	1	1	1	11	0	0	1	1	1	1	1	1
6	1	1	1	11	0	0	1	1	1	1	1	1
7	1	1	1	1	1	1	5	3	3	1	1	1
8	1	1	1	1	1	1	11	0	0	1	1	1
9	1	1	1	1	1	1	11	0	0	1	1	1
10	1	1	1	1	1	1	1	1	1	5	3	3
11	1	1	1	1	1	1	1	1	1	11	0	0
12	1	1	1	1	1	1	1	1	1	11	0	0

The entries $a_{ij} = 20\, g_{ij}$, i.e., to get routing frequencies divide entries by 20, e.g.:

$g_{5,4} = 11/20$
$g_{7,8} = 3/20$
$g_{10,4} = 1/20$

If switch i, $i = 1, 4, 7, 10$, is down, then it follows from (11) that the traffic loss is

$$300+2\frac{11}{20}10+6\frac{1}{20}10+3\frac{1}{20}300=359 \text{ kbits/s}.$$

If switch i, $i = 2, 3, 5, 6, 8, 9, 11, 12$, is down, then the traffic loss is

$$10+\frac{3}{20}300+3\frac{1}{20}300+6\frac{1}{20}10=103 \text{ kbits/s}.$$

Finally, if the hub is down ($i = 13$), use (12); we find that we lose

$$4\frac{9}{20}300+2\frac{9}{20}10=576 \text{ kbits/s}$$

of the traffic.

In this example, we have $B_i = 0$ for $i = 1, \cdots, 12$. By exploiting the symmetrical configuration of the network, we can now write the total loss rate as

$$\text{AR}=359(P_1+P_4+P_7+P_{10})+103(P_2+P_3+P_5+P_6+P_8+P_9+P_{11}+P_{12})+576P_{13}=30.219.$$

Thus, the expected loss rate of data traffic in the long run is 30.219 kbits/s. Since the total traffic TT = 4 (300) + 8 (10) = 1280, we find that the traffic efficiency (TE) in this network is TE = 1 − AR/TT = 0.976.

Example 2: Consider a network structure such as the one shown in Fig. 6(b). All relevant parameters will remain the same as in the previous example, and thus, the steady-state probabilities will also remain the same. Note, however, that if node 3 fails it will cut off the traffic to and from node 1 as well. Thus, the indirect loss of traffic due to the node 3 failure is

$$B_3=(12/20)(300)+(11/20)(10)+(3/20)(300)+(6/20)(10)=234.5.$$

By indirect traffic we mean the traffic which does not initiate or terminate in node 3 but must pass through node 3 in order to reach its destination node. By symmetry $B_3 = B_6 = B_9 = B_{12} = 234.5$. By substituting B_i's into (13), we find that

$$\text{AR}=48.789 \quad \text{and} \quad \text{TE}=0.962.$$

The increase in the expected loss rate is about 64 percent in comparison with previous network configuration.

Example 3: Consider a network configuration such as the one shown in Fig. 6(c) with all the parameters the same as before. The indirect loss of traffic associated with a failure of node 1, 4, 7, or 10 is

$$B_1=B_4=B_7=B_{10}=2\cdot(9/20)(10)+3\cdot(2/20)(300)+6\cdot(2/20)(10)=105.$$

This gives AR = 33.915 and TE = 0.9735. We may notice, perhaps with surprise, that this network layout is better than the layout of Example 2. This is because nodes 1, 4, 7, and 10 are responsible for the bulk of the traffic and are not cut off from the network by failure of another link switch, as in Example 2. For instance, the failure of node 1 causes relatively small transit traffic loss for nodes 2 and 3. On the other hand, failure of node 3 in Example 2 will interrupt a major flow of traffic to and from node 1.

Summary

In this paper we discussed some issues pertaining to evaluation of reliability and/or performance of communication and computer networks.

Based on regenerative and alternating processes, it has been shown how a joint performance measure incorporating both reliability and network performance can be established. Using "traffic rate" as a performance criterion, the method has been successfully applied to assess the performance of burst switching networks and to evaluate alternative network scenarios.

The choice of this performance measure is clearly critical for this application. Other performance measures (e.g., traffic delay, call blocking probabilities, etc.) will be treated in subsequent reports.

The proposed method is very general and adapts easily to various network architectures, environments, and performance criteria.

Acknowledgment

I would like to thank J. Gechter, P. O'Reilly, and A. W. Pierce of GTE Labs, and Prof. J. Keilson of the University of Rochester for many valuable discussions and comments during the preparation of this paper.

References

[1] M. O. Ball, "Computing network reliability," *Oper. Res.*, vol. 17, pp. 823–838, July–Aug. 1979.

[2] ——, "An overview of the computational complexity of network reliability analysis," College Bus. Management, Univ. Maryland, College Park, Working Paper Series MS/S 84-010, 1984.

[3] P. Erdos and A. Renyi, "On random graphs, I," *Publ. Math.*, vol. 6, pp. 290–297, 1959.

[4] ——, "On the strength of connectedness of a random graph," *Acta Math., Acad. Sci. Hung.*, vol. 12, pp. 261–267, 1961.

[5] A. K. Kel'mans, "Some problems of network reliability analysis," *Automat. Remote Contr.*, vol. 26, pp. 564–573, 1965.

[6] ——, "Connectivity of a probabilistic network," *Automat. Remote Contr.*, vol. 28, pp. 444–460, 1967.

[7] V. Bonaventura *et al.*, "Service availability of communication networks," in *Proc. Nat. Telecommun. Conf.*, 1980, pp. 15.2.1–15.2.6.

[8] J. F. Meyer, "On evaluating the performability of degradable computing systems," *IEEE Trans. Comput.*, vol. C-29, pp. 720–731, Aug. 1980.

[9] ——, "Closed-form solutions of performability," *IEEE Trans. Comput.*, vol. C-31, pp. 648–657, July 1982.

[10] V. O. K. Li and J. A. Silvester, "Performance analysis of networks with unreliable components," *IEEE Trans. Commun.*, vol. COM-32, pp. 1105–1110, Oct. 1984.

[11] S. R. Amstutz, "Burst switching—An introduction," *IEEE Commun. Mag.*, vol. 21, pp. 36–42, Nov. 1983.

[12] S. M. Ross, *Stochastic Processes*. New York: Wiley, 1983.

[13] J. Keilson, "A simple algorithm for contract acceptance," *Opsearch*, vol. 7, pp. 157–166, 1970.

[14] F. Haselton, "A PCM frame switching concept leading to burst switching network architecture," *IEEE Commun. Mag.*, vol. 21, pp. 13–19, Sept. 1983.

[15] J. F. Haughney, "Application of burst-switching technology to the defense communications system," *IEEE Commun. Mag.*, vol. 22, pp. 15–21, Oct. 1984.

★

Peter Kubat received the Graduate Diploma in mathematics from Charles University, Prague, Czechoslovakia, in 1969, and the D.Sc. degree in management science from the Technion—Israel Institute of Technology, Haifa, in 1976.

From 1976 to 1978 he visited the Department of Mathematics, University of Toronto, Toronto, Ont., Canada, and from 1978 to 1984 he was a faculty member at the Graduate School of Management, University of Rochester, Rochester, NY. In 1984, he joined the Network Architecture Department of the Telecommunications Research Laboratory, GTE Laboratories Inc., Waltham, MA, and is currently engaged in the investigation of reliability for communication networks and fault-tolerant systems. His research interests include applied stochastic processes, operations management, reliability, communication networks, and mathematical models of computer systems. He has published over 30 papers and technical reports.

Dr. Kubat is a member of the Operations Research Society of America and The Institute of Management Sciences.

On a Class of Integrated Performance/Reliability Models based on Queueing Networks

Olaf Schoen

Institut für Informatik
Technische Universität München

Abstract

Important interdependencies exist between the flow of jobs and the reliability of computing systems. This paper describes a method based on queueing networks for obtaining integrated performance/reliability models which consider these interdependencies.
The underlying queueing networks of BCMP type are briefly described. Errors and failures are classified by their duration and their origin. This leads to failure mechanisms, which allow the description of failure rates increasing with workload, even by second order. By this important reliability measures can be calculated depending on the load. Furthermore the influence of failures on the task flow can be modeled by extending a BCMP task flow model by including recovery of affected jobs. Some of the failure mechanisms lead to extended models which are no longer BCMP tractable. An iterative method to evaluate these models is described. Finally the restrictions of our method are discussed.

1. Introduction

The general procedure in the analysis of systems is to approach aspects of performance and reliability separately. There is, however, an implicit relationship in the tools applied to their evaluation, i.e. simulation and analytical techniques based on Markovian processes are used in both areas (see [1]!). More importantly, there are strong and important interdependencies between the reliability and performance of a system (Fig. 1).
The conventional approach considers the task flow as a function of the system and its load. The reliability depends on the structure of the system and on spontaneous fault processes resulting in errors (and failures) of the system. An error is an incorrect state of a component of the system. A more detailed analysis must consider that system errors may also be generated by the load, e.g. software components might be damaged by false access. System errors lead to incorrect task execution only if the component in which the error occurred is used by a task. An incorrect task execution is called failure. In section 2 we will describe a load-dependent reliability model which considers these effects. From user perspective we consider failures most important and use the term transient failure, for example, as a failure due to a transient error.
Failures reduce the net throughput and require recovery and rerun of tasks. Thus the tasks affected by errors have longer sojourn times, and as by rerun of tasks the system utilization increases, even the sojourn times of non-affected tasks are extended. Phenomena like these can be modeled without diffi-

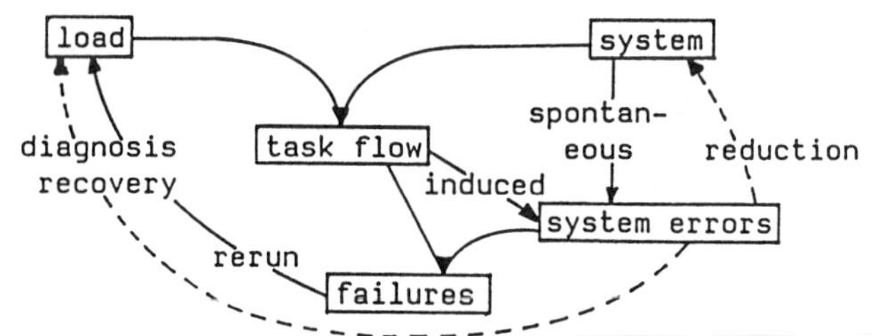

Figure 1: Interdependencies of performance and reliability, (illustrates a simultaneous influence)

culty, as we will show in section 3.
Modeling countermeasures in fault-tolerant systems is more difficult. When failures (and errors) are detected, they initiate countermeasures which increase the load on the system through diagnosis, organizing reconfiguration, and rerun times. Permanent errors require the restauration of the error-free system (e.g. replacing damaged software by error-free copies). More often, however, errors require reconfiguration by the exclusion of defective resources.
There are some recent studies concerning interdependencies of performance and reliability and performance/reliability modeling. [2] and [3] are concerned with statistical analyses of the influence of system load on system failures and reveal a significant second order effect.
[4] - [7] develop and compare some probability distributions in order to describe failure processes due to transient hardware errors and software errors. A doubly stochastic cyclostationary Poisson-distribution in which the dependency of failures on some load indicators is explicitely considered describes the failure processes most exactly.
There are a lot of studies ([8] - [18]) concerning performance/reliability ("performability", see [13]) of Gracefully Degrading Systems (GDS). Usually this is done by estimating the performance of each configuration of the system and by considering these configurations as states of a Markovian process with transitions due to failures (and repairs). These studies do not consider other interdependencies between performance and reliability than that one implemented in GDS, except [11] where failure rates are modeled by second order functions of system load.
A first approach to integrated performance/reliability models is described in [19] where the distribution of the sojourn time of a particular task is calculated depending on load and failures (but constant failure rates). In [20] load-dependent values

for MTTF are estimated from a load model. In this model failures are load-dependent and lead to additional requests due to recovery actions. Reliability-dependent performance values cannot be calculated from this model.

Up to now, a general approach for really integrated performance/reliability models cannot be found in literature. This paper (a completely revised version of [21]; a description of the models can also be found in [22]) describes an approach which is based on a well established instrument for performance evaluation: BCMP queueing networks ([23]). In principle, of course, we could construct any integrated model by Markovian processes; we prefer, however, the more tractable BCMP models.

A BCMP net consists of a number N of nodes (service centers) and tasks (often called customers) in R classes. A task of class r that completes service at node i will next require service at node j in class s with a fixed transition probability $p_{i,r;j,s}$. The flow of tasks in this node×class-space can be decomposed into ergodic subchains; a subchain defines for example a type of job running on the modeled system. Tasks cannot change their subchain. Subchains can be open (external Poisson arrivals) or closed (fixed number of tasks in the subchain).

There are four types of nodes including queueing disciplines first-come-first-served (FCFS), processor sharing, infinite server (no waiting) and preemptive-resume last-come-first-served. The service time distributions must have rational Laplace transforms and may be distinct for each class at a node (except FCFS-nodes).

If all the node utilizations are less than one an equilibrium state of the underlying Markovian process exists which has product form. There are efficient algorithms to compute performance values as utilization, throughput, and sojourn time (see [24], e.g.).

For our models we denote a pair (node, class) by the term locus. Furthermore we admit only one service station for any node; thus at most one task can be served at any time at each node.

Example

We shall demonstrate the scope of our models with a major example. We consider a computing system with three nodes, a CPU, a system base disk (SD), and a data base disk (DD), and four classes of tasks. The system is subject to two task streams (open subchains), a batch stream and a transaction stream. The tasks may move in the node×class space as it is shown by fig. 2. The values at the arcs are the transition probabilities and the values in the circles are the mean service demands at the loci. The queueing discipline is processor sharing at the CPU and FCFS at the disks. We consider the system under two different loads given by the job arrival rates λ and leading to sojourn times Y:

λ (1/s)		batch	trans
load	1	0.01	0.2
	2	0.0125	0.5

Y (s)		batch	trans
load	1	95.2	2.3
	2	302.8	5.9

Table 1 shows the utilizations of the loci and nodes of the model under both of the loads. So far our BCMP example model.

2. Modeling load-dependent reliability

We now analyze failures which depend on the task

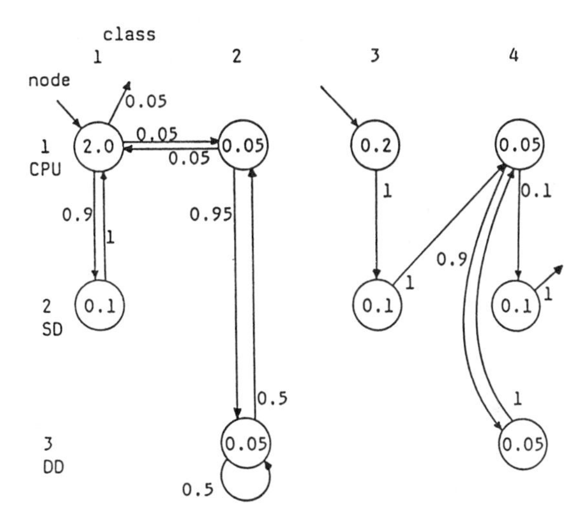

Figure 2: Job flow in the node×class space of the example system

node	load$_1$: class 1	2	3	4	total node	load$_2$: class 1	2	3	4	total node
CPU	0.4	0.01	0.04	0.1	0.55	0.5	0.013	0.1	0.25	0.863
SD	0.018	-	0.02	0.02	0.058	0.023	-	0.05	0.05	0.123
DD	-	0.019	-	0.09	0.109	-	0.024	-	0.225	0.249

Table 1: utilizations of loci and nodes of the example model

flow. We distinguish service dependent and sojourn time dependent failures. Failures depending on the quality and duration of computing phases, storage accesses, and input/output operations belong to the first category; interference causing damage to data stored in main memory, however, belongs to the second category, as this effect will depend on the sojourn time (here: storage time). We begin our analysis by service-dependent failures.

Errors may be transient or permanent. We assume for our models that permanent errors are cleared as soon as they have led to a failure (an incorrect execution of a task). With this assumption, we respect that by BCMP nets only stationary task flow patterns can be evaluated.

Errors may come into existence spontaneously. This is to say that their development and existence is independent of the task execution; but the effect of the error, i.e. the failure, always depends on a task. Spontaneous errors cause spontaneous failures.

Errors also may be induced by tasks, i.e. the execution of the task is the cause of the error. In this case it remains to be distinguished whether the error affects the locus currently held by the causer; if so he himself is subject to the failure, and the error is called self-induced. Otherwise, the causer may induce an error at different loci; we will call these errors cross-induced.

So far we have a classification of failures due to the duration of the underlying errors (transient, permanent) and the cause of their development (spontaneous, self-induced, cross-induced). Every failure (process) will affect tasks at a certain set of loci which is restricted to loci corresponding to the same node of the system. In the case of cross-induced failures one has to specify additionally the set of loci (corresponding to one node) at which tasks cause the errors leading to the failures.

We assign a potential failure rate to every failure, i.e. the maximum rate at which the failure can occur. They are called potential rates since failures can occur only in as much as the loci of the causers and victims of failures are actually occupied by tasks.

The resulting failure rate is derived from a potential failure rate according to the probability that the loci of the victim (and the causer) are occupied. Simplifying, we assume independence of the occupancies. Since the service stations, by assumption, may have at most one task in service at a time we use the utilization as the probability of being occupied.

In the case of self-induced failures, we simply multiply the potential failure rate by the utilization of the locus, since errors are produced with this rate and these necessarily induce a failure of the same locus, regardless of whether the error is transient or permanent.

In the case of spontaneous failures, a potential rate assigned to the locus will generate failures in as much as the locus is occupied by a task, given that the underlying errors are transient. If, however, permanent errors of a set of loci L (corresponding to one node) are to be modeled the errors will lead to a failure in any case. A single locus l $\in L$ is subject to such errors corresponding to its share in the utilization of the set of loci, namely u_l/u_L, u_L the overall utilization of the set of loci.

In the case of cross-induced transient failures, the utilization of the causing as well as of the affected locus is of concern, since a failure depends on the simultaneous occupancy of both loci. In the case of cross-induced permanent failures one must consider - similar to spontaneous permanent failures - that any locus l is affected proportional to its share in the utilization of the whole set of loci L subject to this type of error.

We get the following expressions for the resulting failure rates (denoted by ρ) of a single locus l out of a set of affected loci L:

	transient	permanent
spontaneous	$u_l \cdot \pi_L$	$\frac{u_l}{u_L} \cdot \pi_L$
self-induced	$u_l \cdot \pi_L$	$u_l \cdot \pi_L$
cross-induced	$u_l \cdot (u_K \cdot \pi_{K,L})$	$\frac{u_l}{u_L} \cdot (u_K \cdot \pi_{K,L})$

(1)

where l: locus affected by the failure
L: set of loci affected by the failure
K: set of loci causing the failure
u: utilization
π: potential failure rate.

It is remarkable that the resulting failure rates for cross-induced transient failures increase by second order in case of increasing utilizations!

We now consider those failures whose frequency depends on the sojourn time of the affected jobs. One example is destruction of stored data in a storage medium; this may be effected either by spontaneous or by cross-induced error processes, whereas self-induced errors always can be assigned to some service and therefore can be modeled as above. For the special case of transient defects in storage media we have failure probabilities

$$p_S = Y \cdot n \cdot \pi_S \quad \text{for spontaneous errors, and} \tag{2}$$
$$p_S = Y \cdot n \cdot u_K \cdot \pi_{K,S} \quad \text{for cross-induced errors,}$$

if $p_S \ll 1$, where

p_S : probability that the stored data are corrupted
Y : sojourn time in the storage
n : mean number of storage units used by the job
π_S : potential rate of spontaneous transient errors per unit
u_K : utilization of the set of loci which cause the errors
$\pi_{K,S}$: potential storage error rate caused by jobs at loci of a set K, per storage unit.

Since the sojourn times will increase more than linearly with utilizations, these failures will also increase more than linearly to job arrival rates.

Summary:

Our load-dependent reliability model consists of
- a BCMP performance model which describes the flow of jobs in the system,
- a set of service-dependent failures; every failure is described by
  - its type within the 6 failure mechanisms
  - its potential rate
  - the set of loci which are affected by this failure
  - the set of loci which cause this failure
- sojourn time-dependent failures, described by
  - its type (spontaneous, cross-induced)
  - its potential rate per storage unit

  in the case of cross-induced failures:
  - the set of loci which cause this failure.

Solution of the load-dependent reliability model

The solution procedure for our load-dependent reliability model is very simple:
1. Evaluate the BCMP model in order to estimate the utilization of the loci and nodes of the model;
2. For each failure and each affected locus compute the resulting failure rates according (1) or (2), respectively. Then one can arbitrarily sum the resulting rates for each failure, for each locus, for each node, for each subchain, and for the whole system, respectively.

These failure rates depend on the load, of course. The inverse of the failure rate is the mean time between failures, MTBM. We also can calculate the risk per job

$$p_{job} = P\{\text{job is affected}\} = \frac{\Sigma \rho_{job}}{\lambda_{job}} \quad \text{if } \Sigma\rho_{job} \ll \lambda_{job}$$

where $\Sigma \rho_{job}$: the overall failure rate of the job stream,
λ_{job} : the job stream arrival rate.

We could refine the risk calculation for an individual job by using the individual service times instead of the (mean) utilization.

We could also derive the (load dependent) system availability. To accomplish this, we assume that for any failure there is a mean recovery time specific to this failure, during which the system is not productive, and that there are no further effects of the failure. In this case, with t_f being the mean recovery time after failure f, we have the availability of the system

$$a = 1 - \sum_{f \in F} \rho_f \cdot t_f \qquad \text{if } 1 - a \ll 1,$$

where the summation is over all resulting failure rates ρ_f, $f \in M$. Again, we have neglected that the system might be affected by more than one failure at a time.

Example

We now augment our example model to demonstrate the representability of load-dependent failure rates. We assume (fictive) service-dependent failures with parameters as shown in table 2. Examples for these failures are:

1. recording/sensing errors due to dust particles on the system base disk.
2. IC failures in the CPU.
3. floating point errors.
4. a software error in the routine used at locus (1,1) damages the job status.
5. interference in the case of a simultaneous data transfer from/to the data base disk and the system base disk.
6. the computation phase at locus (1,4) damages the disk driver routine of the system disk.

No.	type	affected loci	causing loci	pot. rate (1/1000h)
1	spontaneous transient	(2,1),(2,3),(2,4)		20.0
2	spontaneous permanent	(1,1),(1,2),(1,3),(1,4)		1.0
3	self-induced transient	(1,3)		40.0
4	self-induced permanent	(1,1)		10.0
5	cross-induced transient	(2,1),(2,3),(2,4)	(3,2),(3,4)	150.0
6	cross-induced permanent	(2,1),(2,3),(2,4)	(1,4)	12.0

Table 2: Parameters of the failures in the example model

The resulting failure rate under $load_1$ for the disk driver error (failure 6) at locus (2,1), for example, is according (1)

$$\rho_{6,(2,1)} = \frac{u_{(2,1)}}{u_{SD}} \cdot (u_{(1,4)} \cdot \pi_6) = 0.37 \cdot 10^{-3} h^{-1}.$$

Table 3 shows the resulting failure rates for each failure and each affected locus under both of the loads.

Additionally we assume sojourn time dependent spontaneous failures in the main memory with a rate of 0.5/1000h per 100 KByte. Batch jobs may occupy 150 KByte and transaction jobs 100 KByte. Then we get the probability p_S that the stored data of one job are corrupted, according (2) and the values of sojourn times:

p_S		batch	trans
load	1	$0.20\ 10^{-4}$	$0.32\ 10^{-6}$
	2	$0.63\ 10^{-4}$	$0.82\ 10^{-6}$

To obtain failure rates for failures in main memory we can multiply the values p_S by the throughput of the subchains:

$$\rho_S = p_{S,job} \cdot \lambda_{job}.$$

So we get:

ρ_S(1/1000h)		batch	trans
load	1	0.72	0.23
	2	2.84	1.48

Respecting service-dependent failures and failures of main memory we get the following overall failure rates:

ρ(1/1000h)		batch	trans	system
load	1	6.49	4.37	10.86
	2	10.315	14.065	24.38

and MTBM values:

MTBM (h)		batch	trans	system
load	1	154.1	228.8	92.1
	2	96.9	71.1	41.0

The risk per job is:

p		batch	trans
load	1	$1.80\ 10^{-4}$	$6.07\ 10^{-6}$
	2	$2.29\ 10^{-4}$	$7.81\ 10^{-6}$

The risk of a job to suffer a failure increases with increasing load because the resulting failure rates increase more than proportionally to the load. So far our example.

No.	node	$load_1$: class 1	2	3	4	total fail.	$load_2$: class 1	2	3	4	total fail.
1	SD	0.36	-	0.4	0.4	1.16	0.46	-	1.0	1.0	2.46
2	CPU	0.73	0.02	0.07	0.1	1.0	0.58	0.015	0.115	0.29	1.0
3	CPU	-	-	1.6	-	1.6	-	-	4.0	-	4.0
4	CPU	4.0	-	-	-	4.0	5.0	-	-	-	5.0
5	SD	0.29	-	0.33	0.33	0.95	0.86	-	1.87	1.87	4.6
6	SD	0.37	-	0.415	0.415	1.2	0.56	-	1.22	1.22	3.0
	Σ=	5.77		4.14		9.91	7.475		12.585		20.06

Table 3: Resulting failure rates of the example model (1/1000h)

3. Modeling failure-dependent load

In section 2 we considered constant recovery times for each failure with no further effects on the task flow. We shall now consider the effect of diagnosis, recovery, and rerun of a single task that was affected by a failure at a locus.

In our node×class space, we will model this effect by transitions into recovery subnets consisting of additional loci which represent diagnosis and recovery of tasks. For this reason we will describe how to estimate the transition probabilities of additional transitions due to failures and how to correct the former transition probabilities out of a locus at which further transitions are inserted. We will consider restrictions for the subnets representing diagnosis and recovery of affected tasks and show how the tasks can continue their execution after passing through these recovery subnets.

The failure probability - i.e. the probability that a task suffers a failure during the expected service time at a locus - can depend either on static or on dynamic model parameters. In the first case we get constant failure probabilities fitting in the BCMP restrictions. This applies to self-induced and spontaneous transient failures because their failure frequencies depend only on the utilization of the affected locus and the potential failure rate. Assuming that the number of failures of a specific type f at one locus l in a time interval is Poisson-distributed we get

$$p_f = P(\{\text{at least one failure at } l \text{ in service time } B_l\})$$

$$p_f = \sum_{j=1}^{\infty} \frac{(\pi \bullet B_1)^j}{j!} \bullet e^{-(\pi \bullet B_1)} \approx \pi \bullet B_1, \text{ for } (\pi \bullet B_1) \ll 1. \quad (3)$$

$(\pi \bullet B_1)$ is a good approximation for the failure probability p_f, because the probability for more than one failure during service time B_1 is very small; in a time interval T we then have the expected number of failures

$$E[\text{number of failures in } T] = \frac{T}{B_1} \bullet u_1 \bullet (\pi \bullet B_1) = u_1 \bullet \pi \bullet T$$

and we get the resulting failure rate $(u_1 \bullet \pi)$.

For the other failure mechanisms we cannot estimate constant failure probabilities, because in these cases they depend on the utilizations of loci other than the affected one or on sojourn times, respectively. In general we get the resulting failure rate ρ as a function of some BCMP-computable performance values. Then we can estimate the failure probabilities p_f to:

$$p_f = \frac{\rho}{u_1} \bullet B_1. \quad (4)$$

These transition probabilities violate the BCMP assumptions but we can evaluate these models by an iterative approach (see below).

So far we have shown how to estimate the transition probabilities due to failures. Since we have added transitions to some loci we have to correct the former transition probabilities such that all transition probabilities from one locus sum to one. This is approximated by summing all failure probabilities at one locus (set F_1) and multiplying the former transition probabilities p_{1k} by the complement of this sum with respect to one:

$$p'_{1k} = p_{1k} \bullet (1 - \sum_{f \in F_1} p_f).$$

This is a good approximation if the failure probabilities are very small compared to one, so that the high order products are almost zero.

Recovery subnets may be of arbitrary complexity; they can consist of zero or more additional loci. If no loci are added the countermeasure due to a failure is represented only by a transition to an arbitrary locus in the same subchain as the affected locus. Additional loci may either correspond to already existing nodes of the system, requiring addition of new classes, or they can correspond to additional nodes e.g. a service processor. In every case it seems to be convenient to provide new classes for recovery subnets to warrant clarity. The mean service times and the service time distributions at the added loci must fit into the rules of BCMP nets.

After an affected task has run through diagnosis and recovery it must return to normal execution. There are four variants for continuation of an affected task:

1. The task continues its execution as if it had not suffered a failure, i.e. it continues from the last locus of the recovery according to the former transitions at the locus where the failure occurred.
2. The task returns to the locus where the failure occurred, i.e. it repeats the computing phase in which it was interrupted. This can be used to model e.g. a CPU retry.
3. The task continues its execution at an arbitrary locus of its subchain. By this we can model a (partial) rerun of the task.
4. The task leaves the system, i.e. the recovery failed and a continuation of the task is not possible. In this case we can estimate the share of tasks which quit due to failures and obtain the lost throughput. This is evident for open subchains but for closed subchains we can insert a transition to that locus where the throughput is measured and estimate the throughput of this transition to get the lost throughput.

These four variants can be used arbitrarily combined.

Example

Returning to our example system, we now introduce recovery actions by additional loci and transitions (see fig. 3). The floating point error may lead into a programmed floating point routine, which is represented by locus (1,5). In this case, the transition probability depends only on static model parameters and is given by (3):

$$q = p_{(1,3);(1,5)} = \pi_3 \bullet E[B_{(1,3)}] = 2.22\ 10^{-6}$$

If by an effect from locus (1,4) the system base disk driver is damaged, the system may recover via a CPU routine which fetches a fresh copy of the driver from the data base disk. Transitions to this recovery take place - with a certain probability - whenever an access to the system base disk is tried; this happens in loci (2,1), (2,3), and (2,4). So we define loci (1,6), (1,7), and (1,8) for the CPU routine and (3,6), (3,7), and (3,8) for the access to the data base disk. Note that the triple recovery representation assures the return to the locus of

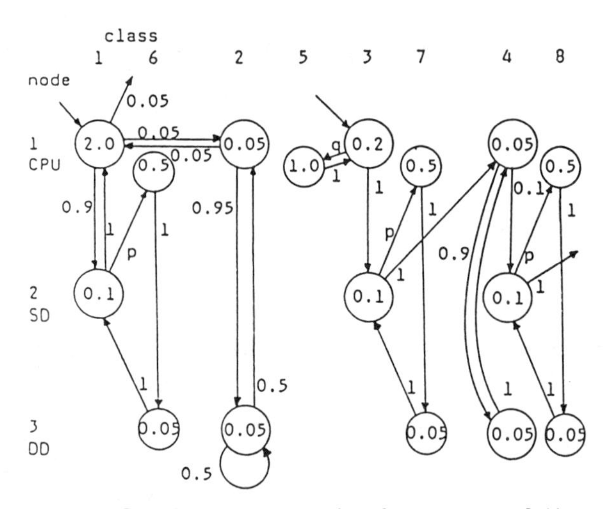

Figure 3: Job flow in the node×class space of the example system after insertion of recovery subnets

origin. The transition probability $p_{(2,1);(1,6)}$, for example, is according (4) for $load_1$

$$p = p_{(2,1);(1,6)} = \frac{\rho_{6,(2,1)}}{u_{(2,1)}} \bullet E[B_{(2,1)}] = 0.57\ 10^{-6}$$

and for $load_2$ $= 0.68\ 10^{-6}$

So we have an example of a transition probability which depends on the job flow and which therefore violates the BCMP assumptions. Figure 3 shows the nodexclass-space of our example model after insertion of these recoveries (The failure probabilities are denoted by q and p, respectively; the correction of the former probabilities is neglected in the figure). The additional load due to recovery from these two failures is too small to have a significant influence on performance values. So far our example.

By an iterative approach (see [22]) one can compute the state probabilities (and the utilizations, throughputs, sojourn times ...) even for models with transition probabilities which depend on BCMP-computable performance values:

```
   {let PR be the transition probabilities into the
   recovery loci}
PR := 0;
repeat evaluate BCMP net;
       compute transition probabilities PR' into the
       recovery;
       modify former transition probabilities
       accordingly;
       exchange PR, PR'
until  |PR - PR'| < ε
```

This algorithm converges rapidly due to the small transition probabilities and the small additional load induced by failures. Evaluations of modified example models have confirmed this. In all cases less than three iterations were necessary.

4. Conclusion

We have a method for modeling and evaluating load-dependent failures and can compute all necessary reliability measures. We can extend this to recovery depending on load-dependent failures. This recovery, however, is restricted to freezing the status during recovery or to a recovery that affects the progress of exactly one job and fits into the BCMP traffic rules, i.e. makes no priorized use of resources. We cannot model recovery actions originated by a single failure which resets a certain class of jobs, nor can we model a recovery which reconfigures the system, i.e. in our model: withdraws some loci, changes transition probabilities, adds further loci, displaces jobs. Approximately, however, the latter case can become tractable in a hyper space, consisting of the finite set of reconfiguration states, each as tractable as our basic model, and inter-reconfiguration-state transitions governed by load-dependent effects as shown above.

References

[1] K.S. Trivedi: *Probability & Statistics with Reliability, Queueing and Computer Science Applications*, Prentice Hall, Englewood Cliffs, 1982.

[2] R.K. Iyer, D.J. Rosetti: *A Statistical Load Dependency Model for CPU Errors at SLAC*, FTCS-12, IEEE, 1982, pp. 363-372.

[3] R.K. Iyer, D.J. Rosetti: *Effect of System Workload on Operating System Reliability: A Study on IBM 3081*, IEEE Trans. Software Eng., vol. SE-11, no. 12, Dec.85, pp. 1438-1448.

[4] S.R. McConnel, D.P. Siewiorek, M.M. Tsao: *The Measurement and Analysis of Transient Errors in Digital Computer Systems*, FTCS-9, IEEE, 1979, pp. 67-70.

[5] X. Castillo: *Workload, Performance, and Reliability of Digital Computing Systems*, FTCS-11, IEEE, 1981, pp. 84-89.

[6] X. Castillo, D.P. Siewiorek: *A Workload Dependent Software Reliability Prediction Model*, FTCS-12, IEEE, 1982, pp. 279-286.

[7] X. Castillo, S.R. McConnel, D.P. Siewiorek: *Derivation and Calibration of a Transient Error Reliability Model*, IEEE Trans. Comp., vol. C-31, no. 7, July 1982, pp. 658-671.

[8] M.D. Beaudry: *Performance-Related Reliability Measures for Computing Systems*, IEEE Trans. Comp., vol. C-27, no. 6, June 1978, pp. 540-547

[9] F.A. Gay, M.L. Ketelsen: *Performance Evaluation for Gracefully Degrading Systems*, FTCS-9, IEEE, 1979, pp. 51-58.

[10] R. Huslende: *A Combined Evaluation of Performance and Reliability for Degradable Systems*, ACM-SIGMETRICS-Conf. on Measurement and Modeling of Comp., Sept. 1981, pp. 157-164.

[11] J.A. Munarin: *Dynamic Workload Model for Performance/Reliability Analysis of Gracefully Degrading Systems*, FTCS-13, IEEE, 1983, pp. 290-295.

[12] J. Arlat, J.C. Laprie: *Performance-Related Dependability Evaluation of Supercomputer Systems*, FTCS-13, IEEE, 1983, pp. 276-283.

[13] J.F. Meyer: *On Evaluating the Performability of Degradable Computing Systems*, IEEE Trans. Comp., vol. C-29, no. 8, Aug. 1980, pp.720-731.

[14] J.F. Meyer, D.G. Furchtgott, L.T. Wu: *Performability Evaluation of the SIFT Computer*, FTCS-9, IEEE, 1979, pp. 43-50.

[15] J.F. Meyer: *Closed-form Solutions of Performability*, FTCS-11, IEEE, 1981, pp. 66-71.

[16] D.G. Furchtgott, J.F. Meyer: *A Performability Solution Method for Degradable Nonrepairable Systems*, IEEE Trans. Comp., vol. C-33, no. 6, June 1984, pp. 550-554.

[17] I. Mitrani, P.J.B. King: *Multiserver Systems Subject to Breakdowns: An Empirical Study*, IEEE Trans. Comp., vol. C-32, no. 1, Jan. 1983, pp. 96-98.

[18] H. Hecht: *Effectiveness Measures for Distributed Systems*, Proc. IEEE Symp. on Reliability in Distributed Software and Database Systems, Pittsburg PA, July 1981, pp. 185-188.

[19] X. Castillo, D.P. Siewiorek: *A Performance-Reliability Model for Computing Systems*, FTCS-10, IEEE, 1980, pp. 187-192.

[20] A. Hac: *A Performance Considered Model for System Reliability*, FTCS-13, IEEE, 1983, pp. 139-143.

[21] E. Jessen, O. Schoen: *On a Class of BCMP-Consistent Integrated Performance/Reliability Models*, Report des Inst. für Informatik der Techn. Univers. München, TUM-I8502, Febr. 1985.

[22] O. Schoen: *Verkehrs/Zuverlässigkeits-Modelle auf der Basis von BCMP-Netzen*, Bericht Nr. 108 des Fachber. Informatik der Univers. Hamburg, FBI-HH-B-108/84, 1984.

[23] F. Baskett, K.M. Chandy, R.R. Muntz, F.G. Palacios: *Open, Closed, and Mixed Networks of Queues with Different Classes of Customers*, Journal ACM, vol. 22, no.2, April 1975, pp. 248-260.

[24] S.S. Lavenberg: *Computer Performance Modeling Handbook*, Academic Press, New York, 1983.

Reprinted from *IEEE Transactions on Computers*, Volume 37, Number 4, April 1988, pages 406–417.

Performability Analysis: Measures, an Algorithm, and a Case Study

R. M. SMITH, KISHOR S. TRIVEDI, SENIOR MEMBER, IEEE, AND A. V. RAMESH

Abstract—**Multiprocessor systems can provide higher performance and higher reliability/availability than single-processor systems. In order to properly assess the effectiveness of multiprocessor systems, measures that combine performance and reliability are needed. We describe the behavior of the multiprocessor system as a continuous-time Markov chain and associate a reward rate (performance measure) with each state. We evaluate the distribution of performability for analytical models of a multiprocessor system using a new polynomial-time algorithm that obtains the distribution of performability for repairable, as well as nonrepairable, systems with heterogeneous components with a substantial speedup over earlier work. Numerical results indicate that distributions of cumulative performance measures over finite intervals reveal behavior of multiprocessor systems not indicated by either steady-state or expected values alone.**

Index Terms—**Availability, fault tolerance, Markov models, Markov reward models, multiprocessor performance, performability, reliability.**

I. Introduction

THE proliferation of fault-tolerant multiple processor systems has given rise to the need to develop composite reliability and performance measures. For this purpose, Meyer [20] developed a conceptual framework of *performability*. In this paper, we consider performability models based on Markov Reward Models (MRM's). We obtain a variety of performability measures on several models of a multiprocessor system to illustrate the effect of different fault-tolerant mechanisms on the ability of the system to complete useful work in a finite time interval. In the course of this study, we show that the distribution of accumulated reward illuminates effects that are not detected by steady-state values, instantaneous measures, or expected values of cumulative measures. Hence, the performability distribution provides new insight on the behavior of multiprocessor computer systems. We describe a new $O(n^3)$ algorithm for the computation of the distribution of accumulated reward in a finite utilization interval where n is the number of states in the MRM.

The evolution of the system through configurations with different sets of operational components is represented by a continuous-time Markov chain (CTMC) which we refer to as a structure-state process. The set of rewards associated with the states of a structure-state process are referred to as the reward structure. Together the structure-state process and the reward structure determine a Markov reward model (MRM). Because the time-scale of the performance-related events (e.g., instruction execution, job service) is at least two orders of magnitude less than the time-scale of the reliability-related events (i.e., component failure, component repair), steady-state values of performance models are used to specify the performance levels or reward rates for each structure state.

We analyze several MRM's of a multiprocessor system with 16 processors, 16 memories, and a crossbar switch. In the Appendix, we describe an improved algorithm to obtain the performability distributions from MRM's with n structure-states that provides an $O(n)$ speedup over the earlier algorithm in [19]. The algorithm may be applied to MRM's constructed for repairable or nonrepairable systems. We demonstrate the use of our algorithm on a problem of moderate size. Previously published results on performability distributions for finite time intervals have been carried out only on very small problems. With the multiprocessor system, we examine the effect of different modeling assumptions on a number of measures including the distribution of accumulated reward.

The freedom to modify the structure-state process as well as the reward structure allows the modeler to represent a wide variety of situations. In the performability domain, there are two extremes. First we may have a structure-state process with only a single state and a possibly complex performance model to generate the reward associated with the single state. A "pure" performance model that ignores failure and repair but considers memory contention overestimates the ability of the system to complete useful work. On the other extreme, a "pure" availability model ignores different levels of performance (other than operational or failed). A model that takes into account both aspects of system behavior by a combined performability measure is more appropriate for the evaluation of computer systems that may undergo a graceful degradation of performance. After completing the Introduction, we describe the multiprocessor system in Section II. In Section III, we present results for MRM's of the multiprocessor system. In the Appendix, we describe and analyze the computational cost of the algorithm used to determine the distribution of accumulated reward for cyclic or acyclic MRM's.

A. Notation

The evolution of the system in time is represented by the finite-state stochastic process $\{Z(t), t \geq 0\}$, which characterizes the dynamics of the system structure and environmental

Manuscript received April 15, 1987; revised November 15, 1987. This work was supported by the U.S. Air Force Office of Scientific Research under Grant AFOSR-84-0132, the U.S. Army Research Office under Grant DAAG29-84-K-0045, and under DOD Research Instrumentation Grant DAAL03-87-G-0066.

The authors are with the Department of Computer Science, Duke University, Durham, NC 27706.

IEEE Log Number 8719361.

influences. $Z(t) \in S = \{1, 2, \cdots, n\}$ is the structure-state of the system at time t. The holding times in the structure-states are exponentially distributed, and hence $Z(t)$ is a homogeneous CTMC. Even in situations where the holding times are generally distributed, they may often be acceptably approximated using a finite number of exponential phases [14]. We let q_{ij} be the transition rate from state i to state j and $Q = [q_{ij}]$ be the n by n generator matrix where

$$q_{ii} = -\sum_{j=1, j\neq i}^{n} q_{ij}.$$

Let $p_i(t)$ denote Prob $[Z(t) = i]$, the probability that the system is in state i at time t. The column vector $\boldsymbol{p}(t)$ of the state probabilities may be computed by solving a matrix differential equation [23]

$$\frac{d}{dt}\boldsymbol{p}(t) = Q^T\boldsymbol{p}(t). \tag{1}$$

The steady-state probability vector $\boldsymbol{\pi}$ of the Markov chain is the solution for the linear system

$$Q^T\boldsymbol{\pi} = 0, \quad \sum_i \pi_i = 1.$$

Let r_i be the reward rate (or the performance level) associated with structure-state i; then the vector $\boldsymbol{r}$ defines the reward structure. The reward rate of the system at time t is defined to be $X(t) = r_{Z(t)}$. We let $Y(t)$ be the accumulated reward until time t, that is, the area under the $X(t)$ curve

$$Y(t) = \int_0^t X(\tau)\, d\tau.$$

Consequently, by interpreting rewards as performance levels, we see that the distribution of accumulated reward is at the heart of characterizing systems that evolve through states with different reward rates (e.g., performance levels). In Fig. 1 we depict a Markov reward model with a three-state CTMC for the structure-state process and a simple reward structure, the transition matrix of the CTMC, as well as sample paths for the stochastic processes $Z(t)$, $X(t)$, and $Y(t)$. Note that a given sample path of $Z(t)$ determines a unique sample path for $X(t)$ and $Y(t)$.

We denote the distribution of accumulated reward by time t evaluated at x as

$$\mathcal{Y}(x, t) \equiv \text{Prob}\,[Y(t) \leq x].$$

A fundamental question about any system is simply, "What is the probability of completing a given amount of useful work within a specified time interval?" The answer is provided by the complement of the above distribution:

$$\mathcal{Y}^c(x, t) \equiv \text{Prob}\,[Y(t) > x].$$

The time-averaged accumulated reward, its distribution, and

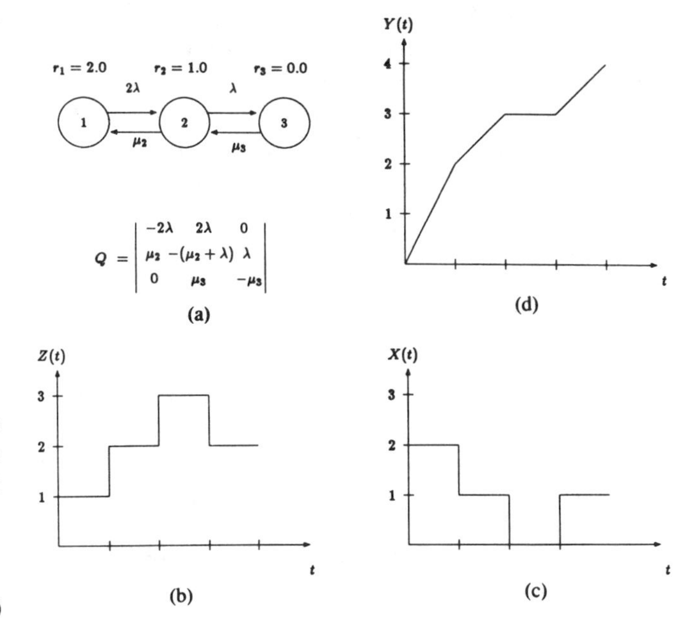

Fig. 1. Three-state Markov reward model with sample paths of $Z(t)$, $X(t)$, and $Y(t)$ processes. (a) Markov reward model. (b) $Z(t)$ path. (c) $X(t)$ path. (d) $Y(t)$ path.

its complementary distribution are denoted as

$$W(t) \equiv \frac{Y(t)}{t} \equiv \frac{1}{t}\int_0^t X(\tau)\, d\tau,$$

$$\mathcal{W}(x, t) \equiv \text{Prob}\,[W(t) \leq x] \text{ and}$$

$$\mathcal{W}^c(x, t) \equiv \text{Prob}\,[W(t) > x].$$

A special case of $W(t)$ is obtained when we assign a reward rate 1 to operational states and zero to nonoperational states. In this case, $W(t)$ is known as the *interval availability* $A_I(t)$. The complementary distributions explicitly answer the questions of the modeler and are easily obtained from the results for $\mathcal{Y}(x, t)$ in the Appendix. To complete our notation, we note that we have assumed a distinguished initial state. To explicity indicate this dependence on the initial state, we will use a subscript on cumulative and time-averaged random variables and their distributions. For example, $W_i(t)$ denotes the time-averaged accumulated reward for the interval $(0, t)$ given that the initial state is i (i.e., $Z(0) = i$).

The ability to complete a given amount of work with probability one is a property of some Markov reward models. An MRM is said to have the *completion property* if does not have a reachable closed set C of states such that $r_i = 0$ for all $i \in C$. As an example of an MRM with the completion property, consider Fig. 1(a) with all parameters greater than zero. Since the probability of remaining in structure state 3 for all but a finite amount of time in an infinite time interval is zero and structure states 1 and 2 have nonzero reward, any finite amount of reward will be accumulated if the time interval is long enough. Because descriptions of fault-tolerant systems almost always include "failed" (zero reward) structure states, we will refer to MRM's of fault-tolerant systems that take repair actions from all "failed" structure states as MRM's with completion. The completion property is a useful distinction because it indicates the most appropriate measures for a model. MRM's with the completion property are appropriately described with $\mathcal{W}(x, t)$, while models without it are readily described with $\mathcal{Y}(x, t)$. Those Markov models without the completion property will be referred to as MRM's with *imperfect repair.* An MRM in which operational states are assigned reward rate 1 and nonoperational states are assigned rate 0 are called *availability models.* In an availability model, if we further require that all nonoperational states are absorbing, than we have a *reliability model.*

B. Previous Work

Early attempts to evaluate fault-tolerant computer systems were restricted to transient analysis of the CTMC describing the evolution of the system over time. The immediate result relating the transient probability to the probability of the system operating at a specified reward level,

$$\text{Prob}\,[X(t)=r] = \sum_{\{j|r_j=r\}} \text{Prob}\,[Z(t)=j] = \sum_{\{j|r_j=r\}} p_j(t),$$

was exploited by Huslende [15] and Wu [28].

Gracefully degrading systems provide useful computation by reconfiguring to adjust to the failure of one or more components. Beaudry used the notation of *computation availability* which in our notation is the expected reward rate at time t:

$$E[X(t)] = \boldsymbol{r}^T\boldsymbol{p}(t) = \sum_i r_i p_i(t),$$

and its limiting value

$$\lim_{t\to\infty} E[X(t)] = \boldsymbol{r}^T\boldsymbol{\pi} = \sum_i r_i\pi_i.$$

These two quantities are generalizations of instantaneous and steady-state availability, respectively. Huslende considered performance reliability by assuming a minimum performance threshold:

$$R(\text{threshold}, t) \equiv \text{Prob}\,[X(\tau) \geq \text{threshold}, \forall \tau \leq t];$$

a generalization of reliability.

Under general assumptions about the stochastic process $\{Z(t), t \geq 0\}$ and the reward structure $\boldsymbol{r}$, Howard [13] studied the expected accumulated reward $E[Y(t)]$ for finite intervals of time and the expected time-averaged accumulated reward over an infinite time interval. It is interesting to note that the limit $t \to \infty$ of the expected value of $X(t)$ and $W(t)$ are equal.

$$\lim_{t\to\infty} E[W(t)] = \sum r_i\pi_i = \lim_{t\to\infty} E[X(t)].$$

With our notation we can express $E[Y(t)]$ as

$$E[Y(t)] = E\left[\int_0^t X(\tau)\,d\tau\right] = \int_0^t E[X(\tau)]\,d\tau = \sum_i r_i \int_0^t p_i(\tau)\,d\tau.$$

To compute $E[Y(t)]$ we define $L_i(t) = \int_0^t p_i(\tau)\,d\tau$ to be an element of $\boldsymbol{L}(t)$ and derive a system of ordinary differential equations for $\boldsymbol{L}(t)$ by integrating (1).

$$\frac{d}{dt}\boldsymbol{L}(t) = Q^T\boldsymbol{L}(t) + \boldsymbol{p}(0).$$

Solutions are readily calculated using methods similar to those used to solve (1). Often we are interested in the behavior of $Y(t)$ far from the mean (as is the case when a system is required to deliver a specific reward with high probability), and in this case the central moments do not provide accurate information. Consequently measures that provide a more detailed look at system behavior are needed.

Recently, considerable attention has been given to the problem of evaluating the distribution of accumulated reward, $\mathcal{Y}(x, t)$. The problem is more easily solved if the distribution of accumulated reward is to be evaluated over an infinite time interval. Beaudry [1] has shown that the distribution of accumulated reward until system failure ($\mathcal{Y}(x, \infty)$) for a system with imperfect repair can be obtained as the time-to-failure distribution of an associated CTMC obtained by simply dividing the rates of transitions leaving a given state i by r_i.

For finite time intervals, Meyer [21] obtained the distribution of accumulated reward in acyclic Markov reward models (no loops in the structure-state CTMC) with r_i being a monotonic function of the state labeling. A direct approach that numerically integrated the convolution equations in the time domain for acyclic models was developed and implemented by Furchtgott and Meyer [9]. The computational complexity is exponential in the number of states so the applicability of the direct time-domain approach is limited to problems with a few states over a short time interval. Subsequently, Goyal and Tantawi [10] developed an $O(n^3)$ algorithm to compute the distribution of accumulated reward in general acyclic structure-state processes with monotonic reward rates. Ciciani and Grassi [3] and Donatiello and Iyer [6] proposed algorithms that do not require the rewards to be monotonic.

MRM's that have cyclic structure-state CTMC's are more difficult. By using the central limit theorem, it can be shown that the asymptotic distribution of the accumulated reward over a time interval $(0, t)$ for t sufficiently large is normally distributed with mean $\lim_{\tau\to\infty} E[X(\tau)]$ multiplied by t and variance $\alpha\sqrt{t}$. Computational methods to determine $\lim_{\tau\to\infty} E[X(\tau)]$ and α may be found in Hordijk *et al.* [12].

Iyer *et al.* [16] describe a recursive technique for computing moments of the distribution of accumulated reward for cyclic MRM's. With the moments in hand, bounds on the distribution of accumulated reward are available. As noted earlier, because the central moments describe the behavior of the distribution about the mean, the bounds are often too loose to be helpful at the extremes, which are often of interest. The difficulties are similar to those one faces extrapolating the value of a continuous function a distance away from a point where all the derivatives are known.

More recently, Goyal, Tantawi, and Trivedi [11] formulated the interval availability problem (a special instance of $W(t)$, for a reward structure with reward rates $r_i = 1$ if state i

TABLE I
MEASURES AND THEIR CHARACTERISTICS

Measure	Notation	Common Model Family	Cumulative or Instantaneous Measure	Steady State or Transient	Distribution or Moment
$p_i(t)$	$P[Z(t) = i]$	av	$Z(t)$:I	T	pmf
π_i	$\lim_{t\to\infty} P[Z(t) = i]$	av	$Z(t)$:I	S	pmf
$A(t)$	$\Sigma_{i\in up}\, p_i(t)$	av	$X(t)$:I	T	M
$A(\infty)$	$\lim_{t\to\infty} A(t)$	av	$X(t)$:I	S	M
Reliability	$P[X(\tau) \geq 1, \forall\tau \geq t]$	rel	$X(t)$:I	T	cdf
$E[X(t)]$	$E[X(t)]$	*all*	$X(t)$:I	T	M
$E[Y(t)]$	$E[Y(t)]$	Imp-rep	$Y(t)$:C	T	M
$\mathcal{Y}(x, t)$	$\mathcal{Y}(x, t)$	Imp-rep	$Y(t)$:C	T	cdf
$\mathcal{Y}(x, \infty)$	$\lim_{t\to\infty} \mathcal{Y}(x, t)$	Imp-rep	$Y(t)$:C	S	cdf
$P[A_I(t) \leq x]$	$P[W(t) \leq x]$	av	$W(t)$:C	T	cdf
$E[W(t)]$	$E[W(t)]$	compl	$W(t)$:C	T	M
$E[W(\infty)]$	$E[W(\infty)]$	compl	$W(t)$:C	S	M
$\mathcal{W}(x, t)$	$\mathcal{W}(x, t)$	compl	$W(t)$:C	T	cdf
$\mathcal{W}(x, \infty)$	$\lim_{t\to\infty} \mathcal{W}(x, t)$	compl	$W(t)$:C	S	cdf

is operational and zero else) as a system of first-order partial differential equations. The randomization technique has also been applied to the interval availability problem by de Souza e Silva and Gail [8].

Puri [22] derived a linear system in the double Laplace transform of the distribution of accumulated reward for a general CTMC and arbitrary reward structure. The numerical solution of the double transform system was proposed in [19]. In the Appendix, we present an improved $O(n^3)$ algorithm to evaluate the distribution of accumulated reward for cyclic and acyclic MRM's with n states. Note that the $O(n)$ speedup over our previous algorithm [19] makes considerably larger MRM's solvable in practice.

In Table I we present the measures that we use to examine the behavior of the example multiprocessor system. We group the measures by the random variables used in their definition. Each measure's properties are then indicated. The properties that we indicate are whether the quantity measured is instantaneous or cumulative, steady state or transient. We also indicate in Table I whether the measure is a distribution or a central moment. We use a column in Table I for each measure to indicate the model families each measure is typically applied to. We use rel, av, Imp-rep, and compl as abbreviations for the reliability, availability, imperfect repair, and completion families, respectively.

Measures used to characterize the behavior of Markov reward models of the multiprocessor system with imperfect repair (without the completion property) are the reliability $R(t)$, the distribution of accumulated reward (performability) over a finite interval $\mathcal{Y}(x, t)$, and $\mathcal{Y}(x, \infty) \equiv \lim_{t\to\infty} \mathcal{Y}(x, t)$. On models with the completion property we use $\mathcal{W}(x, t)$, and $\mathcal{W}(x, \infty) \equiv \lim_{t\to\infty} \mathcal{W}(x, t)$. The effect of changes in the structure-state process, the reward structure, and utilization interval on these measures of performability for MRM's of the multiprocessor system are investigated in the next two sections.

II. Multiprocessor Model Description

We begin with a basic Markov reward model of the multiprocessor system and then indicate a set of changes in the structure-state process and reward structure. The measures obtained for the various models of the multiprocessor system are listed in Table I. In the following section, each graph plots measures for a sequence of illustrative models.

Determining the way changes in the reward structure and the structure-state process affect measures of interest is crucial to using MRM's effectively in the system design process. Efforts to change system behavior in a favorable way must use the appropriate model and measure or they will be ineffective. For example, consider adding a repair facility to a high-reliability nonrepairable system (failure rate of $\sim 10^{-5}$). The steady-state behavior will change radically. However, if the utilization interval is short (~ 10 h) then the repair facility will not substantially change the availability over the 10 h interval. We wish to indicate some situations where the distribution of accumulated reward or its time average will indicate behavior not captured by other measures. We briefly describe the types of failure and repair behavior of the multiprocessor system modeled with structure-state processes. The system consists of 16 processors, 16 memories, and an interconnection network (i.e., crossbar switch) that allows a processor to access any memory. Since the system we analyze is similar to the Carnegie-Mellon multiprocessor system, C.mmp, we use the failure data from that system. Siewiorek in [24] determined the failure rates per hour for the components to be

	Processor	Memory	Switch
Failure Rates:	$\lambda = 0.0000689$	$\gamma = 0.0002241$	$\delta = 0.0002024$.

Viewing the network as a single switch and modeling the system at the processor–memory–switch (PMS) level, we see that the interconnection network is essential for system operation. It is also clear that a minimum number of processors and memories are necessary for the system to be operational. We follow Siewiorek's choice of four processors, four memories, and one interconnection network (switch) as the minimal operating configuration required for handling a task. Each state is specified by a triple (i, j, k) indicating the number of operational processors, memories, and networks, respectively. We augment the states with a nonoperational state F. Events that decrease the number of operational components are associated with failure, and events that increase the number of operational elements are associated with repair. We assume that failures do not occur when the system is not operational. When a component of the multiprocessor system fails, a recovery action must be taken (e.g., shutting down a failed processor, so that it does not fill memories with spurious data), or the whole system will fail and enter state F. The probability that the recovery action is successfully completed is known as the coverage.

We consider two kinds of repair actions, global repair which restores the system to state (16, 16, 1) with rate $\mu = 0.2$ per hour from state F and local repair, which can be thought of as a repair person beginning to fix a component of the system as soon as a component failure occurs. Our model of local repair assumes that there is only one repair person for each component type. We let the local repair rates per hour be

	Processor	Memory	Switch
Local Repair Rates:	$\nu = 2.0$	$\eta = 1.0$	$\epsilon = 0.5$.

A further refinement of the structure-state process can be made with respect to the interconnection network. Siewiorek in [25] notes that the C.mmp interconnection network is actually implemented as a set of 16 fan-out switches for each processor and memory port. In this case, the failure rate of the interconnection system with respect to some operational configuration $(i, j, 1)$ is simply $\delta(i, j) = (i + j) \times$ (fan-out switch failure rate + line failure rate). Since the cause of a failure is uniformly distributed over the fan-out switches and their lines, we will simply let the failure rate associated with each fan-out switch and line pair be 1/32 of the lumped failure rate of the switch. Thus, $\delta(10, 10) = 20 \times 0.000006325 = 0.0001265$. We are pessimistic in that we assume that the failure of one fan-out switch and line brings the system to a nonopertional state (i.e., $(i, j, 0)$). The single or "lumped" network with failure rate δ is more pessimistic than the "distributed" network with failure rate $\delta(i, j)$. A Markov model of the structure-state process for the C.mmp system with a "lumped" network and global repair has 170 states.

The two variations of the structure-state process we consider for the failure transitions are imperfect coverage (i.e., leakage to state F), and the network failure rate ("lumped" or "distributed"). Local or global repair actions are the two kinds of repair strategies investigated. The substantial increase in model complexity that results from adding a local repair capability is evident in Fig. 2, which depicts the structure-state process of a model with a "lumped" interconnection network, local and global repair, and imperfect coverage (365 states). The lower plane in Fig. 2 contains the set of states where component exhaustion has occurred. Most of the states (169) in the lower level are the result of the interconnection network failing. Thirteen states represent system failure due to the exhaustion of operational memories and thirteen more states represent system failure due to the exhaustion of operational processors. The local repair models will include both local and global repair. When we speak of a model with only global repair, we set all local repair rates (ν, η, ϵ) in Fig. 2 to zero and merge all nonoperational states with state F. The structure-state processes of the MRM's of the multiprocessor system thus can be characterized by their failure type (coverage), interconnection network type ("lumped" or "distributed"), and repair type (global or local).

It remains to present the reward structures we use to characterize the performance behavior of the multiprocessor system when it is in a given structure state. The simplest reward structure is obtained by dividing the structure states into two classes, operational and nonoperational, and assigning the reward rate 1.0 to the operational states and 0 to the rest. A more accurate measure of system performance is more closely related to the system's ability to do useful work. Because memory is the slowest resource in the C.mmp system, the effectiveness of the system is limited by the number of available memories. Thus, if there are more memories than processors, performance will still be limited by the memory bandwidth needed by the processors, while if there are more processors than memories, the performance will be limited by the number of memories. A simple capacity-based performance model of an operational structure-state $(i, j, 1)$ is to let the associated reward rate be $\min\{i, j\}$. This performance model is optimistic because it does not consider processors contending for the memories.

When we consider contention for the memories, we use a model developed by Bhandarkar [2] to obtain the average number of busy memories or memory bandwidth. Bhandarkar found the average number of busy memories, and hence the reward rate in an operational state $(i, j, 1)$, to be

$$r_{i,j,1} = m(1-(1-1/m)^l), \qquad (2)$$

where $l = \min\{i, j\}$ and $m = \max\{i, j\}$. We assign a zero reward rate to each nonoperational state. Hence, in addition to a variety of structure-state processes, we also have three reward structures of interest, the availability-based reward structure (0, 1), the capacity-based reward structure ($\min\{i, j\}$), and the contention-based reward structure [see (2)].

The initial state of the system in all our models will be (16, 16, 1) except in Section III-C where $p(0)$, the initial state probability vector, is equal to the steady-state probability vector $\boldsymbol{\pi}$. The effect of changes in utilization-interval length, structure-state process, and reward structure for the multiprocessor MRM's are examined in the next section.

III. Multiprocessor Performability Results

A. The Effects of Coverage and Utilization Interval on $E[X(t)]$ and $E[W(t)]$, Functions of $p(t)$.

First, we use a sequence of models that illustrate the way the completion property affects $E[X(t)]$ and $E[W(t)]$ as a function of time in Figs. 3 and 4, respectively. In both Figs. 3 and 4, we use our contention-based performance model to obtain the reward structure. $E[X(t)]$ is the expected instantaneous reward at time t and has been called the *computation availability* in [1]. This measure answers the question, "What is the expected performance of the system at time t?" $E[W(t)]$ is the expected time-averaged accumulated reward over the interval $(0, t)$. $E[W(t)]$ answers the question, "What is the time-averaged performance of the system over the

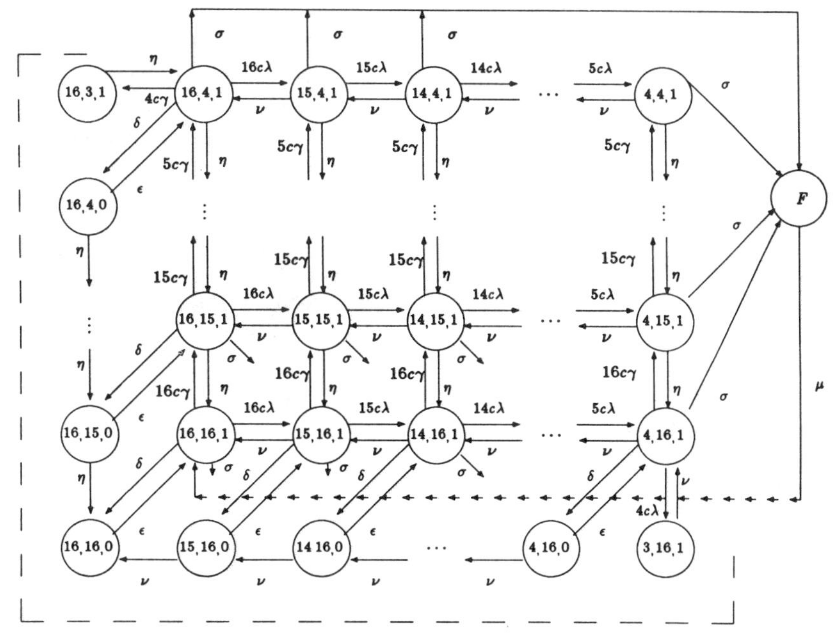

Fig. 2. Markov chain for the multiprocessor system (each state (i, j, k) has a transition with rate $\sigma = ((1 - c)(i\lambda + j\gamma))$ to state F).

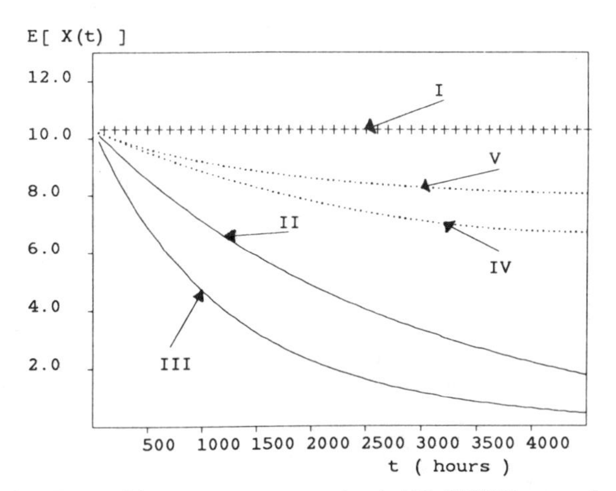

Fig. 3. Expected instantaneous memory bandwidth $E[X(t)]$ versus time t.

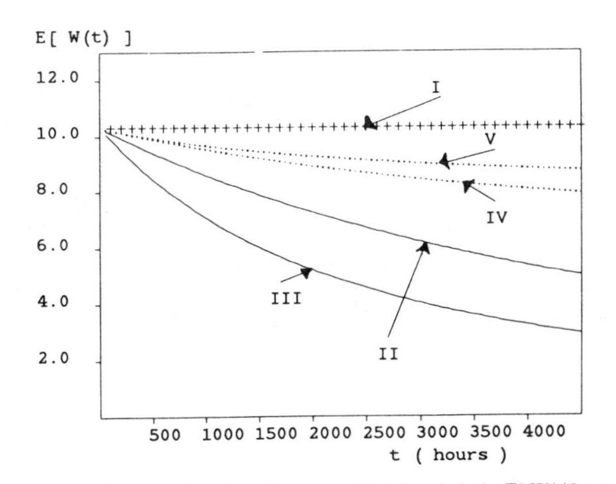

Fig. 4. Time-averaged expected accumulated bandwidth $E[W(t)]$ versus time t.

interval $(0, t)$?'' In Figs. 3 and 4, we let curve I be a ''pure'' performance model of the state (16, 16, 1). The ''pure'' performance model does not have any failures so the system performance is independent of time. With memory contention but no failure, the reward rate is 10.303 and both $E[X(t)]$ and $E[W(t)]$ are 10.303 for all time t. In curve II, only component failures occur ($c = 1$ for coverage), and we see that the expected performance level has been halved at time $t = 2000$. At time $t = 2000$ in Fig. 4, the expected time-averaged accumulated reward has decreased by only one quarter because $E[W(t)]$ is the time average of $E[X(t)]$ over $(0, t)$. Thus, $E[W(t)]$ is relatively insensitive, for large t, to the state of the system at a particular instant, $\tau < t$. Both Figs. 3 and 4 show the importance of the completion property. Models with the completion property (curves I, IV, and V) strongly dominate those without it (curves II and III) indicating the value of global repair for long utilization intervals.

In curves III and V, the coverage is reduced to 0.9. Curve III like II has no repair, and the expected performance level of curve III deteriorates more rapidly than that of curve II. Curve IV has only component failures ($c = 1$), and global repair as well. Consequently, the expected performance level of curve IV is much improved over that of curve II, especially for large t. One might expect curve IV to dominate curve V, which uses the same model as curve IV with $c = 0.9$, just as curve II dominates curve III. However, for large t it is better on the average to experience a coverage failure and rapidly return to the highest reward state (16, 16, 1) rather than spend a long interval in the relatively low reward states before returning to structure state (16, 16, 1). Both of these measures indicate the importance of global repair for longer time intervals.

Unfortunately, $E[X(t)]$ and $E[W(t)]$ do not address the likelihood of completing a given amount of work in a specified interval. $E[W(t)]$ merely gives an indication of the average behavior over a utilization interval. We use $\mathcal{Y}^c(x, t)$ to

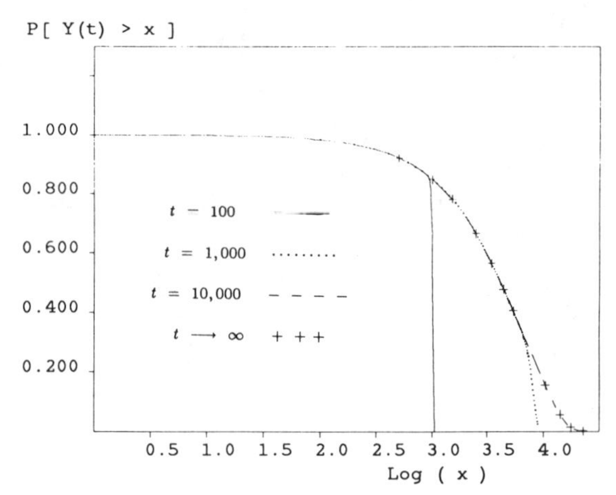

Fig. 5. Complementary distribution of accumulated bandwidth $\mathcal{Y}^c(x, t)$ for different t values.

examine the behavior of a nonrepairable system over different length utilization intervals in the next section.

B. The Effect of Utilization Interval on $\mathcal{Y}^c(x, t)$ for Nonrepairable Models

We consider a model of the C.mmp system with a "lumped" interconnection network, $c = 0.90$ for the coverage, and no repair in Fig. 5. The CTMC of the structure-state process is depicted in Fig. 2 with all repair rates, μ, ν, η, ϵ set to zero. The reward structure is based on the contention-based performance model. Curves I, II, III, and IV plot the value of $\mathcal{Y}^c(x, t)$ for t = 100, 1000, 10 000, and ∞, respectively.

Loosely speaking, $\mathcal{Y}^c(x, t)$ answers the question, "What is the probability that x units of work is completed by time t?" Because the model does not have the completion property, $\mathcal{Y}^c(x, t)$ is substantially less than 1.0 for moderate amounts of accumulated reward even if $t \to \infty$. It is interesting to note that $\mathcal{Y}^c(x, t)$ for moderate t only falls below $\lim_{t\to\infty} \mathcal{Y}(x, t)$ as $x \to \sim 9t$.

The nonrepairable system performs near its asymptotic limit, $\mathcal{Y}^c(x, \infty)$ for moderate t. However, systems that satisfy the completion property will complete any finite amount of work in an arbitrary long utilization interval. When comparing different systems for the same utilization interval, $\mathcal{Y}^c(x, t)$ is quite satisfactory, whether the system satisfies the completion property or not. If we wish to compare the behavior of systems that satisfy the completion property over different utilization intervals, then we need to normalize the curves of the different complementary distributions of accumulated reward so that they can be compared over the same interval. The natural approach is to time average the accumulated reward and use $W(t)$ as the random variable rather than $Y(t)$. In the next section, we examine the behavior of a system that satisfies the completion property over different utilization intervals. The results are rather surprising.

C. The Effect of Utilization Interval on $\mathcal{W}^c(x, t)$ for Models with the Completion Property

As noted in Section III-A, both $E[X(t)]$ and $E[W(t)]$ are functions of the instantaneous probability vector $\boldsymbol{p}(t)$. If we let the initial probability vector $\boldsymbol{p}(0)$ of the system equal the steady-state probability vector $\boldsymbol{\pi}$, then neither $E[X(t)]$ nor $E[W(t)]$ will change since then $\boldsymbol{p}(t) = \boldsymbol{\pi}$ for all t. We show the presence of behavior not detected by these measures in Fig. 6. In Fig. 6 we use a structure-state process with global repair and coverage = 0.95 to model the failure and repair activity of the C.mmp system and the contention-based performance model to obtain the reward structure. The measure $\mathcal{W}^c(x, t)$ can be used to answer the question, "What is the probability that the reward accumulated in the interval $(0, t)$ is at least xt?"

We examine the distribution of time-averaged memory bandwidth (performance) for utilization intervals of length 10, 100, 1000, and 10 000 in curves I, II, III, and IV of Fig. 6. We indicate the steady-state expected reward rate, $\Sigma_i \pi_i r_i$, with a vertical line labeled $V(+\ +\ +)$. We can see the way the curve smooths out and approaches a jump at the steady-state, time-averaged reward rate as t increases. The dynamic behavior of the system in steady state is indicated in Fig. 6. Measures such as $E[X(t)]$ and $E[W(t)]$ are unable to capture the steady-state system dynamics since both these measures are invariant with respect to time for the Markov reward model with $\boldsymbol{p}(0) = \boldsymbol{\pi}$.

D. The Effect of Reward Structure and Model "Family" on $\mathcal{W}^c(x, t)$ for Models with the Completion Property

Insight into the way the structure-state process and the reward structure affect the ability of the multiprocessor system to complete a fixed amount of work in a given time interval $(0, t)$ is obtained from the complementary distribution of time-averaged accumulated reward. We plot the complementary distribution of time-averaged accumulated reward (in this case the time-averaged memory bandwidth) for a basic Markov reward model with a "lumped" interconnection network, perfect coverage ($c = 1$), and global repair. We use "pure" performance models to provide an optimistic upper bound for MRM's comparing the capacity-based and contention-based reward structures resulting from the different performance assumptions about the way memory is accessed. We examine the distribution of $W(t)$ and the distribution of the interval availability $A_I(t)$, in Fig. 7. First we consider the system without failure and repair in curves I and II. The result of this modeling assumption is that no degradation of performance takes place and the state of the system is always (16, 16, 1). Consequently, curves I and II of the complementary distribution of time-averaged accumulated reward are step functions. If we ignore memory contention, then there are 16 processors and 16 memories and the memory bandwidth is 16. It follows that the system performance level (reward rate) is constant and $W(t) = 16$. Curve I in Fig. 7 depicts this unit step form of the complementary distribution of time-averaged accumulated reward. For curve II, we assume that there is contention at the memories. The result of modeling the contention is to lower the ability of the system to deliver useful work. Therefore, the step for curve II occurs at a smaller value of accumulated reward per unit time than the step for curve I. We use the work of Bhandarkar [2] to estimate the effect of contention on the performance of state (16, 16, 1). Hence, with memory

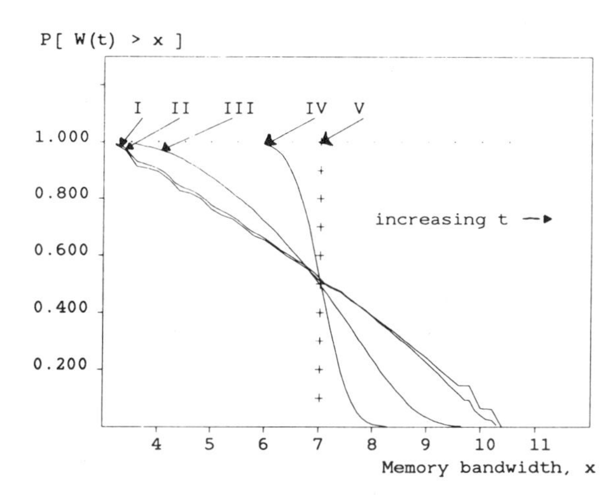

Fig. 6. Complementary distribution of time-averaged accumulated bandwidth $\mathcal{W}^c(x, t)$ for different t values.

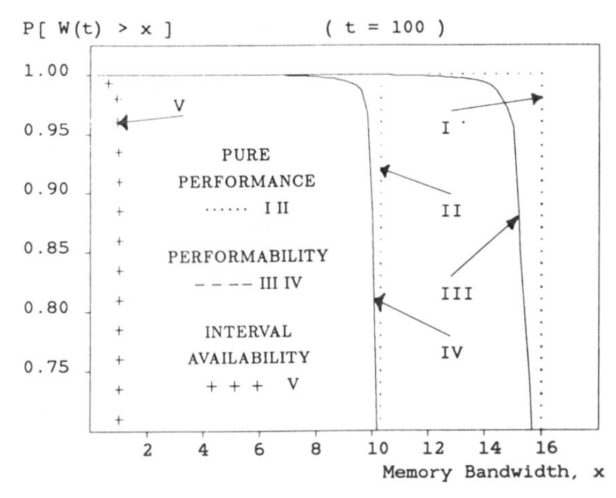

Fig. 7. Complementary distribution of time-averaged accumulated bandwidth $\mathcal{W}^c(x, t)$ for different reward structures.

contention but no failure, the reward rate is 10.303 for all time t. Thus, in curve II the complementary distribution of $W(t)$ is a unit step, though at 10.303 instead of 16 as in curve I.

In curves III and IV, we examine the effect of modeling failure and repair on the complementary distribution of time-averaged accumulated memory bandwidth. For curve III, we assume there is no memory contention and use the capacity-based performance model for each structure state in Fig. 2 (assuming no local repair and merging all nonoperational states in F). The performance (reward rate or memory bandwidth) of operational state $(i, j, 1)$ is set to $\min\{i, j\}$. The gap between curve I and curve III reflects the fact that when failure and repair are taken into account, memory bandwidth varies with time thus lowering the time-averaged accumulated memory bandwidth. Without the occurrence of failures, the memory bandwidth stays constant at 16. Another way of stating the situation is to say that curve III will asymptotically approach curve I as the maximum of all the failure rates tends to zero. Curve IV has a similar relationship to curve II. In curve IV, we use our most detailed performance model and take into account failure and repair. Thus, the performance level (reward rate) of each operational state $(i, j, 1)$ is determined by (2). Because the performance degradation due to component failure is smaller with Bhandarkar's performance estimates than with the capacity-based performance estimates, curve IV more closely approaches curve II than curve III approaches curve I. The relationship of the four curves discussed indicates that the performance model assumptions show an upper limit of the system's ability to complete work. The magnitude of the failure and repair rates effect the rate at which the complementary distribution of time-averaged memory bandwidth declines below the step function defined by the performance model.

We see that "pure" performance models overestimate the ability of the system to complete useful work. For example, using curve IV, we see that the probability that the time-average memory bandwidth is greater than or equal to 9.5 is 0.989, whereas using the "pure" performance-based model of curve II, this probability is 1. It is also true that "pure" failure/repair (availability) models in which the reward rates for operational states are set to 1.0 and nonoperational states are assigned reward 0.0 underestimate the ability of a system to complete useful work when the performance levels are scaled in such a way that the minimum reward operation state has a reward rate ≥ 1.0. Using this reward structure, $W(t)$ is the interval availability $A_I(t)$. To complete the set of reward structures considered for performability models of the multiprocessor system, with curve V we display the complementary distribution of interval availability. We see that curve V is nearly a step function at 1.0 because only a network failure will cause the system to immediately enter state F (13 processor or 13 memory failures must occur before the system will enter state F).

E. The Effect of Coverage and Utilization Interval on $\mathcal{W}^c(x, t)$ for Models with the Completion Property

In this section we continue examining a model of the multiprocessor system with a "lumped" interconnection network, global repair, and different coverage values. We will use the most accurate performance model, namely Bhandarkar's, to obtain the reward structure for the operational states. In Fig. 8 we show the effect of coverage and of the observation period on the chosen measure of effectiveness. As $t \rightarrow 0$, independent of c, $W(t)$ approaches the "pure" performance behavior shown in curve I, a step at 10.303. Curves II, III, and IV (c = 1.0, 0.95, and 0.90, respectively) show that for larger observation intervals (t = 100), the higher coverage curves dominate the lower coverage curves illustrating the effect of coverage on the complementary distribution of time-averaged accumulated reward. In curves V–VII of Fig. 8, we plot the steady-state computation availability, $\lim_{t\rightarrow\infty} E[W(t)] = \Sigma_i r_i \pi_i$, for the different coverage values where π_i is the steady-state probability of being in state i. We can see that as the length of the utilization interval increases, the probability of accumulating a given amount of reward becomes more pessimistic. One cause of this effect is that repair takes place only when the whole system has become inoperable and the failure rates are small enough to make the occurrence of more than one failure in a relatively small interval (100 h) extremely unlikely. States in which a significant number of failures have occurred become more likely as time passes. Allowing repair only when the

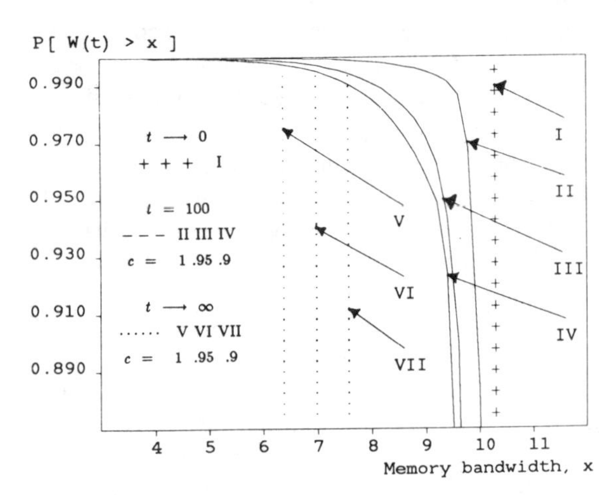

Fig. 8. Complementary distribution of time-averaged accumulated bandwidth $\mathcal{W}^c(x, t)$ for different coverage values.

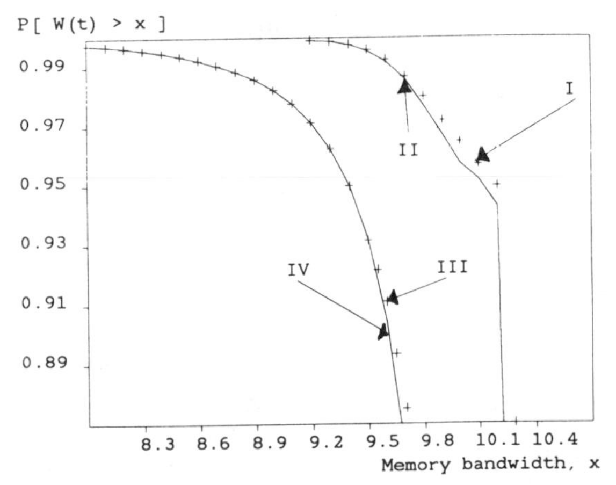

Fig. 9. Complementary distribution of time-averaged accumulated bandwidth $\mathcal{W}^c(x, t)$ for different networks and different repair policies.

system has failed yields the relative position of curves V–VII (c = 1.0, 0.95, and 0.90, respectively). As the coverage probability decreases, the steady-state computation availability actually increases!

The anomalous behavior of the steady-state computation availability is caused by several factors: the disparity in the reward rates of the operational states, the relatively large global repair rate in relation to failure rates, and the assumption that the global repair rate is independent of the number of failed components. If we set the reward rates for all the operational states to be 1.0 and 0 otherwise, then the availability ($\Sigma_i r_i \pi_i$) decreases as the coverage decreases (the anomaly disappears). Similarly, the anomaly disappears if we make the global repair rate comparable to the failure rates or make it dependent on the actual number of components that have failed. Also, local repair causes the anomaly to disappear. The point is that extrapolating from steady-state values and expected values can be misleading.

F. The Effect of Interconnection Network Type and Repair Capabilities on $\mathcal{W}^c(x, t)$ for Models with the Completion Property

We examine the effect of adding a local repair facility for each component type to the multiprocessor system in this section. In Fig. 9 we obtain $\mathcal{W}^c(x, t)$ for two pairs of models with a utilization interval of 100 h. In curve I we plot $\mathcal{W}^c(x, t)$ for a model of the multiprocessor system with both global and local repair, c = 0.90 for coverage and a distributed interconnection network (+ + +). The model used to obtain curve II (solid) is the same as that used for curve I except for a "lumped" instead of distributed interconnection network. The effect of the slightly lower failure rate of the distributed interconnection network is small but discernible.

Curve III is the plot of $\mathcal{W}^c(x, t)$ for the same model as curve I without local repair. Replacing the distributed interconnection network in curve III with the slightly more failure prone "lumped" interconnection network model produces curve IV. The difference between curves III and IV is also quite small. The effect of the local repair facility is sizable for time-averaged memory bandwidth requirements greater than 9.2. This result indicates the value of local repair for the multiprocessor system over even moderately sized utilization intervals. Another way of expressing the situation is to observe that as the time-averaged workload requirement increases, the size of the utilization interval where the local repair facility will substantially increase $\mathcal{W}^c(x, t)$ becomes smaller. Roughly speaking, we can conclude that local repair is worthwhile for systems expected to operate at nearly full capacity (maximum reward rate), even if the utilization interval is of only moderate size.

IV. Conclusion

The ability to determine the distribution of accumulated reward and its time average for moderate size problems is a recent development. We presented a systematic study of a complex multiprocessor system and an $O(n^3)$ algorithm for the computation of the distribution of accumulated reward of general Markov reward models.

The study of Markov reward models of the multiprocessor system points to a number of interesting facts about different performability measures. The first three examples indicate that instantaneous measures do not show the dynamic behavior of the system while $\mathcal{W}^c(x, t)$ does. The next two examples show that steady-state values are deceptive in some circumstances, and the final example indicates the importance a local repair facility may have on the distribution of performability for moderate-size utilization intervals. Furthermore, the study indicates how changes in the failure/repair behavior of the system such as the interconnection network failure rate, repair strategy, and coverage probability affect the complementary distribution of accumulated reward. We also examine the way changing the reward structure and utilization interval affects the distribution of time-averaged accumulated reward. Thus, some inadequacies of steady-state values and expected values are illustrated and an examination of how changes in Markov reward models effect the performability distribution is made. The new algorithm presented in the paper can thus aid the system designer in exploring detailed dynamic behavior of multiprocessor systems.

Appendix
An Algorithm and its Analysis

In this Appendix, we detail the double-transform inversion method for the distribution of accumulated reward. We begin with

$$\mathcal{Y}_i(x, t) \equiv P[Y(t) \le x | Z(0) = i].$$

First, we apply the LST, (i.e., $\int_0^\infty e^{-ux}\, d\mathcal{Y}_i(x, t)$), signified by $\sim$, to $\mathcal{Y}_i(x, t)$ with respect to the work requirement x (transform variable u), and then apply the Laplace transform signified by * with respect to time t (transform variable s). The following linear system has been derived for $\mathcal{Y}^{\sim *}(u, s)$ in [22] and [18].

$$(sI + uR - Q)\mathcal{Y}^{\sim *}(u, s) = e. \tag{3}$$

The matrix of reward rates is $R = \text{diag}\,[r_1, r_2, \cdots, r_i, \cdots, r_n]$, Q is the generator matrix of the CTMC, and e is a column vector of size n with all elements equal to 1.

Using Cramer's rule, we can see that $\mathcal{Y}_i^{\sim *}(u, s)$ is a rational function in s. Hence, it has a partial fraction expansion

$$\mathcal{Y}_i^{\sim *}(u, s) = \sum_{j=1}^{d} \sum_{k=1}^{m_j} a_{ijk}(u)(s - \lambda_j(u))^{-k} \tag{4}$$

where the $\lambda_j(u)$, 1, 2, $\cdots$, j, $\cdots$, d are the d distinct eigenvalues of $[Q - uR]$, each with algebraic multiplicity m_j. The QR algorithm [27] is used to numerically determine eigenvalues of $[Q - uR]$ in $O(n^3)$ time. Using (4) we can invert analytically with respect to s and obtain

$$\mathcal{Y}_i^{\sim}(u, t) = \sum_{j=1}^{d} \sum_{k=1}^{m_j} \frac{a_{ijk}(u)}{(k-1)!} t^{k-1} e^{\lambda_j(u)t}.$$

We define the column vector A_i by

$$A_i = \begin{vmatrix} a_{i11} \\ a_{i12} \\ \vdots \\ a_{i1m_1} \\ \vdots \\ a_{idm_d} \end{vmatrix}$$

and the row vector $E^T(s)$ by

$$E^T(s) = |(s - \lambda_1(u))^{-1} \cdots (s - \lambda_1(u))^{-m_1} \cdots (s - \lambda_d(u))^{-m_d}|.$$

Now (4) can be written in vector notation as

$$\mathcal{Y}_i^{\sim *}(u, s) = E^T(s) A_i.$$

In order to determine A_i, we need n linearly independent equations. For this purpose we choose n distinct values of s, denoted by $s_1, s_2, \cdots, s_n$ sufficiently separated from the eigenvalues of $[Q - uR]$. The matrix E is constructed from the eigenvalues of $[Q - uR]$ and the n values of s.

$$E = \begin{vmatrix} E^T(s_1) \\ E^T(s_2) \\ \vdots \\ E^T(s_n) \end{vmatrix}.$$

A way of choosing the n values of s so that E has a reasonably small condition number is to choose s_j so that the jth element of $E^T(s_j) \sim 1.0$. This causes the diagonal elements of E to be reasonably large, although it does not guarantee E is nonsingular. If E is found to be singular, then a new s value can easily be chosen at the time. We then solve the linear system

$$EA_i = \mathcal{Y}_i^{\sim *}(u, s) = \begin{vmatrix} \mathcal{Y}_i^{\sim *}(u, s_1) \\ \mathcal{Y}_i^{\sim *}(u, s_2) \\ \vdots \\ \mathcal{Y}_i^{\sim *}(u, s_n) \end{vmatrix} \tag{5}$$

for the unknown vector A_i once the right-hand side has been determined. Since the problems we consider are of small size (for the solution of a linear system) we use a direct method (LU factorization) on E. The $O(n^3)$ LU factorization must be done only once and an $O(n^2)$ backsolve must be done for each of the n possible right-hand sides. Thus, we may solve $EA_i = \mathcal{Y}_i^{\sim *}(u, s)$ for all n values of i for a cost that is $O(n^3)$. The practical implication of this fact is that the $\mathcal{Y}^{\sim}(u, t)$ may be obtained at $O(n^3)$ cost if the n different right-hand sides can be obtained at an $O(n^3)$ cost as well.

Since we are interested in the n vectors of partial fraction coefficients A_i, $1 \le i \le n$, let us define

$A = [A_1 A_2 \cdots A_n]$ and

$$Y = [\mathcal{Y}_1^{\sim *}(u, s)\mathcal{Y}_2^{\sim *}(u, s) \cdots \mathcal{Y}_n^{\sim *}(u, s)].$$

Hence, the problem of determining the n^2 partial fraction coefficients can be written in matrix form as

$$EA = Y. \tag{6}$$

Because of the economical way direct methods handle multiple right-hand sides we need only address the problem of determining all the $\mathcal{Y}_i^{\sim *}(u, s)$ that make up Y in $O(n^3)$ time.

We first transform the linear system (3) to a simpler one. Any matrix can be put into upper Hessenberg form using a sequence of Householder unitary similarity transformations [27]. Therefore, we can write

$$U^\dagger = U^{-1} \text{ and } U^\dagger(Q - uR)U = H,$$

where $\dagger$ denotes conjugate transpose and H is an upper Hessenberg matrix. By making the transformation $U^\dagger \mathcal{Y}^{\sim *}(u, s) = \mathcal{M}^{\sim *}(u, s)$ the linear system (3) can be rewritten as

$$(sI - H)\mathcal{M}^{\sim *}(u, s) = U^\dagger e$$

This upper Hessenberg linear system requires only $O(n^2)$ time to determine $\mathcal{M}^{\sim *}(u, s)$. $\mathcal{Y}^{\sim *}(u, s)$ can be regained from $\mathcal{M}^{\sim *}(u, s)$ by a matrix vector product

$$\mathcal{Y}^{\sim *}(u, s) = U\mathcal{M}^{\sim *}(u, s)$$

which also costs $O(n^2)$. An important observation here is that the required sequence of unitary transformations (U) and the matrix H are already available from the QR algorithm that solves the eigenvalue problem for $(Q - uR)$ and hence does not add any cost. Thus, the complexity to obtain $\mathcal{Y}^{\sim *}(u, s)$ is now only $O(n^2)$ for every value of s for each u.

It remains to invert $\mathcal{Y}_i^{\sim}(u, t)$ with respect to u. A number of methods to numerically invert the Laplace transform have been developed. Orthogonal polynomials [26] and Fourier series [4], [5] have been the most commonly used tools for inverting the Laplace transform. To avoid unnecessary notational complexity, we define $V(u) \equiv \mathcal{Y}_i^{\sim}(u, t)/u \equiv \mathcal{Y}_i^{*}(u, t)$ and to follow standard notation let $i = \sqrt{-1}$ in the next two equations. We employ the following method to numerically obtain $v(x)$, the inverse Laplace transform of $V(u)$ using the well-known complex version formula

$$v(x) = \int_{a-i\infty}^{a+i\infty} e^{ux} V(u)\, du = \frac{e^{ax}}{\pi} \int_0^{\infty} \Re\{V(u)e^{iwx}\}\, dw$$

where $u = a + iw$. If the above integral is now discretized using the trapezoidal rule with step size π/T, the following Fourier series approximation $\hat{v}(x)$, of period $2T$, is obtained

$$\hat{v}(x) = \frac{e^{ax}}{T}\left[\frac{V(a)}{2} + \sum_{k=1}^{\infty}\left\{\Re\left(V\left(a + \frac{k\pi i}{T}\right)\right)\cos\left(\frac{k\pi x}{T}\right) - \Im\left(V\left(a + \frac{k\pi i}{T}\right)\right)\sin\left(\frac{k\pi x}{T}\right)\right\}\right]. \quad (7)$$

The discretization error declines exponentially as aT increases [7]

$$\hat{v}(x) - v(x) = \sum_{k=1}^{\infty} e^{-2kaT} v(2kT + x); \qquad 0 \le x \le 2T.$$

Since $v(x) \le 1.0\ \forall x$, the discretization error is easily made very small. Therefore, the bulk of the error in the numerical inversion procedure accrues from truncating the Fourier series. The Fourier series exhibits characteristically slow convergence. However, acceleration methods that allow accurate estimates of the series from the first m terms are known. We use the quotient-difference algorithm of Rutishauser with a remainder estimate suggested by DeHoog *et al.* in [5] to accelerate the convergence of the Fourier series. Cooley *et al.* in [4] use the cosine transform to approximate a series very similar to (7), and Jagerman [17] obtains an expression similar in form to (7) by considering the generating function of a sequence of functionals that converge in the limit to $v(x)$. Because of the $O(n^3)$ cost of computing each function value, the method that reliably yields accurate results with the fewest evaluations is best. We have been pleased with the results obtained when the Fourier series is evaluated to the first m = 80 terms with the DeHoog remainder estimate (even when the desired distribution has jumps at various values of x). The structure of the overall algorithm is as follows.

A: Determine $\mathcal{Y}^{\sim}(u, t)$
for(m values of u) {
  determine the eigenvalues of $(uR - Q)$ — $O(n^3)$
  for(d unique eigenvalues of $(uR - Q)$) {
    solve transformed Hessenberg system — $O(n^2)$
  }
  evaluate partial fraction coefficients — $O(n^3)$
}

TABLE II
COMPUTATION TIME AND FLOPS WITH DIFFERENT VALUES OF n

	$n \rightarrow 4$	10	40	170	365
flops	3.2×10^5	4.0×10^6	1.8×10^8	1.1×10^{10}	1.0×10^{11}
time	6 s	15 s	320 s	3.1 h	25 h

B: Numerical Laplace Transform Inversion
for(n states) {
  for(p desired values of t) {
    for(m values of u) {
      sum partial fraction coefficients
        to evaluate $V(u)$ — $O(n)$
    }
    for(q values of x) {
      sum Fourier series approximation
        to evaluate $v(x)$ — $O(m)$
    }
  }
}

In the worst case, the inner loop of phase *A* of the computation is executed $O(n)$ times. Since each iteration of the inner loop has a computational cost of $O(n^2)$, phase *A* has a computational complexity of $O(mn^3)$. The computational cost of phase *B* is primarily a function of the p different values of time t at which $\mathcal{Y}(x, t)$ is to be evaluated and the m terms in the Fourier series approximation. Phase *B* has a computational complexity of $O(pmn(n + q))$. Therefore, phase *A* comprises the principal computational burden of the algorithm. The total computational effort to obtain $\mathcal{Y}(x, t)$ for q values of x at each of p values of t is $O(pmn(n + q) + mn^3)$. The practical implication is that once the computationally expensive phase *A* has been used to determine $\mathcal{Y}^{\sim}(u, t)$, evaluating $\mathcal{Y}(x, t)$ at other (x, t) points can be done very cheaply.

Often the constants that are brushed under the rug by the $O(\ )$ notation are important. The computational cost of the algorithm is approximately $16m(1 + \alpha)n^3$ where α is a difficulty factor for the QR algorithm that depends on the spectrum of $(Q - uR)$. Since for most matrices $1 \le \alpha \le 2$, the computational cost should be between $32mn^3$ and $48mn^3$. We present in Table II the operation counts (flops) and approximate computation times for determining $\mathcal{Y}_i(x, t)$ on a CONVEX C-1 XP. The operation count values are the median of a small sample, and the time values are the maximum of the same small sample. The order estimates of the previous paragraph indicate the importance of n to the asymptotic behavior of the computation time. Consequently, we fix the number of terms in the Fourier series expansion m at 80 and the number of time values p at 1. The number of values of x (amounts of accumulated reward) is fixed at 100. These are typical values we used to examine $\mathcal{Y}_i(x, t)$ for the examples in this paper. The effect of the $pmn(n + q)$ term is unimportant when $p < n$ and $pq < n^2$. Therefore, the increase in flops becomes approximately n^3 for values of $n \ge 6$. The computation time for the larger state problems is approximate because the jobs were run at a low priority and the CONVEX C-1 was not dedicated to solving these problems.

References

[1] M. Beaudry, "Performance related reliability for computer systems," *IEEE Trans. Comput.*, vol. C-27, pp. 540–547, June 1984.

[2] D. Bhandarkar, "Analysis of memory interference in multiprocessors," *IEEE Trans. Comput.*, vol. C-24, pp. 897–908, Nov. 1975.

[3] B. Ciciani and V. Grassi, "Performability evaluation of fault-tolerant satellite systems," *IEEE Trans. Commun.*, vol. COM-35, pp. 403–409, Apr. 1987.

[4] J. Cooley, A. Lewis, and P. Welch, "The fast Fourier transform: Programming considerations in the calculation of the sine, cosine and Laplace transforms," *J. Sound. Vib.*, vol. 12, pp. 315–337, 1970.

[5] F. DeHoog, J. Knight, and A. N. Stokes, "An improved method for numerical inversion of Laplace transforms," *SIAM J. Sci. Stat. Comput.*, vol. 3, no. 3, pp. 357–366, 1983.

[6] L. Donatiello and B. Iyer, "Analysis of a composite performance reliability measure for fault-tolerant systems," *J. Ass. Comput. Mach.*, vol. 34, no. 1, pp. 179–199, 1987.

[7] F. Durbin, "Numerical inversion of Laplace transforms: An efficient improvement to Dubner and Abate's method," *Comput. J.*, vol. 17, pp. 371–376, 1974.

[8] E. de Souza e Silva and R. Gail, "Calculating cumulative operational time distributions of repairable computer systems," *IEEE Trans. Comput.*, vol. C-35, pp. 322–332, Apr. 1986.

[9] D. Furchtgott and J. Meyer, "A performability solution method for degradable nonrepairable systems," *IEEE Trans. Comput.*, vol. C-33, pp. 550–554, June 1984.

[10] A. Goyal and A. Tantawi, "Evaluation of performability for degradable computer systems," *IEEE Trans. Comput.*, vol. C-36, pp. 738–744, June 1987.

[11] A. Goyal, A. N. Tantawi, and K. S. Trivedi, "A measure of guaranteed availability," IBM Res. Rep. RC 11341, Aug. 1985.

[12] A. Hordijk, D. Iglehart, and R. Schassberger, "Discrete time methods for simulating continuous time Markov chains," *Adv. Appl. Prob.*, vol. 8, pp. 772–788, 1976.

[13] R. A. Howard, *Dynamic Probabilistic Systems, Vol. II: Semi-Markov and Decision Processes.* New York: Wiley, 1971.

[14] M. C. Hsueh, R. K. Iyer, and K. S. Trivedi, "Performability modeling based on real data: A case study," *IEEE Trans. Comput.*, this issue, pp. 478–484.

[15] R. Huslende, "A combined evaluation of performance and reliability for degradable systems," in *Proc. ACM/SIGMETRICS Conf. Measurement Modeling Comput. Syst.*, 1981, pp. 157–164.

[16] B. R. Iyer, L. Donatiello, and P. Heidelberger, "Analysis of performability for stochastic models of fault-tolerant systems," *IEEE Trans. Comput.*, vol. C-35, pp. 902–907, Oct. 1986.

[17] D. L. Jagerman, "An inversion technique for the Laplace transforms," *Bell. Syst. Tech. J.*, vol. 61, pp. 1995–2002, 1982.

[18] V. Kulkarni, V. Nicola, and K. Trivedi, "On modeling the performance and reliability of multi-mode computer systems," *J. Syst. Software*, vol. 6, pp. 175–183, May 1986.

[19] V. Kulkarni, V. Nicola, R. Smith, and K. Trivedi, "Numerical evaluation of performability and job completion time in repairable fault-tolerant systems," in *FTCS Proc.*, vol. 16, 1986.

[20] J. Meyer, "On evaluating the performability of degradable computer systems," *IEEE Trans. Comput.*, vol. C-29, pp. 720–731, Aug. 1980.

[21] ——, "Closed-form solutions of performability," *IEEE Trans. Comput.*, pp. 648–657, July 1982.

[22] P. S. Puri, "A method for studying the integral functionals of stochastic processes with applications: I. The Markov chain case," *J. Appl. Prob.*, vol. 8, pp. 331–343, 1971.

[23] A. R. Reibman and K. S. Trivedi, "Numerical transient analysis of Markov dependability models," *Comput. Oper. Res.*, vol. 15, pp. 19–36, 1988.

[24] D. P. Siewiorek, "Multiprocessors: Reliability modeling and graceful degradation," in *Proc. Infotech State of the Art Conf. Syst. Reliability,* Infotech Int., London, 1977.

[25] D. P. Siewiorek, V. Kini, R. Joobbani, and H. Bellis, "A case study of C.mmp, Cm*, and C.vmp: Part II—Predicting and calibrating reliability of multiprocessor systems," *Proc. IEEE,* vol. 66, pp. 1200–1220, Oct. 1978.

[26] W. T. Weeks, "Numerical inversion of Laplace transforms using Laguerre functions," *J. Ass. Comput. Mach.*, vol. 13, pp. 419–426, July 1966.

[27] J. H. Wilkinson and C. Reinsch, *Handbook for Automatic Computation, Vol. II: Linear Algebra.* Berlin, Germany: Springer-Verlag, 1971.

[28] L. T. Wu, "Operational modes for the evaluation of degradable computing systems," in *Proc. ACM/SIGMETRICS Conf. Measurement Modeling Comput. Syst.*, 1982, pp. 179–185.

R. M. Smith, photograph and biography not available at the time of publication.

Kishor S. Trivedi (M'86–SM'87) received the B.Tech. degree from the Indian Institute of Technology, Bombay, and the M.S. and Ph.D. degrees in computer science from the University of Illinois, Urbana-Champaign.

He is the author of text on *Probability and Statistics with Reliability, Queueing and Computer Science Applications* (Englewood Cliffs, NJ: Prentice-Hall). His research interests are in computing system reliability and performance evaluation. Currently, he is a Professor of Computer Science and Electrical Engineering, Duke University, Durham, NC. He has served as a Principal Investigator on various AFOSR, ARO, Burroughs, IBM, NASA, NIH, and NSF funded projects and as a consultant to industry and research laboratories.

Dr. Trivedi was an Editor of the IEEE TRANSACTIONS ON COMPUTERS from 1983 to 1987. He is an editor of the *Journal of Parallel and Distributed Computing*.

A. V. Ramesh received the B.Tech. degree from the Indian Institute of Technology, Madras, and the M.S. degree in structural engineering from Duke University, Durham, NC.

His research interests are in applied mechanics, numerical analysis, and system theory. Currently, he is in the Ph.D. program in structural engineering at Duke University.

Chapter 6: Conclusion

In preceding chapters, we discussed reliability techniques for distributed systems. Issues raised therein correspond to most real-world problems. One important aspect is to find out whether a given task could be successfully executed, which is possible only if the system still contains an adequate amount of healthy resources. This translates into testing whether certain minimum operational resources exist that are still connected for appropriate coordination. This problem, known as *task-based reliability*, has become crucial to mission-critical applications. The first paper in Subchapter 6.1 addresses this issue in various circuit-switched, network-based multiprocessors.

Another important characteristic of distributed computing systems is localized information processing by a subset of the processing elements. A piece of information may be processed repeatedly within a local subsystem before it changes locality. For example, a seat assignment file in an airline reservation system is heavily used, first, at the airport where passengers are preparing to board. Then, as the plane flies to another airport, the location of the file's use changes. To maximize the efficiency of the system, the file should be accessible at the locality of first use and able to migrate as the need changes. The second paper computes the reliability of transaction-based systems in which multiple copies of files are distributed throughout the system.

The third paper discusses the allocation issue, commonly termed *file allocation problem* (FAP). The basic file placement problem is allocating multiple copies of a single file. It is assumed that each node can access average query and update rates, that each query accesses only one copy of the file, and that all copies must be accessed by an update. Queries, updates, and data storage are represented as system costs, which must be minimized. The problem addresses the most basic attributes of query, update, and storage costs and omits operational features, such as the delay of return traffic and system reliability. The solution to a FAP can be formulated as a linear integer program. For large problems, heuristics offers "reasonable" polynomial time-search strategies, but such strategies do not guarantee precision. They are generally interactive algorithms that generate feasible solutions. A decision algorithm then decides whether and how to improve the solution. Sometimes, sample cases are also simulated to generate near-minimal solutions.

Subchapter 6.2 addresses the problem of *software reliability*, which is viewed as the probability that a given program will operate correctly in a specified environment for a specified duration. With this view in mind, several models have been proposed for estimating the reliability of programs. These models can be broadly categorized as software reliability growth models and statistical models. The former attempt to predict the reliability of a program on the basis of its error history, while the latter estimate it by determining the response (success or failure) of a program to a random sample of test cases, without correcting any errors that may be discovered during the process. A survey of these models appears in Goel (1985), which discusses the appropriateness of the assumptions underlying these models as well as their applicability during the software development cycle. It also gives a step-by-step procedure for developing a software reliability model from available failure data. The other two papers included in this subchapter describe status, perspective, and modeling tools, including an N-version programming approach for software reliability.

Finally, Subchapter 6.3 considers case studies in which operational reliability and availability of digital switching system DX200 and VAXcluster are described. A paper on MAFT architecture for distributed fault tolerance presents another relevant notion toward reliability consciousness. The final paper considers a methodology used in the reliability analysis of a call-handling database system and presents an excellent case study in the reliability area.

EH0300-4/90/0000/0229$01.00 © 1990 IEEE

A RELIABILITY PREDICTOR FOR MIN-CONNECTED MULTIPROCESSOR SYSTEMS

John J. Macaluso, Chita R. Das, and Woei Lin

Computer Engineering Program
Department of Electrical Engineering
The Pennsylvania State University
University Park, PA 16802

ABSTRACT

In the world of cost-effective supercomputing, the use of a multistage interconnection network (MIN) as a means of connecting many processing elements to many memory modules is widespread. Whenever such a system is used in a critical environment, reliability becomes an important issue. Up to this point reliability evaluation methods for multiprocessor systems have been *ad hoc*, that is, designed for, and applicable to, only one or a few types of topology.

This paper presents the first automated simulation package with the ability to perform the reliability simulation of MIN-connected systems. The program is automated in that the required MIN topology is built by the program. The user need only specify the type of MIN and other system characteristics. The underlying strategy of the program is to find the system reliability from the system reachability matrix, which is built by a search procedure requiring $O(NS(N))$ time, where $S(N)$ is the number of switches in an $(N \times N)$ MIN. The package was used to simulate the reliabilities of many topologies proposed in the literature. Some results are presented and used for a comparison of the systems.

1. INTRODUCTION

Multiprocessor systems using multistage interconnection networks (MINs) have been an active area of research for more than a decade. A plethora of different MIN topologies have been proposed to provide communications among N processors (PEs) and N memory modules (MMs). These MINs are generally designed with stages of $(n \times m)$ switching elements (SEs), where n and m are small integers such as 2, 3, or 4. A good body of literature on MINs can be found in [1], and a survey of some fault-tolerant multipath MINs is reported in [2].

The novelty of a multiprocessor lies in its ability to provide high computing power with assured reliability. Reliability becomes important especially when the system is used in a critical application. While performance analyses of the MIN-based systems have been carried out extensively along with their design, relatively little attention has been paid to the reliability issues. Work on fault-tolerant MINs has been mostly confined to finding alternate paths between source and destination sets.

In the past, research pertaining to the reliability evaluation of MINs has addressed either full connectivity without degradation [3] or terminal reliability [4]. A couple of papers have addressed the complete reliability of MIN-based systems considering the failure of PEs, MMs, and SEs [5], [6]. However these works are restricted in a sense that either the system size is limited or the evaluation technique is not applicable to all types of MINs. Recently a combinatorial approach for reliability evaluation of multiprocessors using (4×4) SEs is given in [7]. This analysis is applicable to only unique-path MINs with (4×4) SEs.

As more and more fault-tolerant MINs are proposed, it is essential to develop a methodology for characterizing and comparing one system with another from the reliability standpoint. This type of unified evaluation technique will solve two purposes. First, the reliability of any existing or new MIN-based system can be predicted. Second, depending on the implementation requirements, a cost-effective MIN can be selected. The survey work in [2] has compared the fault-tolerant property of various multipath MINs. However, the work is not complete in a sense that the usual evaluation criterion such as the reliability/availability issue is not addressed. In this paper we are concerned with developing a unified reliability evaluation technique for various types of MIN-based systems.

Analytical evaluation of system reliability considering the degradation of PEs, MMs, and SEs is very difficult due to the NP-hardness of the problem [8]. Therefore, all the analytical evaluation techniques have been restricted to mostly unique-path MINs. As we are interested in analyzing and comparing different multipath strategies, an analytical approach seems almost impossible. Hence, simulation is used as the evaluation tool. This paper presents the first proposed automated simulation package with the ability to perform the reliability evaluation of MIN-connected multiprocessor systems.

The package takes from the user an input file containing the specifications of the topology and the details of the type of analysis desired. The user is freed from the interconnection details because the program has the ability to automatically build the proper topology. A search algorithm is employed during the course of the simulation to find the connectivity between PEs and MMs in the presence of component failures. While the size of the system is not limited by the program, the host machine environment and simulation time may be limiting factors. The reliability model used in this paper is known as task-based reliability [9], where a system remains operational as long as a task can be executed on it. Results of (16×16) and (64×64) systems using the

Faye A. Briggs, *Proceedings of the 1988 International Conference on Parallel Processing*, University Park and London: The Pennsylvania State University Press, 1988, pages 392–399.

following topologies are analyzed in the paper with and without system cost factor involved.

The topologies considered are (2×2) baseline [10], an extra-stage baseline, the (4×4) butterfly [11], the chained MIN [12], [13], the F network [14], the merged delta network (MDN) [15], the inverse augmented data manipulator (IADM) [16], and the interconnection network designed for reliable architectures (INDRA network) [4]. Although this selection covers almost the whole spectrum of MINs proposed in the literature, the program is not limited to only these topologies. It also has the ability to include virtually any MIN-based system. We do not fully describe each of these systems; more complete system descriptions can be found in the literature cited. Since the topologies chosen are representative of those surveyed in [2], this work could be considered a follow-up or extension of the work presented there.

In Section 2 we present an overview of the topologies considered. The simulation techniques are explained in detail in Section 3, including algorithm time complexities. Section 4 gives the results of the system simulations and offers a comparison between them. Concluding remarks are given in Section 5.

2. SYSTEMS SURVEY

This section briefly surveys the different MIN-connected multiprocessor systems listed in the introduction to this paper. We consider a tightly coupled multiprocessor environment where the PEs and MMs are connected through the MIN. The difference between the systems lies solely in the type of interconnection network used for communications among the processors and memories, and therefore each system will be described by its MIN.

2.1 Unique-Path MINs

A unique-path MIN provides only one path between each processor and each of the memory units. The advantage of unique-path networks lies in the simplicity of their implementation. The network uses uncomplicated selector or crossbar switches that require no look-ahead capability (i.e., they need not be independently cognizant of the conditions of the other components in the system). Each switch, if operational, merely routes an input to the proper output link depending upon the value of the request tag bit corresponding to the switch's stage. This simplicity results in fewer internal components and consequently a high average switch reliability relative to more complicated switches, such as those used in the multiple-path systems described later.

The disadvantage of unique-path networks lies in the fact that if any of the switches along a desired path fails, the entire path is eliminated, and the requesting module is unable to access the requested module. However, if a strategy exists whereby another path can be found to act as a detour around the failed element, this fault may be tolerated. These extra paths can be provided at the expense either of redundant passes through the network [17], [18] or of additional hardware; we will consider the latter.

The baseline MIN is one example of an unique-path network. The topology of an (8×8) baseline MIN is shown in Fig. 1. The baseline network consists of $n = \log_2 N$ switch stages, each containing $N/2$ (2×2) crossbar switches. The baseline MIN was chosen to represent other unique-path MINs proven to be topologically equivalent to the baseline by Wu and Feng [10]. We use the baseline in our explanations because of its simplicity of representation.

Another unique-path network, comprised of (4×4) switches, is the butterfly MIN. The BBN Butterfly Parallel Processor^{TM} is a commercially available system with up to 256 processors [11]. Because of the additional links in each switch, the butterfly MIN has only $\log_4 N$ stages each consisting of only $N/4$ switches. Thus, the communications delay of a butterfly MIN is only $O(\log_4 N)$, whereas the baseline delay is $O(\log_2 N)$.

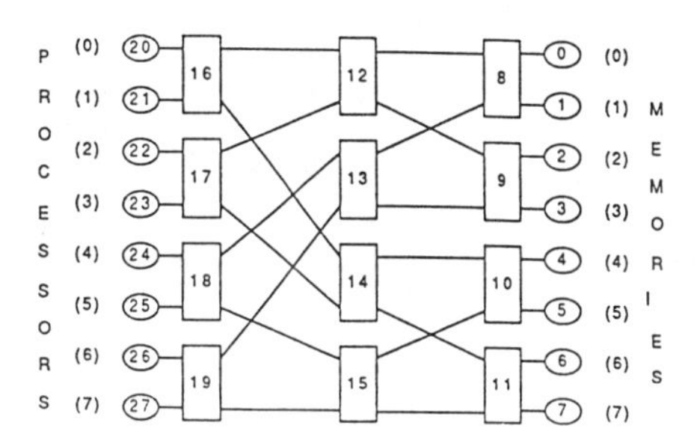

Fig. 1. A (8×8) baseline system.

2.2 Path Redundancy for Unique-path MINs

It may be possible to provide redundant paths to an unique-path MIN by at least four methods. The first involves the addition of an extra stage of switches to the input of the MIN. This is done by duplicating the MIN input stage along with its output link interconnection pattern; this extra stage is inserted between the processors and the previous input stage. This method is examined in this research by the addition of an extra stage to both the baseline and the butterfly MINs. A second redundancy method adds chaining links to each switch, partitions all the switches, and then chains together the switches in the same partition. The chained baseline MIN was examined as an example of this method. A third redundancy strategy consists of replicating r times a network consisting of $(r \times r)$ switches (e.g., the baseline MIN consists of (2×2) crossbar switches, therefore 2 network copies would be provided). This INDRA technique is examined for the baseline network. Finally, the fourth technique crosslinks c copies of an $(\frac{N}{c} \times \frac{N}{c})$ unique path network. This technique can be applied to any of the topologically equivalent unique-path MINs, but traditionally, it is employed using the delta topology to form the MDN.

2.3 Inherently Path-redundant MINs

Anticipating the need for redundant paths in a MIN, some topologies have been designed to provide these multiple paths without a need for further modification. One example, the IADM, uses $(\log_2 N) + 1$ switch stages, each consisting of N (3×3) switches. Another example, the F network, uses $\log_2 N$ switch stages, each with N (4×4) switches. These two networks show most clearly the truism that statically redundant paths require redundant hardware.

3. SIMULATION TECHNIQUES

In this research, it is assumed that the MIN of each of the systems considered has an input side and an output side, and all communications between modules are carried out in one pass from an input position to an output position. Using the system of Fig. 1 as an example, communications between processor 0 (node 20) and processor 3 (node 23) would follow the node path $20 \rightarrow 16 \rightarrow 12 \rightarrow 9 \rightarrow 3$ and not $20 \rightarrow 16 \rightarrow 12 \rightarrow 17 \rightarrow 23$. This example also illustrates the equivalence of input and output positions.

Depending upon the implementation, each of the topologies can be operated under either a circuit-switched or a packet-switched communications protocol. Under circuit switching, a physical link is established between two modules, and is used for transmissions in both directions. In contrast, packet switching is an asynchronous simplex protocol where information packets are exchanged via the network. The program has the ability to simulate either of these protocols.

To use the program, the user need merely edit an already-existing input file. The information in the input file includes the following.

a. The system size.
b. The system type (from a menu list).
c. The communications protocol.
d. The failure rates of PEs, MMs, and SEs.
e. Whether each PE is assigned a local MM.
f. The number of copies of the MIN (for INDRA case).
g. The output file name.

At the beginning of the simulation, this information is read into the program. The program then calls the appropriate procedure to build the desired topology. The ability to build the topology automatically (described later) is especially important in the case of large systems, when hand entry of the interconnection pattern becomes very difficult.

The simulator determines the system condition, whether operational (up) or failed (down), from **R**, the $(N \times N)$ system reachability matrix. **R** describes the connectivity between modules in the following way: if at least one path exists between processor p and memory m, then the matrix element $R[p,m] = 1$, otherwise $R[p,m] = 0$. In an unfailed system (at system startup), $\mathbf{R}=\mathbf{1}$. As components fail, the degree of system degradation can be determined from **R**. If a system is defined as being up if it has at least i processors and j memories all being both operational and completely connected to each other, then the system condition can be determined by examining **R** to ascertain whether a submatrix of order at least $(i \times j)$ and with elements all of value 1 can be found in **R**.

The simulator is based upon the following concept. The reliability evaluation of any system is dependent upon its reachability matrix. Therefore, if a program can be developed to find system reliability from **R**, and if any MIN-based system can be reduced to its reachability matrix, then the reliability of any MIN-based system can be simulated.

The algorithm for simulating system reliability from **R** is as described in [5] and will not be detailed here. The remainder of this section will explain the program with respect to internal system representation and the characterization of a search-traversal algorithm capable of finding **R**. The serial version of the search algorithm is given, and possibilities for a parallel implementation are explained.

3.1 System Representation

The topology of each network is represented by certain constants and arrays as follows (all parenthesized examples correspond to the system of Fig. 1). The system constant a is the total number of PEs, MMs, and SEs present in the system (e.g., $a = 2N + \frac{N}{2}\log_2 N = 28$); the system constant $offset$ $(= a - N)$ is the vertex number of processor 0 (e.g., $offset = 20$); finally, the system constant b is the maximum number of output links per node present in the system (e.g., $b = 2$). The vertices are numbered from 0 to $a-1$ in column-major order beginning with the vertex corresponding to memory 0 and ending with that of processor $N-1$. The output links (if any) are numbered from 0 to $b-1$ from top to bottom. The $(a \times b)$ matrix, **T**, describes the interconnection pattern of each topology as follows: the element $T[i,j]$ is the component connected to output link j of component i, where $i \in \{0..a-1\}$ and $j \in \{0..b-1\}$ (e.g., $T[13,0] = 8$). By convention, if $T[i,j] = -1$, then output link j does not exist for component i (e.g., $T[3,1] = -1$). The one-dimensional boolean array **living** represents the system component failure condition as follows: for all $i = 1 \cdots a-1$, **if** $living[i]$ **then** component i is operational **else** i is failed.

Since all the systems can be represented by this scheme, the user need only indicate the network topology and size. The program calculates the system constants and calls the appropriate procedure to build **T**.

These representations are needed because the procedure which finds the reachability matrix of any system is a search-traversal algorithm. A search-traversal strategy is necessary since the conventional methods of finding **R** reported in [5] do not work in the case of multipath systems such as those surveyed in [2].

3.2 A Serial Algorithm for Finding R

As described in [19], a MIN-based multiprocessor system can be represented by a directed graph. Since this is true, it follows that an $(N \times N)$ system can also be conceptualized as a grove of N search trees, where the root of each tree corresponds to an individual processor vertex, its leaves represent the memory units, and its shape depends upon the network topology. The search tree for the unfailed system of Fig. 1 can be seen in Fig. 2(a). The effect of a component failure will be to prune the tree at the appropriate position(s) held by that component in the tree. For example, if the switch corresponding to node 12 of Fig. 1 fails, the resulting search tree is as seen in Fig. 2(b).

The problem, then, of finding the connectivity between any processor p and any memory m at a given time reduces to: is m present in the search tree of p? This can be accomplished by initializing **R** to **0**, and then performing a reverse preorder traversal of the search tree of p, setting to 1 the appropriate elements

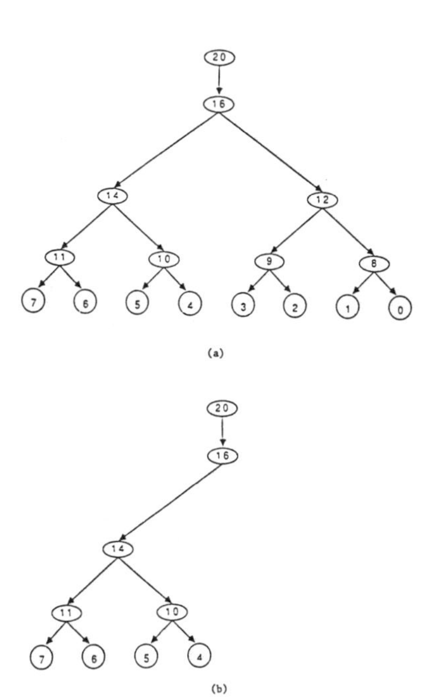

Fig. 2: Search tree for processor 0 of Fig. 1. (a) For an unfailed system. (b) Pruned after failure of node 12.

of **R** each time a leaf vertex (memory) is reached. The termination of the traversal search of the tree of processor p results in the completion of row p of **R**. Therefore, N such searches will complete the entire reachability matrix. The *Search* algorithm of Fig. 3 performs the search of one tree, and the *Reach* algorithm of Fig. 4 uses *Search* to build the reachability matrix, **R**.

The stack of *Search* is initially empty. At the beginning of the search, the *source* vertex is pushed onto the stack. The following steps are repeated while the stack is not empty.

1. A *node* is popped from the stack and checked if it is a leaf; if so, then the proper element of **R** is set to 1.
2. The processed *node* is tagged "visited."
3. The unvisited neighbors of *node* are each pushed onto the stack in birthright order.

Birthright order is from leftmost child to rightmost child in the tree representation. For example, birthright order for vertex 16 of Fig. 1 is vertex 14 and then vertex 12. This order ensures the depth-first traversal desired. The marking of the processed vertices as "visited" serves two purposes.

1. The algorithm is kept from infinitely traveling around any closed loops possibly inherent in a topology (such as the very visible loops of the chained baseline MIN structure).
2. A mechanism is provided by which future consideration of a vertex once processed is denied. This helps to reduce search time by ensuring that each vertex is processed only once, thereby providing an extra stage of pruning in systems whose search trees have vertices holding multiple positions. It is important to note that this pruning mechanism makes this algorithm suitable only for finding whether at least one path exists to each memory—not for finding the number of paths to each memory.

```
procedure  Search (source : integer)
begin
   Mark each vertex "not visited";
   Zero row source - offset of R;
   Reset the local stack;
   Push source onto the stack;
    while stack is not empty  do  begin
      node := stack pop;
      if node < N  then  begin
        (* node is a memory *)
        R[source - offset, node] := 1
      end;  (* if *)
     Mark node "visited";
      for each of node's children i  do
      begin
        if (i is  not visited)
        and ((i < N)
        or (i is living))  then  begin
          Push i onto stack;
        end;  (* if *)
      end;  (* for *)
    end;  (* while *)
end; (* procedure *)
```

Fig. 3. Tree-search algorithm.

```
procedure Reach;
    procedure Search;
begin
L1: for p := 0  to N - 1  do begin
        source := p + offset;
        Search (source);
     end;  (* for *)
     if packet-switched protocol  then
     begin (* packet adjust *)
L2:     for p := 0  to N - 1  do begin
          for m := p + 1  to N - 1  do
          begin
            if R[p, m] = 0
            or R[m, p] = 0  then
            begin
              R[p, m] := 0;
              R[m, p] := 0;
            end;  (* if *)
          end;  (* for *)
        end;  (* for *)
     end;  (* if *)
L3: for each processor p  do begin
        if p is failed  then
          zero row p of  R;
     end;  (* for *)
L4: for each memory m  do begin
        if m is failed  then
          zero column m of  R;
     end;  (* for *)
end; (* procedure *)
```

Fig. 4. Algorithm for finding **R**.

3.3 Algorithm Time Complexities

It is important that the time complexities of *Search* and *Reach* be calculated since they are used frequently in the simulation program. In the calculations that follow, the function $S(N)$ gives the number of switches in the MIN as a function of N (e.g., $S(N) = \frac{N}{2}\log_2 N$) for a $(N \times N)$ baseline MIN). The relative growths of N and $S(N)$ are topology dependent. In the case of the F network, for example, $S(N) = N \log_2 N > N$ always. However, in the separate example of the butterfly network, $S(N) = \frac{N}{4}\log_4 N$, which is greater than N only for large N (i.e., $N > 256$). Since the asymptotic time complexities concern systems with large N, it is assumed in the derivations below that in all cases $S(N) > N$.

3.3.1 Time Complexity of the *Search* Procedure

The *Search* procedure of Fig. 3 consists of two parts: the nonsearch statements (those before the **while** loop) and the statements of the search loop (those making up the **while** loop). The time complexity of the nonsearch statements is $O(S(N))$ because the dominating term is the Mark statement since it initializes all $2N + S(N)$ vertices, and $S(N)$ grows faster than N. The search statements consist of: three assignment statements, each with constant time complexity, and a **for** loop of $O(b) = O(1)$ time complexity since the maximum number of links per node, b, remains constant as the size of the system grows. Thus the statements within the **while** loop are all of constant time complexity, and the time complexity of the entire loop is governed by the maximum number of iterations of the loop as follows. The marking as visited of each processed node ensures that each vertex is considered only once during each tree traversal. Since the maximum number of vertices that a single search can consider is the $S(N)$ switches plus the N memories plus the processor root, the time complexity of a search is $O(S(N))$. Therefore, the time complexity of procedure *Search* is

$$T_S(N) = O(S(N)). \quad (1)$$

3.3.2 Time Complexity of the *Reach* Procedure

Procedure *Reach* can be seen to consist of four loops labeled $L1$ through $L4$ in Fig. 4. Loop $L1$ performs N successive calls on *Search*, and therefore has an $O(NS(N))$ time complexity. Loops $L3$ and $L4$ each perform N iterations of constant-time conditional assignment statements, so each has an $O(N)$ time complexity. Loop $L2$ performs constant-time conditional assignments as many times as there are elements of **R** above the main diagonal, or $\frac{1}{2}(N^2 - N)$. Since L1 is the dominant loop, the time complexity of *Reach* is

$$T_R(N) = O(NS(N)). \quad (2)$$

3.4 A Parallel Algorithm for Finding R

Time requirements for the searches can be lessened even further by performing them in parallel. The parallelism of the *Search* algorithm is apparent; each search is completely independent of every other search (i.e., there are no data dependencies between the searches). This parallelism can be easily exploited on an array of P processors, where $P = 2^n$ for positive integer n. If $N > P$, the technique of loop concurrentization [20] could be used without much difficulty by partitioning the rows of **R** (searches) and assigning a different partition to each processor. For ease of explanation, however, we will consider the case where $P = N$, where each of the processors would be assigned a different row of **R**.

When a packet-switched protocol is being simulated, the execution of the packet-adjustment statements in loop L2 of Fig. 4 introduces data dependencies, and communications between processors becomes necessary. Specifically, each processor adjusts its row of **R**, but only the elements to the right of the main diagonal. Each of these elements $R[i,j]$, where $i > j$, is compared to the element $R[j,i]$ symmetrical to it with respect to the main diagonal. But before a processor i can properly examine an $R[j,i]$ value, it must receive a signal from processor j that the search of row j is complete.

The blocks labeled L1 through L4 of Fig. 4 can each be reduced by a factor of N if the procedure is implemented in parallel. In this case, the dominant execution sequence would be the *Search* procedure of L1. Therefore, if procedure *Reach* is performed in parallel with $P = N$, the time complexity is $O(S(N))$ by Eq. 1.

4. RESULTS AND DISCUSSION

This section compares the selected topologies with respect to their simulated reliabilities. In addition, since the design emphasis on MINs is inspired by a need for cost-effective communication networks, the systems were compared with respect to the ratio of system reliability to system cost (or reliability-to-cost ratio, RCR). The program also has the ability to simulate a system with any given coverage factor, C [21]. Results were obtained for systems of different sizes under both circuit- and packet-switched protocols and with different values of C. However only selected outputs are presented here.

4.1 Elements of the Comparison

For the comparisons to be valid, the processors as well as the memories were assumed to be homogeneous within each system, and the same processor and memory failure statistics were used in each type of system. In this way, each network differed from the others only in its type of MIN.

The reliability of each system is directly related to the mean failure rates of the individual elements making up the interconnection network. The mean failure rates for processors, λ_p, and memories, λ_m, were each taken to be one per 10^4 hours. Since the systems differ only in the type of MIN used, any change in λ_p or λ_m will affect all of the systems equally. Therefore, the comparison depends upon the failure rate of the MIN, which depends upon the failure rates of the individual switching elements. In the absence of any practical failure data, switch failure rates were calculated using the MIL-HDBK-217B reliability model for metal-oxide-semiconductor integrated circuits [22]. Details of the assumed switch design and failure-statistic calculations can be found in [23]. The program considers the MIN to consist of three sets of switches: the input bank, the output bank, and the banks between the input and the output banks. Then the characteristic switch failure rate is assigned to the switches of each set.

The network cost factor is the sum of the costs of all the individual switches comprising the network. The cost of each switch is calculated as a function of the number of its input links, n, and the number of its output links, m, using the equation of the cost function, $C(n,m)$, of Eq. 3.

$$C(n,m) = \begin{cases} nm & \text{for a crossbar switch;} \\ n+m & \text{for a selector switch.} \end{cases} \quad (3)$$

Eq. 4 calculates C_N, the cost of a MIN consisting of x different types of switch, each type i having a cost C_i and a population N_i. The network cost factors are then calculated by dividing each network cost by the minimum cost of all the networks of the same size (in this case the baseline).

$$C_N = \sum_{i=1}^{x} N_i C_i \quad (4)$$

4.2 System Reliability Comparison

The reliability curve for a task requiring 50% of the total number of PEs and 50% of the total number of MMs was obtained for each of the systems.

Figs. 5 and 6 contain the (16 $\times$ 16) system curves for a circuit-switched protocol and a coverage factor of C = 1.0 and C = 0.8 respectively. The difference between the curves of Fig. 5 is noticeable, however when the system's ability to reconfigure itself is relatively weak as in Fig. 6, the fault-tolerant scheme has less of an effect on reliability. In fact, with a coverage factor of 0.8, the topology of the MIN seems to have almost no effect at all on reliability; the curves are almost indistinguishable from each other. This observation follows intuition: a topologically inherent fault-tolerance scheme is of benefit only if the maintenance processor is able to utilize it.

The reliability curves for (64 $\times$ 64) systems under a circuit-switched protocol with C = 1.0 are shown in Fig. 7. As expected, the INDRA, F network, chained MIN, MDN, and extra-stage MINs give high reliability compared to the unique-path MINs. Also, as the system size increased, the reliability gain of multipath networks becomes more pronounced.

Probably the most surprising curve, however, is that of the IADM system. It does not seem to agree with intuition that it would have a reliability consistently below all the others, including the unique-path systems. However, upon further inspection, the reasons become clear. The IADM has multiple paths, but they are not evenly distributed between all processor-memory pairs. For example, there is only one path between processor i and memory j when $i = j$. However, probably the most significant reason for the low reliability is that the IADM contains $N(\log_2(N)+1)$ switches—many more than the $\frac{N}{2}\log_2 N$ switches of the baseline or the $\frac{N}{4}\log_4 N$ switches in the butterfly. If the failure rate of an individual switch is λ_s, then the failure rate of the switches in the system is given by $\sigma\lambda_s$, where σ is the number of active switches at any time. Clearly, if σ is very large (as in the IADM), then the switches will fail much more frequently than if σ is small (as in the baseline, or especially the butterfly). Therefore, the combination of unevenly distributed paths and quick-failing switches makes the IADM a less reliable system compared to the others.

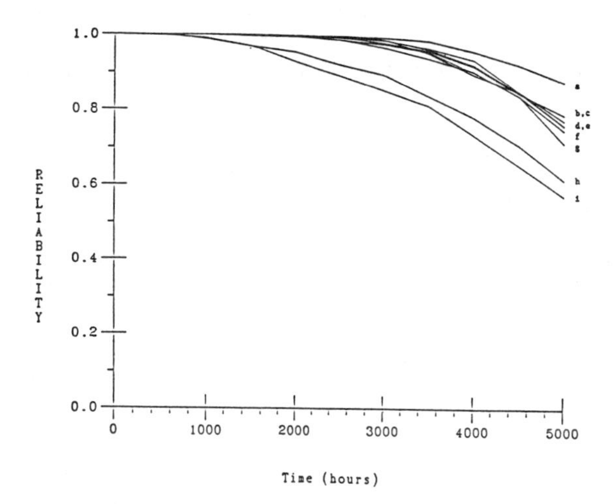

Fig. 5.

$R(t)$ for (16 $\times$ 16) circuit-switched systems with C = 1.0.
(a) INDRA.
(b) F network.
(c) MDN.
(d) Baseline w/ extra stage.
(e) Chained baseline.
(f) Butterfly.
(g) Butterfly w/ extra stage.
(h) Baseline.
(i) IADM.

4.3 System RCR Comparison

When the cost of a system is to be considered along with its reliability, a useful measure is the reliability-to-cost ratio (RCR), i.e., the reliability is divided by the system cost factor. This ratio serves as a comparison between the networks surveyed in Section 2. We observed that for smaller (16$\times$16) systems, the RCR puts the unique-path butterfly and baseline MIN systems at the top of the ranking. The increased reliability of the more complicated multiple-path MIN systems does not compensate for the extra system cost. However, as the size of the system grows, the curves for the unique-path systems fall below some of those with multiple-paths, as seen for a (64 $\times$ 64) system in Fig. 8. In these larger systems, the extra cost begins to be of some benefit, especially the addition of an extra stage to the baseline which takes the number one spot. When cost is considered, the F network falls from the upper positions of the $R(t)$ curves to occupy the lower two positions along with the IADM system in the RCR curves.

4.4 Summary of System Comparisons

From the examination of the curves of Figs. 5 through 8, the following system evaluation is offered.

These observations are based upon the particular switch failure calculated as described above.

The best overall reliability is offered by the extra-stage baseline MIN. The reliability of this topology ranked in the top four. Its value is most clearly seen, however, in the RCR comparison of (64 × 64) systems, where it ranks in the number one spot. This indicates that for large systems where cost is a consideration, the best fault tolerance technique is the addition of an extra stage on a baseline (or topologically equivalent) system. The IADM system was the least reliable of the systems compared due to the reasons mentioned earlier.

Although the simulation of large systems takes a lot of computer time, a comparison of Figs. 5 and 7 shows that the effect of the MIN topology on reliability has a greater effect as the size of the system grows. This indicates rather strongly that the algorithms described in this report should be implemented on a large-grain parallel processor.

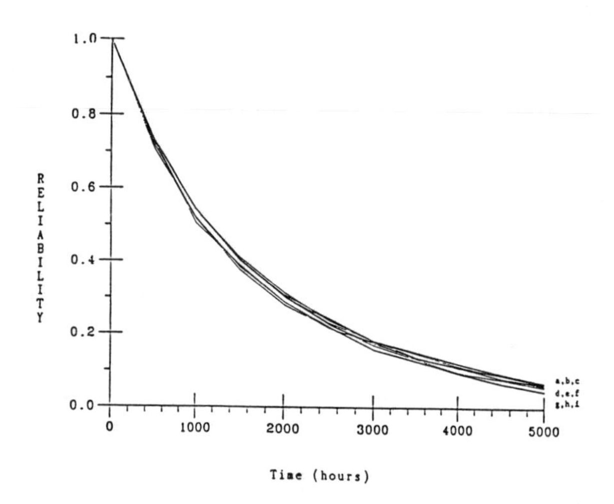

Fig. 6.

$R(t)$ for (16 × 16) circuit-switched systems with $C = 0.8$.
(a) Chained baseline.
(b) Baseline w/ extra stage.
(c) INDRA.
(d) Butterfly w/ extra stage.
(e) Butterfly.
(f) MDN.
(g) Baseline.
(h) F network.
(i) IADM.

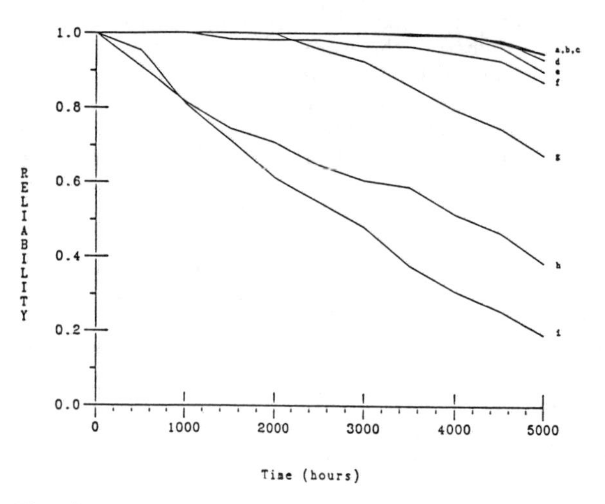

Fig. 7.

$R(t)$ for (64 × 64) circuit-switched systems with $C = 1.0$.
(a) Chained baseline.
(b) INDRA.
(c) Baseline w/ extra stage.
(d) MDN.
(e) Butterfly w/ extra stage.
(f) F network.
(g) Butterfly.
(h) Baseline.
(i) IADM.

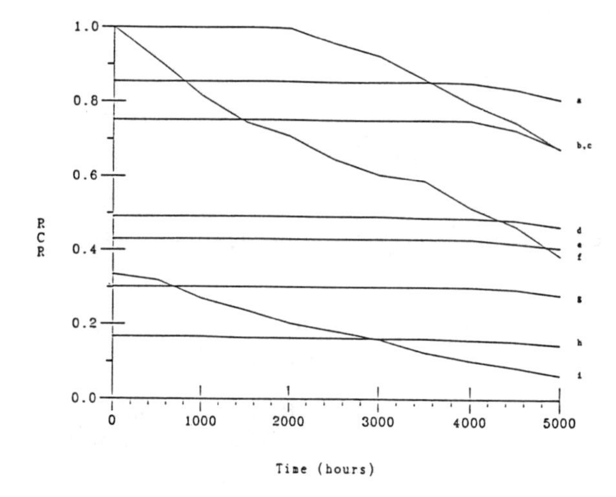

Fig. 8.

RCR for (64 × 64) circuit-switched systems with $C = 1.0$.
(a) Baseline w/ extra stage.
(b) Butterfly w/ extra stage.
(c) Butterfly.
(d) Chained baseline.
(e) INDRA.
(f) Baseline.
(g) MDN.
(h) F network.
(i) IADM.

5. CONCLUSIONS

This paper reports the first automated package with the ability to simulate the system reliability of virtually any MIN-based multiprocessor system. The program accepts the type of MIN and various failure rates from the user. It builds the MIN automatically and stores the interconnection pattern in a matrix. A traversal-search algorithm is used to find the reachability matrix of the system with random faults. The reachability matrix in turn is used in the calculation of the system reliability. The unified approach of this package makes possible system reliability predictions as well as system comparisons with respect to reliability issues.

The package in its present form provides the framework around which many other features can be built. For example, one important extension could be the ability to measure system performance in the presence of faults. Addition of this performance predictor to the reliability model will give the program the capability to predict performance-related reliability measures. Another addition could be the capability to predict the coverage factor of the system from a model of the individual processors and the maintenance processor. In this way, the coverage factor, shown in this research to be so important to system reliability, could be calculated rather than estimated and provided by the user.

REFERENCES

[1] C. -L. Wu and T. -Y. Feng, "A tutorial on interconnection networks for parallel and distributed processing," IEEE, 1984.

[2] G. B. Adams III, D. P. Agrawal, and H. J. Siegel, "A survey and comparison of fault-tolerant multistage interconnection networks," *IEEE Comput.*, vol. 20, pp. 14–27, June 1987.

[3] J. T. Blake and K. S. Trivedi, "Multistage interconnection network reliability," submitted to *IEEE Trans. Comput.*, 1988.

[4] C. S. Raghavendra and A. Varma, "INDRA: a class of interconnection networks with redundant paths," in *1984 Real-Time Systems Symp.*, Computer Society Press, Silver Spring, MD, 1984, pp. 153–164.

[5] C. R. Das and L. N. Bhuyan, "Reliability simulation of multiprocessor systems," in *Proc. 1985 Int. Conf. Parallel Processing*, pp. 591–598.

[6] J. T. Blake and K. S. Trivedi, "Comparing three interconnection networks embedded in a multiprocessor system," Tech Report, Duke Univ. Oct. 1987.

[7] L. Tien and C. R. Das, "Reliability evaluation of butterfly network based multiprocessor systems," submitted to *8th Int. Conf. on Distributed Comput. Systems*, June 1988.

[8] M. O. Ball, "Complexity of network reliability computation," *Networks*, vol. 10, pp. 153–165, 1980.

[9] A. D. Ingle and D. P. Siewiorek, "Reliability models for multiprocessor systems with and without periodic maintenance," in *Proc. 7th Annu. Int. Conf. FTC*, Los Angeles, CA, June 1977, pp. 3–9.

[10] C. -L. Wu and T. -Y. Feng, "On a class of multistage interconnection networks," *IEEE Trans. Comput.*, vol. C-29, pp. 694–702, Aug. 1980.

[11] B. Thomas, "Overview of the Butterfly parallel processor," BBN Laboratories Incorporated, Aug 1985.

[12] V. Kumar, and S. M. Reddy, "Augmented shuffle-exchange multistage interconnection networks," *IEEE Comput.*, vol. 20, pp. 30–40, June 1987.

[13] N. -F. Tzeng, P. -C. Yew, and C. -Q. Zhu, "The performance of a fault-tolerant multistage interconnection network," in *Proc. 1985 Int. Conf. Parallel Processing*, pp. 458–465.

[14] L. Ciminiera and A. Serra, "A connecting network with fault tolerance capabilities," *IEEE Trans. Comput.*, vol. C-35, pp. 578–580, June 1986.

[15] S. M. Reddy and V. Kumar, "On fault-tolerant multistage interconnection networks," in *1984 Int. Conf. Parallel Processing*, pp. 637–648.

[16] R. J. McMillen, Jr., and H. J. Siegel, "Performance and fault tolerance improvements in the inverse augmented data manipulator network," in *9th Symp. Comp. Arch.*, Apr. 1982, pp. 63–72.

[17] J. P. Shen and J. P. Hayes, "Fault-tolerance of a class of connecting networks," in *7th Int. Symp. Comput. Architecture*, May 1980, pp. 61–71.

[18] J. P. Shen and J. P. Hayes, "Fault-tolerance of dynamic full-access interconnection networks," *IEEE Trans. Comput.*, vol. C-33, pp. 241–248, Mar. 1984.

[19] D. P. Agrawal, "Graph theoretical analysis and design of multistage interconnection networks," *IEEE Trans. Comput.*, vol. C-32, pp. 637–648, July 1983.

[20] D. A. Padua and M. J. Wolfe, "Advanced compiler optimizations for supercomputers," *Commun. ACM*, vol. 29, pp. 1184–1201, 1984.

[21] T. F. Arnold, "The concept of coverage and its effect on the reliability model of a repairable system," *IEEE Trans. Comput.*, vol. C-22, pp. 251–254, Mar. 1973.

[22] D.P. Siewiorek and R. Swarz, *The Theory and Practice of Reliable System Design*, Bedford, MA: Digital Press, 1982.

[23] J. J. Macaluso, "On the reliability evaluation of multistage interconnection network based multiprocessor systems," M.S. Thesis, The Pennsylvania State University, 1988.

Reprinted from *IEEE Transactions on Selected Areas in Communications*, Volume 6, Number 8, October 1988, pages 1393–1400.

On Computer Communication Network Reliability Under Program Execution Constraints

ANUP KUMAR, STUDENT MEMBER, IEEE, SURESH RAI, SENIOR MEMBER, IEEE, AND DHARMA P. AGRAWAL, FELLOW, IEEE

***Abstract*—In complex computing environments, such as banking systems, travel agency systems, and power control systems, robustness is an important criterion that determines the quality of software. The ability of the software to handle hardware errors and erroneous inputs is defined as *software robustness*, and system software reliability is a useful measure of it. This paper introduces a new technique to compute software system reliability (SSR). Our method, called FARE (Fast Algorithm for Reliability Evaluation), does not require *a priori* knowledge of multiterminal connections (MC's) for computing the reliability expression. This paper attempts to solve the problem of N-version programming by using the FARE approach. Examples illustrate the technique.**

I. Introduction

IN recent years, computer communication networks (CCN's), such as local area networks, metropolitan area networks, and wide area networks, have become increasingly popular because they offer high fault tolerance, potential for distributed processing, and better reliability than other processing environments [1]. In a graph model of a typical CCN, nodes consist of processing elements, memory units, data files, and programs as CCN resources. These resources interconnect via communication links that enable a program to run by using resources at other nodes. For successful execution of a program at a node, other nodes having necessary data files along with the associated communication links must be operational. The banking system in Fig. 1 shows an example. If bank location 1 generates (executes) a query (program) that requires data files from bank locations 2 and 3, the nodes corresponding to bank locations 2 and 3 and the associated communication links must be all operational. This scenario introduces software system reliability, which is the probability that all programs with distributed data files may run successfully in spite of some faults occurring in processing elements (PE's) and in communication links.

Manuscript received January 5, 1988; revised March 23, 1988. Part of this paper was presented at INFOCOM 1988, New Orleans, LA, March 29-31, 1988. This work was supported in part by the Army Research Office under Contract DAAG-29-85-K-0236.

A. Kumar is with the Computer Systems Laboratory, Electrical and Computer Engineering Department, North Carolina State University, Raleigh, NC 27695.

S. Rai is with the University of Roorkee, Roorkee, India, on leave at the Computer Systems Laboratory, Electrical and Computer Engineering Department, North Carolina State University, Raleigh, NC 27695.

D. P. Agrawal is with the Computer Systems Laboratory, Electrical and Computer Engineering Department, North Carolina State University, Raleigh, NC 27695 on sabbatical leave at AIRMICS, Atlanta, GA.

IEEE Log Number 8821433.

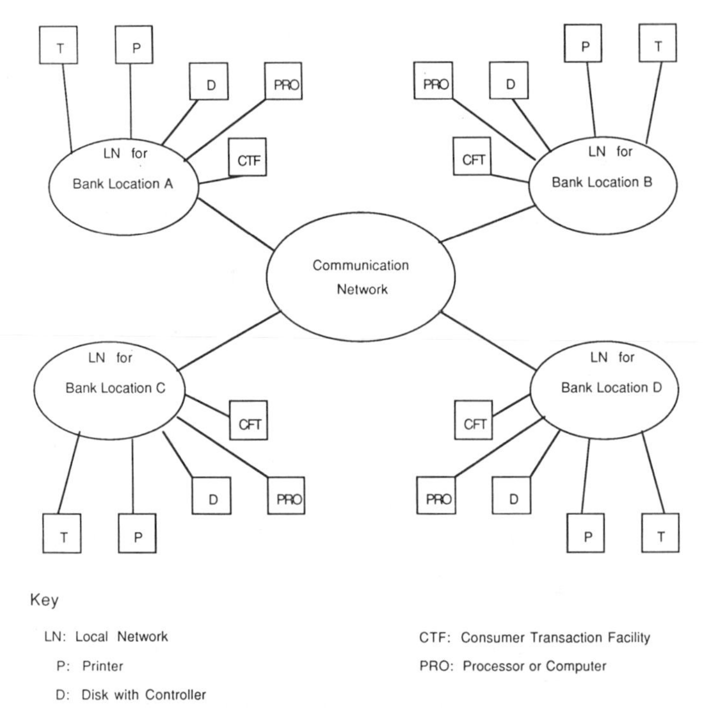

Fig. 1. A distributed banking system.

System software reliability (SSR), in a way, measures software robustness, an important parameter that determines the quality of software [5], [7]. A robust system must have high software system reliability [7]. A software system is robust when the output of a hardware error, or of an input error related to a given application, is inversely proportional to the probability of that error's appearance in the application. If an error is expected to occur with probability p_e and will result in cost C to the user, then the system is robust for all such errors if [7]

$$C \sim 1/p_e. \tag{1}$$

Equation (1) clearly indicates that the errors occurring frequently should have minor effect on a given application. The concept of robustness gives the distinction between software reliability and software system reliability. SSR differs from software reliability in that it does not need the assumption of correct input, fault resilient CCN nodes, and fault free communication links. The ability of software to cope with these faults is, obviously, its robustness property.

For distributed processing, many reliability indexes are proposed: source-to-multiple-terminal (SMT) reliability [12], survivability index [9], computer network reliability [2], and multiterminal reliability [6]. SMT and computer network reliability are popular and are reasonable measures for distributed system performance. The survivability index requires execution time that is very large, even for relatively small networks [10]. The multiterminal reliability algorithm works well as long as there are not too many alternate paths between source and destination pairs.

This paper introduces a method FARE (Fast Algorithm for Reliability Evaluation) for computing SSR. FARE is a one-step method and does not require *a priori* knowledge of multiterminal connections for computing the software system reliability expression. Generating multiterminal connections is the first and an important step with other methods [8], [10]. FARE avoids the complexity involved in enumerating the connections. It processes a 2-D array, derived from a connection matrix, using extended conservative policy as expansion procedure. Then, the concept of termination node vector (TNV) along with 0-SUBstitute and 1-SUBstitute operations produces SSR. These terms are explained in the text. FARE includes a software package. It uses "depth first search" [3] strategy for tree generation, and it stores information only for links that are further required for SSR expression. Therefore, memory use is low because FARE never stores the complete tree for the procedure. FARE evaluates the software system reliability when multiple copies of a program need to be executed simultaneously at different nodes. Such a situation is desirable with N-version programming, and where consistency tests on multiple-copy transfers are used for enhanced reliability [13].

The organization of this paper is as follows. Section II details two practical situations where our method is applicable. Section III describes and illustrates FARE in several examples, including N-version programming. Section IV compares computer execution time and computer storage requirement for FARE to existing methods [8], [10]. Concluding remarks are in Section V.

II. Background

This section details two applications that are useful in explaining the concept of FARE and also the importance of calculating software system reliability. The first application deals with a banking system, and the second deals with a travel agency system.

Banking System: Fig. 1 shows the block-level arrangement for a banking system [14]. The system is a two-level hierarchical network, the lower level local network feeds the needs of each banking location, while the higher level communication network provides connection between all the local networks. Two kinds of processing occur in this environment: bank-level processing and interbank processing. Bank-level processing includes deposits and withdrawals done with the help of a local network (LN). Interbank processing uses a communication network along with LN's, and includes out-of-bank deposits, withdrawals, and higher management reports. A local network can have one or more processors, terminals, disks with controllers, and consumer transaction facilities. Each local network disk stores some or all of the following information:

consumer accounts file (CAF)
interest and exchange rates file (IXF)
automated teller machine accounts file (TAF)
inventory control file of travellers check (ICF)
administrative aids file (ADF)
teller cash control file (TCF).

Abbreviated file names are included in parentheses. A subscript i on file name corresponds to the node i where the file is residing. Thus, $(CAF)_4$ is at node 4.

Fig. 2 shows this system graph theoretic modeling. A node represents any bank location (which is) its local network, and the links show the communication network. Fig. 2 assumes a node has two kinds of entities: 1) queries (programs) generated (executed) by a terminal or CTF, and 2) data files, printers, and so on. These entities are indicated on a node under the title "files available" (FA). Data files are indicated by their names as previously mentioned. Printers (PTR) are included in FA because sometimes a report must print at a remote banking center. The treatment for a printer in the algorithm is the same as for a data file.

A node in Fig. 1 may have all or some of the previously mentioned files. Table I lists these files in its column titled "files available." $P1$ in Table I indicates the query (program) to be executed at node A. This query represents a management report generation (MRG) for all deposits at all the bank location. Files needed for this MRG query are indicated by "files needed" (FNP) and include CAF's at all bank locations. This can be represented logically as: the execution of $P1$ needs CAF_A AND CAF_B AND CAF_C AND CAF_D. This kind of file combination gives rise to the AND-type multiterminal connections. In this combination, all the required files are needed at the source node; here the source is node A. Similarly, $P2$ indicates the query (program) to be executed at note B. This query represents a report on the latest interest and exchange rates. Files needed for this situation include IXF at any node as shown by FNP. Logically, $P2$ is executed at node B if the files needed are IXF_A OR IXF_C OR IXF_D. This type of file combination constitutes the OR-type multiterminal connection. In this scheme, any one of the ORed files are required at the source node.

A complicated situation could involve a mixed combination of AND-type and OR-type connections to process a query. This combination may be termed as a mixed-type multiterminal connection. *Note*: Many more queries (programs) exist in a practical banking system; we have, however, considered examples of only two queries for illustration purposes.

Travel Agency System: This application considers the management of a travel agency network located at different cities. To simplify the discussion, each city has only

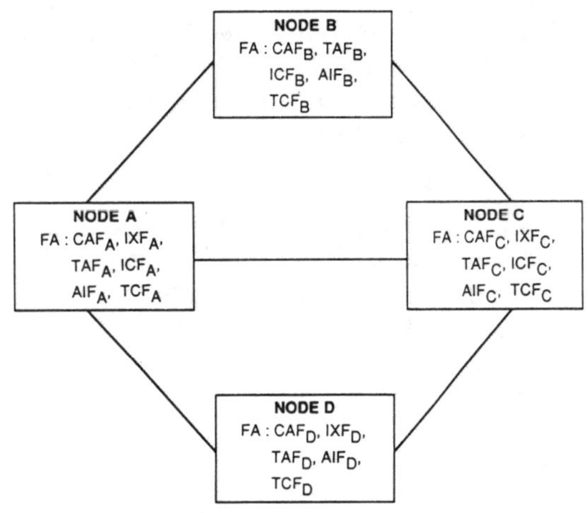

Fig. 2. Graph model for a distributed banking system.

TABLE I
DISTRIBUTION OF PROGRAMS, DATA FILES, AND FILES NEEDED FOR PROGRAM EXECUTION IN THE BANKING SYSTEM OF FIG. 1

Node	Files Available	Set of Programs	Files Needed to Execute a Program
A	CAF_A, IXF_A, TAF_A, ICF_A, AIF_A, TCF_A	P1	$FNP1 = \{CAF_A, CAF_B, CAF_C, CAF_D\}$ $FNP2 = \{IXF_A, IXF_C, IXF_D\}$
B	CAF_B, TAF_B, ICF_B, AIF_B, TCF_B	P2	
C	CAF_C, IXF_C, TAF_C, ICF_C, AIF_C, TCF_C		
D	CAF_D, IXF_D, TAF_D, AIF_D, TCF_D		

one agent. Fig. 3 shows a possible distributed network of the agency [4]. The basic function of each agent is to book hotel accommodations and reserve seats on any airline for their customers. Each agent has a local system and can communicate with others through a communication network. Each city (travel agency) has the following four files:

vacant rooms files (VRF)
seats available file (SAF)
rooms occupied files (ROF)
seats booked file (SBF).

This system can take care of following functions in local processing mode:

- reserve seats from a city by some carrier and reserve rooms in a city
- update the database for seats sold and rooms reserved
- respond to all requests concerning hotel reservations and airlines bookings.

In the remote processing mode, two or more agents communicate to reserve seats on an airline or to book hotel accommodations. For example, an agent in a city A has a request for a roundtrip reservation from A to A, passing through cities B, C, and D. In this situation, remote processing is required to book rooms and airline seats at cities B, C, and D. An initial solution starts at the request site. The reservation at A is done as a local request

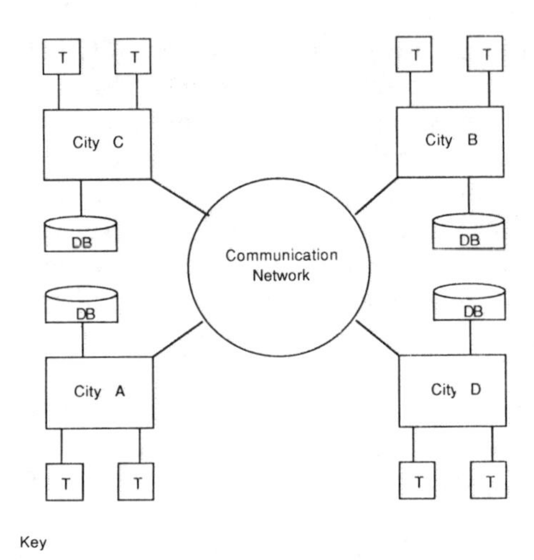

Fig. 3. A distributed travel agency system.

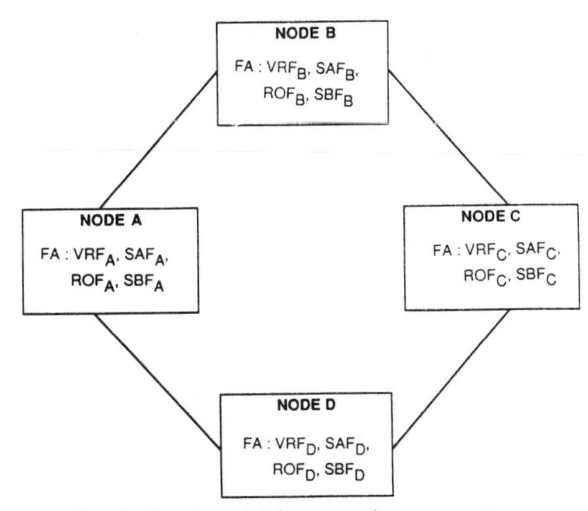

Fig. 4. Graph model for a travel agency system.

by using local processing mode, and the reservations for B, C, and D are done with the help of remote processing through the communication network by checking the SAF's at cities B, C, and D.

This system can be modeled in terms of graph (Fig. 4). The files available on each local node are shown under the title "FA" in Fig. 4. Suppose an agent has a request (program) $P1$ at city A for a roundtrip reservation through B, C, and D. Then he has to have access to SAF_B AND SAF_C AND SAF_D. This kind of file combination necessitates an AND-type multiterminal connection. An OR-type multiterminal connection is required if a report (request or program) $P2$, at node B, has to be generated, which gives the number of seats booked from city A to city B. The agent may do this by using SBF_A OR SBF_B. A more complex situation can have a combination of the AND-type and OR-type multiterminal connection, termed a mixed-type connection.

III. ALGORITHM DEVELOPMENT

This section first defines relevant terms to help explain the algorithm. Steps of the algorithm and examples illustrating the technique follow.

Definition 1: In the distributed computing environments previously discussed, three possible types of multiterminal connection (MC's) requirements exists. Grnarov and Gerla [6] have also shown the need for such requirements in distributed processing environments. These requirements are AND-type MC, OR-type MC, and mixed-type MC. The AND-type MC exists when the transaction of files, residing at different nodes, is simultaneously required for the execution of a program at the source node. The OR-type MC exists when a program to be executed at source node requires transaction of any one of the files residing on nodes $(n_1, n_2, n_3, \cdots, n_m)$. Here the assumption is that the program is executable with the help of any one of these files. Nodes, other than the source node, are terminal nodes (TN's). In mixed-type MC any combination of OR-type and AND-type is possible.

Definition 2: 0-SUBstitute operation takes care of a 0-link where a 0 is replaced in connection matrix (C). This is equivalent to removing a link from the network [15].

Definition 3: 1-SUBstitute operation is performed on a 1-link. This is equivalent to the contraction/fusion concept in graph theoretic terms [15]. 1-SUB operation reduces the size of 2-D array (explained in algorithm) on which it is operated. The algorithm for 1-SUB follows.

```
INPUT:  connection matrix
STEP 1: replace 1 − link(s) by zeros
STEP 2: for all columns corresponding to link(s) = 1
        do
           find all row(s) r having entry = 1
             for each row r do
               row 1 ← row 1 OR row r
               delete row(s) r
             od
        od
STEP 3: delete column(s) corresponding to link(s) =
        1.
```

Definition 4: A link x_i is said to be $0(1)$-link if and only if $X_i = 0(1)$; otherwise, it is a variable link.

Definition 5: This algorithm uses a decomposition policy called extended conservative policy (ECP). The decomposition of C or 2-D array means to expand along different links, and, therefore, perform 1-SUB and 0-SUB operations to get a reduction in size of 2-D array. These events are generated using ECP. Assume that $(r_1, r_2, r_3, \cdots, r_N)$ are the variables in the first row of 2-D array. Various N combinations in this scheme follow:

$$\begin{array}{ll} r_1 & 1 \\ 0\ r_2 & 0\ 1 \\ 0\ 0\ r_3 & 0\ 0\ 1 \\ \vdots & \vdots \\ 0\ 0\ 0\ 0 \ldots r_N & 0\ 0\ 0\ 0 \ldots 1 \end{array}$$

Definition 6: Node vector $(\mathrm{NV})_i$ of an edge x_i is defined as vector where $1(0)$ indicates nodes that are connected (disconnected) by edge x_i. In Table II, the node vector x_3 is [101000].

TABLE II
DISTRIBUTION OF PROGRAMS, DATA FILES, AND FILES NEEDED FOR PROGRAM EXECUTION (FIG. 5)

Node	Files Available	Set of Programs	Files Needed to Execute a Program
1	F1,F5	P1	
2	F1,F2	P4	FNP1 = {F1,F2,F3}
3	F3,F4	P2,P3	FNP2 = {F2,F4,F6}
4	F2,F5	P2,P3	FNP3 = {F1,F3,F5}
5	F3,F6	P4	FNP4 = {F1,F2,F4,F6}
6	F1,F4	P1	

Definition 7: Terminal node vector (TNV) can be obtained from the following:

1) set of source and terminal nodes
2) type of multiterminal connection sought.

For example, in Table IV, terminal node vectors are calculated for $P2$ to be executed at node 3. $P2$ requires files $F2$, $F4$, and $F6$, but $F4$ is already at node 3. The remaining files required are $F2$ and $F6$. These files are available at nodes 2, 4, and 5, respectively. The type of connection can be (2 OR 4) AND 5. As the source node is 3, the terminal node vectors can be written as [d1ld1d] and [dd111d], where "d" denotes "don't-care" condition.

This method continuously monitors the node vector while decomposing the 2-D array by using ECP. Whenever an edge variable is considered, its node vector updates the previous node vector. The updating is done by performing a simple OR operation between two vectors

$$(\mathrm{NV})_{\mathrm{new}} \leftarrow (\mathrm{NV})_{\mathrm{old}} \text{ OR } (\mathrm{NV})_i .$$

To explain the procedure involved, Fig. 6 solves an example. The 2-D array at the root has [000000] as the node vector associated with it. After updating with respect to x_1, the new node vector is

$$\begin{aligned}(\mathrm{NV})_{\mathrm{new}} &= [000000] \text{ OR } [110000] \\ &= [110000].\end{aligned}$$

The following assumptions are made to simplify the algorithm.

1) The distributed system is modeled by an undirected or directed graph G where nodes represent processing elements and edges represent communication links.

2) Graph G is free of self-loop.

3) The failure of one edge is S-independent of the failures of other links.

4) The edge can be in working (good) or faulty (bad) state.

This algorithm does not require *a priori* information

about multiterminal connections and gives the reliability expression directly. To help terminate the decomposition of a 2-D array in subsequent steps, the algorithm specifies a terminating rule: "It involves matching of NV with TNV if the two vectors match then stop decomposing a 2-D array otherwise continue decomposing."

Algorithm

INPUT: [Adjacency list, node vectors, terminating node vectors]

STEP 1: a) Generate connection matrix from given adjacency list
- b) If source node $(j)=1$ then
  - being
    - i) replace link variable in column 1 by zero
    - ii) add unity matrix to it
  - end
  - else
  - begin
    - i) row 1 $\leftrightarrow$ row j
    - ii) column 1 $\leftrightarrow$ column j
    - iii) replace link variable in column 1 by zero
    - iv) add unity matrix to it
    - Call it a 2-D array
  - end.

STEP 2: repeat
- a) suppose there are N links in row 1 of 2-D array POSCOM = store various events E_r using ECP
- b) repeat
  - i) for an E_r use 1-SUB operation with uncomplemented links in E_r (use 0-SUB for complemented links) to decompose 2-D array.
  - ii) update the (NV) for that E_r
  - until all events in POSCOM are exhausted
- until all decomposed 2-D arrays satisfy terminating rule

STEP 3: a) to get software system reliability expression paths from root to all the leaves are traversed.
- b) make the following substitution to get software system reliability expression

$$p_j \leftarrow X_j$$
$$q_j \leftarrow X_j'$$

where $'$ represents Boolean complement

To illustrate the algorithm, we have solved two examples. Fig. 5 shows the distributed system for which the SSR calculation is carried out. Tables II and III give parameters such as file distribution, TNV's, NV's, and also files required by the programs for their execution. Fig. 6 shows the details of our first example in which program 1 is executed at node 1. The software system reliability expression is given as

$$p1p3 + p1q3p4 + q1p3p4 + q1p3q4p5p6 + p1q3q4p2p6 + q1p3q4p5q6p8p7 + p1q3q4p2q6p7p8.$$

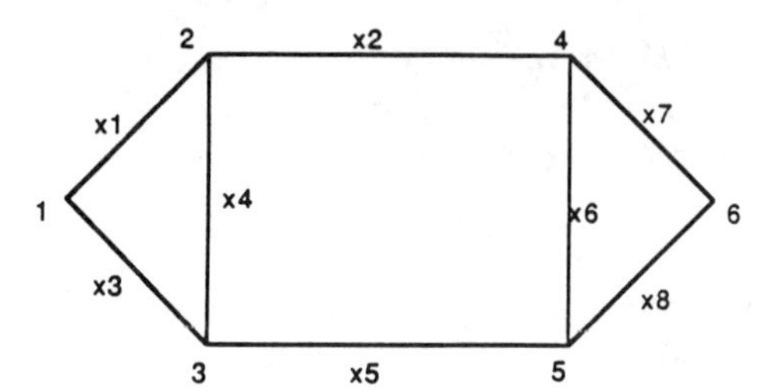

Fig. 5. Topology of a distributed system.

TABLE III
NODE VECTORS AND TERMINAL NODE VECTORS FOR EXECUTION OF PROGRAM 1 ON NODE 1

Link	Node Vector	Terminal Node Vector
1	110000	
2	010100	
3	101000	111ddd
4	011000	11dd1d
5	001010	1d11dd
6	000110	
7	000101	1dd11d
8	000011	

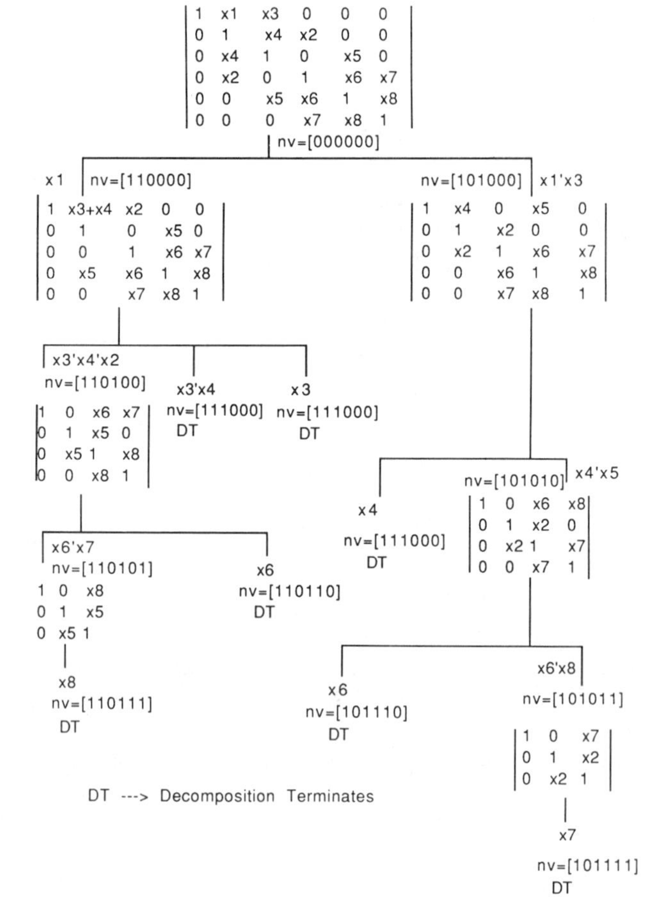

Fig. 6. Results for FARE using DSC of Fig. 5, Tables II and III.

TABLE IV
NODE VECTORS AND TERMINAL NODE VECTORS FOR EXECUTION OF PROGRAM 2 ON NODE 3

Link	Node Vector	Terminal Node Vector
1	110000	
2	010100	
3	101000	
4	011000	d11d1d
5	001010	dd111d
6	000110	
7	000101	
8	000011	

In another example, program $P2$ is executed at node 3. The node vectors and terminal node vectors are in Table IV. The software system reliability expression is as follows:

$$(p4(p3(p2(p5 + q5p6 + q5q6p7(p8)) + q2p5) + q3p1(p2(p5 + q5p6 + q5q6p7(p8)) + q2p5) + q3q1p2(p5 + q5p6 + q5q6p7(p8)) + q3q1q2p5) + q4p3(p1(p2(p5 + q5p6 + q5q6p7(p8)) + q2p5) + q1p5(p6 + q6p8(p7))) + q4q3p5(p6 + q6p8(p7))).$$

The previous algorithm can also solve the problem of N-version programming, but deriving the TNV's takes care of all the source nodes on which the program is to be executed simultaneously [11]. For example, in Table III, the TNV's are [111ddd], [11dd1d], [1d11dd], and [1dd11d] with source node as 1. Now when $P1$ is executed at nodes 1 and 6 simultaneously, then the TNV's are [111ddd], [11dd1d], [1d11dd], [1dd11d], [d11dd1], [d1dd11], [dd11d1], and [ddd111]. The software system reliability expression can be generated exactly the same way as in the previous examples.

$$p1p3 + p1p4q3 + p1p2q6q4q8 + p1p6p7q2q3q4q8 + p3p4q1 + p3p5p6q1q4q7 + p2p4p7q1q3 + p2p7p8q1q3q4 + p2p6p7q1q3q4q8 + p7p8q1q2q3 + p6p7q1q2q3q8 + p6p8q1q3q5q7 + p4p5p8q1q3q7 + p5p6p8q1q3q4q7.$$

IV. COMPARISON

This section compares the FARE algorithm to the one in [10]. We have studied the execution time and memory requirement of N-version programming cases for both algorithms to show how the previously mentioned parameters vary according to the changes in the topology of the communication network. The comparison of execution times for these two algorithms, in the case when one program runs only at one node at a time, is in [8]. Here, we run a different set of programs and a file distribution over various topologies, starting from a simple loop to a completely connected graph. These topologies are in Fig. 7. File distribution, program distribution, and files needed for execution are in Table II. There are four graphs for four programs. Each curve in a graph is drawn for a program executed on all six topologies (indicated by number of links in that topology). In Fig. 8, the execution time of a program is plotted against the number of links in various topologies. As the number of links increases, both algorithms take longer time. However, FARE performs better than the algorithm in [10]. It is because of the following reason: algorithm [10], in worst case, can generate as many as $(n - 1)^{(e-1)}$ intermediate trees, where n denotes number of nodes and e is the maximum in-degree of a node in the graph. This is not the case with FARE because it uses ECP, which prevents explosion during the expansion of intermediate nodes. The number of new nodes created during the expansion is proportional to the maximum in-degree of a node in the graph. Regarding average memory requirements, Table V gives the numerical values for both methods, that is, FARE and the algorithm in [10]. These values are generated by using the 'Time' command on a VAX-780 system and a rigorous analysis needs to be done.

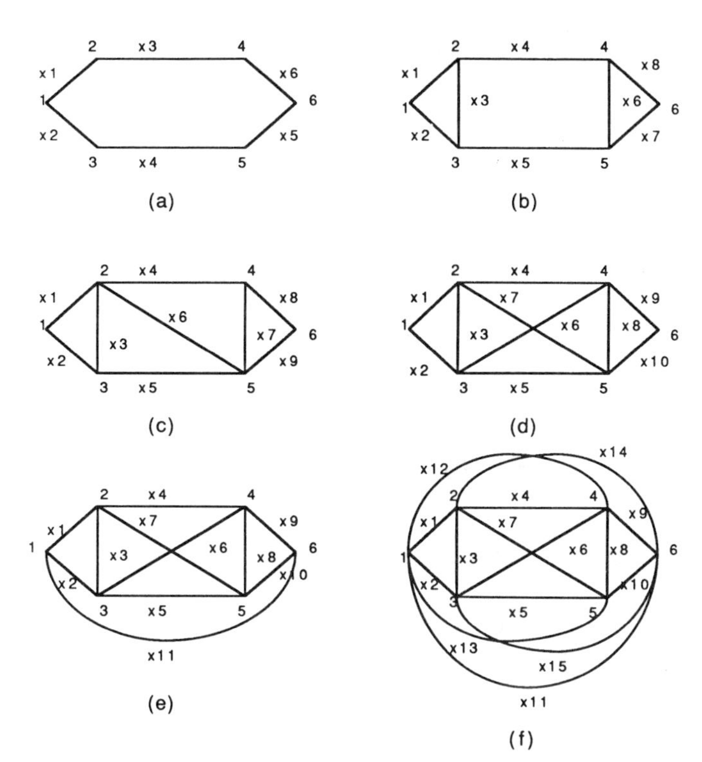

Fig. 7. Topology variants of distributed system.

V. CONCLUSIONS

This paper defines a method called FARE to calculate software system reliability that, in turn, gives the measure of software robustness. The advantage of FARE is that it directly computes the software system reliability expression without *a priori* knowledge of multiterminal connec-

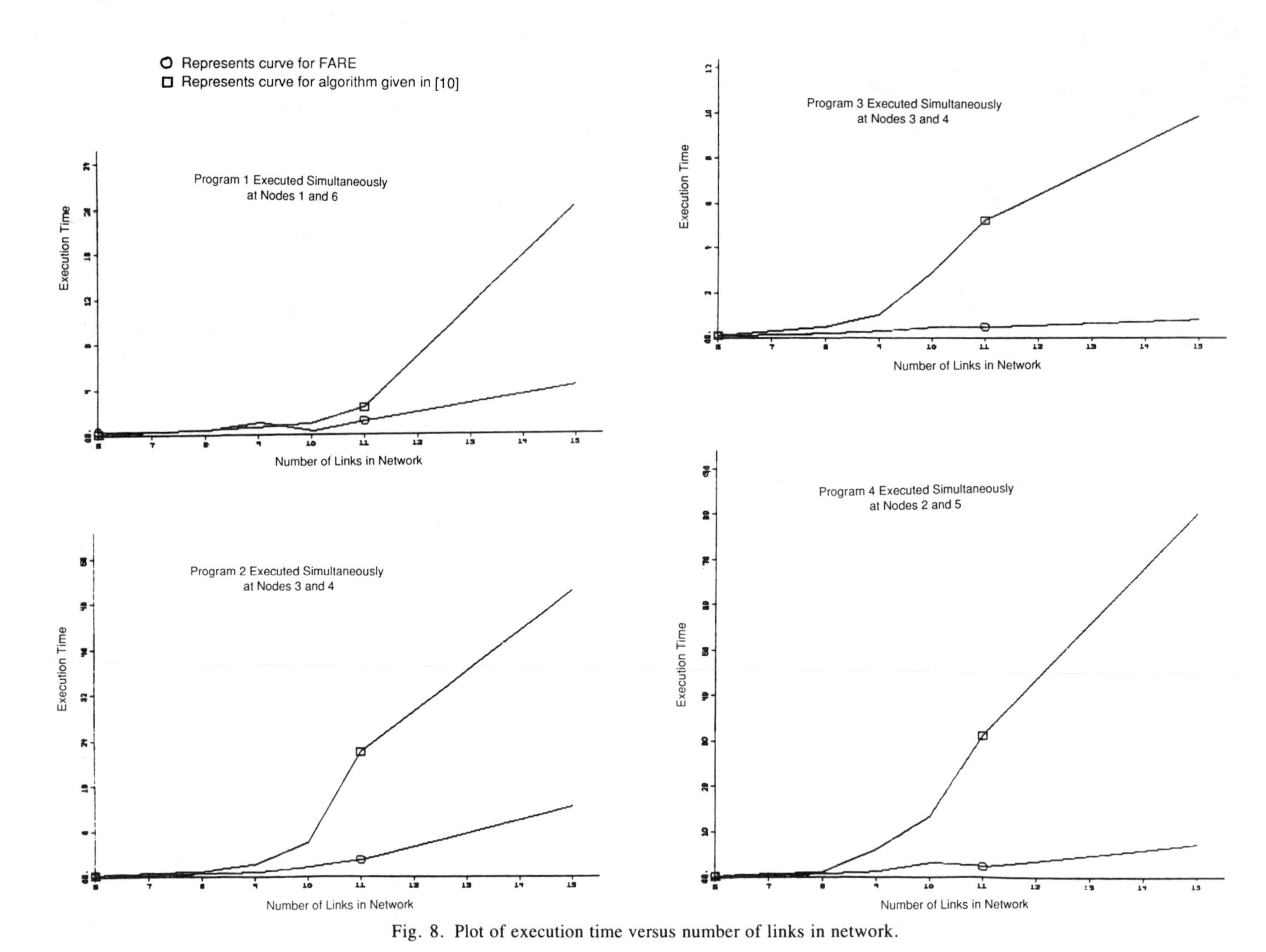

Fig. 8. Plot of execution time versus number of links in network.

TABLE V
AVERAGE (FOR DIFFERENT TOPOLOGIES) MEMORY REQUIREMENT FOR BOTH ALGORITHMS (THIS SET IS FOR PROGRAM 1 EXECUTED SIMULTANEOUSLY AT NODES 1 AND 6)

Topologies	Algorithm in [8]	FARE
a	44 K	32 K
b	45 K	30 K
c	46 K	35 K
d	42 K	32 K
e	47 K	34 K
f	48 K	36 K

tions, as is necessary in [10]. It is possible to calculate the SSR of fairly large distributed systems with this method because of its small execution time and low memory requirement. The graphs in Fig. 8 show the results of running 48 sets of data to study the variation in execution time as topology varies for both methods. They show that FARE performs better than the algorithm in [10].

REFERENCES

[1] D. P. Agrawal, *Advanced Computer Architecture*. Washington, DC: IEEE Computer Society Press, 1986.

[2] K. K. Aggarwal and S. Rai, "Reliability evaluation in computer communication networks," *IEEE Trans. Rel.*, vol. R-30, pp. 32–35, Apr. 1981.

[3] A. V. Aho, J. E. Hopcroft, and J. D. Ullman, *The Design and Analysis of Computer Algorithms*. Reading, MA: Addison-Wesley, 1974.

[4] CORNAFION, *Distributed Computer System*. Amsterdam, The Netherlands: Elsevier Science, 1974.

[5] R. Fairly, *Software Engineering Concept*. Englewood Cliffs, NJ: Prentice-Hall, 1979.

[6] A. Grnarov and M. Gerla, "Multiterminal reliability analysis of distributed processing system," in *Proc. 1981 Int. Conf. Parallel Processing*, Aug. 1981, pp. 79–86.

[7] H. Kopetz, *Software Reliability*. New York: Springer-Verlag, 1979.

[8] A. Kumar, S. Rai, and D. P. Agrawal, "Reliability evaluation algorithms for distributed systems," presented at INFOCOM, 1988.

[9] R. W. Merwin and M. Mirherkerk, "Derivation and use of survivability criterion for DDP systems," in *Proc. Nat. Comput. Conf.*, May 1980, pp. 139–146.

[10] V. K. Prasannakumar, S. Hariri, and C. S. Raghavendra, "Distributed program reliability analysis," *IEEE Trans. Software Eng.*, vol. SE-12, pp. 42–50, Jan. 1986.

[11] S. Rai and D. P. Agrawal, "Reliability of program execution in a distributed system," *IEEE Trans. Software Eng.*, submitted for publication.

[12] A. Satyanarayana and J. N. Hagstrom, "New algorithm for reliability analysis of multiterminal networks," *IEEE Trans. Rel.*, vol. R-30, pp. 325–333, Oct. 1981.

[13] R. K. Scott, J. W. Gault, and D. F. McAllister, "Fault tolerant software reliability modelling," *IEEE Trans. Software Eng.*, vol. SE-13, pp. 582–592, May 1987.

[14] D. A. Sheppard, "Standard for banking communication system," *IEEE Computer*, pp. 92–95, Nov. 1987.

[15] M. N. S. Swamy and K. Tulsiraman, *Graphs, Networks and Algorithms*. New York: Wiley, 1981.

Anup Kumar (S'88) was born in Bareilly, India, on June 22, 1963. He received the B.E. degree in computer science from M.N.R. Engineering College, Allahabad, India, in 1983 and is currently working toward the Ph.D. degree at North Carolina State University, Raleigh.

His research interests include distributed system reliability, fault diagnosis, and parallel processing.

Suresh Rai (SM'85) was born in Meharur, India, on July 9, 1951. He received the graduate degree from Benaras Hindu University in 1972, the M.E. degree from the University of Roorkee in 1974, and the Ph.D. degree from Kurukshetra University in 1980.

He joined the Regional Engineering College at Kurukshetra in 1974 and worked there for six years. After a brief stay at M.M.M. Engineering College, Gorakhpur, he transferred to the University of Roorkee in 1981 as an Associate Professor. He is currently on leave at the School of Engineering at North Carolina State University. He has taught and researched in the areas of reliability engineering, fault diagnosis, and parallel processing. He is a coauthor of the book *Waveshaping and Digital Circuits*.

Dharma P. Agrawal (M'74–SM'79–F'87) received the B.E. degree from Ravishankar University, Raipur, India, in 1966, the M.S. degree from the University of Roorkee in 1968, and the D.Sc. degree from the Federal Institute of Technology, Lausanne, Switzerland, in 1975.

He is a Professor in the Electrical and Computer Engineering Department of North Carolina State University, Raleigh. His research interests include both software and hardware aspects of parallel and distributed processing, computer architecture, reliability, and testing. He is the Group Leader for the 24-node B-HIVE multicomputer project currently being implemented at NCSU.

Dr. Agrawal worked as Program Chairman for the 13th International Symposium on Computer Architecture. His tutorial text, *Advanced Computer Architecture*, has recently been published by the IEEE Computer Society Press, and he has offered tutorials on this subject in conjunction with several conferences. He has been awarded a "Certificate of Appreciation" by the IEEE Computer Society for his contributions to the Publications Board. He has been Co-guest Editor for IEEE TRANSACTIONS ON COMPUTERS, and is an Editor of *Computer Magazine* and the *Journal on Parallel and Distributed Computing*. He has also been a Distinguished Visitor of the Computer Society. He is a member of ACM and SIAM.

Reprinted from *The Proceedings of INFOCOM '87*, 1987, pages 72-76.

AN ALGORITHM FOR OPTIMAL FILE ALLOCATION IN DISTRIBUTED COMPUTING SYSTEMS*

Chiu Ge-Ming and C. S. Raghavendra

Department of Electrical Engineering-Systems
University of Southern California
Los Angeles, CA 90089-0781

ABSTRACT

Optimal allocation of multiple copies of files in distributed computing systems is studied in this paper. In our model, a set of tasks that are preassigned to computing nodes are to be executed with minimum communication cost to access the needed files. Each task would require access to several files and some of which may reside at remote sites. Several copies of each file exists in the system and the redundancy factor is small compared to the number of computing nodes. An efficient iterative algorithm is presented for solving this allocation problem and the approach is illustrated by an example.

I. Introduction

A distributed computer system (DCS) is considered as a collection of computer nodes that are interconnected by a communication network. Information and data can be exchanged via the network between nodes. Each node has its own processing power and private storage. One major advantage of a DCS is obviously the capability of sharing resources such as processing capability, program segments and data files. In particular, the sharing of data files reduces storage space needed to facilitate accesses to a public file. Distributed database is a typical application of file sharing in a DCS. The reliability of the file system is improved as a result of this. However, these advantages are not obtained without associated overheads. Access communication costs and distributed file management are examples of the overheads incurred.

One of the problems regarding file sharing is to allocate copies of files so that some performance measures be optimized. File allocation problem (FAP) has been actively investigated by many researchers [2, 3, 4, 5, 9, 10]. A lot of work has been done on single file allocation problem [2, 12, 13]. In [2], the objective of the problem is to allocate copies of a file to processing nodes so that the sum of file storage cost, query communication cost, and update communication cost is minimized. The relationship between data files and programs that use them is considered in [12]. Problems regarding multiple files were studied along with the considerations of other performance constraints [3, 4, 5, 11]. Network design issues has also been incorporated in the FAP [3, 10, 11]. In [9], a data file is fragmented and distributed among all the processing nodes so that a total cost is minimized. A good survey on FAP can be found in [7].

In this paper, we study a different type of file allocation problem. We assume that a set of tasks in a DCS have already been allocated to processing nodes. Each of these tasks need to access some subset of a file set. Each file in this file set has a fixed number of copies. The problem is to distribute this set of data files so that total communication cost incurred is minimized. In this paper, we only consider query accesses but not update accesses in calculating communication cost. Good examples of this type of DCS are library systems and public information systems where the number of query accesses is much larger than that of update accesses. The number of copies that each file can have may be determined based on other considerations such as reliability issue and budgetary reason etc. This is not an NP-complete problem. It can be solved using exhaustive combinatorial enumeration within polynomial time. However, as the number of copies of each file increases, the complexity of this problem goes up. We will present a different algorithm which is more efficient as number of copies of each file increases. This algorithm is good for the allocation of files whose accesses tend to localize, i.e., processing nodes that access these files are close in terms of communication cost.

This paper is organized as follows. In the next section, a general model of our target system is presented. An optimization algorithm is developed in section III followed by an illustrative example in section IV. The paper is concluded with section V. In Appendix I a subalgorithm needed to make our main algorithm complete is presented.

* This research is supported by the NSF Grant No. DCI-8452003 and a grant from AT&T Information Systems.

II. Model

Our model of a distributed system considered here consists of N processing nodes interconnected by a communication network. A set of T tasks are executed in this system and each task has been designated to a predetermined processing node. Each task demands query services of some subset of a file set. Each file can only have a limited number of copies. The problem is to find the allocation of the set of files to processing nodes which minimizes the total communication cost incurred by query accesses. A task is assumed to always access that copy of file that incurred the minimum communication cost.

Let $C = \left[C_{ij}\right]$ be an N×N matrix with C_{ij} being the communication cost from node i to node j and back. The communication cost to itself is assumed to be zero; i.e., $C_{ii}=0$ for all i.

$\mathcal{T}$ = a set of T tasks, $\left\{ t_1, t_2, \ldots, t_T \right\}$.

$\mathcal{M}$ = a set of M distinct files accessed by tasks in $\mathcal{T}$.

$D = \left[D_{ij}\right]$ be a T×N binary matrix indicating the designation of each task to its processing node.

$$D_{ij} = \begin{cases} 1 & \text{if task } i \text{ is executed on node } j \\ 0 & \text{otherwise} \end{cases}$$

$R = \left[R_{ij}\right]$ be a T×M binary matrix indicating the files demand by tasks in $\mathcal{T}$.

$$R_{ij} = \begin{cases} 1 & \text{if task } i \text{ needs access to file } j \\ 0 & \text{otherwise} \end{cases}$$

The above definitions serve to characterize the distributed system targeted. What's missing are the units of query traffic to each file from a task and the frequencies at which tasks are initiated. As will be described later, these parameters can be included in our model with very little effort. In the following analysis, however, these parameters are assumed to be equal to 1 and be identical among all traffics and tasks. Let m be the maximum number of copies of any file in the system which appears as a constraint in our problem. If we represent an allocation of files by an M×N matrix A with

$$A_{ij} = \begin{cases} 1 & \text{if file } i \text{ has a copy in node } j \\ 0 & \text{otherwise} \end{cases}$$

the associated communication cost C(A) is then given by

$$C(A) = \sum_{i=1}^{T}\sum_{j=1}^{M} R_{ij} \left[\sum_{k=1}^{N} D_{ik} \cdot \min_{l, A_{jl}=1} C_{kl} \right] \qquad (1)$$

We can restate our problem as finding an allocation matrix A^* which gives minimal communication cost $C(A^*)$ under the constraint that

$$\sum_{j=1}^{N} A_{ij} \leq m, \quad i=1,2,\ldots,M \qquad (2)$$

The following example illustrates the problem. If we have

$$C = \begin{bmatrix} 0.0 & 4.3 & 7.2 & 8.1 & 2.2 \\ 5.0 & 0.0 & 2.3 & 7.4 & 3.0 \\ 6.1 & 3.0 & 0.0 & 5.4 & 7.2 \\ 10.0 & 6.1 & 6.0 & 0.0 & 8.0 \\ 4.1 & 4.3 & 5.3 & 6.0 & 0.0 \end{bmatrix}$$

$$T = \left\{ t_1, t_2, t_3, t_4 \right\}$$

$$M = \left\{ f_1 \right\}$$

$$D = \begin{bmatrix} 1 & 0 & 0 & 0 & 0 \\ 0 & 0 & 1 & 0 & 0 \\ 0 & 0 & 0 & 1 & 0 \\ 0 & 0 & 1 & 0 & 0 \end{bmatrix}$$

$$R = \begin{bmatrix} 1 \\ 1 \\ 1 \\ 0 \end{bmatrix}$$

and m = 2.

The optimal allocation matrix will be

$$A^* = \begin{bmatrix} 1 & 0 & 0 & 1 & 0 \end{bmatrix}$$

and $C(A^*) = 0 + 5.4 + 0 = 5.4$.

Since we assume, from Eq.(1), the access to a file is independent from access to another file, the multiple file allocation problem can be decomposed into several individual file allocation problems [7, 14]. The algorithm presented in the next section will then target at allocation of a single file.

III. Optimization Algorithm

Consider the allocation of some file x such that the total communication cost is minimized while limiting the number of copies to m. Let B denote the multiset of processing nodes where those tasks in $\mathcal{T}$ which need to access file x are executed. (The set B is a multiset since there may be more than one task running on the same processing node). Form a matrix C^a by concatenating those rows of C corresponding to distinct elements of B as follows:

$$C^a = \left[C_j \mid j \in B' \right]$$

where B' is the set of distinct elements of B and C_j is the j-th row of C multiplied by its corresponding multiplicity. For any set of nodes A' with one copy of file x in each node, the communication cost C(A') can be expressed as follows:

$$C(A') = \sum_{i=1}^{|B'|} \min_{j \in A'} C^a_{ij} \qquad (3)$$

The problem is reduced to find the set A^* with m or less nodes which minimizes Eq.(3).

The approach taken in our algorithm is to start with an absolute optimal distribution disregarding the constraint of the m-limited copies. It iterates by forwarding along the $|B'|$ chains of node indices of increasing order formed from rows of B' until certain constraints are met. The details of the algorithm is presented below.

Algorithm:

1. Construct a matrix C^a from C as described above.
2. Each row of C^a is sorted individually with respect to the communication costs to create a chain of node indexes of increasing costs. Also created is a new matrix C^d of $|B'| \times (N-1)$ whose elements are the differences of communication costs between two successive nodes in each chain.
3. Initialization:
   3.1 UPBUND = 0; the upper bound of total communication cost in the current iteration is initialized to 0.

3.2 INCRMENT(i) = C^d_{i1}, for i = 1,2,..., $|B'|$; this parameter indicates the increase in cost in case the task(s) associated with i-th row has to access file x from next node in the order.

3.3 Go to step 5.

4. Iteration:

4.1 Update UPBUND as follows:

$$\text{UPBUND} = \text{UPBUND} + \min_{i=1,2,\ldots,|B'|} \text{INCRMENT}(i)$$

4.2 Replace the corresponding smallest INCRMENT(i), as found in step 4.1, by the next element in the same row i of C^d. If the row was exhausted, insert infinitity.

5. Constraint Checking:

5.1 Construct A_i's as follows:

$$A_i = \left\{ j \mid C^a_{ij} \leq \text{UPBUND} \right\}, \qquad \text{for } i = 1,2,\ldots,|B'|$$

Each A_i contains all nodes that can possibly have a copy of file x to be accessed by task(s) in node i such that the communication costs are no greater than UPBUND.

5.2 Perform constraint checking on A_i's ; this is to find out if there is any combination of nodes, one from each A_i, forming an allocation that meets the m-limited constraint. An efficient constraint checking algorithm can be found in Appendix I.

5.3 If the result of step 5.2 is negative, go back to step 4; otherwise go to next step.

5.4 Find out if there are any allocations giving total communication cost less or equal to UPBUND.

5.5 If the result of step 5.4 is negative, go back to step 4; otherwise go to next step.

6. Stop:

The allocation associated with the smallest total cost is the optimal solution and the algorithm stops.

The performance of this algorithm is dependent upon the nature of the problems given. What is important is that it provides an alternative way of approaching an optimal solution for this type of file allocation problem. In particular, with other parameters fixed, its efficiency with respect to time complexity improves as the value of m (the number of copies allowed) increases. This may just be opposite for pure combinatorial method which uses exhausitive search through all $\binom{N}{m}$ possible file asssignments. The number of combinations increases as the number of copies allowed increases. On the other hand, the number of processing nodes of those tasks accessing file x ($|B'|$ in our case) directly affects the complexity of this algorithm. An ideal case for this algorithm will be the one with small $|B'|$/N ratio. Many distributed systems fall in this category. Undoubtedly, an efficient constraint checking algorithm is essential to the success of this approach. If the traffic units and the accessing frequencies by various tasks are considered, we can simply include them in the matrix C^a by multiplying with relevant factors and the same algorithm will apply.

IV. An Illustrative Example

Let us assume that we have a distributed system consisting of 8 processing nodes. A set of 4 tasks are to access a file f_1 and are designated to processing nodes as shown in matrix D below. The unit communication costs between each pair of nodes are given by matrix C. We are to allocate copies of file f_1 so that total communication cost is minimized. In our example, two different constraints, namely m=3 and m=2, will be considered.

$$C = \begin{bmatrix} 0.0 & 3.4 & 7.8 & 0.4 & 10.3 & 8.1 & 10.5 & 4.2 \\ 4.3 & 0.0 & 5.5 & 6.2 & 0.8 & 3.2 & 2.5 & 5.1 \\ 4.8 & 5.4 & 0.0 & 2.2 & 8.3 & 10.1 & 1.8 & 0.4 \\ 1.8 & 9.3 & 6.9 & 0.0 & 0.8 & 4.5 & 3.2 & 1.7 \\ 3.0 & 4.2 & 1.2 & 5.3 & 0.0 & 4.1 & 0.7 & 7.2 \\ 9.8 & 4.2 & 3.2 & 1.0 & 4.9 & 0.0 & 7.9 & 9.9 \\ 2.1 & 7.3 & 11.0 & 5.3 & 1.0 & 4.3 & 0.0 & 9.8 \\ 2.5 & 8.1 & 4.5 & 1.7 & 3.0 & 0.6 & 0.6 & 0.0 \end{bmatrix}$$

$$D = \begin{bmatrix} 1 & 0 & 0 & 0 & 0 & 0 & 0 & 0 \\ 0 & 0 & 0 & 1 & 0 & 0 & 0 & 0 \\ 0 & 0 & 0 & 0 & 1 & 0 & 0 & 0 \\ 0 & 0 & 0 & 0 & 0 & 0 & 0 & 1 \end{bmatrix}$$

The sets B and B' are then obtained.

$$B = B' = \{ 1, 4, 5, 8\}.$$

Matrix C^a is therefore constructed below.

$$C^a = \begin{bmatrix} 0.0 & 3.4 & 7.8 & 0.4 & 10.3 & 8.1 & 10.5 & 4.2 \\ 1.8 & 9.3 & 6.9 & 0.0 & 0.8 & 4.5 & 3.2 & 1.7 \\ 3.0 & 4.2 & 1.2 & 5.3 & 0.0 & 4.1 & 0.7 & 7.2 \\ 2.5 & 8.1 & 4.5 & 1.7 & 3.0 & 0.6 & 1.2 & 0.0 \end{bmatrix}$$

Sorting of C^a creates four chains of increasing order and a difference matrix C^d as shown below:

$$1 \to 4 \to 2 \to 8 \to 3 \to 6 \to 5 \to 7$$
$$4 \to 5 \to 8 \to 1 \to 7 \to 6 \to 3 \to 2$$
$$5 \to 7 \to 3 \to 1 \to 6 \to 2 \to 4 \to 8$$
$$8 \to 6 \to 7 \to 4 \to 1 \to 5 \to 3 \to 2$$

$$C^d = \begin{bmatrix} 0.4 & 3.0 & 0.8 & 3.6 & 0.3 & 2.2 & 0.2 \\ 0.8 & 0.9 & 0.1 & 1.4 & 1.3 & 2.4 & 2.4 \\ 0.7 & 0.5 & 1.8 & 1.1 & 0.1 & 1.1 & 1.9 \\ 0.6 & 0.6 & 0.5 & 0.8 & 0.5 & 1.5 & 3.6 \end{bmatrix}$$

The rest of the algorithm is to forward along the chains to locate the optimal solution. The UPBUND and INCRMENT(i), $i \epsilon \{1,2,3,4\}$, are set in steps 3.1 and 3.2.

UPBUND = 0
INCRMENT(1) = 0.4
INCRMENT(2) = 0.8
INCRMENT(3) = 0.7
INCRMENT(4) = 0.6

At this time UPBUND holds a minimum value of 0 that any allocation can ever possibly have. The A_i's, as generated in step 5.1, will be

$$A_1 = \{1\}$$
$$A_2 = \{4\}$$
$$A_3 = \{5\}$$
$$A_4 = \{8\}$$

Notice that each A_i holds all nodes that can possibly be accessed by node B' (i) while keeping total communication cost less or equal to UPBUND. In other words, if there is a set of nodes, each containing a copy of f_1, that provides a total communication cost no greater than UPBUND, it must cover all A_i's.

Obviously, no set of 3 or less nodes (therefore, 2 or less nodes) can possibly cover all A_i's as a result of step 5.2 and algorithm goes back to step 4. The UPBUND is then updated as follows:

$$UPBUND = 0 + \min(0.4, 0.8, 0.7, 0.6) = 0.4$$

and, hence,

$$INCRMENT(1) = c_{12}^{4} = 3.0$$

A new set of A_i's are generated as a result of the update of UPBUND:

$$A_1 = \{1,4\}$$
$$A_2 = \{4\}$$
$$A_3 = \{5\}$$
$$A_4 = \{8\}$$

Constraint checking reveals that the set, {4,5,8} is the only one that can cover A_1 through A_4 with less than 4 copies. In addition, the total communication cost incurred is 0.4 which is less or equal to UPBUND. Therefore, in the case of m=3, the optimal solution is to store a copy of file f_1 in nodes 4, 5, and 8 respectively. However, for the case of m=2, the algorithm has to continue. This demonstrates the statement made previously regarding the effect of the value of m.

Following the same algorithm, an optimal solution for the case of m=2 is found to be the set of {4,5} which gives a total cost of 2.1. This solution is obtained in the fourth iteration.

V. Conclusions

In this paper, a file allocation problem that has a hard constraint on the number of copies for each file is studied. An optimization algorithm for allocating limited copies of files to processing nodes so that the total communication cost is minimized is developed. The approach offers attractive features that does not exist in a purely combinatorial approach.

Other extensions to this problem are possible. One of them is to also consider the constraint on the number of files that can exist at each node. In this case, a suboptimal allocation can be obtained by deleting nodes that have been assigned enough files and apply the same algorithm to the remaining files without these nodes. The allocation of an individual file will depend upon allocation of other files.

References

[1] P. Bay, A. Thomasian "Optimal Program and Data Locations in Computer Networks", Commun. ACM 20, 5(May 1977), 315-322.

[2] R. G. Casey, "Allocation of copies of a file in an information network", Proc. AFIPS 1972 Spring Jt. Computer Conf., AFIPS Press, Arlington, VA, 251-257.

[3] P. P. S. Chen, J. Akoka, "Optimal Design of Distributed Information Systems", IEEE Trans. Computers, C-29, 12(DEC. 1980), 1068-1080.

[4] W. W. Chu, "Optimal File Allocation in a Multiple Computer System", IEEE Trans. Computers, C-18, 10(Oct. 1969), 885-889.

[5] W. W. Chu, "Optimal File Allocation in a Computer Network", Computer-Communication Systems, N. Abramson and F.F. Kuo, Eds., Prentice-Hall, Englewood Cliffs, N.J., 1973, 82-94.

[6] E. G. Coffman Jr., et. al., "Optimization of the Number of Copies in a Distributed Data Base", IEEE Trans. Software Engineering, SE-7, 1(Jan. 1981), 78-84.

[7] L. W. Dowdy, D. V. Foster, "Comparative Models of the File Assignment Problem", ACM Computing Surveys 14, 2(June 1982), 287-313.

[8] K. P. Eswaran, "Placement of Records in a File and FIle Allocation in a Computer Network", Information Processing 74, IFIPS, 1974, 304-307.

[9] J. F. Kurose, R. Simba, "A Microeconomic Approach to Optimal File Allocation", 6-th Int. Conf. on Dist. Comp. Sys., (Cambridge, MA., May 1986), 28-35.

[10] L. J. Laning, "File Allocation in a Distributed Computer Communication Network", IEEE Trans. Computers, C-32, 3(March 1983), 232-244.

[11] S. Mahmoud, J. S. Riordon, "Optimal Allocation of Resources in Distributed Information Networks", ACM Trans. Database Syst. 1, 1(March 1976), 66-78.

[12] H. L. Morgan, K. D. Levin, "Optimal Program and Data Locations in Computer Networks", Commun. ACM 20, 5(May 1977), 315-322.

[13] C. V. Ramamoorthy, B. W. Wah, "The Isomorphism of Simple File Allocation", IEEE Trans. Computers, C-32, 3(March 1983), 221-232.

[14] B. W. Wah, "File Placement on Distributed Computer Systems", IEEE Computer Jan. 1984, 23-32.

Appendix I:

The efficiency of a constraint checking algorithm is critical to the performance of our optimization algorithm. For purpose of clarity, the constraint checking subproblem is restated below. Given a set of l A_i's determine if there are sets of m or less nodes which may cover all A_i's. Here, A_i is a set of processing nodes where copies of targeted file can possibly be stored. An A_i is said to be covered by a set S if $A_i \cap S \neq \phi$.

The algorithm presented below iterate upon every A_i until all A_i's are exhausted or no set can possibly cover all A_i's with no more than m nodes. Before we further discuss the algorithm, following definitions are needed.

Definition: 1. X(j) is the number of A_i's containing node j, j=1,2,...,N.

Definition: 2. Mx(i) denotes the i-th greatest X(j), i=1,2,...,m.

Observation reveals that the m-limited constraint is not met as long as following inequality condition exists:

$$\sum_{j=1}^{m} Mx(j) < l.$$

This condition facilitate constraint checking during the early stage of iterations. Detailed algorithm is given below with the help of Pascal programming notations.

Algorithm:

1. If $\sum_{j=1}^{m} Mx(j) < l$ then ($S = \phi$ and go to step 4).
2. Select the A_i with smallest $|A_i|$ and construct a set S of sets of nodes as follows:

$$S = \left\{ (p) \;\middle|\; p \in A_i \right\}$$

3. Iterate upon the other A_i's:
   Do while (more A_i) and $S \neq \phi$;
   3.1 pick an A_k with smallest $|A_k|$ among the unprocessed A_i's.
   3.2 for (every element s of S) do;
   if $|s| <$ m then

$$S = S \cup \left\{ \bigcup_{q \in A_k} (s \cup q) \right\} - s$$

   else ($|s| =$ m)
   if $s \cap A_k = \phi$ then

$$S = S - s$$

   end.
4. Return:
   If $S = \phi$ then "constraint not met"
   else "S contains all possible sets of nodes that cover A_i's and meet m-limited constraint"

Software Reliability—Status and Perspectives

C. V. RAMAMOORTHY, FELLOW, IEEE, AND FAROKH B. BASTANI, MEMBER, IEEE

***Abstract*—It is essential to assess the reliability of digital computer systems used for critical real-time control applications (e.g., nuclear power plant safety control systems). This involves the assessment of the design correctness of the combined hardware/software system as well as the reliability of the hardware. In this paper we survey methods of determining the design correctness of systems as applied to computer programs.**

Automated program proving techniques are still not practical for realistic programs. Manual proofs are lengthy, tedious, and error-prone. Software reliability provides a measure of confidence in the operational correctness of the software. Since the early 1970's several software reliability models have been proposed. We classify and discuss these models using the concepts of residual error size and the testing process used. We also discuss methods of estimating the correctness of the program and the adequacy of the set of test cases used.

These methods are directly applicable to assessing the design correctness of the total integrated hardware/software system which ultimately could include large complex distributed processing systems.

***Index Terms*—Correctness probability, error-counting models, error seeding, error size, evaluation of test cases, nonerror-counting models, software fault, software reliability models, testing and debugging phase, testing process, validation phase.**

I. Introduction

IN ORDER TO pave the way for the use of digital computers for critical applications, like nuclear power plant safety control systems, it must be shown that the computer system meets the specified reliability constraints. The theoretical basis for methods of estimating the reliability of the hardware is well developed [6]. Since the early 1970's, several models have been proposed for estimating software reliability and some related parameters, such as mean time to failure (MTTF), residual error content, and other measures of confidence in the software. However, software reliability theory is still rudimentary. In fact, the basis of most software reliability models is often viewed skeptically [57], [114]. Since there is no physical *deterioration* or random malfunction in software, it is preferable to prove that the software meets (or does not meet) its requirements specification. However, current program verification techniques cannot cope with the size and complexity of software for real-time applications [23]. Similarly, exhaustive testing is ruled out by the large number of possible inputs. Besides, testing is limited by other factors, namely, the difficulty in verifying the output corresponding to an input and the lack of realistic inputs (e.g., in a missile defense system).

Most software reliability models attempt to estimate the reliability of the software based on its error history, either during its debugging phase (when errors which are detected are corrected) or during its validation phase (when errors are *not* corrected). For example, an operating system which crashes more often than another has a lower reliability. The difficulty lies in the quantification of this measure.

In the following we discuss some of the software reliability models proposed so far. We attempt to give a physical feel for the models, their assumptions, and the type of data required, and some comments on their applicability to real projects. In Section II we give a definition of software reliability. The existing reliability models are classified in Section III. Section IV discusses models applicable during the debugging phase (these are called "reliability growth models"). We introduce the concepts of error size and testing process and use these to discuss the Jelinski-Moranda models [69], the Shooman model [98], the Schick-Wolverton model [92], the Musa model [75], the Littlewood model [56], the Littlewood-Verrall model [53], and an input domain based stochastic model [85]. We also discuss a general framework for these growth models. In Section V we first discuss the Nelson model [109]. Then we introduce the concepts of equivalence classes and continuity in the input domain. These are used for estimating the correctness probability of the program [82]. In Section VI we briefly discuss the application of error seeding to estimating the efficiency of different testing strategies, the reliability of the set of test cases used (i.e., a measure of how well the software has been tested), and the correctness probability of the program.

The following references discuss other reliability models and their applications: [14], [21], [68], [86], [101], [106].

II. Definition

Software reliability has been defined as the probability that a *software fault* which causes *deviation* from required output by more than *specified tolerances*, in a *specified environment*, does not occur during a *specified exposure period* [109]. Thus, the software need be correct only for inputs for which it is designed (*specified environment*). Also, if the output is correct within the specified tolerances in spite of some error, then the error is ignored. This may happen in the evaluation of complicated floating point expressions where many approximations are used (e.g., polynomial approximations for COSINE, SINE, etc.).

Manuscript received August 25, 1980; revised December 21, 1981. This work was supported by the U.S. Air Force Office of Scientific Research under Contract F49620-79-C-0173.

C. V. Ramamoorthy is with the Department of Electrical Engineering and Computer Science and the Electronics Research Laboratory, University of California, Berkeley, CA 94720.

F. B. Bastani is with the Department of Computer Science, University of Houston, Houston, TX 77004.

It is possible that a failure may be due to errors in the compiler, operating system, microcode, or even the hardware. These failures are ignored in estimating the reliability of the application program. However, the estimation of the overall system reliability will include the correctness of the supporting software and the reliability of the hardware.

The *exposure period* should be independent of extraneous factors like machine execution time, programming environment, etc. For many applications, the appropriate unit of *exposure period* is a *run* corresponding to the selection of a point from the input domain (*specified environment*) of the program. However, for some programs (e.g., an operating system) it is difficult to determine what constitutes a "run." In such cases the unit of *exposure period* may be the calendar or CPU time. Thus, we have

$$R(i) = \text{reliability over } i \text{ runs} = P\{\text{no failure over } i \text{ runs}\} \quad (1)$$

or

$$R(t) = \text{reliability over } t \text{ seconds} = P\{\text{no failure in interval } [0, t]\}. \quad (2)$$

($P\{E\}$ denotes the probability of the event E.)

Definition (1) leads to an intuitive measure of software reliability. Assuming that inputs are selected independently according to some probability distribution function, we have

$$R(i) = [R(1)]^i = (R)^i$$

where $R \equiv R(1)$. We can define the reliability R as follows:

$$R = 1 - \lim_{n \to \infty} \frac{n_f}{n}$$

where

n = number of runs,

n_f = number of failures in n runs.

This is the *operational definition* of software reliability. We can estimate the reliability of a program by observing the outcomes (success/failure) of a number of runs under its operating environment. If we observe n_f failures out of n runs, the estimate of R, denoted by $\hat{R}$, is

$$\hat{R} = 1 - \frac{n_f}{n}.$$

This method of estimating R is the basis of the Nelson model [109].

III. Classification of Software Reliability Models

Fig. 1 shows a classification of some of the existing software reliability models. The classification scheme is based primarily on the phase of software life-cycle during which the model is applicable. The main feature of a model serves as a subclassification. Most of the existing models can be used during the software testing and debugging phase, validation phase, or operational phase. There are several models which do not yield a reliability figure but which measure some parameters useful in evaluating a given software system, for example, the number of remaining errors, test reliability, or confidence in the program. These models are classified as models which yield some measure of the correctness of the program.

A. Testing and Debugging Phase

During this phase the implemented software is tested and debugged. It is often assumed that the correction of errors does not introduce any new errors. Hence, the reliability of the program increases, and therefore the models used during this phase are also called reliability growth models.

The error-counting models estimate both the number of errors remaining in the program as well as its reliability. Deterministic, Bayesian, and Markov error-counting models have been proposed. The deterministic models assume that if the model parameters are known, then the correction of an error results in a known increase in reliability. This category includes the Jelinski-Moranda [47], Shooman [24], [95], [96], Musa [75], and Schick-Wolverton [92] models. The general Poisson model [4] is a generalization of these four models. The Bayesian model due to Littlewood [56] models the (usual) case where larger errors are detected earlier than smaller errors. All the preceding models neglect the time required to correct an error. This aspect is modeled as a Markov process by Trivedi and Shooman [108]. The model also yields an estimate of the availability of the software system.

The number of errors remaining in the program is useful in estimating the maintenance cost. However, with these models it is difficult to incorporate the case where new errors may be introduced in the program as a result of imperfect debugging. Further, the reliability estimate is unstable if the estimate of the number of remaining errors is low [33], [60].

The nonerror-counting models only estimate the reliability of the software. The Jelinski-Moranda geometric de-eutrophication model [69] and a simple model used in the Halden project [21] are deterministic models in this category. The stochastic models consider the situation where different errors have different effects on the failure rate of the program. The correction of an error results in a stochastic increase in the reliability. A stochastic model based on the nature of the input domain of the program is developed in [85].

The deterministic and the stochastic models assume that the reliability is unchanged during the interval between consecutive error corrections. While this is true in an absolute sense, the reliability estimate as *perceived* by the person testing the program, increases as the number of successful consecutive runs increases. This situation is modeled by the Bayesian growth models. The model due to Littlewood and Verrall [53] and the Mixed-Gamma model developed in [85] are stochastic and Bayesian.

The models described above treat the program as a black box. That is, the reliability is estimated without regard to the structure of the program. The validity of their assumption usually increases as the size of the program increases. Since programs for critical control systems may be of medium size

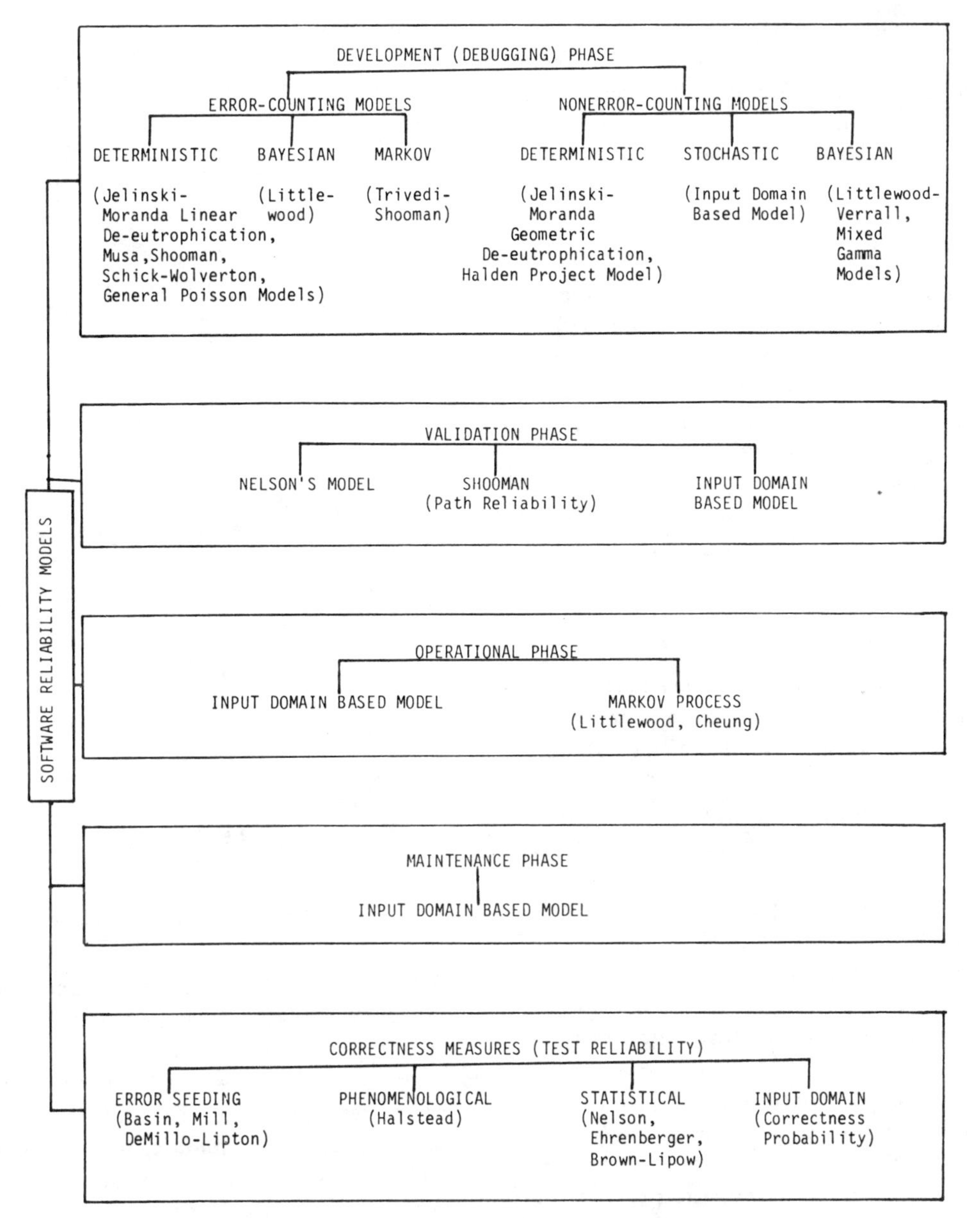

Fig. 1. Classification of software reliability models.

only, these models are mainly used to obtain a preliminary estimate of the software reliability.

B. Validation Phase

Software developed for critical applications, like air-traffic control, must be shown to have a high reliability prior to actual use. At the end of the development phase, the software is subjected to a large amount of testing in order to estimate the reliability. Errors found during this phase are not corrected. In fact, if errors are discovered the software may be rejected.

The Nelson model [109] is based on statistical principles. The software is tested with test cases having the same distribution as the actual operating environment. The operational definition (Section II) is used to obtain the reliability estimate.

The only disadvantage of the Nelson model is that a large amount of test cases are required in order to have a high confidence in the reliability estimate. The approach developed in [82] reduces the number of test cases required by exploiting the nature of the input domain of the program. An important feature of this model is that the testing need not be random—*any* type of test-selection strategy can be used.

C. Operational Phase

In most cases the successive inputs to the program are not independent. For example, in process control systems the sensor inputs are correlated in time due to physical constraints. As a first approximation, the operating environment of a program can be viewed as consisting of a number of different distributions. For a certain (random) period the input is selected using one distribution and then a transition is made to another distribution. Littlewood [55] and Cheung [18] have modeled the input distribution selection mechanism as a Markov process.

The input domain based model developed in [82] incorporates the feature that the outcome of a particular run may depend on the outcomes of the previous runs. For example, if a new test case is identical to some earlier, successful test case, then its outcome will be successful with certainty. Thus, in a sense, the software is being continuously validated during the operational phase.

D. Maintenance Phase

During the maintenance phase the possible activities are: error correction, addition of new features, and improvements in algorithms. Any of these activities can perturb the reliability of the system. The new reliability can be estimated using the models for the validation phase. However, it may be possible to estimate the change in the reliability using fewer test cases by ensuring that the original features have not been altered. We do not know of any existing software reliability models applicable during this phase. The input domain based approach is discussed in [7].

E. Correctness Measures

Software for critical applications must have a reliability estimate of 1. In these cases the confidence in the estimate is very important. We have classified all methods which provide some measure of how well the software has been tested under the "correctness measures" category.

Error seeding [92] and program mutation [22] directly measure the test reliability [38]—all seeded errors and all nonequivalent mutations should be detected by the test cases.

The software science (phenomenological) approach due to Halstead [40] gives an empirical prediction of the error content of the software based on the number of operators and operands. If the predicted error content is high, then more testing is required.

The statistical approach developed at TRW [109] yields a confidence in the reliability estimate based on statistical sampling theory. A method of estimating the *representativeness* of the test cases, that is, whether the test cases are adequately representative of the operating environment, is discussed in [15]. Ehrenberger has developed statistical models based on the coverage of the input domain, program functions, and program structures [14].

An input domain based model for directly estimating the correctness probability of a program is developed in [82]. An uncertainty measure based on fuzzy subsets of the input domain is discussed in [84].

In this section we have classified many software reliability models without describing them in detail. References [14], [21], [92], and [106] contain a detailed survey of most of the models. Some of the models are discussed in the next two sections.

IV. Software Reliability Growth Models

During the debugging phase the software is tested and all errors which are detected are corrected. Assuming that no new errors are introduced, the reliability increases. Hence, the models used for assessing software reliability based on its error history are called reliability growth models. Similar growth models have been developed and used for assessing the reliability of hardware design [25], [31].

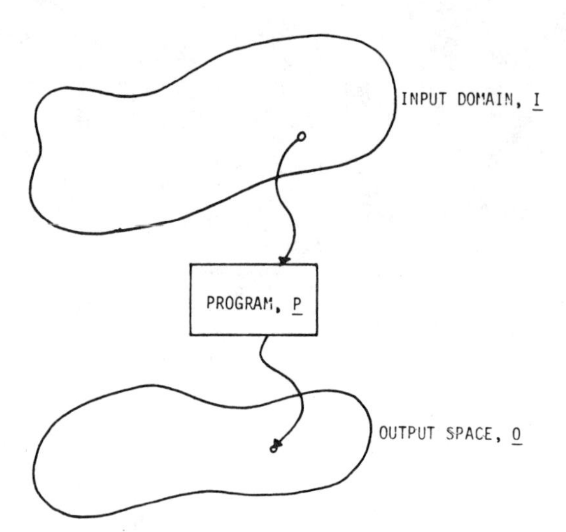

Fig. 2. Functional view of a program.

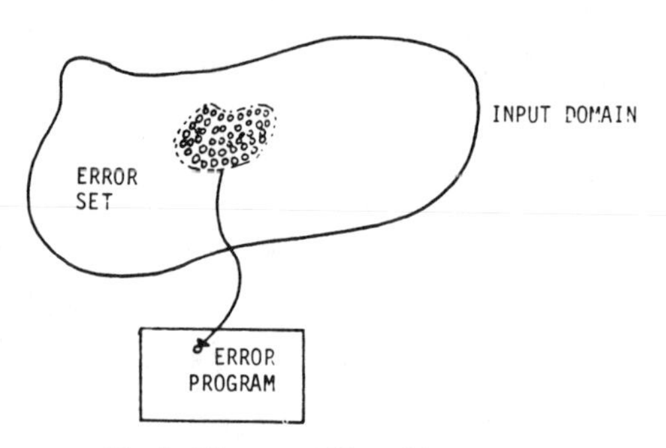

Fig. 3. Elements affected by an error.

In this section we first develop a general framework for software reliability growth models using the concepts of error sizes and testing process. Then we discuss two classes of these models, namely, error-counting models and nonerror-counting models. The validation of some of the models and their application in other aspects of software development are also discussed.

A. Error Sizes

A program P maps its input domain I into its output space O, as shown in Fig. 2. Each element in I is mapped to a unique element in O if we assume that the state variables (i.e., output variables whose values are used during the next run, as in process control software) are considered a part of both I and O.

Fig. 3 shows the elements in the input domain that are affected by an error in the program. That is, if there are no other errors and if this error is corrected, then the program will be correct for all inputs.

The software reliability models used during the development phase are intimately concerned with the size of an error. This is defined as follows.

Definition: The size of an error is the probability that an element selected from I according to the test case selection criterion results in failure due to that error.

An error is easy if it has a large size (i.e., if it affects many input elements) since then it will be easily detected. Similar'y, an error is subtle if it has a small size since then it is relatively more difficult to detect the error. The size of an error depends on the way the inputs are selected. Good test case selection strategies, like boundary value testing, path testing, and range testing, magnify the size of an error since they exercise error-prone constructs. Likewise, the observed (effective) error size is lower if the test cases are randomly chosen from the input domain.

We can generalize the notion of "error size" by basing it on the different methods of observing programs. For example, an error has a large size *visually* if it can be easily detected by code reading. Similarly, an error is difficult to detect by code review if it has a small size (e.g., only when one character is missing).

There is an important relationship between the total size of the remaining errors in a program Ve_r and its operational reliability $R(1)$, as defined in Section II:

$$Ve_r = 1 - R(1),$$

i.e.,

$$Ve_r = \lim_{n \to \infty} \frac{n_f}{n}$$

where n_f is the number of failures in n runs. Thus, the size of the remaining errors can be estimated as follows:

$$\hat{Ve}_r \approx \frac{n_f}{n}.$$

The development phase is assumed to consist of the following cycle:

1) the program is tested until an error is found,
2) the error is corrected and step 1) is repeated.

As we have noted above, the error history of a program depends on the testing strategy employed, so that the reliability models must consider the testing process used.

B. Testing Process

Fig. 4 illustrates a simple example where the error history is strongly dependent on the testing process (see also [94]). Assume that the program has three paths which partition the input domain into three disjoint subsets [Fig. 4(a)]. The shaded regions in the figure correspond to set of elements affected by the errors in the program. Some errors affect elements in more than one path. The errors can also overlap, that is, an input can be affected by more than one error.

Fig. 4(b) illustrates a possible error history when each input is considered as equally likely. Initially errors are frequently detected. As these are corrected, the interval between error detection increases since fewer errors remain. Fig. 4(c) shows the case where a path is tested "well" before testing another path. Thus, whenever a switch is made to a new path the error detection rate increases. Fig. 4(c) also corresponds to the case where different test case selection strategies are used. In the beginning test cases are selected randomly (equally likely) since the probability of detecting errors is high. Later, more elaborate strategies like boundary value testing are used. These exercise error-prone constructs and thus have a higher probability of detecting errors.

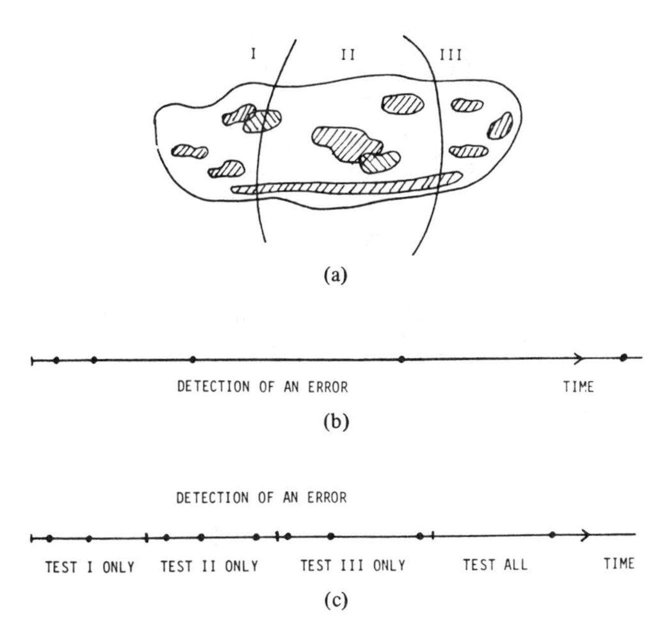

Fig. 4. Illustrating the effect of the testing process on the error history. (a) Input domain with initial errors. (b) Random (equally likely) testing. (c) Path testing.

C. General Growth Model

The major assumption of all software reliability growth models is the following.

Assumption: Inputs are selected randomly and independently from the input domain according to the operational distribution.

This is a very strong assumption and will not hold in general, especially so in the case of process control software where successive inputs are correlated in time during system operation. For example, if an input corresponds to a temperature reading then it cannot change rapidly. To complicate the issue further, most process control software systems maintain a history of the input variables. The input to the program is not only the current sensor inputs, but also their history. This further reduces the validity of the above assumption. The assumption is necessary in order to keep the analysis simple. However, in the following we relax it as follows.

Assumption: Inputs are selected randomly and independently from the input domain according to some probability distribution (which can change with time).

This means that the effective error size varies with time even though the program is not changed. This permits a straightforward modeling of the testing process. Let

j = number of failures experienced,
k = number of runs since the jth failure,
$T_j(k)$ = testing process for the kth run after j failures,
$V_j(k)$ = size of residual errors for the kth run after j failures; this can be random.

Now

$$P\{\text{success on } k\text{th run} \mid j \text{ failures}\} = 1 - V_j(k) = 1 - f(T_j(k))\lambda_j$$

where

λ_j = error size under operational inputs; this can be a random variable; $0 \leqslant \lambda_j \leqslant 1$;

$f(T_j(k))$ = severity of testing process relative to operational distribution; $0 \leq f(T_j(k)) \leq 1/\lambda_j$.

Hence

$$R_j(k \mid \lambda_j) = P\{\text{no failure over } k \text{ runs} \mid \lambda_j\}$$

$$= \prod_{i=1}^{k} P\{\text{no failure on the } i\text{th run} \mid \lambda_j\}$$

since successive test cases have independent failure probability. Hence

$$R_j(k \mid \lambda_j) = \prod_{i=1}^{k} [1 - f(T_j(i))\lambda_j],$$

i.e.,

$$R_j(k) = E_{\lambda_j}\left[\prod_{i=1}^{k} [1 - f(T_j(i))\lambda_j]\right] \tag{3}$$

where $E_{\lambda_j}[\cdot]$ is the expectation over λ_j.

For many types of software, e.g., operating systems and real-time process control systems, it is difficult to identify "runs" or there may be a very large number of such runs if we assume that each cycle constitutes a run. In these cases, it is simpler to work in continuous time. The above relation becomes

$$R_j(t) = E_{\lambda_j}[e^{-\lambda_j \int_0^t f(T_j(s))\,ds}] \tag{4}$$

where

λ_j = failure rate after jth failure; $0 \leq \lambda_j \leq \infty$,

$T_j(s)$ = testing process at time s after jth failure,

$f(T_j(s))$ = severity of testing process relative to operational distribution; $0 \leq f(T_j(s)) \leq \infty$.

Remarks:

1) As we have noted above, $f(T_j(\cdot))$ is the severity of the testing process relative to the operational distribution, where the testing severity is the ratio of the probability that a run based on the test case selection strategy detects an error to the probability that a failure occurs on a run selected according to the operational distribution. Obviously, during the operational phase, $f(T_j(\cdot)) = 1$. Thus, a testing strategy which simulates the operational input would be ideal from a reliability modeling point of view since it would simplify the model considerably. But this conflicts with our need to improve the software by detecting as many errors as possible by employing nonrandom testing strategies such as path testing, function testing, and boundary value testing [59]. In general, it is difficult to determine the severity of these test cases and most models assume that $f(T_j(\cdot)) = 1$. However, for some testing strategies we can attempt to quantify $f(T_j(\cdot))$. For example, in function testing the severity increases as we switch to untested functions since these are more likely to contain errors than functions which have already been tested.

2) In the continuous case, the time is the CPU time.

3) The software reliability models discussed below can be applied (in principle) to any type of software. However, their validity increases as the size of the software and the number of programmers involved increases.

4) This process is a type of doubly stochastic process; these processes were originally studied by Cox in 1955 [19].

1) Error-Counting Models: The models discuss in this section attempt to estimate the software reliability in terms of the estimated number of errors remaining in the program. The Jelinski-Moranda [47] was the first error-counting model. The Shooman model [24], [95] underwent some changes [96], [98] and is now similar to the Jelinski-Moranda model. The Schick-Wolverton model [92] extended the Jelinski-Moranda model by incorporating a factor representing the severity of the test cases. The Musa model [75] is equivalent to the Jelinski-Moranda model. However, it is better developed and is the first model to insist on execution time. Recently, Littlewood [56] developed a model where the failure rate of successive errors is stochastically decreasing, unlike the previous models which assume that all errors have the same failure rate.

a) General Poisson Model: The general Poisson model (GPM) is discussed in [4]. It generalizes the Jelinski-Moranda linear de-eutrophication model [47], [69], the Shooman model [24], [95], [96], [98], and the Schick-Wolverton model [91], [92]. The key parts of the Musa model are also generalized by this model.

The GPM model assumes that [see (4)]

$$f(T_j(s)) = \alpha s^{\alpha-1}$$

$$\lambda_j = (N - M_j)\varphi$$

where

N = number of errors originally present,

M_j = number of errors corrected *before* the jth failure and *after* the $(j-1)$th failure,

α, φ = constants

$$\therefore R_j(t) = e^{-\varphi(N-M_j)t^{\alpha}}.$$

Assumptions: The assumptions of the GPM model are as follows:

1) consecutive inputs have independent failure probabilities,

2) all errors have the same disjoint failure rate φ,

3) the severity of the testing process is proportional to a power of the elapsed CPU time,

4) no new errors are introduced.

Discussion: Assumption 1) has already been discussed above. Assumption 2) is a major drawback of these models [57]: earlier errors are likely to have a larger failure rate since they are detected more easily. Assumption 3) depends to a large extent on the testing strategy used. Intuitively, as time increases, the severity of the testing increases [92]. Assumption 4) is not true in general and can lead to invalid estimates [4]. Musa [75] partly overcomes this by estimating the total number of errors to be eventually detected.

The GPM-based models have been applied to several projects with mixed results [4], [21], [41], [66], [69], [75], [78], [92], [101], [102]. Unfortunately, with the exception of the Musa model [41], [75], [78], the time used is the calendar time, partly due to the lack of data. Musa [78] has also directly validated the assumptions of his model using several sets of data.

Application: As discussed in the derivation of (4), the execution time should be used. Thus, the data required are the CPU time between failures and the number of errors corrected at each time. The validity of the assumptions increases as the size of the programs (in terms of code length and the number of programmers) increases. If the number of errors discovered is small, then the maximum likelihood estimate of the parameters is unstable [33]. Littlewood and Verrall [60] have derived general conditions which the error data must satisfy in order for the parameters of the Jelinski-Moranda linear de-eutrophication model to have finite values. The MLE's can be computed by solving the following equations for $\hat{N}, \hat{\alpha}, \hat{\varphi}$:

$$\sum_{i=1}^{n} \frac{1}{\hat{N} - M_{i-1}} - \sum_{i=1}^{n} \hat{\varphi} t_i^{\hat{\alpha}} = 0;$$

$$\frac{n}{\hat{\alpha}} + \sum_{i=1}^{n} \log t_i - \sum_{i=1}^{n} \hat{\varphi}(\hat{N} - M_{i-1}) t_i^{\hat{\alpha}} \log t_i = 0;$$

$$\frac{n}{\hat{\varphi}} - \sum_{i=1}^{n} (\hat{N} - M_{i-1}) t_i^{\hat{\alpha}} = 0$$

where

M_i = number of errors corrected before the ith failure,

t_i = time (CPU) between the $(i - 1)$th and the ith failures,

n = number of failures.

These are discussed further in [4].

b) Littlewood's Model: The GPM is a *deterministic* reliability growth model. This is shown in Fig. 5(a) where after each error correction the residual failure rate (λ_j) decreases by a known amount. We have already noted that different errors have different failure rates. Thus, a better approach is to model the failure rate as a random variable. This results in Fig. 5(b). Here the change in the failure rate is random. Also, the failure rate is constant during the interval between error corrections. However, if one adopts a Bayesian viewpoint, then the failure rate, *as perceived by the tester*, is a varying quantity even when no changes are made to the software: the fact that there has been no failure makes us more confident in the program. This is shown in Fig. 5(c). This issue is discussed in detail by Littlewood in [56] and [57].

In Littlewood's model [56] $f(T_j(s)) = 1$ (i.e., the testing is assumed to reflect the operational environment) and λ_j is a random variable. The *prior* distribution of the failure rate φ of each error is assumed to be the gamma distribution with parameters α and β. The *posterior* distribution is computed as follows:

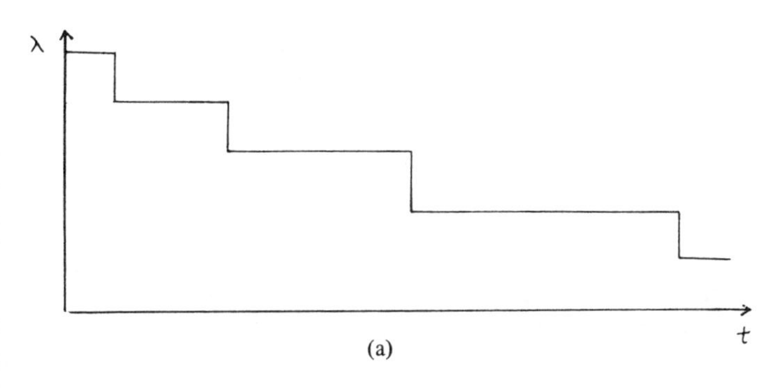

(a)

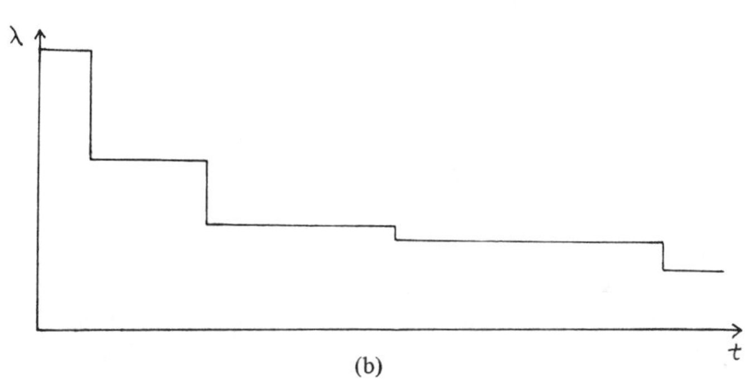

(b)

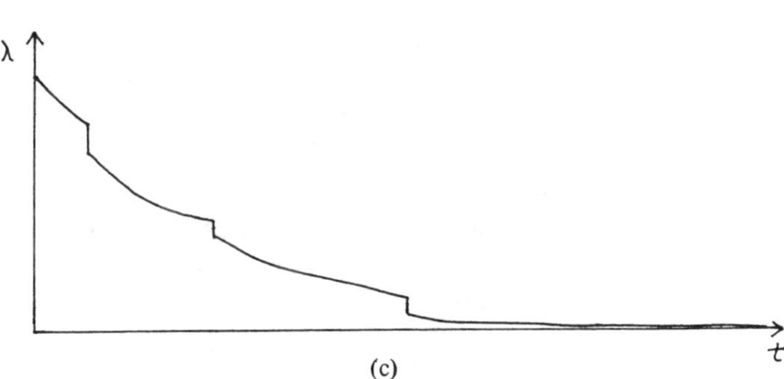

(c)

Fig. 5. Modeling of failure rates. (a) Deterministic. (b) Stochastic. (c) Bayesian.

Thus, the *posterior* distribution is gamma $(\alpha, \beta + t)$. If M_j errors have been corrected, then assuming that each error constitutes an independent error process, λ_j has the gamma $((N - M_j)\alpha, \beta + t)$ distribution where N is the original number of errors in the software.

Therefore, from (4),

$$R_j(t) = \int_0^{\infty} \frac{e^{-lt}(\beta + s_j)^{(N-M_j)\alpha} l^{(N-M_j)\alpha - 1} e^{-(\beta + s_j)l}}{\Gamma((N - M_j)\alpha)} \, dl = \left[\frac{\beta + s_j}{\beta + s_j + t}\right]^{(N-M_j)\alpha}$$

$$\text{pdf}\{\varphi = l \mid \text{no failure due to this error during time } t\} = \frac{\text{pdf}\{\text{no failure due to this error during time } t \mid \varphi = l\} \, \text{pdf}\{\varphi = l\}}{\int_0^{\infty} \text{pdf}\{\text{no failure due to this error during time } t \mid \varphi = l\} \, \text{pdf}\{\varphi = l\} \, dl}$$

$$= \frac{e^{-lt} \dfrac{\beta^{\alpha} l^{\alpha - 1} e^{-\beta l}}{\Gamma(\alpha)}}{\displaystyle\int_0^{\infty} e^{-lt} \frac{\beta^{\alpha} l^{\alpha-1} e^{-\beta l}}{\Gamma(\alpha)} \, dl} = \frac{(\beta + t)^{\alpha} l^{\alpha - 1} e^{-(\beta + t)l}}{\Gamma(\alpha)}.$$

where

s_j = time of the jth failure,

t = elapsed time since jth failure.

Thus, $R_j(t)$ is a *Pareto* distribution which appears to be more suitable for software systems since it permits very large error free intervals unlike the *exponential* distribution [60].

Assumptions: The assumptions of Littlewood's model are as follows:

1) consecutive inputs have independent failure probabilities,

2) at any time, the failure rates of the errors remaining in the program are independently identically distributed random variables; in particular, the distribution is assumed to be the gamma distribution,

3) the program failure rate is the sum of the individual failure rates,

4) the input process simulates the operational environment,

5) no new errors are introduced.

Discussion: One major weakness is assumptions 2) and 3) taken together. This is not true in general. For example, consider the extreme case where the failure rates of the errors have the constant distribution c, and the errors affect the *same* elements in the input domain. Then the program failure rate is also c and the removal of each error, except the last error, leaves the failure rate unchanged.

A safer approach would be to assume that the errors are disjoint. This would become more reasonable as the size of the software increases. However, the analysis is likely to be complicated since we can no longer assume that the errors have independent failure rates. The data required are the CPU time between errors and the number of errors corrected. Additional aspects of the model are discussed in [56].

2) Nonerror-Counting Models: The models discussed in this section do not consider the number of errors remaining in the program. Instead they consider the effects of the remaining errors. This permits a simple modeling of the possibility that new errors are introduced while correcting an error.

a) Jelinski-Moranda Geometric De-Eutrophication Model: This model was first proposed by Moranda [69], [73]. A generalization of the model is

$$f(T_j(s)) = \alpha s^{\alpha - 1}$$

$$\lambda_j = KD^{j-1}$$

where K, D, α are constants to be estimated. Therefore,

$$R_j(t) = e^{-KD^{j-1}t^{\alpha}}.$$

Thus, this is a GPM with geometrically decreasing failure rate. It models the case where consecutive errors have decreasing sizes. This is more realistic than the assumption that all errors have the same size (as in the GPM models discussed in Section IV-C.1a). An application of this model is discussed in [69]. An interesting observation is that the estimate of the parameters of this model may exist even in cases where those of the linear de-eutrophication model do not exist, i.e., fail to converge [106], [21].

b) Input Domain-Based Stochastic Model: This model is discussed in detail in [85]. In (3) let

$$f(T_j(k)) = 1$$

$$\lambda_j = \text{a random variable.}$$

Thus, the testing process is assumed to be identical to the operational environment. Let Δ_j be the size of the jth error. Then

$$\Delta_j = \lambda_{j-1} - \lambda_j.$$

Intuitively, errors which are caught later have a smaller size than those which are caught earlier. However, this is true only in a probabilistic sense. This can be modeled by requiring that

$$\Delta_j \underset{st}{\leqslant} \Delta_{j-1}$$

where $X \leqslant_{st} Y$ means that X is *stochastically* smaller than Y. In order to model a variety of situations, we assume that $\Delta_j \sim \lambda_{j-1} X$, where $X \sim F$, and F is assumed to be piecewise continuous. ($X \sim F$ means that the random variable X has the distribution F.)

A special case is $F \equiv \beta(r, s), r \geqslant 1, s \geqslant 1$. The β (beta) distribution is discussed in [12] and [44] and is often used in the study of hardware and software reliability [109]. Assuming that initially there is an error present for any input (i.e., $\lambda_0 = 1$), we get

$$E[\lambda_j] = a^j, \quad \text{where } a = \frac{s}{r+s}$$

$$\text{MTTF}_j = \left[\frac{r+s-1}{s-1}\right]^j \approx \frac{1}{a^j} \quad \text{for } s \gg 1.$$

We can attempt to predict the value of the constant "a." For example, if we assume that the size of the next error is symmetrically distributed between 0 and λ_{j-1}, then $r = s$, so that $a = 1/2$.

Thus, the simple model permits the prediction of the model parameters and the modeling of various debugging situations.

Assumptions: The assumptions of the model are as follows:

1) successive inputs have independent failure probability,

2) an assumption is made regarding the distribution of the change in the residual error size after each correction,

3) an assumption is made regarding the initial error size,

4) the testing process is assumed to be the same as the operating environment; however, this can be easily relaxed.

Discussion: The key assumption is the second one. It is possible to validate this assumption by estimating the size of each error detected and corrected and assuming that the remaining errors are disjoint from the corrected errors. The size of each error can be estimated using the sampling technique discussed in Section V.

It is difficult to estimate the parameters of F based on the error history of the program. Attempts to estimate the MLE of the parameters by trial and error can fail if k is large since then $\binom{k}{i}$ becomes very large. A practical approach is to compute a series of approximations to $R_j(k)$. For example, the first approximation is

$$P_j(k) = P\left\{\frac{k \text{ successful runs before failure} \mid j \text{ errors detected}}{\text{corrected}}\right\}$$
$$= E[(1 - \lambda_j)^k \lambda_j]$$
$$\approx (1 - E[\lambda_j])^k E[\lambda_j]$$
$$= (1 - a^j)^k a^j.$$

This is similar to the Jelinski–Moranda geometric de-eutrophication model [69]. The MLE of a is determined by solving the following equation for $\hat{a}$:

$$\sum_{i=1}^{j} \frac{n_i i \hat{a}^i}{1 - \hat{a}^i} = \frac{j(j+1)}{2}$$

where n_i is the number of successful runs between the $(i - 1)$th and the ith errors.

Application: In this section we discuss the application of this model to the error data derived from the OECD Halden reactor project [21] and the EPRI (Electrical Power Research Institute) project [62]. Both the Halden and the EPRI projects involved research on the development methodology for critical software for nuclear power plant safety control systems. A major problem is the validation of the software and the assessment of its reliability.

Fig. 6 shows the application of the first approximation of the stochastic model to the Halden project data. Fig. 6(a) shows the estimate of the constant "a" and the predicted MTTF's. From a comparison of the predicted and actual MTTF we cannot conclude much regarding the validity of the model. However, we make two important observations. First, the estimate of "a" shows rapid convergence, so that by the jackknife technique of Mosteller and Tukey [74] we can conclude that the fit is reasonably good, i.e., additional data do not change the model parameters much. Second, from Fig. 6(b) we see that the error data lie largely within the 90 percent upper and lower confidence bounds. The fit is relatively good considering the large fluctuations in the actual data.

For the EPRI project, two programs PROGA and PROGB were developed based on the same design. (Another program has been developed based on the same requirement specification, but the error data for it is not yet available.) Both the programs were tested with the same set of test cases. However, the testing was not random—functional testing was used. We assume that

$$f(n\text{th function}) = n$$

where the "nth function" denotes the nth *distinct* function to be tested since the start of testing. This seems valid since the size of errors is magnified if the $n - 1$ (reasonably) debugged functions are excluded from testing.

We consider the mixed function testing process, i.e., a function is tested a certain number of times, then a transition is made to another function (which may have been already tested), and so on. The MLE of "a" is determined by solving the following equation for $\hat{a}$:

$$\frac{j(j+1)}{2} = \sum_{i=1}^{j} \sum_{k=1}^{n_i} \frac{i n_{ik} F_{ik} \hat{a}^i}{1 - F_{ik} \hat{a}^i}$$

i = ERROR NO.	$\frac{1}{a}$	MTTF = $[1/a]^i$	ACTUAL MTTF
0	-	-	1
1	2.0	2.0	1
2	1.59	2.5281	2
3	1.508	3.43338	10
4	1.676	7.89035	6
5	1.606	10.6838	1
6	1.5015	11.4591	198
7	1.8436	72.3883	203
8	1.874651	152.509	3598
9	2.16678	1052.77	49
10	2.115301	1793.58	1611
11	2.090145	3326.14	9986
12	2.105023	7569.93	-

(a)

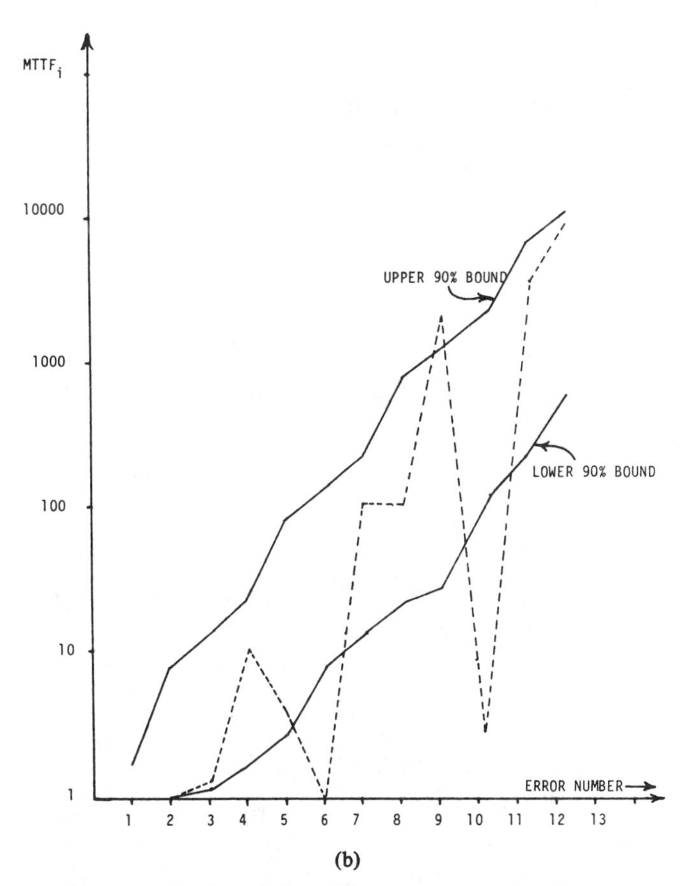

(b)

Fig. 6. (a) Application of the geometric model to the Halden project error data. (b) Upper/lower 90 percent bounds for the time between failures.

where

j = the total number of errors detected,

n_i = the number of transitions from function to function between the detection of the $(i-1)$th and the ith errors,

n_{ik} = the number of times the kth function (in the sequence between the detection of the $(i-1)$th and the ith errors) is tested,

$F_{ik} = n$, if the kth function (in the sequence between the detection of the $(i-1)$th and the ith errors) is the nth *distinct* function to be tested.

Fig. 7(a) shows the application of the model to PROGA, yielding $R(1) = 0.99993$ and $\text{MTTF}_6 = 13903$. The application of the model to the data for PROGB is shown in Fig. 7(b). Here $R(1) = 0.99990$ and $\text{MTTF}_5 = 12896$.

Further discussions of the model and its application to project management appear in [85].

c) Littlewood–Verrall Model: This model is discussed in [53]. In (4) let

$$f(T_j(s)) = 1;$$

$$\lambda_j = \text{a random variable.}$$

An innovative feature of this model is the direct modeling of the case where the programmer may introduce new errors while correcting an error. This is achieved by assuming that

$$\lambda_j \underset{st}{\leqslant} \lambda_{j-1}.$$

Let $g(\lambda_j, j)$ be the probability density function (pdf) of λ_j. Then

$$R_j(t) = \int_0^\infty e^{-\lambda_j t} g(\lambda_j, j)\, d\lambda_j.$$

Let $h(t \mid \lambda_j)$ be the pdf of time to failure given λ_j. Then

$$h(t \mid \lambda_j) = \frac{d}{dt}\,[R_j(t \mid \lambda_j)] = \lambda_j e^{-\lambda_j t}.$$

Littlewood and Verrall have adopted the Bayesian approach, i.e., the pdf of λ_j is updated as additional data are gathered. Let α be the parameters of g; let $P_0(\alpha)$ be the *prior* pdf of α and $P_1(\alpha)$ be the *posterior* pdf of α. Then (see [53])

$$P_1(\alpha) \propto \left\{ \prod_{i=1}^{n} \int h(t_i \mid l)\, g(l, i \mid \alpha)\, dl \right\} P_0(\alpha)$$

where t_i is the length of the ith interval. Therefore,

$$R_j(t_{n+1}) = \int_0^\infty e^{-\lambda_j t_{n+1}} \int g(\lambda_j, j \mid \alpha)\, P_1(\alpha)\, d\alpha\, d\lambda_j.$$

The particular distribution assumed for g in [53] is the gamma distribution. This can model a variety of different situations.

Assumptions: The following are assumed:

1) successive inputs have independent failure probability,
2) the failure rate is stochastically decreasing,
3) the distribution of λ_j, denoted by $g(\lambda_j, j)$ is assumed,

j	n_j	n_{ji}	f_{ji}	$\frac{1}{a}$
1	1	1	1	2.0000
2	1	20	1	4.0000
3	1	20	2	3.6973
4	1	40	2	3.3828
5	2	200	1	3.4863
		200	2	3.9073
6	5	1000	2	
		120	3	
		300	4	
		600	5	
		1	6	
$R(1) = 0.99993$; $\text{MTTF}_6 = 13903$				

(a)

j	n_j	n_{ji}	f_{ji}	$\frac{1}{a}$
1	1	1	1	2.0000
2	1	20	1	4.0000
3	2	200	1	5.2366
		1	2	
4	2	40	2	4.5700
		20	3	
5	3	1000	3	4.8426
		120	4	
		1	5	
6	3	300	5	-
		600	6	
		8	7	
$R(1) = 0.99990$; $\text{MTTF}_5 = 12896$				

(b)

Fig. 7. (a) Application of the stochastic model to the EPRI project data–PROGA. (b) Application of the stochastic model to the EPRI project data–PROGB.

4) the *prior* distribution of parameters of g is assumed,
5) the testing process is similar to the operational environment.

Discussion: Although the model is moderately general, it has not been widely used. Some applications can be found in [53] and [61]. The model cannot utilize additional information on the relation between λ_j and λ_{j-1}. For example, it cannot model the case where it is *assured* that no new errors are introduced, i.e., $\lambda_j \leq \lambda_{j-1}$. Also, $g(\lambda_j, j)$ includes a growth parameter which is ad hoc. Further, from the results indicated in [61] it appears as though the distribution of the time to next failure is too broad to yield a satisfactory indication of the current status of the software. A sharp distribution is preferred if the reliability indication is to be used in determining a stopping time for software testing.

D. Summary

We can view λ as a random walk process in the interval $(0, e)$. Each time the program is changed (due to error corrections or other modifications) λ changes. In the formulation of (3) and (4), λ_j denotes the *state* of λ after the jth change to the program. Let Z_j denote the time between failures after the jth change. Z_j is a random variable whose distribution depends on λ_j. In all the above continuous (discrete) time models, we have assumed this distribution to be the exponential (geometric) distribution with parameter λ_j, provided that $f(T_j(\cdot)) = 1$. We do not know anything about the random walk process of λ other than a sample of time between failures. Hence, one approach is to construct a model for λ and fit the parameters of the model to the sample data. Then we assume that the future behavior of λ can be predicted from the behavior of the model.

Some of the models for λ which have been developed are as follows.

General Poisson Model [4]: The set of possible states are $(0, e/N, 2e/N, \cdots, e)$; $\lambda_j = (N - j)e/N$; the parameters are e, N; there is a finite number of possible states.

Geometric De-Eutrophication Model [69]: The set of possible states are $(e, ed, ed^2, ed^3, \cdots)$, where $d < 1$; $\lambda_j = ed^j$; the parameters are e, d; there is an infinite (although countable) number of states.

Stochastic (Input Domain) Model: The state space is continuous over the interval $(0, e)$; $\lambda_j = \lambda_{j-1} + \Delta_j$, where $\Delta_j \sim \lambda_{j-1}X$, $X \sim \beta(r, s)$; the parameters are r and s; note the $e = 1$ in (3).

An alternative approach is the Bayesian approach advocated by Littlewood [57]. Here we postulate a *prior* distribution for each of $\lambda_1, \lambda_2, \cdots, \lambda_j$. Then based on the sample data, we compute the *posterior* distribution of λ_{j+1}. Some additional discussions appear in [85].

In this section we have discussed some of the theoretical issues of modeling the software reliability growth process. Additional discussions can be found in [50], [54], [57], [67], [70], [71], [97], [103], and [107]. Many other software reliability growth models have been proposed. Some of these are discussed in [21], [37], [89], [92], [93], and [107]. The data requirements of these models have been investigated in [26]. Trivedi and Shooman [108] have developed a model which includes the debugging and error correction time. Reliability growth models have been used for other purposes. Some examples are: code reading [48], estimating the stopping time for testing a program [33], [34], [77], [85], estimating the number of executable paths given that the inputs are selected randomly [72], and estimation of hardware reliability growth [73].

In the following section we discuss models which estimate the reliability of the program based on the results of testing performed in its validation phase. In this phase no changes are made to the software even if new errors are discovered.

V. Validation Phase

As we have already noted, the validity of most software reliability growth models increases with the size of the software since personnel peculiarities are averaged out. However, software used for highly critical, real-time purposes, as in nuclear power plant safety control systems, are of medium size (approximately a few thousand lines of some high level Language code). The programs must be shown to possess very high reliability. Ideally, we would like to prove that the software meets its specification. However, existing proof techniques cannot tackle programs of this size and complexity. For this reason, after the software has been debugged to the satisfaction of the developer(s) it enters a validation phase. During this phase the software is thoroughly tested, specifically for estimating its reliability. If any errors are discovered during this phase, they are *not* corrected. In fact, the software may be rejected if even a single new error is discovered. In this section we discuss methods of estimating software reliability during the validation phase. We first discuss Nelson's method [63], [79], [109]. Then we develop a model for estimating the reliability and the correctness probability of the program based on its input domain.

A. The Nelson Model

This model [63], [109] is based on the definition given in Section II. Test cases are selected randomly according to the operational distribution. Then

$$\hat{R}(1) = 1 - \frac{n_f}{n}$$

where

n = total number of test cases,

n_f = number of failures out of these n runs.

Fig. 8 shows the size of the remaining errors Ve_r in the program when inputs are selected according to the operational distribution. Ve_r is an unknown quantity which is related to the reliability $R(1)$. Specifically

$$Ve_r = 1 - R(1).$$

In Section IV-C we observed that if inputs are selected independently from the input domain, then

$$\text{number of runs to next failure} \sim \text{geometric}\ (Ve_r).$$

Hence, if n_f is the number of failures out of n runs, the MLE of Ve_r is

$$\hat{Ve}_r = \frac{n_f}{n}.$$

This is true since no changes are made to the program even if errors are detected, so that Ve_r remains constant. This is the difference with the models discussed in Section IV which consider the case where Ve_r changes due to debugging actions.

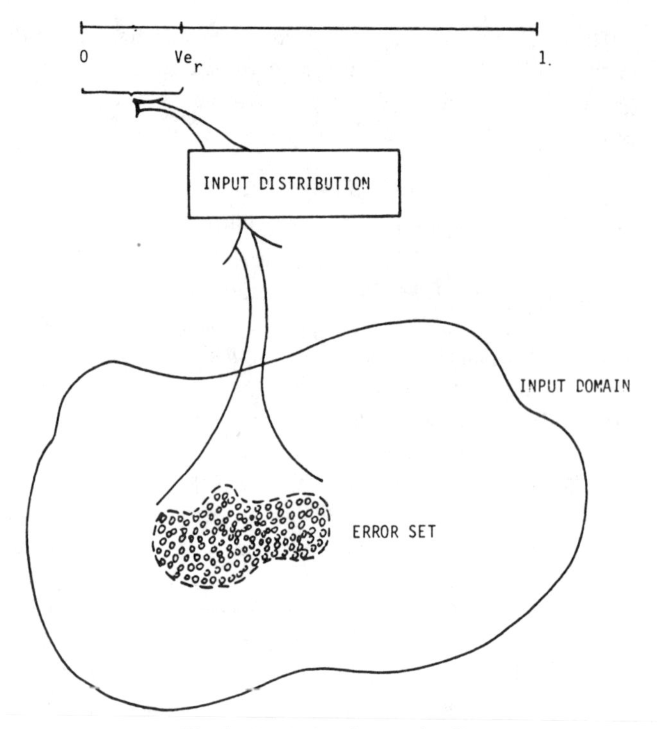

Fig. 8. Operational error size Ve_r.

The Nelson model is the only model whose theoretical foundations are sound. However, it suffers from a number of practical drawbacks.

1) In order to have a high confidence in the reliability estimate, a large number of test cases must be used.

2) It does not take into account "continuity" in the input domain. For example, if the program is correct for a test case, then it is likely that it is correct for all test cases executing the same sequence of statements.

3) It assumes random sampling of the input domain. Thus, it cannot take advantage of testing strategies which have a higher probability of detecting errors, e.g., boundary value testing, etc. Further, for most real-time control systems the successive inputs are *correlated* if the inputs are sensor readings of physical quantities, like temperature, which cannot change rapidly. In these cases we cannot perform random testing.

4) It does not consider any complexity measure of the program, e.g., number of paths, statements, etc. Generally, a complex program should be tested more that a simple program for the same confidence in the reliability estimate.

In order to overcome these drawbacks, the model has been extended [79] as follows.

The input domain is divided into several equivalence classes. The division can be based on paths or some other criteria when the number of paths is too large (e.g., program subfunctions). It is assumed that there is some continuity over an equivalence class, i.e., if the program executes correctly for an input from the jth equivalence class, then it will execute correctly for any randomly selected input from the same equivalence class with probability $1 - \epsilon_j$, where $\epsilon_j \ll 1$. Then

$$R(1) = \sum_{j=1}^{m} P_j(1 - \epsilon_j)$$

where

m = number of equivalence classes,

P_j = probability of selecting an input from the jth equivalence class during actual operation.

Discussion: This model is a big improvement over the original form. Some comments are as follows.

1) The assignment of values to ϵ_j is ad hoc; no theoretical justification is given for the assignment [79],

2) The model uses only one type of complexity measure, namely, number of paths, functions, etc. However, it does not consider the relative complexity of each path, function, etc.

Many other interesting aspects of the Nelson model are discussed in [109].

B. Input Domain Based Model

This model is discussed in detail in [82]. It removes most of the objections to the Nelson model. The price is the increased complexity of the model. However, it is perfectly suited for medium size programs such as those for real-time control systems.

In Section II we defined the operational reliability of a program by the following equation:

$$R = 1 - \lim_{n \to \infty} \frac{n_f}{n}$$

where n_f is the number of failures in n runs. This is the time domain definition of software reliability. In Section IV we discussed the concept of error size and found that

$$\hat{R} = 1 - \hat{Ve}_r$$

where $\hat{Ve}_r$ is the estimated remaining error size. This is the input domain definition of software reliability. $\hat{Ve}_r$ can be determined by testing the program and locating and estimating the size of errors found. An obvious advantage of this approach is that it permits *any* testing strategy to be used. An accompanying parameter is the correctness probability of the program. This requires the concept of probabilistic equivalence classes defined as follows: E is a probabilistic equivalence class if $E \subseteq I$, where I is the input domain of the program P, and P is correct for all elements in E, with probability $P\{X_1, \cdots, X_d\}$, if P is correct for each $X_i \in E, i = 1, \cdots, d$. Then $P\{I|X\}$ is the correctness probability of P based on the set of test cases X. (Obviously, the program must be correct for each element in X.) Probabilistic equivalence classes are derived from the requirements specification and the program source code in order to minimize control flow errors. A suggested selection criterion [82] is as follows.

Let E be a probabilistic equivalence class; $X \in E$ if an error in the program which affects any element in E *can* affect X, and *vice versa*. The results of this classification scheme are as follows:

1) it includes all paths without loops since distinct paths differ in at least one statement,

2) multiple conditions are treated separately since an error in one condition need not affect the other conditions,

3) loops are restricted to a finite number of repetitions.

In order to further minimize control flow errors, these classes should be intersected with classes derived from the requirements specification [112].

Finally, we can estimate the correctness probability of the program using the continuity assumption, i.e., closely related points in the input domain are "correlated" with respect to the implementation of a function. This is true in general for algebraic programs where errors usually affect an interval of nearby points. These regions correspond to high probability equivalence classes, such as those formed on the basis of program paths.

At first we assume that the program has a single input variable. We further assume that the correctness probability of a point depends at most on the correctness of its neighbors. If the equivalence class is discrete, say $(x_1, x_2, \cdots, x_n)$, then this means that

$$P\{x_i \text{ correct} \mid x_1, \cdots, x_{i-1}, x_{i+1}, \cdots, x_n \text{ correct}\}$$
$$= P\{x_i \text{ correct} \mid x_{i-1}, x_{i+1} \text{ correct}\}.$$

This leads to a mathematically tractable derivation of the correctness probability.

Let $E_i = [a, a+v]$ denote a continuous equivalence class of a program P, selected using the above selection criterion. Fig. 9 shows the interpretation of the nearest neighbor dependency assumption for E_i. Fig. 9(a) illustrates the case where a single test case is used. It means that the *distribution* of a probabilistic equivalence class containing the test case is the exponential distribution. In principle, any other distribution can be considered. However, the derivation of the correctness probability may be intractable. Fig. 9(b) considers the case where two test cases are used. The region between these two points is validated by both of them.

In general, it can be shown that

$$P\{E_i \text{ is correct} \mid 1 \text{ test case}\} = e^{-\lambda V}$$

$$P\{E_i \text{ is correct} \mid n \text{ test cases having successive distances } x_j, j = 1, \cdots, n-1\} = e^{-\lambda V} \prod_{j=1}^{n-1} \left[\frac{2}{1 + e^{-\lambda x_j}}\right].$$

Here λ is a parameter of the equivalence class. In particular, the mean length of E_i is $1/\lambda$. λ can range from 0 to ∞. A value of 0 means that E_i is an equivalence class with finite degree. (The degree of an equivalence class is the number of distinct test cases which completely validate it.) For example, $f(x) = constant$ can be verified by one test case.

The above analysis can be easily generalized to m dimensions by assuming independent behavior along each coordinate. However, for equivalence classes of finite degrees we can adopt a more satisfactory approach. If D is the degree of the equivalence class then it can be shown that

$$\lambda V \approx \frac{D-1}{N}$$

where N is the number of elements in the class (due to the finite word length of the computer used). This means that more testing should be performed when using computers of small word lengths, which is intuitive.

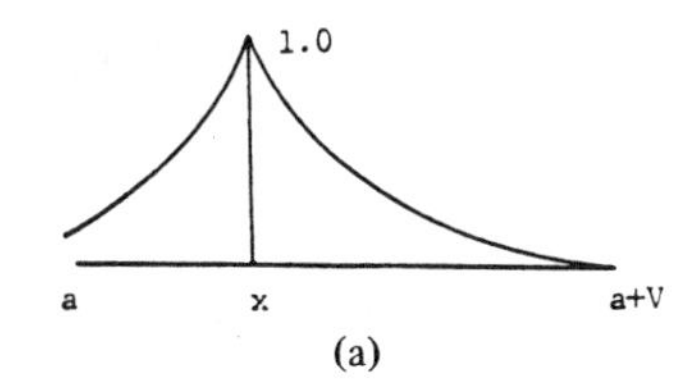

(a)

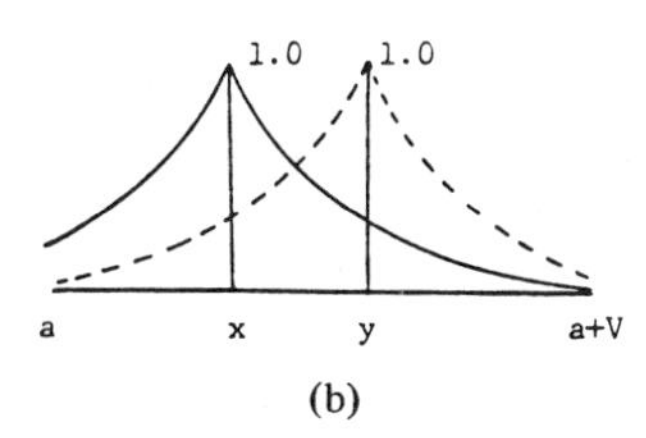

(b)

Fig. 9. Interpretation for continuous equivalence classes. (a) Single test case. (b) Two test cases.

D can be related to the testing complexity of the program. For example, for a program computing a polynomial

$$a_0 + a_1 x + a_2 x^2 + \cdots + a_n x^n.$$

D is $(n+1)$ since $(n+1)$ distinct test cases completely validate it. More generally, for algebraic programs we can estimate D by viewing each expression as a group of multinomials and function calls. Thus

$$x^3 z + y \cos\left(x^2 y + \frac{xw}{z}\right)$$

can be written as

$$a_1 = xw$$
$$a_2 = z$$
$$a_3 = \frac{a_1}{a_2}$$
$$a_4 = x^2 y + a_3$$
$$a_5 = \cos(a_4)$$
$$a_6 = x^3 z + y a_5.$$

Each subexpression is either a multinomial or a function call. (Division and exponential are considered as functions with two input parameters.) For a function call with n input parameters, the complexity is defined to be n + the complexity of the function. If the function has been validated, its complexity is 1. For a multinomial in k variables having highest degree n, the complexity is defined to be

$$\prod_{i=1}^{k} \left[\frac{n+i}{i}\right].$$

This is equal to the maximum number of terms in a multinomial in k variables having highest degree n. When k is large, this complexity measure will be very large. In these cases the complexity measure can be taken to be equal to the number of terms in the expression. This is acceptable since code reading would detect any extra or missing terms. Thus, in the above example, the complexity is $6 + 2 + 3 + 20 + 2 + 70 = 103$. For such general expression, we *estimate* D by equating it to the testing complexity of the equivalence class. Thus

$$\lambda V \approx \frac{\hat{D} - 1}{N}.$$

Assumption: The distribution of the equivalence classes has to be assumed; this includes prior estimation of the parameters of the distribution.

Discussion: The advantages of this model are as follows.

1) *Any* test case selection strategy can be used; this will minimize the testing effort since we can choose test cases which exercise error-prone constructs.

2) It does not assume random sampling.

3) It takes into account the complexity of the program: a simple program is tested less than a complicated program for the same correctness probability. The model also yields the optimal testing strategy to be used. Specifically, for algebraic programs the test cases should be spread out over the input domain for higher correctness probability.

The disadvantages of the model are as follows.

1) It is relatively expensive to determine the equivalence classes and their complexity.

2) The assumption has to be justified by experience.

C. Summary

The models discussed in this section are especially attractive for medium size programs whose reliability cannot be accurately estimated by using reliability growth models. These models also have the advantage of considering the structure of the program. This enables the joint use of program proving and testing in order to validate the program and assess its reliability [62]. Another model which considers the program structure is discussed in [100].

The correctness probability measure discussed above is an example of criteria which determine how well the program has been tested. A similar criterion based on fuzzy set theory is discussed in [7].

As evident from the discussions in Section III, no specific software reliability model has found wide acceptance. This is partly due to the cost involved in gathering failure data and partly because of the difficulty in modeling the testing process. In the following, we outline a method combining well-established proof procedures with software reliability estimation methods. It is particularly suitable for critical control systems.

1) During the testing and debugging phase two different software reliability growth models are used, primarily for determining the stopping time. This is a measure of the amount of testing needed in order to reach a desired reliability goal. The latter is stringent since software for critical systems may be rejected if even a single error is found after this phase.

2) During the validation phase the equivalence classes are determined based on program paths using the selection criterion discussed in Section V-B. Boundary value and range testing are performed in order to ensure that the classes are properly chosen.

3) If the path corresponding to each equivalence class can be verified (e.g., by using symbolic execution [83]), then the correctness probability of the class is 1.

4) If the correctness of the path cannot be verified, then the degree of the equivalence class is estimated. Next, as many test cases as necessary are used in order to achieve a desired correctness probability estimate. It can be shown that for a given number of test cases, the correctness probability estimate is maximized if the test cases are spread evenly over the equivalence class [82].

In the next section we discuss an experimental approach in determining the correctness probability of the program. This can be used as a verification of the theoretical estimate.

VI. Error Seeding

The method of "error seeding" is a powerful experimental approach to evaluating software development tools and techniques. It was originally used by Mills [36], [92] to estimate the number of errors remaining in the program. In Section VI-A we show that error seeding provides an independent estimate of the correctness probability of a program. Section VI-B discusses its application to evaluating the efficiency of different testing techniques and the reliability of different sets of test cases.

A. Correctness Probability

The program is seeded with m errors (one at a time), and for each error all the test cases are run until the error is detected or the set of test cases is exhausted. Then

correctness probability based on input set X

$$= P\{I \mid X\}$$

$$= P\{\text{no error in program} \mid \text{program works for } X\}$$

$$= P\{\text{any error in the program is detected by } X\}$$

$$= \frac{m_d}{m}$$

where m_d is the number of seeded errors detected by the set of test cases X. The size of the errors seeded in the program provides a natural confidence level in the resulting estimate of the correctness probability.

The error seeding approach is not always practical since the program may have to be tested $m|X|$ times and this can be very large. For example, if $m = 1000$, $|X| = 10^6$, then the program may have to be tested 10^9 times! Thus, a model for estimating the correctness probability—such as developed in Section V-B—is essential. Another factor in favor of the theoretical model is the auxiliary derivations regarding optimal testing, dependency of the correctness probability on the program complexity, machine word length, etc. [82]. These results are reasonable. In particular, the spread of test cases over the input domain has been suggested as a good testing practice by Howden and is used as a criteria of how well the program has been tested by Ehrenberger [14].

B. Evaluation of Test Cases

An important application of error seeding is to the evaluation of different testing strategies and sets of test cases. We first discuss the efficiency of a testing strategy and then we discuss the reliability of test cases.

1) Testing Efficiency: The efficiency of a test case selection strategy is defined as the probability that test cases selected randomly using the strategy will detect an error, i.e., it is the effective error size viewed by the testing strategy. The efficiency of a test case depends on various factors, especially the number of errors remaining in the program and the type of the errors. If we consider the *relative* efficiency of different testing strategies, then the primary factor is the type of errors in the program. For example, initially static analysis techniques can detect many errors, but their efficiency falls as syntax errors, missing initialization, and other similar errors are eliminated. Similarly, some errors are difficult to detect by testing or dynamic analysis, although they are easily detected by code reivew. An example is missing control flow type of errors which are extremely difficult to detect by testing the code, since the missing code may affect a very small part of the input domain, i.e., have a very small error size. However, code review and static analysis can expose these errors. Further studies on the nature and frequency of software errors can be found in [2], [27], [90], and [99].

The efficiency of a testing strategy can be estimated by error seeding. We seed errors in the program according to various types and sizes and then apply different testing strategies. The results indicate the efficiency of the strategies and the types of errors for which they are mainly effective.

Experiment: The following experiment was performed in the EPRI project [62]. Sixteen errors were seeded in the program. One was caught during code review. Eight were caught using data flow analysis tools. The errors which escaped detection by these techniques were of the type of missing assignment statement and bad expressions. These were detected by testing.

2) Reliability of Test Cases: A data selection criterion was originally defined by Goodenough and Gerhart [38] to be reliable if it ensured the selection of tests that are consistent in their ability to reveal errors. Here we define the reliability of a set of test cases, *irrespective of the test selection strategy*, to be some measure of the confidence in the correctness of the program if it works for the given test cases. That is, the reliability of test cases is a measure of our belief in the *a posteriori* correctness of the software. The correctness probability (Section IV-B) and correctness possibility [7] are measures of the reliability of test cases. Other measures are: 1) the coverage of program structure, its functions, and its input domain [14]; these are due to Ehrenberger; 2) representativeness of the set of test cases [15]; and 3) statistical confidence in the reliability estimate [109]. Also, software complexity measures can be used to guide the amount of testing [17], [29], [40], [64].

An interesting experimental approach is program mutation due to DeMillo and Lipton [22]. However, this technique is expensive since the number of mutations is combinatorially explosive for programs of realistic sizes. The majority of the mutations have large error sizes and these are easily detected. A practical solution is to seed the program with errors, such that the size of the errors is controlled. We now discuss one such experiment.

Experiment: The error seeding technique was used to assess the reliability of the set of test cases used in the first phase of the EPRI project [62]. Eight errors were seeded. Errors in expressions were detected by almost all the test cases, implying that arithmetic errors usually have a large size. However, three errors escaped detection. These were of the boundary value and missing control flow type of errors. This indicates that further test cases exercising the ranges of the variables and boundary conditions are necessary.

VII. Conclusion

We have addressed only a few topics in software reliability. The bulk of this paper is devoted to software reliability growth models. We have developed a general framework for these models using the concepts of error size and testing process. We distinguished between error-counting and nonerror-counting models. If only the reliability estimate is required, then the nonerror-counting models are preferable since it is easy to extend them to include situations where new errors are introduced into the software as a result of changes. Error-counting models should be used when an estimate of the number of remaining errors is needed. This may be required if resources have to be allocated for the maintenance phase (assuming that the average resource per error correction is known). It is also possible to estimate the number of errors remaining by using error seeding techniques. (Another approach is to develop a random walk model for the number of errors remaining in the program.)

Finally, we briefly discussed the Nelson model and its extension, and the input domain based model for the validation phase. We also mentioned the error seeding technique in evaluating testing techniques, sets of test cases, and the correctness probability of programs.

Some aspects which we have not discussed here are: 1) operational reliability [18], [51], [55], [58], 2) reliability estimate after perturbation during the maintenance phase (discussion on maintenance (growth dynamics) can be found in [8], [9], [10], [30], [52], [104], and [105]), 3) the combination of hardware and software reliability estimates to get the overall system reliability estimate [16], [49], [50], [107], the nature and frequency of software errors [2], [27], [90], [99], 4) methods of achieving reliable software: there are several references on this aspect, some of which are [3], [5], [11], [28], [32], [35], [39], [42], [43], [45], [65], [80], [81], [83], [87], and [88], 5) other metrics, besides software reliability, which reflect the quality of software systems are discussed in [13], [20], [110], and [111], 6) finally, there have been a few case studies on the efficacy of these techniques [1], [21], [46], [76], [113].

Acknowledgment

The authors wish to thank B. Littlewood, E. Nelson, G. Schick, and R. Wolverton for their careful review of an earlier draft of the paper. Their detailed comments have greatly improved the quality of the paper.

References

[1] J. Abe, K. Sakamura, and H. Aiso, "An analysis of software project failure," in *Proc. 4th Int. Conf. Software Eng.*, Munich, Germany, Sept. 1979, pp. 378–385.

[2] F. Akiyama, "An example of software system debugging," in

Proc. IFIP 1971. Amsterdam, The Netherlands: North-Holland, 1972, pp. 353-359.

[3] P. G. Anderson, "Redundancy techniques for software quality," in *Proc. Annu. Rel. and Maintainability Symp.*, 1978, pp. 86-93.

[4] J. E. Angus, R. E. Schafer, and A. Sukert, "Software reliability model validation," in *Proc. Annu. Rel. and Maintainability Symp.*, San Francisco, CA, Jan. 1980, pp. 191-199.

[5] R. G. Babb, II and L. L. Trip, "An engineering framework for software standards," in *Proc. Annu. Rel. and Maintainability Symp.*, San Francisco, CA, Jan. 1980, pp. 214-219.

[6] R. E. Barlow and F. Proschan, *Statistical Theory of Reliability and Life Testing.* New York: Holt, Rinehart and Winston, 1975.

[7] F. B. Bastani, "An input domain based theory of software reliability and its application," Ph.D. dissertation, Dep. Elec. Eng. Comput. Sci., Univ. of California, Berkeley, 1980.

[8] L. A. Belady and M. M. Lehmann, "An introduction to growth dynamics," in *Statistical Computer Performance Evaluation,* W. Freiberger, Ed. New York: Academic, 1972, pp. 503-511.

[9] —, "A model of large program development," *IBM Syst. J.*, vol. 15, no. 3, pp. 225-252, 1976.

[10] —, "The characteristic of large systems," IBM Tech. Rep. RC6785 (#28969), Sept. 13, 1977.

[11] W. R. Bezanson, "Reliable software through requirements definition using data abstraction," in *Microelectronics and Reliability*, vol. 17. New York: Pergamon, 1978, pp. 85-92.

[12] P. J. Bickel and K. A. Doksum, *Mathematical Statistics: Basic Ideas and Selected Topics.* San Francisco: Holden-Day, 1977.

[13] B. W. Boehm, J. R. Brown, and M. Lipow, "Quantitative evaluation of software quality," in *Proc. 2nd Int. Conf. Software Eng.*, San Francisco, CA, Oct. 1976, pp. 592-605.

[14] S. Bologna and W. Ehrenberger, "Applicability of statistical models for reactor safety software verification," Rep., 1978.

[15] J. R. Brown and M. Lipow, "Testing for software reliability," in *Proc. 1975 Int. Conf. Rel. Software*, Los Angeles, CA, Apr. 1975, pp. 518-527.

[16] W. L. Bunce, "Hardware and software: An analytical approach," in *Proc. Annu. Rel. and Maintainability Symp.*, San Francisco, CA, Jan. 1980, pp. 209-213.

[17] N. Chapin, "A measure of software complexity," in *Proc. 1979 Nat. Comput. Conf.*, AFIPS, New York, June 1979, pp. 995-1002.

[18] R. C. Cheung, "A user-oriented software reliability model," in *Proc. COMPSAC 1978*, Chicago, IL., Nov. 1978, pp. 565-570.

[19] D. R. Cox and P.A.W. Lewis, *The Statistical Analysis of Series of Events.* London: Methuen, 1966.

[20] B. Curtis, S. B. Sheppard, and P. Milliman, "Third time charm: Stronger prediction of programmer performance by software complexity metrics," in *Proc. 4th Int. Conf. Software Eng.*, Munich, Germany, Sept. 1979, pp. 356-360.

[21] G. Dahll and J. Lahti, "Investigation of methods for production and verification of computer programmes with high requirements for reliability," OECD Halden Reactor Project, Preliminary Rep., 1978.

[22] R. A. DeMillo, R. J. Lipton, and F. G. Sayward, "Hints on test data selection: Help for the practicing programmer," *Computer*, pp. 34-41, Apr. 1978.

[23] R. A. Demillo, R. J. Lipton, and A. J. Perlis, "Social processes and proofs of theorems and programs," *Commun. Ass. Comput. Mach.*, vol. 22, pp. 271-280, May 1979.

[24] J. C. Dickson *et al.*, "Quantitative analysis of software reliability," in *Proc. Annu. Rel. and Maintainability Symp.*, San Francisco, CA, 1972, pp. 148-157.

[25] J. T. Duane, "Learning curve approach to reliability monitoring," *IEEE Trans. Aerosp.*, vol. 2, pp. 563-566, 1964.

[26] L. Duvall *et al.*, "Data needs for software reliability modelling," in *Proc. Annu. Rel. and Maintainability Symp.*, San Francisco, CA, Jan. 1980, pp. 200-208.

[27] A. Endres, "An analysis of errors and their causes in system programs," in *Proc. Int. Conf. on Rel. Software*, Los Angeles, CA, 1975, pp. 327-336.

[28] M. E. Fagan, "Design and code inspections to reduce errors in program development," *IBM Syst. J.*, vol. 15, no. 3, pp. 182-211, 1976.

[29] A. R. Feuer and E. B. Fowlkes, "Relating computer program maintainability to software measures," in *Proc. 1979 Nat. Comput. Conf.*, AFIPS, New York, June 1979, pp. 1003-1012

[30] —, "Some results from an empirical study of computer software," in *Proc. 4th Int. Conf. Software Eng.*, Munich, Germany, Sept. 1979, pp. 351-355.

[31] J. M. Finkelstein, "Starting and limiting values for reliability growth," *IEEE Trans. Rel.*, vol. R-28, pp. 111-114, June 1979.

[32] K. F. Fischer and M. G. Walker, "Improved software reliability through requirements verification," *IEEE Trans. Rel.*, vol. R-28, pp. 233-240, Aug. 1979.

[33] E. H. Forman and N. D. Singpurwalla, "An empirical stopping rule for debugging and testing computer software," *J. Amer. Stat. Ass.*, vol. 72, pp. 750-757, Dec. 1977.

[34] —, "Optimal time intervals for testing hypotheses on computer software errors," *IEEE Trans. Rel.*, vol. R-28, pp. 250-253, Aug. 1979.

[35] J. D. Gannon and J. J. Horning, "The impact of language design on the production of reliable software," in *Proc. Int. Conf. Rel. Software*, Los Angeles, CA, 1975, pp. 10-22.

[36] E. Girard and J.-C. Rault, "A programming technique for software reliability," in *Rec. 1973 IEEE Symp. Comput. Software Rel.*, New York, May 1973, pp. 44-50.

[37] A. L. Goel and K. Okumoto, "Time-dependent error-detection model for software reliability and other performance measures," *IEEE Trans. Rel.*, vol. R-28, pp. 206-211, Aug. 1979.

[38] J. B. Goodenough and S. L. Gerhart, "Toward a theory of test data selection," *IEEE Trans. Software Eng.*, vol. SE-1, pp. 156-173, June 1975.

[39] S. J. Greenspan and C. L. McGowan, "Structuring software development for reliability," in *Microelectronics and Reliability*, vol. 17. New York: Pergamon, 1978, pp. 75-84.

[40] M. H. Halstead, *Elements of Software Science.* New York: Elsevier, 1977.

[41] P. A. Hamilton and J. D. Musa, "Measuring reliability of computation center software," in *Proc. 3rd Int. Conf. on Software Eng.*, Atlanta, GA, 1978, pp. 29-36.

[42] H. Hecht, "Fault-tolerant software for real-time applications," *Comput. Surveys*, vol. 8, pp. 391-407, Dec. 1976.

[43] —, "Fault-tolerant software," *IEEE Trans. Rel.*, vol. R-28, pp. 227-232, Aug. 1979.

[44] P. G. Hoel, S. C. Port, and C. J. Stone, *Introduction to Probability Theory.* Boston: Houghton Mifflin, 1971.

[45] P. P. Howley, Jr., "Software quality assurance for reliable software," in *Proc. Annu. Rel. and Maintainability Symp.*, 1978, pp. 73-78.

[46] D. R. Jeffrey and M. J. Lawrence, "An inter-organizational comparison of programmer productivity," in *Proc. 4th Int. Conf. Software Eng.*, Munich, Germany, Sept. 1979, pp. 369-377.

[47] Z. Jelinski and P. Moranda, "Software reliability research," in *Statistical Computer Performance Evaluation*, W. Freiberger, Ed. New York: Academic, 1972, pp. 465-484.

[48] —, "Applications of a probability-based model to code reading," in *Rec. 1973 IEEE Symp. Comput. Software Rel.*, New York, May 1973, pp. 78-81.

[49] R. H. Keegan and R. C. Howard, "Approximation method for estimating meaningful parameters for a software-controlled electromechanical system," in *Proc. Annu. Rel. and Maintainability Symp.*, 1976, pp. 434-439.

[50] M. B. Kline, "Software and hardware R & M: What are the differences?," in *Proc. Annu. Rel. and Maintainability Symp.*, San Francisco, CA, Jan. 1980, pp. 179-185.

[51] H. Kopetz, "Software redundancy in real-time systems," in *Proc. IFIP Congress 1974.* Amsterdam, The Netherlands: North-Holland, 1974, pp. 182-186.

[52] M. M. Lehman and F. N. Parr, "Program evolution and its impact on software engineering," in *Proc. 2nd Int. Conf. Software Eng.*, San Francisco, CA, Oct, 1976, pp. 350-357.

[53] B. Littlewood and J. L. Verrall, "A Bayesian reliability growth model for computer software," *J. Roy. Stat. Soc.*, vol. 22, no. 3, pp. 332-346, 1973; also in *Rec. 1973 IEEE Symp. Comput. Software Rel.*, New York, May 1973, pp. 70-77.

[54] B. Littlewood, "MTBF is meaningless in software reliability," *IEEE Trans. Rel.*, vol. R-24, p. 82, Apr. 1975.

[55] —, "A reliability model for Markov structured software," in *Proc. 1975 Int. Conf. Rel. Software*, Los Angeles, CA, pp. 204-207.

[56] —, "A Bayesian differential debugging model for software reliability," Dep. Math., City Univ., London, England, June

1979; also in *Proc. COMPSAC 1980*, Chicago, IL, 1980, pp. 511–519.
[57] —, "How to measure software reliability and how not to . . . ," *IEEE Trans. Rel.*, vol. R-28, pp. 103–110, June 1979; also in *Proc. 3rd Int. Conf. on Software Eng.*, Atlanta, GA, May 1978, pp. 37–45.
[58] —, "Software reliability model for modular program structure," *IEEE Trans. Rel.*, vol. R-28, pp. 241–246, Aug. 1979.
[59] —, private communication.
[60] B. Littlewood and J. L. Verrall, "On the likelihood function of a debugging model for computer software reliability," Dep. Math., City Univ., London, England, 1980.
[61] B. Littlewood, "Theories of software reliability: How good are they and how can they be improved?," *IEEE Trans. Software Eng.*, vol. SE-6, pp. 489–500, Sept. 1980.
[62] A. B. Long *et al.*, "A methodology for the development and validation of critical software for nuclear power plants," in *Proc. 1st Int. Conf. on Software Appl.*, Nov. 1977.
[63] W. H. MacWilliams, "Reliability of large real-time control software systems," in *Rec. 1973 IEEE Symp. Comput. Software Rel.*, New York, May 1973, pp. 1–6.
[64] T. J. McCabe, "A complexity measure," *IEEE Trans. Software Eng.*, vol. SE-2, pp. 308–320, Dec. 1976.
[65] H. D. Mills, "On the development of large reliable software," in *Rec. IEEE Symp. Comput. Software Rel.*, New York, May 1973, pp. 155–159.
[66] I. Miyamoto, "Software reliability in on-line real-time environment," in *Proc. Int. Conf. Rel. Software*, Los Angeles, CA, 1975, pp. 194–203.
[67] —, "Toward an effective software reliability evaluation," in *Proc. 3rd Int. Conf. on Software Eng.*, Atlanta, GA, 1978, pp. 46–55.
[68] S. N. Mohanty, "Models and measurements for quality assessment of software," *Comput. Surveys*, vol. 11, pp. 251–275, Sept. 1979.
[69] P. B. Moranda, "Prediction of software reliability during debugging," in *Proc. 1975 Annu. Rel. and Maintainability Symp.*, Washington, DC, Jan. 1975, pp. 327–332.
[70] —, "The (sad) status of; (unapproached) limits to; and (manifold) alternatives for software measurement techniques," in *Proc. COMPCON 1978*, San Francisco, CA, Mar. 1978, pp. 353–354.
[71] —, "Software reliability revisited," *Computer*, vol. 11, pp. 92–94, April. 1978.
[72] —, "Limits to program testing with random number inputs," in *Proc. COMPSAC 1978*, Chicago, IL, Nov. 1978, pp. 521–526.
[73] —, "Event-altered rate models for general reliability analysis," *IEEE Trans. Rel.*, vol. R-28, pp. 376–381, Dec. 1979.
[74] F. Mosteller and J. W. Tukey, *Data Analysis and Regression: A Second Course in Statistics.* Reading, MA: Addison-Wesley, 1977.
[75] J. D. Musa, "A theory of software reliability and its applications," *IEEE Trans. Software Eng.*, vol. SE-1, pp. 312–327, Sept. 1975.
[76] —, "An exploratory experiment with *foreign* debugging of programs," in *Proc. Symp. Comput. Software Eng.*, New York, Apr. 1976, pp. 499–511.
[77] —, "Software reliability measures applied to system engineering," in *Proc. 1979 Nat. Comput. Conf.*, AFIPS, New York, June 1979, pp. 941–946.
[78] —, "Validity of execution-time theory of software reliability," *IEEE Trans. Rel.*, vol. R-28, pp. 181–191, Aug. 1979.
[79] E. Nelson, "Estimating software reliability from test data," in *Microelectronics and Reliability*, vol. 17. New York: Pergamon, 1978, pp. 67–74.
[80] D. L. Parnas, "Information distribution aspects of design methodology," in *Proc. IFIP 1971.* Amsterdam, The Netherlands: North-Holland, 1972, pp. 339–344.
[81] —, "The influence of software structure on reliability," in *Proc. Int. Conf. Rel. Software*, Los Angeles, CA, 1975, pp. 358–362.
[82] C. V. Ramamoorthy and F. B. Bastani, "An input domain based approach to the quantitative estimation of software reliability," in *Proc. Taipei Seminar on Software Eng.*, Taipei, Taiwan, 1979.
[83] C. V. Ramamoorthy *et al.*, "A systematic approach to the development and validation of critical software for nuclear power plants," in *Proc. Taipei Seminar on Software Eng.*, Taipei, Taiwan, 1979.
[84] C. V. Ramamoorthy and F. B. Bastani, "Practical considerations for the development of process control software," presented at *INTERKAMA 1980*, Düsseldorf, West Germany, Oct. 1980.
[85] —, "Modelling of the software reliability growth process," in *Proc. COMPSAC 1980*, Chicago, IL, 1980, pp. 161–169.
[86] J.-C. Rault, "An approach towards reliable software," in *Proc. 4th Int. Conf. on Software Eng.*, Munich, Germany, Sept. 1979, pp. 220–230.
[87] D. J. Reifer, "A new assurance technology for computer software," in *Proc. Annu. Rel. and Maintainability Symp.*, 1976, pp. 446–451.
[88] —, "Software failure modes and effect analysis," *IEEE Trans. Rel.*, vol. R-28, pp. 247–249, Aug. 1979.
[89] H. Remus and S. Zilles, "Prediction and management of program quality," in *Proc. 4th Int. Conf. Software Eng.*, Munich, Germany, Sept. 1979, pp. 341–350.
[90] R. J. Rubey, "Quantitative aspects of software validation," in *Proc. Int. Conf. on Rel. Software*, Los Angeles, CA, 1975, pp. 246–251.
[91] G. J. Schick and R. W. Wolverton, "Achieving reliability in large scale software systems," in *Proc. 1974 Rel. and Maintainability Symp.*, Los Angeles, CA, Jan. 1974, pp. 302–319.
[92] —, "An analysis of competing software reliability models," *IEEE Trans. Software Eng.*, vol. SE-4, pp. 104–120, Mar. 1978.
[93] N. F. Schneidewind, "An approach to software reliability prediction and quality control," in *Proc. 1972 Fall Joint Comput. Conf.*, AFIPS, vol. 41, pp. 837–847.
[94] —, "Analysis of error processes in computer software," in *Proc. Int. Conf. on Rel. Software*, Los Angeles, CA, 1975, pp. 337–346.
[95] M. L. Shooman, "Probability models for software reliability prediction," in *Statistical Computer Performance Evaluation*, W. Freiberger, Ed. New York: Academic, 1972, pp. 485–502.
[96] —, "Operational testing and software reliability estimation during program development," in *Rec. 1973 IEEE Symp. Comput. Software Rel.*, New York, May 1973, pp. 51–57.
[97] M. L. Shooman and S. J. Amster, "Software reliability: An overview," in *Reliability and Fault Tree Analysis*, Barlow, Fussel, and Singpurwalla, Eds. Philadelphia, SIAM, 1975, pp. 655–685.
[98] M. L. Shooman, "Software reliability: Measurement and models," in *Proc. 1975 Annu. Rel. and Maintainability Symp.*, Washington, DC, Jan. 1975, pp. 485–491.
[99] M. L. Shooman and M. I. Bolsky, "Types, distribution, and test and correction times for programming errors," in *Proc. Int. Conf. Rel. Software*, Los Angeles, CA, 1975, pp. 347–357.
[100] M. L. Shooman, "Structured models for software reliability prediction," in *Proc. 2nd Int. Conf. Software Eng.*, San Francisco, CA, 1976, pp. 268–280.
[101] A. N. Sukert, "A four-project empirical study of software error prediction models," in *Proc. COMPSAC 1978*, Chicago, IL, Nov. 1978, pp. 577–582.
[102] —, "Empirical validation of three error prediction models," *IEEE Trans. Rel.*, vol. R-28, pp. 199–205, Aug. 1979.
[103] A. N. Sukert and A. L. Goel, "A guidebook for software reliability assessment," in *Proc. Annu. Rel. and Maintainability Symp.*, San Francisco, CA, Jan. 1980, pp. 186–190.
[104] E. B. Swanson, "The dimensions of maintenance," in *Proc. 2nd Int. Conf. Software Eng.*, San Francisco, CA, Oct. 1976, pp. 492–497.
[105] —, "On the user-requisite variety of computer application software," *IEEE Trans. Rel.*, vol. R-28, pp. 221–226, Aug. 1979.
[106] J. Tal, "Development and evaluation of software reliability estimators," Dep. Elec. Eng., Univ. of Utah, Salt Lake City, UTEC SR 77-013, 1976.
[107] W. E. Thomson and P. O. Chelson, "On the specification and testing of software reliabiity," in *Proc. Annu. Rel. and Maintainability Symp.*, San Francisco, CA, Jan. 1980, pp. 379–383.
[108] A. K. Trivedi and M. L. Shooman, "A many-state Markov model for the estimation and prediction of computer software performance parameters," in *Proc. 1975 Int. Conf. Rel. Software*, Los Angeles, CA, pp. 208–220.
[109] TRW Defense and Space Systems Group, "Software reliability study," Redondo Beach, CA, Rep. 76-2260.1-9-5, 1976.

[110] G. F. Walters and J. A. McCall, "The development of metrics for software R & M," in *Proc. Annu. Rel. and Maintainability Symp.*, 1978, pp. 78–85.

[111] —, "Software quality metrics for life-cycle cost-reduction," *IEEE Trans. Rel.*, vol. R-28, pp. 212–220, Aug. 1979.

[112] E. J. Weyuker and T. J. Ostrand, "Theories of program testing and the application of revealing subdomains," in *Dig. Workshop on Software Testing and Test Documentation*, FL, Dec. 1978, pp. 1–18.

[113] B. B. White, "Program standards help software maintainability," in *Proc. Annu. Rel. and Maintainability Symp.*, 1978, pp. 94–98.

[114] M. V. Zelkowitz, "Perspectives on software engineering," *Comput. Surveys*, vol. 10, pp. 197–216, June 1978.

C. V. Ramamoorthy (M'57–SM'76–F'78) received the undergraduate degrees in physics and technology from the University of Madras, Madras, India, the M.S. degree and the professional degree of Mechanical Engineer, both from the University of California, Berkeley, and the M.A. and Ph.D. degrees in applied mathematics and computer theory from Harvard University, Cambridge, MA.

He was associated with Honeywell's Electronic Data Processing Division from 1956 to 1971, last as Senior Staff Scientist. He was a Professor in the Department of Electrical Engineering and Computer Sciences at the University of Texas, Austin. Currently, he is a Professor in the Department of Electrical Engineering and Computer Sciences, University of California, Berkeley.

Dr. Ramamoorthy was Chairman of the Education Committee of the IEEE Computer Society and Chairman of the Committee to develop E.C.P.D. Accreditation Guidelines for Computer Science and Engineering Degree Programs. He also was the Chairman of the AFIPS Education Committee, a member of the Science and Technology Advisory Group of the U.S. Air Force, and a member of the Technology Advisory Panel of Ballistic Missile Defense (U.S. Army). Currently, he is Vice President of the IEEE Computer Society for Educational Activities.

Farokh B. Bastani (M'82) received the B.Tech. degree in electrical engineering from the Indian Institute of Technology, Bombay, India, in 1977, and the M.S. and Ph.D. degrees in electrical engineering and computer science from the University of California, Berkeley, in 1978 and 1980, respectively.

He joined the University of Houston, Houston, TX, in 1980, where he is currently Assistant Professor of Computer Science. His research interests include developing a design methodology and quality assessment techniques for large-scale computer systems.

Reprinted from *IEEE Transactions on Software Engineering*, Volume SE-11, Number 12, December 1985, pages 1411–1423.

Software Reliability Models: Assumptions, Limitations, and Applicability

AMRIT L. GOEL, MEMBER, IEEE

***Abstract*—A number of analytical models have been proposed during the past 15 years for assessing the reliability of a software system. In this paper we present an overview of the key modeling approaches, provide a critical analysis of the underlying assumptions, and assess the limitations and applicability of these models during the software development cycle. We also propose a step-by-step procedure for fitting a model and illustrate it via an analysis of failure data from a medium-sized real-time command and control software system.**

***Index Terms*—Estimation, failure count models, fault seeding, input domain models, model fitting, NHPP, software reliability, times between failures.**

Introduction and Background

An important quality attribute of a computer system is the degree to which it can be relied upon to perform its intended function. Evaluation, prediction, and improvement of this attribute have been of concern to designers and users of computers from the early days of their evolution. Until the late 1960's, attention was almost solely on the hardware related performance of the system. In the early 1970's, software also became a matter of concern, primarily due to a continuing increase in the cost of software relative to hardware, in both the development and the operational phases of the system.

Software is essentially an instrument for transforming a discrete set of inputs into a discrete set of outputs. It comprises of a set of coded statements whose function may be to evaluate an expression and store the result in a temporary or permanent location, decide which statement to execute next, or to perform input/output operations.

Since, to a large extent, software is produced by humans, the finished product is often imperfect. It is imperfect in the sense that a discrepancy exists between what the software can do versus what the user or the computing environment wants it to do. The computing environment refers to the physical machine, operating system, compiler and translator, utilities, etc. These discrepancies are what we call software faults. Basically, software faults can be attributed to an ignorance of the user requirements, ignorance of the rules of the computing environment, and to poor communication of software requirements between the user and the programmer or poor documentation of the software by the programmer. Even if we know that software contains faults, we generally do not know their exact identity.

Currently, there are two approaches available for indicating the existence of software faults, viz. program proving, and program testing. Program proving is formal and mathematical while program testing is more practical and heuristic. The approach taken in program proving is to construct a finite sequence of logical statements ending in the statement, usually the output specification statement, to be proved. Each of the logical statements is an axiom or is a statement derived from earlier statements by the application of an inference rule. Program proving by using inference rules is known as the inductive assertion method. This method was mainly advocated by Floyd, Hoare, Dijkstra, and recently Reynolds [39]. Other work on program proving is on the symbolic execution method. This method is the basis of some automatic program verifiers. Despite the formalism and mathematical exactness of program proving, it is still an imperfect tool for verifying program correctness. Gerhart and Yelowitz [10] showed several programs which were proved to be correct but still contained faults. However, the faults were due to failures in defining what exactly to prove and were not failures of the machanics of the proof itself.

Program testing is the symbolic or physical execution of a set of test cases with the intent of exposing embedded faults in the program. Like program proving, program testing remains an imperfect tool for assuring program correctness. A given testing strategy may be good for exposing certain kinds of faults but not for all possible kinds of faults in a program. An advantage of testing is that it can provide useful information about a program's actual behavior in its intended computing environment, while proving is limited to conclusions about the program's behavior in a postulated environment.

In practice neither proving nor testing can guarantee complete confidence in the correctness of a program. Each has its advantages and limitations and should not be viewed as competing tools. They are, in fact, complementary methods for decreasing the likelihood of program failure.

Due to the imperfectness of these approaches in assuring a correct program, a metric is needed which reflects the degree of program correctness and which can be used

Manuscript received February 4, 1985; revised July 31, 1985 and September 30, 1985. This work was supported in part by Rome Air Development Center, GAFB, and by the Computer Applications and Software Engineering (CASE) Center at Syracuse University.

The author is with the Department of Electrical and Computer Engineering and the School of Computer and Information Science, Syracuse University, Syracuse, NY 13244.

in planning and controlling additional resources needed for enhancing software quality. One such quantifiable metric of quality that is commonly used in software engineering practice is software reliability. This measure has attracted considerable attention during the last 15 years and continues to be employed as a useful metric. A commonly used approach for measuring software reliability is via an analytical model whose parameters are generally estimated from available data on software failures. Reliability and other relevant measures are then computed from the fitted model.

Even though such models have been in use for some time, the realism of many of the underlying assumptions and the applicability of these models for assessing software reliability continue to be questioned. It is the purpose of this paper to evaluate the current state-of-the-art related to this issue. Specifically, the key modeling approaches are briefly discussed and a critical analysis of their underlying assumptions, limitations, and applicability during the software development cycle is presented.

It should be pointed out that the emphasis of this paper is on software reliability modeling approaches and several related but important issues are only briefly mentioned. Examples of such issues are the practical and theoretical difficulties of parametric estimation, statistical properties of estimators, unification of models via generalized formulations or via, say, a Bayesian interpretation, validation and comparison of models, and determination of optimum release time. For a discussion of these issues, the reader is referred to Goel [19].

The term software reliability is discussed in Section II along with a classification of the various modeling approaches. The key models are briefly described in Sections III, IV, and V. An assessment of the main assumptions underlying the models is presented in Section VI and the applicability of these models during the software development cycle is discussed in Section VII. A step-by-step procedure for fitting a model is given in Section VIII and is illustrated via an analysis of software failure data from a medium-sized command and control system. A summary of some related work and concluding remarks are presented in Section IX.

II. Meaning and Measurement of Software Reliability

There are a number of views as to what software reliability is and how it should be quantified. Some people believe that this measure should be binary in nature so that an imperfect program would have zero reliability while a perfect one would have a reliability value of one. This view parallels that of program proving whereby the program is either correct or incorrect. Others, however, feel that software reliability should be defined as the relative frequency of the times that the program works as intended by the user. This view is similar to that taken in testing where a percentage of the successful cases is used as a measure of program quality.

According to the latter viewpoint, software reliability is a probabilistic measure and can be defined as the probability that software faults do not cause a failure during a specified exposure period in a specified use environment. The probabilistic nature of this measure is due to the uncertainty in the usage of the various software functions and the specified exposure period here may mean a single run, a number of runs, or time expressed in calendar or execution time units. To illustrate this view of software reliability, suppose that a user executes a software product several times according to its usage profile and finds that the results are acceptable 95 percent of the time. Then the software is said to be 95 percent reliable for that user.

A more precise definition of software reliability which captures the points mentioned above is as follows [30]. Let F be a class of faults, defined arbitrarily, and T be a measure of relevant time, the units of which are dictated by the application at hand. Then the reliability of the software package with respect to the class of faults F and with respect to the metric T, is the probability that no fault of the class occurs during the execution of the program for a prespecified period of relevant time.

Assuming that software reliability can somehow be measured, a logical question is what purpose does it serve. Software reliability is a useful measure in planning and controlling resources during the development process so that high quality software can be developed. It is also a useful measure for giving the user confidence about software correctness. Planning and controlling the testing resources via the software reliability measure can be done by balancing the additional cost of testing and the corresponding improvement in software reliability. As more and more faults are exposed by the testing and verification process, the additional cost of exposing the remaining faults generally rises very quickly. Thus, there is a point beyond which continuation of testing to further improve the quality of software can be justified only if such improvement is cost effective. An objective measure like software reliability can be used to study such a tradeoff.

Current approaches for measuring software reliability basically parallel those used for hardware reliability assessment with appropriate modifications to account for the inherent differences between software and hardware. For example, hardware exhibits mixtures of decreasing and increasing failure rates. The decreasing failure rate is seen due to the fact that, as test or use time on the hardware system accumulates, failures, most likely due to design errors, are encountered and their causes are fixed. The increasing failure rate is primarily due to hardware component wearout or aging. There is no such thing as wearout in software. It is true that software may become obsolete because of changes in the user and computing environment, but once we modify the software to reflect these changes, we no longer talk of the same software but of an enhanced or a modified version. Like hardware, software exhibits a decreasing failure rate (improvement in quality) as the usage time on the system accumulates and faults, say, due to design and coding, are fixed. It should also be noted that an assessed value of the software reliability measure is always relative to a given use environment. Two users exercising two different sets of paths in the same software are likely to have different values of software reliability.

A number of analytical models have been proposed to address the problem of software reliability measurement. These approaches are based mainly on the failure history of software and can be classified according to the nature of the failure process studied as indicated below.

Times Between Failures Models: In this class of models the process under study is the time between failures. The most common approach is to assume that the time between, say, the $(i - 1)$st and the ith failures, follows a distribution whose parameters depend on the number of faults remaining in the program during this interval. Estimates of the prarameters are obtained from the observed values of times between failures and estimates of software reliability, mean time to next failure, etc., are then obtained from the fitted model. Another approach is to treat the failure times as realizations of a stochastic process and use an appropriate time-series model to describe the underlying failure process.

Failure Count Models: The interest of this class of models is in the number of faults or failures in specified time intervals rather than in times between failures. The failure counts are assumed to follow a known stochastic process with a time dependent discrete or continuous failure rate. Parameters of the failure rate can be estimated from the observed values of failure counts or from failure times. Estimates of software reliability, mean tme to next failure, etc., can again be obtained from the relevant equations.

Fault Seeding Models: The basic approach in this class of models is to "seed" a known number of faults in a program which is assumed to have an unknown number of indigenous faults. The program is tested and the observed number of seeded and indigenous faults are counted. From these, an estimate of the fault content of the program prior to seeding is obtained and used to assess software reliability and other relevant measures.

Input Domain Based Models: The basic approach taken here is to generate a set of test cases from an input distribution which is assumed to be representative of the operational usage of the program. Because of the difficulty in obtaining this distribution, the input domain is partitioned into a set of equivalence classes, each of which is usually associated with a program path. An estimate of program reliability is obtained from the failures observed during physical or symbolic execution of the test cases sampled from the input domain.

III. Times Between Failures Models

This is one of the earliest classes of models proposed for software reliability assessment. When interest is in modeling times between failures, it is expected that the successive failure times will get longer as faults are removed from the software system. For a given set of observed values, this may not be exactly so due to the fact that failure times are random variables and observed values are subject to statistical fluctuations.

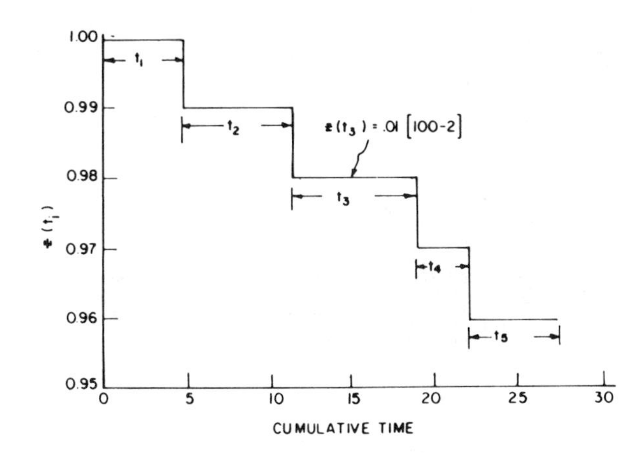

Fig. 1. A typical plot of $Z(t_i)$ for the JM model ($N = 100$, $\phi = 0.01$).

A number of models have been proposed to describe such failures. Let a random variable T_i denote the time between the $(i - 1)$st and the ith failures. Basically, the models assume that T_i follows a known distribution whose parameters depend on the number of faults remaining in the system after the $(i - 1)$st failure. The assumed distribution is supposed to reflect the improvement in software quality as faults are detected and removed from the system. The key models in this class are described below.

Jelinski and Moranda (JM) De-Eutrophication Model

This is one of the earliest and probably the most commonly used model for assessing software reliability [20]. It assumes that there are N software faults at the start of testing, each is independent of others and is equally likely to cause a failure during testing. A detected fault is removed with certainty in a negligible time and no new faults are introduced during the debugging process. The software failure rate, or the hazard function, at any time is assumed to be proportional to the current fault content of the program. In other words, the hazard function during t_i, the time between the $(i - 1)$st and ith failures, is given by

$$Z(t_i) = \phi[N - (i - 1)],$$

where ϕ is a proportionality constant. Note that this hazard function is constant between failures but decreases in steps of size ϕ following the removal of each fault. A typical plot of the hazard function for $N = 100$ and $\phi = 0.01$ is shown in Fig. 1.

A variation of the above model was proposed by Moranda [29] to describe testing situations where faults are not removed until the occurrence of a fatal one at which time the accumulated group of faults is removed. In such a situation, the hazard function after a restart can be assumed to be a fraction of the rate which attained when the system crashed. For this model, called the geometric de-eutrophication model, the hazard function during the ith testing interval is given by

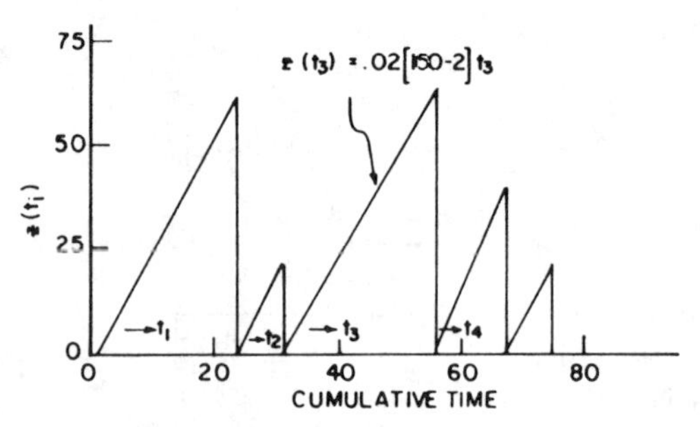

Fig. 2. A typical plot of the hazard function for the SW model (N = 150, ϕ = 0.02).

$$Z(t_i) = Dk^{i-1},$$

where D is the fault detection rate during the first interval and k is a constant ($0 < k < 1$).

Schick and Wolverton (SW) Model

This model is based on the same assumptions as the JM model except that the hazard function is assumed to be proportional to the current fault content of the program as well as to the time elapsed since the last failure [40] is given by

$$Z(t_i) = \phi\{(N - (i - 1))\} t_i$$

where the various quantities are as defined above. Note that in some papers t_i has been taken to be the cumulative time from the beginning of testing. That interpretation of t_i seems to be inconsistent with the interpretation in the original paper, see, e.g., Goel [15].

We note that the above hazard rate is linear with time within each failure interval, returns to zero at the occurrence of a failure and increases linearly again but at a reduced slope, the decrease in slope being proportional to ϕ. A typical behavior of $Z(t_i)$ for $N = 150$ and $\phi = 0.02$ is shown in Fig. 2.

A modification of the above model was proposed in [41] whereby the hazard function is assumed to be parabolic in test time and is given by

$$Z(t_i) = \phi[N - (i - 1)] (-at_i^2 + bt_i + c)$$

where a, b, c are constants and the other quantities are as defined above. This function consists of two components. The first is basically the hazard function of the JM model and the superimposition of the second term indicates that the likelihood of a failure occurring increases rapidly as the test time accumulates within a testing interval. At failure times ($t_i = 0$), the hazard function is proportional to that of the JM model.

Goel and Okumoto Imperfect Debugging Model

The above models assume that the faults are removed with certainty when detected. However, in practice [47] that is not always the case. To overcome this limitation, Goel and Okumoto [11], [13] proposed an imperfect debugging model which is basically an extension of the JM model. In this model, the number of faults in the system at time t, $X(t)$, is treated as a Markov process whose transition probabilities are governed by the probability of imperfect debugging. Times between the transitions of $X(t)$ are taken to be exponentially distributed with rates dependent on the current fault content of the system. The hazard function during the interval between the $(i - 1)$st and the ith failures is given by

$$Z(t_i) = [N - p(i - i)]\lambda.$$

where N is the initial fault content of the system, p is the probability of imperfect debugging, and λ is the failure rate per fault.

Littlewood–Verrall Bayesian Model

Littlewood and Verall [25], [26] took a different approach to the development of a model for times between failures. They argued that software reliability should not be specified in terms of the number of errors in the program. Specifically, in their model, the times between failures are assumed to follow an exponential distribution but the parameter of this distribution is treated as a random variable with a gamma distribution, viz.

$$f(t_i|\lambda_i) = \lambda_i \, e^{-\lambda_i t_i}$$

and

$$f(\lambda_i|\alpha, \psi(i)) = \frac{[\psi(i)]^\alpha \, \lambda_i^{\alpha-1} \, e^{-\psi(i)\lambda_i}}{\Gamma\alpha}$$

In the above, $\psi(i)$ describes the quality of the programmer and the difficulty of the programming task. It is claimed that the failure phenomena in different environments can be explained by this model by taking different forms for the parameter $\psi(i)$.

IV. Fault Count Models

This class of models is concerned with modeling the number of failures seen or faults detected in given testing intervals. As faults are removed from the system, it is expected that the observed number of failures per unit time will decrease. If this is so, then the cumulative number of failures versus time curve will eventually level off. Note that time here can be calander time, CPU time, number of test cases run or some other relevant metric. In this setup, the time intervals may be fixed *a priori* and the observed number of failures in each interval is treated as a random variable.

Several models have been suggested to describe such failure phenomena. The basic idea behind most of these models is that of a Poisson distribution whose parameter takes different forms for different models. It should be noted that Poisson distribution has been found to be an excellent model in many fields of application where interest is in the number of occurrences.

One of the earliest models in this category was proposed by Shooman [43]. Taking a somewhat similar approach, Musa [31] later proposed another failure count model based on execution time. Schneidewind [42] took a differ-

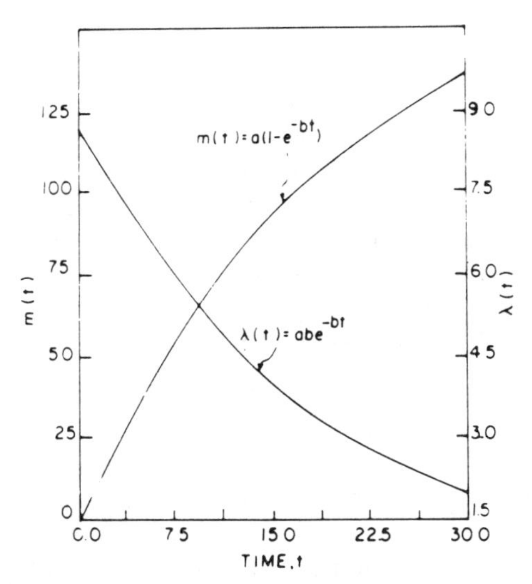

Fig. 3. A typical plot of the $m(t)$ and $\lambda(t)$ functions for the Goel–Okumoto NHPP model (a = 175, b = 0.05).

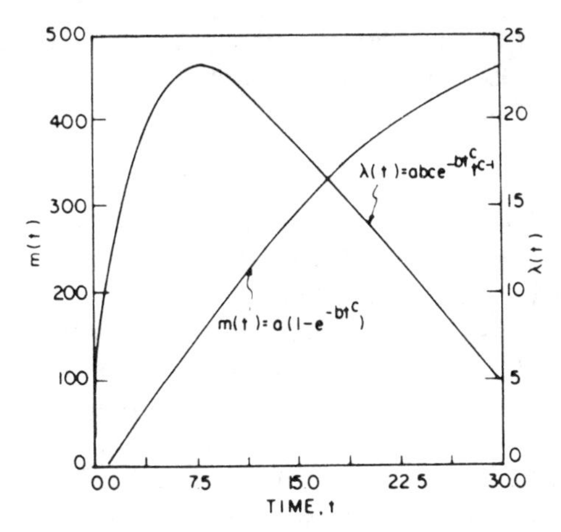

Fig. 4. A typical plot of the $m(t)$ and $\lambda(t)$ functions for the Goel generalized NHPP model (a = 500, b = 0.015, c = 1.5).

ent approach and studied the fault counts over a series of time intervals. Goel and Okumoto [11] introduced a time dependent failure rate of the underlying Poisson process and developed the necessary analytical details of the models. A generalization of this model was proposed by Goel [16]. These and some other models in this class are described below.

Goel–Okumoto Nonhomogeneous Poission Process Model

In this model Goel and Okumoto [12] assumed that a software system is subject to failures at random times caused by faults present in the system. Letting $N(t)$ be the cumulative number of failures observed by time t, they proposed that $N(t)$ can be modeled as a nonhomogeneous Poisson process, i.e., as a Poisson process with a time dependent failure rate. Based on their study of actual failure data from many systems, they proposed the following form of the model

$$P\{N(t) = y\} = \frac{(m(t))^y}{y!} e^{-m(t)}, \quad y = 0, 1, 2, \cdots$$

where

$$m(t) = a(1 - e^{-bt}),$$

and

$$\lambda(t) \equiv m'(t) = abe^{-bt}.$$

Here $m(t)$ is the expected number of failures observed by time t and $\lambda(t)$ is the failure rate. Typical plots of the $m(t)$ and $\lambda(t)$ functions are shown in Fig. 3.

In this model a is the expected number of failures to be observed eventually and b is the fault detection rate per fault. It should be noted that here the number of faults to be detected is treated as a random variable whose observed value depends on the test and other environmental factors. This is a fundamental departure from the other models which treat the number of faults to be a fixed unknown constant.

In some environments a different form of the $m(t)$ function might be more suitable than the one given above, see, e.g., Ohba [36] and Yamada *et al.* [48].

Using a somewhat different approach than described above, Schneidewind [42] had earlier studied the number of faults detected during a time interval and failure counts over a series of time intervals. He assumed that the failure process is a nonhomogeneous Poisson process with an exponentially decaying intensity function given by

$$d(i) = \alpha e^{-\beta i}, \ \alpha, \beta > 0, \ i = 1, 2, \cdots$$

where α and β are the parameters of the model.

Goel Generalized Nonhomogeneous Poisson Process Model

Most of the times between failures and failure count models assume that a software system exhibits a decreasing failure rate pattern during testing. In other words, they assume that software quality continues to improve as testing progresses. In practice, it has been observed that in many testing situations, the failure rate first increases and then decreases. In order to model this increasing/decreasing failure rate process, Goel [16], [17] proposed the following generalization of the Goel–Okumoto NHPP model.

$$P\{N(t) = y\} = \frac{(m(t))^y}{y!} e^{-m(t)}, \qquad y = 0, 1, 2, \cdots,$$

$$m(t) = a(1 - e^{-bt^c}),$$

where a is expected number of faults to be eventually detected, and b and c are constants that reflect the quality of testing. The failure rate for the model is given by

$$\lambda(t) \equiv m'(t) = abc\, e^{-bt^c} t^{c-1}.$$

Typical plots of the $m(t)$ and $\lambda(t)$ functions are shown in Fig. 4.

Musa Execution Time Model

In this model Musa [31] makes assumptions that are similar to those of the JM model except that the process modelled is the number of failures in specified execution time intervals. The hazard function for this model is given by

$$z(\tau) = \phi f(N - n_c)$$

where τ is the execution time utilized in executing the program up to the present, f is the linear execution frequency (average instruction execution rate divided by the number of instructions in the program), ϕ is a proportionality constant, which is a fault exposure ratio that relates fault exposure frequency to the linear execution frequency, and n_c is the number of faults corrected during $(0, \tau)$.

One of the main features of this model is that it explicitly emphasizes the dependence of the hazard function on execution time. Musa also provides a systematic approach for converting the model so that it can be applicable for calendar time as well.

Shooman Exponential Model

This model is essentially similar to the JM model. For this model the hazard function [43], [44] is of the following form

$$z(t) = k\left[\frac{N}{I} - n_c(\tau)\right]$$

where t is the operating time of the system measured from its initial activation, I is the total number of instructions in the program, τ is the debugging time since the start of system integration, $n_c(\tau)$ is the total number of faults corrected during τ, normalized with respect to I, and k is a proportionality constant.

Generalized Poisson Model

This is a variation of the NHPP model of Goel and Okumoto and assumes a mean value function [1] of the following form.

$$m(t_i) = \phi(N - M_{i-1})\, t_i^{\alpha}$$

where M_{i-1} is the total number of faults removed up to the end of the $(i - 1)$st debugging interval, ϕ is a constant of proportionality, and α is a constant used to rescale time t_i.

IBM Binomial and Poisson Models

In these models Brooks and Motley [6] consider the fault detection process during software testing to be a discrete process, following a binomial or a Poisson distribution. The software system is assumed to be developed and tested incrementally. They claim that both models can be applied at the module or the system level.

Musa–Okumoto Logarithmic Poisson Execution Time Model

In this model [33] the observed number of failures by some time τ is assumed to be a NHPP, similar to the Goel–Okumoto model, but with a mean value function which is a function of τ, viz.

$$\mu(\tau) = \frac{1}{\theta} \cdot \ln(\lambda_0 \theta \tau + 1),$$

where λ_0 and θ represent the initial failure intensity and the rate of reduction in the normalized failure intensity per failure, respectively. This model is also closely related to Moranda's geometric de-eutrophication model [29] and can be viewed as a continuous version of this model.

V. Fault Seeding and Input Domain Based Models

In this section we give a brief description of a few time-independent models that have been proposed for assessing software reliability. As mentioned earlier, the two approaches proposed for this class of models are fault seeding and input domain analysis.

In fault seeding models, a known number of faults is seeded (planted) in the program. After testing, the numbers of exposed seeded and indigenous faults are counted. Using combinatorics and maximum likelihood estimation, the number of indigenous faults in the program and the reliability of the software can be estimated.

The basic approach in the input domain based models is to generate a set of test cases from an input (operational) distribution. Because of the difficulty in estimating the input distribution, the various models in this group partition the input domain into a set of equivalence classes. An equivalence class is usually associated with a program path. The reliability measure is calculated from the number of failures observed during symbolic or physical execution of the sampled test cases.

Mills Seeding Model

The most popular and most basic fault seeding model is Mills' Hypergeometric model [27]. This model requires that a number of known faults be randomly seeded in the program to be tested. The program is then tested for some amount of time. The number of original indigenous faults can be estimated from the numbers of indigenous and seeded faults uncovered during the test by using the hypergeometric distribution. The procedure adopted in this model is similar to the one used for estimating population of fish in a pond or for estimating wildlife. These models are also referred to as tagging models since a given fault is tagged as seeded or indigenous.

Lipow [23] modified this problem by taking into consideration the probability of finding a fault, of either kind, in any test of the software. Then, for statistically independent tests, the probability of finding given numbers of indigenous and seeded faults can be calculated. In another modification, Basin [2] suggested a two stage procedure with the use of two programmers which can be used to estimate the number of indigenous faults in the program.

Nelson Model

In this input domain based model [35], the reliability of the software is measured by running the software for a sample of n inputs. The n inputs are randomly chosen from the input domain set $E = (E_i : i = 1, \cdots, N)$ where each E_i is the set of data values needed to make a run. The random sampling of n inputs is done according to a probability distribution P_i; the set $(P_i : i = 1, \cdots N)$ is the operational profile or simply the user input distribution. If n_e is the number of inputs that resulted in execution failures, then an unbiased estimate of software reliability $\hat{R}_1$ is $\{1 - (n_e/n)\}$. However, it may be the case that the test set used during the verification phase may not be representative of the expected operational usage. Brown and Lipow [7] suggested an alternative formula for $\hat{R}$ which is

$$\hat{R}_2 = 1 - \sum_{j=1}^{N} \left(\frac{f_j}{n_j}\right) p(E_j)$$

where n_j is the number of runs sampled from input subdomain E_j and f_j is the number of failures observed out of n_j runs.

The main difference between Nelson's $\hat{R}_1$ and Brown and Lipow's $\hat{R}_2$ is that the former explicitly incorporates the usage distribution or the test case distribution while the latter implicitly assumes that the accomplished testing is representative of the expected usage distribution. Both models assume prior knowledge of the operational usage distribution.

Ramamoorthy and Bastani Model

In this input domain based model, the authors are concerned with the reliability of critical, real-time, process control programs. In such systems no failures should be detected during the reliability estimation phase, so that the reliability estimate is one. Hence, the important metric of concern is the confidence in the reliability estimate. This model provides an estimate of the conditional probability that the program is correct for all possible inputs given that it is correct for a specified set of inputs. The basic assumption is that the outcome of each test case provides at least some stochastic information about the behavior of the program for other points which are close to the test point. The specific model is discussed in [3], [38]. A main result of this model is

$$P\,\{\text{program is correct for all points in } [a, a+V] \mid \text{it is correct for test cases having successive distances } x_j, j = 1, \cdots, n-1\} = e^{-\lambda V} \prod_{j=1}^{n-1} \left[\frac{2}{1 + e^{-\lambda x_j}}\right],$$

where λ is a parameter which is deduced from some measure of the complexity of the source code.

Unlike other sampling models, this approach allows any test case selection strategy to be used. Hence, the testing effort can be minimized by choosing test cases which exercise error-prone constructs. However, the model concerning the parameter λ needs to be validated experimentally.

A related model based on fuzzy set theory is discussed in [4].

VI. Model Assumptions and Limitations

In this section we evaluate the implications of the various assumptions underlying the models described above. The main purpose of the following discussion is to focus attention on the framework within which the existing models have been developed. The applicability of such models during the software development cycle will be discussed in the next section.

Before proceeding further, it is helpful to note that a precise, unambiguous statement of the underlying assumptions is necessary to develop a mathematical model. The physical process being modeled, the software failure phenomenon in our case, can hardly be expected to be so precise. It is, therefore, necessary to have a clear understanding of the statement as well as the intent of an assumption.

In the following discussion, the assumptions are evaluated one at a time. Not all of the assumptions discussed here are relevant to any given model but, as a totality, they provide an insight into the kind of limitations imposed by them on the use of the software reliability models. It should be pointed out that the arguments presented here are not likely to be universally applicable because the software development process is very environment dependent. What holds true in one environment may not be true in another. Because of this, even assumptions that seem reasonable, e.g., during the testing of one function or system, may not hold true in subsequent testing of the same function or system. The ultimate decision about the appropriateness of the underlying assumptions and the applicability of the models will have to be made by the user of a model. What is presented here should be helpful in determining whether the assumptions associated with a given model are representative of the user's development environment and in deciding which model, if any, to use.

Times Between Failures Are Independent

This assumption is used in all times between failure models and requires that successive failure times be independent of each other. In general, this would be the case if successive test cases were chosen randomly. However, testing, especially functional testing, is not based on independent test cases, so that the test process is not likely to be random. The time, or the additional number of test cases, to the next failure may very well depend on the nature or time of the previous fault. If a critical fault is uncovered, the tester may decide to intensify the testing process and look for more potential critical faults. This in turn may mean shorter time to the next failure. Although strict adherence to this assumption is unlikely, care should be taken in ensuring some degree of independence in data points.

A Detected Fault Is Immediately Corrected

The models that require this assumption assume that the software system goes through a purification process as testing uncovers faults. An argument can be made that this assumption is at least implicitly satisfied in many testing situations. Sometimes, when a fault is encountered, testing can proceed without removing it. In that case, the future fault detection process can be assumed to behave as if the fault had been physically removed. If, however, the fault is in the path that must be tested further, this assumption would be satisfied only if the fault is removed prior to proceeding with the remainder of the test bucket or if new test cases were generated to get around it.

No New Faults Are Introduced During the Fault Removal Process

The purpose of this assumption is to ensure that the modeled failure process does have a monotonic pattern. That is, the subsequent faults are exposed from a system that has less faults than before. In general, this may not be true if faults are debugged after each occurrence because other paths may be affected during debugging, leading to additional faults in the system. It is generally considered to be a restrictive assumption in reliability models. The only way to strictly satisfy this is to ensure that the correction process does not introduce new faults. If, however, the additional faults introduced constitute a very small fraction of the fault population, the practical effect on model results would be minimal.

Failure Rate Decreases with Test Time

This assumption implies that the software gets better with testing in a statistical sense. This seems to be a reasonable assumption in most cases and can be justified as follows. As testing proceeds, faults are detected. They are either removed before testing continues or they are not removed and testing is shifted to other parts of the program. In the former case, the subsequent failure rate decreases explicitly. In the latter case, the failure rate (based upon the entire program) decreases implicitly since a smaller portion of the code is subjected to subsequent testing.

Failure Rate Is Proportional to the Number of Remaining Faults

This assumption implies that each remaining fault has the same chance of being detected in a given testing interval between failures. It is a reasonable assumption if the test cases are chosen to ensure equal probability of executing all portions of the code. However, if one set of paths is executed more thoroughly than another, more faults in the former are likely to be detected than in the latter. Faults residing in the unreachable or never tested portion of the code will obviously have a low, or zero, probability of being detected.

Reliability Is a Function of the Number of Remaining Faults

This assumption implies that all remaining faults are equally likely to appear during the operation of a system and is used when reliability estimates are based on the number of remaining faults. If the usage is uniform, then this is clearly a reasonable assumption. If, however, some portions are more likely to be executed than others, this assumption will not hold. However, the reliability of the system can be recomputed by incorporating information about differences in usage. In other words, a reliability measure conditioned on usage rather than based on the number of remaining faults, would be more suitable. If, however, such information is not available, the assumption of uniform usage is the only reasonable one to make. In that case, the estimated reliability value should be interpreted with caution.

Time Is Used as a Basis for Failure Rate

Most models use time as a basis for determining changes in failure rate. This usage assumes that testing effort is proportional to either calendar time or execution time. Also, time is generally easy to measure and most testing records are kept in terms of time. Another argument in favor of this measure is that time tends to smooth out differences in test effort.

If, however, testing is not proportional to time, the models are equally valid for any other relevant measure. Some examples of such measures are lines of code tested, number of functions tested, and number of test cases executed.

Failure Rate Increases Between Failures

This assumption implies that the likelihood of finding a fault increases as the testing time increases within a given failure interval. This would be a justifiable assumption if software were assumed to be subject to wearout within the interval. But, generally, this is not the case with software systems. Another situation where such an assumption might be justifiable is where testing intensity increases within the interval in the same fashion as does the failure rate.

Testing Is Representative of the Operational Usage

This assumption is necessary when a reliability estimate based on testing is projected into the operational phase. It is used primarily in input domain based models. The times between failures and fault count models would also need this assumption if they are used to assess operational reliability.

Test cases are generally chosen to ensure that the functional requirements of the system are correctly met. A given user of the system, however, may not use the functions in the same proportion as done during testing. In that case, testing will not reflect the operational usage. If information about usage pattern is available, testing effort can be modified to be representative of the use profile.

VII. Applicability of Software Reliability Models

In this section we suggest the classes of models that might be applicable in various phases of the software development process. Some of the general comments made in the beginning of Section VI about the importance and

TABLE I
LIST OF KEY ASSUMPTIONS BY MODEL CATEGORY

Times Between Failures (TBF) Models
- Independent times between failures.
- Equal probability of the exposure of each fault.
- Embedded faults are independent of each other.
- Faults are removed after each occurrence.
- No new faults introduced during correction, i.e., perfect fault removal.

Fault Count (FC) Models
- Testing intervals are independent of each other.
- Testing during intervals is reasonably homogeneous.
- Numbers of faults detected during nonoverlapping intervals are independent of each other.

Fault Seeding (FS) Models
- Seeded faults are randomly distributed in the program.
- Indigeneous and seeded faults have equal probabilities of being detected.

Input Domain Based (IDB) Models
- Input profile distribution is known.
- Random testing is used.
- Input domain can be partitioned into equivalent classes.

interpretation of assumptions are also applicable to the discussion here. In particular, note that a precise statement of assumptions is necessary for modeling even though the development process being modeled is extremely unlikely to be that precise. A partial explanation for this apparent inconsistency lies in the fact that a model is, simply, an attempt to summarize the complexity of the real process in order to understand it and possibly control it. In order to be useful, a software reliability model, thus, has to be simple and cannot capture in detail every facet of the modeled failure process. A realization of such constrants imposed on a mathematical model would be helpful in choosing one which can adequately represent the environment within a given development phase.

In the following discussion, we consider the four classes of software reliability models (see Table I) and assess their applicability during the design, unit testing, integration testing, and operational phases.

Design Phase

During the design phase, faults may be detected visually or by other formal or informal procedures. Existing software reliability models are not applicable during this phase because the test cases needed to expose faults as required by fault seeding and input domain based models do not exist, and the failure history required by time dependent models is not available.

Unit Testing

The typical environment during module coding and unit testing phase is such that the test cases generated from the module input domain do not form a representative sample of the operational usage distribution. Further, times between exposures of module faults are not random since the test strategy employed may not be random testing. In fact, test cases are usually executed in a deterministic fashion.

Given these conditions, it seems that the fault seeding models are applicable provided it can be assumed that the indigenous and seeded fault have equal probabilities of being detected. However, a difficulty could arise if the programmer is also the tester in this phase. The input domain based models seem to be applicable, except that matching the test profile to operational usage distribution could be difficult. Due to these difficulties, such models, although applicable, may not be usable.

The time dependent models, especially the time between failures models, do not seem to be applicable in this environment since the independent times between failures assumption is seriously violated.

Integration Testing

A typical environment during integration testing is that the modules are integrated into partial or whole systems and test cases are generated to verify the correctness of the integrated system. Test cases for this purpose may be generated randomly from an input distribution or may be generated deterministically using a reliable test strategy, the latter being probably more effective. The exposed faults are corrected and there is a strong possibility that the removal of exposed faults may introduce new faults.

Under such testing conditions, fault seeding models are theoretically applicable since we still have the luxury of seeding faults into the system. Input domain based models based on an explicit test profile distribution are also applicable. The difficulty in applying them at this point is the very large number of logic paths generated by the whole system.

If deterministic testing (e.g., boundary value analysis, path testing) is used, times between failures models may not be appropriate because of the violation of the independence of interfailure times assumption. Fault count models may be applicable if sets of test cases are independent of each other, even if the tests within a set are chosen deterministically. This is so because in such models the system failure rate is assumed to decrease as a result of executing a set of test cases and not at every failure.

If random testing is performed according to an assumed input profile distribution, then most of the existing software reliability models are applicable. Input domain based models, if used, should utilize a test profile distribution which is statistically equivalent to the operational profile distribution. Fault seeding models are applicable likewise, since faults can be seeded and the equal probability of fault detection assumption may not be seriously violated. This is due to the random nature of the test generation process.

Times between failures and failure count models are most applicable with random testing. The next question could be about choosing a specific model from a given class. Some people prefer to try a couple of these models on the same failure history and then choose one. However, because of different underlying assumptions of these models, there are subtle distinctions among them. Therefore, as far as possible, the choice of a specific model should be based on the development environment considerations.

Acceptance Testing

During acceptance testing, inputs based on operational usage are generated to verify software acceptability. In this phase, seeding of faults is not practical and the exposed faults are not usually immediately corrected. The fault seeding and times between failures models are thus not applicable. Many other considerations here are similar to those of intergration testing so that the fault count and input domain based models are generally applicable.

Operational Phase

When the reliability of the software as perceived by the developer or the "friendly users" is already acceptable, the software is released for operational use. During the operational phase, the user inputs may not be random. This is because the user may use the same software function or path on a routine basis. Inputs may also be correlated (e.g., in real-time systems), thus losing their randomness. Furthermore, faults are not always immediately corrected. In this environment, fault-count models are likely to be most applicable and could be used for monitoring software failure rate or for determining the optimum time for installing a new release.

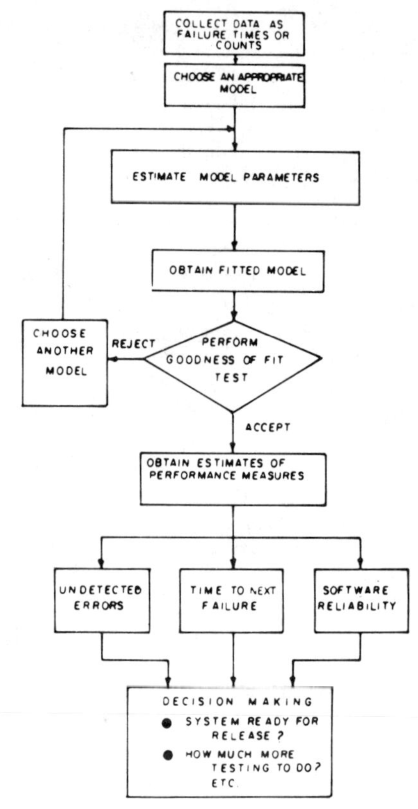

Fig. 5. Flowchart for software reliability modeling and decision making.

VIII. Development and Use of a Model

A step-by-step procedure for fitting a model to software failure data is presented in this section. The procedure is illustrated via analyses of data from a medium size, real-time command and control system. The use of the fitted model for computing reliability and other performance measures, as well as for decision-making, is also explained.

Modeling Procedure

The various steps of the model fitting and decision making process are shown in Fig. 5 and are described below.

Step 1—Study Software Failure Data: The models discussed in this paper require that failure data be available. For most of the models, such data should be in the form of either times between failures or as failure counts. The first step in developing a model is to carefully study such data in order to gain an insight into the nature of the process being modeled.

It is highly desirable to plot the data as a function of, say, calendar time, execution time, or number of test cases executed. The objective of such plots is to try to determine the appropriate variables to use in the model. For example, based on the data and information about the development environment, a model of failures versus unique test cases run may be more important and relevant than a model of failures versus, say, calendar time. Sometimes it is desirable to model several such combinations and then use the fitted models for answering a variety of questions about the failure process. Occasionally, it may be necessary to normalize the data to, for example, account for changes in system size during testing.

Step 2—Choose a Reliability Model: The next step is to choose an appropriate model based upon an understanding of the testing process and of the assumptions underlying the models discussed earlier. The data and plots from Step 1 are likely to be very helpful in this process. A check about the "goodness" of the chosen model for the data at hand will be made in Step 5.

Step 3—Obtain Estimates of Model Parameters: Different methods are generally required depending upon the nature of available data. The most commonly used one is the method of maximum likelihood because it has very good statistical properties. However, sometimes, the method of least squares or some other method may be preferred.

Step 4—Obtain the Fitted Model: The fitted model is obtained by substituting the estimated values of the parameters in the chosen model. At this stage, we have a fitted model based on the available failure data and the chosen model form.

Step 5—Perform Goodness-of-Fit Test: Before proceeding further, it is advisable to conduct the Kolmogorov–Smirnov, or some other suitable goodness-of-fit test to check the model fit. If the model fits, i.e., if it is a satisfactory descriptor of the observed failure process, we can move ahead. However, if the model does not fit, we have to collect additional data or seek a better, more appropriate model. There is no easy answer to either how much data to collect or how to look for a better model. Decisions on these issues are very much problem dependent and require a clear understanding of the models and the software development environment.

Step 6—Obtain Estimates of Performance Measures: At this stage, we can compute various quantitative measures to assess the performance of the software system. Some useful measures are shown in Fig. 5. Confidence bounds can also be obtained for these measures to evaluate the degree of uncertainty in the computed values of the performance measures.

Step 7—Decision Making: The ultimate objective of developing a model is to use it for making some decisions about the software system, e.g., whether to release the system or continue testing. Such decisions are made at this stage of the modeling process based on the information developed in the previous steps.

An Example of Software Reliability Modeling

We now employ the above procedure to illustrate the development of a software reliability model based on failure data from a real-time, command and control system. The delivered number of object instructions for this system was 21 700 and it was developed by Bell Laboratories. The data were reported by Musa [32] and represent the failures observed during system testing for 25 hours of CPU time.

For purposes of this illustration, we employ the NHPP model of Goel and Okumoto [12]. We do so because of its simplicity and applicability over a wide range of testing situations as also noted by Misra [28], who successfully used this model to predict the number of remaining faults in a space shuttle software subsystem.

Step 1: The original data were reported as times between failures. To overcome the possible lack of independence among these values, we summarized the data into numbers of failures per hour of execution time. The summarized data are given in Table II. A plot of the hourly data is shown in Fig. 6 and a plot of $N(t)$, the cumulative number of failures by t, is shown in Fig. 7. Some other plots shown in Fig. 7 will be discussed later.

Step 2: A study of the data in Table II and of the plot in Fig. 6 indicates that the failure rate (number of failures per hour) seems to be decreasing with test time. This means that an NHPP with a mean value function $m(t) = a(1 - e^{-bt})$ should be a reasonable model to describe the failure process.

Step 3: For the above model, two parameters, a and b, are to be estimated from the failure data. We chose to use the method of maximum likelihood for this purpose [14], [16]. The estimated values for the two parameters are $\hat{a} = 142.32$ and $\hat{b} = 0.1246$. Recal that $\hat{a}$ is an estimate of the expected total number of faults likely to be detected and $\hat{b}$ represents the number of faults detected per fault per hour.

Step 4: The fitted model based on the data of Table II and the parameters estimated in Step 3 is

$$\hat{m}(t) = 142.32\ (1 - e^{-0.1246t})$$

TABLE II

FAILURES IN 1 HOUR (EXECUTION TIME) INTERVALS AND CUMULATIVE FAILURES

Hour	Number of Failures	Cumulative Failures
1	27	27
2	16	43
3	11	54
4	10	64
5	11	75
6	7	82
7	2	84
8	5	89
9	3	92
10	1	93
11	4	97
12	7	104
13	2	106
14	5	111
15	5	116
16	6	122
17	0	122
18	5	127
19	1	128
20	1	129
21	2	131
22	1	132
23	2	134
24	1	135
25	1	136

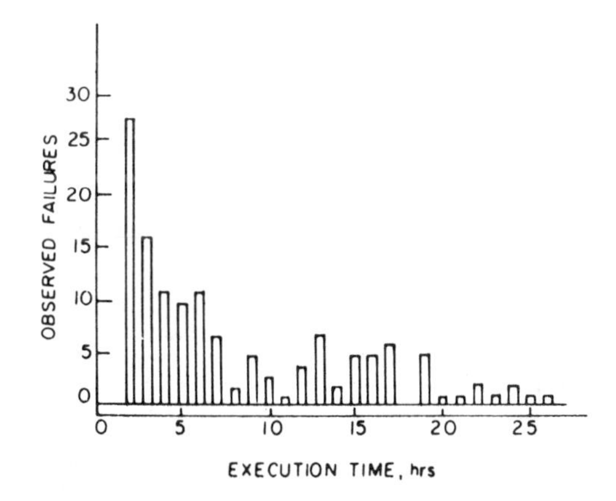

Fig. 6. Plot of the number of failures per hour.

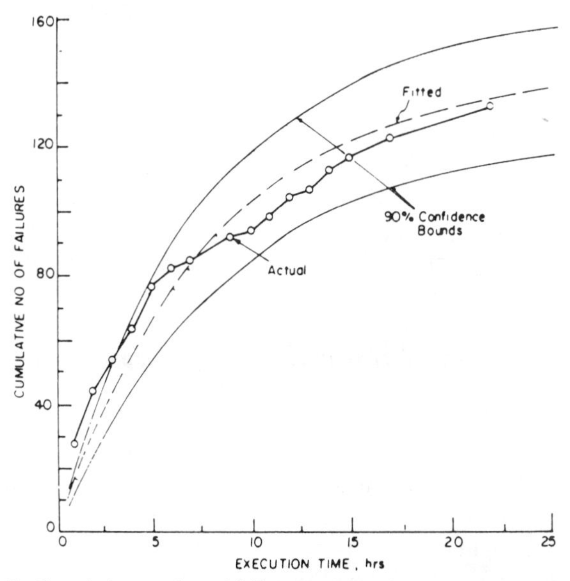

Fig. 7. Cumulative number of failures as a function of execution time and confidence bounds.

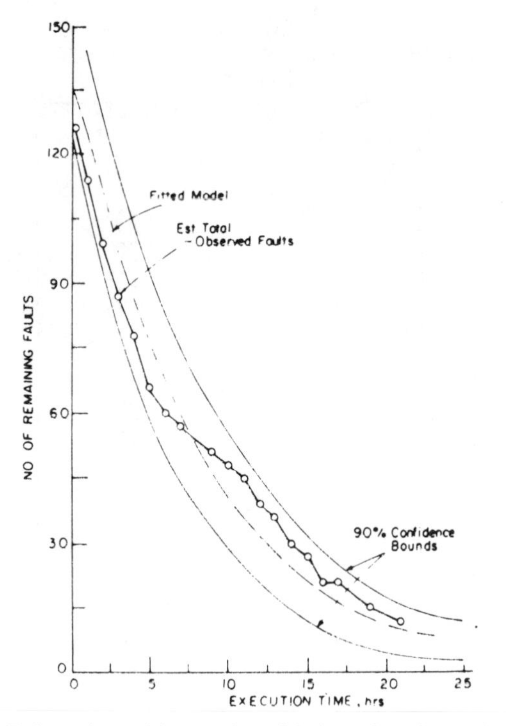

Fig. 8. Estimated remaining number of faults and confidence bounds.

and

$$\hat{\lambda}(t) = 17.73 \cdot e^{-0.1246t}.$$

Step 5: In this case, we used the Kolmogorov–Smirnov goodness-of-fit test for checking the adequacy of the model. For details of this test, see Goel [17]. Basically, the test provides a statistical comparison between the actual data and the model chosen in Step 2. The fitted model in Step 4 passed this test so that it could be considered to be a good descriptor of the data in Table II. The plots in Fig. 7 also provide a visual check of the goodness-of-fit of the model.

Step 6: For illustration purposes, we computed only one performance measure, the expected number of remaining faults, at various testing times. A plot of these values is shown in Fig. 8.

Plots of the confidence bounds for the expected cumulative number of failures, and the expected number of remaining faults are also shown in Figs. 7 and 8, respectively. A study of these plots indicates that the chosen NHPP model provides an excellent fit to the data and can be used for purposes of describing the failure behavior as well as for prediction of the future failure process. The information available from this can be used for planning, scheduling, and other management decisions as indicated below.

Step 7: The model developed above can be used for answering a variety of questions about the failure process and for determining the additional test effort required until the system is ready for release. This type of information can be sought at various points of time and one does not have to wait until the end of testing. For illustrative purposes suppose that failure data through only 16 hours of testing are available, and a total of 122 failures (see Table II) have been observed. Based on these data, the fitted model is

$$\hat{m}(t) = 138.37\ (1 - e^{-0.133t}).$$

An estimate of the remaining number of faults is 16.37 with a 90 percent confidence interval of (4.64–28.11). Also, the estimated one hour ahead reliability is 0.165 and the corresponding 90 percent confidence interval is (0.019–0.310).

Now, suppose that a decision to release software for operational use is to be based on the number of remaining faults. Specifically, suppose that we would release the system if the expected number of remaining faults is less than or equal to 10. In the above analysis we saw that the best estimate of this quantity at present is 16.37, which means that we should continue testing in the hope that additional faults can be detected and removed. If we were to carry on a similar analysis after each additional hour of testing, the expected number of remaining faults after 20 hours would be 9.85 so that the above release criterion would be met.

The above simple example was meant to illustrate the kind of information that can be obtained from a software reliability model. In practice, determination of release time, additional testing effort, etc. are based on much more elaborate considerations than remaining faults. The results from models such as the ones developed here can be used as inputs into the decision-making process.

IX. Concluding Remarks

In this paper, we have provided a review and an evaluation of software reliability models. Four classes of analytical models, along with their underlying assumptions and limitations, were described. The use and applicability of such models during software development and operational phases was also discussed. A methodology for developing a model from failure data was proposed and illustrated via an example. The objective was to provide a model user an insight into the usefulness and limitations of such models that would be helpful in determining which model, if any, to use in a given software development environment.

The material presented here primarily dealt with the development and use of a software reliability model. Several related, but important aspects [19] that were not addressed are model unification issues [22], optimum release time determination [9], parametric estimation problems, comparisons of models [1], [15], [41], [46], and alternate approaches to such models [8], [21].

It should be noted that the above analytical models are primarily useful in estimating and monitoring software reliability, viewed as a measure of software quality. Since they treat the software product and the development process as a blackbox, they cannot be explicitly used for assessing the role of various tools and techniques in determining software quality. For this purpose, more detailed models will be needed that explicitly incorporate information about the software system and the development process.

Acknowledgment

P. Valdes and A. Deb provided valuable help on an earlier version of this paper. Constructive comments by P. Moranda and F. Bastani were very helpful in improving the quality of the paper. The author is grateful to all of them.

References

[1] J. E. Angus, R. E. Schafer, and A. Sukert, "Software reliability model validation," in *Proc. Annu. Reliability and Maintainability Symp.*, San Francisco, CA, Jan. 1980, pp. 191-193.

[2] S. L. Basin, "Estimation of software error rate via capture-recapture sampling," Science Applications, Inc., Palo Alto, CA, 1974.

[3] F. B. Bastani, "An input domain based theory of software reliability and its application," Ph.D. dissertation, Univ. California, Berkeley, 1980.

[4] —, "On the uncertainty in the correctness of computer programs," *IEEE Trans. Software Eng.*, vol. SE-11, pp. 857-864, Sept. 1985.

[5] B. W. Boehm, J. R. Brown, M. Lipow, "Quantitative evaluation of software quality," in *Proc. 2nd Int. Conf. Software Eng.*, San Francisco, CA, Oct. 1976, pp. 592-605.

[6] W. D. Brooks and R. W. Motley, "Analysis of discrete software reliability models," Rep. RADC-TR-80-84, Apr. 1980.

[7] J. R. Brown and M. Lipow, "Testing for software reliability," in *Proc. Int. Conf. Reliable Software*, Los Angeles, CA, Apr. 1975, pp. 518-527.

[8] L. H. Crow and N. D. Singpurwalla, "An empirically developed Fourier series model for describing software failures," *IEEE Trans. Rel.*, vol. R-33, pp. 175-183, June 1984.

[9] E. H. Forman and N. D. Singpurwalla, "An empirical stopping rule for debugging and testing computer software," *J. Amer. Statist. Ass.*, vol. 72, no. 360, pp. 750-757, 1977.

[10] S. Gerhart and L. Yelowitz, "Observations of fallibility in applications of modern programming methodologies," *IEEE Trans. Software Eng.*, vol. SE-2, pp. 195-207, May 1976.

[11] A. L. Goel and K. Okumoto, "An analysis of recurrent software failures in a real-time control system," in *Proc. ACM Annu. Tech. Conf.*, ACM, Washington, DC, 1978, pp. 496-500.

[12] —, "A time dependent error detection rate model for software reliability and other performance measures," *IEEE Trans. Rel.*, vol. R-28, pp. 206-211, 1979.

[13] —, "A Markovian model for reliability and other performance measures of software systems," in *Proc. Nat. Comput Conf.*, New York, vol. 48, 1979, pp. 769-774.

[14] A. L. Goel, "A software error detection model with applications," *J. Syst. Software*, vol. 1, pp. 243-249, 1980.

[15] —, "A summary of the discussion on an analysis of competing software reliability models," *IEEE Trans. Software Eng.*, vol. SE-6, pp. 501-502, 1980.

[16] —, "A guidebook for software reliability assessment," Rep. RADC-TR-83-176, Aug. 1983.

[17] —, "Software reliability modelling and estimation techniques," Rep. RADC-TR-82-263, Oct. 1982.

[18] A. L. Goel, V. R. Basili, and P. M. Valdes, "When and how to use a software reliability model," in *Proc. 7th Software Eng. Workshop*, NASA/GSFC, Greenbelt, MD, Nov. 1983.

[19] A. L. Goel, "Software reliability modeling and related topics: A survey," Dep. of Elec. and Comput. Eng., Syracuse Univ., Syracuse, NY, Tech. Rep., Oct. 1985.

[20] Z. Jelinski and P. Moranda, "Software reliability research," In *Statistical Computer Performance Evaluation*, W. Freiberger, Ed. New York: Academic, 1972, pp. 465-484.

[21] W. Kremer, "Birth-death and bug counting," *IEEE Trans. Rel.*, vol. R-32, pp. 27-47, Apr. 1983.

[22] N. Langberg and N. D. Singpurwalla, "Unification of some software reliability models via the Bayesian approach," *SIAM J. Sci. Statist. Comput.*, vol. 6., pp. 781-790, July 1985.

[23] M. Lipow, "Estimation of software package residual errors," TRW, Redondo Beach, CA, Software Series Rep. TRW-SS-72-09, 1972.

[24] —, "Maximum likelihood estimation of parameters of a software time-to-failure distribution," TRW, Redondo Beach, CA, Systems Group Rep. 2260.1.9.-73B-15, 1972.

[25] B. Littlewood and J. L. Verrall, "A Bayesian reliability growth model for computer software," *Appl. Statist.*, vol. 22, pp. 332-346, 1973.

[26] B. Littlewood, "Theories of software reliability: How good are they and how can they be improved?" *IEEE Trans. Software Eng.*, vol. SE-6 pp. 489-500, 1980.

[27] H. D. Mills, "On the statistical validation of computer programs," IBM Federal Syst. Div., Gaithersburg, MD, Rep. 72-6015, 1972.

[28] P. N. Misra, "Software reliability analysis," *IBM Syst. J.*, vol. 22, no. 3, pp. 262-270, 1983.

[29] P. B. Moranda, "Prediction of software reliability during debugging," in *Proc. Annu. Reliability and Maintainability Symp.*, Washington, DC, Jan. 1975, pp. 327-332.

[30] —, private communication, 1982.

[31] J. D. Musa, "A theory of software reliability and its application," *IEEE Trans. Software Eng.*, vol. SE-1, pp. 312-327, 1971.

[32] —, "Software Reliability Data," DACS, RADC, New York, 1980.

[33] J. D. Musa and K. Okumoto, "A logarithmic Poisson execution time model for software reliability measurement," in *Proc. 7th Int. Conf. Software Eng.*, Orlando, FL, Mar. 1983, pp. 230-237.

[34] G. J. Myers, *Software Reliability, Principles and Practices*. New York: Wiley, 1976.

[35] E. Nelson, "Estimating software reliability from test data," *Microelectron. Rel.*, vol. 17, pp. 67-74, 1978.

[36] M. Ohba, "Software reliability analysis models," *IBM J. Res. Develop.*, vol. 28, pp. 428-443, July 1984.

[37] K. Okumoto and A. L. Goel, "Availability and other performance measures of software systems under imperfect maintenance," in *Proc. COMPSAC*, Chicago IL, Nov. 1978, pp. 66-71.

[38] C. V. Ramamoorthy and F. B. Bastani, "Software reliability: Status and perspectives," *IEEE Trans. Software Eng.*, vol. SE-8, pp. 359-371, July 1982.

[39] J. Reynolds, *The Craft of Programming*. Englewood Cliffs, NJ: Prentice-Hall, 1981.

[40] G. J. Schick and R. W. Wolverton, "Assessment of software reliability," presented at the 11th Annu. Meeting German Oper. Res. Soc., DGOR, Hamburg, Germany; also in *Proc. Oper. Res.*, Physica-Verlag, Wirzberg-Wien, 1973, pp. 395-422.

[41] G. J. Schick and R. W. Wolverton, "An analysis of computing software reliability models," *IEEE Trans. Software Eng.*, vol. SE-4, pp. 104-120, 1978.

[42] N. F. Schneidewind, "Analysis of error processes in computer software," in *Proc. Int. Conf. Reliable Software*, Los Angeles, CA, Apr. 1975, pp. 337-346.

[43] M. L. Shooman, "Probabilistic models for software reliability prediction," in *Statistical Computer Performance Evaluation*, W. Freiberger, Ed. New York: Academic, 1972, pp. 485-502.

[44] —, "Software reliability measurement and models," in *Proc. Annu. Reliability and Maintainability Symp.*, Washington, DC, Jan. 1975, pp. 485-491.

[45] —, "Structural models for software reliability and prediction," in *Proc. 2nd Int. Conf. Software Eng.*, San Francisco, CA, Oct. 1976, pp. 268-273.

[46] A. N. Sukert, "An investigation of software reliability models," in *Proc. Annu. Reliability and Maintainability Symp.*, Philadelphia, PA, Jan. 1977, pp. 478-484.

[47] T. A. Thayer, M. Lipow, and E. C. Nelson, "Software reliability study," Rep. RADC-TR-76-238, Aug. 1976.

[48] S. Yamada, M. Ohba, and S. Osaki, "S-shaped reliability growth modeling for software error detection," *IEEE Trans. Rel.*, vol. R-32, pp. 475-478, Dec. 1983.

Amrit L. Goel (M'75), for a photograph and biography, see this issue, p. 1410.

Reprinted from *IEEE Transactions on Software Engineering*, Volume SE-13, Number 5, May 1987, pages 582–592.

Fault-Tolerant Software Reliability Modeling

R. KEITH SCOTT, JAMES W. GAULT, MEMBER, IEEE, AND DAVID F. McALLISTER, MEMBER, IEEE

Abstract—**In situations in which computers are used to manage life-critical situations, software errors that could arise due to inadequate or incomplete testing cannot be tolerated. This paper examines three methods of creating fault-tolerant software systems, Recovery Block, *N*-Version Programming, and Consensus Recovery Block, and it presents reliability models for each. The models are used to show that one method, the Consensus Recovery Block, is more reliable than the other two.**

The results of an experiment used to validate the models are presented. It is demonstrated that, for highly reliable acceptance tests, the Consensus Recovery Block system gave the highest reliability. In all cases, the Consensus Recovery Block and Recovery Block systems were better than the *N*-Version Programming systems.

A simple cost model that shows the relative costs of increasing software reliability using the three fault-tolerant methods is presented.

Index Terms—**Consensus recovery block, fault-tolerant software, *N*-version programming, recovery block.**

I. Introduction

VARIOUS methods of increasing software reliability by attempting to make software fault-intolerant have been proposed. Hecht [1] notes that all such attempts are inadequate for the following reasons:

1) There are no indications when validation, testing, or reviewing is complete.
2) No existing methods *guarantee* correct program execution.
3) The methods tend to be costly.
4) The execution of a "perfect" program may be affected by bad data or environmental abnormalities.
5) There are no guidelines for limited program verification after a change or fix has been applied.

In a similar fashion, Keirstead [2] observes:

> "The principal methods for examining software products are formal testing and proof of correctness; of these techniques, formal testing is the most widely accepted. It is well known, however, that testing can only establish the presence of errors but cannot assure their absence. Testing also relies ultimately on manual skills to identify test cases and to evaluate results, and this can often have serious consequences on the reliable assessment of software."

Manuscript received March 30, 1984.

R. K. Scott is with IBM Corporation, Research Triangle Park, NC 27709.

J. W. Gault is with the U.S. Army European Research Office, London, England.

D. F. McAllister is with the Department of Computer Science, North Carolina State University, Raleigh, NC 27650.

IEEE Log Number 8613883.

Therefore, attempts to guarantee highly reliable software by invoking software fault-intolerant methods are futile.

An alternate technique is to assume that a program will always retain a residual number of design faults regardless of the amount of testing to which it is subjected. With this assumption, the software must be designed to supply "correct" results even if faults are encountered. The software is then said to be "fault-tolerant."

II. Recovery Block

One approach to software fault-tolerance is the "Recovery Block" [3], [4]. As the name implies, the basic goal is to detect a software fault in a program, recover the machine state at the time the faulty program was entered, and execute another program that performs the same function as the faulty program. A number of independently designed programs that perform the same function must be developed. It is assumed that if the programs are developed independently, the probability of common design faults among the different versions is negligibly small. Therefore if one program version supplies an incorrect output due to some inherent design fault, another version should be able to supply correct output since, by assumption, it would not contain the same error-producing fault. The versions are ranked based on some criteria. This ranking is a reflection of the "graceful degradation" in the versions [5]. When the given task is requested, only the "best" program is executed. If an acceptance test detects erroneous output, then the "next best" one is executed, etc., until an acceptable output is obtained. If all versions are deemed faulty, an error is posted.

A. Recovery Block Reliability Model

There have been a number of reliability models proposed for Recovery Block programming. Hecht [1] and Grnarov *et al.* [6] have proposed models based on correlated errors. Wei [7] proposed a simple model based on acceptance test errors and exhaustion of versions. Soneriu's model [8] views any redundant software system as a collection of independent models. At any point in time, some of the members of the collection are "up" and others are "down." System reliability is then based on the number of members in the up state.

1) Independent Recovery Block Reliability Model: In our approach, there are four distinct types of errors that can occur in a Recovery Block structure. A Type 1 error occurs when any program version produces an incorrect result, but the acceptance test labels the result as correct. In a Type 2 error, the final version produces correct results, but the acceptance test erroneously determines that the results are incorrect. A Type 3 error occurs when the recovery mechanism cannot successfully recover the input state of the previous version in preparation for executing another version, or it could not successfully invoke the next version. Finally, a Type 4 error occurs when the last version produces incorrect results and the acceptance test judges that the results are incorrect. Fig. 1 depicts the event tree [9] of a Recovery Block structure with two versions, one primary and one backup. The paths terminating in the four error types are defined. Fig. 1 can be used to determine the probability of a given error. First, consider the following events and their associated probabilities:

$P(C_i)$ = Probability of version i executing correctly.
$P(I_i)$ = Probability of version i executing incorrectly $= 1 - P(C_i)$.
$P(C_R)$ = Probability of successful recovery.
$P(I_R)$ = Probability of unsuccessful recovery $= 1 - P(C_R)$.
$P(A_I)$ = Probability of accepting an incorrect result.
$P(R_I)$ = Probability of rejecting an incorrect result $= 1 - P(A_I)$.
$P(R_C)$ = Probability of rejecting a correct result.
$P(A_C)$ = Probability of accepting a correct result $= 1 - P(R_C)$.
$P_{RB}(E_K, n)$ = Probability of a type k error given n versions.

By using Fig. 1, the probability of a given error type for an $n = 2$ Recovery Block System can be found by multiplying together the probabilities encountered while tracing a path from the top event, executing the primary version, to a node terminated by the error number. (Since the versions are independently developed, it is assumed that they are statistically independent. This allows the probability of a joint event to be calculated by multiplying together the probabilities of the individual events.) We then sum these products over all such paths for a given error type. Using this procedure, it can be seen that

$$P_{RB}(E_1, 2) = P(I_1)\,P(A_I) + P(C_1)\,P(R_C)\,P(C_R)\,P(I_2)\,P(A_I) + P(I_1)\,P(R_I)\,P(C_R)\,P(I_2)\,P(A_I).$$

$$P_{RB}(E_2, 2) = P(C_1)\,P(R_C)\,P(C_R)\,P(C_2)\,P(R_C) + P(I_1)\,P(R_I)\,P(C_R)\,P(C_2)\,P(R_C).$$

$$P_{RB}(E_3, 2) = P(C_1)\,P(R_C)\,P(I_R) + P(I_1)\,P(R_I)\,P(I_R).$$

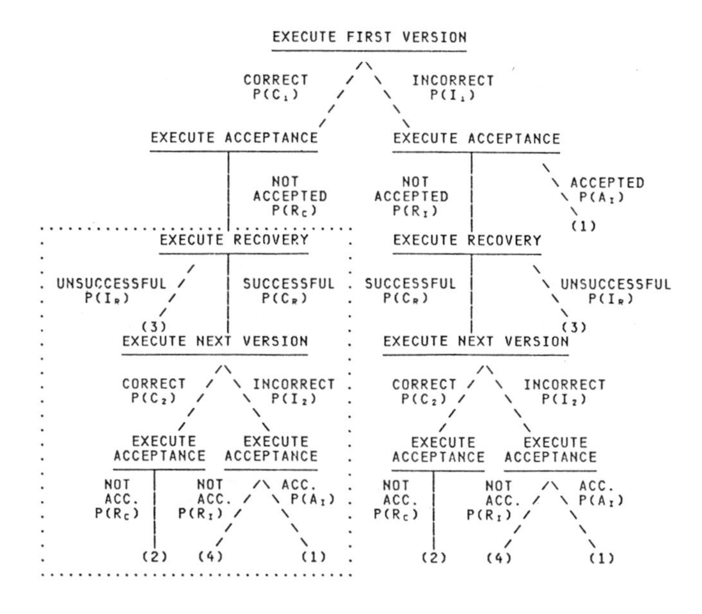

Fig. 1. Error Types for an $n = 2$ Recovery Block.

$$P_{RB}(E_4, 2) = P(C_1)\,P(R_C)\,P(C_R)\,P(I_2)\,P(R_I) + P(I_1)\,P(R_I)\,P(C_R)\,P(I_2)\,P(R_I).$$

The probability of any error is:

$$R_{RB}(E, 2) = P_{RB}(E_1, 2) + P_{RB}(E_2, 2) + P_{RB}(E_3, 2) + P_{RB}(E_4, 2).$$

This gives a system reliability of:

$$R_{RB} = 1 - P_{RB}(E, 2).$$

To extend the tree to $n = 3$, simply remove the terminating nodes labeled (2) and (4) at the bottom of the tree in Fig. 1 and replace each by a repetition of the tree structure enclosed in the dotted box with $P(C_2)$ and $P(I_2)$ becoming $P(C_3)$ and $P(I_3)$. This procedure can be extended to any value of n. Applying this technique yields the following recurrence equations for the four error types:

Type 1 Error

Define $P_{RB}(E_1, 0) = 0$ and $P_{RB}(E_1, 1) = P(I_1)\,P(A_I)$. Then for $n \geq 2$:

$$P_{RB}(E_1, n) = P_{RB}(E_1, n-1) + \left[P_{RB}(E_1, n-1) - P_{RB}(E_1, n-2)\right]\left(\frac{P(C_R)\,P(I_n)}{P(I_{n-1})}\right) \cdot \left[P(C_{n-1})\,P(R_C) + P(I_{n-1})\,P(R_I)\right]. \quad (1)$$

Type 2 Error

Define $P_{RB}(E_2, 1) = P(C_1)\, P(R_C)$. Then for $n \geq 2$:

$$P_{RB}(E_2, n) = P_{RB}(E_2, n-1)\, P(C_R) \cdot P(C_n)\left(P(R_C) + \frac{P(I_{n-1})\, P(R_I)}{P(C_{n-1})}\right). \tag{2}$$

Type 3 Error

Define $P_{RB}(E_3, 1) = 0$ and $P_{RB}(E_3, 2) = P(C_1)\, P(R_C)\, P(I_R) + P(I_1)\, P(R_I)\, P(I_R)$. Then for $n \geq 3$:

$$P_{RB}(E_3, n) = P_{RB}(E_3, n-1) + \left[P_{RB}(E_3, n-1) - P_{RB}(E_3, n-2)\right] P(C_R) \cdot \left[P(C_{n-1})\, P(R_C) + P(I_n)\, P(R_I)\right]. \tag{3}$$

Type 4 Error

Define $P_{RB}(E_4, 1) = P(I_1)\, P(R_I)$. Then for $n \geq 2$:

$$P_{RB}(E_4, n) = P_{RB}(E_4, n-1)\, P(C_R) \cdot P(I_n)\left(P(R_I) + \frac{P(C_{n-1})\, P(R_C)}{P(I_{n-1})}\right). \tag{4}$$

System Failure

$$P_{RB}(E, n) = P_{RB}(E_1, n) + P_{RB}(E_2, n) + P_{RB}(E_3, n) + P_{RB}(E_4, n). \tag{5}$$

And of course, system reliability is:

$$R_{RB} = 1 - P_{RB}(E, n). \tag{6}$$

a) Closed Form Solutions for the Independent Model: Equations (1)-(4) can be simplified to facilitate the study of the effects of parameter changes on system reliability, such as increasing the number of versions, etc. First, assume that the probability of accepting an incorrect answer is the same as the probability of rejecting a correct answer; i.e., $P(A_I) = P(R_C) = P(A)$ and $P(A_C) = P(R_I) = 1 - P(A)$. Also, assume that the probability of correct state recovery between the versions' execution is 1, $P(C_R) = 1$ and $P(I_R) = 0$. Finally, assume that all versions have the same reliability, $P(I_i) = P(I)$ and $P(C_i) = P(C)$. Then, for $0 < P(A) < 1$ and $0 < P(I) < 1$, the equations for the four error types become

$$P_{RB}(E_2, n) = \left[P(A) - P(A)\, P(I)\right] \cdot \left[P(A) - 2P(A)P(I) + P(I)\right]^{(n-1)}, \tag{8}$$

$$P_{RB}(E_3, n) = 0, \tag{9}$$

and

$$P_{RB}(E_4, n) = \left[P(I) - P(A)\, P(I)\right] \cdot \left[P(A) - 2P(A)\, P(I) + P(I)\right]^{(n-1)}. \tag{10}$$

Consider the following theorem:

Theorem 1: If $0 < P(I) < 1$ and $0 < P(A) < 1$, then $0 < [P(A) - 2P(A)\, P(I) + P(I)] < 1$.

Proof: If $P(I) \in (0, 0.5)$, then $1 - 2P(I) > 0$ and

$$\frac{-P(I)}{1 - 2P(I)} < 0 < P(A) < 1 < \frac{1 - P(I)}{1 - 2P(I)}.$$

Multiplying by $1 - 2P(I)$ gives

$$-P(I) < P(A)(1 - 2P(I)) < 1 - P(I)$$

or

$$0 < P(A) - 2P(A)\, P(I) + P(I) < 1.$$

If $P(I) \in (0.5, 1)$, then $1 - 2P(I) < 0$ and

$$\frac{1 - P(I)}{1 - 2P(I)} < 0 < P(A) < 1 < \frac{-P(I)}{1 - 2P(I)}.$$

Multiplying by $1 - 2P(I)$ establishes the result.

Finally, if $P(I) = 0.5$, the result is immediate. This establishes the theorem.

Using Theorem 1 and taking limits as $n \to \infty$, (7)-(10) become:

$$\lim_{n\to\infty} P_{RB}(E_1, n) = \frac{P(A)\, P(I)}{1 - [P(A) - 2P(A)\, P(I) + P(I)]}, \tag{7'}$$

$$\lim_{n\to\infty} P_{RB}(E_2, n) = 0, \tag{8'}$$

$$\lim_{n\to\infty} P_{RB}(E_3, n) = 0, \tag{9'}$$

and

$$\lim_{n\to\infty} P_{RB}(E_4, n) = 0. \tag{10'}$$

It can be seen that, theoretically, increasing the number of versions n can drive the probability of a Type 2, Type

$$P_{RB}(E_1, n) = P(A)\, P(I) \sum_{j=1}^{n} [P(A) - 2P(A)\, P(I) + P(I)]^{(j-1)} = \frac{P(A)\, P(I)\left[1 - [P(A) - 2P(A)\, P(I) + P(I)]^n\right]}{1 - [P(A) - 2P(A)\, P(I) + P(I)]}, \tag{7}$$

3, and Type 4 error to zero. The probability of a Type 1 error cannot be eliminated by increasing the number of versions. Asymptotically, it approaches a value determined by the reliability of the component versions and the acceptance test.

2) Dependent Recovery Block Reliability Model: In the previous sections, it was assumed that since program versions were independently developed, they would also be statistically independent. This assumption allowed the probability of any joint event to be calculated by multiplying together the probabilities of the individual events that comprise it. In reality, the assumption of statistical independence may not hold [10]–[13]. In such cases, Fig. 1 can still be used to compute the probability of the four error types. For an $n = 3$ Recovery Block system, the probability of a Type 1 error is the sum of the following joint probabilities:

$$\begin{aligned} P_{RB}(E_1, 3) = {} & P(I_1A_I) + P(C_1R_CC_RI_2A_I) \\ & + P(I_1R_IC_RI_2A_I) \\ & + P(C_1R_CC_RC_2R_CC_RI_3A_I) \\ & + P(C_1R_CC_RI_2R_IC_RI_3A_I) \\ & + P(I_1R_IC_RC_2R_CC_RI_3A_I) \\ & + P(I_1R_IC_RI_2R_IC_RI_3A_I). \end{aligned} \tag{11}$$

Similarly, the probabilities of the other error types are as follows:

$$\begin{aligned} P_{RB}(E_2, 3) = {} & P(C_1R_CC_RC_2R_CC_RC_3R_C) \\ & + P(I_1R_IC_RC_2R_CC_RC_3R_C) \\ & + P(C_1R_CC_RI_2R_IC_RC_3R_C) \\ & + P(I_1R_IC_RI_2R_IC_RC_3R_C) \end{aligned} \tag{12}$$

$$\begin{aligned} P_{RB}(E_3, 3) = {} & P(C_1R_CI_R) + P(I_1R_II_R) \\ & + P(C_1R_CC_RC_2R_CI_R) + P(C_1R_CC_RI_2R_II_R) \\ & + P(I_1R_IC_RC_2R_CI_R) + P(I_1R_IC_RI_2R_II_R) \end{aligned} \tag{13}$$

$$\begin{aligned} P_{RB}(E_4, 3) = {} & P(C_1R_CC_RC_2R_CC_RI_3R_I) \\ & + P(I_1R_IC_RC_2R_CC_RI_3R_I) \\ & + P(C_1R_CC_RI_2R_IC_RI_3R_I) \\ & + P(I_1R_IC_RI_2R_IC_RI_3R_I). \end{aligned} \tag{14}$$

Error probabilities for Recovery Block systems with $n > 3$ can be found by extending Fig. 1 as suggested in Section II-A-1.

III. *N*-Version Programming

Another prevalent method of software fault-tolerance is the *N*-Version Programming method [4]. In this method, n programmers ($n > 1$) are asked to independently develop and debug a program, each working from a common specification. The n programs are placed under the control of a supervisor program that dispatches all versions of the program with the necessary input when the specified task is required. Upon completion, the outputs of the n programs are compared. Since all versions of the program were independently developed, it is assumed that the probability of a common software error is zero. This implies that the probability of agreement on an incorrect output is negligibly small. Therefore, if at least two program versions agree on an output, that output is assumed to be "correct."

A. N-Version Programming Reliability Model

There have been a number of reliability models proposed for *N*-Version Programming. Some depend on probabilities of correlated errors [6] while others assume version independence and rely on majority voting schemes [7]. Also, Soneriu's [8] model for redundant software could be adapted to *N*-Version Programming.

1) Independent N-Version Programming Reliability Model: In our model of *N*-Version Programming, there are three types of errors that can occur. A Type 1 error occurs when all outputs disagree. A Type 2 Error occurs when an incorrect output occurs more than once. Finally, a Type 3 Error occurs when there is an error in the voting procedure. The probability of a system error is then the sum of the three probabilities of a Type 1, 2, or 3 Error.

If it is assumed that the probabilities of Type 2 and Type 3 errors are zero, then the software system resembles a 2 of n redundant hardware system [14]. The system reliability is then the probability of at least two versions reaching similar, correct conclusions upon execution. For example, consider an *N*-Version Programming system composed of three independent modules. It is assumed that there exists one and only one correct result for a given set of inputs. Let $P(I_i)$ = probability of version i executing incorrectly, and $P(C_i) = 1 - P(I_i)$ = probability of version i executing correctly. Then the probability of a system error $P_{NVP}(E, n)$ is P(Type 1 Error). In this three-version programming system, the probability of at least two versions executing incorrectly becomes:

$$\begin{aligned} P_{NVP}(E, 3) & = P(I_1)\,P(I_2)\,P(I_3) + P(C_1)\,P(I_2)\,P(I_3) \\ & \quad + P(I_1)\,P(C_2)\,P(I_3) \\ & \quad + P(I_1)\,P(I_2)\,P(C_3) \\ & = P(I_1)\,P(I_2) + P(I_1)\,P(I_3) + P(I_2)\,P(I_3) \\ & \quad - 2P(I_1)\,P(I_2)\,P(I_3). \end{aligned} \tag{15}$$

Therefore, the reliability of a 2 of 3 *N*-Version Programming system is:

$$R_{NVP} = 1 - P_{NVP}(E, 3). \tag{16}$$

If each version has the same reliability $P(C)$, the reliability in a 2 of n voting system is [14]:

$$R_{NVP} = \sum_{j=2}^{n} \binom{n}{j} P(C)^j P(I)^{(n-j)} \tag{17}$$

where $P(I) = 1 - P(C)$.

2) Dependent N-Version Programming Reliability Model: As with a Recovery Block, a system reliability model for an N-Version Programming system can be found if the system is composed of dependent versions. Recall that the probability of an error is:

$$P_{NVP}(E, n) = P_{NVP}(E_1, n) + P_{NVP}(E_2, n) + P_{NVP}(E_3, n). \tag{18}$$

For a dependent N-Version Programming, composed of three dependent versions we have

$$P_{NVP}(E_1, 3) = P(I_1 I_2 I_3) + P(C_1 I_2 I_3) + P(I_1 C_2 I_3) + P(I_1 I_2 C_3). \tag{19}$$

If the versions are dependent, the probability of a Type 2 error, the multiple occurrence of an incorrect answer, cannot be assumed to be zero. We will continue to assume the probability of a Type 3 error is zero. Therefore, the probability of an error in a dependent N-Version Programming system is:

$$P_{NVP}(E, n) = P_{NVP}(E_1, n) + P_{NVP}(E_2, n) - P_{NVP}(E_1, n) \cap P_{NVP}(E_2, n). \tag{20}$$

IV. Consensus Recovery Block

The Consensus Recovery Block [10], [15] is a software fault-tolerant technique that combines aspects of Recovery Blocks and N-Version Programming. This new, hybrid technique attempts to overcome serious problems in the two established fault-tolerant techniques.

The acceptance test is the most crucial component of a Recovery Block, yet there are no guidelines for developing such a test. Once designed, there is no methodology for determining if the acceptance test is adequate for its intended purpose. Acceptance tests are also prone to design faults just as are the programs for which they are required to verify. Finally, no general guidelines exist for the placement and implementation of Recovery Blocks in a software system. The Consensus Recovery Block technique reduces the importance of the acceptance test in a fault-tolerant software system.

N-Version Programming may not be appropriate in applications in which multiple correct outputs are possible. Even the "same" output computed by different program versions may not agree "exactly" due to roundoff errors and other constraints imposed by the physical limitations of digital computers. In these situations, an N-Version Programming system may erroneously conclude that it could not produce a correct output even though multiple correct outputs may have occurred. System reliability is therefore degraded. The Consensus Recovery Block offers reduces the effects of this N-Version Programming limitation.

A. Description of the Consensus Recovery Block

The Consensus Recovery Block requires the development of n independent versions of a program, an acceptance test, and a voting procedure. The versions are ranked based on a set of criteria as in a conventional Recovery Block system. Upon invocation of the Consensus Recovery Block, all versions execute and submit their outputs to a voting procedure. Since it assumed that there are no common faults, if two or more of the versions agree on one output, that output is designated as correct. If there is no agreement, that is, the versions supply incorrect outputs or multiple correct outputs, then a modified Recovery Block is entered. The output of the "best" version is examined by the acceptance test. If that output is judged acceptable, it is treated as a correct output. If, on the other hand, the output is not accepted, the next best version's output is subjected to the acceptance test. This process continues until an acceptable output is found, or the n outputs are exhausted. Notice that there is no requirement for input state recovery since all versions execute in a "parallel" fashion as in an N-Version Programming system, not in a "serial" fashion as in a conventional Recovery Block.

B. Independent Consensus Recovery Block Reliability Model

The n program versions are executed, and the outputs are checked to see if at least two agree. If there is agreement, a Type 5 error occurs if the consensus output is incorrect. Under the assumption of independence, we will assume that the probability of a Type 5 error is zero. If there is no agreement, the outputs are sequentially subjected to the acceptance test in accordance with the Recovery Block scheme. The system is now susceptible to the four error types described in Section II-A-1 with the exception of a Type 3 error. Since no recovery is performed, $P(\text{Type 3 error}) = 0$. This implies that $P(C_R) = 1$. Fig. 2 shows a flowchart for a Consensus Recovery Block.

Using Fig. 2, it can be seen that the probability of a system error in a Consensus Recovery Block with n versions is:

$$\begin{aligned} P_{CRB}(E, n) &= P(G)\, P(G_I) + P(D)\, P_{RB}(E, n \mid P(C_R) = 1) \\ &= P_{CRB}(E_5, n) + P(D)\, P_{RB}(E, n \mid P(C_R) = 1) \end{aligned} \tag{21}$$

where

$$P(G) = P(2 \text{ or more outputs agree}).$$

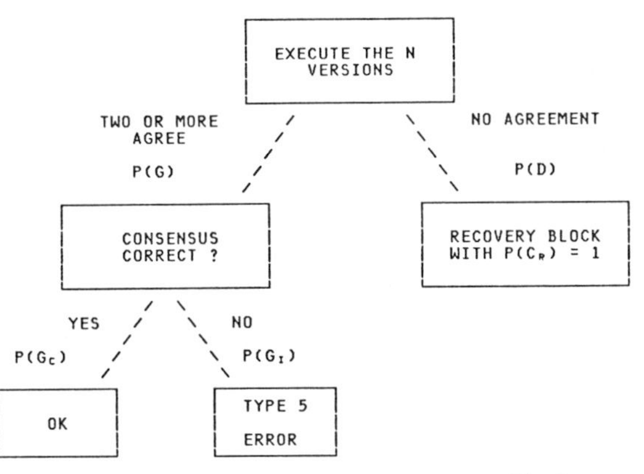

Fig. 2. Flowchart of a Consensus Recovery Block.

$P(G_I) = P(A$ recurring output is incorrect).

$P(G_C) = P(A$ recurring output is correct) $= 1 - P(G_I)$.

$P(D) = P$(All outputs are different).

$P_{RB}(E, n \mid P(C_R) = 1) = P$(Error in Recovery Block given perfect recovery).

Then, assuming $P(G_I) = 0$,

$$P_{CRB}(E, n) = P(D)\left[P_{RB}(E_1, n) + P_{RB}(E_2, n) + P_{RB}(E_4, n)\right] \quad (22)$$

where $P(E_1, n)$, $P(E_2, n)$, and $P(E_4, n)$ are defined by (1), (2), and (4), respectively, with $P(C_R) = 1$.

From the Law of Total Probability [9]:

$$P(D) = P(D \mid U)\, P(U) + P(D \mid N)\, P(N) \quad (23)$$

where event U denotes the existence of one and only one correct output, and N denotes its complement—multiple correct outputs, and $P(U) = 1 - P(N)$. It can be seen that $P(D)$ is the probability of failure of an N-Version Programming system. Therefore, the probability of a system error in a Consensus Recovery Block is:

$$P_{CRB}(E, n) = \left[P(D \mid U)\, P(U) + P(D \mid N)\, P(N)\right] * \left[P_{RB}(E_1, n) + P_{RB}(E_2, n) + P_{RB}(E_4, n)\right] \quad (24)$$

for $P(C_R) = 1$.

And as always, system reliability is

$$R_{CRB} = 1 - P_{CRB}(E, n). \quad (25)$$

Expressions for the probability of a Type 1, Type 2, and Type 4 error in a Consensus Recovery Block can be found by multiplying (1), (2), and (4), respectively, by $P(D)$.

1) Closed Form Solutions for the Independent Model: Since the equations for the probabilities of the errors in a Consensus Recovery Block are the error equations for a Recovery Block multiplied by a constant, $P(D)$, closed form solutions can be found. As in Section II-A-1-a, assume that the probability of accepting an incorrect answer is the same as the probability of rejecting a correct answer; i.e., $P(A_I) = P(R_C) = P(A)$ and $P(A_C) = P(R_I) = 1 - P(A)$. Also, assume that the probability of correct state recovery between the versions' execution is 1, $P(C_R) = 1$ and $P(I_R) = 0$. Finally, assume that all versions have the same reliability, $P(I_i) = P(I)$ and $P(C_i) = P(C)$. Then, for $0 < P(A) < 1$ and $0 < P(I) < 1$:

$$P_{CRB}(E_1, n) = P(D)\frac{P(A)\,P(I)\left[1 - \left[P(A) - 2P(A)\,P(I) + P(I)\right]^n\right]}{1 - \left[P(A) - 2P(A)\,P(I) + P(I)\right]}, \quad (26)$$

$$P_{CRB}(E_2, n) = P(D)\left[P(A) - P(A)\,P(I)\right]\left[P(A) - 2P(A)\,P(I) + P(I)\right]^{(n-1)}, \quad (27)$$

$$P_{CRB}(E_3, n) = 0, \quad (28)$$

and

$$P_{CRB}(E_4, n) = P(D)\left[P(I) - P(A)\,P(I)\right]\left[P(A) - 2P(A)\,P(I) + P(I)\right]^{(n-1)}. \quad (29)$$

Using Theorem 1, (26)–(29) become:

$$\lim_{n \to \infty} P_{CRB}(E_1, n) = \frac{P(D)\,P(A)\,P(I)}{1 - \left[P(A) - 2P(A)\,P(I) + P(I)\right]}, \quad (26')$$

$$\lim_{n \to \infty} P_{CRB}(E_2, n) = 0, \quad (27')$$

$$\lim_{n \to \infty} P_{CRB}(E_3, n) = 0, \quad (28')$$

and

$$\lim_{n \to \infty} P_{CRB}(E_4, n) = 0. \quad (29')$$

a) Consensus Recovery Block Theorems: Using the independent reliability model presented here, three theorems describing the behavior of a Consensus Recovery Block can be established.

Theorem 2: If $0 < P(D) < 1$, then a Consensus Recovery Block system is more reliable than a Recovery Block system with the same components.

Proof: From the definition of system reliability, it suffices to show that:

$$P_{CRB}(E, n) < P_{RB}(E, n).$$

However, (21) states that the probability of an error in a Consensus Recovery Block is the product of $P(D)$ and the probability of an error in a Recovery Block. By assumption $P(D) < 1$, hence the result follows immediately. (Notice that the above probabilities of error are equal only in the unlikely event that $P(D) = 0$ or 1.)

Theorem 3: If the probability of a Recovery Block error is less than 1, then a Consensus Recovery Block system is more reliable than a *N*-Version Programming system composed of the same components; i.e.

$$P_{CRB}(E, n) < P_{NVP}(E, n).$$

Proof: Again from (21):

$$P_{CRB}(E, n) = P(D)\, P_{RB}(E, n).$$

As stated in Section IV-B, $P(D)$ is the probability of error in an *N*-Version Programming system, therefore

$$P_{CRB}(E, n) = P_{NVP}(E, n)\, P_{RB}(E, n).$$

Since $P_{RB}(E, n) < 1$, the result follows.

Theorem 4: If $P(D) < 1$ and state recovery is certain, then the reliability of a Consensus Recovery Block system is less sensitive to acceptance test errors than a Recovery Block system composed of the same components. That is

$$\frac{\partial P_{CRB}(E, n)}{\partial P(A)} < \frac{\partial P_{RB}(E, n)}{\partial P(A)}.$$

Proof: Using (7)–(10) and (22), and taking the partial derivatives with respect to $P(A)$, yields

$$\frac{\partial P_{CRB}(E, n)}{\partial P(A)} = P(D)\, \frac{\partial P_{RB}(E, n)}{\partial P(A)}.$$

and the result follows.

2) Dependent Consensus Recovery Block Reliability Model: Equations for the probabilities of errors in a Consensus Recovery Block can be found for systems composed of dependent versions. Using Fig. 2, it can be seen that the probability of a Type 1 error in a Consensus Recovery Block is the joint probability of event D and a Type 1 error in a conventional Recovery Block, i.e.

$$P_{CRB}(E_1, n) = P(DRB(E_1, n))$$

then, for $n = 3$:

$$\begin{aligned} P_{CRB}(E_1, 3) = {} & P(DI_1A_I) + P(DC_1R_CC_RI_2A_I) \\ & + P(DI_1R_IC_RI_2A_I) \\ & + P(DC_1R_CC_RC_2R_CC_RI_3A_I) \\ & + P(DC_1R_CC_RI_2R_IC_RI_3A_I) \\ & + P(DI_1R_IC_RC_2R_CC_RI_3A_I) \\ & + P(DI_1R_IC_RI_2R_IC_RI_3A_I). \end{aligned} \quad (30)$$

Similarly, the probabilities of the other error types are as follows:

$$\begin{aligned} P_{CRB}(E_2, 3) = {} & P(DC_1R_CC_RC_2R_CC_RC_3R_C) \\ & + P(DI_1R_IC_RC_2R_CC_RC_3R_C) \\ & + P(DC_1R_CC_RI_2R_IC_RC_3R_C) \\ & + P(DI_1R_IC_RI_2R_IC_RC_3R_C) \end{aligned} \quad (31)$$

$$\begin{aligned} P_{CRB}(E_3, 3) = {} & P(DC_1R_CI_R) + P(DI_1R_II_R) \\ & + P(DC_1R_CC_RC_2R_CI_R) \\ & + P(DC_1R_CC_RI_2R_II_R) \\ & + P(DI_1R_IC_RC_2R_CI_R)\, P(DI_1R_IC_RI_2R_II_R) \end{aligned} \quad (32)$$

and

$$\begin{aligned} P_{CRB}(E_4, 3) = {} & P(DC_1R_CC_RC_2R_CC_RI_3R_I) \\ & + P(DI_1R_IC_RC_2R_CC_RI_3R_I) \\ & + P(DC_1R_CC_RI_2R_IC_RI_3R_I) \\ & + P(DI_1R_IC_RI_2R_IC_RI_3R_I). \end{aligned} \quad (33)$$

For a Consensus Recovery Block composed of dependent components, it is not possible to assume that the probability of a Type 5 error, agreement on an incorrect output, is zero. Therefore,

$$\begin{aligned} P_{CRB}(E, n) = {} & P_{CRB}(E_1, n) + P_{CRB}(E_2, n) \\ & + P_{CRB}(E_4, n) + P_{CRB}(E_5, n) \end{aligned} \quad (34)$$

V. Experimental Validation of the Models

An experiment was conducted at North Carolina State University to validate the six reliability models presented in Sections II, III, and IV [10], [11], [13].

A. Description of the Experiment

Sixty-five members of an undergraduate computer science class were given an assignment to write a Pascal program that would manage a hypothetical national package shipment system. (See [10] for a complete description of the problem.) The students were to read an input file that defined a shipment network composed of nodes interconnected by transportation links. The input file supplied a number for each node, its geographical coordinates, and the numbers of the other nodes to which it was connected. A request file was to be interrogated to determine package shipment requirements. The students were asked to route a package through the network from the given originating node to the desired destination node, using interconnecting nodes and links as needed. Each node had a specified delay attributed to time needed to receive, handle, and if necessary, ship a package. It was required that a package reach its destination within ten days after a shipment request was made. To make the problem more difficult, an event file had to be monitored that changed the delay time of the nodes, disconnected or re-established links, and crashed or re-established nodes. The students were required to maintain a simulation clock that started at 0 days at the beginning of the program and was to be updated in discrete steps as events and requests were encountered relative to the clock.

For output, the students were to supply the simulation time at which requests occurred, and the nodes through which the package traveled while proceeding to its destination. The time at which a package left a node was also listed.

The 65 programs were collected and tested with some sample input files. It was determined that 16 of the programs could react to the sample inputs without experiencing any type of fatal execution error. A network of 50 nodes and appropriate interconnecting links was created. The network was initialized so that 4 percent of the nodes and links, selected at random, would be down. The remaining nodes were assigned randomly selected delay times. An event file was built by randomly selecting one of five possible events and applying this event to a randomly selected node or link as necessary. The interevent time was also randomly selected. The request file was built of randomly selected origin and destination nodes, with the interrequest times being fixed to avoid overlapping requests. The requests were generated until a set of 100 requests that could be successfully handled by the "best" program was found. This set of 100 inputs constituted our input test set.

1) Reliabilities of the Individual Programs: In order to use the reliability models, the reliabilities of the individual programs that comprised the fault-tolerant systems had to be estimated. One pool of 50 test cases was used to assess the reliability of the individual programs, and the remaining 50 were used to test the behavior of the fault-tolerant system of interest. The test cases were randomly assigned to each of the two pools so that each pool got the same (as nearly as possible) number of test cases of roughly equal difficulty. This approach to test selection solved our problem of biased estimation. See Wiggs [13] for a detailed analysis of test case selection based on input difficulty.

Each program was supplied with the network definition file, the event file, and a request file composed of the 50 test cases. The number of failures out of 50 test cases was used to estimate the program reliability. Program reliability ranged from 1.00 to 0.16 with an average of 0.68 and a median of 0.74.

2) Limitations of the Experiment: This experiment, even though it yielded many useful observations, suffered from a number of deficiencies. First, the solutions to any given input were not unique. The programs often found multiple correct ways of routing a package through the network. This fact biased our results in favor of the Recovery Block and the Consensus Recovery Block. Also, the programs did not always reach a determinant state of correctness or incorrectness. An arbitrarily limit was imposed on the amount of time used to deliver a package to its destination. This meant that some programs could have been pursuing valid, albeit inefficient, algorithms for package routing, only to have the simulation clock expire before the package arrived at the designated node. By definition, such outputs were labeled "incorrect."

3) Acceptance Test: In addition to program versions, Recovery Block and Consensus Recovery Block systems, require an acceptance test. For this problem, the acceptance test was a defined as a program that checked a program's outputs to see if:

1) The package routing began and ended at the proper nodes.
2) Routing was completed in the required time.
3) Nodes used for transportation were physically connected by links.
4) The correct delay time was used in the elapsed time calculations.
5) Transmission occurred at the proper time relative to the simulation clock.
6) The end-to-end transmission time was calculated correctly.
7) The nodes used by the program were "up" at the time required by the program.
8) The links used by the program were "up" at the time required by the program.

In all likelihood, such a simplistic acceptance test could be programmed without error. The acceptance test was therefore programmed to randomly make the "wrong" decisions at a specified frequency.

4) Model Verification: In order to study the effectiveness of the Recovery Block models, the 16 programs were randomly grouped into 50 unique combinations of 3 programs. The programs in each triplet were ordered by decreasing reliability, and each triplet was regarded as a three-version Recovery Block. The expected reliability for each Recovery Block was calculated using various values of acceptance test reliability. Each triplet and acceptance test was stimulated with the second set of 50 input test cases. The actual system reliability was noted, and the predicted system reliability was compared to the actual reliability.

As in the Recovery Block study, the 50 triplets in our experiment were treated as *N*-Version Programming systems. The reliabilities estimated from the first pool of 50 test cases were used to predict the reliability of the fault-tolerant systems. The second pool of 50 test cases was used to estimate the actual reliability of the *N*-Version Programming systems. The two reliabilities were compared to see how successfully the models predicted system reliability.

As before, our 50 triplets were treated as Consensus Recovery Block systems. The reliabilities estimated from the first 50 test cases were used to predict the reliability of the fault-tolerant systems. The last 50 test cases were used to measure the actual reliability of the Consensus Recovery Block systems. The two reliabilities were compared to see how successful the modeling had been.

If the null hypothesis is that a given model could accurately predict system reliability, the mean of the 50 differences between the actual reliabilities and the predicted reliabilities should be zero. A student t-test was performed to test this null hypothesis. Complete results of the experiment for various confidence levels are presented in [10] and [11].

B. Conclusions from the Experiment

The Recovery Block part of this experiment rendered three important observations. First, even through the 16 programs were developed independently, there was still some correlation among the probabilities of error which caused the independent models to consistently over estimate reliability. This was not due to algorithmic similarities; the problem was designed so that many diverse algorithms could be conceived with which to solve it. The dependence was related to a "difficulty factor." If one program found a particular test case difficult and gave a wrong answer, then the probability was nonzero that other programs also had trouble finding a correct answer, even though the programs used different algorithms. The difficulty factor in software reliability estimation is discussed in detail in Wiggs [13]. Second, the experiment showed that the proposed dependent Recovery Block reliability model could indeed be used to predict system reliability if the necessary joint probabilities can be found. In our experiment, the null hypothesis accepted with a 99 percent confidence level. Finally, it was demonstrated that the Recovery Block method of fault-tolerance can be used to improve the reliability of a software system. The amount of improvement depends on the reliability of the acceptance test, but even acceptance tests of poor quality (0.75) can provide reliability improvements.

The first observation of the *N*-Version Programming portion of the experiment was that occurrences of multiple incorrect outputs could be predicted accurately, but their occurrence did not affect the success of reliability prediction. Second, none of the four models accurately predicted the system reliability because of the possibility of multiple correct answers. The definition of "success" in an *N*-Version Programming system is the agreement on a (assumably) correct output. The programs provided a number of different correct solutions to each input test case, but an *N*-Version Programming system does not recognize an output as correct unless it occurs more than once. This limitation of *N*-Version Programming also explains the final observation. An *N*-Version Programming system used in situations that can have multiple correct outputs may not provide fault-tolerance. In fact, as in the case of our experiment, the reliability of an *N*-Version Programming system may be *less* than that of the programs that comprise it.

The following conclusions were drawn for the Consensus Recovery Block systems. First, the independent models did not accurately predict the system reliability, but the dependent models did so to a 95 percent confidence level. The reason for this is the inherent dependence created by the difficulty of the input test cases as explained earlier. Also, it was demonstrated that, at least for reasonably good acceptance tests, the Consensus Recovery Block system does offer a measure of fault-tolerance. However, in systems with poor acceptance tests, the reliability of the Consensus Recovery Block system may in fact be less than that of the individual programs that comprise it.

The last question investigated was whether or not a Consensus Recovery Block system is more reliable than a Recovery Block system or an *N*-Version Programming system composed of the same programs. It was found that Consensus Recovery Block systems were more reliable than corresponding *N*-Version Programming systems. The Consensus Recovery Block systems were more reliable than corresponding Recovery Block systems only when a highly reliable acceptance test was used. This is because in a Consensus Recovery Block system, reliability is degraded by both acceptance test errors and the occurrence of multiple correct outputs. A Recovery Block is only affected by the acceptance test reliability.

VI. Using the Models

In this section, use of the models is demonstrated with two examples. First, we develop a simple model to compare the relative costs of the systems described in this paper. *N*-Version Programming is becoming a popular fault-tolerant tool. Our claim is that adding an acceptance test to the process, thus forming a Consensus Recovery Block system, is usually worth the additional expense since an acceptance test is normally much cheaper to develop than the component versions and provides a significant improvement in reliability.

In the second example, the independent models are used to show the effects of acceptance test reliability on system reliability for the three methods of software fault-tolerance. This example shows the point at which a Recovery Block or Consensus Recovery Block system becomes more reliable than a corresponding *N*-Version Programming system.

In both examples, we will restrict our comparisons to three-version systems.

A. Comparison to Improving a Single Version

Suppose, for example, that a program has been shown to have a reliability of 0.98, and its specification requires a reliability of at least 0.99. We wish to determine the most cost effective way to achieve the required reliability increase.

In a three version Recovery Block system, if the original program constitutes one of the versions, then (1)–(6) show that two additional programs of reliability 0.80 and 0.70 combined with an acceptance test program with a reliability of 0.93 would give a system reliability of 0.99. (Note: This triplet of program reliabilities is just one of many that could be used to achieve the required system reliability.) Similarly, (15) and (16) can be used to show that for an *N*-Version Programming system composed of three versions, programs of reliability 0.98, 0.93, and 0.91 are needed to achieve a system reliability of 0.99. Finally, (22) and (25) require two additional versions both with reliabilities of 0.60 and an acceptance test with re-

ACTION	VERSION RELIABILITY			ACCEP. TEST RELI.	SYSTEM RELI.	COSTS (B)			
	$P(C_1)$	$P(C_2)$	$P(C_3)$			0.50	0.60	0.70	0.80
IMPROVE	0.98				0.9800	2.9289	5.3925	9.6564	16.9455
NVP	0.98	0.93	0.91		0.9908	7.1130	9.1719	11.8287	15.2575
RB	0.98	0.80	0.70	0.93	0.9900	5.9516	7.1514	8.6246	10.4405
CRB	0.98	0.60	0.60	0.83	0.9905	4.3750	4.9135	5.5267	6.2263

Fig. 3. Costs of achieving a 0.99 reliable system for $A = 0.5$.

liability 0.83 to create a three version Consensus Recovery Block with a reliability of 0.99.

In order to assess the cost of the additional programming efforts needed to achieve the reliability goal, a very simple cost model will be used. Assume that the cost of producing a given program is a function of the desired reliability. That is:

$$C(R) = \frac{A}{(1 - R)^B} \quad \text{for } 0 \leq R < 1 \qquad (35)$$

where C is the cost, R is the required reliability, B is a constant that incorporates the costs of producing a program such as personnel expenses, computer resource expenses, etc., and A is a constant related to the complexity of the program. (For the purposes of this example, the units of A will be left undefined.) Assume that for this hypothetical programming environment, $B = 0.5$, and all versions of the program have the same complexity $A = 1$. Then, the cost of improving the single program's reliability to 0.99 is:

$$C(R) = \frac{1}{(1 - 0.99)^{0.5}} - \frac{1}{(1 - 0.98)^{0.5}} = 2.9289$$

and the cost of writing two additional programs for an N-Version Programming is:[1]

$$C(R) = \frac{1}{(1 - 0.93)^{0.5}} - \frac{1}{(1 - 0.91)^{0.5}} = 7.1130.$$

If we assume that the acceptance test is one-half as complex as a program version, then the cost of producing a Recovery Block system is:

$$C(R) = \frac{1}{(1 - 0.80)^{0.5}} - \frac{1}{(1 - 0.70)^{0.5}} - \frac{0.5}{(1 - 0.93)^{0.5}} = 5.9516.$$

[1]It has been assumed that the cost of producing N versions of a program is N times the cost of producing 1 version. Clearly, this is a worst case assumption. Some resources, such as testing and debug efforts, system requirements, etc., can surely be shared. Therefore, the cost of producing N versions is $\leq N$ times the cost of producing just one.

For a Consensus Recovery Block the cost is:

$$C(R) = \frac{1}{(1 - 0.60)^{0.5}} - \frac{1}{(1 - 0.60)^{0.5}} - \frac{0.5}{(1 - 0.83)^{0.5}} = 4.3750.$$

Fig. 3 summarizes the results for $B = 0.5$, 0.6, 0.7, and 0.8. In this hypothetical situation, if programming costs are low, $B = 0.5$, it is more cost effective to continue the test and debugging process on one version and improve its reliability to the required level than to implement any form of fault-tolerance. As programming costs increase, $B = 0.6$, it would be less expensive to create a Consensus Recovery Block than to continue work on the single version. However, improving the single version is less expensive than developing either a Recovery Block or N-Version Programming system. In an expensive programming environment, $B = 0.7$, it would be better to form either a Recovery Block or Consensus Recovery Block. Finally, at $B = 0.8$, any of the three forms of fault-tolerance is cheaper than improving the original program. Notice that in all cases, the Consensus Recovery Block is cheaper than the Recovery Block which in turn is cheaper than N-Version Programming.

B. Relative Comparisons

The example in Fig. 4 compares the system reliability of Recovery Block, N-Version Programming, and Consensus Recovery Block systems composed of the same versions. It demonstrates that by expending the resources to produce an acceptance test, significant reliability increases can be realized over an N-Version Programming system.

In this example, (6), (16), and (25) are used to compute system reliabilities for fault-tolerant systems composed of three independent program versions, each with a reliability of 0.90. It is assumed that there exists one and only one correct output for each set of inputs. The reliability of the acceptance test is varied from 0.60 to 0.99. Notice that the reliability of the Recovery Block increases as a function of the acceptance test reliability and surpasses the reliability of the N-Version Programming system at about 0.80. Also note that even for poor acceptance tests,

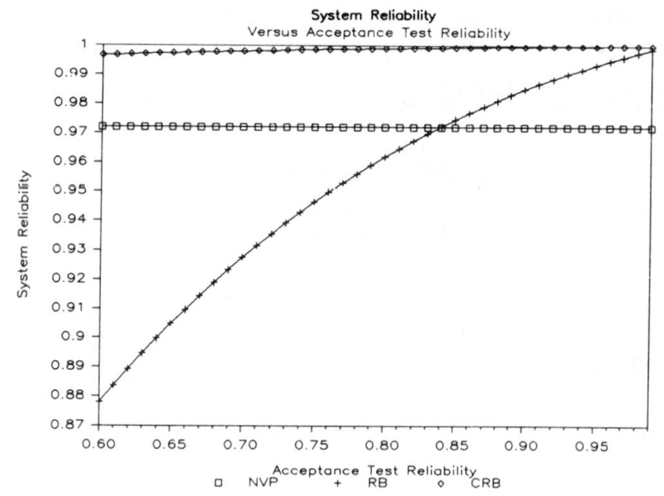

Fig. 4. System reliability comparison.

the reliability of the Consensus Recovery Block is much greater than that of the Recovery Block or N-Version Programming system.

VII. Summary and Conclusions

It is not possible to *guarantee* the reliability of a program by validation and testing. If extremely reliable programs are required, some form of software fault-tolerance must be used. This paper examines three methods of achieving fault-tolerance and proposes reliability models for each. Since it is common practice to assume statistical independence among program versions in fault-tolerant systems, independent reliability models are presented. However, experiments have shown that some dependence does exist [10]–[13], and models assuming statistical dependence are also presented.

The proposed models are used to demonstrate the superiority of one of the methods, the Consensus Recovery Block, over the other two. The models prove that the Consensus Recovery Block is less susceptible to acceptance test errors than the Recovery Block and is, in general, more reliable than a Recovery Block composed of the same programs. It also is shown that the Consensus Recovery Block is more reliable than an N-Version Programming system composed of the same programs. This particularly true when the occurrence of multiple correct outputs is a possibility. A simple cost model is used to demonstrate the use of the models for relative cost comparisons in achieving a given reliability.

References

[1] H. Hecht, "Fault-tolerant software for real-time applications," *Comput. Surveys*, vol. 8, no. 4, pp. 391–407, Dec. 1976.

[2] R. E. Keirstead, "On the feasibility of software certification," Stanford Res. Inst., Menlo Park, CA, SRI Project 2383, 1975.

[3] B. Randell, "System structure for software fault tolerance," *IEEE Trans. Software Eng.*, vol. SE-1, pp. 220–231, June 1975.

[4] T. Anderson and P. A. Lee, *Fault Tolerance*. Englewood Cliffs, NJ: Prentice-Hall International, 1981.

[5] P. A. Lee, "A reconsideration of the recovery block scheme," Comput. Lab., Univ. Newcastle upon Tyne, Tech. Rep. 119, 1978.

[6] A Grnarov, J. Arlat, and A. Avizienis, "On the performance of fault-tolerance strategies," in *Proc. FTCS-10, Tenth Annu. Int. Symp. Fault-Tolerant Comput.*, 1980, pp. 251–253.

[7] A. Yu-Wu Wei, "Real-time programming with fault-tolerance," Ph.D. dissertation, Univ. Illinois, Urbana-Champaign, 1981.

[8] M. Doru Soneriu," A methodology for the design and analysis of fault-tolerant operating systems," Ph.D. dissertation, Illinois Inst. Technol., Chicago, 1981.

[9] K. S. Trivedi, *Probability Statistics with Reliability, Queuing, and Computer Science Applications*. Englewood Cliffs, NJ: Prentice-Hall, 1982.

[10] R. K. Scott, "Data domain modeling of fault-tolerant software reliability," Ph.D. dissertation, North Carolina State Univ., Raleigh, NC, 1983.

[11] R. K. Scott, J. W. Gault, D. F. McAllister, and J. Wiggs, "Experimental verification of six fault-tolerant software reliability models," in *Proc. FTCS-14, Fourteenth Annu. Int. Symp. Fault-Tolerant Comput.*, 1984.

[12] —, "Investigation of version dependence in fault-tolerant software," in *Proc. Avionics Panel Spring 1984 Meeting Design for Tactical Avionics Maintainability*, 1984.

[13] J. Wiggs, "Version dependency in fault-tolerant software," Master's thesis, North Carolina State Univ., Raleigh, NC, 1984.

[14] B. S. Dhillon and C. Singh, *Engineering Reliability*. New York: Wiley, 1981.

[15] R. K. Scott, J. W. Gault, and D. F. McAllister, "The consensus recovery block," in *Proc. Total Systems Reliability Symp.*, 1983, pp. 74–85.

R. Keith Scott received the B.A. degree in physics from the University of North Carolina in 1975 and the M.S.E. degree in electrical engineering from North Carolina A&T State University in 1977.

In 1977, he joined International Business Machines at Research Triangle Park, NC. Taking advantage of IBM's Graduate Work Study program, he received the Ph.D. degree in electrical and computer engineering from North Carolina State University in 1983. He is currently Manager of the Department of Controller Microcode Development for IBM, Research Triangle Park.

James W. Gault (S'70–M'70) received the B.S. degree from Colorado State University Fort Collins, in 1962 and the M.S. and Ph.D. degrees from the University of Iowa, Iowa City, in 1969.

He worked for Collins Radio Company from 1962 to 1967 and taught in the Department of Electrical Engineering of the North Carolina State University from 1969 to 1982. He is presently the Director of the U.S. Army European Research Office in London, England.

David McAllister (M'84) is Professor and Graduate Administrator in the Department of Computer Studies at North Carolina State University. His areas of interest include software reliability and computer graphics.

Dr. McAllister is a member of the Association for Computing Machinery.

Operational Reliability of the DX 200 Switching System

Tapani Purho; Nokia Telecommunications; Helsinki

Zekrollah Aflatuni; Nokia Telecommunications; Helsinki

Jouni Soitinaho; Nokia Telecommunications; Helsinki

Keywords: Availability Performance, Reliability Performance, Maintainability, Field Data, Reliability Improvement

Abstract

Availability performance of telecommunication services is one of the most important factors when the quality of service of a telecommunication network is assessed. This paper summarizes the operational objectives and field results of availability performance of the DX 200 Digital Switching System widely used in the telephone network of the Finnish PTT and private telephone companies.

Analysis of the field results from the point of view of quality and reliability improvement is also presented.

Introduction

The DX 200 is a switching system for a wide range of applications, from the very small local exchanges to network and transit exchanges. The number of subscribers may vary from 60 to 40 000, and the number of trunks, from 30 to 7000.

The DX 200 switching system is developed and manufactured by Nokia Telecommunications. Nokia Company is the largest private company in Finland employing over 30 000 people. Electronics is the most important branch of the company.

The principal structure of the DX 200 exchange is presented in figure 1. The subscribers are connected to the subscriber modules and the telephone call is transmitted through two switches, the subscriber stage switch and group stage switch, to exchange terminals and from there to the transmission network.

The distributed control of the DX 200 system is based on the microcomputers linked together via messagebus.

The operation and maintenance personnel has access to the exchange through video display units connected to the operation and maintenance computer. Operation and maintenance procedures can also be performed from the centralized operation and maintenance center, where from all the exchanges of a certain geographical area can be supervised.

From the reliability point of view all the essential units having effect on more than 64 subscribers or on one PCM transmission line have been provided with redundancy. Three redundancy principles have been applied: hot stand by, load sharing and (n + 1) redundancy.

The mechanical construction of an exchange is based on racks, subracks and plug-in units. One thousand subscribers can be connected to one rack. One subscriber line card has connections for sixteen subscribers. A typical processor unit is of the size of one subrack comprising about twenty plug-in units.

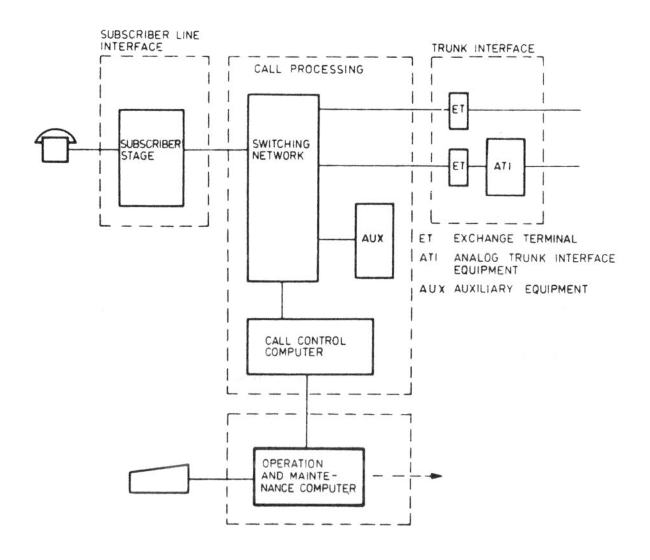

Figure 1. Basic structure of the DX 200 system.

Availability concept of the DX 200 system

For reliability management of the DX 200 system the availability concept of CCITT G.106 (Ref. 1) has been adopted. The availability performance factors of the DX 200 system are presented in figure 2.

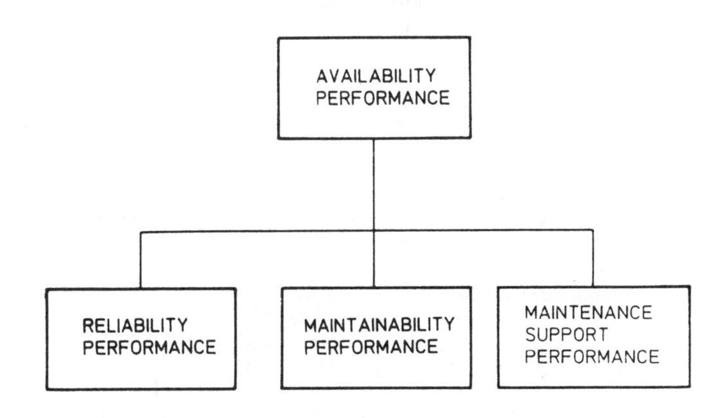

Figure 2. Dependability concepts.

The following measures for each availability performance factor are calculated during the operation of the exchanges:

1. Availability performance:

- unavailability of exchange
- unavailability of system
- unavailability of average subscriber
- unavailability of transmission line

2. Reliability performance:

- failure rate of exchange
- failure rate of system

3. Maintainability performance:

- active repair time
- accuracy of automatic fault localization

4. Maintenance support performance:

- administrative + logistic delay
- probability of spare part shortage

The objective values and calculated results for most of the measures are presented in this paper.

The analysis of these measures also renders very useful data for other system management purposes of the company. Such data is also presented here.

The availability concept provides the starting point for all the availability related activities during the life cycle of the system. The activities and the reliability management have been presented in earlier papers (Refs. 2-3). Especially the failure data collecting procedure (Ref. 4) was developed to meet the requirement of calculating the actual values for the availability performance related measures when the exchanges are in operation.

Definitions and assumptions

1. Relevant failures

Any software or hardware failure except for the failures belonging to one of the following classes is classified as relevant.

- failures due to operation and maintenance personnel's unauthorised actions or neglect
- failures due to events and actions during transport and installation outside contractor's control and responsibility
- failures due to physical, thermal, chemical or electrical environment outside the environmental specification limits
- failures due to other failures not being repaired within 4 hours (applied to redundancy situations or failure combinations)
- secondary failures
- intermittent failures (one relevant failure is counted when the failure is found)
- failures due to changes in the maintenance and redundancy structure or other reasons than repair of a relevant failure
- failures due to the fact that notified amendments have not been introduced into the exchange
- failures due to work and equipment not covered by the evaluation.

2. Time concept

The time period, starting from the system testing is divided into the periods shown in figure 3.

Data presented here concerns the total time period for software analysis and the operation period for hardware analysis.

3. Availability performance

In calculating exchange availability it is assumed that exchange unavailability occurs under one the following conditions:

Terminal exchange:	More than 50 % of the subscriber lines are out of service
Transit exchange:	More than 50 % of the trunks of the exchange are out of service
Tandem exchange:	More than 50 % of the equivalent subscriber lines are out of service.

4. Operational availability

Operational availability A is defined as:

A = Accumulated up-time/accumulated up-time + accumulated down-time)

Mean operational availability for a DX 200 exchange is calculated considering a system comprising of N exchanges in operation for a time period of one year.

$$DTs = \sum_{i=1}^{n} DTi \qquad (2)$$

$$DTi = \frac{dTi \times ni}{N} \qquad (3)$$

where

DTs = avarage system down-time/year (hours)

DTi = exchange i weighted down-time/year

dTi = exchange i down-time/year assuming that the exchange has been operating for at least one year

ni = exchange i equivalent subscriber lines

N = $\sum n_i$ = total number of equivalent subscriber lines under observation

Mean operational availability is therefore calculated as

$$A = 1 - \frac{DTs}{T}$$

T = observation time (one year)

Availability performance of the DX 200 switching system

For calculating the values of availability performance related measures, the field data and configuration management system was developed for the DX 200 exchanges. In this system, the relevant data concerning all the exchanges in operation is kept and updated within the company. From reliability point of view the field data system is a tool for estimating and verifying the availability performance of the DX 200 exchanges (Refs. 5-6).

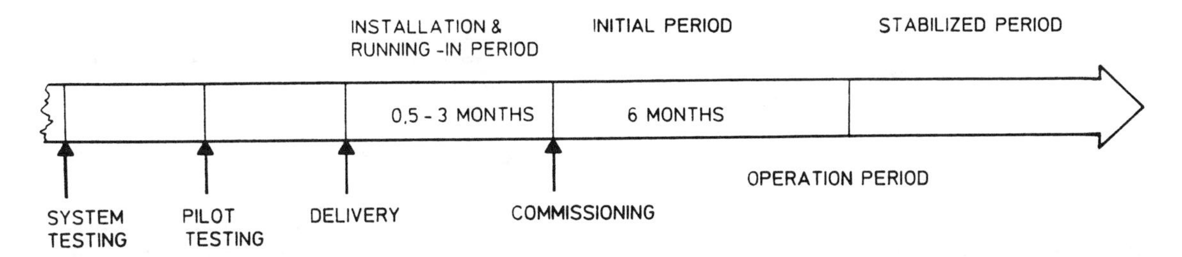

Figure 3. Periods of time starting from the system testing.

The operational results presented here are based on field data collected from over 200 exchanges during the period of 1983 - 1986. Most of the exchanges are unmanned and without airconditioning.

1. Availability performance measures

Mean operational and intrinsic availability for the DX 200 system during the years 1984 to 1986 is shown in figure 4.

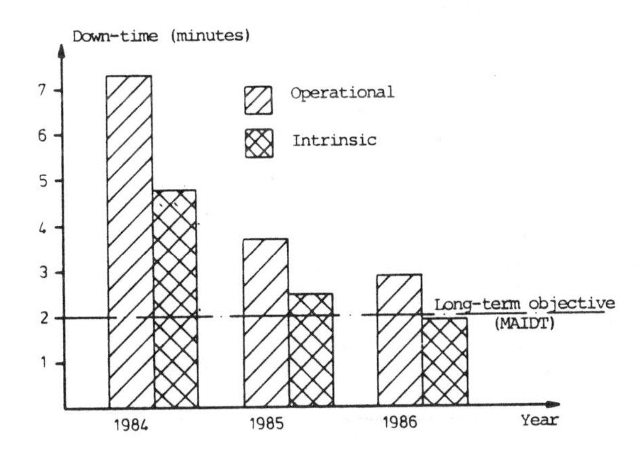

Figure 4. Mean operational and intrinsic down-time.

It is shown that the long term mean accumulated intrinsic down-time (MAIDT) objective assessed for the system was achieved in 1986.

Average down-time for a single subscriber (considering all the failures during the year 1985) is 29 min/year.

2. Reliability performance measures

The reliability performance measure of the DX 200 system is expressed as the number of faults/100 subscriber lines per year. Figure 5 illustrates the trend of hardware faults (total and relevant faults for the period 1984 to 1986.)

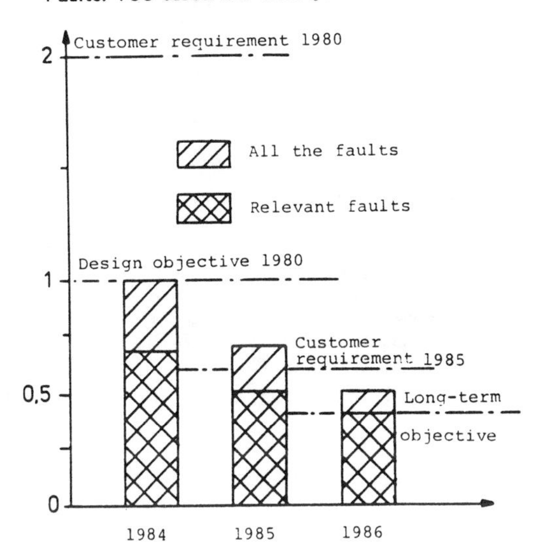

Figure 5. Hardware failure rate.

For reliability performance in 1980, a design objective was set in the reliability and maintainability program of the DX 200 switching system. The objective is based on the international requirements of CCITT and national requirements of customers.

Because the objective assessed in 1980 was reached in 1984, a new objective was set in 1985. At the same time the customer requirement was tightened up due to good experiences from the first exchanges in operation.

Distribution of the hardware failures is further demonstrated in figure 6 for each year using the following failure causes.

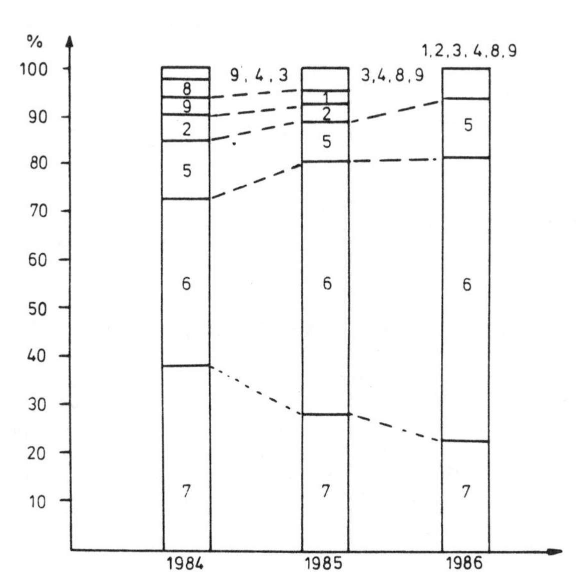

Figure 6. Distribution of hardware failures

1. Design
2. Manufacturing
3. Installation
4. Operator
5. External
6. Component
7. Unknown
8. Secondary faults
9. Organizational

Referring to figure 6 it can be noticed that as the number of failures due to "design", "manufacturing" and "unknown" reasons is decreasing, the percentage of component failures is increasing relatively, which is a good trend towards an ideal situation where 100 % of all the failures are component failures.

"Unknown" failures still constitute a noticeable part of the total failures. Software originated faults, insufficient fault localization capabilities, operator's errors and intermittent failures of some components can be mentioned as some of the reasons for the malfunction.

3. Maintainability performance

Maintainability performance is characterized by the active repair time of hardware failure.

Lognormal plot of active repair time data during the period of 1984 to 1986 is shown in figure 7. From the plot, the median value for active repair time is found to be 25 minutes.

Customer requirement for the median time to repair is 30 minutes.

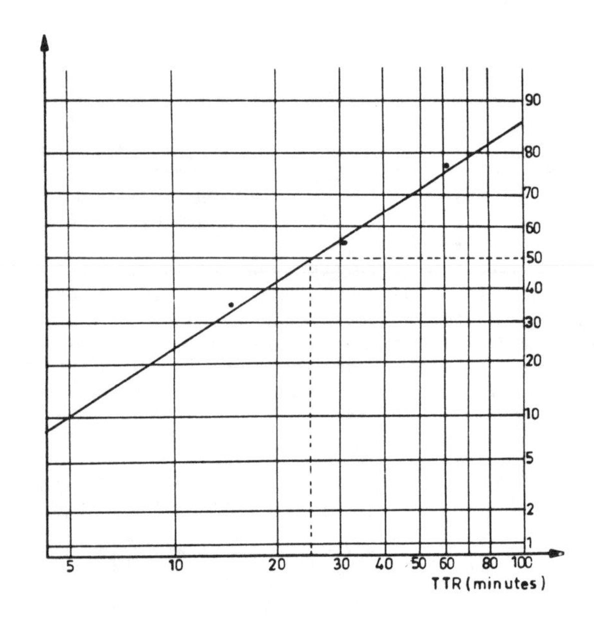

Figure 7. Lognormal plot of active TTR data of the system during the period of 1984 to 1986.

4. Availability performance report

Availability performances of all the DX 200 exchanges in operation are continuously monitored by means of special reporting. An example of availability performance report of one of our customers is presented in figure 8. These reports are delivered to the customers according to the maintenance contract.

AVAILABILITY PERFORMANCE REPORT 9.6.1986

Reporting period 1.1.85 - 31.12.85 Operating hours 1361520

TAMPERE TELEPHONE COMPANY

Number of equivalent subscriber lines 33344

1. Exchange faults	Hardware	Software	Documentation	Total
Number of fault reports	193			
Number of relevant faults	134			

2. Exchange availability

Percent	99.999
Operational unavailability (min/year)	4.8

3. Exchange failure rate/100 equivalent subscriber lines/year

All fault reports	0.70
Relevant faults	0.49

4. Distribution of hardware faults

Design	1.55	Operator	0.00	Not known	30.57
Manufacturing	2.07	External	4.66	Secondary fault	2.07
Installation	0.52	Component	52.40	Organizational	0.00

5. Distribution of active recovery and repair time

Median value 28 min

1-15 min	21.76	16-30 min	9.84	31-60 min	12.95
61-120 min	14.51	Over 120 min	8.81	Not known	32.12

6. Distribution of causes of faults

Total break down	1.55
Subscriber modules out of use	2.07
PCM lines out of use	5.18
No traffic disturbances	26.42
Loss of accounting	0.00
Unknown reason	3.11
Others	61.66

Figure 8. Availibility Performance Report

Analysis of field data

From the point of view of Nokia Telecommunications, the more detailed analysis of the field data provides an effective means to improve the in-house process of developing, manufacturing and installing the DX 200 system. In this chapter, the data is analysed from the point of view of software and hardware development as well as component engineering.

1. Software analysis

Software quality can be described by means of many different methods and metrics depending on the viewpoint. One of the most important demands for a quality measure is that it can be followed as a function of time.

The continuous development of the DX 200 switching system has been organized into successive system releases. As the different system releases include different numbers of new functions the software quality has to be monitored according to each system release.

This paper next presents some error data of the DX 200 software. Firstly, a presentation according to the system releases is given, and then according to the programs. The data is based on error reports connected since October 1983.

Figures 9 and 10 illustrate the error count of DX 200 operation and maintenance software for system releases 'A' and 'B' as a function of time. Software errors are classified according to their effect on the operation of the system. A major error may have an effect on the traffic handling capacity, or it may prevent some operation and maintenance functions from being performed succesfully. Minor errors have less serious consequences and their effect can usually be avoided.

Figure 9. Error rate curve for system release 'A'.

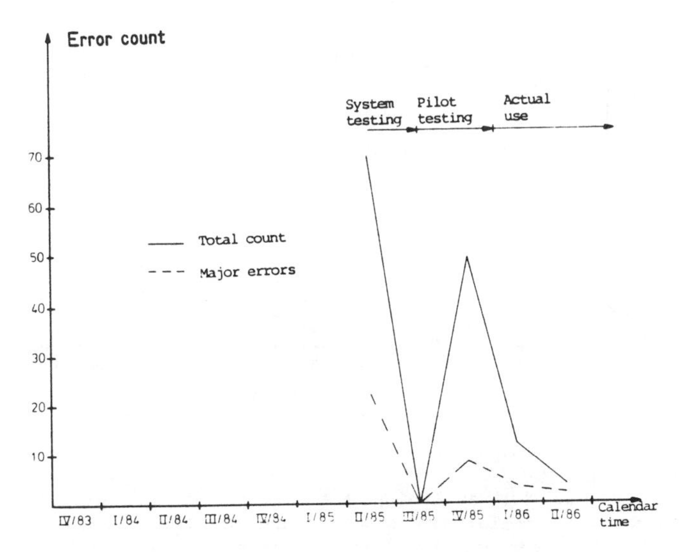

Figure 10. Error rate curve for system release 'B'.

The curves consist of three parts corresponding to the following phases: system testing, pilot testing and actual use.

Every new system release undergoes a system testing performed by a specified test group. Because the system is then working like in actual use it is reasonable to take the system testing phase into consideration here. As can be seen in figure 9, the errors found in the system testing are the reasons for the starting high segment of each curve.

After the system testing, the pilot testing phase begins. This can be seen quite clearly in both curves. Pilot testing has been found effective to complete the system testing by revealing errors which can be detected in real network only.

Delivery of a system release into a new environment, as well as correcting of errors in working exchanges, cause new peaks in the graphs. Those peaks are lower and lower within the same system release. The absolute error count of one system release cannot be compared with that of another system release because of the different number of new functions incorporated in them.

The analysis can be continued at the program level. The basic component of the DX 200 software is a program which is executed under the real time operating system. The DX 200 system includes over 300 programs and over one million code lines of high level language.

The development of the DX 200 is a continuous process where new programs are designed and existing programs maintained at the same time. Maintenance means correction of errors and implementation of new functions into existing programs. Sometimes a whole program must be redefined as the demands increase. In this environment the field quality data of programs can be used as a development aid. The field data system provides an excellent tool for handling the large amount of data at the program level.

The quality measures of programs are based on the error count. Many different measures can be defined according to the cause and the effect of the error. The quality of different programs can be compared only if the measure takes the size of the program into consideration. One simple way is to divide the error count by the number of code lines for getting the error density.

Figure 11 illustrates the error density of the DX 200 operation and maintenance programs of one system level. The system testing and the pilot testing phases are included in the analysis. The errors are not classified in any respect but all errors are regarded as equal. As some errors cannot be pointed to one single program, they are counted separately and included in the error count of all the programs.

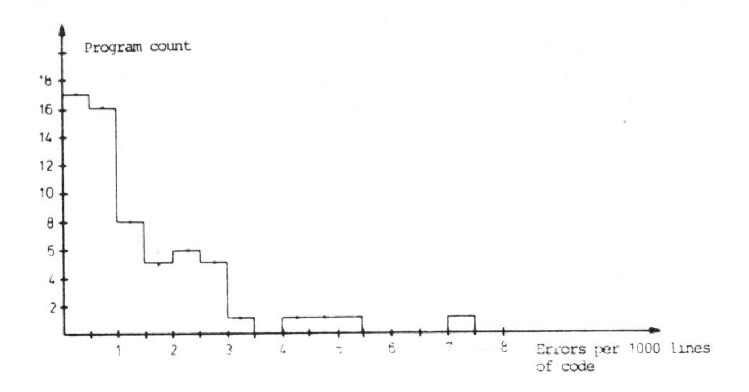

Figure 11. Error densities of programs.

The measures can be used to indicate potential quality problems in some of the programs. Those problems have to be analyzed more precisely and the analysis results can then be used to prevent similar problems from arising in the future.

2. Hardware analysis

For finding out the reasons for improvement of reliability performance of the DX 200 system as shown in figure 5 some hardware design modifications were made.

The dependency between the failure rate of the plug-in units and the number of modifications made to the units were analysed.

The results are presented in figure 12. The following designations are used:

λ_R: ratio of the observed and the predicted failure rates

N_{PU}: number of plug-in unit types belonging to one category of λ_R

N_M : number of modifications made to the plug-in units of one category

N_A : avarage number of modifications / plug-in unit type.

The observed failure rates of the units were calculated from the field data of the years -83, -84 and -85. In the calculations only the faults belonging to the category of component fault (see figure 6) were taken into account.

The analysis is composed of 30 different unit types categorised into four groups according to their λ_R. The number of modifications for each unit was separately calculated during the period -83 to -85.

Referring to figure 12, a conclusion can be drawn that the greater the observed failure rate of the unit is the more changes the unit type has experienced.

Because some of the modifications have been implemented during -85, further improvement on the reliability performance of the DX 200 system is expected in the near future.

λ_R	N_{PU}	N_M	N_A
1	7	14	2.0
1...2	7	15	2.1
2...4	9	22	2.4
4	4	12	3.0

Figure 12. Average number (N_A) of modifications as a function of relative failure rate (λ_R)

Although the dependence of the hardware failure rate from the hardware modifications is rather clear, other reasons for quality improvement were searched.

In figure 13, the distribution of the failure rate of all plug-in units of the DX 200 system in operation is presented. As can be seen, the plug-in units having a failure rate below 2×10^{-6}/h compose 80 % of all the faults.

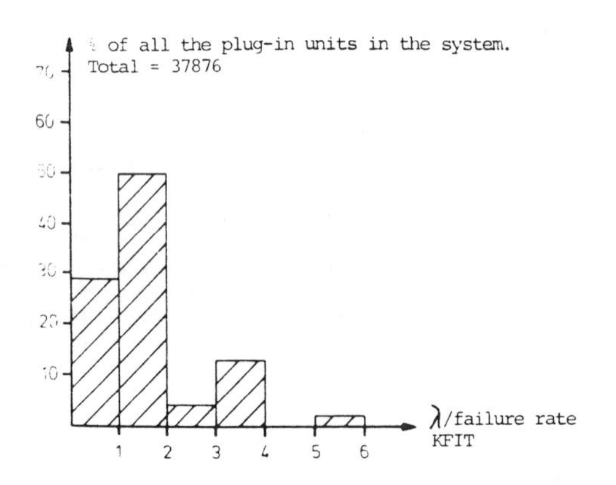

Figure 13. Plot of plug-in unit failure rates during 1985.

Figures 14 and 15 show the reliability improvement on plug-in unit level. It can be seen that the reliability prediction made in 1980 was quite pessimistic, with about 35 % of the plug-in unit types being predicted to have a failure rate greater than 5×10^{-6}/h. The field results of the year 1984, as shown in figure 14, indicate that only 24 % of the plug-in unit types have a failure rate exceeding 5×10^{-6}/h. The results of the year 1985 are even better as can be seen from figure 14. As a consequence of the field experience of the years 1984 and 1985, a new reliability prediction was made (figure 15) which expected that no plug-in unit type will have a failure rate exceeding 5×10^{-6}/h and that 96 % of all plug-in unit types will have a failure rate below 2×10^{-6}/h. This is expected to be achieved by the end of 1987.

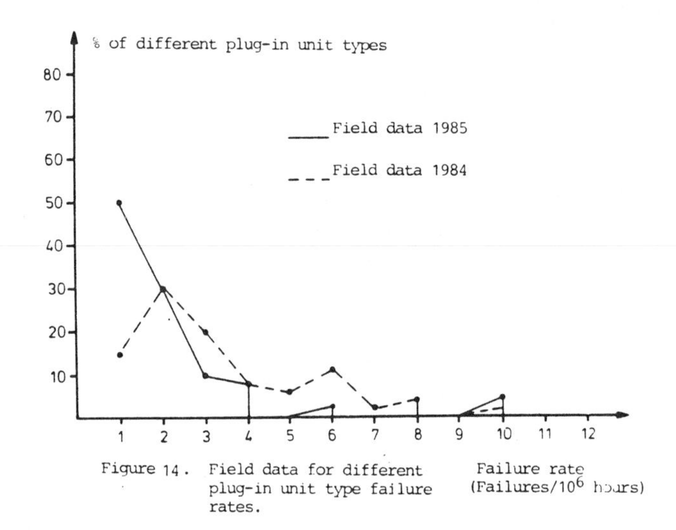

Figure 14. Field data for different plug-in unit type failure rates.

Failure rate (Failures/10^6 hours)

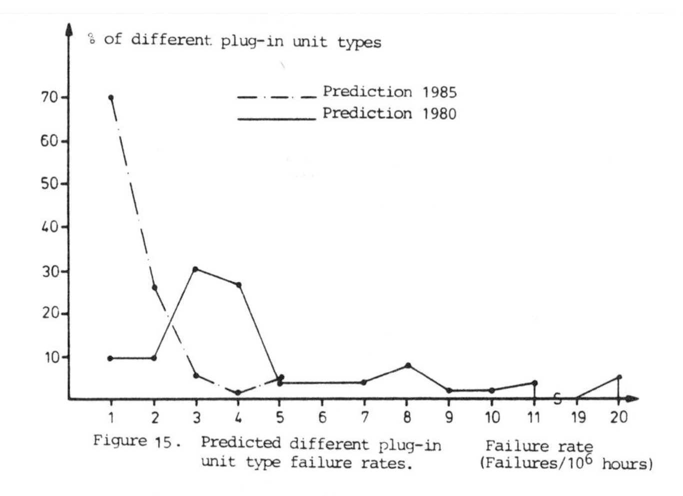

Figure 15. Predicted different plug-in unit type failure rates.

Failure rate (Failures/10^6 hours)

One measure for analysing the testability of plug-in units is to follow the repair time of faulty units in the repair center. Such an analysis is performed by dividing the plug-in units into 6 categories (A to F) according to the level of active MTTR for each plug-in unit type. The histogram of MTTR is shown in figure 16. Active MTTR for an average plug-in unit is calculated to be 47 minutes.

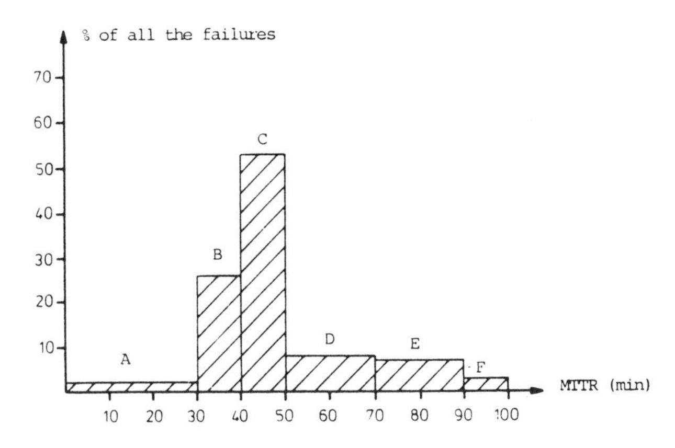

Figure 16. Histogram of active MTTR of plug-in units, period 1985.

Component type	Failure rate (Fits)			
	Prediction 1980	Field data 1984 60 % confidence	* Prediction 1985 60 % confidence	Field data 1986 60 % confidence
Bipolar LS SSI MSI	 10 50	 9 14	 6.0 7.0	 7 7
Bipolar S SSI MSI	 20 80	 10 -	 6 11	 4 -
CMOS digital (MSI)	-	16	17	6
Memory MOS SRAM 4K DRAM 16K DRAM 64K	 300 500 -	 28 70 -	 24 82 110	 22 20 17
Memory Bipolar RAM 4K	200	64	36	12
Microprocessor 8085 8086	 1000 2000	 350 +660	 200 300	 59 +606
Interface Bipolar SSI MSI	 50 100	 9 14	 5 8	 8 32
Linear OP. AMP Double	200	11	8	8
Transistor NPN PNP	 10 30	 12 17	 5 11	 4 13
Euroconnector Relay	20 100	18 4	20 9	3 10

Figure 17. Failure rate for various component technologies used in the DX 200 system.

* Prediction applicable to 1987
\+ Component hours is too small

3. Component analysis

One of the reasons for the pessimistic reliability prediction of the DX 200 system in 1980 was the lack of field experience of the electronic components used in the system. It should be noticed especially, that there was very little experience about the use of memory and processor circuits. In the estimation of the failure rate of these components the models of the MIL HDBK 217 and some other similar models were used.

Figure 17 shows failure rate data for various component technologies. The failure rates are calculated using the 60 % upper confidence limit. The environmental temperature is measured to be in the range of 30 to 40°C. As can be seen, the field failure rates for memories and processor circuits are much lower than the predicted values in 1980. Due to this, all the plug-in units containing these circuits have much lower failure rates than it was predicted. Since the reliability prediction of 1985 is based on the field data collected so far, it is highly feasible that the DX 200 switching system will reach the long-term objective shown in figure 5.

Conclusion

The availability performance of the DX 200 switching system has been demonstrated. Some improvement in the availability performance as well as in the reliability performance during the operating years of 1984 to -85 and 1986 has been demonstrated. One reason for this is that some design modifications in software and hardware have been implemented, and another main reason is the continuous reliability growth of the new component technology. As a result of this improvement the amount of maintenance work in the operating exchanges has decreased to one fourth of the work required by the old crossbar and relay technology.

The analysis of the field data has led to remarkable improvements in the design, manufacturing and installation of the DX 200 Switching System.

Because of rather pessimistic reliability models used in predicting LSI circuit reliability, Nokia has developed a model of its own (Ref. 7) taking into account the constant reliability growth of LSI circuits.

References

1. CCITT Recommentation G.106,"Terms and Definitions Related to Quality of Service, Availability and Reliability", 1984.

2. J. Anttila, P. Jääskeläinen, "Reliability and Maintainability Programme for a Telecommunication Microprocessor System", International Conference on Reliability and Maintainability, Paris, 1978.

3. J. Anttila, K. Olkkola, "Reliability Programme - A Systematic and Practical Approach to the Reliability of Telecommunication Systems", Nordic SINTOM-Conference: Reliability and Safety, Hurdal Norway, 1980.

4. P. Hämäläinen, T. Purho, "Failure Data Collecting of the DX 100 and DX 200 Telephone Exchanges", International Conference on Reliability and Maintainability, Lannion France, 1980.

5. T. Purho, "The field Data and Configuration Management System of the DX 200 Digital Switching Exchanges", Relectronics Symposium 1985, Budapest.

6. J. Anttila, T. Purho, "Availability Performance of the DX 200 and DX 100 Digital Switching Systems in operation", Relectronics Symposium 1983, Budapest.

7. Sauli Palo, "Reliability Prediction of Microcircuits, Microelectron", Reliab., vol. 23, no. 2. pp. 283 - 294, 1983.

Biographies

Tapani Purho
Nokia Telecommunications
P.O.BOX 33
SF - 02601 Espoo
FINLAND

Mr. Tapani Purho is the Quality Assurance Manager in the Development and Engineering Division of Nokia Telecommunications. He is responsible for the quality assurance of hardware and software development as well as the quality approval of the electronic components used in the DX 200 system. Prior to his current duty he worked as a Reliability Engineer in the DX 200 project. Mr. Purho received his MSc degree in 1977 in electrical engineering from the Helsinki University of Technology. Mr. Purho has been a member of international CCITT and CECC working groups and national working groups of IEC organisation since 1980.

Zekrollah Aflatuni
Nokia Telecommunications
P.O.BOX 33
SF - 02601 Espoo
FINLAND

Mr. Zekrollah Aflatuni is currently serving as Reliaability Manager at Nokia Telecommunications, Development and Engineering Division. His main responsibility is reliablity assurance of the DX 200 Digital Switching System. Prior to this he was the Quality Control Manager of Mobira Company involved mainly in mobile equipment quality assurance. Mr. Aflatuni received his BSc degree in electrical engineering from Middle East Technical University in 1969 and the degree of Lisenciate of Technology from Helsinki University of Technology in 1973. He has been involved in reliability and quality assurance for 12 years.

Jouni Soitinaho
Nokia Telecommunications
P.O.BOX 33
SF - 02601 Espoo
FINLAND

Mr. Jouni Soitinaho is a Quality Engineer in Development and Engineering Division of Nokia Telecommunications. He received his MSc degree in electrical engineering from Helsinki University of Technology in 1985. From 1982 until 1984 he was involved in the reliability planning of DX 200. After that he continued his work with the DX 200 project being responsible for the software quality assurance. His main activities include developing and utilizing software design standards and practices as well as controlling quality of software development.

Reprinted with permission from *Digital Technical Journal*, Number 5, September 1987, pages 69–79.

Edward E. Balkovich
Prashant Bhabhalia
William R. Dunnington
Thomas F. Weyant

VAXcluster Availability Modeling

VAXcluster systems use redundant hardware—processors, interconnects, and storage elements—and software to achieve high system availability. No special hardware or software is required. A simple, first-order availability model is used to illustrate how this redundancy improves availability. Four VAXcluster configurations are analyzed to show that redundancy decreases system unavailability by two orders of magnitude. Decomposition techniques were used to develop these first-order availability models, which were then analyzed using "textbook" reliability analysis techniques. More complex configurations and models of broader classes of faults will require the support of more sophisticated modeling tools.

An increasing number of specialized computer systems are being dedicated to tasks that are critical to the success of an organization. For example, in the financial services industry or in manufacturing, it must be possible to access a computing system to deliver a service or to manufacture a product. Any loss of access to the computing system adversely impacts business. The ability to access a computing system when it is needed (commonly referred to as availability) is becoming an important metric used to select such computer systems. Obviously, high availability also improves the quality of service provided by general-purpose computing systems, such as those providing timesharing services.

VAXcluster systems provide high availability.[1] They can be configured so that there is no single point of failure. Each cluster is a multiple-computer system, built from standard hardware and software elements. VAXcluster systems can be expanded in increments to provide the computing power, data resources, and storage capabilities typically associated with mainframe systems.

Although these systems are not fault tolerant, they can detect, isolate, and recover from faults in their processor, interconnect, and storage subsystems. (Fault tolerance generally implies that a recovery from a fault is completely invisible to an application.) While VAXcluster systems can detect, isolate, and recover from faults, the recovery from some types of faults impacts the applications and their design. For example, a VAXcluster system will retry an I/O operation if a fault is detected in either the interconnect or storage subsystems.

The integrity of the I/O operation is ensured by the operating system. If a processor fails, however, the computations hosted by it are lost. A user must start a new session on another (available) processor. The user must depend on an application, not the operating system, to recover the state of the computation to the point at which the fault occurred. For example, a journal file can be used to recover an editing session or database transaction. In this case, the integrity of the computation is assured by the application, not by the operating system.

This paper documents a study using simple first-order models to show how the inherent redundancy of VAXcluster systems is used to achieve high availability. Although more sophisticated models are possible, the models used in this study were sufficient to illustrate the main points. It is assumed that the reader is familiar with the basic technical concepts of VAXcluster systems presented in our companion papers.[2,3] It is not assumed that the reader is familiar with the standard methods of analyzing availability used to illustrate the points of this study.

VAXcluster Structure

Figure 1 illustrates a simple VAXcluster system with terminals connected to the system via a LAT server. Either processor is accessible through that server, and dual-ported disks are accessible through either Hierarchical Storage Controller (HSC). The HSC devices and the processors are

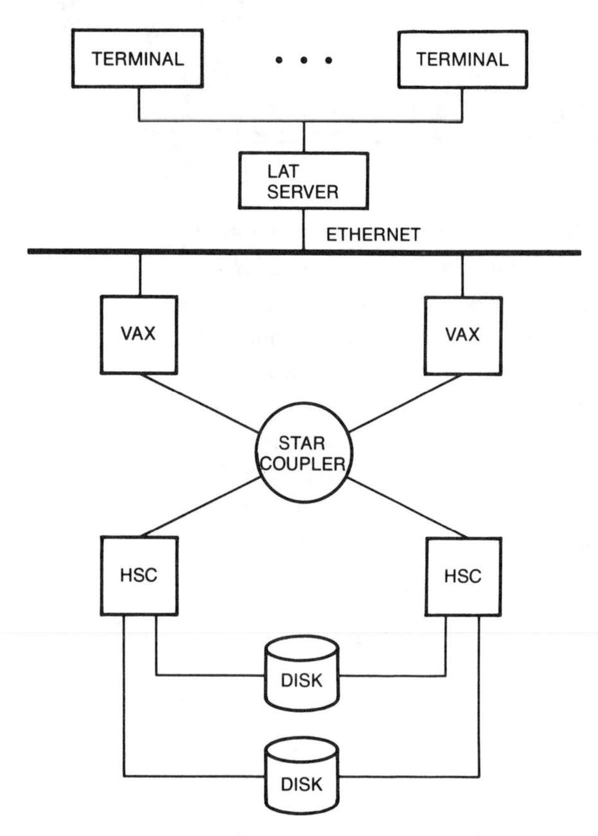

Figure 1 Simple VAXcluster Configuration

connected by a Star Coupler, a passive device offering two independent datapaths between each node of the system. Multiple disks are used to shadow a volume of information. This simple system illustrates all the basic forms of redundancy in VAXcluster systems.

Processor Failures

If a processor or its Computer Interconnect (CI) adapter fails, all computations in progress on that processor will be lost. The processor and the adapter can detect some types of faults and inform the VAXcluster system of them immediately. Other types of faults are detected by the other VAXcluster processors by way of time-outs.

When other processors detect a fault in a processor or its adapter, they reconfigure themselves to remove the failed processor from the cluster. The reconfiguration times depend on the number of locks in the system and on the number of I/O devices in the configuration. The average reconfiguration time after a processor failure is a small number of seconds.[4] After the reconfiguration is complete, the user can begin a new session on the remaining processor. Appropriately constructed applications, such as those employing journaling, can then be recovered to the point of the failure.

Interconnect Failures

The Star Coupler, a passive device, has a negligible failure rate compared with the other elements. The individual CI paths attached to a single adapter have active elements, however, and the failure rates for those paths must be considered.

If a single path fails, the CI adapter will retry the transmission on the redundant path. The retry is invisible to both the processor and the HSC device using the adapter.

If both paths fail, neither the processor nor the HSC device attached to the adapter can communicate with other elements of the VAXcluster configuration. The effect is similar to a processor or HSC failure. However, other processors and HSC devices can continue to communicate with each other.

Hierarchical Storage Controller Failures

HSC failures are managed by the VAX processors. The HSC device can detect some faults and inform the cluster about them immediately. Other types of faults are detected by the VAX processors and the disks by time-outs. When a fault is detected in an HSC device, the VAX processors will retry any I/O operations in progress by using the redundant HSC device. An HSC failure is invisible to the process issuing the QIO operation. The times required to reconfigure the system after an HSC failure depend on the number of outstanding I/O operations, the number of I/O devices, and the use of volume shadowing. The average time is typically a small number of seconds.

Volume shadow sets, hosted by an HSC device, must be reconstructed if that device fails. Although the shadow set is available during reconstruction, this process involves additional I/O that competes with user requests to read or write to the volume shadow set.

Disk Failures

HSC devices detect disk failures. Volume shadowing allows an HSC device to retry a failed I/O operation using another member of the volume shadow set. The failure of a disk in a shadow set is invisible to the process issuing the QIO operation. When a fault is detected, the volume shadow set will be reconfigured to remove the failed volume. Once again, the average time required to reconfigure the shadow set after a disk failure is a small number of seconds.

VAXcluster Configurations Considered

Modeling Procedure

This paper focuses on the availability modeling of four simple VAXcluster configurations. The goals of the study were to

- Demonstrate the sensitivity of different reliability and availability parameters
- Demonstrate how different types of redundancy improve VAXcluster availability

These goals were achieved by first modeling the availability of a baseline configuration consisting of a VAX processor, an HSC storage controller, and a disk drive. Each element in the configuration represented a single point of failure. Next, redundancy in the form of a second VAX processor was added to the baseline configuration to create a second configuration. Another HSC storage controller was then added to create a third configuration. Finally, a disk drive and volume shadowing were added to create a fourth and fully redundant configuration. These four simple configurations were used to study the principal forms of redundancy in a VAXcluster system.

Referring to Figure 1, the configurations considered here consisted of VAX processors, a Star Coupler, HSC storage controllers, and disk drives; they did not include the Ethernet, the LAT server, or the user terminals.

Baseline Configuration — Model 1

The baseline configuration, Figure 2, consisted of a VAX processor, an HSC storage controller, and a disk drive. The processor and the storage controller were connected by way of a Star Coupler whose failure rate is negligible compared to that of the other elements. Figure 2 also shows the configuration diagram translated into a reliability block diagram in which the series positioning of each element represents a single point of failure for the configuration.

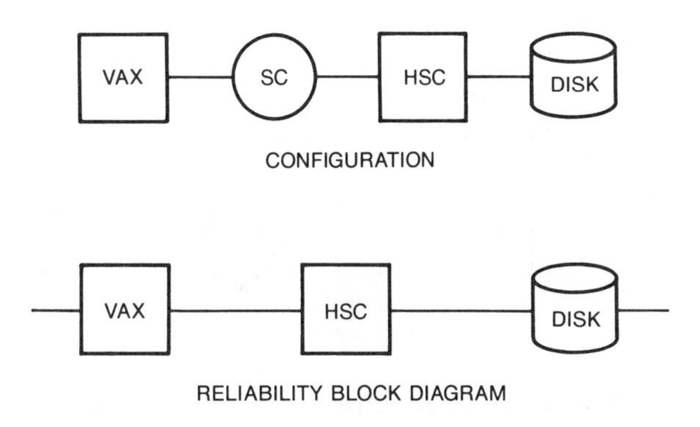

Figure 2 Baseline Configuration (Model 1)

Redundant Processor Configuration — Model 2

The second configuration considered in the study, Figure 3, added redundancy in the form of a second VAX processor. The failure of either processor or its CI adapter requires a failover process to the redundant processor with its associated VAXcluster reconfiguration activities. These activities usually complete in a matter of seconds.

In the reliability block diagram for the hardware model, the redundant VAX processors are shown in parallel because both must fail for the configuration to fail. However, the HSC device

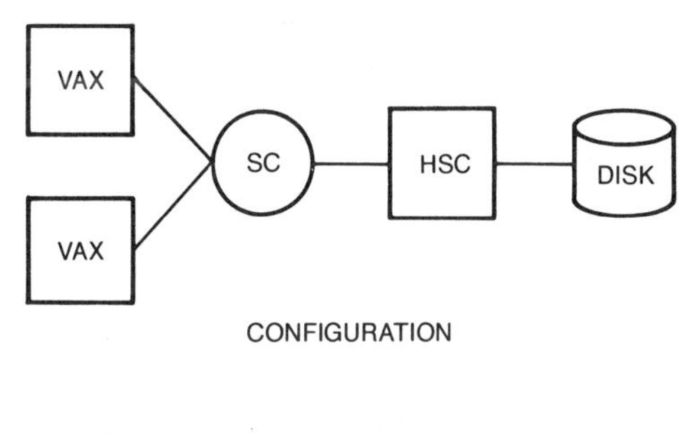

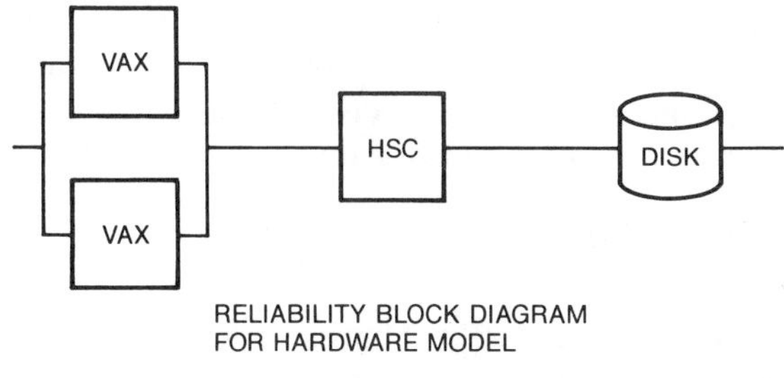

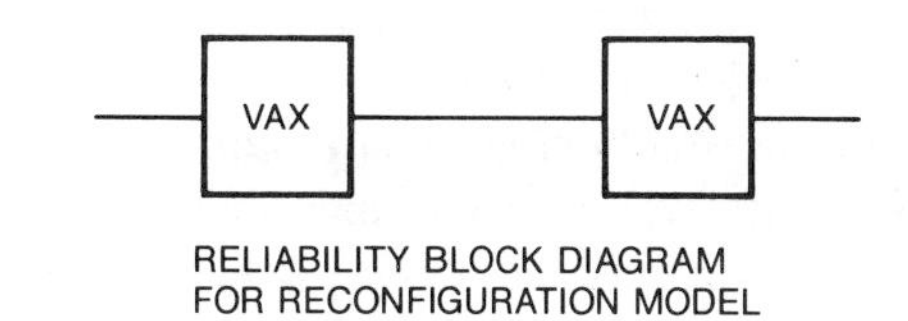

Figure 3 Configuration with Redundant Processor (Model 2)

and the disk drive are still shown as single points of failure.

If either processor fails, the VAXcluster system will undergo a reconfiguration. Depending on the user application, the system may be unavailable during the failover process.[5] This condition is represented in the reliability block diagram by the two VAX processors in series.

Similarly, the reconfiguration operation is repeated when a repaired VAX processor is reestablished in the VAXcluster system. Again, depending on the user application, the system may be unavailable until the reconfiguration completes. Since either VAX processor could fail, the reliability block diagram is again valid for this condition.

Redundant Storage Controller Configuration — Model 3

In the third configuration, Figure 4, additional redundancy in the form of a second HSC storage controller was added to the Model 2 configuration, which already had a redundant VAX processor. Now the failure of either a VAX processor or an HSC storage controller requires a failover process to either the redundant processor or the controller with the associated VAXcluster reconfiguration activities.

When a repaired HSC storage controller is re-established in a VAXcluster system, there is no reconfiguration operation. Instead, the HSC device is placed in "warm stand-by" redundancy. That is, the device is not actively re-established in the VAXcluster system unless the other HSC device fails. This situation contrasts with that of the active redundancy of the VAX processor, which is immediately reconfigured back into operation as soon as it is repaired.

Fully Redundant Configuration — Model 4

A fourth configuration, Figure 5, added further redundancy in the form of a second disk drive and volume shadowing to the Model 3 configuration, which already had a redundant VAX processor and HSC storage controller.

In volume shadowing, write commands are applied to all available volumes in the shadow set. Read commands are accomplished using any available volume. A fault in a disk causes it to be removed from the shadow set. A repaired volume is merged back into a shadow set by first copying the data from an available volume as a background activity. Only upon becoming identical to existing members of the set will the repaired volume again become an available member of the shadow set.

A detailed description and analysis of the Model 4 configuration is given later.

Modeling Approach

Several formal definitions are needed to quantify VAXcluster availability.

Availability is the proportion of time that service is available from a VAXcluster system to perform a user application.

It is important to remember that this definition of availability is a general one. As the nature of the application, the size of the VAXcluster configuration, and the amount of redundancy change, availability can be defined in more complex

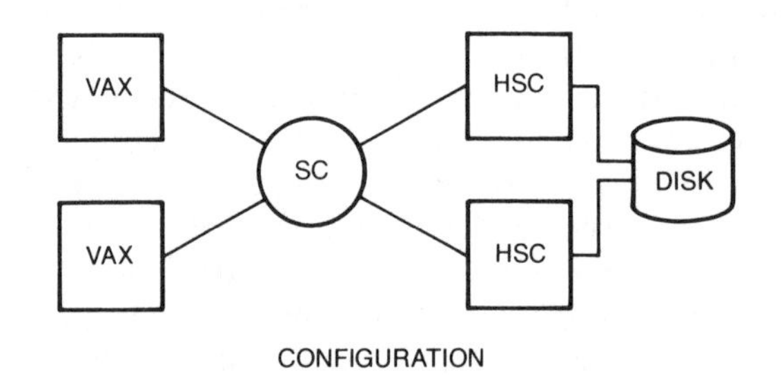

CONFIGURATION

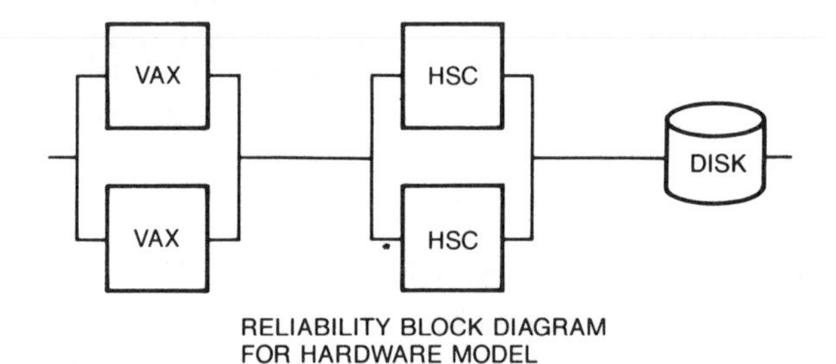

RELIABILITY BLOCK DIAGRAM FOR HARDWARE MODEL

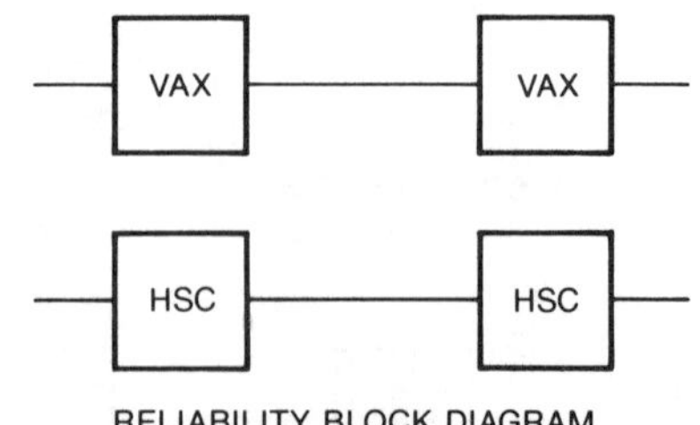

RELIABILITY BLOCK DIAGRAM FOR RECONFIGURATION MODEL

Figure 4 Configuration with Redundant Processor and Storage Controller (Model 3)

ways. For the configurations used in this study, at least one of each type of element must be running for the VAXcluster system to be operational.

Unavailability is the proportion of time that service is interrupted and that a VAXcluster system cannot perform a user application.

In this study, the related metric of downtime in minutes per year will be used rather than the system unavailability.

Reconfiguration time is the time taken to initially detect a failed element and remove it from the VAXcluster system. For a failed VAX processor, this time also includes the time taken later to re-establish the repaired element's membership in the cluster.

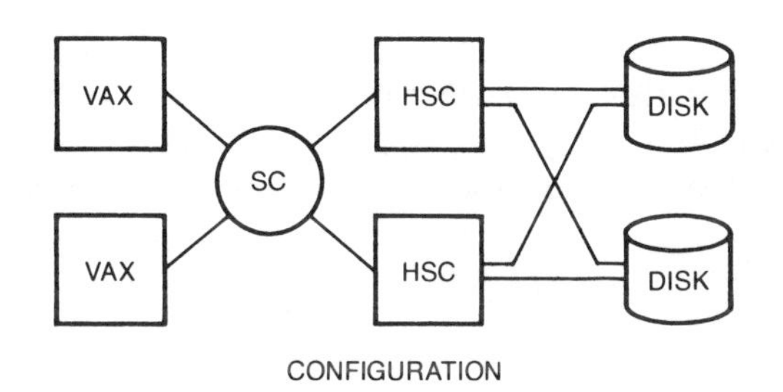

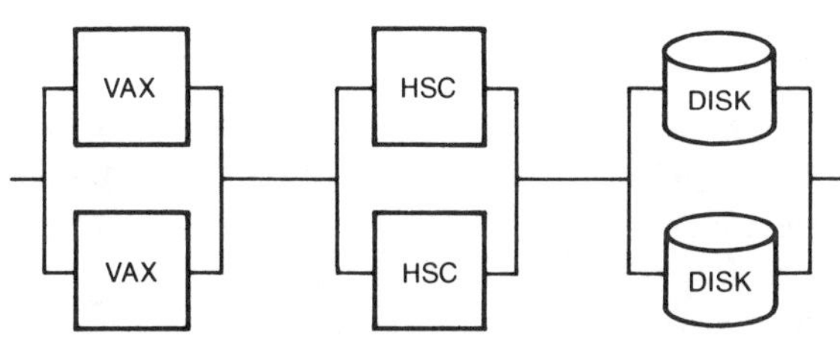

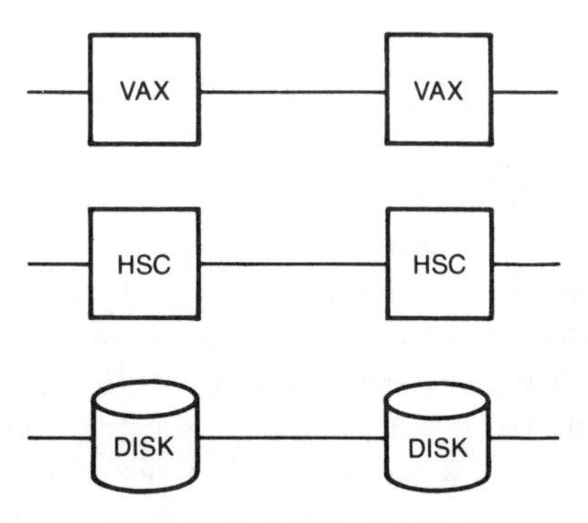

Figure 5 Configuration of Fully Redundant System (Model 4)

Note that the HSC device employs "warm stand-by" redundancy and therefore does not have any significant reconfiguration time associated with re-establishing membership in the cluster.

VAXcluster reconfiguration activities usually complete in a matter of seconds; however, in extremely rare cases, much longer times are possible.

Overview

The most common approach to modeling complex systems consists of structurally dividing a system into smaller subsystems, such as processors, controllers, and disks.[6] The availability of each subsystem is then analyzed separately, and the individual subsystem solutions are combined to obtain the system solution. One important assumption must be made to achieve a solution: the behavior of each subsystem must be independent from that of any other subsystem.

Furthermore, a decomposition technique can be applied to certain behaviors that cause system outages due to failures in redundant subsystems. In these cases, the recovery to an operational system happens quickly. Similar behavior is also present when the failed subsystem is repaired and is ready to rejoin the system to make it a fully configured system. This type of decomposition is called behavioral decomposition.

With this approach to structural and behavioral decomposition, hardware failures and VAXcluster reconfigurations are modeled separately. Such a decomposition allows the model to analyze both VAXcluster reconfigurations and complete system failures due to hardware failures. It also allows the model to analyze the sensitivity of system availability to each factor.

In this study, availability modeling captured the following factors:

- Hard failures requiring a repair call
- VAXcluster reconfigurations during which the VAXcluster system was assumed to be unavailable in this analysis
- Response time for maintenance personnel
- Time-to-repair

The following factors were not considered (except for the impact of reconfigurations due to hardware failures):

- Intermittent failures
- Transient failures

- Quorum disks
- Operational errors
- Software errors

The following modeling parameters were used:

- The mean time-between-failures (MTBF) and mean time-to-repair (MTTR) of each of the following elements:
  - VAX processor
  - HSC storage controller
  - Disk drive
- VAXcluster reconfiguration times caused by
  - VAX processor failure
  - Re-establishment of the repaired VAX processor into the VAXcluster configuration
  - HSC storage controller failure
  - Disk drive failure
- Response time for maintenance

The remainder of this section describes in detail the modeling of the fourth configuration (Model 4).

Analysis of Hardware Failure

Consider the structural decomposition of the VAXcluster configuration. Three subsystems were connected in series, each consisting of two elements in parallel. At least one element in each subsystem had to be operational for the VAXcluster system to be operational. The hardware reliability block diagram is shown in Figure 5.

Repairable systems are those for which an automatic or manual repair can be made if an element fails. Assume that each element is subject to failure and has its own repair facility.[7] If the time-to-failure of element i is exponentially distributed with failure rate λ_i, and the time-to-repair of element i is exponentially distributed with repair rate μ_i, the instantaneous availability can be obtained by the following equation:

$$A_i(t) = \frac{\mu_i}{\lambda_i+\mu_i} + \frac{\lambda_i}{\lambda_i+\mu_i} e^{-(\lambda_i+\mu_i)t}$$

As t approaches infinity, $A_i(t)$ approaches the steady-state availability and A_i equals $\mu_i/(\lambda_i+\mu_i)$.

The steady-state availability of a single element is given by the following equation:

$$A = \mu/(\lambda+\mu)$$

in which λ is the failure rate of the element and μ is the repair rate of the element. The time-to-failure and the time-to-repair are assumed to be exponentially distributed.

The steady-state availability of two elements in parallel is[8]

$$A = 1-(1-A_1)(1-A_2)$$

In Model 4, the elements in each subsystem are two VAX processors, or two HSC storage controllers, or two disk drives. Using the equation above, the availability of the processor subsystem, A_p, can be expressed as

$$A_p = 1-\left(\frac{\lambda_p}{\lambda_p+\mu_p}\right)^2$$

Similarly, the availability of the HSC storage controller subsystem, A_b, and the availability of the disk drive subsystem, A_r, can be expressed as

$$A_b = 1-\left(\frac{\lambda_b}{\lambda_b+\mu_b}\right)^2$$

and

$$A_r = 1-\left(\frac{\lambda_r}{\lambda_r+\mu_r}\right)^2$$

The aggregate availability of the VAXcluster system is

$$A_s = A_p \times A_b \times A_r$$

For exponentially distributed times, the failure rate, λ, is $1/MTBF$ and the repair rate, μ, is $1/MTTR$.

Analysis of Reconfiguration Times

Next, consider the behavioral decomposition caused by the reconfiguration that occurs when one element in a subsystem fails and an automatic failover to a second (redundant) element takes place. During this process, a reconfiguration occurs when a failed element leaves the VAXcluster system. For processors only, another reconfiguration occurs when a repaired processor later rejoins the VAXcluster system. Depending on the user application, the VAXcluster system may be unavailable to perform user applications during these reconfigurations.

For example, consider the following time line:

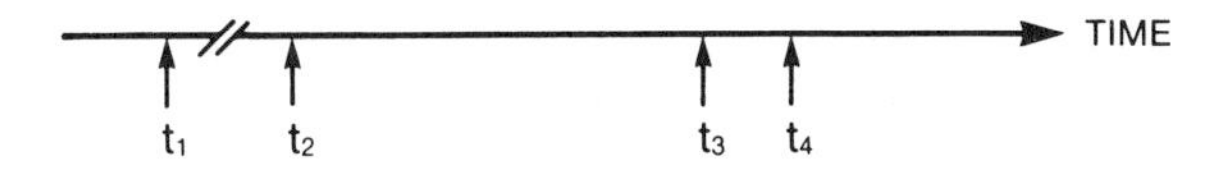

Figure 6

Time t_1 to t_2 is the VAXcluster reconfiguration time for a failed VAX processor to be detected and removed from the VAXcluster membership. Time t_2 to t_3 is the repair time for the failed hardware element. Time t_3 to t_4 is the time for the repaired VAX processor to be re-established in the VAXcluster membership.

Figure 5 includes the reliability block diagram representing the VAXcluster reconfiguration behavior of the Model 4 configuration. Each subsystem is shown as two elements in series. If any single element is not operational, the subsystem can be unavailable due to a VAXcluster reconfiguration.

For two elements in series, the availability is[8]

$$A = A_1 \times A_2$$

In model 4, the elements in each subsystem are two VAX processors, or two HSC storage controllers, or two disk drives.

Applying the equation above for elements in series, the availability of the processor subsystem, A_p, is

$$A_p = \left\{\frac{\mu_p}{(\lambda_p + \mu_p)}\right\}^2$$

Note that for the VAX processor, the rate μ_p is the reciprocal of the sum of the times t_1 to t_2 and t_3 to t_4.

Similarly, the availability of the HSC storage controller subsystem, A_h, and the availability of the disk drive subsystem, A_r, is

$$A_h = \left\{\frac{\mu_h}{(\lambda_h + \mu_h)}\right\}^2$$

and

$$A_r = \left\{\frac{\mu_r}{(\lambda_r + \mu_r)}\right\}^2$$

The aggregate availability of the VAXcluster system is

$$A_s = A_p \times A_h \times A_r$$

Assuming an operation running 24 hours a day, 365 days per year, the downtime equals $(1 - A_s) \times 525{,}600$ minutes per year. This figure is the downtime caused only by reconfigurations. The total downtime is the sum of the downtime caused by hardware failures and the downtime caused by VAXcluster reconfigurations.

Extensions to the Models

The simple models considered in this study can be extended in several dimensions.

The complexity of the configurations can be increased either by adding more VAXcluster elements or by extending the bounds of the models to include the Ethernet and its attachments. A complex configuration could include multiple clusters and multiple Ethernet segments. More complex definitions of availability are needed as the configurations increase in complexity. These definitions range from the single-user view to a measure of system productivity.

Only permanent (hard) hardware failures are considered in this study. Intermittent and transient hardware and software failures, as well as operational errors, can be added as extensions to future models. The downtime allocation reported in the literature typically attributes about one third of the total to each of the hardware, software, and operator-induced failures.[9] This result includes the effectiveness of system recovery that can be hardware based, software based, or both. Certain insidious failures can result in ineffective recovery, even in the presence of hardware or software redundancies. The term "fault coverage" represents the joint probability of fault detection and successful failover to a redundant element. A fault-coverage factor of one is assumed in this study.

This study also assumes that the subsystems of VAX processors, HSC storage controllers, and disk drives are independent. Relaxing this assumption adds to the complexity of the modeling approach. Similarly, a simplistic maintenance strategy is assumed in which each cluster element has its own repair facility.

The extensions described above add more realism to the modeling approach at the expense of added complexity in both model formulation and solution technique. Moreover, the textbook formulae used in this study are limiting and often inappropriate.

Markov modeling is a particularly useful analytic technique for formulating and solving these complex models.[7] Simulation is an alternative but computationally less efficient technique.

Another valuable industry-wide tool is the Symbolic Hierarchical Automatic Reliability and Performance Evaluator (SHARPE) software.[10] SHARPE's hierarchical feature allows complex subsystem models to be combined into a system model for efficient solution. SHARPE also employs state-of-the-art matrix-solving routines to solve large and often ill-conditioned problems arising from the Markov model formulation of these complex configurations.

Results and Conclusions

This section discusses the results of this study in detail.

The Impact of Initial Redundancy

In Model 1, no redundancy exists in the system.

In Model 2, the redundancy of the additional VAX processor reduces the total downtime to 16 percent of the downtime in Model 1.

In Model 3, the redundancy of an additional VAX processor and an HSC storage controller reduces the total downtime to almost 7 percent of the downtime in Model 1.

In Model 4, the total redundancy of an additional VAX processor, an HSC storage controller, and a disk drive reduces the total downtime to slightly under 1 percent of the downtime in Model 1.

These results show that redundancy does work to increase the availability of the system. Figure 7 shows the effect on total downtime as different forms of redundancy are introduced. A fully redundant configuration reduces system downtime by two orders of magnitude.

VAXcluster Reconfiguration Downtime

Figure 8 is an expanded view of the decrease in total downtime for the three models that include redundancy. It also shows the contribution of VAXcluster reconfigurations to total downtime. Here the typical duration of reconfiguration is used. Since Model 1 has no redundancy, the VAXcluster reconfiguration downtime is zero.

Impact of Increased Frequency of Reconfigurations

Since the previous results considered the frequency of reconfigurations equal to that of hardware failures, it was necessary to study the impact of an increased frequency of reconfigurations on downtime.

Figure 9 shows the linear relationship between reconfiguration downtime and an increase in the frequency of reconfigurations. It also shows the trend in the reconfiguration downtime as the duration of reconfiguration is first varied to three and then to six times the typical value. As shown, the key to reduced downtime is keeping the duration and the frequency of reconfigurations as low as practical. High-reliability hardware is a major factor in keeping the frequency of reconfigurations low.

Contribution of Individual VAXcluster Elements

This study also examined how much downtime an individual VAXcluster element contributes toward the total downtime.

Figure 10 shows the contribution of each element (CPU, HSC, and disk) toward the total downtime for Model 4. At a given MTBF, the VAX processor contributed 82 percent of the total

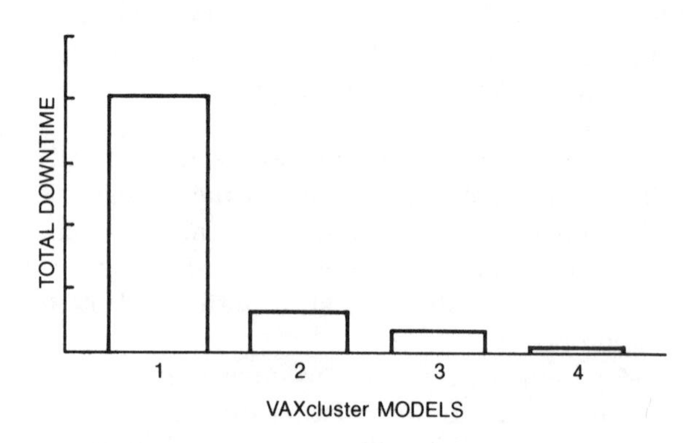

Figure 7 Impact of Initial Redundancy

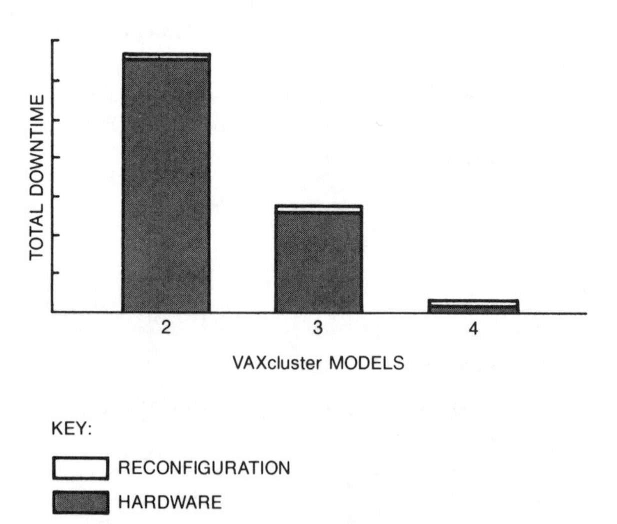

Figure 8 Total System Downtime by Model

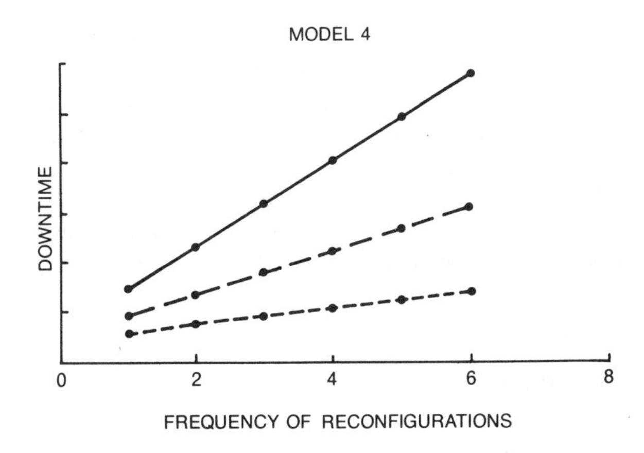

Figure 9 Reconfiguration Downtime by Frequency of Reconfigurations

downtime. When the MTBF of that particular VAX processor was improved, its contribution dropped to 57 percent.

Typical VAXcluster configurations would generally include more than the two disks used in this study. Having more disks would change the contribution of the disk subsystem to the system unavailability. (Analyzing the impact of additional disks is outside the scope of this paper.)

The reliability improvement in the MTBF of the VAX processor decreased both the hardware and the reconfiguration downtime. Figure 11 shows a decrease of approximately 58 percent in total downtime.

Hardware Downtime versus Response Time

This study included a response time for maintenance for each call as part of the recovery time. If an on-site maintenance person were available, the response time would be eliminated, thus speeding the recovery of a failed element. When this strategy is considered, the hardware downtime drops by almost 60 percent. Figure 12 shows this reduction as applied to Model 4.

The N of M Redundancy Case

The results given so far have been for (1 of 1) and (1 of 2) configurations of VAX processors, storage controllers, and disks. In this section, the hardware downtime results for VAX processors are generalized to the (N of M) redundancy case. The assumption is that N processors are required for capacity and M processors represent $M-N$ redundancy. The steady-state availability is defined as the probability of at least (N of M) processors working. The cluster is assumed to be unavailable when less than N processors are working. Note that, depending on the configuration and application, clusters with less than N working could be considered as partially available. The case of the partially available cluster is not considered here.

The (N of M) availability, as defined above, is

$$Availability_{(N\ of\ M)} = \sum_{i=0}^{M-N} \frac{M!}{i!(M-i)!} \left(\frac{\mu}{\mu+\lambda}\right)^{M-i} \left(1-\frac{\mu}{\mu+\lambda}\right)^{i}$$

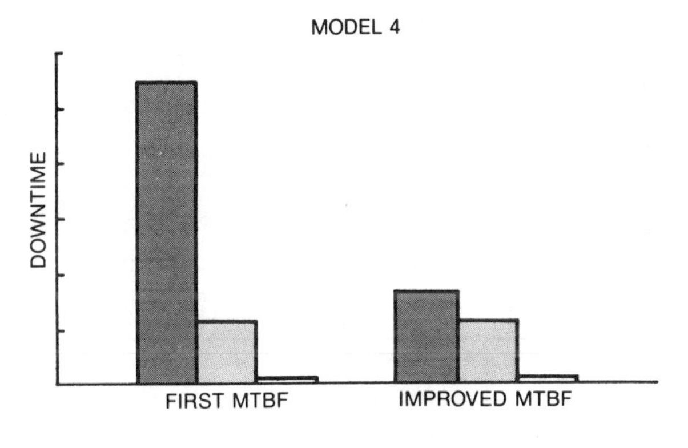

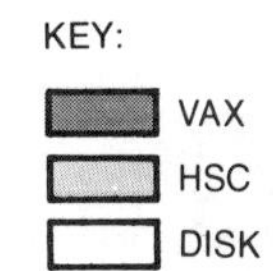

Figure 10 Contributions of Individual VAXcluster Elements to Downtime

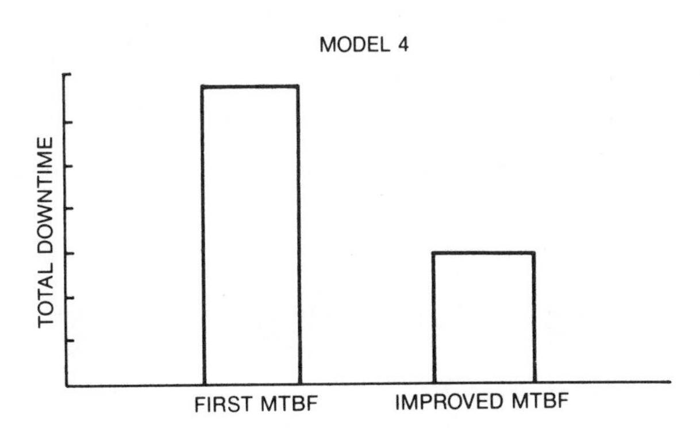

Figure 11 Total System Downtime by VAX Processor MTBF

An application of the (N of M) availability expression for VAX processors is shown in Figure 13. The number of VAX processors required to run applications to capacity was set to 1, 2, 3, and 4. The values for M were set to $N+0$, $N+1$, and $N+2$. High availability is typically measured in values much greater than 0.99. Therefore, to distinguish the variation in availability, the origin in Figure 13 is not zero but much greater than 0.9. With no redundancy ($M=N+0$), availability decreases with an increase in the number of processors. That decrease occurs because more CPUs must be available to deliver the application, bringing about a greater likelihood of failure and outage. This result is shown in the graph by the downward trend of the "$N+0$" bars. Adding a single redundant CPU ($M=N+1$) greatly improves system availability. Adding a second redundant CPU ($M=N+2$) has little additional effect on availability. The additional improvement is not visible on the graph, even with the expanded vertical scale. It can therefore be assumed that "$N+1$" redundancy is sufficient for most applications.

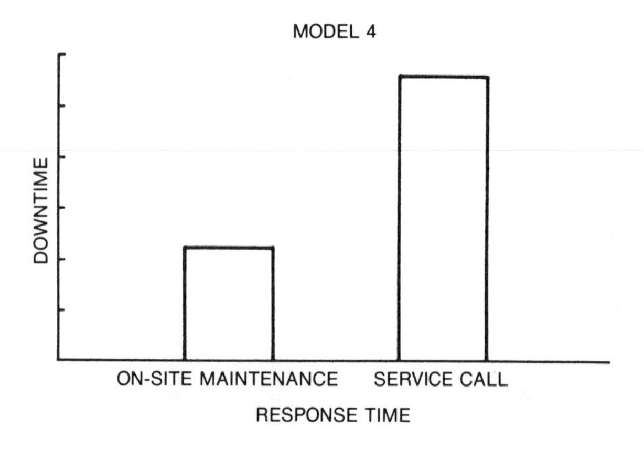

Figure 12 Hardware Downtime versus Response Time

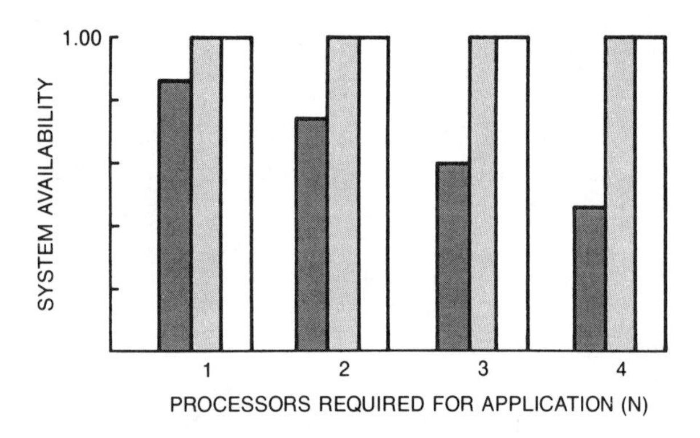

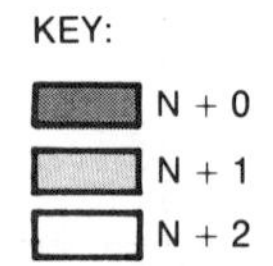

Figure 13 The (N of M) VAX Processor Redundancy Case

Summary

VAXcluster systems achieve high availability by eliminating single points of failure with redundant hardware. Redundancy is introduced at the level of standard processors, interconnects, storage elements, and software. No special-purpose hardware or software is required. The same hardware and software could be used to construct a less available uniprocessor system without volume shadowing.

The simple analytic models of VAXcluster availability developed in this study show that redundancy yields dramatic improvements in system availability for the system configuration shown in Figure 1. The average downtime of the system is reduced by nearly two orders of magnitude from that of a similar uniprocessor system without volume shadowing.

Because they can be expanded incrementally, VAXcluster systems requiring a minimum number of N processors to achieve a performance goal can achieve significant improvements in availability with the addition of a single redundant processor. There is no requirement to fully replicate all the original N processors.

The system configurations analyzed in this study are simple ones designed to illustrate the most important concepts of VAXcluster systems. The downtime of a more complex VAXcluster configuration, with many additional processors, HSC devices, and disk drives, changes system downtime in complex ways. In general, additional redundant hardware causes multiple hardware failures to become less of a factor. When faults do occur, however, time is required to reconfigure the system. Some applications may view these small reconfiguration times as a source of system downtime. In such cases, additional hardware increases both the frequency of reconfigurations and their contribution to system downtime. Continuing efforts to improve hardware reliability are particularly important to reduce the downtime due to multiple hardware failures and the frequency of reconfigurations that might be counted as downtime by an application.

The analysis used in this study uses structural and behavioral decompositions of systems. Structural decomposition is the most common approach to modeling complex systems. However, this approach assumes that each subsystem behaves independently. For the systems and phenomena considered in this study, recovery to an operational state happens quickly following a system reconfiguration caused by a fault in a redundant subsystem. Similar behavior is also present when a failed VAX processor subsystem is repaired and is ready to rejoin the system.

These modeling approaches were applied to the VAXcluster system, which was considered to be repairable. Structural decomposition was used to model the hardware failures of each VAX processor, HSC device, and disk drive in the system. Behavioral decomposition was used separately to model the reconfiguration times.

Notes and References

1. This paper is limited to CI-based VAXcluster systems. Local Area VAXcluster systems, implemented with Ethernet, are not considered in this analysis. The reader should be aware that there are significant configuration differences between CI-based VAXcluster systems and Local Area VAXcluster systems that lead to important differences in system availability.

2. N. Kronenberg, H. Levy, W. Strecker, and R. Merewood, "The VAXcluster Concept: An Overview of a Distributed System," *Digital Technical Journal* (September 1987, this issue): 7–21.

3. *VAXcluster Systems Handbook* (Bedford: Digital Equipment Corporation, Order No. EB-28858-46, 1986).

4. E. Los, S. Snaman, S. Szeto, and D. Thiel, Corrections to "Cluster State Transitions," *VAXcluster Systems Quorum*, vol. 2, issue 3 (Digital Equipment Corporation, February 1987): addendum.

5. During reconfiguration, significant processor resources are used to reconstruct the lock manager database. Some real-time applications may view the reconfiguration time as a system outage.

6. S. Bavuso et al., *Dependability Analysis of Typical Fault-Tolerant Architectures Using HARP*, CS-1986-18.

7. K. Trivedi, *Probability and Statistics with Reliability, Queuing and Computer Science Applications* (Englewood Cliffs: Prentice Hall, 1982).

8. P. O'Connor, *Practical Reliability Engineering* (Chichester: John Wiley & Sons, Ltd., 1985).

9. D. Siewiorek and R. Swarz, *The Theory and Practice of Reliable System Design* (Bedford: Digital Press, 1982).

10. R. Sahner and K. Trivedi, *SHARPE: Symbolic Hierarchical Automatic Reliability and Performance Evaluator,* (Durham: Duke University Department of Computer Science, September 1986).

Reprinted from *IEEE Transactions on Computers*, Volume 37, Number 4, April 1988, pages 398–405.

The MAFT Architecture for Distributed Fault Tolerance

ROGER M. KIECKHAFER, MEMBER, IEEE, CHRIS J. WALTER, MEMBER, IEEE, ALAN M. FINN, MEMBER, IEEE, AND PHILIP M. THAMBIDURAI, MEMBER, IEEE

Abstract—**This paper describes the Multicomputer Architecture for Fault-Tolerance (MAFT), a distributed system designed to provide extremely reliable computation in real-time control systems. MAFT is based on the physical and functional partitioning of executive functions from application functions. The implementation of the executive functions in a special-purpose hardware processor allows the fault-tolerance functions to be transparent to the application programs and minimizes overhead. Byzantine Agreement and Approximate Agreement algorithms are employed for critical system parameters. MAFT supports the use of multiversion hardware and software to tolerate built-in or generic faults. Graceful degradation and restoration of the application workload is permitted in response to the exclusion and readmission of nodes, respectively.**

Index Terms—**Approximate agreement, distributed systems, fault tolerance, flight control, interactive consistency, real-time systems.**

I. Introduction

THE computerization of life-critical control systems has placed extreme safety requirements on real-time computing systems. Perhaps the most stringent requirement to date is that proposed for flight-critical control functions in advanced commercial transport aircraft. The failure probability of such systems is required to be about 10^{-10}/h; this is approximately three orders of magnitude more stringent than the corresponding requirement for military aircraft [1]. In those few systems designed to specifically address this extreme level of dependability, fault-tolerance is achieved through modular redundancy and the voting of data from replicated tasks [2]–[4].

Many problems unique to extremely reliable real-time systems were first addressed by the Software Implemented Fault-Tolerance (SIFT) distributed computer system [2], [9], leading to some fundamental results in fault-tolerance theory [8]. However, performance measurements have shown that SIFT executive functions can consume up to 80 percent of the system throughput [5]. This result demonstrates that the computational overhead of maintaining fault tolerance can seriously detract from the system resources available for application functions. A contemporary alternative to the software intensive approach of SIFT is the Fault Tolerant Multiprocessor (FTMP) [3]. FTMP provides hardware assistance for several system executive functions, such as voting and synchronization. Still, studies have shown that up to 60 percent of the system throughput can be consumed by FTMP executive functions [10]. A descendent of FTMP is the Fault Tolerant Processor (FTP) of the Advanced Information Processing System (AIPS) program [11]. FTP has shown much higher efficiencies than either SIFT or FTMP [4]. Whereas SIFT and FTMP are multiprocessor systems, FTP functions as a uniprocessor, employing redundant processing channels solely to provide fault tolerance. The effective processing power of FTP is thus limited to that of a single processor.

This paper describes a system called the Multicomputer Architecture for Fault-Tolerance (MAFT) [7], a distributed computer system designed to combine extreme reliability with high performance in a real-time environment, which is the result of an extended research effort in ultrareliable fault-tolerant systems [6].

Johnston *et al.* [1] have tabulated the system requirements for a variety of aerospace-related control applications. These requirements can be partitioned into two categories: performance and dependability. For a commercial flight-control system, the performance requirements are characterized by control loop frequencies (iteration rates) of up to 200 Hz, instruction throughputs of up to 5.5 million instructions per second (MIPS), I/0 rates of up to one million bits per second (MBPS), and a transport lag (input to output delay) as short as 5 ms. These numbers are representative of the applications that MAFT was meant to support, and are taken as minimum performance objectives for the system. The major dependability objective of MAFT is the mission safety specification that the probability of system failure be approximately 10^{-10}/h over a 10 h mission, without repairs. To meet this objective, MAFT must tolerate subtle and highly improbable faults, such as multiple coincident faults, malicious or "Byzantine" faults, and built-in or "generic" faults.

Section II presents an overview of the MAFT architecture. Section III, comprising the bulk of the text, describes how the executive functions are implemented in MAFT. It also

Manuscript received June 8, 1987; revised November 23, 1987.

R. M. Kieckhafer was with the Bendix Aerospace Technology Center of the Allied Signal Corporation, Columbia, MD, 21045. He is now with the Department of Computer Science, University of Nebraska, Lincoln, NE 68588-0115.

C. J. Walter is with the Bendix Aerospace Technology Center of the Allied Signal Corporation, Columbia, MD 21045.

A. M. Finn was with the Bendix Aerospace Technology Center of the Allied Signal Corporation, Columbia, MD 21045. He is now with the United Technologies Research Center, East Hartford, CT 06108.

P. M. Thambidurai is with the Bendix Aerospace Technology Center of the Allied Signal Corporation and the Department of Electrical Engineering, Duke University, Durham, NC 27705.

IEEE Log Number 8719330.

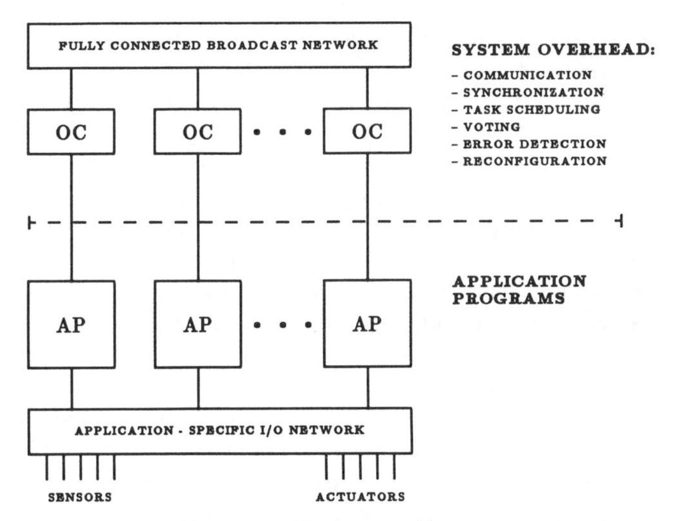

Fig. 1. MAFT system architecture.

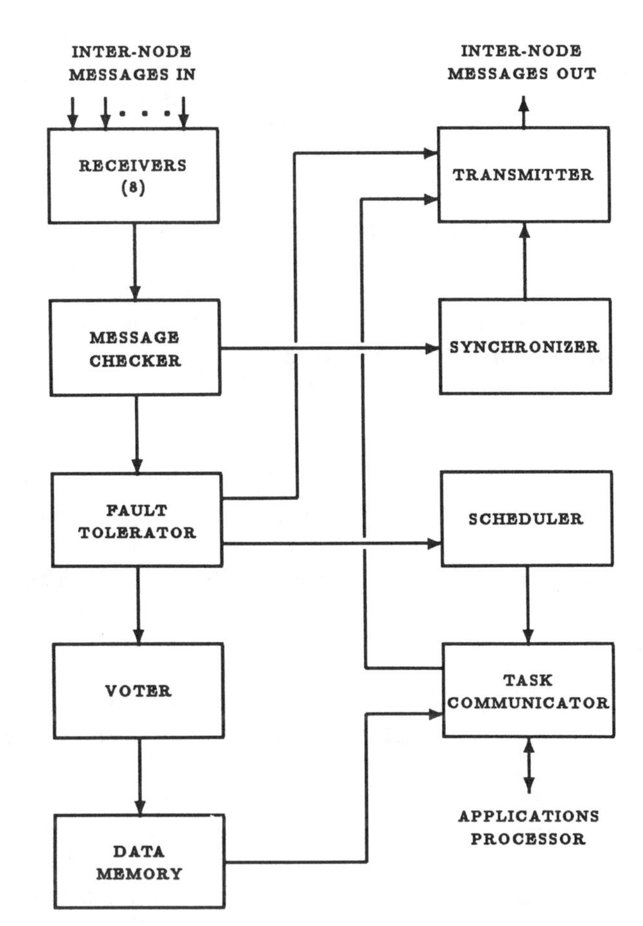

Fig. 2. Operations Controller block diagram.

discusses the capabilities and limitations of the current implementation of the architecture.

II. MAFT Architecture

A MAFT system consists of several semi-autonomous computers (nodes) connected by a broadcast bus network, as shown in Fig. 1. Each node is partitioned into two separate processors called the operations controller (OC) and the application processor (AP), respectively. The OC is a hardware-intensive data-driven processor designed to handle the vast majority of the system's executive functions. These functions include internode communication and synchronization, data voting, error detection, task scheduling, and system reconfiguration.

The OC/AP partitioning leaves the AP free to execute the application programs. Typical application functions performed by the AP include reading sensors, performing control law computations, and sending commands to actuators. While the OC is a specific device common to all MAFT systems, the AP may be any processor appropriate to a given application. The AP operating system may be extremely simple, since overall management of the distributed system is performed by the OC.

A problem facing any distributed system is maintaining agreement between nonfaulty nodes in the presence of faults. Various parameters may require exact or Byzantine Agreement [8], or Approximate Agreement [13]. A variety of solutions to the distributed agreement problem exist. The earliest solution employs hardware interstages in the communication paths [14], as used in FTP. Others make constraining assumptions about the behavior of the communication system [20], [21]. In MAFT, it is assumed that any error is possible, no matter how malicious. Therefore, interactive consistency [15] and convergent voting [13] algorithms are applied where required to maintain Byzantine Agreement and Approximate Agreement, respectively.

It is difficult to quantify the probability that generic faults exist in any given system. However, software fault-tolerance studies have shown that the existence of generic faults cannot be ignored [12]. To deal with generic faults, MAFT permits the use of design diversity in both software and hardware. This feature complicates the problem of detecting faults and managing the system. For example, truly dissimilar copies of a given task will most likely have slightly different execution times. Similarly, differences in such factors as numerical stability, arithmetic precision, and timing can cause slight differences in the various copies of a given data value. Unlike SIFT, FTMP, and FTP, equality among nonfaulty copies of a particular data value is not required in MAFT. Also, unlike FTMP and FTP, copies of a task are not required to run in exact synchrony.

Two MAFT prototypes are currently being implemented. The first is a demonstrator in which the OC's are implemented in standard 7400 series TTL technology. To date, four of six planned nodes have been assembled and operated as a system. In the second prototype, the OC's are implemented in a set of seven 2 μm CMOS chips. The current realization of the architecture will support up to eight nodes, providing sufficient redundancy to tolerate multiple coincident faults.

III. OC Functions

The OC is a special purpose data-driven processor composed of several semi-autonomous subsystems. These subsystems, illustrated in Fig. 2, perform the executive functions listed in Fig. 1.

Communication

Each OC has two communication functions. It must communicate with all other OC's in the system, and with its own AP. An OC communicates with the other OC's through formatted messages transmitted over its dedicated serial broadcast link. The effective bandwidth in the demonstrator implementation is about one MBPS per node. Dedicated receivers in the receiver subsystem monitor each link and receive all messages broadcast by the OC associated with that link. The Message Checker subsystem subjects each received message to a variety of physical and logical checks. Messages deemed to be correct are passed to the other subsystems, as appropriate, for further processing.

Messages are classified as data, scheduling, synchronization, or error-management messages. A data message is broadcast by an OC whenever it receives a computed data value from its own AP. There are two types of scheduling messages: the task completed/started (CS) message, transmitted by an OC whenever its own AP completes an application task, and the task interactive consistency (TIC) message, transmitted by all nodes at predefined intervals. The synchronization or system state (SS) message is also transmitted by all nodes at predefined intervals. There are two types of error-management messages: the error report (ERR) message transmitted by a node when it detects an error committed by any node, and the base penalty count (BPC) message, transmitted by all nodes at predetermined intervals. The BPC message informs the recently powered-up node of the error status of all nodes in the system.

The contents of all internode messages are protected by a strict link protocol and an error control code (ECC). However, the fault tolerance of MAFT is not predicated upon the reliable delivery of all messages. MAFT relies on modular redundancy and distributed agreement algorithms to mask the effects of corrupted messages. While the ECC and link protocols cannot guarantee the detection of individual corrupted messages, they do probabilistically preclude the persistent undetectable corruption of multiple messages.

An OC communicates with its AP through an asynchronous parallel interface located in the Task Communicator subsystem. The interface appears as a simple input/output device to the AP. This approach eliminates the need for special purpose interfacing hardware or software in the AP. Upon completion of a task, the AP informs the OC. The OC responds with the next task to be started. The AP immediately starts the new task while the OC broadcasts a CS message to all OC's. Voted data required by the task are provided by the OC from the data memory as requested by the AP. Similarly, the OC receives from the AP any output data which require voting. Each data value is broadcast to all OC's in a data message as soon as it is received from the AP.

Synchronization

There are two major synchronization functions in MAFT; the first pertains to steady-state operation, the second to startup. The steady-state algorithm is responsible for maintaining synchronization of the nodes in the "operating set," which is defined as the subset of the system nodes that are synchronized to each other and considered to be nonfaulty. The startup algorithm has two modes: "cold start" mode for the initial synchronization of the system, and "warm start" mode for the synchronization of a node to an existing "operating set."

Steady State: The steady-state algorithm is similar to that used by SIFT [9]. Loose, frame-based synchronization is achieved through the exchange of system state (SS) messages whose transmission implicitly denotes the local clock time. One iteration of the steady-state algorithm entails the broadcast of two versions of the SS message. For the first message, each node counts a nominal number of local clock ticks and then broadcasts a "presync" SS message. Each node timestamps these presync messages on receipt. Since all nodes use the same nominal count parameter, each local timestamp is a function of the sending node's initial synchronization skew, the sending node's local clock rate, and the message transmission delays across the link. Each node computes an error estimate as the difference between the timestamp for its own presync message and its voted value for all the presync timestamps. The error estimate is used to adjust the number of clock ticks counted before sending the second message. For the second message, each node counts the locally adjusted nominal number of clock ticks and then broadcasts a "sync" SS message. Voted timestamp values are produced by the fault-tolerant voting algorithm described in the data management and voting section. The correctness and accuracy of this algorithm may be derived by regarding it as a hardware implementation of Srikanth and Toueg's optimal clock synchronization algorithm [17].

The accuracy of steady-state synchronization depends on several parameters: the length of the synchronization interval, clock drift, and message delivery delay. In a worst-case flight-control configuration, the nodes are physically distributed throughout a large aircraft. Given this configuration, the skew between any two nonfaulty nodes will not exceed 18 μs in the current OC implementation, even in the presence of a malicious fault.

Startup: In previous systems, guaranteed automatic startup was not considered necessary [16]. However, reliable "cold start" may be required to recover from catastrophic midmission events, such as a severe lightning strike or multiple power supply interruptions. Reliable "warm start" is required for graceful readmission of nodes previously excluded for transient faults.

For cold start, all nonfaulty nodes converge to and agree upon a single operating set. Each node synchronizes temporally by iteratively timestamping, in overlapping buffers, the receipt of SS messages for a given period of time. The period is chosen such that one iteration of the synchronization algorithm by any operational, nonfaulty node will be heard. Each node's synchronization interval is determined from these data and voting produces a target synchronization time. Only bounded corrections which shorten the second synchronization message interval are allowed in converging to the target time. This is an implementation of a nonterminating Approximate Agreement algorithm [13].

Termination of the cold start algorithm is achieved by Byzantine Agreement. In the interval preceeding the presync SS message, each node records a bit vector denoting which other nodes are "in sync with" (ISW) it. The ISW vectors are broadcast in the sync SS message. If all the nodes were started at approximately the same time, with no faults, it would be sufficient to determine the largest clique from the received ISW vectors. An operating set would be formed when enough nodes were in the clique. However, variations in message delivery time or the presence of faults may produce asymmetric ISW vectors (node i in sync with node j, but not vice versa). Therefore, the received ISW vectors are rebroadcast in the presync SS message. An interactive consistency algorithm is used to compute consistent ISW vectors, and reduce them to a potential operating set vector. When there are max $[(3f+1), (\lfloor N/2 \rfloor + 1)]$ nodes in the potential operating set vector (where f is the maximum number of maliciously faulty nodes, and N is the number of nodes in the system), reconfiguration to that operating set may be initiated.

During warm start, an operating set already exists. A starting node announces its intention to join the operating nodes by beginning to broadcast SS messages. If the starting node detects at least max $[(3f + 1), (\lfloor N/2 \rfloor + 1)]$ operating nodes, it converges to its voted value of the operating nodes' timestamps. When the starting node has synchronized itself with the operating nodes, the operating nodes may reconfigure to include it in the operating set.

Data Management and Voting

Each OC stores a copy of all shared application data values in its own data memory. This scheme provides N-way redundant storage of all data, and makes it immediately available to any node. The OC handles the management and voting of application data in a manner transparent to the AP.

Each data message is tagged with a data identification descriptor (DID) which uniquely labels the data contained in the message. Upon receipt of a data message from any node, each OC performs reasonableness checks to filter out data which are outside of predefined maximum and minimum limits. Data which pass the reasonableness checks are forwarded to the voter subsystem and trigger a new vote on the values with that DID. The voter does not wait for the arrival of all expected copies of a DID. Rather, it performs an "on-the-fly" vote using the new copy and any previously received copies. If one copy of the task generating the data value should fail or "hang," the voted value of the remaining copies is still available.

A dynamic deviance check verifies that all inputs used in the voting process are within a specified window of the voted value. The size of the deviance window is individually defined for each DID. The voter performs the deviance check every voting cycle and identifies the source node for each copy which fails the check. Deviance checking and on-the-fly voting permit truly dissimilar versions to be employed in both software and hardware.

Two algorithms are available for voting on application data. The first is the familiar "median select" (MS) algorithm which selects the center value for any odd number of inputs and averages the two central values for an even number of inputs. The second algorithm is the "mean of the medial extremes" (MME). These algorithms are just two variations in the more general Fault-Tolerant Midpoint voting strategy [13]. This strategy discards the μ most extreme values from either end of a sorted set of N inputs, and computes the mean of the two remaining extremal values. Computationally, the MS and MME algorithms vary only in their choice of μ. MS uses $\mu = \lfloor (N - 1)/2 \rfloor$, whereas MME used $\mu = \lfloor (N - 1)/3 \rfloor$.

The two algorithms offer different fault-tolerance properties. MS is the more robust algorithm since it discards more extremal values than MME. However, MME is convergent in the presence of up to μ erroneous copies regardless of the nature of those errors. Dolev [13] has shown that MME has a guaranteed convergence rate of 1/2. Conversely, a malicious error can prevent MS from converging. The application designer selects which of the two algorithms is to be used for each DID.

Application Task Scheduling

In MAFT, the application software is broken into tasks—indivisible blocks of code which must be executed without interruption on a single AP. As viewed by the OC, each task has several properties: iteration frequency, relative priority, desired redundancy, and intertask dependencies.

Dependencies between tasks may include concurrent forks and joins (AND-FORKS and AND-JOINS) or conditional branches (OR-FORKS and OR-JOINS). The MAFT scheduler subsystem treats all tasks as periodic. Nonperiodic behavior is obtained through conditional branching. MAFT supports tasks of varying frequencies within the constraints of a binary frequency distribution, i.e., all frequencies are 2^{-j} times the highest frequency, where j is a nonnegative integer.

The iteration period of a task is the reciprocal of its frequency. The shortest iteration period in the MAFT hierarchy is the "atomic period," so called because it is indivisible with respect to iteration periods. The boundaries between atomic periods coincide with the transmission of an SS message. While no single task may be repeated more than once per atomic period, several shorter tasks may be scheduled and executed on each node during the same atomic period. Conversely, a low-frequency task may run for several atomic periods.

The longest permissible iteration period is referred to as the "master period," and corresponds to the period of the lowest frequency task in the workload. The current OC implementation allows up to 1024 atomic periods per master period, permitting tasks of up to ten distinct frequencies to execute concurrently.

Scheduler Operation: The scheduling strategy selected for MAFT is a fault-tolerant variation of a deterministic priority-list algorithm. In this approach, each task is assigned a unique priority number. When a node becomes available, the list of tasks is scanned in order of decreasing priority. The first task encountered which is ready for execution is selected. In MAFT, the assignment of tasks to nodes is determined by the task reconfiguration process and is static for any given operating set. A separate priority list is thus maintained for each node.

The scheduling function is fully replicated so that the scheduler of each node selects tasks for every node in the system. The selection for its own node is used to dispatch tasks for its own application processor. The selections for all other nodes are used to monitor the actions of the other schedulers.

When an AP completes its currently assigned task, it passes a branch condition (BC) value to its OC, to be used by the schedulers in resolving OR-FORK dependencies. The OC immediately broadcasts a task completed/started (CS) message, containing the received BC.

To maintain Byzantine Agreement on scheduling data, the scheduler employs an Interactive Consistency algorithm. This algorithm requires synchronized transmission rounds in which each node rebroadcasts the data received in the previous round. The number of rebroadcasts required to guarantee resolution of the disagreement is equal to the maximum number of simultaneous malicious faults to be tolerated [15]. MAFT incorporates one round of rebroadcast to deal with a single malicious fault.

Synchronization of the rebroadcasts is accomplished by defining a "subatomic" period whose boundaries are marked by the simultaneous transmission of task interactive consistency (TIC) messages by all nodes. An integer number of subatomic periods constitutes one atomic period. The TIC message contains two bytes, reflecting the scheduling activity of all nodes during the previous subatomic period. The first byte is a task completed (TC) byte indicating those nodes from which a CS message was received. The second byte is a BC byte indicating the branch condition contained in each received CS message. After the TIC messages have been received, each node executes an Interactive Consistency Algorithm on the contents of the TIC messages. Thus, all nodes reach Byzantine Agreement on the existence and content of each CS message transmitted.

Task execution timers synchronized to the subatomic period boundaries monitor the execution time of each task. If the execution time of a task is shorter than a predetermined minimum or longer than a predetermined maximum, then an execution timer error is reported. In addition, a task sequence error is reported whenever a node fails to execute tasks in the sequence expected by the schedulers.

Scheduler Performance: To ensure consistency among schedulers, all data structure modifications are based on the agreed upon contents of TIC messages, rather than CS messages. The "confirmation delay" created by waiting for TIC messages imposes a potential performance penalty in the release of successor tasks. If the task precedence graph defined by the intertask dependencies consists of a single path of dependent tasks, then the AP's will be idle while completions are confirmed. However, if the graph contains parallel paths, then independent tasks may be executed while dependencies are resolved. In the type of real-time workloads expected for MAFT applications, there is a significant amount of available parallelism, indicating that high-AP utilization can be maintained.

Fig. 3 shows the precedence graph for part of an actual workload currently being developed for a particular MAFT-based flight control system [22]. This graph concurrently processes the control laws for each of the three axes of rotation, along with a system monitoring function. All four paths converge by an AND-JOIN to a validity checking function,

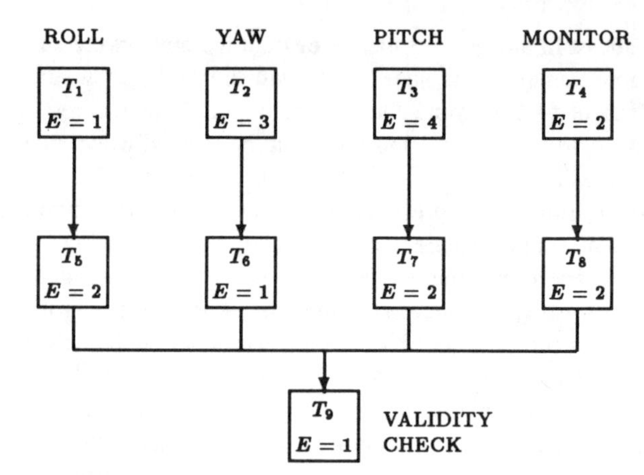

Fig. 3. Sample workload fragment.

<table>
<tr><td>PERIOD</td><td>1</td><td>2</td><td>3</td><td>4</td><td>5</td><td>6</td><td>7</td><td>8</td><td>9</td><td>10</td><td>11</td><td>12</td></tr>
<tr><td>GROUP 1</td><td>1</td><td colspan="4">3</td><td colspan="2">5</td><td colspan="2">7</td><td colspan="2">ϕ</td><td>9</td></tr>
<tr><td>GROUP 2</td><td colspan="3">2</td><td colspan="2">4</td><td>6</td><td>ϕ</td><td colspan="2">8</td><td colspan="2">ϕ</td><td>9</td></tr>
</table>

Fig. 4. Full six-node system standard Gantt chart.

task 9. The maximum execution time of each task, E, is listed in units of subatomic periods.

Assume that this workload requires triple redundancy in all tasks except task 9. Task 9 must be executed by all nodes in the operating set, and therefore has a variable redundancy of N, rather than a fixed redundancy of three. Fig. 4 is a standard Gantt chart [19] of the workload on a six-node system, with tasks 1, 3, 5, 7, and 9 assigned to nodes 1, 2, and 3 (Group 1), and tasks 2, 4, 6, 8, and 9 assigned to nodes 4, 5, and 6 (Group 2). The null task ϕ indicates processor time unavailable to this precedence graph due to confirmation delay.

The Gantt chart illustrates the effect of confirmation delay on processor utilization. During subatomic period 7, the AP's of nodes 4, 5, and 6 are idle because tasks 7 and 8 cannot be released until the completion confirmation of tasks 3 and 4, respectively. Similarly, all processors are idle during subatomic periods 10 and 11 because task 9 cannot be released until completion of tasks 7 and 8 is confirmed. This schedule utilizes 19 of the 24 available processor periods, for an overall processor utilization of 79 percent.

The minimum length of the subatomic period in the current OC implementation is 400 μs. For a full eight-node system, a minimum of seven subatomic periods is required per atomic period, yielding a maximum control loop frequency of 357 Hz. The flight-control workload above is designed for a 500 μs subatomic period. With this workload, the six-node system can support an atomic period as short as 6 ms for a frequency of 167 Hz.

Task Reconfiguration

In MAFT, task reconfiguration refers to the process of redistributing the application workload to account for changes in the system operating set. The overall objectives of task reconfiguration are to provide graceful degradation of application functions as resources are lost, and graceful restoration of those functions as resources are restored. The task reconfiguration process produces an eligibility table, indicating those nodes to which each task may be assigned. The actual selection and execution of tasks from the eligibility table are controlled by the normal scheduling process at run time. A new value for the system operating set is periodically computed by the fault tolerator subsystem and compared to the previous operating set. If the two sets differ, then task reconfiguration is initiated. Byzantine Agreement is assured by an Interactive Consistency algorithm performed on the operating set value before it is used.

A detailed description of the task reconfiguration algorithm is presented in [18]. Briefly, the algorithm employs three independent processes, combining their results to determine the eligibility of each task. Each process is path independent, and therefore reversible. The *Global Task Activation Process* activates or deactivates individual tasks to account for changes in the overall system capability. This process is realized by defining a set of relevant nodes and an initial activation status (either active or inactive) for each task. When the number of relevant nodes remaining in the operating set reaches a predetermined threshold, the activation status of the task is complemented. The *Task Reallocation Process* reallocates tasks among the operating nodes of the system to maintain the desired redundancy of each task. This process is realized by assigning tasks to nodes in a predetermined order of preference. If the redundancy of a task is R, then the task will be executed by the R most preferred nodes for that task in the current operating set. The *Task–Node Status Matching Process* allows tasks to be prohibited from executing on individual nodes based upon each node's operational status (i.e., included or excluded). For example, it may be desirable to prohibit output tasks from executing on an excluded node, assigning instead a comprehensive set of diagnostic tasks.

The ability to maintain functionality while the system degrades depends on the assignment of the node preferences for each task. A simple algorithm for defining preferences has been applied to the example workload of Fig. 3. This algorithm starts with the full six-node system, then attempts to equalize the workload as N, the number of nodes in the operating set, is reduced. For each value of N, preferences are assigned in order of decreasing task execution time. The preference orderings produced by this algorithm are shown in Table I. For each task i, the table lists its execution time E_i in subatomic periods, its required redundancy $R_i(N)$, and the system nodes in order of decreasing preference.

The effectiveness of task reconfiguration for the example workload of Fig. 3 is shown in Table II for all values of N from the minimal three-node system through the full six-node system. $W(N)$ is the total processing time required by the workload, computed as the summation of $E_i R_i(N)$ over all i. The value $t_0(N) = W(N)/N$ is the ideal (shortest) execution time for the workload. The value t is the actual execution time observed for each operating set. The AP utilization is then $t_0(N)/t$ for each case.

Table II shows that AP utilization is never less than 70 percent for any operating set. In the minimal three-node system, the workload requires 20 subatomic periods. Given

TABLE I
TASK-TO-NODE PREFERENCE ORDERING

i	E_i	$R_i(N)$	Node Preference Order
1	1	3	1 2 3 5 4 6
2	3	3	4 5 6 1 3 2
3	4	3	1 2 3 4 6 5
4	2	3	4 5 6 2 1 3
5	2	3	1 2 3 5 4 6
6	1	3	4 5 6 2 3 1
7	2	3	1 2 3 6 5 4
8	2	3	4 5 6 3 2 1
9	1	N	1 2 3 4 5 6

TABLE II
MULTIPROCESSING EFFICIENCY

N	$W(N)$	$t_o(N)$	t	% AP Utilization
3	54	18.00	20	90
4	55	13.75	17-18	76-81
5	56	11.20	15-16	70-75
6	57	9.50	12	79

the 500 μs subatomic period length, the minimum atomic period is 10 ms for a frequency of 100 Hz. Deactivation or replacement of less critical tasks can help to maintain loop frequency in severely degraded systems. For example, if the maximum permissible atomic period for this application is 18 subatomic periods (9 ms), then the three-node system exceeds its deadline by two subatomic periods. Deactivation of the four-period long monitoring function (tasks 7 and 8) when $N = 3$, and activation of a simpler two-period long monitoring function would permit the control functions to meet their deadlines.

Error Handling

An OC contains error detection mechanisms which continuously monitor the behavior of all nodes as revealed by their message traffic. These errors are reported in one or more of 31 error flags contained in an ERR message. The OC also uses a penalty counting mechanism to communicate the overall health of the node. A base penalty count (BPC) is maintained which indicates the current value of the accrued penalties for every node. An incremental penalty count (IPC) is also maintained for each node, containing a proposed penalty assessment for the node based on error detections during the current atomic period. Whenever an error is detected, a penalty weight, unique to the triggered detection mechanism, is added to the IPC. The value of the penalty weight for each detection mechanism is set by the application designer to reflect the relative severity of the type of error detected.

At the beginning of every atomic period, each node broadcasts an error report (ERR) message regarding each node in the system. The ERR message contains the accused node's identity, error flags, BPC, and IPC. Upon receipt of a round of ERR messages, each node votes on the contents of the error flag, BPC, and IPC fields. An updated BPC value is calculated for each node by adding its voted IPC to its voted BPC value. In this way, Byzantine Agreement on both the IPC and BPC counts is guaranteed. (The message which contained the original error is the first round broadcast of an interactive consistency algorithm; the ERR message constitutes the rebroadcast.)

Once Byzantine Agreement is reached, the BPC value of an accused node is updated and compared to a predefined exclusion threshold. If the value of the BPC exceeds the threshold, then each OC will recommend exclusion of the accused node from the operating set. The recommendation is made through a "next operating set" field contained in the next SS message. Upon receipt of the SS messages, each node votes on the contents of the next operating set fields in accordance with the Interactive Consistency algorithm, and takes the voted value as the new operating set of the system. Thus, Byzantine Agreement is also guaranteed on the inclusion or exclusion of each node in the operating set.

Reconfiguration of the application tasks may be initiated immediately or may be delayed until the beginning of the next master period, at the application designer's discretion. The excluded node is immediately prohibited from participation in any voting or decision making processes, masking its errors until reconfiguration is completed.

A faulty node will continue to accrue penalties, even after it has been excluded. If the node was excluded for a transient fault, it will subsequently exhibit correct behavior, and will *not* accrue penalties. At the beginning of each master period, the BPC of each node is decremented by a specified amount; this amount may be zero for permanent exclusion. When the BPC of an excluded node falls below a predetermined readmission threshold, each OC recommends readmission via the same sequence used to decide on exclusion.

Under normal circumstances, error detection mechanisms will not be exercised for extended periods of time, resulting in extremely long latency times for faults within the detection mechanisms themselves. Therefore, a system level self-test mechanism has been implemented in the OC. At a predetermined time, one node, selected on a rotating basis, broadcasts a stream of erroneous bits designed to exercise specific error detection mechanisms. All nonfaulty nodes respond by broadcasting error reports on the received errors. Addition of the IPC to the BPC is suspended to prevent exclusion of the originating node. Any node containing a faulty error detection mechanism is identified by the fact that its error report differs from the consensus. Using the consensus as the basis for correctness decouples the system self-test mechanism from the actual contents of the erroneous message stream. Thus, an error in the test messages generated by a specific source node does not affect the accuracy of the diagnosis. Rotation of source node duties ensures full test converage as long as at least one node is capable of correctly generating the self-test message stream.

IV. Summary

The MAFT system has been designed to provide high performance and extreme reliability across a broad spectrum of real-time applications. To satisfy these goals, MAFT relies on the functional and physical partitioning of system executive functions from application functions. The executive functions are performed by hardware-intensive data-driven operations controllers (OC) while application functions are assigned to general-purpose application processors (AP).

MAFT meets or exceeds its stated performance objectives by providing control loop frequencies in excess of 350 Hz, and bus bandwidths of about one MBPS. Since MAFT supports multiprocessing and permits a wide choice of AP's, the throughput of 5.5 MIPS is not difficult to attain. MAFT employs Interactive Consistency and Convergent Voting algorithms to maintain Byzantine Agreement and Approximate Agreement, respectively, on critical system and application parameters. Specific MAFT features such as on-the-fly voting and deviance checking support the use of dissimilar hardware and software in the system. Two different prototypes have been implemented and are currently undergoing testing.

Acknowledgment

The MAFT architecture is the result of long-term research and development efforts involving several divisions of the Allied Signal Corporation. The authors wish to thank all those who have contributed to MAFT and to the production of this paper, particularly M. C. McElvany. The example workload, due to F. Herman, is also acknowledged.

References

[1] M. W. Johnston *et al.*, "AIPS system requirements (revision 1)," CSDL-C-5738, Charles Stark Draper Lab., Inc., Cambridge, MA, Aug. 1983.

[2] J. H. Wensley *et al.*, "SIFT: Design and analysis of a fault-tolerant computer for aircraft control," *Proc. IEEE*, vol. 66, Oct. 1978.

[3] A. L. Hopkins *et al.*, "FTMP—A highly reliable fault-tolerant multiprocessor for aircraft," *Proc. IEEE*, vol. 66, Oct. 1978.

[4] T. B. Smith, "Fault-tolerant processor concepts and operation," in *Proc. Fourteenth IEEE Fault-Tolerant Comput. Symp.*, June 1984.

[5] D. L. Palumbo and R. W. Butler, "Measurement of SIFT operating system overhead," NASA Tech. Memo. 86322, 1985.

[6] A. Whiteside *et al.*, "Fault-tolerant multicomputer system for control applications," in *Proc. Eleventh IEEE Fault-Tolerant Comput. Symp.*, June 1981.

[7] C. J. Walter *et al.*, "MAFT: A multicomputer architecture for fault-tolerance in real-time control systems," in *Proc. IEEE Real-Time Syst. Symp.*, Dec. 1985.

[8] L. Lamport, R. Shostak, and M. Pease, "The Byzantine generals problem," *ACM TOPLAS*, vol. 4, pp. 382–401, July 1982.

[9] J. Goldberg *et al.*, "Development and analysis of the software implemented fault-tolerance (SIFT) computer," Final Rep. NASA Contract NASA-CR-172146, Feb. 1984.

[10] E. W. Czeck, D. P. Siewiorek, and Z. Segall, "Fault free performance validation of a fault-tolerant multiprocessor: Baseline and synthetic workload measurements," Carnegie Mellon Univ., Dep. Comput. Sci., CMU-CS-85-117.

[11] ——, "Advanced information processing system (AIPS) system requirements (revision 1)," Rep. CSDL-C-5709, Charles Stark Draper Lab., Inc., Cambridge, MA, Oct. 1984.

[12] J. C. Knight *et al.*, "A large scale experiment in N-version programming," in *Proc. Fifteenth IEEE Fault-Tolerant Comput. Symp.*, June 1985, pp. 135–139.

[13] D. Dolev *et al.*, "Reaching approximate agreement in the presence of faults," in *Proc. Third Symp. Reliability Distributed Software Database Syst.*, Oct. 1983.

[14] D. Davies and J. Wakerly, "Synchronization and matching in redundant systems," *IEEE Trans. Comput.*, vol. C-27, pp. 531–539, June 1978.

[15] M. Pease, R. Shostak, and L. Lamport, "Reaching agreement in the presence of faults," *J. Ass. Comput. Mach.*, vol. 27, pp. 228–234, Apr. 1980.

[16] R. W. Butler, "An assessment of the real-time application capabilities of the SIFT computer system," NASA Tech. Memo. 84432, Apr. 1982.

[17] T. K. Srikanth and S. Toueg, "Optimal clock synchronization," *J. Ass. Comput. Mach.*, vol. 34, July 1987.

[18] R. M. Kieckhafer, "Task reconfiguration in a distributed real-time system," in *Proc. Eighth IEEE Real-Time Syst. Symp.*, Dec. 1987.

[19] G. K. Manacher, "Production and stabilization of real-time task schedules," *J. Ass. Comput. Mach.*, vol. 14, July 1967.

[20] O. Babaoglu and R. Drummond, "Streets of Byzantium: Network architectures for fast reliable broadcasts," *IEEE Trans. Software Eng.*, vol. SE-11, pp. 546–554, June 1985.

[21] K. Perry, "Randomized Byzantine agreement," in *Proc. Fourth Symp. Rel. Distributed Software Database Syst.*, Silver Springs, MD, Oct. 1984, pp. 107–118.

[22] D. P. Gluch and M. J. Paul, "Fault-tolerance in distributed digital fly-by-wire flight control systems," in *Proc. AIAA/IEEE Seventh Digital Avion. Syst. Conf.*, Oct. 13–16, 1986.

Roger M. Kieckhafer (S'81–M'83) received the B.S. degree in nuclear engineering from the University of Wisconsin, Madison, in 1974 and the M.S. and Ph.D. degrees in electrical engineering from Cornell University, Ithaca, NY, in 1982 and 1983, respectively.

From 1983 to 1987 he was with the Bendix Aerospace Technology Center of the Allied Signal Corporation, Columbia, MD, as a member of the Technical Staff. While there, he was engaged in research on fault-tolerant distributed computer systems, and was one of the designers of the Multicomputer Architecture for Fault-Tolerance (MAFT). He is currently with the University of Nebraska, Lincoln, as an Assistant Professor in the Department of Computer Science. His research interests include computer architecture, distributed systems, and fault- tolerance.

Dr. Kieckhafer is a member of the Association for Computing Machinery and the IEEE Computer Society.

Chris J. Walter (M'75) received the B.S. degree in electrical engineering from the University of Notre Dame, Notre Dame, IN, in 1975 and the M.S. degree in computer science and electrical engineering from the University of Michigan, Dearborn, in 1978 and is completing the D.Sc. degree in computer science at the George Washington University, Washington, DC.

Since 1980 he has been a member of the Technical Staff at the Bendix Aerospace Technology Center of the Allied Signal Corporation working in the design of fault-tolerant computer architectures and leads the System Integration and Reliability Group. His research interests include fault tolerance, computer architecture, and distributed systems. He is currently involved in the development of a highly architecture for artificial intelligence applications.

Mr. Walter is a member of the IEEE Technical Committee on Fault Tolerance.

Alan M. Finn (S'80–M'83) received B.S. degrees in mathematics and electrical engineering from Rensselaer Polytechnic Institute, Troy, NY, in 1977. In 1979 he matriculated at Cornell University, Ithaca, NY. Between 1977 and 1979 he was employed by IBM, Poughkeepsie, NY, where he worked in Advanced Processor Development for the 308x series of machines and received the Ph.D. degree in 1983.

From 1983 to 1986, he was employed by the Bendix Aerospace Technology Center of the Allied Signal Corporation where he worked on the MAFT fault-tolerant real time control system for commerical aircraft. He is currently employed by the United Technologies Research Center and is working on fault-tolerance and signal processing. His research interests include computer architecture, fault tolerance, signal processing, and finite field mathematics.

Dr. Finn is a member of the IEEE Computer Society and the ACM.

Philip M. Thambidurai (M'87) received the B.S. degree in physics from Geneva College, Beaver Falls, PA, in 1978, and the M.S. degree in nuclear physics from North Carolina State University, Raleigh, in 1981.

Since 1985 he has been a member of the Technical Staff at the Bendix Aerospace Technology Center of the Allied Signal Corporation. He is currently involved in research and development related to future versions of the MAFT system. His research interests include fault tolerance, computer architecture, real-time systems, and modeling. He is also working towards the Ph.D. in electrical engineering at Duke University, Durham, NC.

Mr. Thambidurai is a member of the IEEE Computer Society.

Reprinted from *IEEE Transactions on Reliability*, Volume 38, Number 1, April 1989, pages 153–158.

Reliability Issues with Multiprocessor Distributed Database Systems: A Case Study

Chi-Ming Chen, Member IEEE
Bell Communications Research, Red Bank
Jose D. Ortiz, Member IEEE
Bell Communications Research, Red Bank

Key Words — **Mated architecture, Telecommunications, Multiprocessor, Downtime, Uncoverage.**

Reader Aids —
Purpose: Advance reliability analysis techniques
Special math needed for explanations: None
Special math needed to use: None
Results useful to: Reliability analysts, network planners, and design engineers

Summary & Conclusions — **Database systems are becoming an integral part of Bell Operating Companies' voice and data networks. This paper presents the results and the methodology used in the reliability analysis of a call-handling database system. This paper also identifies those areas where the reliability modeling methods and analysis techniques need to be extended to cover all causes of outages, especially when dealing with new architectures, such as the Next Generation Switch.**

The analysis indicates that for this architecture, node configurations with one redundant processor meet the hardware-allocated downtime objective. However, some configurations do not meet this objective when hardware failure rates, repair times, and other parameters such as uncoverage are increased. Some of these parameters can be controlled by the maintenance policies of the support organization, but the critical ones are mainly defined by the reliability and quality programs that equipment suppliers have in place. Reliability and quality analyses can be used to address the many areas in a supplier's program that affect product reliability. Because of the new directions that switching systems are taking, studies to extend the current reliability and quality generic requirements and analysis capabilities to these multiprocessor architectures and the Next Generation Switch are under way. The goal is to handle failures due to causes other than hardware more effectively.

1. INTRODUCTION

Database systems are becoming an integral part of the voice and data networks of the Bell Operating Companies ("Bell Operating Company" refers to any Bell Operating Company divested from AT&T). Equipment used in these networks must meet high standards of reliability and quality (R&Q) generally associated with telecommunication products, but historically not with database systems. However, recent advances in computer architectures, especially the parallel computer architectures, are affecting the design of new database systems. Multiprocessor system architectures are part of the parallel computer architectures proposed to implement reliable services. Multiprocessor systems evolved from single processor systems (with and without duplication) with the following three major motivations:

- increased throughput
- increased availability
- reduced response time[1].

The goals of the multiprocessor database system that was analyzed included these motivations among many other requirements.

Bell Communications Research (Bellcore) publishes generic requirements to inform the industry of Bellcore's view of the R&Q criteria covering major systems used in the networks of Bell Operating Companies. Bellcore also conducts R&Q analyses of selected telecommunications products. An R&Q analysis can include several of the following areas: system reliability, physical design and component reliability practices, software reliability and quality, product support systems, quality programs, and manufacturing processes [2, 3]. Bellcore also studies the development of new methods for advancing the R&Q technology.

Sections 2-4 of this paper are devoted to the reliability analysis of an illustrative multiprocessor database system architecture. The discussion of this architecture includes system description, reliability objectives, and analysis methodology. Section 5 discusses several areas of enhancement on current reliability modeling and analysis capabilities.

2. EXAMPLE-SYSTEM DESCRIPTION

The example database system can provide call handling services such as the 800 service (toll-free numbers) application. The processing functions and database reside in one pair of nodes, each of which normally handles about one-half of the traffic. However, under certain conditions, such as a major failure of one node, the remaining one is capable of handling 100 percent of the traffic. This is known as a mated architecture (see figure 1). The interface between the voice and data networks and the database services is provided by a signaling network with multiple connections to the voice and data network and to the database nodes.

Each node consists of a multiprocessor architecture with access to high-capacity disk units containing programs and data (see figure 2). Each processor or central processing unit (CPU) has a front-end unit used to interface with the signaling network and has a system disk. The number of processors provided in a node corresponds to

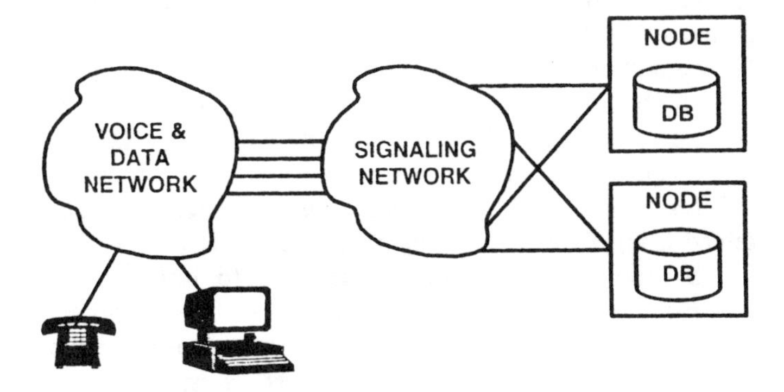

Fig. 1. Sample Mated Architecture

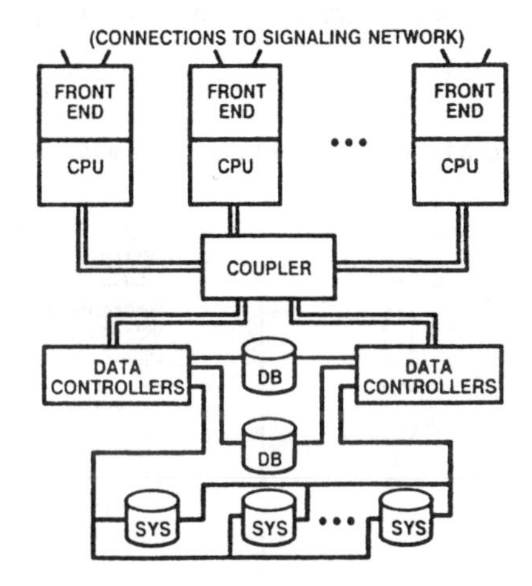

Fig. 2. Block Diagram of Node

the quantity required to handle all the anticipated traffic, with one additional processor included for redundancy. This makes it possible for one node, with a failed processor and with its mate down for any reason, to continue to provide service without interruption or degradation for an extended period. There are 3-5 processors in a node, with a corresponding number of system disks and 2 additional database (DB) disks. Access to the disks is via a coupler and data controllers; the latter are used to maximize access performance. In this example each processor handles service requests independently.

The example architecture has several characteristics that impact on the reliability of the system. The most important is that more than one-half of the processors within a node must be active for the node to remain operational. The minimum number of processors required for an active node is defined as the quorum. This is intended as a safeguard to avoid corruption of data resulting from groups of processors forming independent entities within a node and simultaneously accessing the databases—an event that could occur under certain failure conditions due to the design of the operating system. Other characteristics of this architecture are that at least one data controller and one database disk must be available for system operation.

3. RELIABILITY OBJECTIVES

The reliability objectives for the call-handling database specify a mated architecture with a system downtime of 3 minutes per year. This corresponds to an equivalent system downtime of 20.9 hours per year for a single node, if failures of the nodes are statistically independent. This means that each node can be down due to failures or other reasons for 20.9 hours, but that the overlap is only 3 minutes per year including scheduled downtime. Generally, objectives for network switching systems and applications do not provide an allowance for scheduled downtime. As a result, any known downtime is generally accounted for in the analysis by using tighter objectives or by including representative failure modes. Sources of such scheduled downtime include, but are not limited to, common operation and support activities, preventive maintenance, and software updates. For the example system no activities requiring scheduled downtime were identified.

4. ANALYSIS METHODOLOGY

The methodology for reliability analysis is outlined in the RQSSGR [3]:

a. Identify the failure modes and their system effect.
b. Allocate to hardware a fraction of downtime.
c. Generate Markov reliability models.
d. Estimate hardware failure rates, repair times and other parameters.
e. Solve the models and generate downtime predictions.
f. Conduct sensitivity analysis.

These steps are further discussed below.

4.1 *Failure Modes*

During the analysis, no failure mode was identified in the design of this architecture that could result in a failure of both nodes in the mated pair. As a result, the analysis concentrated on the reliability performance of each node and how it related to the derived objective of 20.9 hours (section 3).

The major failure modes identified for each node are multiple processor failures (below-quorum), node initialization, uncoverage failures, under-capacity, and failures of the disk community including the data controller and coupler. Node initializations are required whenever a processor leaves or joins the node—to keep track of the quorum among other things. Uncoverage is defined as the probability that a single hardware failure in the node causes a node outage. Uncoverage failures are caused by design errors or unforeseen system response to failures. Under-capacity failures can occur when a node has insufficient processing capacity to provide acceptable service. In this environment, minimum acceptable service is measured by the time required to process service requests. The following set of events results in under-capacity failures: one node in the mated pair is down, several processors are disabled in the active node (beyond the provided redundancy), and the traffic level corresponds to that anticipated for a high traffic period. The assumptions used to estimate under-capacity downtime are conservative (overestimate downtime) because the node may exercise network control functions to handle traffic overload conditions. By denying service to a fraction of service requests handled by the network, the node can effectively continue to provide adequate service to the remaining traffic. For analysis purposes, the under-capacity results provide an upper bound measure of this failure mode.

4.2 *Downtime Allocation*

This analysis concentrates on the effect of hardware failures on downtime, although it includes operational considerations. However, the downtime objective covers all possible sources of failures in the system: hardware, software, procedural, or others. To compare the downtime estimates obtained from the reliability models, which are hardware based, an allocation of 20 percent of the objective is made to hardware-caused failures. A 20 percent allocation corresponds to an allowable hardware downtime for one node of about 250 minutes per year (20 percent of 20.9 hours per year).

This allocation corresponds to the guidelines specified for network switching equipment in the LSSGR. The LATA (Local Access and Transport Area) Switching Systems Generic Requirements (LSSGR) document is a Bellcore compilation of Bellcore's views of generic requirements for switching systems based on the typical needs of Bell Operating Companies [4]. The LSSGR serves as a guide for the analysis of new switching systems. Section 12 addresses the issues pertaining to reliability. This technical reference as well as others listed in this paper is available from the Customer Service organization in Bellcore.

4.3 *Reliability Models and Analysis Parameters*

Markov models are a commonly used reliability modeling technique [5]. These models consist of state transition diagrams that are used to identify the critical functional states, such as operational, failed, and degraded, of the hardware covered by each failure mode. The diagrams also show the transitions among the various states accounting for the designed redundancy, recovery times and strategies, and other parameters that represent the system architecture. Based on the diagram structure and on assumptions for the parameter values (eg, constant failure and repair rates), well-defined methods (stochastic theory) exist for solving such models (determining the steady-state transition probabilities) and mechanized tools are documented in the literature that facilitate such calculations. Although the recovery (restore) rates have travel and on-site repair components, for the illustrative model in this analysis, a simplifying assumption is made that travel and on-site repair times are combined. This results in slightly higher downtime estimates, but appreciably reduces the modeling complexities required to handle states with multiple processor failures. The downtime estimates from the models were obtained using Bellcore's mechanized tools.

Several Markov models were used to estimate the downtime due to the major failure modes of this architecture. Figure 3 shows one of the models used to estimate the effect of the below-quorum and uncoverage failure modes in the processor community. Similar models were used to estimate the downtime due to node initializations and failures in the disk community (disks and controllers). Figure 3 also describes the states and parameters used in this model.

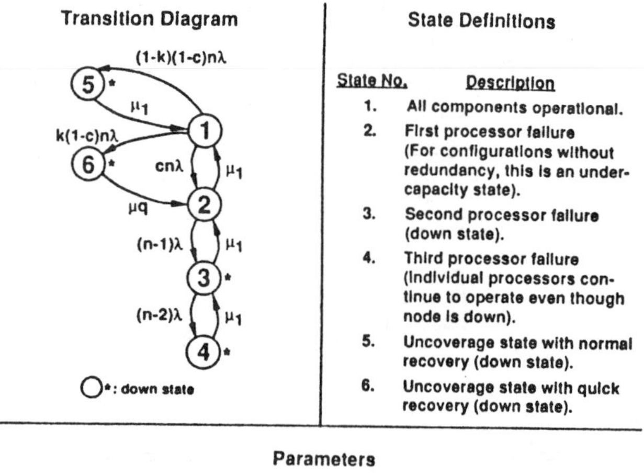

Fig. 3. Sample Markov Model for Processor Community

The mean-time-between-failures (MTBF) and mean-time-to-repair (MTTR) values used in this example are:

Component	MTBF (hours)	MTTR (hours)
Processor	1 200	4.2
Processor Initialization	2 900	0.0083
Data Controller	5 000	4.0
Disk	10 000	4.0
Coupler	730 000	5.5

To account for uncoverage, 1 percent of all single hardware faults are assumed to result in an uncoverage failure.

However, two methods of recovery from uncoverage failures are possible, quick and nominal repairs. Quick repairs (assumed to take 30 minutes) occur when the system operator is able to reconfigure the node around the fault from a maintenance station (local or remote). After this quick repair, although the node is not down, the processor that failed must be repaired. For the analysis, 90 percent of uncoverage faults are assumed in this category. The remaining uncoverage events require a typical repair with its associated nominal repair time.

4.4 *Analysis Results*

Figure 4 summarizes the hardware downtime for several configurations as estimated from the reliability models. For example, the configuration with 3 processors for capaci*y and no redundancy (3 + 0) has an estimated downtime of over 400 minutes per year and exceeds the hardware allocated objective of 250 minutes per year. Thus, node configurations without redundancy exceed the objective. The unusual effect (larger downtime for greater redundancy) observed for configurations 2 + 1 and 2 + 2 is explained below.

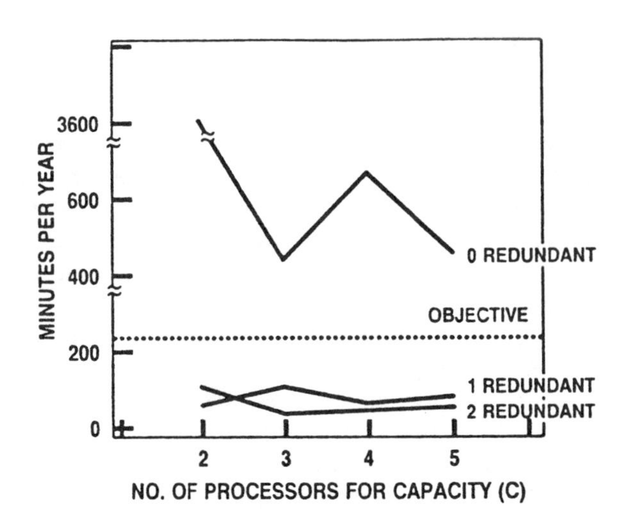

Fig. 4. Predicted Hardware Downtime

Of interest is the contribution that the failure modes have on system downtime. Figure 5 shows this effect for several node configurations. Except for configuration 2 + 0, no configuration without redundancy meets the objective—because of the under-capacity failure mode. For configurations with 4 processors or fewer, the below-quorum failure mode is dominant (excluding under-capacity). Configurations with more than 4 processors owe most of their downtime to uncoverage and node initializations (again excluding under-capacity). The effect of the disk community remains unchanged for all configurations.

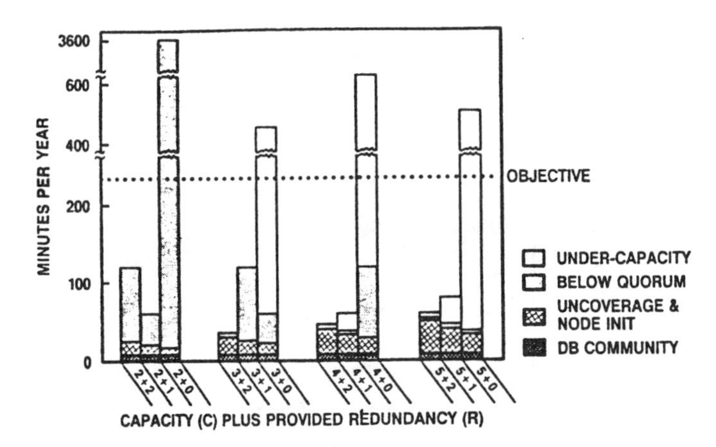

Fig. 5. Hardware Downtime by Failure Mode

The unusual effect for configurations 2 + 1 and 2 + 2 are explained by the quorum requirement of the node. By adding a redundant processor to configuration 2 + 1, the number of failures that brings the node down (2 failures) remains unchanged. This is because for configuration 2 + 2, the quorum increases by 1 to a value of 3. However, configuration 2 + 2 has more processors that can fail, and therefore the higher downtime for the below quorum category. The same effect can be observed for configuration 3 + 0 and 3 + 1, with the latter having a larger below-quorum downtime, but the effect is overshadowed by the under-capacity downtime of configuration 3 + 0. This quorum "quirk" becomes less noticeable for larger configurations.

The effect that failure modes have on system downtime can be summarized as follows. With increasing redundancy the downtime for under-capacity and below-quorum failures decreases. At the same time, the downtime for uncoverage and node initialization increases. The effect of the disk community is the same for all configurations. Figure 6 shows the net effect where the composite downtime gradually decreases with increasing redundancy until a minimum is reached at which uncoverage and initializations become dominant. This figure outlines the practical limitations of the benefits of increased redundancy for this particular architecture.

4.5 *Sensitivity Analysis*

The effect on system downtime of changing the values of several model parameters is discussed here. Figure 7 shows the effect that the failure rate has on system downtime. Configurations 3 + 1 and 4 + 1 both meet the objective for the nominal failure rate (factor of 1). But when the failure rate doubles, configuration 3 + 1 does not meet the objective. Because failure rates are a critical parameter, accurate estimates should be obtained for the equipment by using well based methods such as those provided for telecommunication products in [6]. Increases in failure rates can occur if equipments are purchased from suppliers other than the one analyzed, or if the manufacturing technology lacks control or is new.

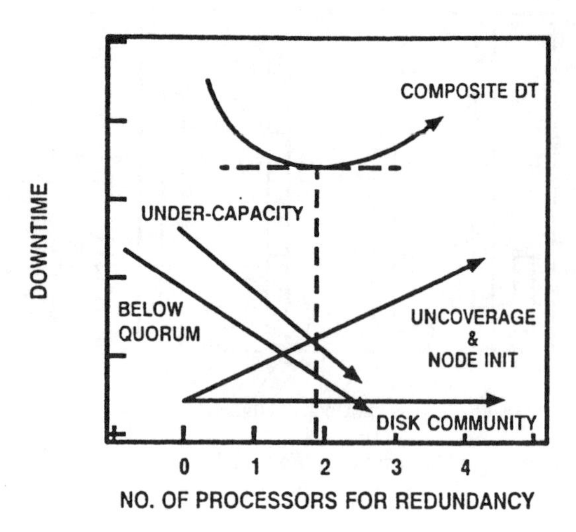

Fig. 6. Hardware Downtime Conceptual Model

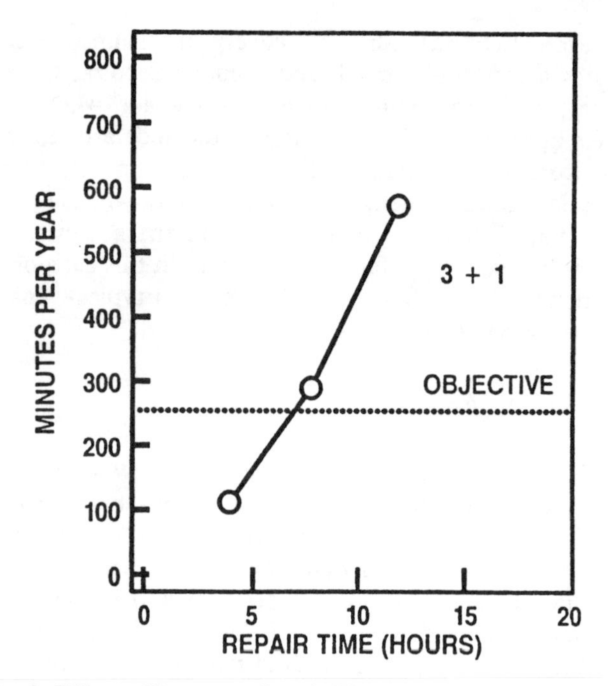

Fig. 8. Effect of Processor Repair Time

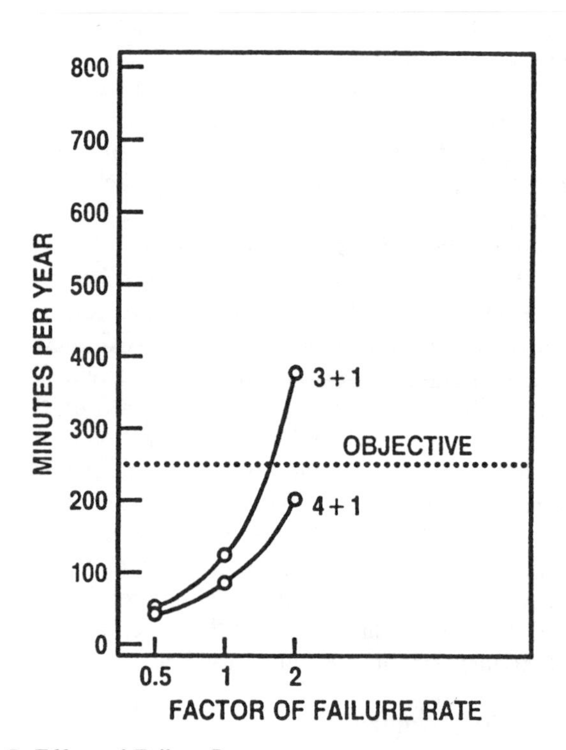

Fig. 7. Effect of Failure Rate

Figure 8 shows the effect that repair times have on downtime for configuration 3 + 1. When repair times double, the downtime does not meet the objective. Repair times are generally defined by the maintenance and diagnostic capabilities of the system. The availability of spares and the experience of the maintenance work-force also impact on system repair times. The maintenance organization has some control on the repair time components due to sparing and maintenance policies, but equipment maintainability is a function of the technology and design practices used by the supplier.

Another important parameter is uncoverage. For this analysis, the nominal uncoverage is 1 percent. The uncoverage value is varied from 0.1 percent to 10 percent for two values of the proportion of quick recoveries: 90 percent, the nominal value, and 50 percent, a lower more pessimistic value. Figure 9 shows the effect on the hardware downtime that the uncoverage parameters have on configuration 4 + 1. The objective is met for uncoverage values less than 3 percent, regardless of the proportion of quick recoveries and almost to 10 percent for the nominal proportion of quick recoveries. However, for uncoverage values greater than 3.5 percent the objective is not met for the 50 percent quick recovery proportion.

Uncoverage in a system reflects how well the redundancy schemes and recovery strategies have been implemented by the system designer. Extensive reliability testing, including systematic fault insertion under normal and stressed operating conditions, can be used to estimate the uncoverage parameters. Field-performance and maintenance-personnel training play an important role to estimate uncoverage values. It can be presumed that, for newly deployed systems, the uncoverage parameters fall in the more pessimistic ranges.

5. AREAS OF ENHANCEMENT

Many stored program control (SPC) switching systems employ a duplexed architecture (single processor with hot standby) to provide redundancy for high reliability. The example multiprocessor database system represents an important point of departure from duplex architectures to provide highly reliable services. Probably, similar computer architectures will be deployed in the networks and operation support systems of Bell Operating Companies.

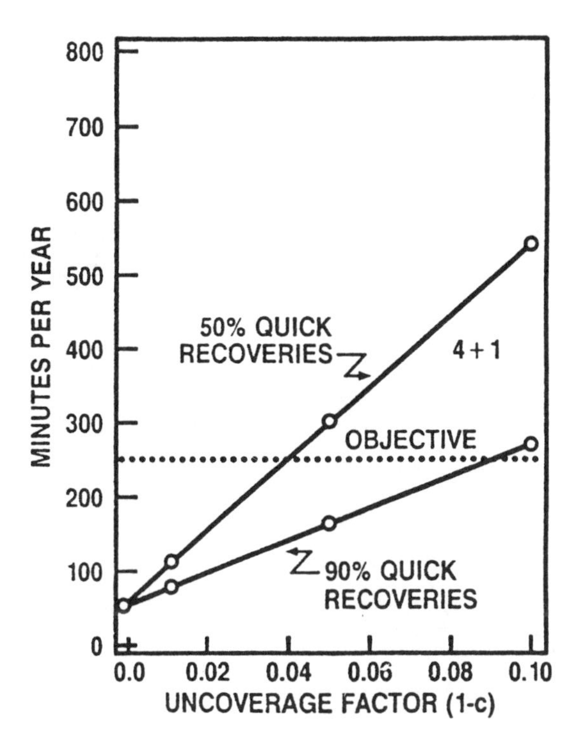

Fig. 9. Effect of Uncoverage on Downtime

The 20 percent downtime allocated to failures due to hardware causes, was as specified in the LSSGR. These guidelines are substantiated by extensive data from the SPC switching systems using duplex redundancy schemes. For different system architectures such as multiprocessor systems, the allocation scheme requires further study.

Beyond the complication of a new architecture and the downtime allocation schemes, other reasons exist to extend the current reliability modeling and analysis capabilities. Current techniques are most effective with hardware, but hardware devices are becoming more reliable, so that the task of meeting hardware system objectives is becoming increasingly less burdensome for system designers. Also, the experience with SPC switching systems shows that software and recovery deficiencies and procedural errors account for over two-thirds of the downtime [7]. Within Bellcore, studies of fault-tolerant computing techniques and human factors affecting equipment reliability are being conducted to investigate possible means of reducing the failures due to these causes.

Recently a list of areas requiring attention in the development of R&Q requirements for the Next Generation Switch (NGS) was proposed [8]. Distributed systems are likely to become the future switching architecture rather than those using the centralized arrangement. The focus of reliability concern then shifts from system outages to partial losses or degradation of service. The partial outage issue is being considered in a revision of the LSSGR.

REFERENCES

[1] P. C. Patton, "Multiprocessor: Architecture and applications", *IEEE Computer*, vol 18, 1985 Jun, pp 29-40.

[2] G. G. Brush, R. J. Ferrise, P. A. Link, T. C. Tweedie, "The Bellcore reliability and quality analysis process", *IEEE Global Telecommunications Conf.*, 1985 Dec, pp 30.1.1-30.1.5.

[3] Bellcore, "Reliability and quality switching systems generic requirements (RQSSGR)", TR-TSY-000284, Issue 1, 1986 Nov.

[4] Bellcore, "LATA switching systems generic requirements (LSSGR)", TR-TSY-000064, Issue 2, 1987 Jul.

[5] J. F. Kitchin, "Approximate Markov modeling of systems with redundancy and non-exponential repair", *IEEE J. Select. Areas Communication*, vol SAC-4, 1986 Oct, pp 1133-1137.

[6] Bellcore, "Reliability prediction procedure for electronic equipment", TR-TSY-000332, Issue 1, 1985 Sep.

[7] Syed R. Ali, "Analysis of total outage data for stored program control switching systems", *IEEE J. Select. Areas Communication*, vol SAC-4, 1986 Oct, pp 1044-1046.

[8] C. L. Davis, Jr., R. J. Ferrise, "An agenda for the reliability and quality of the next generation switch", *Int'l Switching Symp.*, 1987 Mar, pp C5.3.1-C5.3.6.

AUTHORS

Dr. Chi-Ming Chen; Bell Communications Research; 331 Newman Springs Rd., 3J211; Red Bank, New Jersey 07701 USA.

Chi-Ming Chen (S'80, M'85) was born in Taiwan, Republic of China, in 1949. He received the BS and MS degrees in Physics from National Tsing Hua University, in 1971 and 1973, respectively. He received the MS degree in Computer Science from the Pennsylvania State University in 1981 and the PhD degree in Computer and Information Science from the University of Pennsylvania in 1985. Since then, he has been a Member of the Technical Staff in the Switching Analysis and Reliability Technology Center of Bellcore. His areas of interests include network reliability, performance modeling, and fault-tolerant computing. He is also participating in the T1Q1 standards work.

Jose D. Ortiz; Bell Communications Research; 331 Newman Springs Rd., 2X-305; Red Bank, New Jersey 07701 USA.

Jose D. Ortiz (M'87) was born in Havana, Cuba, in 1952. He received the BS in Systems Engineering and the ME in Industrial Engineering from the University of Florida in 1974 and 1976, respectively. Prior to divestiture, he was a Member of the Technical Staff in Bell Labs in the Quality Assurance Center with various product analysis and studies responsibilities. After divestiture, in his capacity as an MTS in the Switching Analysis and Reliability Technology Center of Bellcore, his responsibilities entailed conducting reliability and quality analyses of network switching products. He is a district manager in the same organization with responsibility for conducting technical analyses of packet switching systems.

Manuscript TR88-355 received 1988 March 11; revised 1988 June 27.

IEEE Log Number 25469

◄ TR ►

Bibliography

Journals

[A.1] *IEEE Transactions on Computers*

[A.2] *IEEE Transactions on Software Engineering*

[A.3] *IEEE Computer*

[A.4] *IEEE Transactions on Reliability*

[A.5] *IEEE Transactions on Reliability,* Special Issue on Reliability of Parallel and Distributed Computing Networks, April 1989.

[A.6] *IEEE Transactions on Communications.*

[A.7] *IEEE Journal on Selected Areas in Communication*

[A.8] *IEEE Transactions on Circuit Theory*

[A.9] *SIAM Journal on Computing*

[A.10] *SIAM Journal on Algebraic and Discrete Methods*

[A.11] *SIAM Journal on Applied Mathematics*

[A.12] *Journal of Graph Theory*

[A.13] *Operations Research*

[A.14] *Network*

[A.15] *Reliability Engineering* (U.K.)

[A.16] *Microelectronics and Reliability* (U.K.)

Books

[B.1] M. Hall, Jr., *Combinatorial Theory,* Blaisdell Publishing, Waltham, Mass., 1967.

[B.2] J. Abadie (Ed.), *Integer and Nonlinear Programming,* North-Holland/American Elsevier, N.Y., 1970.

[B.3] G.S.G. Beveridge and R.S. Schechter, *Optimization: Theory and Practice,* McGraw-Hill, N.Y., 1970.

[B.4] D.M. Himmelblau, *Applied Nonlinear Programming,* McGraw-Hill, N.Y., 1972.

[B.5] H.A. Taha, *Integer Programming: Theory, Applications and Computations,* Academic Press, N.Y., 1975.

[B.6] D.I.A. Cohen, *Combinatorial Theory,* John Wiley and Sons, N.Y., 1978.

[B.7] F.A. Tillman, C.L. Hwang, and W. Kuo, *Optimization of Systems Reliability,* Marcel Dekker, Inc., N.Y., 1980.

[B.8] E. Bannai and T. Ito, *Algebraic Combinatorics and Association Schemes,* Benjamin-Cummings, Redwood City, Calif., 1984.

[B.9] B. Bollobas, *Combinatorics,* Cambridge University Press, Cambridge, England, 1986.

[B.10] F. Harary, *Graph Theory,* Addison-Wesley Pub. Co., Reading, Mass., 1969.

[B.11] W. Mayeda, *Graph Theory,* John Wiley and Sons, N.Y., 1972.

[B.12] C. Berge, *Graphs and Hypergraphs,* North-Holland, Amsterdam, The Netherlands, 1973.

[B.13] N. Deo, *Graph Theory with Application to Engineering and Computer Science,* Prentice Hall, Englewood Cliffs, N.J., 1974.

[B.14] A.V. Aho, J.E. Hopcroft, and J.D. Ullman, *The Design and Analysis of Computer Algorithms,* Addison-Wesley Pub. Co., Reading, Mass., 1974.

[B.15] M.R. Garey and D.S. Johnson, *Computers and Intractability: A Guide to the Theory of NP-Completeness,* W.H. Freeman, N.Y., 1979.

[B.16] L. Kleinrock, *Queuing System: Vol. 1 Theory,* John Wiley and Sons, N.Y., 1975.

[B.17] H. Kobayashi, *Modeling and Analysis: An Introduction to System Performance Evaluation Methodology,* Addison-Wesley Pub. Co., Reading, Mass., 1978.

[B.18] U.N. Bhatt, *Elements of Applied Stochastic Processes,* John Wiley and Sons, N.Y., 1984.

[B.19] V.E. Benes, *Mathematical Theory of Connecting Networks and Telephone Traffic,* Academic Press, Orlando, Fla., 1965.

[B.20] F.L. Bauer, *Software Engineering—An Advanced Course,* Springer-Verlag, N.Y., 1977.

[B.21] J.L. Peterson, *Petrinet Theory and the Modeling of Systems,* Prentice Hall, Englewood Cliffs, N.J., 1981.

[B.22] S. Cerri and G. Pelagatti, *Distributed Data Principles and Systems,* McGraw-Hill, N.Y., 1984.

[B.23] M. Ajmone Marsam, G. Balbo, and G. Conte, *Performance Models of Multiprocessor Systems,* MIT Press, Cambridge, Mass., 1986.

[B.24] W. Reisig, *Petri Nets: An Introduction,* Springer-Verlag, N.Y., 1985.

[B.25] B. Bhargava (Ed.), *Concurrency and Reliability in Distributed Systems,* Van Nostrand and Reinhold, N.Y., 1987.

[B.26] G.J. Lipovski and M. Malek, *Parallel Computing,* John Wiley and Sons, N.Y., 1987.

[B.27] D. Bertsekas and R. Gallagher, *Data Networks,* Prentice-Hall, Englewood Cliffs, N.J., 1987.

[B.28] M. Schwartz, *Telecommunication Networks: Protocols, Modeling and Analysis,* Addison-Wesley Pub . Co., Reading, Mass., 1987.

[B.29] M.L. Shooman, *Probabilistic Reliability: An Engineering Approach,* McGraw-Hill, N.Y., 1968.

[B.30] [MIL-HDBK-217 B74] *Military Standardization Handbook: Reliability Prediction of Electronic Equipment,* Sept. 1974.

[B.31] R.E. Barlow and F. Proschan, *Statistical Theory of Reliability and Life Testing: Probability Models,* Holt, Rinehart and Winston, N.Y., 1975.

[B.32] R.E. Barlow (Ed.), *Reliability and Fault Tree Analysis: Theoretical and Applied Aspects of System Reliability and Safety Assessment,* Society of Industrial and Applied Mathematics, Philadelphia, Penn., 1975.

[B.33] N.R. Mann, R.E. Shafer, and N.D. Singpururalla, *Methods for Statistical Analysis of Reliability and Life Data,* John Wiley and Sons, N.Y., 1974.

[B.34] O.C. Smith, *Introduction to Reliability in Design,* McGraw-Hill, N.Y., 1976.

[B.35] B.S. Dhillon and C. Singh, *Reliability Engineering,* John Wiley and Sons, N.Y., 1981.

[B.36] E.J. Henley and H. Kumanto, *Reliability Engineering and Risk Assessment,* Prentice-Hall, Englewood Cliffs, N.J., 1981.

[B.37] K.S. Trivedi, *Probability and Statistics with Reliability, Queuing and Computer Science Applications,* Prentice-Hall, Englewood Cliffs, N.J., 1982.

[B.38] D.P. Siewiorek and R.S. Swartz, *The Theory and Practice of Reliable System Design,* Digital Press, Bedford, Mass., 1982.

[B.39] B.S. Dhillon, *Reliability Engineering in Systems Design and Operation,* Van Nostrand and Reinhold, N.Y., 1983.

[B.40] R. Billinton and R.N. Allan, *Reliability Evaluation of Engineering Systems: Concepts and Techniques,* Plenum Press, N.Y., 1983.

[B.41] D.T. O'Conner, *Practical Reliability Engineering,* John Wiley and Sons, N.Y., 1985.

[B.42] C.J. Colbourn, *The Combinatorics of Network Reliability,* Oxford University Press, Oxford, England, 1987.

Research Paper/Reports

Chapter 1: Introduction

[1.1] C.-L. Wu and T.-Y. Feng, "On a Class of Multistage Interconnection Networks," *IEEE Trans. Computers,* Vol. C-29, No. 8, Aug. 1980, pp. 694–702.

[1.2] Y.W. Ng and A. Avizienis, "Unified Reliability Model for Fault Tolerant Computer," *IEEE Trans. Computers,* Vol. C-29, No. 11, Nov. 1980, pp. 1002–1011.

[1.3] L.N. Bhuyan and D.P. Agrawal, "Design and Performance of a General Class of Interconnection Networks," *IEEE Trans. Computers,* Vol. C-30, No. 8, Aug. 1981, pp. 587–590.

[1.4] J.P. Ignizio, D.F. Palmer, and C.M. Murphy, "A Multicriteria Approach to Supersystem Architecture Definition," *IEEE Trans. Computers,* Vol. C-31, No. 5, May 1982, pp. 410–418.

[1.5] C.P. Kruskal and M. Snir, "The Performance of Multistage Interconnection Networks for Multiprocessors," *IEEE Trans. Computers,* Vol. C-32, No. 12, Dec. 1983, pp. 1091–1098.

[1.6] L.N. Bhuyan and D.P. Agrawal, "Generalized Hypercube and Hyperbus Structures for a Computer Network," *IEEE Trans. Computers,* Vol. C-33, No. 4, April 1984, pp. 323–333.

[1.7] K. Padmanabhan, "Fault Tolerance and Performance Improvement in Multiprocessor Interconnection Networks," Ph.D. Dissertation, Univ. of Illinois at Urbana-Champaign, Urbana, Ill., May 1984.

[1.8] J. Spragins, "Limitations of Current Telecommunication Network Reliability Models," *Proc. IEEE GLOBECOM '84,* IEEE, N.Y., 1984.

[1.9] J.P. Shen and J.P. Hayes, "Fault-Tolerance of Dynamic Full Access Interconnection Networks," *IEEE Trans. Computers,* Vol. C-33, No. 3, March 1984, pp. 241–248.

[1.10] D.P. Agrawal, V.K. Janakiram and G.C. Pathak, "Evaluating the Performance of Multicomputer Configurations," *Computer,* Vol. 19, No. 5, May 1986, pp. 23–37.

[1.11] A. Ghafoor, "A Class of Fault-Tolerant Multiprocessor Networks," *IEEE Trans. Reliability,* Vol. R-38, April 1989, pp. 5–15.

[1.12] L. Bhuyan, Q. Yang, and D.P. Agrawal, "Performance of Multiprocessor Interconnection Networks," *Computer,* Vol. 22, No. 2, Feb. 1989, pp. 25–37.

[1.13] P.C. Patton, "Multiprocessor: Architecture and Applications," *Computer,* Vol. 18, No. 6, June 1985, pp. 29–40.

[1.14] J. Spragins, et. al., "State of the Art in Telecommunication Network Reliability Modeling," *Proc. Pacific Computer Communication Symp.,* Seoul, Republic of Korea, Oct. 1985.

[1.15] A. Varma and C.S. Raghavendra, "Performance Analysis of a Redundant Path Interconnection Network," *Technical Report 0285-9,* Dept. of Electrical Engineering Systems, University of Southern California, Los Angeles, Calif., Jan. 1985.

[1.16] M.O. Locks, "Recent Developments in Computing of System Reliability," *IEEE Trans., Reliability,* Vol. R-34, Dec. 1985, pp. 425–436.

[1.17] R.F. Yanney and J.P. Hayes, "Distributed Recovery in Fault-Tolerant Multiprocessor Networks," *IEEE Trans. Computers,* Vol. C-35, No. 10, Oct. 1986, pp. 871–879.

[1.18] J.C. Bermond, et al., "Strategies for Interconnection Networks: Some Methods from Graph Theory," *J. of Parallel and Distributed Computing,* Vol. 3, 1986, pp. 433–449.

[1.19] A. Varma, "Design and Analysis of Reliable Interconnection Networks," Ph.D. Dissertation, Dept. of Electrical Engineering, University of Southern California, Los Angeles, Calif., Jan. 1986.

[1.20] C. Raghavendra and J. Silvester, "A Survey of Multi-Connected Loop Topologies for Local Computer Networks," *Computer Networks and ISDN Systems,* Vol. 11, 1986, pp. 29–42.

[1.21] K. Hwang, "Advanced Parallel Processing with Supercomputer Architectures," *Proc. of IEEE,* Vol. 75, No. 10, Oct. 1987, pp. 1348–1379.

[1.22] G.B. Adams, D.P. Agrawal, and H.J. Siegel, "A Survey and Comparison of Fault-Tolerant Multistage Interconnection Networks," *Computer,* Vol. 20, No. 6, June 1987, pp. 14–27.

Chapter 2: Multiprocessor Systems Reliability

[2.1] T.F. Arnold, "The Concept of Coverage and Its Effect on the Reliability Models of a Repairable System," *IEEE Trans. Computers,* Vol. C-22, No. 3, March 1973, pp. 251–254.

[2.2] I.A. Baqai and T. Lang, "Reliability Aspects of the Illiac IV Computer," *Proc. 1976 Intl. Conf. Parallel Processing,* IEEE Computer Society Press, Wash., D.C., Aug. 1976, pp. 123–131.

[2.3] A.D. Ingle and D.P. Siewiorek, "Reliability Model for Multiprocessor Systems with and without Periodic Maintenance," *Proc. 7th FTCS,* IEEE Computer Society Press, Wash., D.C., June 1977, pp. 3–9.

[2.4] J. Losq, "Effect of Failures on Gracefully Degradable Systems, *Proc., 7th FTCS,* IEEE Computer Society Press, Wash., D.C., June 1977, pp. 29–34.

[2.5] Y.W. Ng and A. Avizienis, "A Reliability Model for Gracefully Degrading and Repairable Fault-Tolerant Systems," *Proc. 7th FTCS,* IEEE Computer Society Press, Wash., D.C., June 1977, pp. 22–28.

[2.6] D.P. Siewiorek, "Multiprocessors: Reliability Modeling and Graceful Degradation," *Infotech State of the Art Conf. on System Reliability,* London, England, 1977, pp. 48–73.

[2.7] I.E. Mahgoub and D.P. Agrawal, "Impact of Cluster Network Failure on the Performance of Cluster-Based Supersystems, *Proc. Intl. Conf. Parallel Processing,* IEEE Computer Society Press, Wash., D.C., Aug. 1986, pp. 743–749.

[2.8] V. Cherkassky and M. Malek, "A Measure of Graceful Degradation in Parallel-Computer Systems," *IEEE Trans., Reliability,* Vol. R-38, April 1989, pp. 76–81.

[2.9] I. Koren and Z. Koren, "On Gracefully Degrading Multiprocessors with Multistage Interconnection Networks, *IEEE Trans. Reliability,* Vol. R-38, April 1989, pp. 82–89.

[2.10] T.C.K. Chou and J.A. Abraham, "Performance/Availability Model of Shared Resource Multiprocessors," *IEEE Trans. Reliability,* Vol. R-29, April 1980, pp. 70–74.

[2.11] V. Cherkassky, E. Opper, and M. Malek, "Reliability and Fault Diagnosis Analysis of Fault-Tolerant Multistage Interconnection Networks," *Proc. 14th FTCS,* IEEE Computer Society Press, Wash., D.C., June 1984.

[2.12] C.S. Raghavendra and D.S. Parker, "Reliability Analysis of an Interconnection Network," *Proc., 4th Distributed Computing Conf.,* IEEE Computer Society Press, Wash., D.C., May 1984, pp. 461–471.

[2.13] R.K. Iyer, Reliability Evaluation of Fault Tolerant Systems: Effect of Variability in Failure Rates,"

IEEE Trans. Computers, Vol. C-33, No. 5, May 1984, pp. 197–200.

[2.14] C.R. Das and L.N. Bhuyan, "Reliability Simulation of Multiprocessor Systems," *Proc., Intl. Conf. Parallel Processing,* IEEE Computer Society Press, Wash., D.C., Aug. 1985, pp. 591–598.

[2.15] C. Botting, S. Rai, and D.P. Agrawal, "Reliability Computation of Multistage Interconnection Networks," *IEEE Trans. Reliability,* Vol. R-38, April 1989, pp. 138–145.

[2.16] B.L. Menezes, R.M. Jenevein, and M. Malek, "Reliability Analysis of KYKLOS Interconnection Networks," *Proc. 6th Distributed Computing Conf.,* IEEE Computer Society Press, Wash., D.C., 1986, pp. 46–53.

[2.17] W. Najjar and J.L. Gaudiot, "Reliability and Performance Modeling of Hypercube Based Multiprocessor," *2nd Intl. Workshop on Applied Mathematics and Performance/Reliability Models for Computer Communication Systems,* Italy, May 1987.

[2.18] V. Cherkassky and M. Malek, "Graceful Degradation of Multiprocessor Systems," *Proc. Intl. Conf. Parallel Processing,* Pennsylvania State University Press, University Park, Penn., Aug. 1987, pp. 885–888.

[2.19] J.T. Blake, A.L. Reibman, and K.S. Trivedi, "Sensitivity Analysis of Reliability and Performance Measures for Multiprocessor Systems," *Proc. ACM SIGMETRICS Conf. on Measurements and Modeling of Computer Systems,* ACM Inc., N.Y., May 1988.

[2.20] J.T. Blake and K.S. Trivedi, "Comparing Reliabilities of Two Fault-Tolerant Interconnection Networks," *Proc. 18th FTCS,* IEEE Computer Society Press, Wash., D.C., June 1988.

[2.21] J. Kim, C.R. Das, W. Lin, and M.J. Thathutaveetil, "Reliability Analysis of Hypercube Multicomputer," *Technical Report,* Computer Engineering Program, Dept. of Electrical Engineering, Pennsylvania State Univ., University Park, Penn., 1988.

[2.22] A. Verma and C.S. Raghavendra, "Reliability Analysis of Redundant-Path Interconnection Networks," *IEEE Trans. Reliability,* Vol. R-38, April 1989, pp. 130–137.

Chapter 3: Multiterminal Reliability Evaluation

[3.1] G. Hilborn, "Measures for Distributed Processing Network Survivability," *Proc. of the 1980 National Computer Conf.*, AFIPS Press, Reston, Va., May 1980.

[3.2] R.E. Merwin and M. Mirhakak, "Derivation and Use of a Survivability Criterion for DDP Systems," *Proc. of the 1980 National Computer Conf.*, AFIPS Press, Reston, Va., May 1980, pp. 139–146.

[3.3] S. Rai, "A Cut-Set Approach to Reliability Evaluation in Communication Networks," *IEEE Trans. Reliability,* Vol. R-31, Dec. 1982, pp. 428–431.

[3.4] J. Kim, C.R. Das, W. Lin, and T. Feng, "Reliability Evaluation of Hypercube Multicomputers," *IEEE Trans. Reliability,* Vol. R-38, April 1989, pp. 121–129.

[3.5] S. Pateras and J. Rajski, "Design of a Self-Reconfiguring Interconnection Network for Fault-Tolerant VLSI Processor Arrays," *IEEE Trans. Reliability,* Vol. R-38, April 1989, pp. 40–50.

[3.6] J. Garcia-Molina, "Reliability Issues for Fully Replicated Distributed Databases," *Computer,* Vol. 16, No. 9, Sept. 1982, pp. 34–42.

[3.7] C.S. Raghavendra and S.V. Makam, "Dynamic Reliability Modeling and Analysis of Computer Networks," *Proc. Intl. Conf. Parallel Processing,* IEEE Computer Society Press, Wash., D.C., Aug. 1983.

[3.8] R.K. Wood, "A Factoring Algorithm Using Polygon-to-Chain Reduction for Computing k-Terminal Network Reliability," *Networks,* Vol. 15, 1985, pp. 173–190.

[3.9] S. Hariri, C.S. Raghavendra, and V.K. Prasanna Kumar, "Reliability Measures for Distributed Processing Systems," *Proc. 6th Intl. Conf. Distributed Computing Systems,* IEEE Computer Society Press, Wash., D.C., May 1986, pp. 564–571.

[3.10] S. Rai and D.P. Agrawal, "Reliability of Program Execution in a Distributed Environment," *IEEE Trans. Software Engineering,* 1989 (communicated).

[3.11] C.S. Raghavendra, V.K. Prasanna Kumar, and S. Hariri, "Reliability Analysis in Distributed Systems," *IEEE Trans. Computers,* Vol. C-37, No. 3, March 1988, pp. 352–358.

Chapter 4: Multimode and Dependent-Failure Analysis

4.1: Multimode Analysis

[4.1] E. Phibbs and S.H. Kuwamamota, "An Efficient Map Method for Processing Multistate Logic Trees," *IEEE Trans. Reliability,* Vol. R-21, May 1972, pp. 93–98.

[4.2] B.S. Dhillon, "The Analysis of the Reliability of Multistage Device Networks," Ph.D. Thesis, Dept. of Industrial Engineering, Univ. of Windsor, Ontario, Canada, 1975.

[4.3] E. El-Neweihi, F. Proschan, and J. Sethuraman, "Multistate Coherent Systems," *J. Applied Probability,* Vol. 15, 1978, pp. 675–688.

[4.4] R.E. Barlow and A.S. Wu, "Coherent Systems with Multistate Components," *Math Oper. Res.,* Vol. 3, 1978, pp. 275-281.

[4.5] S. Ross, "Multivalued State Component Systems," *Ann. Prob.,* Vol. 7, 1979, pp. 379–383.

[4.6] W.S. Griffith, "Multistate Reliability Models," *J. Appl. Prob.,* Vol. 17, 1980, pp. 735–744.

[4.7] X. Janan, "On Multistate System Analysis," *IEEE Trans. Reliability,* Vol. R-34, Oct. 1985, pp. 329–337.

[4.8] I.J. Sauks, "Diagraph Matrix Analysis," *IEEE Trans. Reliability,* Vol. R-34, Dec. 1985, pp. 437-446.

[4.9] S.N. Chiou and V.O.K. Li, "Reliability Analysis of Communication Network with Multimode Components," *IEEE J. on Selected Areas in Comm.,* Vol. SAC-4, Oct. 1986, pp. 1156–1161.

[4.10] C.L. Yang and P. Kubat, "Reliability Analysis and Efficient Computation of Most Probable States for Communication Networks with Multimode Components," GTE Labs, Inc., Waltham, Mass., 1987.

[4.11] J. Yuan, et al., "Reliability Modeling and Evaluation for Networks under Multiple and Fluctuating Operational Conditions," *IEEE Trans. Reliability,* Vol. R-36, Dec. 1987, pp. 557–564.

[4.12] K.V. Le and V.O.K. Li, "Modeling and Analysis of Systems with Multimode Components and Dependent Failures," *IEEE Trans. Reliability,* Vol. R-38, April 1989, pp. 68–75.

4.2: Dependent-Failure Analysis

[4.13] J.D. Spragins, "Dependent Failures in Data Communication Systems," *IEEE Trans. Reliability,* Vol. R-25, Dec. 1977, pp. 1494–1499.

[4.14] J.D. Spragins and J. Assiri, "Communication Network Reliability Calculations with Dependent Failures," in *Proc. IEEE National Telecom. Conf.,* IEEE, N.Y., 1980, pp. 25.2.1–25.2.5.

[4.15] J.D. Spragins, J.D. Markov, M.W. Doss, S.A. Mitchell, and D.C. Squire, "Communication Network Availability Predictions Based on Measurement Data," *Proc. Intl. Test Conf.,* IEEE Computer Society Press, Wash., D.C., 1981.

[4.16] S.N. Pan and J.D. Spragins, "Dependent Failure Reliability Models for Tactical Communication Networks," *Proc. IEEE Intl. Conf. on Communication,* IEEE Computer Society Press, Wash., D.C., 1983, pp. 765–771.

[4.17] Y. Lam and V. Li, "On Reliability Calculation of Network with Dependent Failures," *Proc. IEEE GLOBECOM,* IEEE, N.Y., Dec. 1983, pp. 1499–1503.

[4.18] V.O.K. Li and J.A. Silvester, "Performance Analysis of Networks with Unreliable Components," *IEEE Trans. Computers,* Vol. C-32, No. 10, Oct. 1984, pp. 1105–1110.

[4.19] J. Yuan, "A Conditional Probability Approach to Reliability with Common-Cause Failures," *IEEE Trans. Reliability,* Vol. R-34, April 1985, pp. 38–42.

[4.20] D.R. Shier and J.D. Spragins, "Exact and Approximate Dependent Failure Models of Telecommunications Networks," *Proc. INFOCOM '85,* IEEE Computer Society Press, Wash., D.C., 1985, pp. 200–205.

[4.21] Y.F. La and V.O.K. Li, "Reliability Modeling and Analysis of Communication Networks with Dependent Failures," *ibid.*

[4.22] K.C. Chae and G.M. Clark, "System Reliability in the Presence of Common Cause Failures," *IEEE Trans. Reliability,* Vol. R-35, April 1986, pp. 32–35.

[4.23] C.D. Lai, "Bounds on Reliability of a Coherent System with Positively Correlated Components," *IEEE Trans. Reliability,* Vol. R-35, Dec. 1986, pp. 508–511.

[4.24] W.G. Schneeweiss, "The Failure Frequency of Systems with Dependent Components," *IEEE Trans. Reliability,* Vol. R-35, Dec. 1986, pp. 512–517.

[4.25] Y. Lam and U. Li, "An Improved Algorithm for Performance Analysis of Networks with Unreliable Components," *IEEE Trans. Computers,* Vol. C-34, No. 5, May 1987, pp. 496–497.

Chapter 5: Performability

[5.1] X. Castillo and D.P. Siewiorek, "A Performance-Reliability Model for Computing Systems," *Proc. 10th FTCS,* IEEE Computer Society Press, Wash., D.C., Oct. 1980, pp. 187–192.

[5.2] J.F. Meyer, "Closed Form Solutions for Performability," *IEEE Trans. Computers,* Vol. C-31, No. 7, July 1982, pp. 648–657.

[5.3] J. Arlat and J.C. Laprie, "Performance-Related Dependability Evaluation of Supercomputer Systems," *Proc. 13th FTCS,* IEEE Computer Society Press, Wash., D.C., June 1983, pp. 276–283.

[5.4] J.F. Meyer, "Unified Performance-Reliability Evaluation," *Proc. American Control Conf.,* June 1984.

[5.5] D.G. Furchtgott and J.F. Meyer, "A Performability Solution Method for Degradable Non-Repairable Systems," *IEEE Trans. Computers,* Vol. C-33, No. 6, June 1984, pp. 550–554.

[5.6] M. Smotherman, R.M. Geist, and K.S. Trivedi, "Provably Conservative Approximations to Complex Reliability Models," *IEEE Trans. Computers,* Vol. C-35, No. 4, April 1986.

[5.7] B.R. Iyear, et al., "Analysis of Performability for Stochastic Models of Fault-Tolerant Systems," *IEEE Trans. Computers,* Vol. C-35, No. 10, Oct. 1986, pp. 902–907.

[5.8] L. Donatello and B.R. Iyear, "Analysis of a Composite Performance Reliability Measure for Fault-Tolerant Systems," *J. ACM,* Vol. 34, No. 1, Jan. 1987, pp. 179–199.

[5.9] P. Kubat, "Estimation of Reliability for Communication/Computer Networks; Simulation/Analytic Approach," *Proc. Parallel Processing Conf.,* Pennsylvania State University Press, University Park, Penn., 1987, pp. 608–614.

[5.10] M.C. Hsueh, R.K. Iyer, and K.S. Trivedi, "A Measurement-Based Performability Model for a Multiprocessor System," *Mathematical Computer Performance and Reliability,* North Holland, Amsterdam, The Netherlands, 1987.

Chapter 6: Conclusion

6.1: Task-Based Reliability

[6.1] L.N. Bhuyan and S.R. Das, "Dependability Evaluation of Multicomputer Networks," *Proc. Intl. Conf. Parallel Proc.,* IEEE Computer Society Press, Wash., D.C., Aug. 1985, pp. 591–598.

[6.2] W.W. Chu, et al., "Task Allocation in Distributed Data Processing," *Computer,* Vol. 13, No. 11, Nov. 1980, pp. 57–69.

[6.3] L.W. Dowdy and D.V. Foster, "Comparative Models of the File Assignment Problem," *Computing Surveys,* Vol. 14, June 1982, pp. 287–313.

[6.4] J.A. Bannister and K.S. Trivedi, "Task Allocation in Fault-Tolerant Distributed Systems," *Acta Informatica,* Vol. 20, 1983, pp. 261–281.

[6.5] B.W. Wah, "File Placement on Distributed Computer Systems," *Computer,* Vol. 17, No. 1, Jan. 1984, pp. 23–32.

[6.6] W.W. Chu, et al., "Estimation of Intermodule Communication and Its Application in Distributed Processing Systems," *IEEE Trans. Computers,* Vol. C-33, No. 8, Aug. 1984, pp. 691–699.

[6.7] C.-C. Shen and W.-H. Tsai, "A Graph Matching Approach to Optimal Task Assignment in Distributed Computing Systems Using a Minimax Criterion," *IEEE Trans. Computers,* Vol. C-34, No. 3, March 1985, pp. 197–203.

[6.8] W.W. Chu and L.M.-T. Lan, "Task Allocation and Precedence Relations for Distributed Real-Time Systems," *IEEE Trans. Computers,* Vol. C-36, No. 6, June 1987, pp. 667–679.

[6.9] T.L. Casavant and J.G. Kuhl, "A Taxonomy of Scheduling in General Purpose Distributed Computing System," *IEEE Trans. Software Engineering,* Vol. SE-14, No. 2, Feb. 1988, pp. 141–154.

[6.10] G.C. Pathak and D.P. Agrawal, "Task Division and Multicomputer Systems," *Proc. 5th Intl. Conf. Distributed Computing Systems,* IEEE Computer Society, Wash., D.C., 1984, pp. 273–280.

[6.11] S.M. Shatz and J.-P. Wang, "Models and Algorithms for Reliability-Oriented Task-Allocation in Redundant Distributed Computer Systems," *IEEE Trans. Reliability,* Vol. R-38, April 1989, pp. 16–27.

[6.12] S. Hariri and C.S. Raghavendra, "Distributed Functions Allocation for Reliability and Delay Optimization," *Proc. IEEE/ACM 1986 Fall Joint Computer Conf.,* IEEE Computer Society Press, Wash., D.C., Nov. 1986, pp. 344–352.

[6.13] W.W. Chu and K.K. Leung, "Module Replication and Assignment for Real-Time Distributed Processing Systems," Special Issue of *IEEE Proceedings,* May 1987, pp. 547–562.

[6.14] S.M. Shatz and J.-P. Wang, "Introduction to Distributed Software Engineering," *Computer,* Vol. 20, No. 10, Oct. 1987, pp. 23–31.

[6.15] K.G. Shin, T.-H. Lin, and Y.-H. Lee, "Optimal Checkpointing of Real-Time Tasks," *IEEE Trans. Computers,* Vol. C-36, No. 11, Nov. 1987, pp. 1328–1341.

[6.16] J.-P. Wang and S.M. Shatz, "Task Allocation for Reliability in Distributed Computing Systems," *UIC Technical Report,* Dec. 1987.

[6.17] F. Distante and V. Piuri, "Hill-Climbing Heuristics for Optimal Hardware Dimensioning and Software Allocation in Fault-Tolerant Distributed Systems," *IEEE Trans. Reliability,* Vol. R-38, April 1989, pp. 28–39.

6.2: Software Reliability

[6.18] T.K. Nayak, "Software Reliability: Statistical Modeling and Estimation," *IEEE Trans. Reliability,* Vol. R-35, Dec. 1986, pp. 566–570.

[6.19] L. Chen and A. Avizienis, "N-Version Programming: A Fault-Tolerance Approach to Reliability of Software Operation," *Proc. 8th FTCS,* IEEE Computer Society, Wash., D.C., 1978, pp. 3–9.

[6.20] G.J. Schick and R.W. Wolverton, "An Analysis of Competing Software Reliability Models," *IEEE Trans. Software Engineering,* Vol. SE-4, March 1978, pp. 104–120.

[6.21] J.W. Duran and J.J. Wiorkowski, "Quantifying Software Validity by Sampling," *IEEE Trans. Reliability,* Vol. R-29, June 1980, pp. 141–144.

[6.22] J.E. Angus, "The Application of Software Reliability Models to a Major C^3I System," *Proc. 1984 Annual Reliability and Maintainability Symp.,* IEEE Computer Society, Wash., D.C., Jan. 1984.

[6.23] Special Issue on Software Reliability, *IEEE Trans. on Software Engineering,* Vol. SE-11, No. 12, Dec. 1985

[6.24] J.A. Stankovic, *Reliable Distributed System Software,* IEEE Computer Society Press, Wash., D.C., 1985.

[6.25] *Proc. Symp. Reliability in Distributed Software and Database Systems,* IEEE Computer Society Press, Wash., D.C., 1987.

[6.26] Special Issue on Software Reliability, *IEEE Trans. Software Engineering,* Vol. SE-12, No. 12, Jan. 1986.

[6.27] K.H. Kim, S. Heu, and S.M. Yang, "Performance Analysis of Fault-Tolerant Systems in Parallel Execution of Conversations," *IEEE Trans. Reliability,* Vol. R-38, April 1989, pp. 96–102.

6.3: Case Studies

[6.28] H. Frank, "Survivability Analysis of Command and Control Communication Networks—Part i," *IEEE Trans. Computers,* Vol. C-22. May 1974.

[6.29] H. Frank, "Survivability Analysis of Command and Control Communication Networks—Part ii," *IEEE Trans. Computers,* Vol. C-22, May 1974.

[6.30] G.G. Brush, R.J. Ferrise, P.A. Link, and T.C. Tweedie, "The Bellcore Reliability and Quality Analysis Process," *IEEE Global Telecom. Conf.,* IEEE, N.Y., Dec. 1985, p. 30.1.1.

[6.31] I.A. Geigenhaum, "INTELSAT System Reliability," *Proc. Reliability and Maintainability Symp.,* IEEE, N.Y., 1980, pp. 415–421.

[6.32] R.E. Fleming, et al., "Fault Tolerant C^3I System —A_O, A_I, MTBF Allocation," *Proc. Reliability and Maintainability Symp.,* IEEE, N.Y., 1986, p. 352.

[6.33] O.C. Ibe, R.C. Howe, and K.S. Trivedi, "Approximate Availability Analysis of VAXcluster Systems, *IEEE Trans., Reliability,* Vol. 38, April 1989, pp. 146–152.

IEEE COMPUTER SOCIETY

A member society of the Institute of Electrical and Electronics Engineers, Inc.

Policies of the IEEE Computer Society

Headquarters Office

1730 Massachusetts Avenue, N.W.
Washington, DC 20036-1903
Phone: (202) 371-1012
Telex: 7108250437 IEEE COMPSO

Publications Office

10662 Los Vaqueros Circle
Los Alamitos, CA 90720
Membership and General Information: (714) 821-8380
Publications Orders: (800) 272-6657

European Office

13, Avenue de l'Aquilon
B-1200 Brussels, Belgium
Phone: 32 (2) 770-21-98
Telex: 25387 AWALB

Asian Office

Ooshima Building
2-19-1 Minami-Aoyama, Minato-ku
Tokyo 107, Japan

IEEE Computer Society Press Publications

Monographs: A monograph is a collection of original material assembled as a coherent package. It is typically a treatise on a small area of learning and may include the collection of knowledge gathered over the lifetime of the authors.

Tutorials: A tutorial is a collection of original materials prepared by the editors and reprints of the best articles published in a subject area. They must contain at least five percent original materials (15 to 20 percent original materials is recommended).

Reprint Books: A reprint book is a collection of reprints that are divided into sections with a preface, table of contents, and section introductions that discuss the reprints and why they were selected. It contains less than five percent original material.

Technology Series: The technology series is a collection of anthologies of reprints each with a narrow focus of a subset on a particular discipline.

Submission of proposals: For guidelines on preparing CS Press Books, write Editor-in-Chief, IEEE Computer Society, 1730 Massachusetts Avenue, N.W., Washington, DC 20036-1903 (telephone 202-371-1012).

Purpose

The IEEE Computer Society advances the theory and practice of computer science and engineering, promotes the exchange of technical information among 97,000 members worldwide, and provides a wide range of services to members and nonmembers.

Membership

Members receive the acclaimed monthly magazine *Computer*, discounts, and opportunities to serve (all activities are led by volunteer members). Membership is open to all IEEE members, affiliate society members, and others seriously interested in the computer field.

Publications and Activities

Computer. An authoritative, easy-to-read magazine containing tutorial and in-depth articles on topics across the computerfield, plus news, conferences, calendar, interviews, and new products.

Periodicals. The society publishes six magazines and four research transactions. Refer to membership application or request information as noted above.

Conference Proceedings, Tutorial Texts, Standards Documents. The Computer Society Press publishes more than 100 titles every year.

Standards Working Groups. Over 100 of these groups produce IEEE standards used throughout the industrial world.

Technical Committees. Over 30 TCs publish newsletters, provide interaction with peers in specialty areas, and directly influence standards, conferences, and education.

Conferences/Education. The society holds about 100 conferences each year and sponsors many educational activities, including computing and science accreditation.

Chapters. Regular and student chapters worldwide provide the opportunity to interact with colleagues, hear technical experts, and serve the local professional community.

Ombudsman

Members experiencing problems—magazine delivery, membership status, or unresolved complaints—may write to the ombudsman at the Publications Office.

Other IEEE Computer Society Press Texts

Monographs

Integrating Design and Test: Using CAE Tools for ATE Programming:
Written by K.P. Parker
(ISBN 0-8186-8788-6 (case)); 160 pages

JSP and JSD: The Jackson Approach to Software Development (Second Edition)
Written by J.R. Cameron
(ISBN 0-8186-8858-0 (case)); 560 pages

National Computer Policies
Written by Ben G. Matley and Thomas A. McDannold
(ISBN 0-8186-8784-3 (case)); 192 pages

Physical Level Interfaces and Protocols
Written by Uyless Black
(ISBN 0-8186-8824-6 (case)); 240 pages

Protecting Your Proprietary Rights in the Computer and High Technology Industries
Written by Tobey B. Marzouk, Esq.
(ISBN 0-8186-8754-1 (case)); 224 pages

Tutorials

Ada Programming Language
Edited by S.H. Saib and R.E. Fritz
(ISBN 0-8186-0456-5); 548 pages

Advanced Computer Architecture
Edited by D.P. Agrawal
(ISBN 0-8186-0667-3); 400 pages

Advanced Microprocessors and High-Level Language Computer Architectures
Edited by V. Milutinovic
(ISBN 0-8186-0623-1); 608 pages

Communication and Networking Protocols
Edited by S.S. Lam
(ISBN 0-8186-0582-0); 500 pages

Computer Architecture
Edited by D.D. Gajski, V.M. Milutinovic, H.J. Siegel, and B.P. Furht
(ISBN 0-8186-0704-1); 602 pages

Computer Communications: Architectures, Protocols and Standards (Second Edition)
Edited by William Stallings
(ISBN 0-8186-0790-4); 448 pages

Computer Grahics (2nd Edition)
Edited by J.C. Beatty and K.S. Booth
(ISBN 0-8186-0425-5); 576 pages

Computer Graphics Hardware: Image Generation and Display
Edited by H.K. Reghbati and A.Y.C. Lee
(ISBN 0-8186-0753-X); 384 pages

Computer Grahics: Image Synthesis
Edited by Kenneth Joy, Max Nelson, Charles Grant, and Lansing Hatfield
(ISBN 0-8186-8854-8 (case)); 384 pages

Computer and Network Security
Edited by M.D. Abrams and H.J. Podell
(ISBN 0-8186-0756-4); 448 pages

Computer Networks (4th Edition)
Edited by M.D. Abrams and I.W. Cotton
(ISBN 0-8186-0568-5); 512 pages

Computer Text Recognition and Error Correction
Edited by S.N. Srihari
(ISBN 0-8186-0579-0); 364 pages

Computers for Artificial Intelligence Applications
Edited by B. Wah and G.-J. Li
(ISBN 0-8186-0706-8); 656 pages

Database Management
Edited by J.A. Larson
(ISBN 0-8186-0714-9); 448 pages

Digital Image Processing and Analysis: Volume 1: Digital Image Processing
Edited by R. Chellappa and A.A. Sawchuk
(ISBN 0-8186-0665-7); 736 pages

Digital Image Processing and Analysis: Volume 2: Digital Image Analysis
Edited by R. Chellappa and A.A. Sawchuk
(ISBN 0-8186-0666-5); 670 pages

Digital Private Branch Exchanges (PBXs)
Edited by E.R. Coover
(ISBN 0-8186-0829-3); 400 pages

Distributed Control (2nd Edition)
Edited by R.E. Larson, P.L. McEntire, and J.G. O'Reilly
(ISBN 0-8186-0451-4); 382 pages

Distributed Database Management
Edited by J.A. Larson and S. Rahimi
(ISBN 0-8186-0575-8); 580 pages

Distributed-Software Engineering
Edited by S.M. Shatz and J.-P. Wang
(ISBN 0-8186-8856-4 (case)); 304 pages

DSP-Based Testing of Analog and Mixed-Signal Circuits
Edited by M. Mahoney
(ISBN 0-8186-0785-8); 272 pages

End User Facilities in the 1980's
Edited by J.A. Larson
(ISBN 0-8186-0449-2); 526 pages

Fault-Tolerant Computing
Edited by V.P. Nelson and B.D. Carroll
(ISBN 0-8186-0677-0 (paper) 0-8186-8667-4 (case)); 432 pages

Gallium Arsenide Computer Design
Edited by V.M. Milutinovic and D.A. Fura
(ISBN 0-8184-0795-5); 368 pages

Human Factors in Software Development (Second Edition)
Edited by B. Curtis
(ISBN 0-8186-0577-4); 736 pages

Integrated Services Digital Networks (ISDN) (Second Edition)
Edited by W. Stallings
(ISBN 0-8186-0823-4); 404 pages

Interconnection Networks for Parallel and Distributed Processing
Edited by C.-I. Wu and T.-y. Feng
(ISBN 0-8186-0574-X); 500 pages

Local Network Equipment
Edited by H.A. Freeman and K.J. Thurber
(ISBN 0-8186-0605-3); 384 pages

Local Network Technology (3rd Edition)
Edited by W. Stallings
(ISBN 0-8186-0825-0); 512 pages

Microprogramming and Firmware Engineering
Edited by V. Milutinovic
(ISBN 0-8186-0839-0); 416 pages

Modern Design and Analysis of Discrete-Event Computer Simulations
Edited by E.J. Dudewicz and Z. Karian
(ISBN 0-8186-0597-9); 486 pages

New Paradigms for Software Development
Edited by William Agresti
(ISBN 0-8186-0707-6); 304 pages

Object-Oriented Computing—Volume 1: Concepts
Edited by Gerald E. Peterson
(ISBN 0-8186-0821-8); 214 pages

Object-Oriented Computing—Volume 2: Implementations
Edited by Gerald E. Peterson
(ISBN 0-8186-0822-6); 324 pages

Office Automation Systems (Second Edition)
Edited by H.A. Freeman and K.J. Thurber
(ISBN 0-8186-0711-4); 320 pages

Parallel Architectures for Database Systems
Edited by A.R. Hurson, L.L. Miller, and S.H. Pakzad
(ISBN 0-8186-8838-6 (case)); 478 pages

Programming Productivity: Issues for the Eighties (Second Edition)
Edited by C. Jones
(ISBN 0-8186-0681-9); 472 pages

Recent Advances in Distributed Data Base Management
Edited by C. Mohan
(ISBN 0-8186-0571-5); 500 pages

Reduced Instruction Set Computers
Edited by W. Stallings
(ISBN 0-8186-0713-0); 384 pages

Reliable Distributed System Software
Edited by J.A. Stankovic
(ISBN 0-8186-0570-7); 400 pages

Robotics Tutorial (2nd Edition)
Edited by C.S.G. Lee, R.C. Gonzalez, and K.S. Fu
(ISBN 0-8186-0658-4); 630 pages

Software Design Techniques (4th Edition)
Edited by P. Freeman and A.I. Wasserman
(ISBN 0-8186-0514-0); 736 pages

Software Engineering Project Management
Edited by R. Thayer
(ISBN 0-8186-0751-3); 512 pages

Software Maintenance
Edited by G. Parikh and N. Zvegintzov
(ISBN 0-8186-0002-0); 360 pages

Software Management (3rd Edition)
Edited by D.J. Reifer
(ISBN 0-8186-0678-9); 526 pages

Software-Oriented Computer Architecture
Edited by E. Fernandez and T. Lang
(ISBN 0-8186-0708-4); 376 pages

Software Quality Assurance: A Practical Approach
Edited by T.S. Chow
(ISBN 0-8186-0569-3); 506 pages

Software Restructuring
Edited by R.S. Arnold
(ISBN 0-8186-0680-0); 376 pages

Software Reusability
Edited by Peter Freeman
(ISBN 0-8186-0750-5); 304 pages—

Software Reuse: Emerging Technology
Edited by Will Tracz
(ISBN 0-8186-0846-3); 392 pages

Structured Testing
Edited by T.J. McCabe
(ISBN 0-8186-0452-2); 160 pages

Test Generation for VLSI Chips
Edited by V.D. Agrawal and S.C. Seth
(ISBN 0-8186-8786-X (case)); 416 pages

VLSI Technologies: Through the 80s and Beyond
Edited by D.J. McGreivy and K.A. Pickar
(ISBN 0-8186-0424-7); 346 pages

Reprint Collections

Selected Reprints: Dataflow and Reduction Architectures
Edited by S.S. Thakkar
(ISBN 0-8186-0759-9); 460 pages

Selected Reprints on Logic Design for Testability
Edited by C.C. Timoc
(ISBN 0-8186-0573-1); 324 pages

Selected Reprints: Microprocessors and Microcomputers (3rd Edition)
Edited by J.T. Cain
(ISBN 0-8186-0585-5); 386 pages

Selected Reprints in Software (3rd Edition)
Edited by M.V. Zelkowitz
(ISBN 0-8186-0789-0); 400 pages

Selected Reprints on VLSI Technologies and Computer Graphics
Edited by H. Fuchs
(ISBN 0-8186-0491-3); 490 pages

Technology Series

Artificial Neural Networks: Theoretical Concepts
Edited by V. Vemuri
(ISBN 0-8186-0855-2); 160 pages

Computer-Aided Software Engineering (CASE)
Edited by E.J. Chikofsky
(ISBN 0-8186-1917-1); 132 pages